Circuit and System Theory

Circuit and System Theory

GLADWYN LAGO
University of Missouri-Columbia

LLOYD M. BENNINGFIELD
Wichita State University

JOHN WILEY & SONS
New York Chichester Brisbane Toronto

Library of Congress Cataloging in Publication Data:

Lago, Gladwyn Vaile, 1917–
Circuit and system theory.

Includes index.
1. Electric circuits. 2. System analysis.
3. Electric engineering–Mathematics. 4. Engineering mathematics. I. Benningfield, Lloyd M., joint author.
II. Title.
TK454.L33 621.319′2 79-10878
ISBN 0-471-04927-1

Printed in the United States of America

10 9 8 7 6 5 4 3 2 1

Preface

The scientific community continually faces new problems that are raised by the exposure to an ever-increasing body of knowledge. One general result of this exposure has been that material initially entering a curriculum at an advanced level is selectively moved to lower level courses. In this process, material is condensed, summarized, and generally crammed into already tight programs. As students in such programs reach various levels, they are expected to mature to a level attained a few years ago only by graduate students. With this net result, there is a real necessity for educational institutions to select and feature theories and concepts of fundamental importance that are amenable to a broad range of applications.

In applying this need to technical matters, we find that few subjects are more important than the study of linear system theory. Many problems faced by the scientist are either linear or can be approximated by linear systems. Before students can learn to solve most non-linear problems, they must understand the simpler linear problem.

Although much of the material in this book was taught at the senior or graduate level a few years ago, this book is directed to the junior level audience. We have deliberately and conscientiously tried introducing each new concept with a simple explanation, using carefully chosen examples to bridge the gap between the introductory courses and the material in this book. We are attempting to lead the reader through a wide range of topics and ideas so that he or she will require only a minimum of help from outside sources. The reader or student who finishes this book should be better able to comprehend the current literature in this area.

This book can be used in almost any department of engineering, physics, or applied mathematics. It assumes a knowledge of Kirchhoff's laws and simple a.c. circuit theory. Therefore, this book should meet the needs of the person who, perhaps out of school for a number of years, wishes to review and to keep abreast of recent developments.

The content of this book must be considered first by reviewing the complementarity of classical and transform methods of analysis. The classical methods were once the only known way of solving linear systems. Heaviside operational calculus was developed years ago and succeeded by the Laplace transform methods. The classical methods are sometimes referred to as time-domain methods whereas the transform methods are referred to as frequency-domain methods. Transform methods became so popular that in many quarters there was a movement away from time-domain methods. Later, however, state variable methods were developed, and one method of solving state variable equations is in the time domain. Underlying this shift of emphasis is the availability of the digital computer that can obtain numerical solutions. Because of this shift in direction, time-domain methods are again recognized as being a fundamental part of system theory. Therefore, this book opens with a coverage of classical methods of solving differential equations but also presents transform methods.

Another aspect or point of view relevant to the content of this text is that circuit theory is a much broader subject than is sometimes recognized. Circuit theory has moved to the more abstract method of analysis and synthesis of circuits through the use of the transfer function, which can be represented as a box. The input and output of the box can be thought of as signals. Circuit theory has become concerned with the representation of signals and with the manner in which these signals are transmitted through the circuit.

From this point of view, the transition from circuit theory to system theory is only a change in notation. Some authors feel obliged, from the outset, to mix examples of mechanical systems and electro-mechanical systems with circuit examples. For many cases, it seems preferrable to use simple circuit examples until the matematical tools have been developed and understood. Thereafter, the transition from circuit theory to other physical systems, as discussed in Chapters 7 and 8, is very simple.

There is obviously more material in this book than can be covered in one semester. In fact, the book can be used for a two-semester sequence in the junior year. For a one-semester course, Chapters 7, 8, and 12 can be omitted and some of the material in the latter part of other chapters can be considered briefly. Inclusion of this extra material makes possible the creation of a course suited to a department's specific needs.

The authors wish to acknowledge the generous and helpful suggestions and encouragement from colleagues at both the University of Missouri-Columbia and Wichita State University. So many students have offered helpful suggestions that we cannot mention them by name.

G. V. Lago
L. M. Benningfield

Contents

Circuit and System Theory

Chapter 1

Circuit and System Concepts

1-1. INTRODUCTION

Most persons have a reasonably clear concept of circuits, but an understanding of the extent and nature of systems can develop in a great many directions. There are about as many different concepts of a *system* as there are persons dealing with such topics, for example, as those ranging from business management to body temperature regulation, from a missile trajectory to population density, and from human behavior to stereo tone controls. Perhaps the following definition from *Webster's Third New International Dictionary* indicates why such a diverse collection of things fall under the designation of the term: "A system is a complex unity formed of many diverse parts subject to a common plan or serving a common purpose."

In some sense the totality of our universe, as a unit, would satisfy the above definition, but the very extent of this idea overwhelms the human mind, and it is not really amenable to useful study. There are some people involved with world simulations, or computer analyses of various mathematical models of our world, but rather than pursue such overwhelmingly complex systems, most of us deal with small groups of parts so organized or interrelated as to enable us to accomplish some relatively simple goal. In order to deal with even such smaller groupings, it is necessary to isolate our objectives by determining what quantities affect the system to be studied, and how these effects are introduced, as well as what results from the action of the system on items external to it. It is not always an easy task to accomplish this isolation of the group of parts we desire to study as a system.

As one system example, it is certainly possible, and can be of considerable practical use, to consider the mechanism of body temperature control. Such a system requires the determination of suitable representations for metabolic heat generation within the body; heat transfer from the body core through the muscle and skin, together with heat loss by respiration; and the mechanism by which the body senses and controls its own temperature. In accomplishing such an isolation it is necessary to determine the external effects due to such environmental conditions as air temperature and velocity. It is also necessary to account for such things as the condition of the temperature-sensing mechanism of the body, and the rate and type of food input for production of heat.

The body temperature control mechanism described above is a complex unity that maintains body temperature and so sustains life with its various components all contributing to that effect. Thus, the body-temperature control mechanism qualifies as a system.

Another frequently studied example of a system is the economy of some political unit, such as a city, state, region, or country. Such items as material flow, productivity, and money supply are involved, and the external effects, from and upon surrounding territories, would also need to be considered. Again we have the essential characteristics of a system.

As another example of a unity formed of diverse parts, we might consider an antenna positioning control consisting of a mechanism to move the antenna, a means of inputing the desired antenna position, and the necessary components to relate desired position to actual position; of course a source of energy to operate this group of components is also needed. This again is a system, though its parts form a less complex grouping than in either of the previous examples.

In considering the three examples above, several additional points can be made. We progress from the first to the third by going from (1) a system totally structured without human design, through (2) a system in which humans have considerable effect on the structure, to (3) one in which, so far as concerns its form as a system, man is totally responsible. It is also evident that, basically, certain quantities serve as inputs to the system and certain others serve as outputs. This input-output concept makes it convenient for people dealing with systems to use the *block diagram* of Fig. 1-1.1 to represent the system relationships involved. This block diagram will be fundamental to most of our work with circuits and systems.

1-2. THE PROBLEM OF MODELING

The diagram of Fig. 1-1.1 provides a framework within which we can begin to consider more specifically what inputs are involved, what outputs are appropriate, and how the outputs and inputs are related. From such a representation it is then possible to consider

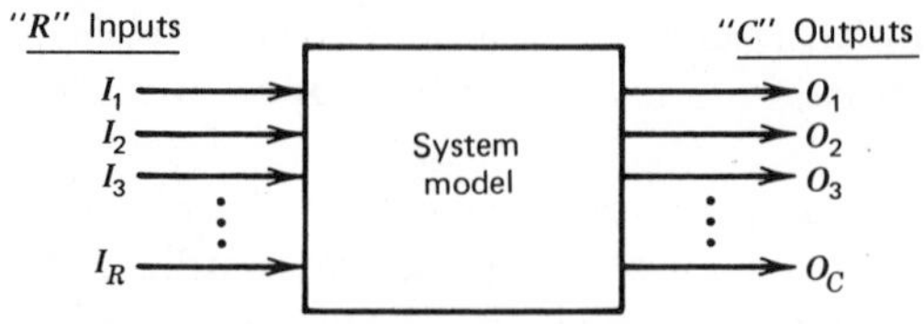

Fig. 1-1.1

the effect of such things as variation of the input and change in system structure or interconnection. Development of the above diagramatic relationship is called the *modeling of a system* and is one of the main steps in studying systems.

In general, the development of a relationship between input and output quantities for a system involves *mathematical approximation* because the true nature of the relationship is not known or can not be represented exactly within reasonable mathematical complexity. One major factor in the development of system models is their ultimate use. The more critical the information to be obtained from the model, the more complex the model generally becomes in order to account more precisely for more aspects of the relationships among its quantities.

The development of suitable mathematical system models applies various laws and principles of physics, chemistry, biology, sociology, and business, establishing algebraic relationships, continuous dynamic relationships that involve differential and/or integral equations, and discrete dynamic relationships that involve difference equations. In this text the reader is introduced to a variety of electrical, mechanical, electro-mechanical, fluidic, thermal, chemical, and socio-economic systems. In general, the models considered are said to be linear, which eliminates many systems from consideration, and requires rather crude approximations of the relationships among the quantities describing others.

In spite of these limitations, linear systems form a very important subset of systems because of the simplicity of their models, the ease with which solutions are obtained, and the generality of the solutions. In essence the concept of linearity relates to the following conditions: if, *first*, changing inputs by a given multiple results in changing outputs by the same multiple, and *second*, the outputs to the sum of two sets of inputs can be found as the sum of the outputs resulting from each set of inputs applied individually, then the system is said to be *linear*.

The development of system models can provide a benefit beyond a specific model under study. Many systems behave analogously or similarly to other totally different systems; in its behavior a thermal or fluidic system may, for example, resemble an electrical circuit containing only resistors and capacitors; and such similar behavior allows someone familiar with one type of system to generalize his knowledge and thereby facilitate his learning about new systems. Insight gained from familiarity with one type of system may be extended to analogous systems.

1-3. OBTAINING THE SOLUTION FROM THE MODEL

The second part of the study of systems relates to the determination of system response to various types of inputs. If a person is left in a freezer for ten hours without food and with only certain clothing as protection, what happens to body temperature as determined from a model of the body's temperature control system? If the supply of money as determined by the Federal Reserve Board is doubled, what happens to the price of various products as determined by some model of the national economy? In the determination of these responses from a model, a solution is obtained.

Depending on the desired use of solutions, many different forms of models and solution techniques may be used. If simple studies of the time variation of a single system variable are adequate, then a solution technique that solves the single pertinent differential equation (or perhaps algebraic or difference equation) is adequate, and more general solution of sets of equations probably results in excessive solution cost or effort. On the other hand, if the more complex variation of a set of system variables is needed, then more general solution methods and models may be needed. This text presents the mathematical tools of the classical methods of solving differential equations, one-sided and two-sided Laplace transforms and Fourier transforms, and the state-variable methods of solving linear system problems.

At one time, classical methods of solution were the only known methods of solving differential equations; therefore, the earlier literature in system theory used the classical methods. Such methods are sometimes referred to as *time-domain methods*.

In approximately the first third of this century, the Heaviside operational calculus method was developed. As originally conceived, Heaviside's method had certain limitations and rested upon a shaky mathematical basis. To overcome these limitations the Laplace transform methods were developed and have been used extensively since the middle of the 1930's. Such transform methods are sometimes referred to as *frequency-domain methods*. As happens when something new comes along, there is a tendency to discard the old, even when it still has much to offer. The result was a movement away from the time-domain methods as the frequency-domain methods became popular.

Since the middle 1950's, state-variable methods which draw upon the time-domain have been popular. Behind the scene in this shift of emphasis is the availability of the digital computer, whose use has made state-variable equations a very useful way of solving system problems. Because of this shift, time-domain methods are again recognized as a fundamental part of system theory.

1-4. THE *n*th-ORDER LINEAR ORDINARY DIFFERENTIAL EQUATIONS (LODE)

As mentioned previously, many systems and circuits can be modeled by linear models involving ordinary differential equations. An example of a linear ordinary (not partial) differential equation is

$$a_n(t)\frac{d^n x(t)}{dt^n} + a_{n-1}(t)\frac{d^{n-1}x(t)}{dt^{n-1}} + \cdots + a_1(t)\frac{dx(t)}{dt} + a_0(t)\,x(t) = f(t) \quad (1\text{-}4.1)$$

where the variable t is the independent variable interpreted as *time*, and $x(t)$ is the dependent variable. Since Eq. 1-4.1 is an ordinary differential equation, involving only one independent variable (no dynamic variation with physical space in this case), the system being described is called a lumped parameter system. If the system has distributed parameters (such as a transmission line), the resulting equation would be a partial differential equation. If $f(t) = 0$, the equation is said to be homogeneous; whereas if $f(t) \neq 0$, the equation is nonhomogeneous. This equation is an nth order equation because n is the highest ordered derivative contained in the equation. This equation is said to be linear, because it is of the first degree in the dependent variable $x(t)$ and in all its derivatives, and the $a_n(t)$ terms are a function of t and not of $x(t)$. Engineers commonly refer to systems described by equations of this sort as time-varying systems. If any of the derivative terms are of higher degree than the first, or any of the a_n terms are functions of $x(t)$ rather than t, the differential equation is said to be non-linear.

If the $a_n(t)$ terms are constants, the linear differential equation is relatively easy to solve. It is fortunate that many physical systems may be modeled by such equations. Another argument for studying such equations is that it is helpful to have an understanding of linear equations before attempting to study the more difficult non-linear equations.

If all a_n terms of Eq. 1-4.1 are constant, the equation becomes an nth-order linear differential equation, with constant coefficients as given by

$$a_n \frac{d^n x}{dt^n} + a_{n-1} \frac{d^{n-1} x}{dt^{n-1}} + \cdots + a_1 \frac{dx}{dt} + a_0 x = f(t) \tag{1-4.2}$$

where as a notational convenience, x replaces $x(t)$. For the most part, we will be studying equations of this sort.

1-5. THE COMPLEMENTARY COMPONENT OF THE SOLUTION OF LODE's

Before discussing the complete solution of Eq. 1-4.1, we will examine the homogeneous equation

$$a_n \frac{d^n x_c}{dt^n} + a_{n-1} \frac{d^{n-1} x_c}{dt^{n-1}} + \cdots + a_1 \frac{dx_c}{dt} + a_0 x_c = 0 \tag{1-5.1}$$

where the dependent variable x is replaced by x_c to indicate that the solution of the homogeneous equation is only one component of the total solution called the complementary component.

The solution x_c of Eq. 1-5.1 is any functional relationship, among the variables and the a_n constants, that satisfies the equation but contains no derivative. People who have been working with such equations for years have found such a basic functional relationship to be

$$x_c(t) = A\,\epsilon^{st} \tag{1-5.2}$$

where A and s are parameters treated as constants for a specific LODE.

To be a solution, $x_c(t)$ must satisfy Eq. 1-5.1; $x_c(t)$ and its derivatives have the form

$$\left.\begin{array}{ll} x_c = A\,\epsilon^{st} & \dfrac{d^2 x_c}{dt^2} = As^2\,\epsilon^{st} \\ & \vdots \\ \dfrac{dx_c}{dt} = As\,\epsilon^{st} & \dfrac{d^n x_c}{dt^n} = As^n\,\epsilon^{st} \end{array}\right\} \tag{1-5.3}$$

Upon substituting into Eq. 1-5.1, a factor $A\,\epsilon^{st}$ appears in each term and can be factored out as

$$(a_n s^n + a_{n-1} s^{n-1} + \cdots + a_1 s + a_0) A\,\epsilon^{st} = 0 \tag{1-5.4}$$

For this equation to equal zero either $A\,\epsilon^{st}$ or the terms inside the bracket must equal zero. The term $A\,\epsilon^{st}$ cannot equal zero (except in very special cases) because this is $x_c(t)$, the desired non-zero complementary solution. Therefore, the expression within the brackets must equal zero, as

$$a_n s^n + a_{n-1} s^{n-1} + \cdots + a_1 s + a_0 = 0 \tag{1-5.5}$$

Equation 1-5.5 is called the *characteristic equation.*

THE CHARACTERISTIC EQUATION

We see the characteristic equation can be obtained in a more direct manner by observing that s is of the same degree in each term of Eq. 1-5.5 as the order of the derivative of the corresponding term of Eq. 1-4.1. As a specific example, the differential equation

$$4\frac{d^3 x_c}{dt^3} + 6\frac{d^2 x_c}{dt^2} + 12\frac{dx_c}{dt} + 30x_c = 0 \tag{1-5.6}$$

has the characteristic equation

$$4s^3 + 6s^2 + 12s + 30 = 0 \tag{1-5.7}$$

FACTORING THE CHARACTERISTIC EQUATION

In order to determine specific values for s for this example, the characteristic equation must be factored. While low-degree equations such as Eq. 1-5.7 can be factored by formula, it is common practice to use calculators or computers and numerical methods to determine factors. The general theory behind factoring such equations is, therefore, more the subject of a book on numerical methods than of this text.

For the present discussion, we assume that Eq. 1-5.5 can be factored with distinct factors, as

$$a_n(s + \alpha_1)(s + \alpha_2)(s + \alpha_3)\cdots(s + \alpha_n) = 0 \tag{1-5.8}$$

where it is noted that there are a number of roots $s_i = -\alpha_i$ equal to the order of the LODE.

ADDITIONAL CONSIDERATIONS OF THE COMPLEMENTARY COMPONENT OF THE SOLUTION

The roots of the characteristic equation lead to such functions as

$$\left.\begin{matrix} x_{c1} = A_1\,\epsilon^{-\alpha_1 t} & x_{c3} = A_3\,\epsilon^{-\alpha_3 t} \\ x_{c2} = A_2\,\epsilon^{-\alpha_2 t} & x_{cn} = A_n\,\epsilon^{-\alpha_n t} \end{matrix}\right\} \tag{1-5.9}$$

each of which satisfy the homogeneous equation, (Eq. 1-5.1). For example, upon substituting $x_{c1} = A_1 \epsilon^{-\alpha_1 t}$, this becomes

$$[a_n(-\alpha_1)^n + a_{n-1}(-\alpha_1)^{n-1} + \cdots + a_1(-\alpha_1) + a_0] A_1 \epsilon^{-\alpha_1 t} = 0 \qquad (1\text{-}5.10)$$

The expression inside the bracket equals zero because it is forced to do so as a part of the method of determining α_1 (as is also true for all α_i).

Precisely the same development can be repeated for x_{c2}, x_{c3}, etc. If each function satisfies the homogeneous differential equation, the homogeneous version of the LODE will also be satisfied by any linear combination of these terms, such as

$$A_1 \epsilon^{-\alpha_1 t} + A_2 \epsilon^{-\alpha_2 t} + A_3 \epsilon^{-\alpha_3 t} + \cdots + A_n \epsilon^{-\alpha_n t} \qquad (1\text{-}5.11)$$

When such an expression is substituted into the homogeneous equation, the result is zero, because the mathematical operations involved over the field of complex numbers are commutative and distributive, resulting in a sum of n expressions such as in the left side of Eq. 1-5.10 (one for each α_i), each of which is zero.

The general solution to the homogeneous differential equation is obtained by adding all n functions of Eqs. 1-5.9, as in

$$x_c = A_1 \epsilon^{-\alpha_1 t} + A_2 \epsilon^{-\alpha_2 t} + \cdots + A_n \epsilon^{-\alpha_n t} \qquad (1\text{-}5.12)$$

1-6. THE PARTICULAR COMPONENT OF THE SOLUTION FOR LODE's

We now turn our attention to the non-homogeneous equation

$$a_n \frac{d^n x}{dt^n} + a_{n-1} \frac{d^{n-1} x}{dt^{n-1}} + \cdots + a_1 \frac{dx}{dt} + a_0 x = f(t) \qquad (1\text{-}6.1)$$

It is apparent that the complementary solution x_c does not satisfy this equation, because of the presumed non-zero $f(t)$ on the right-hand side. What is needed is some particular function of the x (identified by x_p) that satisfies Eq. 1-6.1, as indicated by

$$a_n \frac{d^n x_p}{dt^n} + a_{n-1} \frac{d^{n-1} x_p}{dt^{n-1}} + \cdots + a_1 \frac{dx_p}{dt} + a_0 x_p = f(t) \qquad (1\text{-}6.2)$$

This x_p is called a *particular solution* because it satisfies the differential equation for a particular driving function $f(t)$. We will take up the details of obtaining the x_p for various $f(t)$ through the examples later in the next two chapters.

1-7. THE COMPLETE SOLUTION FOR LODE's

Suppose we substitute $(x = x_p + x_c)$ into the original non-homogeneous Eq. 1-6.1, as in

$$a_n \frac{d^n(x_p + x_c)}{dt^n} + a_{n-1} \frac{d^{n-1}(x_p + x_c)}{dt^{n-1}} + \cdots + a_1 \frac{d(x_p + x_c)}{dt} + a_0(x_p + x_c) = f(t) \quad (1\text{-}7.1)$$

Again, because of the commutative and distributive properties of the mathematical operations involved, this equation can be rewritten

$$a_n \frac{d^n x_p}{dt^n} + a_{n-1} \frac{d^{n-1} x_p}{dt^{n-1}} + \cdots + a_1 \frac{dx_p}{dt} + a_0 x_p + a_n \frac{d^n x_c}{dt^n} + a_{n-1} \frac{d^{n-1} x_c}{dt^{n-1}} + \cdots + a_1 \frac{dx_c}{dt} + a_0 x_c = f(t) \quad (1\text{-}7.2)$$

The sum of the first $n + 1$ terms (those involving x_p) of this equation is equal to $f(t)$ because this is exactly how x_p is defined in Eq. 1-6.2. Therefore, these terms on the left side can be canceled with the $f(t)$ on the right, leaving

$$a_n \frac{d^n x_c}{dt^n} + a_{n-1} \frac{d^{n-1} x_c}{dt^{n-1}} + \cdots + a_1 \frac{dx_c}{dt} + a_0 x_c = 0 \quad (1\text{-}7.3)$$

This is the homogeneous equation whose solution was determined as Eq. 1-5.12. The complete solution to the non-homogeneous equation for distinct roots of the characteristic equation is

$$x = x_p + x_c = x_p + A_1 \epsilon^{-\alpha_1 t} + A_2 \epsilon^{-\alpha_2 t} + \cdots + A_n \epsilon^{-\alpha_n t} \quad (1\text{-}7.4)$$

In summary, x_p is called the ***particular solution*** because it satisfies the non-homogeneous equation for a particular driving function $f(t)$, and x_c is called the *complementary* solution because it complements the x_p to form the complete solution $x = x_p + x_c$. We will go over many of these ideas several times while discussing examples in the following chapters.

Chapter **2**

Systems Described by First-Order Differential Equations

2-1. INTRODUCTION

There is no one way to organize a book of this sort. Some authors feel obliged, from the outset, to mix examples of circuits, mechanical systems, electro-mechanical systems. To us it seems that this procedure has certain disadvantages. New ideas about properties of linear systems, the method of modeling a system, how to solve the resulting equations, get mixed together. Before any one subject is digested, another is introduced.

Also, this omelet method of presenting the material does not recognize the broad nature of circuit theory. Although circuit theory begins with the study of a set of elements connected in a specific manner, it is much more than this. Circuit theory has developed a set of theorems, such as *superposition*, and *reciprocity*, which apply to all linear systems. Circuit theory has moved to more abstract methods of analysis and synthesis, through the use of the transfer functions and block diagrams. The inputs and outputs of the block can be thought of as signals. Circuit theory has become concerned with the mathematical representation of signals, and with the manner in which these signals are transmitted through the blocks. From this point of view, the transition from circuit theory to system theory is only a change in notation.

Although only circuit examples are used through Chapter 6, the presentation is general enough to apply to other systems. Chapters 7 and 8 present a discussion of the modeling of a variety of physical systems. The ideas discussed in the earlier chapters can be immediately applied to these systems.

2-2. CIRCUITS WITH ONE STORAGE ELEMENT: I

In this chapter, we study circuits that lead to first-order differential equations, and in Chapter 3, circuits that lead to higher-order equations. We assume the reader has a rudimentary knowledge of circuit theory. The three circuit elements are shown in Table 2-2.1. The equations in the table are called *volt–ampere relationships*, because they show how v and i are related for each element. Both v and i are functions of time and should more properly be written as $v(t)$ and $i(t)$. However, it is tacitly agreed that v and i represent $v(t)$ and $i(t)$, respectively. In each case, the reference direction of the current, i, is into the plus (+) terminal of the reference voltage, v. The $i(0+)$ and $v(0+)$ initial conditions are positive if $i(0+)$ is in the positive direction of i, and the polarity of $v(0+)$ agrees with that of v.

Example 2-2.1. Figure 2-2.1(a) indicates that an R-C circuit is connected to a constant voltage source by closing a switch, S, at $t = 0$, and (b) conveys this same information in a different form. The $u_{-1}(t)$ symbol in Fig. 2-2.1(b) indicates the unit-step function. If

Table 2-2.1

The Symbols for the Three Elements		
R	L	C
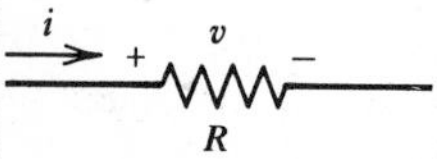	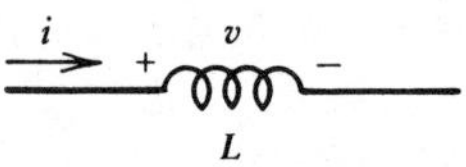	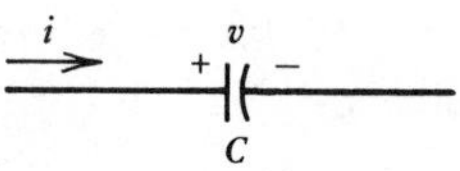
The Mathematical Models for the Elements		
R	L	C
$v = Ri$	$v = L\dfrac{di}{dt}$	$v = S\displaystyle\int_0^t i\,dt + v(0+)$
$i = Gv$	$i = \Gamma\displaystyle\int_0^t v\,dt + i(0+)$	$i = C\dfrac{dv}{dt}$
where $G = \dfrac{1}{R}$	where $\Gamma = \dfrac{1}{L}$	where $C = \dfrac{1}{S}$

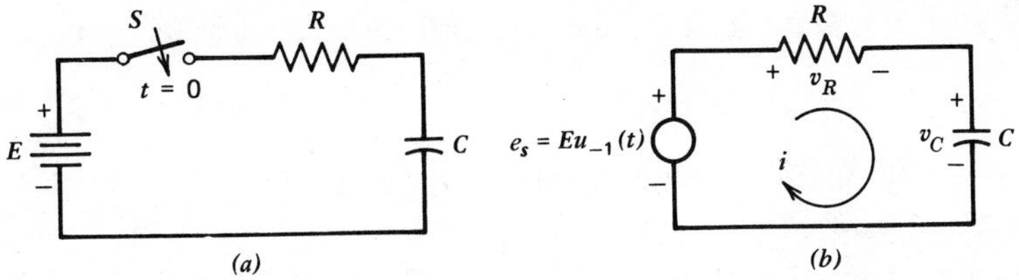

Fig. 2-2.1

the reader is unfamiliar with this notation, or with the family of singularity functions, he should turn to Appendix A.

The current i is added as shown in (b), and the Kirchhoff voltage law equation (KVLE) is written in Eq. 2-2.1; the volt-ampere relationships are added in Eq. 2-2.2, and the more conventional way of writing this is shown in Eq. 2-2.3:

$$Eu_{-1}(t) = v_R + v_C \tag{2-2.1}$$

$$Eu_{-1}(t) = Ri + \frac{1}{C}\int_0^t i\,dt \tag{2-2.2}$$

$$E = Ri + \frac{1}{C}\int_0^t i\,dt \qquad (\text{for } t > 0) \tag{2-2.3}$$

Eq. 2-2.3 is differentiated, treating E as a constant, giving

$$0 = R\frac{di}{dt} + \frac{i}{C} \tag{2-2.4}$$

As discussed in Section 1-4, Eq. 2-2.4 is homogeneous; the particular component of the solution is zero, and the complementary component becomes the complete solution. The steps of Section 1-5 are carried out without discussion:

$$i = A\,\epsilon^{st} \qquad \frac{di}{dt} = As\,\epsilon^{st} \tag{2-2.5) (2-2.6}$$

$$0 = RAs\,\epsilon^{st} + \frac{A\,\epsilon^{st}}{C} = \left(Rs + \frac{1}{C}\right)A\,\epsilon^{st} \tag{2-2.7}$$

$$Rs + \frac{1}{C} = 0 \qquad s_1 = -\frac{1}{RC} \tag{2-2.8) (2-2.9}$$

$$i = A\,\epsilon^{-t/RC} \tag{2-2.10}$$

where A is an arbitrary constant to be determined.

Although there are many methods of determining A, the conventional way is by using initial conditions. The switch S is closed at $t = 0$ and initial current $i(0+)$ is the value of the current at $t = 0+$. Equation 2-2.3 is good for all $t > 0$; so it is good for $t = 0+$, and we write

$$E = R\,i(0+) + \frac{1}{C}\int_{0-}^{0+} i\,dt \qquad (2\text{-}2.11)$$

The limits on the integral contain $t = 0$, the time S is closed. Since i is finite, the integral term is zero, and Eq. 2-2.11 can be solved for $i(0+)$ as

$$i(0+) = \frac{E}{R} \qquad (2\text{-}2.12)$$

At $t = 0+$, Eq. 2-2.10 becomes

$$i(0+) = \frac{E}{R} = A\,\epsilon^{-(0+)/RC} = A \qquad (2\text{-}2.13)$$

and A is determined as E/R. Upon substitution for A, Eq. 2-2.10 is

$$i = \frac{E}{R}\epsilon^{-t/RC} \qquad (2\text{-}2.14)$$

When we look at Eq. 2-2.14, we remember that the equations leading to this are good only for $t > 0$. To include this idea in our answer we write

$$i = \frac{E}{R}\epsilon^{-t/RC}\,u_{-1}(t) \qquad (2\text{-}2.15)$$

Example 2-2.2. We again solve Example 2-2.1 by picking up Eq. 2-2.2 as

$$E\,u_{-1}(t) = Ri + \frac{1}{C}\int_0^t i\,dt \qquad (2\text{-}2.16)$$

We take the derivative of both sides as

$$E\,u_0(t) = R\frac{di}{dt} + \frac{i}{C} \qquad (2\text{-}2.17)$$

Equations 2-2.16 and 2-2.17 are written to be good for all values of time, so the proposed form for the solution must also, and i is written

$$i = A\,\epsilon^{st}\,u_{-1}(t) \qquad (2\text{-}2.18)$$

To find the derivative of i, we must be careful. At $t = 0-$, $i(0-) = 0$; but at $t = 0+$, $i(0+) = A$. Therefore at $t = 0$, i has a discontinuity equal to A, which means the derivative of i contains an impulse at $t = 0$ of magnitude A in addition to the derivative in the usual sense.

$$\frac{di}{dt} = A\,u_0(t) + As\,\epsilon^{st}\,u_{-1}(t) \qquad (2\text{-}2.19)$$

Equations 2-2.18 and 19 are substituted into Eq. 2-2.17 as

$$E\,u_0(t) = AR\,u_0(t) + \left(Rs + \frac{1}{C}\right) A\,\epsilon^{st}\,u_{-1}(t) \tag{2-2.20}$$

The magnitudes of the impulse on both sides must be equal, or

$$E = AR \qquad A = \frac{E}{R} \qquad \text{(2-2.21) (2-2.22)}$$

as found before. Since there is no $u_{-1}(t)$ on the left, the magnitude of the $u_{-1}(t)$ term on the right must equal zero, which again yields

$$Rs + \frac{1}{C} = 0 \tag{2-2.23}$$

from which s can be found. This development has determined both A and s at the same time, and Eq. 2-2.18 can now be written

$$i = \frac{E}{R}\,\epsilon^{-t/RC}\,u_{-1}(t) \tag{2-2.24}$$

One of the main arguments against using classical methods is in the need for a set of auxiliary steps to solve for the arbitrary constants in the solution. This example is included to suggest how the impulse family of functions can be used for this purpose. Although we will use some of these ideas on a few occasions, for the most part we will drop the subject and reflect the more traditional attitude.

Example 2-2.3. The circuit of Fig. 2-2.2 is used with $e_s = E\,u_{-1}(t)$, and the circuit is initially at rest. The direction of the current is added, and the Kirchhoff's voltage law equation (KVLE) is written for $t > 0$

$$E = L\frac{di}{dt} + Ri \tag{2-2.25}$$

This is a non-homogeneous equation; therefore, the solution i contains both a particular component and a complementary component as

$$i = i_p + i_c \tag{2-2.26}$$

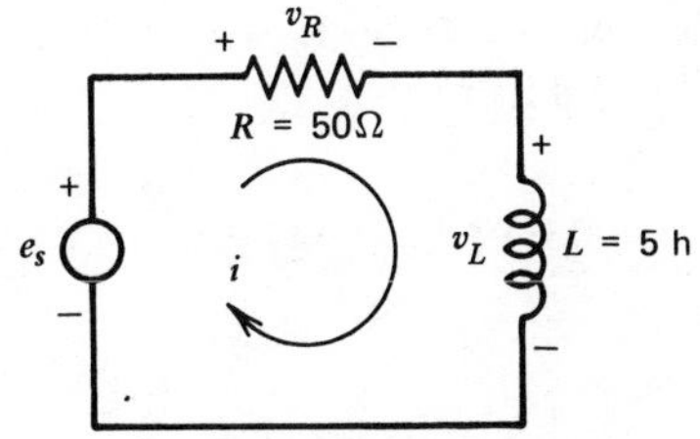

Fig. 2-2.2

The particular component of the solution i_p must satisfy the differential equation for the particular driving function as given by

$$E = L\frac{di_p}{dt} + Ri_p \qquad (2\text{-}2.27)$$

It is reasoned, to satisfy this equation, that i_p is a constant:

$$i_p = K \qquad \frac{di_p}{dt} = 0 \qquad (2\text{-}2.28)$$

Upon substitution, Eq. 2-2.27 becomes, and K is found as

$$E = L(0) + RK \qquad K = i_p = \frac{E}{R} \qquad (2\text{-}2.29)\ (2\text{-}2.30)$$

Since i_p satisfies the differential equation, why is i_p not the complete solution? The answer is that i_p does not satisfy the initial conditions. In this example $i(0+) = 0$, whereas $i_p(0+) = E/R$. Therefore, the complementary component i_c must be added to i_p so that $i = i_p + i_c$ can match the proper initial conditions.

The component i_c is obtained from the homogeneous equation

$$0 = L\frac{di_c}{dt} + Ri_c \qquad (2\text{-}2.31)$$

as discussed in Section 1-5. To review this briefly, $i = i_p + i_c$ is substituted into Eq. 2-2.25, giving

$$E = L\frac{d(E/R + i_c)}{dt} + R\left(\frac{E}{R} + i_c\right) \qquad (2\text{-}2.32)$$

$$E = L\frac{di_c}{dt} + E + Ri_c \qquad (2\text{-}2.33)$$

from which Eq. 2-3.31 is obtained:

$$0 = L\frac{di_c}{dt} + Ri_c \qquad (2\text{-}2.34)$$

When i_p is substituted on the right-hand side of Eq. 2-2.25, it yields an E that cancels the E on the left side because this is precisely how i_p was defined as given by Eq. 2-2.27.

The characteristic equation, and its root are

$$0 = Ls + R \qquad s_1 = -\frac{R}{L} \qquad (2\text{-}2.35)\ (2\text{-}2.36)$$

and the form of i_c is

$$i_c = A\,\epsilon^{-(R/L)t} \qquad (2\text{-}2.37)$$

For the complete solution, i_c is added to i_p as

$$i = \frac{E}{R} + A\,\epsilon^{-(R/L)t} \tag{2-2.38}$$

where A is determined from the appropriate initial conditions.

As stated, E is applied to the R-L circuit initially at rest. By this we mean that before $t = 0$, $i = 0$; or more specifically, $i(0-) = 0$. We can reason that $i(0+)$ also is zero. We can do this by assuming that $i(0+) = K$, which is not zero, and showing that this leads to a contradiction. If $i(0-) = 0$ and $i(0+) = K \neq 0$, then i has a discontinuity at $t = 0$, and di/dt would contain an impulse at $t = 0$. We again direct our attention to Eq. 2-2.25, repeated as

$$E = L\frac{di}{dt} + Ri$$

An impulse on the right side of this equation must be matched by an impulse on the left side. There is no impulse on the left side, therefore we have reached a contradiction, and $i(0+)$ must be zero.

This initial condition is applied to Eq. 2-2.38 to determine A:

$$i(0+) = 0 = \frac{E}{R} + A\,\epsilon^{-R(0+)/L} \qquad A = -\frac{E}{R} \tag{2-2.39) (2-2.40}$$

Upon substitution, Eq. 2-2.38 becomes

$$i = \frac{E}{R} - \frac{E}{R}\epsilon^{-(R/L)t} = \frac{E}{R}(1 - \epsilon^{-(R/L)t}) \tag{2-2.41}$$

To show that this equation has meaning only for positive time, we write

$$i = \frac{E}{R}(1 - \epsilon^{-(R/L)t})\,u_{-1}(t) \tag{2-2.42}$$

Example 2-2.4. The circuit of Fig. 2-2.2 is used again with $e_s = 10\epsilon^{-6t}u_{-1}(t)$ and the circuit is initially at rest. The numerical values are used for R and L to make the work more compact. The KVLE for $t > 0$ is written

$$10\epsilon^{-6t} = 5\frac{di}{dt} + 50i \tag{2-2.43}$$

The i_p is determined from

$$10\epsilon^{-6t} = 5\frac{di_p}{dt} + 50i_p \tag{2-2.44}$$

This equation is examined with the thought as to the form i_p must have to satisfy this equation. Since the left side has a factor ϵ^{-6t}, each term on the right side must contain this factor. This suggests an i_p of the form

$$i_p = K\epsilon^{-6t} \qquad \frac{di_p}{dt} = -6K\epsilon^{-6t} \tag{2-2.45)(2-2.46}$$

These are substituted into Eq. 2-2.44 to yield K as

$$10\epsilon^{-6t} = -30K\epsilon^{-6t} + 50K\epsilon^{-6t} \qquad K = 0.5 \tag{2-2.47)(2-2.48}$$

Finally, i_p is given by

$$i_p = 0.5\epsilon^{-6t} \tag{2-2.49}$$

The i_c is determined from the homogeneous equation

$$0 = 5\frac{di_c}{dt} + 50i_c \tag{2-2.50}$$

This is identical with Eq. 2-2.31, and the solution can be found from

$$i_c = A\epsilon^{-(R/L)t} = A\epsilon^{-10t} \tag{2-2.51}$$

The complete solution is

$$i = i_p + i_c = 0.5\epsilon^{-6t} + A\epsilon^{-10t} \tag{2-2.52}$$

The constant A and the final solution are

$$i(0+) = 0 = 0.5 + A \qquad \text{or} \qquad A = -0.5 \tag{2-2.53}$$

$$i = 0.5(\epsilon^{-6t} - \epsilon^{-10t})u_{-1}(t) \tag{2-2.54}$$

Example 2-2.5. The circuit of Fig. 2-2.2 is used again with $e_s = 10u_0(t)$. If a KVLE is written for $t > 0$, the driving function will not appear, since $10u_0(t)$ occurs at $t = 0$. Therefore, an equation that is good for all time must be written

$$10u_0(t) = 5\frac{di}{dt} + 50i \tag{2-2.55}$$

For Eq. 2-2.55 to hold at $t = 0$, some term on the right side must contain an impulse to balance the impulse of the left side. If we assume that i contains an impulse at $t = 0$

$$i = K u_0(t) \qquad \text{then} \qquad \frac{di}{dt} = K u_{+1}(t) \tag{2-2.56)(2-2.57}$$

We have now reached a contradiction, because there is no $u_{+1}(t)$ term on the left side of the equation. Therefore, i does not contain an impulse, but it does contain a discontinuity at $t = 0$.

We will complete this example using two different methods. First we will use a heuristic type of reasoning that is convenient but possibly not very satisfying; and then we will use a method similar to that in Ex. 2-2.2.

Using the first method, we reason that at $t = 0-$, $i(0-) = 0$, and that at $t = 0+$, i has jumped to some unknown value of K. Therefore, the magnitude of the discontinuity in i at $t = 0$ is K, and di/dt contains a $K\,u_0(t)$ term in addition to the derivative in the usual sense. If we substitute into Eq. 2-2.55 and examine the equation only at $t = 0$, we get

$$10u_0(t) = 5[K\,u_0(t)] + (\text{other finite terms}) \qquad (2\text{-}2.58)$$

The other terms on the right are finite and are negligible compared with the first term. The K is solved as

$$K = 10/5 = 2 = i(0+) \qquad (2\text{-}2.59)$$

For $t > 0$, Eq. 2-2.55 becomes

$$0 = 5\frac{di}{dt} + 50i \qquad (2\text{-}2.60)$$

which we have already solved as Eq. 2-2.51:

$$i = A\,\epsilon^{-10t} \qquad (2\text{-}2.61)$$

Using the $i(0+) = 2$ condition, A is found to be 2, and the final result is

$$i = 2\epsilon^{-10t}\,u_{-1}(t) \qquad (2\text{-}2.62)$$

As we look back over this example we see that the impulse of voltage throws an initial current into the inductor at $t = 0$, producing an $i(0+)$. After $t = 0$, this current follows an exponential decay toward zero.

Using the second method, we start with the fact that we know something about the shape of the current after $t = 0$, which in this case is

$$i = K\,\epsilon^{-10t} \qquad (2\text{-}2.63)$$

In order to substitute into Eq. 2-2.55, we find the derivative of i. Because of the discontinuity in i at $t = 0$, this becomes

$$\frac{di}{dt} = K\,u_0(t) - 10K\,\epsilon^{-10t} \qquad (2\text{-}2.64)$$

Upon substitution, Eq. 2-2.55 becomes

$$10u_0(t) = 5[K\,u_0(t) - 10K\,\epsilon^{-10t}] + 50K\,\epsilon^{-10t} \qquad (2\text{-}2.65)$$

The exponential terms cancel, leaving

$$10u_0(t) = 5K\,u_0(t) \qquad (2\text{-}2.66)$$

from which K is found and is substituted into Eq. 2-2.63.

$$K = 2 \qquad i = 2\epsilon^{-10t} \qquad (2\text{-}2.67)\,(2\text{-}2.68)$$

INTRODUCTION TO THE BLOCK DIAGRAM

In Fig. 1-1.1, a block diagram was shown displaying the input-output relationship for a system. In this particular example, the input is $10u_0(t)$, and the output $2\epsilon^{-10t}$. One of the conventional ways of describing the system itself is to use the response of the system to a unit impulse. We will go into the reasons for this later, but will make use of the idea here. The block diagram for the circuit of Fig. 2-2.2 is shown in Fig. 2-2.3. The response of the circuit to $u_0(t)$ is shown inside the box. Since the actual input is $10u_0(t)$, the output is 10 times $0.2\epsilon^{-10t}$.

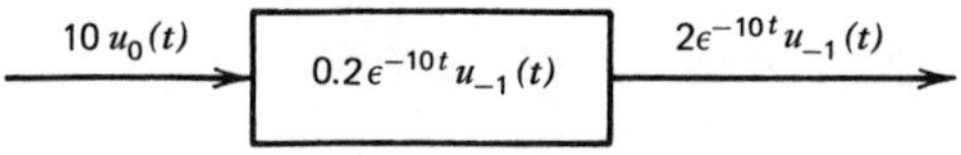

Fig. 2-2.3

Example 2-2.6. The circuit of Fig. 2-2.2 is driven by a voltage $e_s = 10u_{-1}(t)$ and numerical values are used for R and L. Rather than work the problem again, we simply substitute numerical values into Eq. 2-2.42:

$$i = 0.2\,(1 - \epsilon^{-10t})\,u_{-1}(t) \tag{2-2.69}$$

The block diagram for the circuit is shown in Fig. 2-2.4. Note that the system is still described by the unit impulse response.

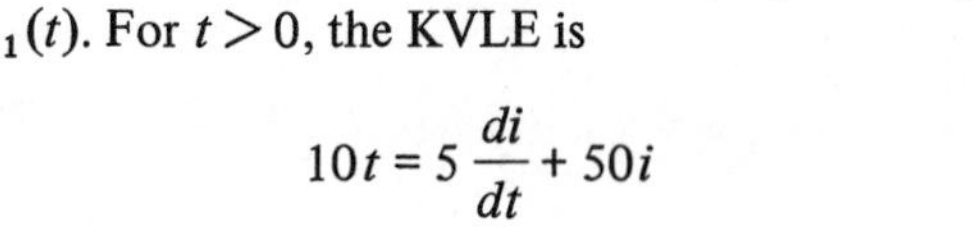

Fig. 2-2.4

Example 2-2.7. The circuit of Fig. 2-2.2 is driven by $e_s = 10u_{-2}(t)$ which can be expressed as $e_s = 10t\,u_{-1}(t)$. For $t > 0$, the KVLE is

$$10t = 5\frac{di}{dt} + 50i \tag{2-2.70}$$

The particular component of the solution is obtained from

$$10t = 5\frac{di_p}{dt} + 50i_p \tag{2-2.71}$$

An examination of this equation displays that the two terms on the right side must sum to $10t$. Therefore, i_p must contain a term Kt. However, this is insufficient, as can be seen from a trial effort by substituting

$$i_p = Kt \qquad \frac{di_p}{dt} = K \tag{2-2.72}$$

into Eq. 2-2.71

$$10t = 5K + 50Kt \tag{2-2.73}$$

If the coefficients of the t terms are matched on both sides of the equation

$$10 = 50K \qquad K = 0.2 \tag{2-2.74}$$

However, if the coefficients of the t^0 terms are matched

$$K = 0 \tag{2-2.75}$$

and a contradiction has been reached.

The i proposed in Eq. 2-2.72 has the correct slope, but must be shifted by an appropriate amount to match the $10t$ curve. A second try is

$$i_p = K_1 t + K_2 \qquad \frac{di_p}{dt} = K_1 \tag{2-2.76}$$

and upon substitution, Eq. 2-2.71 becomes

$$\begin{aligned} 10t &= 5[K_1] + 50[K_1 t + K_2] \\ 10t &= 50K_1 t + [5K_1 + 50K_2] \end{aligned} \tag{2-2.77}$$

Equating coefficients of the t^1 and t^0 on both sides and solving gives

$$K_1 = 0.2 \qquad K_2 = -0.02 \tag{2-2.78}$$

and the desired i_p is

$$i_p = 0.2t - 0.02 \tag{2-2.79}$$

The current $i = i_p + A\epsilon^{-10t}$, and with A determined is

$$i = (0.2t - 0.02 + 0.02\epsilon^{-10t})\, u_{-1}(t) \tag{2-2.80}$$

The block diagram for the circuit is shown in Fig. 2-2.5.

Example 2-2.8. The circuit of Fig. 2-2.2 is driven by $e_s = 10u_{-3}(t)$, which can be written as $e_s = 10t^2/2u_{-1}(t)$. For $t > 0$, the KVLE is

$$5t^2 = 5\frac{di}{dt} + 50i \tag{2-2.81}$$

The component i_p is found from

$$5t^2 = 5\frac{di_p}{dt} + 50i_p \tag{2-2.82}$$

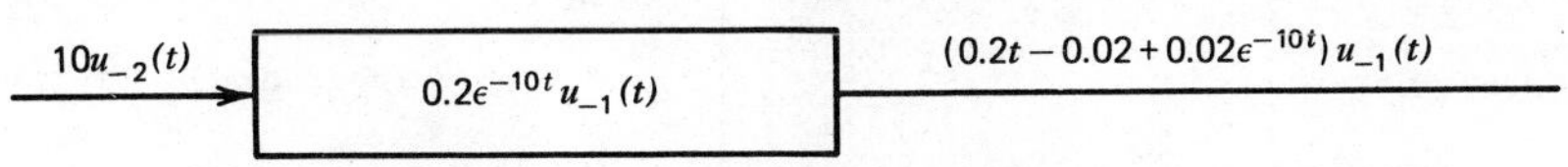

Fig. 2-2.5

The argument developed during the last example can be extended to show that i_p has the form

$$i_p = K_1 t^2 + K_2 t + K_3 \tag{2-2.83}$$

This equation is substituted into Eq. 2-2.82, and the three K's are determined. The i_p and the i_c are added, and the final equation is

$$i = (0.1t^2 - 0.02t + 0.002 - 0.002\epsilon^{-10t})\,u_{-1}(t) \tag{2-2.84}$$

THE INPUT-OUTPUT RELATIONSHIP

We could draw another block diagram for this example, but instead we display the relationships shown in Fig. 2-2.6 in order to bring out another idea.

Input 1 produces *Output 1*; in a similar manner, *Input 2* produces *Output 2*; etc. The impulse response of the system remains the same, because the system is not changed as the input changes.

Let us concentrate our attention on the inputs. *Input 2* is the integral of *Input 1*; *Input 3* is the first integral of *Input 2*, and the second integral of *Input 1*; etc.

Let us next look at the outputs. *Output 2* is the integral of *Output 1*, and to show this we integrate *Output 1* as

$$\int_0^t 2\epsilon^{-10t}\,dt = -0.2\epsilon^{-10t}\big|_0^t = 0.2\,(1 - \epsilon^{-10t})\,u_{-1}(t) \tag{2-2.85}$$

which checks *Output 2*; similarly, *Output 3* is the integral of *Output 2*; etc.

The reason for this correspondence between the integrating of both the input and the output is one of the properties of systems described by linear differential equations with constant coefficients. Later we will learn to find the output response of a system to any

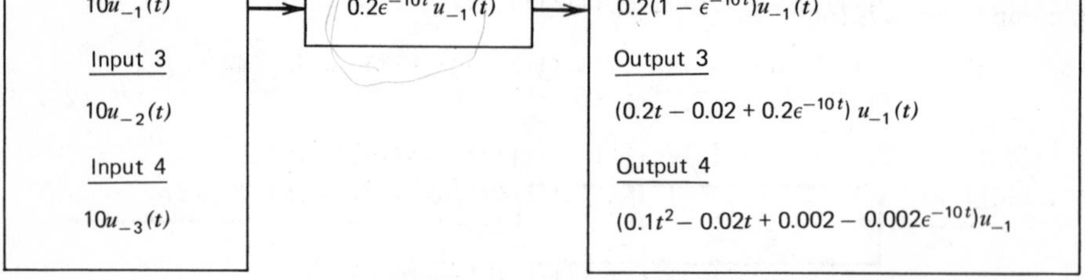

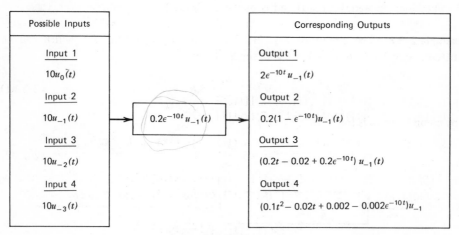

Fig. 2-2.6

input function when the impulse response of the system is known. This will be done by a procedure known as convolution, which is based on an integral.

In the previous discussion, the emphasis has been placed on integration. Obviously, we can move in the opposite direction through differentiation. For example, if *Input 4* produces *Output 4*, then the new input, which is the derivative of *Input 4* will produce a new response, which is the derivative of *Output 4*. To do this, differentiation must be performed in the generalized sense; that is, the derivative of a discontinuity produces an impulse, etc.

If for some reason a person is having difficulty determining the impulse response of a system, he can first find the step function response, and then take the derivative of this result to find the impulse response.

COMMENTARY

In the examples, we started with a specific circuit of a voltage source in series with a resistor and an inductor. The input is the voltage, and the output is a current. The impulse response of this circuit summarizes the input-output relation in an abstract sort of way. Figure 2-2.6 does not show the elements that make up the circuit, nor how the elements are arranged, but it does contain the information about the dynamic behavior of the circuit.

If we generalize our thinking, the impulse response of Fig. 2-2.6 can represent a large number of different circuits. In fact, the impulse response may represent a mechanical system, a hydraulic system, or a thermal system, among many other possibilities. The input may be a force, a pressure, or a heat source, and the output might be velocity, a position, or a temperature.

The theory of circuits goes beyond the study of specific collections of R, L, and C elements arranged in a certain manner. Circuit theory goes from the specific to the general, and develops ways of describing signals applied to a circuit and how these signals change as they progress through the circuit, etc. From this point of view, system theory is an extension of circuit theory. The main difference is in the physical interpretation of the equations.

2-3. CIRCUITS WITH ONE STORAGE ELEMENT: II–ADDITIONAL EXAMPLES

Two examples are included to show how sinusoidal sources can be handled.

Example 2-3.1. The same R–L circuit is again used except this time the driving function is a sinusoid, as shown in Fig. 2-3.1. For $t > 0$, the KVLE is

$$20 \sin (4t + 60^\circ) = 5 \frac{di}{dt} + 50i \tag{2-3.1}$$

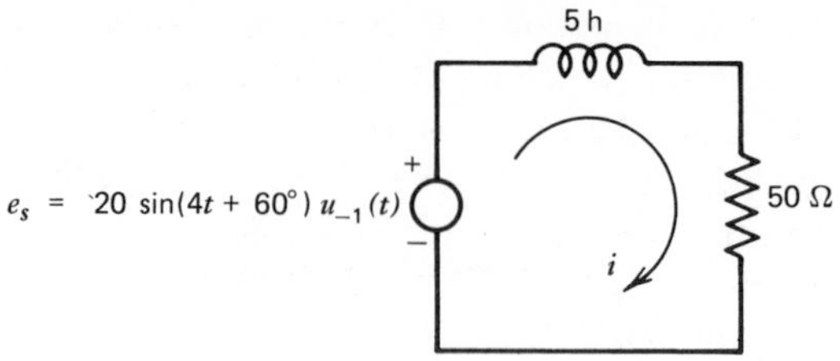

Fig. 2-3.1

The particular component of the solution is determined such that it satisfies the equation

$$20 \sin(4t + 60°) = 5 \frac{di_p}{dt} + 50 i_p \tag{2-3.2}$$

The sinusoid is a very unusual function; it is the only periodic function whose derivative and integral have the same shape as the original function. A function $f(t)$ is periodic if it satisfies the equation

$$f(t) = f(t + T)$$

where T is the period of the function.

The sinusoid has another important property. If two sinusoids of the same frequency (period) are added, the sum is also a sinusoid regardless of the amplitude and phase of the two waves being added.

With these thoughts in mind we return our attention to Eq. 2-3.2, which i_p must satisfy. If i_p is chosen to be a sinusoid, di_p/dt is also a sinsoid. Upon substitution, the two terms on the right are sinusoids; therefore, the sum is a sinusoid and can match the driving function on the left. The i_p is not in phase with the driving function, because there is a phase shift when the two terms on the right are added. With all these considerations in mind, the form for i_p is chosen as

$$i_p = K_1 \cos 4t + K_2 \sin 4t \tag{2-3.3}$$

$$\frac{di_p}{dt} = -4K_1 \sin 4t + 4K_2 \cos 4t \tag{2-3.4}$$

The left side of Eq. 2-3.2 is expanded, and Eqs. 2-3.3 and 4 are substituted into the right side, giving

$$10 \sin 4t + 17.32 \cos 4t = 5[-4K_1 \sin 4t + 4K_2 \cos 4t] + 50[K_1 \cos 4t + K_2 \sin 4t] \tag{2-3.5}$$

This equation is rewritten as

$$10 \sin 4t + 17.32 \cos 4t = (-20K_1 + 50K_2) \sin 4t + (50K_1 + 20K_2) \cos 4t \tag{2-3.6}$$

The corresponding coefficients of sine and cosine terms are equated

$$10 = -20K_1 + 50K_2 \qquad 17.32 = 50K_1 + 20K_2 \tag{2-3.7}$$

These equations are solved for K_1 and K_2

$$K_1 = 0.229 \qquad K_2 = 0.292 \tag{2-3.8}$$

and are substituted back into Eq. 2-3.3

$$i_p = 0.229 \cos 4t + 0.292 \sin 4t \tag{2-3.9}$$

which can finally be written as

$$i_p = 0.371 \sin (4t + 38.2^\circ) \tag{2-3.10}$$

To complete the solution, the particular and the complementary components are added, giving

$$i = i_p + i_c = 0.371 \sin (4t + 38.2^\circ) + A\,\epsilon^{-10t} \tag{2-3.11}$$

at $t = 0+$, $i(0+) = 0$ from which A is found as

$$0 = 0.371 \sin 38.2 + A \qquad \text{or} \qquad A = -0.23 \tag{2-3.12}$$

The final solution is

$$i = [0.371 \sin (4t + 38.2^\circ) - 0.23\epsilon^{-10t}]\, u_{-1}(t) \tag{2-3.13}$$

All of this work to find the particular component should seem very familiar to anyone who has studied circuit theory. The sinusoid is such an important driving function that special techniques have been developed to handle it. These techniques are called a-c circuit theory. We assume that the reader has some knowledge of a-c circuit theory; therefore, it is not our purpose to develop these techniques here, but rather to place them in the proper perspective in reference to what we are doing.

REVIEW OF A-C CIRCUIT METHODS

All a-c circuit methods have been developed as a systematic procedure to solve a differential equation that is driven by a sinusoidal function for the particular component of the solution. The a-c circuit methods are composed of three essential parts:

1. If the driving function (or functions) is given as a time function, the first thing to do is to represent this function as a phasor. We will refer to this operation as taking the direct phasor transform.
2. The proper phasor equations are written and manipulated to find the desired solution as a phasor.
3. The solution phasor is reinterpreted as a time function. We will refer to this operation as taking the inverse phasor transform.

We return to the example at hand to demonstrate these statements. The voltage driving this system is

$$e_s = 20 \sin (4t + 60^\circ) \tag{2-3.14}$$

Part 1. We take the direct phasor transform by representing this voltage with a phasor

$$\mathbf{E} = 20\underline{/60^\circ} \tag{2-3.15}$$

Note we are using the peak value for E, not the usual rms value.

Part 2.

$$Z = R + j\omega L = 50 + j(4)\,5 = 53.8\underline{/21.8^\circ} \tag{2-3.16}$$

$$\mathbf{I} = \frac{\mathbf{E}}{Z} = \frac{20\underline{/60^\circ}}{53.8\underline{/21.8^\circ}} = 0.371\underline{/38.2^\circ} \tag{2-3.17}$$

Part 3. We take the inverse phasor transform and label the current as i_p, since this result is the particular component of the solution

$$i_p = 0.371 \sin(4t + 38.2^\circ) \tag{2-3.18}$$

This result checks the one found previously as Eq. 2-3.10.

We could divide Eq. 2-3.15 by $\sqrt{2}$ to convert the E to an rms value. This would also yield I in Eq. 2-3.17 as an rms value, but when we convert back to peak value in Eq. 2-3.18 we would have to multiply by $\sqrt{2}$ and thus obtain the same result. Since these two operations with the $\sqrt{2}$ cancel, we just avoid the whole subject by staying with peak values all the way through the development.

Example 2-3.2. The same R-L circuit is again used, except this time we will consider a class of driving functions and only talk about the work in general terms. The class of function is indicated in Fig. 2-3.2 as a periodic non-sinusoidal function. If this function satisfies the Dirichlet conditions, it can be expanded into an infinite Fourier series, as shown by

$$e_s = \frac{a_0}{2} + \sum_{n=1}^{\infty} a_n \cos n\,\omega_1 t + \sum_{n=1}^{\infty} b_n \sin n\,\omega_1 t \tag{2-3.19}$$

Where the a's and b's are determined for the function in question. When this function is applied to the circuit, the equation for $t > 0$ is

$$\frac{a_0}{2} + \sum_{n=1}^{\infty} a_n \cos n\,\omega_1 t + \sum_{n=1}^{\infty} b_n \sin n\,\omega_1 t = 5\frac{di}{dt} + 50i \tag{2-3.20}$$

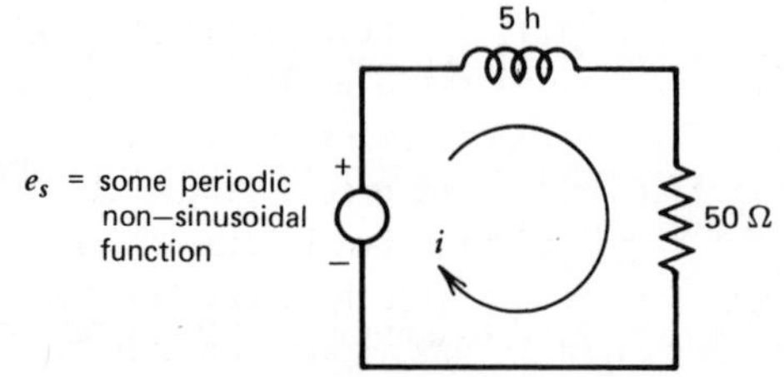

Fig. 2-3.2

The particular component of the solution can now be found as an infinite Fourier series. For each frequency of the applied Fouier series, a separate a-c circuit problem is solved in a manner similar to that shown in the previous example to find the corresponding term in the infinite series of the response. For the complete solution, the complementary component must be added to this, as indicated by

$$i = [i_{p1} + i_{p2} + i_{p3} + \cdots + A\,\epsilon^{-10t}]\,u_{-1}(t) \tag{2-3.21}$$

Obviously no one attempts to solve an infinite number of terms in i_p. Only enough terms are found as needed to satisfy the required accuracy for the specific problem at hand.

2-4. GENERAL COMMENTARY

In the last two sections, a number of examples are presented to demonstrate the procedure of using classical methods to solve linear differential equations with constant coefficients. Although the examples involve circuits that lead to first-order equations, a number of the general properties can be seen from these examples.

In each example, the particular component of the solution satisfies the differential equation for the particular driving function. The complementary component of the solution satisfies the homogeneous equation. The same *R-L* circuit was used in many of these examples to dramatize that the form of the complementary component does not change unless the homogeneous differential equation changes. All the previous examples with the same *R-L* circuit have the solution of the form

$$i = i_p + A\,\epsilon^{-10t} \tag{2-4.1}$$

The i_p varies, and A varies, but the form $i_c = A\,\epsilon^{-10t}$ remains the same.

Table 2-4.1 summarizes some of these thoughts. The first column lists all the driving functions we have used so far, plus a few others. The second column lists the form of each particular component of the solution that accompanies the corresponding driving function. Although the only examples presented thus far are for first-order equations, these forms hold for the general *n*th-order equation. Obviously, these first two columns could be extended, but those items listed suggest how other particular components can be found.

The third column is added to emphasize that the form of the complementary component does not depend on the driving function, but only on the homogeneous equation.

Table 2-4.2 is included to mention that other names are often used as equivalent to particular and complementary components. The first column lists three terms that are used as equivalents. The term *steady-state response* is used by electrical engineers to mean the *particular component*. This term can be used appropriately for *Items 1*, *7*, *8*, and *9* in Table 2-4.1 because all these responses last indefinitely. However, the term seems inappropriate for the items that die out with time. The term *forced response* is used to indicate that the driving function forces this response onto the system, and hence it is the *particular component*.

Table 2-4.1

Item no.	Form of the driving function applied at $t = 0$	Form of the particular component of the solution	Form of the complementary component of the solution
1	A	K	
2	$A\epsilon^{-at}$	$K\epsilon^{-at}$	
3	At	$K_1 t + K_2$	
4	At^2	$K_1 t^2 + K_2 t + K_3$	
5	At^n	$K_1 t^n + K_2 t^{n-1} + \cdots + K_{n+1}$	For a given differential equation the form of i_c remains the same regardless of the form of the driving function.
6	$[A_1 t^n + A_2 t^{n-1} + \cdots + A_n t + A_{n+1}]$	$K_1 t^n + K_2 t^{n-1} + \cdots + K_{n+1}$	
7	$A \sin \omega t$	$B_1 \cos \omega t + B_2 \sin \omega t$	
8	$A \cos \omega t$	$B_1 \cos \omega t + B_2 \sin \omega t$	
9	Nonsinusoidal periodic function	An infinite Fourier series of sinusoidal terms	
10	$A\epsilon^{-at} \sin \omega t$	$\epsilon^{-at}[K_1 \cos \omega t + K_2 \sin \omega t]$	
11	$A\epsilon^{-at} \cos \omega t$	$\epsilon^{-at}[K_1 \cos \omega t + K_2 \sin \omega t]$	
12	$At\epsilon^{-at}$	$\epsilon^{-at}[K_1 t + K_2]$	

Table 2-4.2

Particular component	Complementary component
Steady-state response	Transient response
Forced response	Natural response

The second column also lists three terms that are equivalent. The term *transient response* is used by electrical engineers. The term *natural response* is used to indicate that this is the way the system responds independent of the driving function.

2-5. ONE STORAGE ELEMENT IMBEDDED IN AN ALL-RESISTIVE NETWORK

If a circuit is composed of a number of sources and resistors but has only one storage element, the current or voltage associated with this element can be found with the aid of Thevenin's theorem or Norton's theorem and the equations already derived.

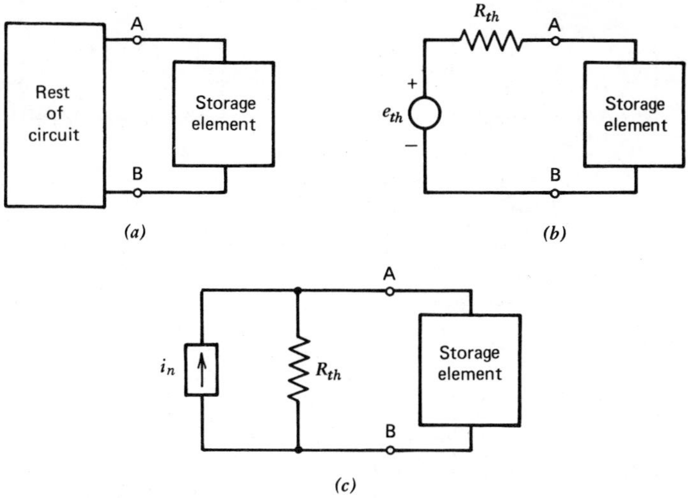

Fig. 2-5.1

The entire circuit can be divided into (1) the storage element and (2) the rest of the circuit, as shown in Fig. 2-5.1(a). When Thevenin's theorem is used, the storage element is removed and the voltage appearing across the open terminals A–B is determined. This voltage is called e_{th}. All the sources are removed, leaving their internal impedance, and the resistance looking back into terminals A–B is determined and is called R_{th}. The rest of the circuit can be replaced by e_{th} in series with R_{th}, and, so far as conditions external to terminals A–B are concerned, the new circuit will yield identical results with the original circuit. Since the storage element is external to the transformation, it may be added as shown in Fig. 2-5.1(b), and the voltage and current associated with the storage element in (b) will be identical with the voltage and current in (a).

In a similar manner, the original circuit of Fig. 2-5.1(a) can be replaced by that shown in (c) through use of Norton's theorem.

Example 2-5.1. The circuit shown in Fig. 2-5.2(a) is in steady-state with the switch S opened, and S is closed at $t = 0$. The desired solution is the current i in the 2-henry inductor.

All the concepts of circuit theory can be used in solving this circuit. For example, the principle of superposition can be used to find $i(0-)$. In Fig. 2-5.2(b) the voltage source is applied, and the current source is removed, leaving its internal impedance, which is infinite. The inductor acts as a short in steady-state, and i_1 (the component of $i(0-)$ due to the voltage source) is

$$i_1 = \frac{12}{6} = 2 \text{ amp} \tag{2-5.1}$$

The component of $i(0-)$ due to the current source is determined from the circuit shown in Fig. 2-5.2(c), where the voltage source is removed, leaving its internal impedance, a short. Due to the symmetry of the circuit, the current of 6 amperes divides into

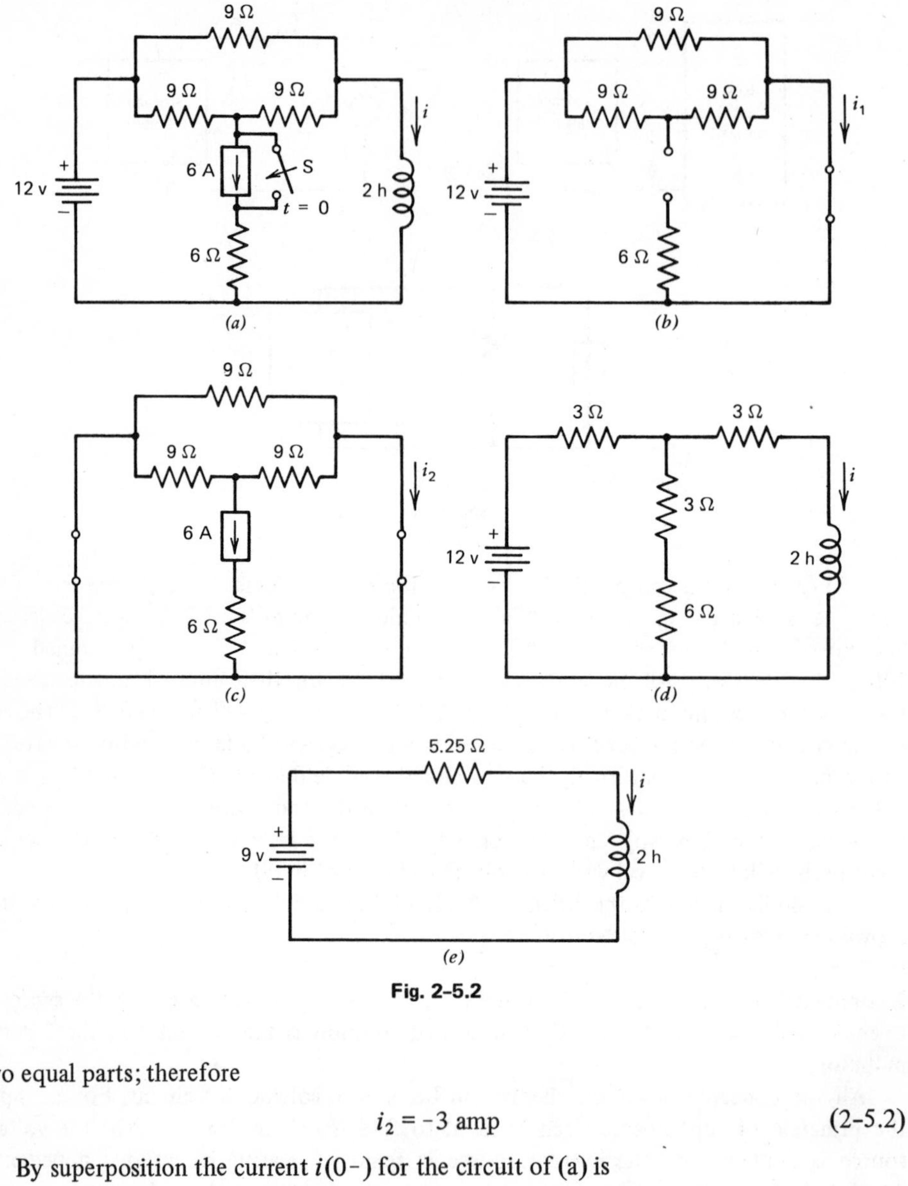

Fig. 2-5.2

two equal parts; therefore

$$i_2 = -3 \text{ amp} \tag{2-5.2}$$

By superposition the current $i(0-)$ for the circuit of (a) is

$$i(0-) = i_1 + i_2 = 2 - 3 = -1 \text{ amp} \tag{2-5.3}$$

Because of the inductor, this becomes $i(0+)$.

After S is closed, the resulting circuit could be solved with three loop currents. However, we replace the Δ of 9-Ω resistors by the equivalent Y of 3-Ω resistors as shown in (d). Next, the inductor is removed, and the battery and resistive part of the circuit is replaced by the Thevenin's theorem, and (e) results. Since the inductor is external to all these transformations, the current i in the circuit of (e) is the same as i in (a). The current has the form

$$i = \frac{9}{5.25} + A\,\epsilon^{-5.25t/2} = 1.72 + A\,\epsilon^{-2.63t} \tag{2-5.4}$$

The A is determined from

$$i(0+) = -1 = 1.72 + A \quad \text{or} \quad A = -2.72 \tag{2-5.5}$$

The desired solution is

$$i = 1.72 - 2.72^{-2.63t} \quad \text{for } t > 0 \tag{2-5.6}$$

2-6. ONE RESISTOR IMBEDDED IN A NETWORK CONTAINING ONLY L OR ONLY C ELEMENTS

If a circuit is composed of any number of sources and storage elements all of one type, but with only one resistor, the current or voltage associated with the resistor can be found with the aid of Thevenin's theorem or Norton's theorems and the equations already derived.

The entire circuit can be divided into the resistive element and the rest of the circuit, as shown in Fig. 2-6.1(a). When Thevenin's theorem is used, the resistor is removed and e_{th}, the voltage appearing across the open terminals A-B, is determined. All the sources are removed, leaving their internal impedance. When looking back into the A-B terminals, this circuit reduces to one equivalent storage element. The resulting Thevenin's theorem circuit is shown in Fig. 2-6.1(b), and the corresponding Norton's theorem circuit in (c).

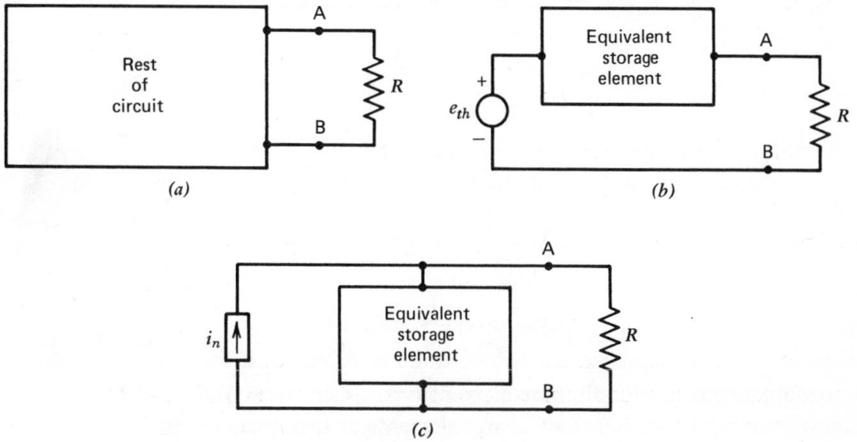

Fig. 2-6.1

Table 2-6.1. Circuits Containing Elements of Only One Kind

Equivalent Value of n Elements in Series		
The R element	The L element	The C element
$R_T = R_1 + R_2 + \cdots + R_n$	$L_T = L_1 + L_2 + \cdots + L_n$	$S_T = S_1 + S_2 + \cdots + S_n$
Equivalent Value of n Elements in Parallel		
The R element	The L element	The C element
$G_T = G_1 + G_2 + \cdots + G_n$	$\Gamma_T = \Gamma_1 + \Gamma_2 + \cdots + \Gamma_n$	$C_T = C_1 + C_2 + \cdots + C_n$
Voltage Division Across Elements in Series (See Fig. 1-6.2a for explanation of notation)		
The R element	The L element	The C element
$v_{R2} = \dfrac{R_2}{R_1 + R_2 + \cdots + R_n} v_t$	$v_{L2} = \dfrac{L_2}{L_1 + L_2 + \cdots + L_n} v_t$	$v_{S2} = \dfrac{S_2}{S_1 + S_2 + \cdots + S_n} v_t$
Current Division Among Elements in Parallel (See Fig. 1-6.2b for Explanation of Notation)		
The R element	The L element	The C element
$i_{R2} = \dfrac{G_2}{G_1 + G_2 + \cdots + G_n} i_t$	$i_{L2} = \dfrac{\Gamma_2}{\Gamma_1 + \Gamma_2 + \cdots + \Gamma_n} i_t$	$i_{C2} = \dfrac{C_2}{C_1 + C_2 + \cdots + C_n} i_t$

Table 2-6.1 is included to summarize how circuits containing only elements of one kind can be manipulated. This table shows that the R, the L, and the S elements behave in an analogous manner, as do the G, the Γ and the C, where

$$G = \frac{1}{R} \quad \Gamma = \frac{1}{L} \quad \text{and} \quad C = \frac{1}{S} \tag{2-6.1}$$

If a voltage v_t is placed across n elements of the same kind in a series as shown in Fig. 2-6.2(a), the voltage across each of the elements has exactly the same wave shape as v_t and differs only in magnitude. The voltage across the second element is used as an example, with the assumption that the elements are all inductors:

$$v_2 = \frac{L_2}{L_1 + L_2 + \cdots + L_n} v_t \tag{2-6.2}$$

The corresponding result for all three element types is shown in Table 2-6.1.

Likewise, if a current i_t is sent into n elements of the same type in parallel, as shown in Fig. 2-6.2(b), the current into each of the elements has exactly the same shape as i_t

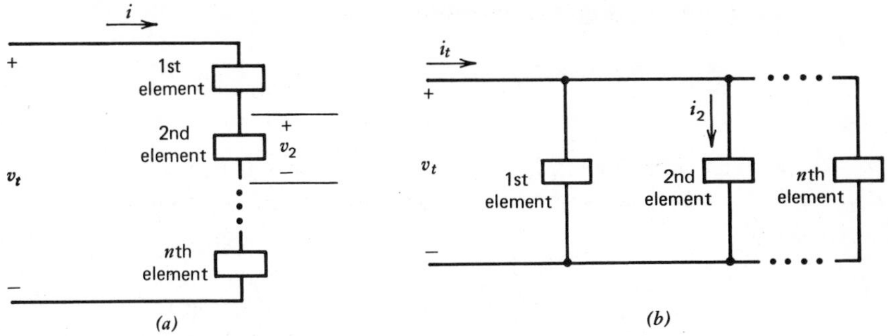

Fig. 2-6.2

and differs only in magnitude. The current into the second element is used as an example, with the assumption that all elements are inductors:

$$i_2 = \frac{\Gamma_2}{\Gamma_1 + \Gamma_2 + \cdots + \Gamma_n} i_t \tag{2-6.3}$$

The corresponding results for all three element types are shown in Table 2-6.1.

If the circuit is too complicated to solve by using series and parallel combination, more general methods, such as writing the proper set of KVLE's or KCLE's are needed.

Example 2-6.1. The desired solution in the circuit of Fig. 2-6.3(a) is the current i_R. The circuit of (a) is a two loop-current problem and could be solved this way. However, Thevenin's theorem reduces this to a one loop-current problem as shown in (b). To use Thevenin's theorem, the resistor is removed, and the voltage e_{th} across the A-B terminals can be found from

$$e_{th} = \frac{S_2}{S_1 + S_2} e_s = \frac{C_1}{C_1 + C_2} e_s = \frac{15}{45} e_s = 3.33 u_{-1}(t) \tag{2-6.4}$$

The Thevenin's theorem capacitance is that looking into the A-B terminals, with e_s replaced by a short:

$$C_{th} = C_1 + C_2 = 45\ \mu f \tag{2-6.5}$$

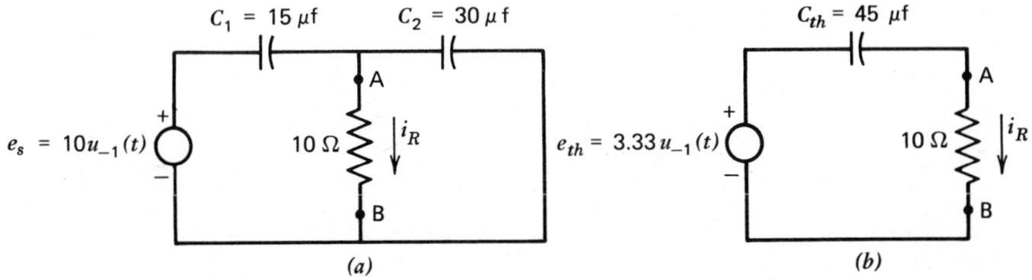

Fig. 2-6.3

We have solved the resulting circuit of (b) before, i_R is

$$i_R = \frac{E}{R} \epsilon^{-(t/RC)} u_{-1}(t) = 0.333\epsilon^{-2,222t} u_{-1}(t) \tag{2-6.6}$$

Example 2-6.2. The desired solution in the circuit of Fig. 2-6.4(a) is the current i_R. To find e_{th}, the 50-Ω resistor is removed in (b), and, as a preliminary step: the voltages e_{L2} and e_{L4} are found.

$$e_{L2} = \frac{L_2}{L_1 + L_2} e_s = \frac{6}{3+6} e_s = 40 \sin(4t + 60^\circ) u_{-1}(t) \tag{2-6.7}$$

$$e_{L4} = \frac{L_4}{L_3 + L_4} e_s = \frac{4.5}{9+4.5} e_s = 20 \sin(4t + 60^\circ) u_{-1}(t) \tag{2-6.8}$$

The voltage e_{th} is

$$e_{th} = e_{L2} - e_{L4} = 20 \sin(4t + 60^\circ) u_{-1}(t) \tag{2-6.9}$$

The Thevenin's theorem inductance is found from the circuit shown in (c) as

$$L_{th} = \frac{L_1 L_2}{L_1 + L_2} + \frac{L_3 L_4}{L_3 + L_4} = 5 \text{ h} \tag{2-6.10}$$

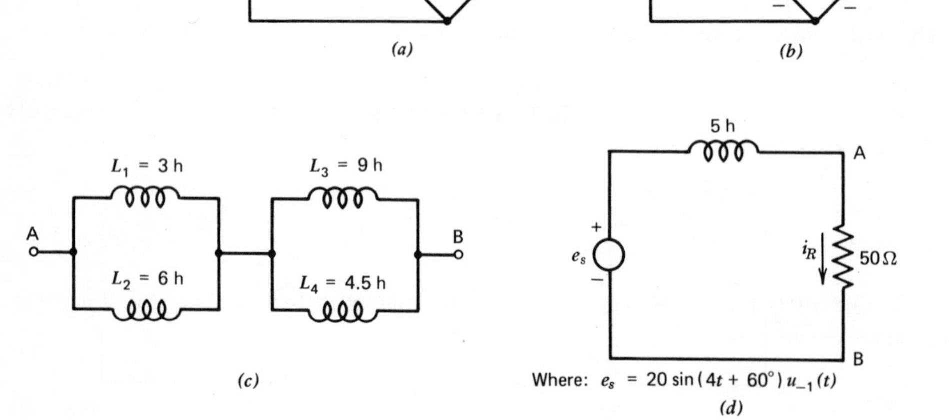

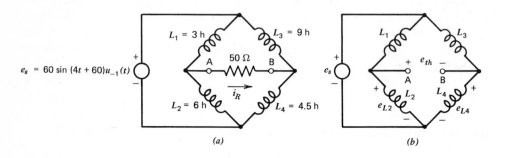

Fig. 2-6.4

The Thevenin's theorem circuit in (d) is exactly the circuit of Ex. 2-3.1. The solution for i_R, found before as Eq. 2-3.13, is

$$i_R = 0.371 \sin(4t + 38.2°) - 0.23\epsilon^{-10t} \tag{2-6.11}$$

2-7. THE GENERAL FORM FOR THE SOLUTION TO A FIRST-ORDER SYSTEM

If a circuit is described by a first-order differential equation, the desired response is either a current or a voltage of the form

$$i = i_p + A\,\epsilon^{-t/t_c} \qquad v = v_p + B\,\epsilon^{-t/t_c} \tag{2-7.1}$$

If the circuit is an R-L type, the time constant t_c is

$$t_c = \frac{L}{R} \tag{2-7.2}$$

or, if it is an R-C type

$$t_c = RC \tag{2-7.3}$$

The A and the B can be determined from the known conditions as

$$\begin{aligned} i(0+) &= i_p(0+) + A \quad \text{or} \quad A = i(0+) - i_p(0+) \\ v(0+) &= v_p(0+) + B \quad \text{or} \quad B = v(0+) - v_p(0+) \\ i &= i_p + [i(0+) - i_p(0+)]\,\epsilon^{-t/t_c} \\ v &= v_p + [v(0+) - v_p(0+)]\,\epsilon^{-t/t_c} \end{aligned} \tag{2-7.4}$$

It is certainly not desirable to learn a subject by substituting into the proper formula, but this type of problem is so simple it is difficult not to remember these forms.

SOLUTION TO THE FIRST-ORDER SYSTEM, USING THE GENERAL FORM

With the use of Eqs. 2-7.4 and a knowledge of Thevenin's theorem, many circuits can be solved without actually drawing the Thevenin's theorem circuit. The next two examples demonstrate this idea.

Example 2-7.1. The circuit shown in Fig. 2-7.1 is in steady-state, with S1 and S2 closed, and both switches are opened at $t = 0$. We wish to find the equations for both v_{C1} and v_{C2} after $t = 0$.

Before $t = 0$ in steady-state, the 300 volts divides into three equal parts and the capacitors maintain the voltages at $t = 0+$, as

$$\begin{aligned} v_{C1}(0+) &= 200 \text{ v} \\ v_{C2}(0+) &= 100 \text{ v} \end{aligned} \tag{2-7.5}$$

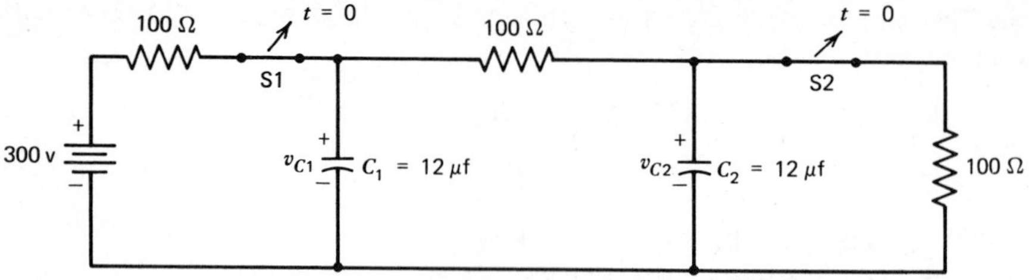

Fig. 2-7.1

The current i_R could be determined by using Thevenin's theorem circuit, and from this v_{C1} and v_{C2} could be determined through integration. However, we find v_{C1} and v_{C2} directly by use of the equation

$$v_C = v_{ss} + [v(0+) - v_{ss}]\,\epsilon^{-t/RC_{th}} \tag{2-7.6}$$

By use of the principle of conservation of charge, v_{ss} is found from

$$C_1\,v_{C1}(0+) + C_2\,v_{C2}(0+) = (C_1 + C_2)\,v_{ss} \tag{2-7.7}$$

where v_{ss} is voltage at $t = \infty$ on both C_1 and C_2. For this example, v_{ss} is

$$v_{ss} = 150\text{ v} \tag{2-7.8}$$

The capacitance C_{th} is made up of C_1 and C_2 in series and has a value

$$C_{th} = 6\,\mu f \tag{2-7.9}$$

When these pieces of information are collected, v_{C1} becomes

$$v_{C1} = 150 + [200 - 150]\,\epsilon^{-t/(100\times 6\times 10^{-6})} = 150 + 50\epsilon^{-1667t} \tag{2-7.10}$$

In a similar manner, v_{C2} is

$$v_{C2} = 150 + [100 - 150]\,\epsilon^{-1667t} = 150 - 50\epsilon^{-1667t} \tag{2-7.11}$$

With a little experience, solutions such as these can almost be written by inspection.

Example 2-7.2. In the circuit of Fig. 2-7.2, S is opened at $t = 0$ and the desired solution is the voltage v across current source. At $t = 0+$, the current in the inductor is zero, and 2

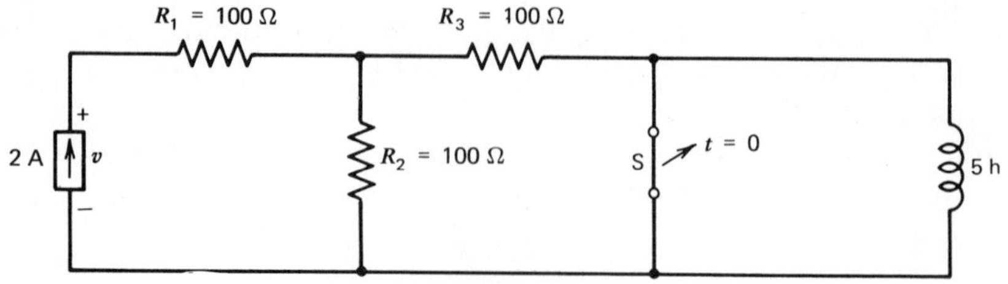

Fig. 2-7.2

amperes exists in R_1 and R_2 in series; therefore, $v(0+)$ is equal to 400 volts. At $t = \infty$, the inductor L behaves as a short, and $v_{ss} = 300$ volts. The R_{th} is R_2 and R_3 in series, or $R_{th} = 200\ \Omega$. The solution for v is

$$v = 300 + [400 - 300]\ \epsilon^{-(200/5)t} = 300 + 100\epsilon^{-40t} \qquad (2\text{-}7.12)$$

2-8. SWITCHING WITHOUT NEED FOR A COMPLEMENTARY COMPONENT OF THE SOLUTION

The particular component of the solution satisfies the differential equation, but since it may not satisfy the initial conditions, the complementary component is needed. Stated another way, if the particular component satisfies the initial condition, there is no need for the complementary component.

Two simple examples are shown in Fig. 2-8.1. In (a), the capacitor has E volts on it before $t = 0$, and when S is closed, the circuit is already in steady-state, hence

$$i = 0 \qquad v_C = E \qquad (2\text{-}8.1)$$

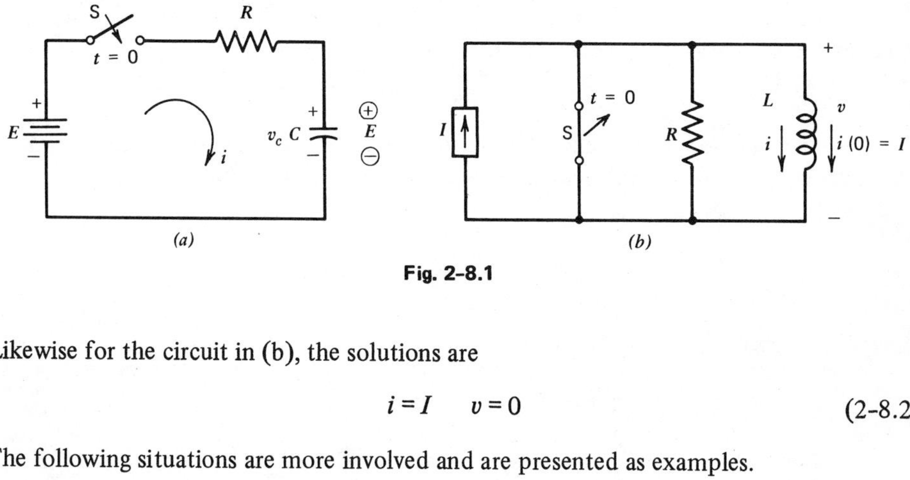

Fig. 2-8.1

Likewise for the circuit in (b), the solutions are

$$i = I \qquad v = 0 \qquad (2\text{-}8.2)$$

The following situations are more involved and are presented as examples.

Example 2-8.1. In the circuit of Fig. 2-8.2, the switch S is thrown to *Position 1* until steady-state is reached, and at $t = 0$, S is thrown to *Position 2*. The desired solution is to find a value of E such that the complementary component of the solution is zero. The form for the solution is

$$i = i_p + [i(0+) - i_p(0+)]\ \epsilon^{-10t} \qquad (2\text{-}8.3)$$

If $i(0+) = i_p(0+)$, then the solution reduces to

$$i = i_p \qquad (2\text{-}8.4)$$

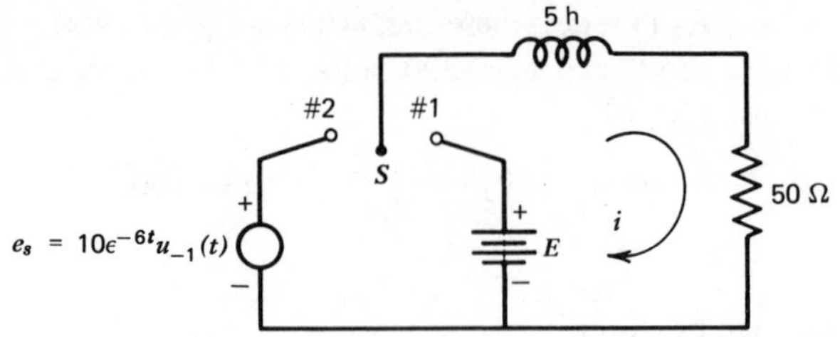

Fig. 2-8.2

If S were thrown directly to *Position 2*, the problem is identical to Ex. 2-2.4, where i_p was found to be

$$i_p = 0.5\epsilon^{-6t} \tag{2-8.5}$$

For the complementary component of the solution to be zero

$$i(0+) = i_p(0+) = 0.5 \tag{2-8.6}$$

Finally, for $i(0+)$ to have this value, E must equal

$$E = Ri(0+) = 50(0.5) = 25 \text{ v} \tag{2-8.7}$$

Example 2-8.2. This example uses the circuit of Fig. 2-8.2 except

$$e_s = 20 \sin (4t + \delta) \tag{2-8.8}$$

The circuit is initially at rest, and S is thrown to *Position 2*. The desired solution is to find two values of δ such that the complementary component of the solution is zero.

After $t = 0$, the equation is

$$e_s = 20 \sin (4t + \delta) = 5\frac{di}{dt} + 50i \tag{2-8.9}$$

The i_p is found by using a-c circuit theory

$$E = 20\underline{/\delta}$$

$$I = \frac{E}{Z} = \frac{20\underline{/\delta}}{50 + j(4)(5)} = 0.371\underline{/\delta - 21.8^\circ} \tag{2-8.10}$$

From which i_p is determined as

$$i_p = 0.371 \sin (4t + \delta - 21.8^\circ) \tag{2-8.11}$$

For $i_c = 0$, then

$$i_p(0+) = i(0+) = 0 = 0.371 \sin (\delta - 21.8^\circ) \tag{2-8.12}$$

and the two values of δ are determined as

$$\begin{aligned} \delta_1 - 21.8^\circ &= 0^\circ \quad \text{or} \quad \delta_1 = 21.8^\circ \\ \delta_2 - 21.8^\circ &= 180^\circ \quad \text{or} \quad \delta_2 = 201.8^\circ \end{aligned} \tag{2-8.13}$$

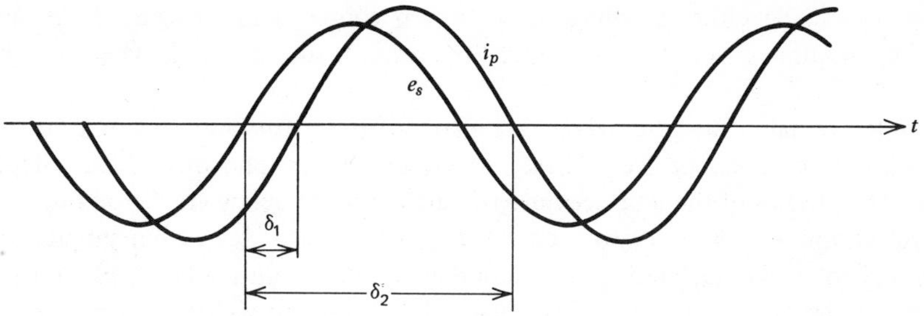

Fig. 2-8.3

If the switch S is closed at one of these points in the cycle, the i_p not only satisfies the differential equation but also satisfies the initial conditions, and there is no need for a complementary component of the solution.

Figure 2-8.3 is presented to help visualize this situation. The applied voltage e_s and the steady-state component of the current i_p are sketched. The values of δ_1 and δ_2 are indicated on this figure.

COMMENTARY

Examples 2-8.1 and 2 featured the inductor. If $i(0+)$ in the inductor is equal to $i_p(0+)$, then the equation

$$i = i_p + [i(0+) - i_p(0+)]\, \epsilon^{-Rt/L} \tag{2-8.14}$$

reduces to

$$i = i_p \tag{2-8.15}$$

Although the capacitor is not considered in these examples, the same situation exists for the *R-C* circuit. The most convenient quantity to work with in this type of circuit is the q (or voltage) on the capacitor. This type of problem is left for exercises at the end of the chapter.

There are a number of situations in which the ideas suggested by these examples are very important. One is when a large inductive load is being connected to a power system. If the load is thrown onto the system at the wrong point in the cycle, a large transient results. If it is thrown onto the system at such a point in the cycle that $i_c = 0$, the load disturbs the system by the minimum amount.

2-9. DEPENDENT SOURCES

The sources that have been used so far are known as *independent sources*, which means that the magnitude of the source does not depend on anything external to the source itself. A *dependent* (or controlled) *source* on the other hand, is a source whose magnitude depends on the current or the voltage at some other point in the circuit. Since a dependent source can be either a voltage or a current source whose magnitude in turn may depend upon either a voltage or a current, the result is four different types of dependent sources.

In order to match the external characteristics of transistors and vacuum tubes, models are developed for these devices. These models usually contain one or more dependent sources. The physics of these devices and the subject of modeling are beyond the scope of this book. Some of the results of dependent sources can be seen from simple examples and do add to the understanding of system theory. If a circuit contains only independent sources, the circuit is said to be passive whereas if it contains dependent sources, the circuit is said to be active. Since the dependent sources feed energy into the circuit from other locations, such as from the power supply, the circuit may become unstable, as discussed in the examples.

Example 2-9.1. The circuit shown in Fig. 2-9.1(a) is at rest when e_s is applied at $t = 0$. The branch currents are indicated by j's. The voltage source shown in series with the 3-Ω resistor is a dependent source in that this voltage depends on the current j_2.

In some earlier examples, the circuit to the left of Points A and B was replaced by the use of Thevenin's theorem and a one-loop-current problem resulted. We cannot do this here, because the current j_2 would be internal to the Thevenin-theorem equivalent circuit, and we must know j_2 to find the voltage of the dependent source. Therefore, we keep the circuit in its original form, and choose i_1 and i_2, as shown in Fig. 2-9.1(b). The two KVLE's for $t > 0$ are

$$10 = 15i_1 + 10i_2 \qquad 10 - Ki_1 = 10i_1 + 13i_2 + 4\frac{di_2}{dt} \tag{2-9.1}$$

The current i_2 is eliminated between these two equations, yielding

$$6\frac{di_1}{dt} + (9.5 - K)\,i_1 = 3 \tag{2-9.2}$$

The characteristic equation is

$$6s + (9.5 - K) = 0 \tag{2-9.3}$$

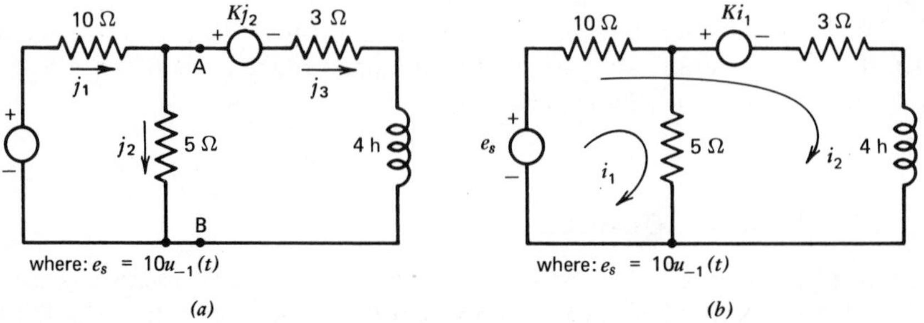

Fig. 2-9.1

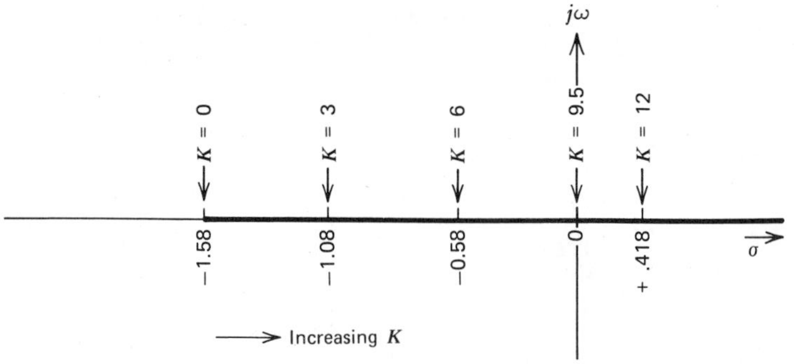

Fig. 2-9.2

and the root is

$$s_1 = -\frac{(9.5 - K)}{6} \tag{2-9.4}$$

Without worrying about the physics of the device, we assume that K can take on a range of values. An examination of the root s_1 given by Eq. 2-9.4 shows that the root depends on the value of K. This idea is portrayed in Fig. 2-9.2. When $K = 0$, the root is located at $s_1 = -1.58$. As K is increased, the root moves to the right along the portion of the real axis shown by the heavy line. This heavy line is called the root-locus. The specific values for $K = 3$ and $K = 6$ are shown. When $K = 9.5$, the root is at the origin, and when $K > 9.5$ the root is in the right half s-plane.

Equation 2-9.2 can be solved for i_1 as

$$i_1 = \frac{3}{9.5 - K} + \frac{10 - 2K}{3(9.5 - K)}\,\epsilon^{-(9.5-K)t/6} \tag{2-9.5}$$

The solution for i_2 is

$$i_2 = \frac{(5 - K)}{(9.5 - K)} + \frac{(K - 5)}{(9.5 - K)}\,\epsilon^{-(9.5-K)t/6} \tag{2-9.6}$$

The two currents for the various values of K are shown in Table 2-9.1.

The equations for $K = 0$ show the way the circuit would respond if the dependent source were not present. For values of K smaller than 9.5, the exponent in the complementary component of the solutions is negative, and hence the complementary component of the current eventually goes to zero.

However, if K is greater than 9.5, this exponent is positive, and the complementary component increases without bound with increasing time. The basic reason for this is that the dependent source is feeding energy into the system faster than it is being dissipated or stored in the circuit. Such a system is said to be unstable because the input to the system no longer has control of the output.

Root-locus methods, as this term is used in the literature, contain many facets not covered in this example. In control system theory, the concept of open-loop and closed-

Table 2-9.1

K	Equation for i_1	Equation for i_2
0	$0.316 + 0.351\epsilon^{-1.58t}$	$0.527 - 0.527\epsilon^{-1.58t}$
3	$0.462 + 0.205\epsilon^{-1.08t}$	$0.308 - 0.308\epsilon^{-1.08t}$
5	0.667	0
9.5	$0.667 + 0.5t$	$-0.75t$
12	$-1.2 + 1.867\epsilon^{+0.418t}$	$2.8 - 2.8\epsilon^{+0.418t}$

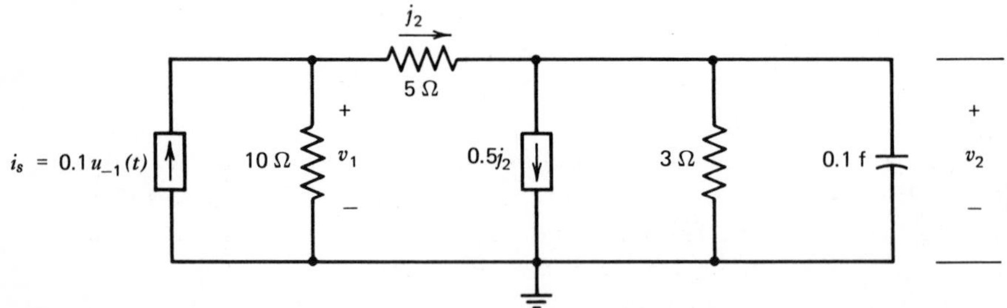

Fig. 2-9.3

loop systems is fundamental. This is discussed briefly in Chapter 5. When root-locus methods are applied to the general subject of control system theory, the roots of a closed-loop system are studied by starting with a knowledge of the corresponding open-loop system and by applying a well-defined set rules. A complete discussion of this subject is beyond the scope of this book.[1]

Example 2-9.2. The circuit as shown in Fig. 2-9.3 is at rest when i_s is applied at $t = 0$. The current source shown in parallel with the 3-Ω resistor is a dependent source in that the current depends on j_2. Node-to-datum voltages are shown on the circuit.

The desired solution is the voltage on the capacitor, which is v_2. The two KCLE's for $t > 0$ are

$$0.1 = \left(\frac{1}{10} + \frac{1}{5}\right) v_1 - \frac{1}{5} v_2$$
$$-0.5j_2 = -\frac{1}{5} v_1 + \left(\frac{1}{5} + \frac{1}{3}\right) v_2 + 0.1 \frac{dv_2}{dt} \tag{2-9.7}$$

[1] See *Control System Theory*, by Lago and Benningfield, The Ronald Press, Chapter 11.

The value of the current j_2

$$j_2 = \frac{(v_1 - v_2)}{5} \tag{2-9.8}$$

is substituted into Eqs. 2-9.7 to yield

$$\begin{aligned} 0.1 &= 0.3v_1 - 0.2v_2 \\ 0 &= -0.1v_1 + 0.433v_2 + 0.1\frac{dv_2}{dt} \end{aligned} \tag{2-9.9}$$

These two equations are solved as

$$0.0333 = 0.366v_2 + 0.1\frac{dv_2}{dt} \tag{2-9.10}$$

The final solution is

$$v_2 = 0.0908 - 0.0908\epsilon^{-3.366t} \tag{2-9.11}$$

This circuit is explored in more detail in the problems.

2-10. NON-LINEAR SYSTEMS

Non-linear systems are much more difficult to analyze than linear systems. Most of the time, special techniques are required to solve a non-linear problem, and the techniques change with the nature of the problem. Quite often numerical methods are used along with the digital computer. The following two examples are presented as an introduction to this important area of study.

Example 2-10.1. Very few non-linear systems can be solved for in closed form. By this is meant a solution that can be written as a function. This circuit example is one of the exceptions.

The circuit to be analyzed is shown in Fig. 2-10.1. The resistive element is non-linear, and the volt-ampere relationship is defined to be

$$v = 5i^2 \tag{2-10.1}$$

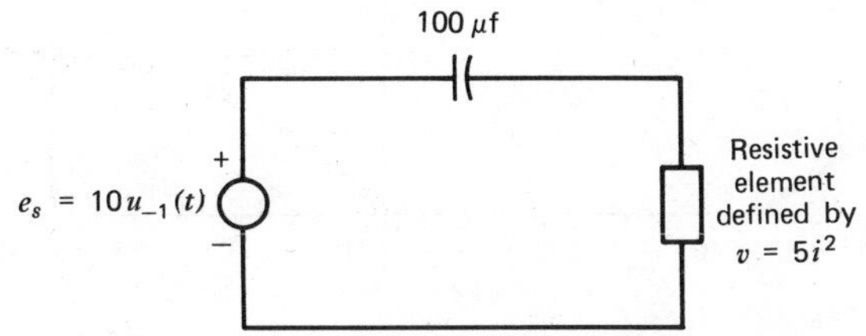

Fig. 2-10.1

The KVLE after $t = 0$ is

$$10 = \frac{1}{10^{-4}} \int_0^t i\,dt + 5i^2 \tag{2-10.2}$$

This equation is differentiated and solved for di as

$$0 = 10^4 i + 10i \frac{di}{dt} \qquad di = -10^3\,dt \qquad \text{(2-10.3) (2-10.4)}$$

Upon integration, the result is

$$i = -10^3 t + A \tag{2-10.5}$$

To find $i(0+)$, we return to Eq. 2-10.2, written at $t = 0+$ as

$$10 = 0 + 5[i(0+)]^2 \tag{2-10.6}$$

and solve for $i(0+)$

$$i(0+) = \sqrt{2} \tag{2-10.7}$$

The final solution is

$$i = -10^3 t + \sqrt{2} \tag{2-10.8}$$

At first glance it seems that as t goes to infinity, the solution goes to negative infinity, but an inspection of the circuit shows that this doesn't make sense. The original equation, Eq. 2-10.2, is correct only until the voltage on the capacitor reaches 10 volts, at which time i becomes zero. Therefore, Eq. 2-10.8 can be set equal to zero, and this time can be determined as

$$t = \sqrt{2} \times 10^{-3}$$

A sketch of i is shown in Fig. 2-10.2(a), and the capacitor voltage in (b).

Example 2-10.2. The circuit to be analyzed is shown in Fig. 2-10.3(a). The resistance is non-linear, and the graph in (b) is a plot of its volt-ampere relationship.

The approximate solution to this problem can be obtained by using linear circuit analysis in the following manner. The curve in (b) can be approximated by a number of

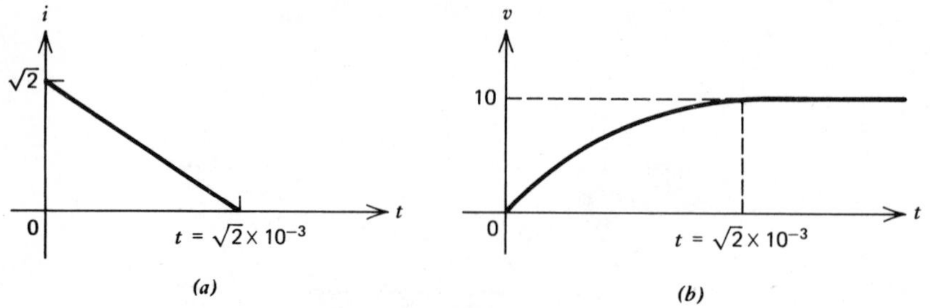

Fig. 2-10.2

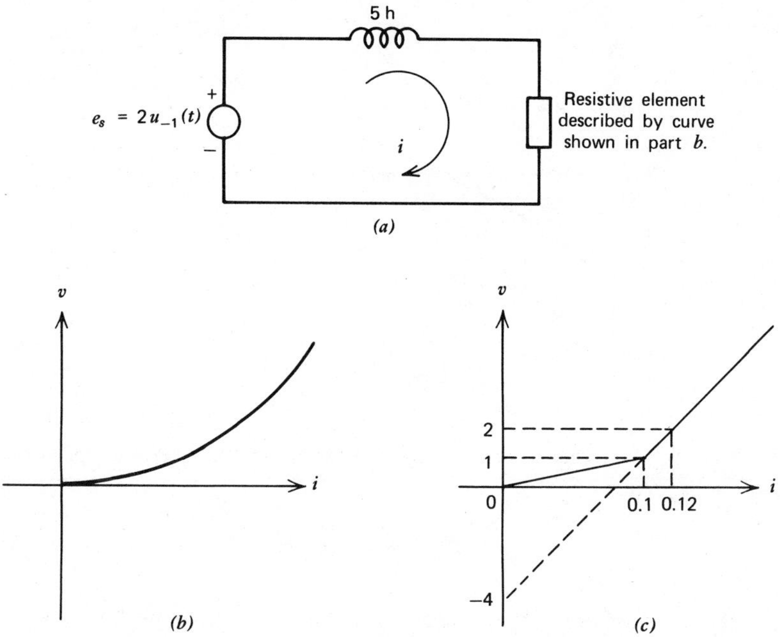

Fig. 2-10.3

segments of straight lines. The more segments taken, the better the approximation, but also the greater the amount of work. To show the method with a minimum of detail, only two segments were used. A certain amount of judgment is required in approximating the curve with two straight lines. The lines chosen are those shown in (c). The straight-line approximation assumes the resistance is constant over that portion of the curve. The value of the resistance is v/i or the slope of the straight line.

When the current i is between $0 < i < 0.1$, R has a value given by

$$R = \frac{v}{i} = \frac{1}{0.1} = 10\ \Omega$$

The differential equation is

$$2 = 5\frac{di}{dt} + 10i \qquad (2\text{-}10.9)$$

The solution to this equation is

$$i = \frac{E}{R}(1 - \epsilon^{-Rt/L}) = 0.2(1 - \epsilon^{-2t}) \qquad (2\text{-}10.10)$$

We find such a time t_1 that $i = 0.1$

$$0.1 = 0.2(1 - \epsilon^{-2t_1}) \quad \text{or} \quad t_1 = 0.347 \qquad (2\text{-}10.11)$$

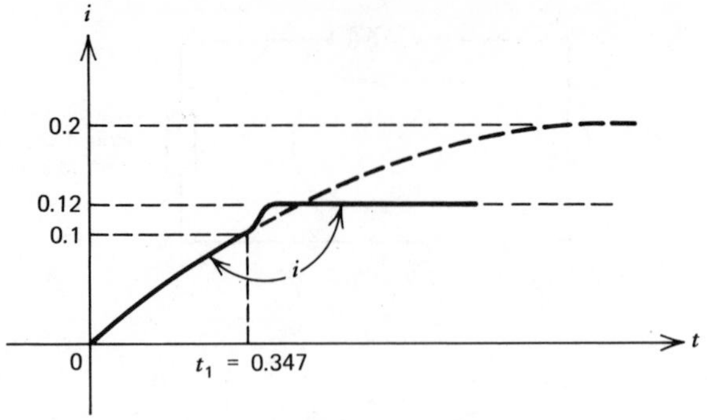

Fig. 2-10.4

After $t = t_1$, we use the second straight segment to approximate the volt-ampere curve for the resistance.

$$R = \frac{\Delta v}{\Delta i} = \frac{1}{0.02} = 50\ \Omega \tag{2-10.12}$$

Since this segment does not pass through the origin, the v-axis intercept is found to be $v = -4$, and the equation for the volt-ampere relationship is

$$v_R = -4 + 50i \tag{2-10.13}$$

The equation for $t > 0.347$ becomes

$$2 = 5\frac{di}{dt} - 4 + 50i \qquad \text{or} \qquad 6 = 5\frac{di}{dt} + 50i \tag{2-10.14}$$

The form for the solution is

$$i = 0.12 + A\,\epsilon^{-10t} \qquad t > 0.347 \tag{2-10.15}$$

At $t = 0.347$, the current $i = 0.1$:

$$0.1 = 0.12 + A\,\epsilon^{-3.47} \qquad \text{or} \qquad A = -0.64 \tag{2-10.16}$$

The two solutions are brought together and written as

$$\begin{aligned} i &= 0.2(1 - \epsilon^{-2t}) \qquad 0 < t < 0.347 \\ i &= 0.12 - 0.64\epsilon^{-10t} \qquad 0.347 < t \end{aligned} \tag{2-10.17}$$

These solutions are sketched in Fig. 2-10.4.

PROBLEMS

2-1 The circuit shown is at rest before $t = 0$. Find the equation for the voltage v; from this find i_R and i_L; and check your work by showing $i_s = i_R + i_L$.

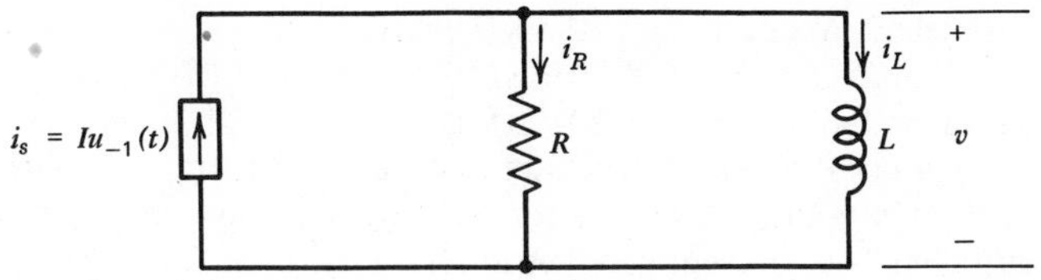

2-2 The circuit shown is at rest before $t = 0$. Write the differential equation and solve in terms of q. From this find i, v_R, and v_C, and check your work by showing $e_s = v_R + v_C$.

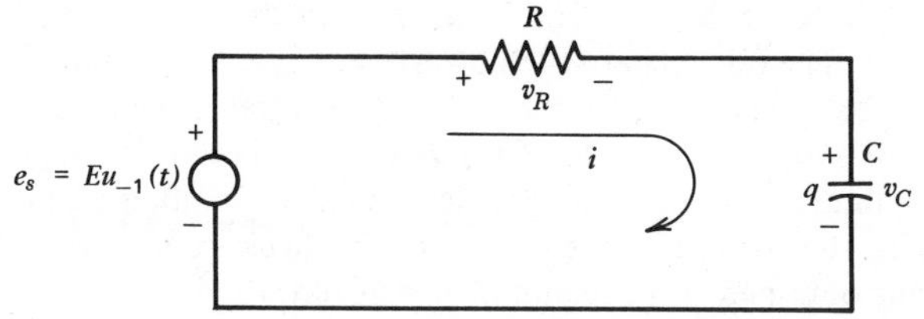

2-3 Given the circuit of Problem 2-1: Write the KCLE in the conventional manner, then define a function λ as $\lambda = \int_0^t v\,dt$ and rewrite the KCLE in terms of λ. Solve for λ; find i_L, v, and i_R; and check your work by showing that $i_s = i_R + i_L$.

2-4 The circuit as shown is at rest before $t = 0$. Write the KCLE, and solve for v. From this solve for i_R, and i_C, and check your work by showing that $i_s = i_R + i_C$.

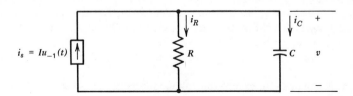

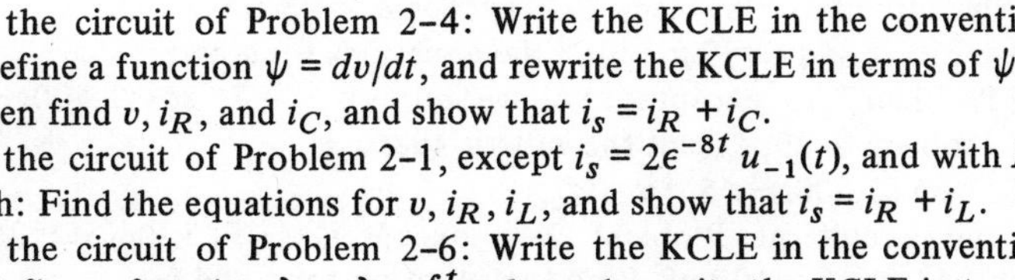

2-5 Given the circuit of Problem 2-4: Write the KCLE in the conventional manner; then define a function $\psi = dv/dt$, and rewrite the KCLE in terms of ψ. Solve for ψ, and then find v, i_R, and i_C, and show that $i_s = i_R + i_C$.

2-6 Given the circuit of Problem 2-1, except $i_s = 2\epsilon^{-8t}\, u_{-1}(t)$, and with $R = 50\ \Omega$ and $L = 5$ h: Find the equations for v, i_R, i_L, and show that $i_s = i_R + i_L$.

2-7 Given the circuit of Problem 2-6: Write the KCLE in the conventional manner; then define a function λ as $\lambda = \int_0^t v\,dt$, and rewrite the KCLE in terms of λ. Solve for λ, and then find i_L, v, and i_R and show that $i_s = i_R + i_L$.

2-8 Given the circuit of Problem 2-4, except $i_s = 2\epsilon^{-8t}\, u_{-1}(t)$, and with $R = 1{,}000\ \Omega$ and $C = 100\ \mu$f: Find the equations for v, i_R, and i_C, and show that $i_s = i_R + i_L$.

2-9 Repeat Problem 2-8, except $i_s = 20u_0(t)$.

2-10 Repeat Problem 2-8, except $i_s = 20u_{-1}(t)$.

2-11 Repeat Problem 2-8, except $i_s = 20u_{-2}(t)$.

2-12 Repeat Problem 2-8, except $i_s = 20u_{-3}(t)$.

2-13 Repeat Problem 2-6, except $i_s = 20u_{-1}(t)$.

2-14 Repeat Problem 2-6, except $i_s = 20u_{-2}(t)$.
2-15 Repeat Problem 2-6, except $i_s = 20u_{-3}(t)$.
2-16 Repeat Problem 2-6, except $i_s = 20u_{-4}(t)$.
2-17 Given the circuit of Problem 2-2, except $e_s = 100\epsilon^{-4t}u_{-1}(t)$, and with $R = 1{,}000\ \Omega$ and $C = 100\ \mu$f: Find the equations for i, v_R, v_C and show that $e_s = v_R + v_C$.
2-18 Repeat Problem 2-17, except $e_s = 100u_{-1}(t)$.
2-19 Repeat Problem 2-17, except $e_s = 100u_{-2}(t)$.
2-20 Repeat Problem 2-17, except $e_s = 100u_{-3}(t)$.
2-21 Repeat Problem 2-17, except $e_s = 100u_{-4}(t)$.
2-22 Develop a figure similar to that of Fig. 2-2.6 for the circuit of Problem 2-8, with the various inputs obtained from the problems as given by: *Input 1* from Problem 2-9; *Input 2* from Problem 2-10; *Input 3* from Problem 2-11; and *Input 4* from Problem 2-12. The outputs are the corresponding voltage v outputs from this same set of problems. Show that *Output 2* is the integral of *Output 1*; that *Output 3* is the integral of *Output 2*; and that *Output 4* is the integral of *Output 3*.
2-23 Extend the figure developed in Problem 2-22 by adding an input $i_s = 20u_{+1}(t)$, and then find the corresponding output v. To check your work, take the derivative of the output obtained from the input $i_s = 20u_0(t)$.
2-24 Repeat Problem 2-22, except use the circuit of Problem 2-6 and the inputs as given by: *Input 1* from Problem 2-13; *Input 2* from Problem 2-14; *Input 3* from Problem 2-15; and *Input 4* from Problem 2-16. Extend this figure by adding an input $20u_0(t)$, and then find the corresponding output v. To check your work, take the derivative of *Output 1*. This answer divided by 20 should be placed in the block showing the unit impulse response of the system.
2-25 Repeat Problem 2-22, except use the circuit of Problem 2-17 and the inputs as given by: *Input 1* from Problem 2-18; *Input 2* from Problem 2-19; *Input 3* from Problem 2-20; and *Input 4* from Problem 2-21. Extend this figure by adding an input $100u_0(t)$. To check, take the derivative of *Output 1*.
2-26 Given the circuit of Fig. 2-2.2, except the voltage source is $e_s = 20 \cos(6t + 20^\circ)\, u_{-1}(t)$: Solve for the current i, using the element values as given.
2-27 Repeat Problem 2-26, except the voltage source is $e_s = 15 \sin(8t - 25^\circ)\, u_{-1}(t)$.
2-28 Repeat Problem 2-26, except the voltage source is $e_s = -25 \cos(10t - 45^\circ)\, u_{-1}(t)$.
2-29 Given the circuit of Problem 2-2, except $e_s = 75 \sin(1{,}000t - 35^\circ)\, u_{-1}(t)$, and with $R = 150\ \Omega$ and $C = 5\ \mu$f: Find the equations for q and i.
2-30 Repeat Problem 2-29, except the voltage source is $e_s = 75 \cos(1{,}500t + 25^\circ)\, u_{-1}(t)$.
2-31 Repeat Problem 2-29, except the voltage source is $e_s = -125 \sin(2{,}000t - 65^\circ)\, u_{-1}(t)$.
2-32 Given the circuit of Problem 2-4, except $i_s = -0.25 \sin(5t - 25^\circ)$ and with $R = 1{,}000\ \Omega$ and $C = 100\ \mu$f: Find the equation for v and i_C.
2-33 Repeat Problem 2-32, except $i_s = +0.25 \cos(10t + 35^\circ)\, u_{-1}(t)$.
2-34 Repeat Problem 2-32, except $i_s = [0.15 \cos 15t + 0.20 \sin 15t]\, u_{-1}(t)$.
2-35 A non-sinusoidal periodic voltage source is applied to the circuit of Problem 2-26. The first three terms of the Fourier series of this wave shape are given as $e_s = [10 + 15 \cos 8t + 10 \cos 16t + \cdots]\, u_{-1}(t)$. Find the current i for this wave shape. Obviously, i_p can be carried out to only three terms.
2-36 Repeat Problem 2-35, except the voltage source is $e_s = [15 - 20 \sin 6t + 12 \sin 12t + \cdots]\, u_{-1}(t)$.

2-37 Repeat Problem 2.35, except the voltage source is $e_s = [5 - (10 \cos 5t) + (7 \cos 15t) + \cdots] u_{-1}(t)$.

2-38 A non-sinusoidal periodic current source is applied to the circuit of Problem 2-32. The first three terms of the Fourier series of this wave shape are given as $i_s = [0.5 - 0.35 \sin 3t - 0.15 \cos 9t + \cdots] u_{-1}(t)$. Find the voltage v for this wave shape. Obviously, v_p can be carried out to only three terms.

2-39 Repeat Problem 2-38, except the current source is $i_s = [0.3 + 0.25 \cos 5t - 0.15 \cos 10t + \cdots] u_{-1}(t)$.

2-40 The following voltage is applied to the circuit of Fig. 2-2.2: $e_s = 10\epsilon^{-2t} \sin 4t\, u_{-1}(t)$. Find the equation for i.

2-41 Repeat Problem 2-40, except $e_s = 10\epsilon^{-3t} \cos 6t\, u_{-1}(t)$.

2-42 Repeat Problem 2-40, except $e_s = \epsilon^{-2t}\,[10 \sin 4t + 20 \cos 4t]\, u_{-1}(t)$.

2-43 Repeat Problem 2-40, except $e_s = 10t\, \epsilon^{-4t}\, u_{-1}(t)$.

2-44 Repeat Problem 2-40, except $e_s = \epsilon^{-2t}[2t + 4]\, u_{-1}(t)$.

2-45 Repeat Problem 2-40, except $e_s = 10t^2\, \epsilon^{-4t}\, u_{-1}(t)$.

2-46 Given the circuit as shown: The switch S is opened until steady-state conditions are reached, and S is closed at $t = 0$. Find the equation for i_L after $t = 0$. Do this by using Thevenin's theorem.

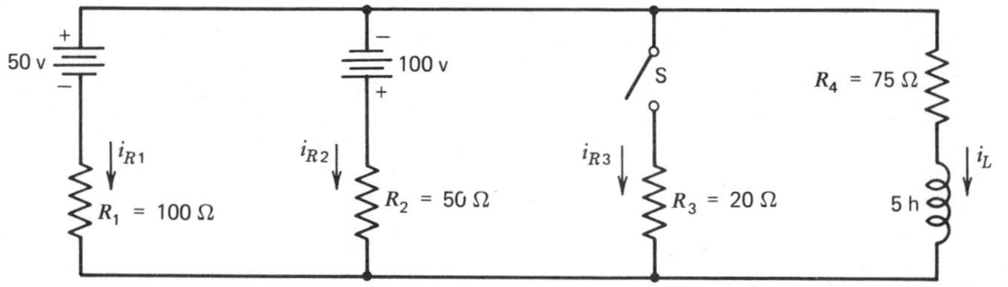

2-47 Repeat Problem 2-46, except use Norton's Theorem.

2-48 Repeat Problem 2-46, except use the methods suggested in Section 2.7.

2-49 As a continuation of Problem 2-46, find the equation for the current i_{R3}. Note that R_3 is internal to the Thevenin's theorem circuit. Return to the original circuit given in Problem 2-46, and find the unknown quantities from the known.

2-50 As a continuation of Problem 2-46, find i_{R3} by the methods suggested in Section 2.7.

2-51 Given the circuit as shown in Problem 2-46: The switch S is closed until steady state conditions are reached, and S is opened at $t = 0$. Find the equation for i_L after $t = 0$. Do this by using Thevenin's theorem.

2-52 Repeat Problem 2-51, except use Norton's Theorem.

2-53 Repeat Problem 2-51, except use the methods suggested in Section 2.7.

2-54 As a continuation of Problem 2-51, find the equation for the current i_{R2}. See Problem 2-49 for a comment.

2-55 As a continuation of Problem 2-51, find i_{R2} by the methods suggested in Section 2.7.

2-56 Find i_R in Example 2-6.1 using Norton's theorem.

2-57 Find i_R in Example 2-6.1 using window currents i_1 and i_2, and then find $i_R = i_1 - i_2$.

2-58 Find i_R in Example 2-6.2 using Norton's Theorem.

2-59 The circuit shown is at rest, and S is opened at $t = 0$. Find the equation for i_R using Thevenin's theorem.

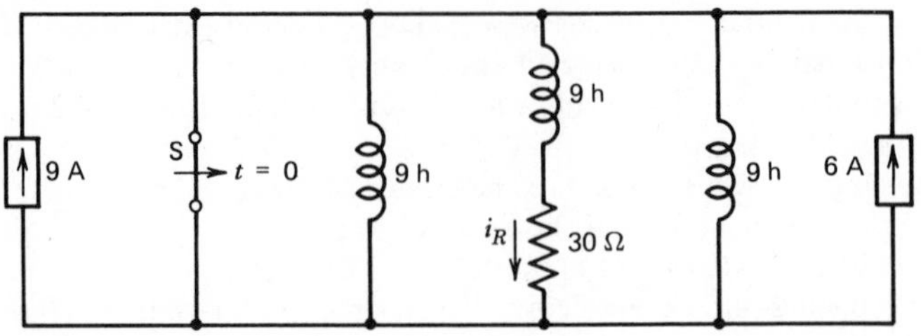

2-60 Repeat Problem 2-59, except find i_R using Norton's Theorem.

2-61 Find i_R in Problem 2-59 by the methods in Section 2.7.

2-62 In the circuit shown, the switch S is thrown to *Position 1* until steady-state is reached; and at $t = 0$, S is thrown to *Position 2*. Find the equations for q and i, and sketch when $E = 25$ volts and $e_s = 75 \sin(1{,}000t - 35^\circ)$.

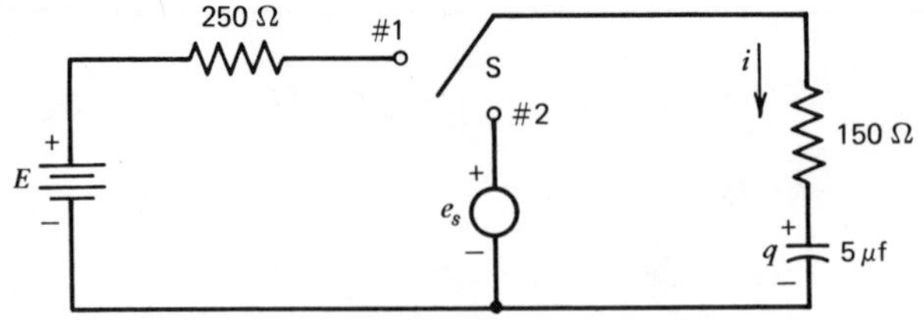

2-63 Given the circuit of Problem 2-62, with $E = 0$ and $e_s = 75 \sin(1{,}000t + \delta)$: Find such two values of δ that the complementary component of the solution is zero.

2-64 Repeat Problem 2-63, except $E = 25$, and $e_s = 75 \sin(1{,}000t + \delta)$.

2-65 For the circuit of Problem 2-62, E is a variable and $e_s = 75 \sin(1{,}000t + \delta)$. Find such maximum value of E that the complementary component of the solution is zero. Find the one value of δ corresponding to this E.

2-66 For the circuit of Problem 2-62, E is a variable and $e_s = 100\epsilon^{-1{,}000t}$. Find such value of E that the complementary component of the solution is zero.

2-67 Repeat Problem 2-66 for $e_s = 100t^2$.

2-68 In the circuit shown, the switch S is thrown to *Position 1* until steady-state is reached, and at $t = 0$, S is thrown to *Position 2*. Find the equations for v and i_L, and sketch when $I = 5$ amp and $i_s = 10 \cos(10t + 50^\circ)$.

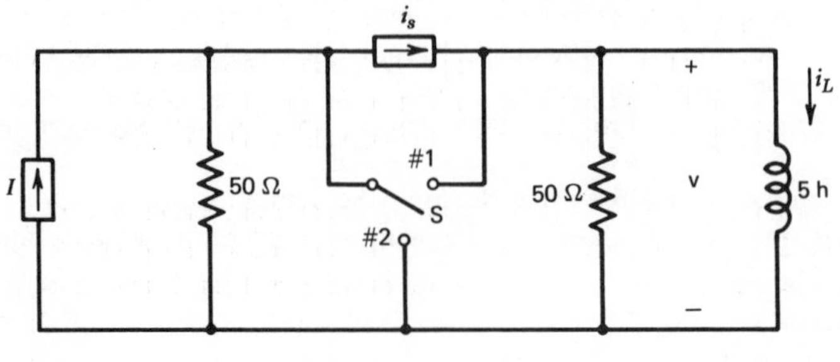

2-69 Given the circuit of Problem 2-68, with $I = 0$ and $i_s = 10 \cos(10t + \delta)$: Find such two values of δ that the complementary component of the solution is zero.

2-70 Repeat Problem 2-69, except $I = 5$ and $i_s = 10 \cos(10t + \delta)$.

2-71 For the circuit of Problem 2-68, I is a variable and $i_s = 10 \cos(10t + \delta)$. Find such maximum value of I that the complementary component of the solution is zero. Find the one value of δ corresponding to this I.

2-72 For the circuit of Problem 2-68, I is a variable and $i_s = 10\epsilon^{-6t}$. Find such value of I that the complementary component of the solution is zero.

2-73 The circuit shown is at rest before $t = 0$. Find i_R after $t = 0$.

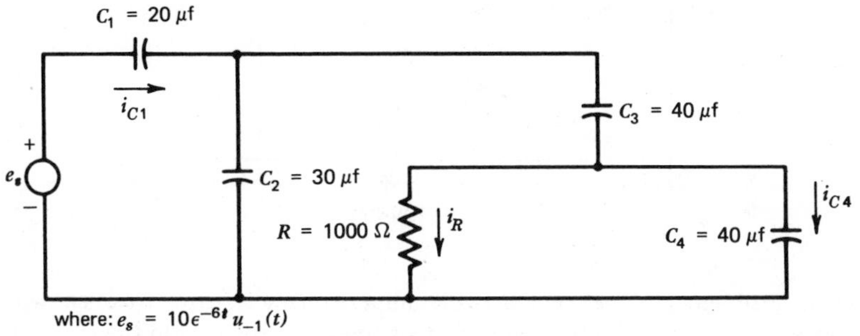

2-74 Repeat Problem 2-73, except find the current i_{C4}.

2-75 The circuit shown is at rest before $t = 0$. Find i_R after $t = 0$.

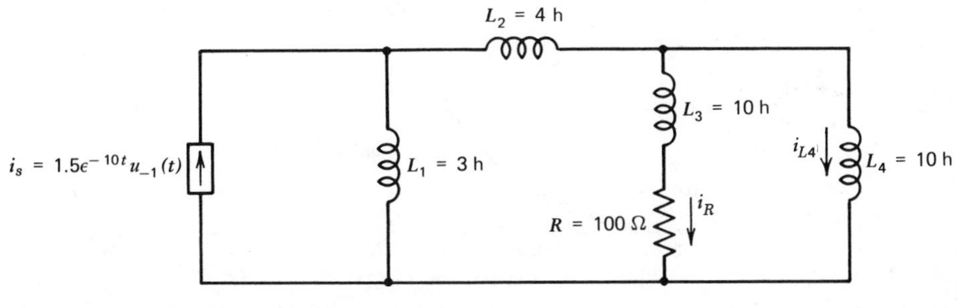

2-76 Repeat Problem 2-75, except find the current i_{L4}.

2-77 In the circuit of Fig. 2-9.3 the dependent source $0.5j_2$ is changed to Kj_2, where K is a positive constant. Discuss the behavior of the circuit as K takes on a range of values. Draw a root-locus to aid in the discussion. Find value of K (denoted K_c) for which the circuit becomes unstable. Set $K = \frac{1}{2} K_c$ and find the equation for v_2 and repeat for $K = 2K_c$.

2-78 Given the circuit of Fig. 2-9.1, except the polarity of the dependent voltage source is reversed. Find the equation for i as a function of K. Analyze and discuss the behavior of circuit as K takes on a range of values. Draw a root-locus to aid in the discussion.

2-79 Given the circuit Fig. 2-9.3, except the dependent current sources is now Kv_1. Find the equation of the voltage v_2 as a function of K. Analyze and discuss the behavior of the circuit as K takes on a range of values. Draw a root-locus to aid in the discussion.

2-80 Repeat Problem 2-79, except reverse the direction of the current in the dependent current source.

2-81 This problem uses the circuit of Fig. 2-10.3, except the volt-ampere characteristic for the non-linear element and the two straight line approximations are as shown. Find the two equations for i, and sketch.

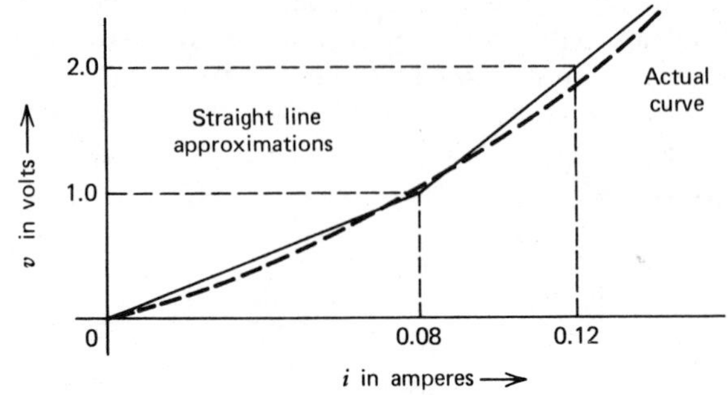

2-82 Repeat Problem 2-81, except the volt-ampere characteristics are as shown and the curve is approximated with three straight lines as shown. Find the three equations for i, and sketch.

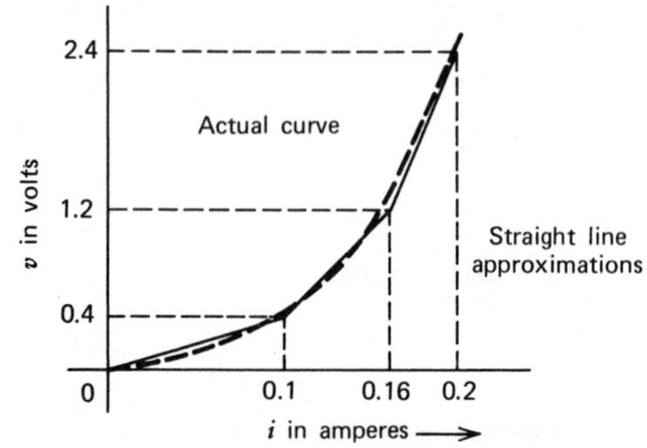

2-83 The volt-ampere characteristics in Problem 2-81 also describes the non-linear resistance in the circuit shown. Find the two equations for v, and sketch.

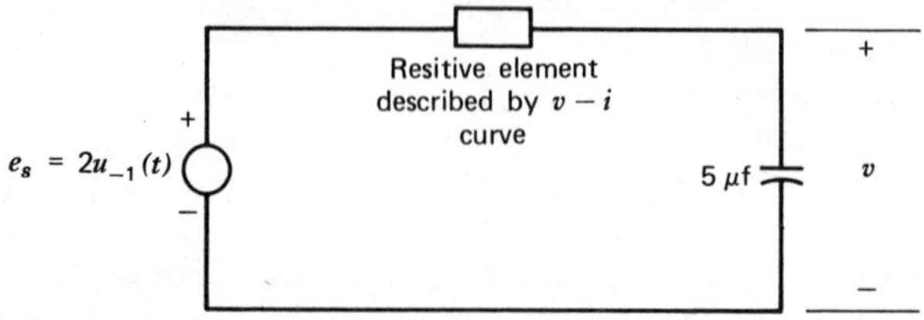

2-84 Same as Problem 2-83, except the volt-amperes characteristics are those shown in Problem 2-82. Find the three equations for v, and sketch.

2-85 Example 2-10.1 is one of the few non-linear circuits for which solutions can be obtained in a closed form. Plot the volt-ampere curve described by the non-linear resistance as given in this example. Approximate this volt-ampere characteristics between $v = 0$ and $v = 10$ volts by using three straight lines. Obtain the equation for i for each of these three parts of the problem, and sketch.

Chapter **3**

Systems Described by Second and Higher-Order Differential Equations

Chapter 3 extends the use of classical methods of solving differential equations as developed in Chapter 2 to systems described by second- and higher-order differential equations.

3-1. THE *R-L-C* SERIES CIRCUIT

The KVLE for the *R-L-C* series circuit of Fig. 3-1.1 for $t > 0$ is

$$e_s = L\frac{di}{dt} + Ri + \frac{1}{C}\int_0^t i\,dt + v(0) \tag{3-1.1}$$

To remove the integral sign, the derivative is taken

$$\frac{de_s}{dt} = L\frac{d^2i}{dt^2} + R\frac{di}{dt} + \frac{i}{C} \tag{3-1.2}$$

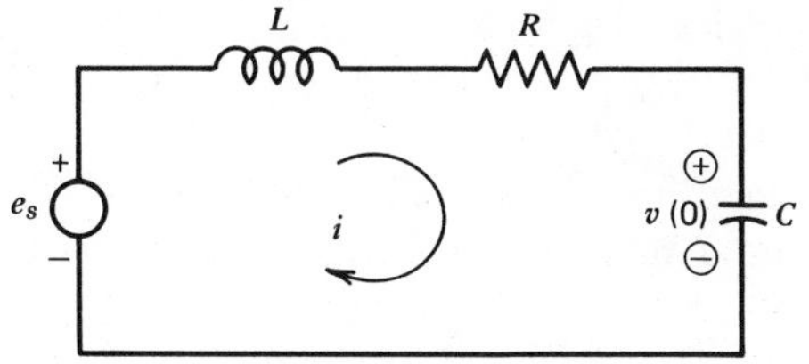

Fig. 3-1.1

Since this is a non-homogeneous equation, the solution is made up of

$$i = i_p + i_c \tag{3-1.3}$$

where i_p must satisfy the equation

$$\frac{de_s}{dt} = L\frac{d^2 i_p}{dt^2} + R\frac{di_p}{dt} + \frac{i_p}{C} \tag{3-1.4}$$

and i_c satisfies

$$0 = L\frac{d^2 i_c}{dt^2} + R\frac{di_c}{dt} + \frac{i_c}{C} \tag{3-1.5}$$

The corresponding characteristic equation is

$$0 = Ls^2 + Rs + \frac{1}{C} \tag{3-1.6}$$

The quadratic formula yields the roots

$$s_1 \,\&\, s_2 = \frac{-R \pm \sqrt{R^2 - (4L/C)}}{2L} = -\frac{R}{2L} \pm \sqrt{\frac{R^2}{4L^2} - \frac{1}{LC}} \tag{3-1.7}$$

For notational convenience, a and b are defined as

$$a = \frac{R}{2L} \qquad b = \sqrt{\frac{R^2}{4L^2} - \frac{1}{LC}} \tag{3-1.8}$$

and with this notation, the roots of the characteristic equation are

$$s_1 \,\&\, s_2 = -a \pm b \tag{3-1.9}$$

The b can take on three different forms. If $R^2/4L^2 > 1/LC$, b is the square root of a positive quantity and is real. If $R^2/4L^2 = 1/LC$, b has the value of zero. If $R^2/4L^2 < 1/LC$, b is the square root of a negative number and is imaginary. This situation is handled by defining β as

$$\beta = \sqrt{\frac{1}{LC} - \frac{R^2}{4L^2}} \tag{3-1.10}$$

When b becomes imaginary, β becomes real

$$b = j\beta \tag{3-1.11}$$

Because of these possibilities, three forms are needed for the solution. At the present time these are called *Case I*, *Case II*, and *Case III*. Later, names will be given to these cases.

Case I. Here, $R^2/4L^2 > 1/LC$, or $R > 2\sqrt{L/C}$ and both roots are real:

$$s_1 = -a + b \qquad s_2 = -a - b \tag{3-1.12}$$

The solution for current is $i = i_p + i_c$, which becomes

$$i = i_p + A_1\, \epsilon^{(-a+b)t} + A_2\, \epsilon^{(-a-b)t} \tag{3-1.13}$$

For a specific problem, i_p is determined, and A_1 and A_2 are evaluated using two initial conditions. This is done shortly in the examples, but for the moment we are interested in the forms of the solution.

As soon as i is determined for this circuit, the voltages v_R, v_L, and v_C can be found. The voltage v_R is Ri, and obviously has the same form as i. The voltage v_L is found from $v_L = L\, di/dt$, and since the derivative of an exponential does not change the exponent, v_L has the same form. A similar statement can also be made for v_C. Hence, all the voltages for this same circuit have the form

$$v = v_p + A_1\, \epsilon^{(-a+b)t} + A_2\, \epsilon^{(-a-b)t} \tag{3-1.14}$$

Of course, the A's between Eqs. 3-1.13 and 14 will be different, because different initial conditions are involved.

Equation 3-1.13 is a good form, which we will use most of the time. As a teaching device, however, we want to learn to put this equation into another form.

The ϵ^{-at} can be factored out of the last two terms of Eq. 3-1.13 as

$$i = i_p + \epsilon^{-at}\,[A_1\, \epsilon^{bt} + A_2\, \epsilon^{-bt}] \tag{3-1.15}$$

The terms ϵ^{+bt} and ϵ^{-bt} can be expanded as[1]

$$i = i_p + \epsilon^{-at}\,[A_1 (\cosh bt + \sinh bt) + A_2 (\cosh bt - \sinh bt)] \tag{3-1.16}$$

$$i = i_p + \epsilon^{-at}\,[(A_1 + A_2) \cosh bt + (A_1 - A_2) \sinh bt] \tag{3-1.17}$$

Since A_1 and A_2 are arbitrary constants, their sum and difference are also constants. For convenience we let

$$B_1 = A_1 + A_2 \qquad \text{and} \qquad B_2 = A_1 - A_2 \tag{3-1.18}$$

The solution can finally be written in the form

$$i = i_p + \epsilon^{-at}\,[B_1 \cosh bt + B_2 \sinh bt] \tag{3-1.19}$$

[1] The hyperbolic functions show the relationship between the underdamped and overdamped cases. Once this relationship is understood, the exponential form of the solution is preferred. A review of hyperbolic functions is in Appendix B.

Any voltage in this circuit can also be put into the form

$$v = v_p + \epsilon^{-at}\,[B_1 \cosh bt + B_2 \sinh bt] \tag{3-1.20}$$

In summary, Eqs. 3-1.13 and 19 are simply two different ways of writing the same equation. For obvious reasons, Eq. 3-1.13 is called the exponential form, and Eq. 3-1.19 the hyperbolic form.

Case II. Here, $R_c^2/4L^2 = 1/LC$, or $R_c = 2\sqrt{L/C}$. This specific value of R is denoted by the subscript C on the R_c. For this case $b = 0$, and hence the two roots are equal.

$$s_1 = s_2 = -a \tag{3-1.21}$$

We can obtain the solution for this case from Eq. 3-1.19 by letting $b = 0$, if this is done properly. When B_2 is determined from the initial conditions, it always contains a b in the denominator, as

$$B_2 = \frac{B_3}{b} \tag{3-1.22}$$

With this notation, Eq. 3-1.19 becomes

$$i = i_p + \epsilon^{-at}\left[B_1 \cosh bt + B_3 \frac{\sinh bt}{b}\right] \tag{3-1.23}$$

The limit is taken as $b \to 0$:

$$i = i_p + \epsilon^{-at}\left\{B_1 \lim_{b\to 0}\,[\cosh bt] + B_3 \lim_{b\to 0}\left[\frac{\sinh bt}{b}\right]\right\} \tag{3-1.24}$$

The $\lim_{b\to 0}\,[\cosh bt] = 1$, based on the definition of cosh bt. However, the other term leads to

$$\lim_{b\to 0}\left[\frac{\sinh bt}{b}\right] = \frac{0}{0} \tag{3-1.25}$$

This indeterminant form is evaluated, using l'Hospital's rule:

$$\lim_{b\to 0}\left[\frac{\sinh bt}{b}\right] = \lim_{b\to 0}\left[\frac{(d \sinh bt)/db}{db/db}\right] = t \tag{3-1.26}$$

The B_3 is an arbitrary constant and is replaced with B_2. Equation 3-1.24 becomes

$$i = i_p + \epsilon^{-at}\,[B_1 + B_2 t] \tag{3-1.27}$$

In a similar manner, all the voltages can be put in the form

$$v = v_p + \epsilon^{-at}\,[B_1 + B_2 t] \tag{3-1.28}$$

Case III. Here, $R < 2\sqrt{L/C}$, and both roots are complex

$$s_1 = -a + j\beta \qquad s_2 = -a - j\beta \tag{3-1.29}$$

The solution for this case is obtained from Eq. 3-1.23 by letting $b = j\beta$:

$$i = i_p + \epsilon^{-at}\left[B_1 \cosh j\beta t + B_3 \frac{\sinh j\beta t}{j\beta}\right] \tag{3-1.30}$$

$$i = i_p + \epsilon^{-at}\left[B_1 \cos \beta t + \frac{B_3 j \sin \beta t}{j\beta}\right] \tag{3-1.31}$$

$$i = i_p + \epsilon^{-at}\,[B_1 \cos \beta t + B_2 \sin \beta t] \tag{3-1.32}$$

In a similar manner, all the voltage terms can be put into the form

$$v = v_p + \epsilon^{-at}\,[B_1 \cos \beta t + B_2 \sin \beta t] \tag{3-1.33}$$

In the limit as $R \to 0$, *Case III* becomes

$$a = \frac{R}{2L} = 0 \qquad \beta = \sqrt{\frac{1}{LC} - \frac{R^2}{4L^2}} = \frac{1}{\sqrt{LC}} \tag{3-1.34}$$

$$i = i_p + \left[B_1 \cos \frac{1}{\sqrt{LC}} t + B_2 \sin \frac{1}{\sqrt{LC}} t\right] \tag{3-1.35}$$

$$v = v_p + \left[B_1 \cos \frac{1}{\sqrt{LC}} t + B_2 \sin \frac{1}{\sqrt{LC}} t\right] \tag{3-1.36}$$

Since the equations for *Case III* contain circular functions, the currents and voltages oscillate, and from now on we refer to this as the oscillatory case. *Case I* does not oscillate, and we refer to this as the overdamped case. *Case II* is the borderline between the oscillatory case and the overdamped case and is referred to as the critically damped case.

3-2. THE *R-L-C* PARALLEL CIRCUIT

The second circuit to be considered is the *R-L-C* parallel circuit shown in Fig. 3-2.1. The KCLE for $t > 0$ is

$$i_s = C\frac{dv}{dt} + Gv + \frac{1}{L}\int_0^t v\,dt + i(0) \tag{3-2.1}$$

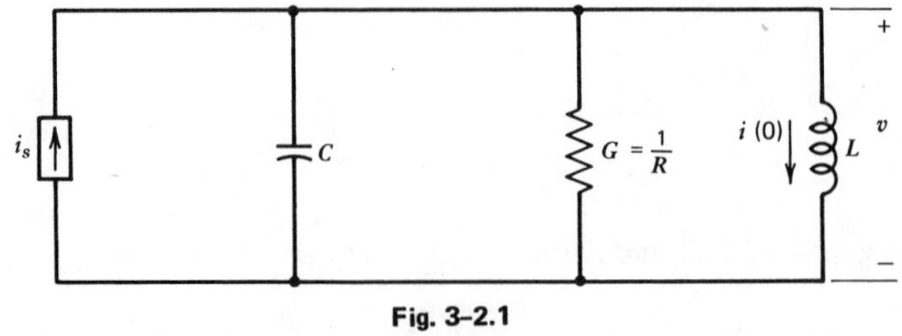

Fig. 3-2.1

A comparison of Eq. 3-2.1 with Eq. 3-1.1 shows the two equations to have identical forms. This, of course, is just one example of the more general concept known as duality. If in Eq. 3-1.1 we replace

$$e_s \longrightarrow i_s \quad L \longrightarrow C \quad R \longrightarrow G \quad C \longrightarrow L \quad i \longrightarrow v \tag{3-2.2}$$

we obtain Eq. 3-2.1. Hence, we can make these same replacements in the first set of solutions to obtain the solutions here.

For the series *R-L-C* circuit, a_s and b_s are defined as

$$a_s = \frac{R}{2L} \qquad b_s = \sqrt{\frac{R^2}{4L^2} - \frac{1}{LC}} \tag{3-2.3}$$

where the subscript s indicates the *R-L-C* elements are in series. For the parallel *R-L-C* circuit, the analogous quantities are

$$a_p = \frac{G}{2C} = \frac{1}{2RC} \qquad b_p = \sqrt{\frac{G^2}{4C^2} - \frac{1}{LC}} \tag{3-2.4}$$

where the subscript p indicates the *R-L-C* elements are in parallel. All the solutions previously listed can be rewritten for this new circuit.

3-3. CIRCUITS WITH TWO STORAGE ELEMENTS

The following examples demonstrate the concepts discussed in the last section.

Example 3-3.1. The circuit shown in Fig. 3-3.1(a) is in steady-state, with S in *Position 1* and at $t = 0$, S is thrown to *Position 2*. The desired solutions after $t = 0$ are the current i in the inductor and the charge q on the capacitor, as shown in (b).
Before $t = 0$, i_{ss} is

$$i_{ss} = \frac{10}{100} = 0.1 \tag{3-3.1}$$

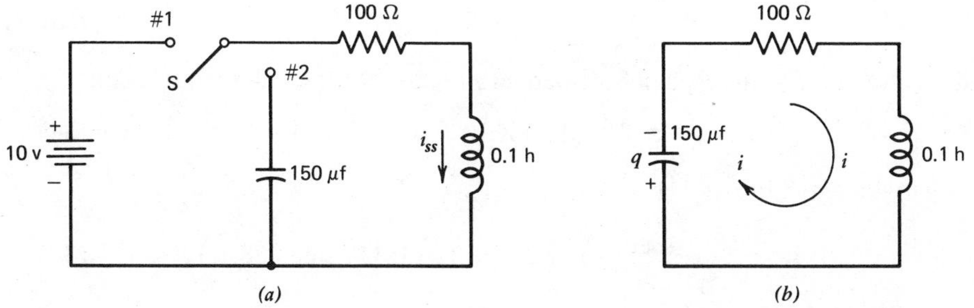

Fig. 3-3.1

After $t = 0$, $i_{ss} = q_{ss} = 0$, and the a and b terms and the roots are

$$a = \frac{R}{2L} = 500 \qquad b = \sqrt{\frac{R^2}{4L^2} - \frac{1}{LC}} = 428.1 \tag{3-3.2}$$

$$\begin{aligned} s_1 &= -a + b = -500 + 428.1 = -71.9 \\ s_2 &= -a - b = -500 - 428.1 = -928.1 \end{aligned} \tag{3-3.3}$$

These roots are for the over-damped case, and the solutions can be written in either the exponential or hyperbolic form. For the current i, these forms are

$$i = 0 + A_1 \epsilon^{-71.9t} + A_2 \epsilon^{-928.1t} \tag{3-3.4}$$

or

$$i = 0 + \epsilon^{-500t} [B_1 \cosh 428.1t + B_2 \sinh 428.1t] \tag{3-3.5}$$

In this example, the hyperbolic form is used, but the same circuit is employed again in Ex. 3-3.4, at which time the exponential form is used.

In order to find the two constants, two initial conditions are needed. At $t = 0+$, the capacitor and inductor will maintain q and i, respectively; therefore

$$q(0+) = 0 \quad \text{and} \quad i(0+) = 0.1 \tag{3-3.6}$$

If we solve for q first, these are precisely the conditions we need.

$$q = 0 + \epsilon^{-500t} [B_1 \cosh 428.1t + B_2 \sinh 428.1t] \tag{3-3.7}$$

At $t = 0+$

$$q(0+) = 0 = B_1 \tag{3-3.8}$$

Upon substitution, Eq. 3-3.7 is

$$q = \epsilon^{-500t} [B_2 \sinh 428.1t] \tag{3-3.9}$$

The derivative is

$$i = \frac{dq}{dt} = \epsilon^{-500t} [B_2 (428.1) \cosh 428.1t] + [B_2 \sinh 428.1t](-500)\, \epsilon^{-500t} \tag{3-3.10}$$

At $t = 0+$

$$i(0+) = 0.1 = 428.1 B_2 \tag{3-3.11}$$

This is solved for B_2, and B_2 is substituted into Eq. 3-3.9 for the desired solution

$$q = \epsilon^{-500t} [2.34 \times 10^{-4} \sinh 428.1t] \tag{3-3.12}$$

One simple way to find i is

$$i = \frac{dq}{dt} = \epsilon^{-500t} [0.1 \cosh 428.1t - 0.117 \sinh 428.1t] \tag{3-3.13}$$

In later examples, we shall put i in its proper form and find the solution for i without first finding q.

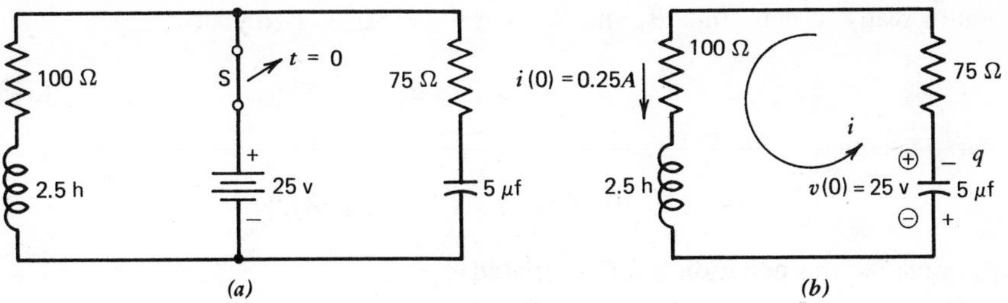

Fig. 3-3.2

Example 3-3.2. The circuit shown in Fig. 3-3.2(a) is in steady-state with S closed, and S is opened at $t = 0$. The desired solutions are the current in the inductor and the q on the capacitor. The circuit after $t = 0$ is shown in (b) with the current direction shown. Before $t = 0$ in steady-state, the current in the inductor is

$$i_{ss} = \frac{E}{R} = \frac{25}{100} = 0.25$$

This becomes $i(0+)$. The voltage on the capacitor is equal to 25 volts, with the polarity shown in ⊕ and ⊖ marks in (b).

The a and b terms and the roots are

$$a = \frac{R}{2L} = 35 \qquad b = \sqrt{\frac{R^2}{4L^2} - \frac{1}{LC}} = j281 \tag{3-3.14}$$

$$s_1 \;\&\; s_2 = -35 \pm j281 \tag{3-3.15}$$

The roots are complex; therefore, the solution is of the oscillatory form.

In the preceding example, we found q first and then determined i as the derivative of q. To be different, the current i is determined first from

$$i = 0 + \epsilon^{-35t}\,[B_1 \cos 281t + B_2 \sin 281t] \tag{3-3.16}$$

To find B_1 and B_2, two initial conditions are needed. From the statement of the problem, $q(0+)$ and $i(0+)$ are known. Since it is more convenient to differentiate Eq. 3-3.16 than it is to integrate it, the initial conditions $i(0+)$ and $(di/dt)\,(0+)$ are preferred. To find $(di/dt)\,(0+)$, the KVLE is written

$$0 = 2.5\frac{di}{dt} + 175i + \frac{1}{5 \times 10^{-6}}\int_0^t i\,dt - 25 \tag{3-3.17}$$

and the known initial conditions are placed in this equation

$$0 = 2.5\frac{di}{dt}(0+) + 175[0.25] + 0 - 25 \tag{3-3.18}$$

This is solved for

$$\frac{di}{dt}(0+) = -7.5 \tag{3-3.19}$$

We are now ready to determine B_1 and B_2. At $t = 0+$, Eq. 3-3.16 yields

$$i(0+) = 0.25 = B_1 \qquad (3\text{-}3.20)$$

Eq. 3-3.16 becomes

$$i = \epsilon^{-35t}\,[0.25 \cos 281t + B_2 \sin 281t] \qquad (3\text{-}3.21)$$

To determine B_2, this equation is differentiated as

$$\frac{di}{dt} = e^{-35t}\,[-0.25\,(281) \sin 281t + B_2\,(281) \cos 281t]$$

$$+ [0.25 \cos 281t + B_2 \sin 281t](-35)\,\epsilon^{-35t} \qquad (3\text{-}3.22)$$

At $t = 0+$, this becomes

$$\frac{di}{dt}(0+) = -7.5 = B_2\,(281) - 35\,(0.25) \qquad (3\text{-}3.23)$$

The B_2 is found as

$$B_2 = 0.00445 \qquad (3\text{-}3.24)$$

and upon substitution, i is

$$i = \epsilon^{-35t}\,[0.25 \cos 281t + 0.00445 \sin 281t] \qquad (3\text{-}3.25)$$

The q can be found from

$$q = \int_0^t i\,dt - 125 \times 10^{-6} = \epsilon^{-35t}\,[-125 \cos 281t + 873 \sin 281t] \times 10^{-6} \qquad (3\text{-}3.26)$$

We suggest you carry out the details of this integration to find q and determine for yourself whether it is easier to differentiate or integrate such equations.

Example 3-3.3. The circuit shown in Fig. 3-3.3(a) is in steady-state with the switch S closed, and S is opened at $t = 0$. The desired solution is the voltage on the capacitor.

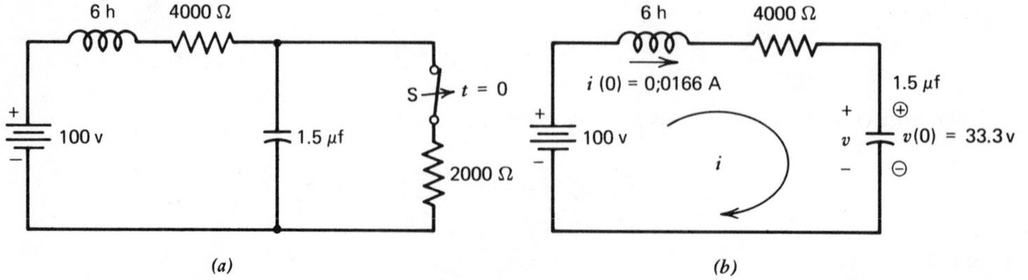

Fig. 3-3.3

Before $t = 0$ in steady-state, the inductor behaves as if it were a short, and the capacitor as an open circuit. The i_{ss} is obtained as

$$i_{ss} = \frac{100}{6{,}000} = 0.0166$$

The voltage on the capacitor is

$$v_{ss} = 0.0166 \times 2{,}000 = 33.3$$

These become the initial conditions. The circuit after $t = 0$ is shown in Fig. 3-3.3(b).

The a and b terms are found as

$$a = \frac{R}{2L} = 333.3 \qquad b = \sqrt{\frac{R^2}{4L^2} - \frac{1}{LC}} = 0 \tag{3-3.27}$$

and the roots of the characteristic equation are

$$s_1 \;\&\; s_2 = -333.3 + 0 \tag{3-3.28}$$

since the roots are real and equal, the circuit is critically damped. The form for the voltage on the capacitor is

$$v = v_p + \epsilon^{-333.3t}\,[B_1 + B_2 t] \tag{3-3.29}$$

By inspection of the circuit, v in steady-state (v_p) is 100 volts:

$$v = 100 + \epsilon^{-333.3t}\,[B_1 + B_2 t] \tag{3-3.30}$$

At $t = 0+$, $v(0+) = 33.3$:

$$v(0+) = 33.3 = 100 + B_1 \qquad \text{or} \qquad B_1 = -66.6 \tag{3-3.31}$$

Upon substituting, v becomes

$$v = 100 + \epsilon^{-333.3t}\,[-66.6 + B_2 t] \tag{3-3.32}$$

Before we can continue, we need the derivative of v at $t = 0+$ as the second initial condition. This is obtained from

$$i = C\frac{dv}{dt} \qquad \text{or} \qquad \frac{dv}{dt} = \frac{i}{C}$$

We know $i(0+)$, and hence can find

$$\frac{dv}{dt}(0+) = \frac{i(0+)}{C} = \frac{0.0166}{1.5 \times 10^{-6}} = 11{,}111 \tag{3-3.33}$$

To continue, Eq. 3-3.32 is differentiated

$$\frac{dv}{dt} = \epsilon^{-333.3t}\,[B_2] + [-66.6 + B_2 t](-333.3)\,\epsilon^{-333.3t} \tag{3-3.34}$$

At $t = 0+$ this becomes

$$\frac{dv}{dt}(0+) = 11{,}111 = B_2 + 22{,}222 \qquad \text{or} \qquad B_2 = -11{,}111 \tag{3-3.35}$$

The final solution is

$$v = 100 + \epsilon^{-333.3t}[-66.6 - 11{,}111t] \tag{3-3.36}$$

The current i can be found as

$$i = C\frac{dv}{dt} = \epsilon^{-333.3t}[0.0166 + 5.55t] \tag{3-3.37}$$

For practice, we suggest that you find i in this manner and to also start with the form

$$i = \epsilon^{-333.3t}[B_1 + B_2 t] \tag{3-3.38}$$

and again determine i.

Example 3-3.4. The circuit shown in Fig. 3-3.4 is at rest when e_s is applied. The desired solutions are the charge q and the voltage v on the capacitor.
The KVLE for $t > 0$ is

$$10t = 0.1\frac{di}{dt} + 100i + \frac{1}{150 \times 10^{-6}}\int_0^t i\,dt \tag{3-3.39}$$

This equation is written in terms of q as

$$10t = 0.1\frac{d^2q}{dt^2} + 100\frac{dq}{dt} + \frac{q}{150 \times 10^{-6}} \tag{3-3.40}$$

The $q = q_p + q_c$, where q_p must satisfy

$$10t = 0.1\frac{d^2q_p}{dt^2} + 100\frac{dq_p}{dt} + \frac{q_p}{150 \times 10^{-6}} \tag{3-3.41}$$

as discussed in Section 2-2, q_p has the form

$$q_p = K_1 t + K_2 \qquad \frac{dq_p}{dt} = K_1 \qquad \frac{d^2q_p}{dt^2} = 0 \tag{3-3.42}$$

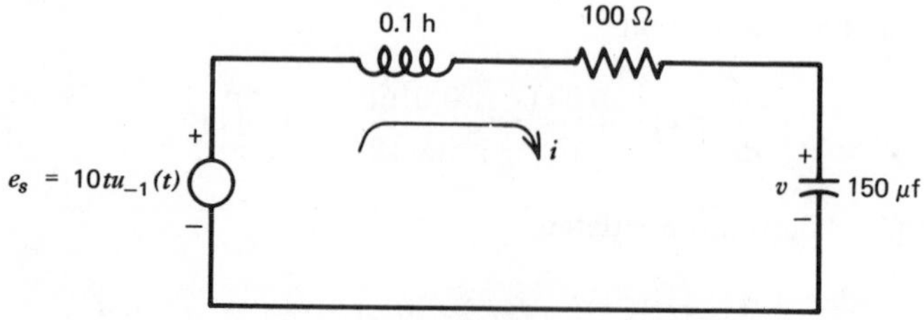

Fig. 3-3.4

Upon substitution, Eq. 3-3.41 is

$$10t = 0 + 100\,[K_1] + \frac{[K_1 t + K_2]}{150 \times 10^{-6}} \tag{3-3.43}$$

Upon equating the corresponding coefficients of t^1 and t^0 on both sides of this equation, the following two equations are obtained.

$$10 = \frac{K_1}{150 \times 10^{-6}} \qquad 0 = 100K_1 + \frac{K_2}{150 \times 10^{-6}} \tag{3-3.44}$$

The K_1 and K_2 are found from these two equations:

$$K_1 = 1.5 \times 10^{-3} \qquad K_2 = -2.25 \times 10^{-5} \tag{3-3.45}$$

and q_p is

$$q_p = 1.5 \times 10^{-3} t - 2.25 \times 10^{-5} \tag{3-3.46}$$

The element values of this circuit are identical with those of Ex. 3-3.1; therefore, the characteristic equation and the roots are the same. The solution is put into exponential form as

$$q = 1.5 \times 10^{-3} t - 22.5 \times 10^{-6} + A_1\,\epsilon^{-71.9t} + A_2\,\epsilon^{-928.1t} \tag{3-3.47}$$

At $t = 0+$

$$q(0+) = 0 = -22.5 \times 10^{-6} + A_1 + A_2 \tag{3-3.48}$$

$$\frac{dq}{dt} = 1.5 \times 10^{3} - 71.9A_1\,\epsilon^{-71.9t} - 928.1A_2\,\epsilon^{-928.1t} \tag{3-3.49}$$

At $t = 0+$

$$\frac{dq}{dt}(0+) = i(0+) = 0 = 1.5 \times 10^{-3} - 71.9A_1 - 928.1A_2 \tag{3-3.50}$$

Equations 3-3.48 and 3-3.50 are solved for

$$A_1 = 22.6 \times 10^{-6} \qquad A_2 = -0.1375 \times 10^{-6} \tag{3-3.51}$$

The solution for q becomes

$$q = 1.5 \times 10^{-3} t - 22.5 \times 10^{-6} + 22.6 \times 10^{-6}\epsilon^{-71.9t} - 0.1375 \times 10^{-6}\epsilon^{-928.1t} \tag{3-3.52}$$

and the voltage on the capacitor is

$$v = \frac{q}{C}$$
$$v = 10t - 0.15 + 0.151\epsilon^{-71.9t} - 0.000917\epsilon^{-928.1t} \tag{3-3.53}$$

Example 3-3.5. The circuit shown in Fig. 3-3.5 is at rest when e_s is applied. The desired solution is the current i.

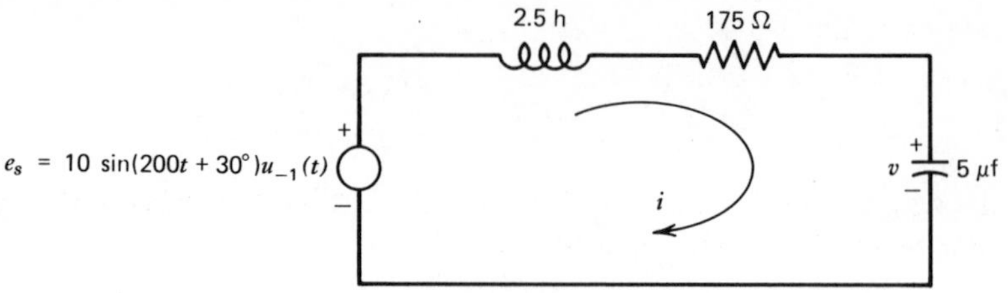

Fig. 3-3.5

The KVLE is

$$10 \sin (200t + 30°) = 2.5 \frac{di}{dt} + 175i + \frac{1}{5 \times 10^{-6}} \int_0^t i\, dt \tag{3-3.54}$$

At $t = 0+$, this equation is used to find $di/dt\,(0+)$:

$$10 \sin (30°) = 2.5 \frac{di}{dt}(0+) + 0 + 0 \qquad \frac{di}{dt}(0+) = 2 \tag{3-3.55}$$

To find the particular component of the current a-c circuit theory is used. The phasor E_s representing the voltage e_s is

$$E_s = 10\underline{/30°} \tag{3-3.56}$$

The impedance Z is found as

$$Z = R + j\left(\omega L - \frac{1}{\omega C}\right) = 529\underline{/-70.7°} \tag{3-3.57}$$

The current as a phasor is calculated to be

$$I = \frac{E_s}{Z} = \frac{10\underline{/30°}}{529\underline{/-70.7°}} = 1.89 \times 10^{-2}\underline{/100.7°} \tag{3-3.58}$$

The i_p is the inverse phasor transform:

$$i_p = 1.89 \times 10^{-2} \sin (200t + 100.7°) \tag{3-3.59}$$

The circuit is identical with that of Ex. 3-3.2; therefore, the roots of the characteristic equation are the same, and the form for i is

$$i = 1.89 \times 10^{-2} \sin (200t + 100.7°) + \epsilon^{-35t}\,[B_1 \cos 281t + B_2 \sin 281t] \tag{3-3.60}$$

Since the work should now be familiar, B_1 and B_2 are found with a minumum of discussion.
At $t = 0+$, $i(0+) = 0$:

$$i(0+) = 0 = 1.89 \times 10^{-2} \sin (100.7°) + B_1 \qquad \text{or} \qquad B_1 = -1.86 \times 10^{-2} \tag{3-3.61}$$

$$i = 1.89 \times 10^{-2} \sin (200t + 100.7°) + \epsilon^{-35t}[-1.86 \times 10^{-2} \cos 281t + B_2 \sin 281t] \tag{3-3.62}$$

$$\frac{di}{dt} = 1.89 \times 10^{-2} (200) \cos (200t + 100.7°)$$
$$+ \epsilon^{-35t} [-1.86 \times 10^{-2}(-281) \sin 281t + B_2 (281) \cos 281t]$$
$$+ [-1.86 \times 10^{-2} \cos 281t + B_2 \sin 281t] (-35) \epsilon^{-35t} \quad (3\text{-}3.63)$$

At $t = 0+$

$$\frac{di}{dt}(0+) = 2 = 1.89 \times 10^{-2} (200) \cos (100.7°) + 281B_2 + 35 (1.86 \times 10^{-2}) \quad (3\text{-}3.64)$$

or

$$B_2 = 0.73 \times 10^{-2}$$

$$i = \{1.89 \sin (200t + 100.7°) + \epsilon^{-35t} [-1.86 \cos 281t + 0.73 \sin 281t]\} \times 10^{-2} \quad (3\text{-}3.65)$$

Example 3-3.6. The circuit shown in Fig. 3-3.6 is at rest when e_s is applied. The desired solution is the current i.
The KVLE is

$$10\epsilon^{-100t} = 6\frac{di}{dt} + 4000i + \frac{1}{C}\int_0^t i\, dt \quad (3\text{-}3.66)$$

which at $t = 0+$ yields

$$10 = 6\frac{di}{dt}(0+) \quad \text{or} \quad \frac{di}{dt}(0+) = 1.667 \quad (3\text{-}3.67)$$

Equation 3-3.66 is differentiated

$$-1{,}000\epsilon^{-100t} = 6\frac{d^2i}{dt^2} + 4{,}000\frac{di}{dt} + \frac{i}{1.5 \times 10^{-6}} \quad (3\text{-}3.68)$$

The particular component of the current is found from the equation

$$-1{,}000\epsilon^{-100t} = 6\frac{d^2i_p}{dt^2} + 4{,}000\frac{di_p}{dt} + \frac{i_p}{1.5 \times 10^{-6}} \quad (3\text{-}3.69)$$

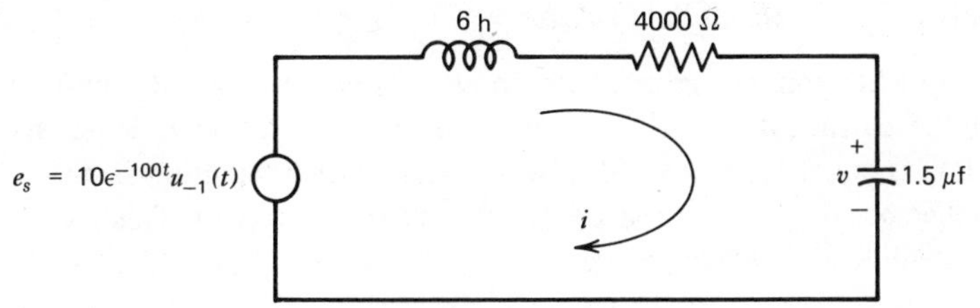

Fig. 3-3.6

The i_p component has the form

$$i_p = K\epsilon^{-100t} \qquad \frac{di_p}{dt} = -100K\epsilon^{-100t}$$
$$\frac{d^2 i_p}{dt^2} = 10{,}000K\epsilon^{-100t} \tag{3-3.70}$$

These forms are substituted into Eq. 3-3.69, and K is found to be

$$K = -3.06 \times 10^{-3} \tag{3-3.71}$$

The element values are exactly those of Example 3-3.3 and the roots of the characteristic equation are

$$s_1 \;\&\; s_2 = -333.3 \tag{3-3.72}$$

The form of the desired solution is

$$i = -3.06 \times 10^{-3}\,\epsilon^{-100t} + \epsilon^{-333.3t}\,[B_1 + B_2 t] \tag{3-3.73}$$

At $t = 0+$

$$i(0+) = 0 = -3.06 \times 10^{-3} + B_1 \qquad B_1 = 3.06 \times 10^{-3} \tag{3-3.74}$$

$$i = -3.06 \times 10^{-3}\,\epsilon^{-100t} + \epsilon^{-333.3t}\,[3.06 \times 10^{-3} + B_2 t] \tag{3-3.75}$$

$$\frac{di}{dt} = 0.306\,\epsilon^{-100t} + \epsilon^{-333.3t}\,[B_2] + [3.06 \times 10^{-3} + B_2 t](-333.3)\,\epsilon^{-333.3t} \tag{3-3.76}$$

At $t = 0+$

$$\frac{di}{dt}(0+) = 1.667 = 0.306 + B_2 - 3.06(333.3) \times 10^{-3} \qquad \text{or} \qquad B_2 = 2.38 \tag{3-3.77}$$

The final solution is

$$i = -3.06 \times 10^{-3}\,\epsilon^{-100t} + \epsilon^{-333.3t}\,[3.06 \times 10^{-3} + 2.38t] \tag{3-3.78}$$

3-4. SWITCHING WITHOUT NEED FOR A COMPLEMENTARY COMPONENT OF THE SOLUTION

The discussion of Section 2-8 is extended to circuits containing both an inductor and a capacitor. If the current in the inductor $i(0+)$ is equal to $i_p(0+)$ and the charge on the capacitor $q(0+)$ is equal to $q_p(0+)$, the two constants of integration in the solution will be zero and the solution will consist of the particular component. This idea is developed through one example and several problems.

Example 3-4.1. We use the circuit of Ex. 3-3.6, which is shown in Fig. 3-3.6. For this example, i_p has already been found in Ex. 3-3.6 as

$$i_p = -3.06 \times 10^{-3}\, \epsilon^{-100t} \tag{3-4.1}$$

By the methods previously discussed, q_p can be found as

$$q_p = 3.06 \times 10^{-5}\, \epsilon^{-100t} \tag{3-4.2}$$

From these, the values of $i(0+)$ is found equal to $i_p(0+)$ as

$$i(0+) = i_p(0+) = -3.06 \times 10^{-3} \tag{3-4.3}$$

and $q(0+)$ is set equal to $q_p(0+)$

$$q(0+) = q_p(0+) = 3.06 \times 10^{-5} \tag{3-4.4}$$

These are the desired quantities such that the complementary component of the solution is zero.

3-5. MORE COMPLICATED CIRCUITS WITH TWO STORAGE ELEMENTS

A group of R-L-C elements may combine into an R-L-C series circuit or an R-L-C parallel circuit. Circuits of this sort are included in the problems. This article considers a group of elements connected in a more general manner. One such circuit is that shown in Fig. 3-5.1.

It is possible to solve more complicated circuits in terms of the general parameters. For example, the characteristic equation for the circuit of Fig. 3-5.1 is

$$L_1L_2s^2 + (RL_1 + RL_2 + R_1L_2 + R_2L_1)\,s + (R_1R + R_2R + R_1R_2) = 0 \tag{3-5.1}$$

This equation depends upon all five elements, and a study of how each element affects the solution is not worth the effort. For these reasons, numerical coefficients are used in the examples that follow.

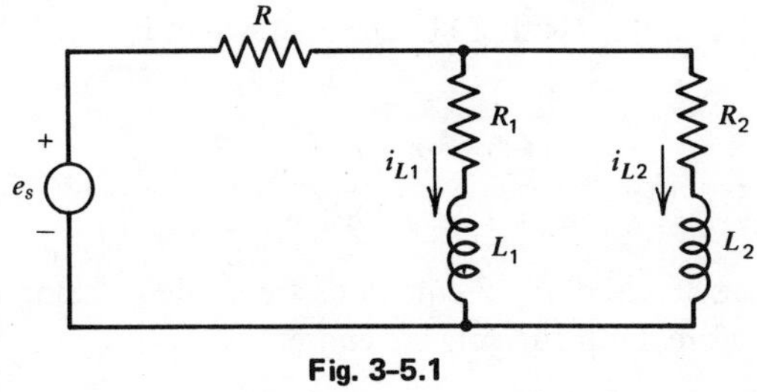

Fig. 3-5.1

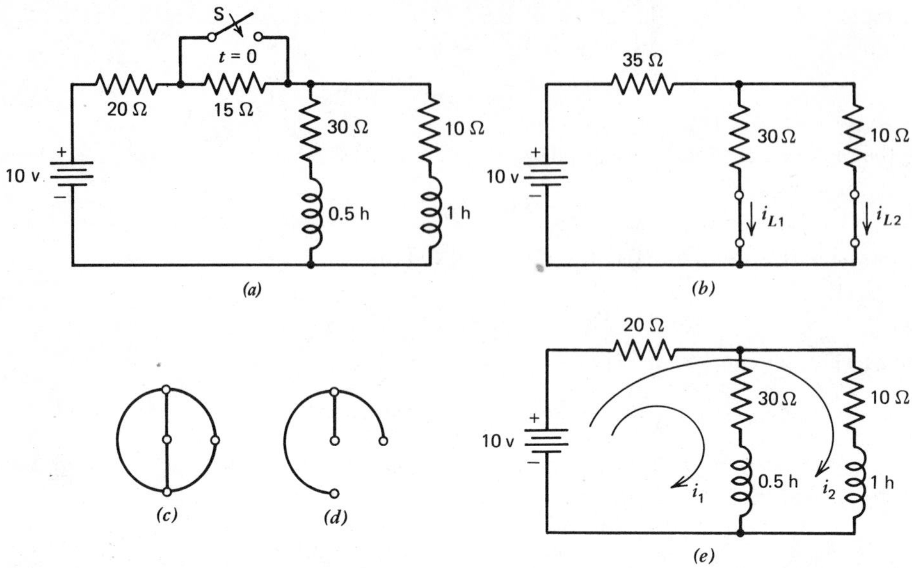

Fig. 3-5.2

Example 3-5.1. In the circuit as shown in Fig. 3-5.2(a), S is opened until steady-state conditions are reached, and S is closed at $t = 0$. The desired solution is the current in the 0.5-henry inductor.

Figure 3-5.2(b) is the equivalent circuit before $t = 0$ in steady-state. The currents i_{L1} and i_{L2} are found as

$$i_{L1} + i_{L2} = 0.235 \tag{3-5.2}$$

$$i_{L1} = \left(\frac{10}{40}\right) 0.235 = 0.0588$$
$$i_{L2} = \left(\frac{30}{40}\right) 0.235 = 0.1762 \tag{3-5.3}$$

When the switch S is closed, these become the initial conditions:

$$i_{L1}(0+) = 0.0588 \qquad i_{L2}(0+) = 0.1762 \tag{3-5.4}$$

The topology of the circuit (with S closed) is shown in Fig. 3-5.2(c), and the following are determined:(2)

$$b = 5 \qquad n = n_t - 1 = 3$$
$$n_t = 4 \qquad l = b - n = 2 \tag{3-5.5}$$

If node-pair voltages are used, $n = 3$ equations are needed; if loop-currents, $l = 2$ equations are needed. Therefore, loop currents are chosen.

2 Appendix C presents a short review of the subject of linear independence among circuit variables.

Since the desired solution is the current in the 0.5-h inductor, we choose a tree so that the 0.5-h inductor is a link, then only one loop current will exist in this inductor. We also let the other inductor be a link. The tree is shown in Fig. 3-5.2(d) and the loop currents in (e).

The KVLE's written around these two loops are

$$10 = 50i_1 + 0.5\frac{di_1}{dt} + 20i_2$$
$$10 = 20i_1 + 30i_2 + \frac{di_2}{dt} \tag{3-5.6}$$

The system is described by two equations that must be solved simultaneously. To solve for i_1, we wish to eliminate i_2 between these two equations. To do this, the first equation is solved for i_2 as

$$i_2 = 0.5 - 2.5i_1 - 0.025\frac{di_1}{dt} \tag{3-5.7}$$

and is differentiated to give

$$\frac{di_2}{dt} = -2.5\frac{di_1}{dt} - 0.025\frac{d^2i_1}{dt^2} \tag{3-5.8}$$

These quantities are substituted into the second equation

$$10 = 20i_1 + 30\left(0.5 - 2.5i_1 - 0.025\frac{di_1}{dt}\right) + \left(-2.5\frac{di_1}{dt} - 0.025\frac{d^2i_1}{dt^2}\right) \tag{3-5.9}$$

The terms are combined as

$$0.025\frac{d^2i_1}{dt^2} + 3.25\frac{di_1}{dt} + 55i_1 = 5 \tag{3-5.10}$$

The particular component of the solution is found as

$$i_{1p} = \frac{5}{55} = 0.091 \tag{3-5.11}$$

The characteristic equation is

$$0.025s^2 + 3.25s + 55 = 0 \tag{3-5.12}$$

and the roots of this equation are

$$s_1 = -20 \qquad s_2 = -110 \tag{3-5.13}$$

As a side comment, we see from these roots that the system is overdamped, which is precisely what should be expected. The only way a circuit can oscillate is for there to be an exchange of energy between a capacitor and an inductor. Since this circuit contains only inductors, it must be overdamped.

The solution for i_1 has the form

$$i_1 = 0.091 + A_1\,\epsilon^{-20t} + A_2\,\epsilon^{-110t} \tag{3-5.14}$$

To find the two arbitrary constants, two initial conditions are needed. Although $i_1(0+) = 0.0588$ and $i_2(0+) = 0.1762$ are known, we need $(di_1/dt)(0+)$. To find this, we return to the first of Eqs. 3-5.6 repeated as

$$10 = 50i_1 + 0.5\frac{di_1}{dt} + 20i_2 \tag{3-5.15}$$

and write this equation at $t = 0+$

$$10 = 50\,(0.0588) + 0.5\frac{di_1}{dt}(0+) + 20\,(0.1762) \tag{3-5.16}$$

From which we find

$$\frac{di_1}{dt}(0+) = 7.072 \tag{3-5.17}$$

We now have the form for i_1, which contains two constants, and we have the two desired initial conditions. The solution from here on is identical to work we have done before.

Equation 3-5.14 at $t = 0+$ yields

$$i_1(0+) = 0.0588 = 0.091 + A_1 + A_2 \tag{3-5.18}$$

$$\frac{di_1}{dt} = -20A_1\,\epsilon^{-20t} - 110A_2\,\epsilon^{-110t} \tag{3-5.19}$$

which at $t = 0+$ yields

$$\frac{di_1}{dt}(0+) = 7.072 = -20A_1 - 110A_2 \tag{3-5.20}$$

The constants A_1 and A_2 are found from Eqs. 3-5.18 and 20 as

$$A_1 = 0.0393 \qquad A_2 = -0.0715 \tag{3-5.21}$$

The solution for i_1 is

$$i_1 = 0.091 + 0.0393\epsilon^{-20t} - 0.0715\epsilon^{-110t} \tag{3-5.22}$$

The current i_2 can be determined as

$$i_2 = 0.273 - 0.0786\epsilon^{-20t} - 0.0179\epsilon^{-110t} \tag{3-5.23}$$

and the voltage v across the 10-Ω resistor as

$$v_{10\Omega} = 2.73 - 0.786\epsilon^{-20t} - 0.179\epsilon^{-110t} \tag{3-5.24}$$

Let us generalize our thinking about systems of this sort by imagining a much more complicated system. All the different currents and the voltages in this system are related through a set of linear differential equations with constant coefficients. One variable can be found from the solutions of the other variables by the appropriate number of differentiations and the recombining of terms, etc. None of these processes changes the exponents in the form of the solution. In other words, all the solutions for the variables in a given system have the same form. You will not find the current in one part of a circuit that is oscillatory, and another current in another part of the same circuit that is overdamped.

With these thoughts in mind, we can write the forms for some of the voltages and currents in this circuit as

$$
\begin{aligned}
i_2 &= 0.273 + A_1\,\epsilon^{-20t} + A_2\,\epsilon^{-110t}\\
v_{10\Omega} &= 2.73 + A_1\,\epsilon^{-20t} + A_2\,\epsilon^{-110t}\\
v_{1\text{h}} &= 0 + A_1\,\epsilon^{-20t} + A_2\,\epsilon^{-110t}\\
v_{0.5\text{h}} &= 0 + A_1\,\epsilon^{-20t} + A_2\,\epsilon^{-110t}\\
v_{30\Omega} &= 2.73 + A_1\,\epsilon^{-20t} + A_2\,\epsilon^{-110t}\\
v_{20\Omega} &= 7.27 + A_1\,\epsilon^{-20t} + A_2\,\epsilon^{-110t}
\end{aligned}
\tag{3-5.25}
$$

As presented here, the use of A_1 and A_2 is simply a way of identifying constants and for each equation, these constants will be evaluated, by use of a different set of initial conditions. The A_1 and A_2 in each equation will be different. The point is that these variables all have the same form.

Example 3-5.2. The circuit as shown in Fig. 3-5.3(a) has the switch S opened before $t = 0$, until steady-state is reached, and S is closed at $t = 0$. The desired solution is the current in the inductor after $t = 0$. The currents i_1 and i_2 are chosen as shown in (c).

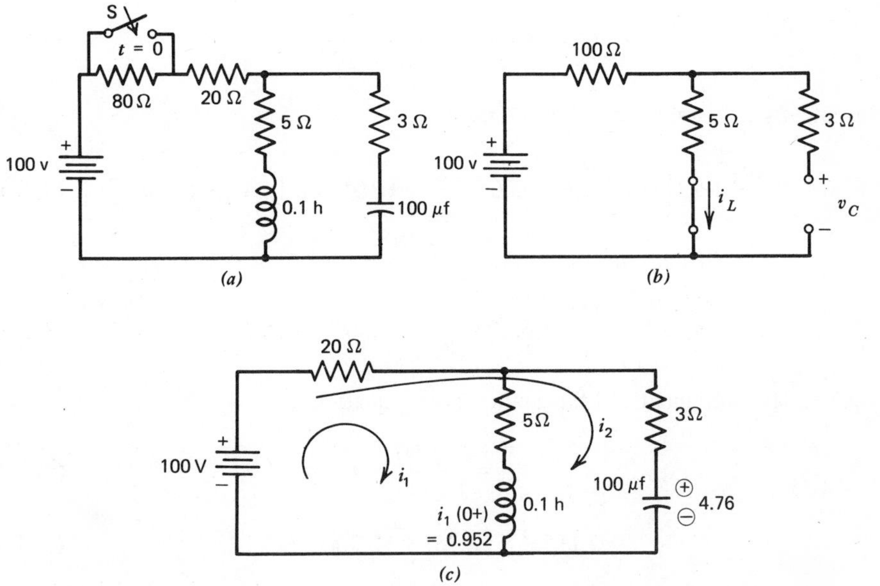

Fig. 3-5.3

The equivalent circuit in steady-state before $t = 0$ is shown in (b). This current in the inductor becomes $i_1(0+)$:

$$i_L = \frac{100}{105} = 0.952 = i_1(0+) \tag{3-5.26}$$

and the voltage on the capacitor becomes $v_C(0+)$:

$$v_C = 5i_L = 5 \times 0.952 = 4.76 = v_C(0+) \tag{3-5.27}$$

The KVLE's are

$$\begin{aligned} 100 &= 25i_1 + 0.1\frac{di_1}{dt} + 20i_2 \\ 100 &= 20i_1 + 23i_2 + \frac{1}{10^{-4}}\int_0^t i_2\,dt + 4.76 \end{aligned} \tag{3-5.28}$$

The first of these is solved for

$$i_2 = 5 - 1.25i_1 - 0.005\frac{di_1}{dt} \tag{3-5.29}$$

and for

$$\frac{di_2}{dt} = -1.25\frac{di_1}{dt} - 0.005\frac{d^2i_1}{dt^2} \tag{3-5.30}$$

The second of Eqs. 3-5.28 is differentiated

$$0 = 20\frac{di_1}{dt} + 23\frac{di_2}{dt} + \frac{i_2}{10^{-4}} \tag{3-5.31}$$

and i_2 and di_2/dt are substituted in the result:

$$0 = 20\frac{di_1}{dt} + 23\left[-1.25\frac{di_1}{dt} - 0.005\frac{d^2i_1}{dt^2}\right] + 10^4\left[5 - 1.25i_1 - 0.005\frac{di_1}{dt}\right] \tag{3-5.32}$$

When terms are collected, this becomes

$$0.115\frac{d^2i_1}{dt^2} + 58.75\frac{di_1}{dt} + 12{,}500i_1 = 50{,}000 \tag{3-5.33}$$

The particular component and the characteristic equation are

$$i_{1p} = \frac{50{,}000}{12{,}500} = 4 \tag{3-5.34}$$

$$0.115s^2 + 58.75s + 12{,}500 = 0 \tag{3-5.35}$$

This equation is factored and the form for i_1 is

$$s_1 \;\&\; s_2 = -255 \pm j209$$
$$i_1 = 4 + \epsilon^{-255t}\,[B_1 \cos 209t + B_2 \sin 209t] \tag{3-5.36}$$

The $i_1(0+)$ is known, but its derivative at $t = 0+$ is not. Upon first inspection of Eqs. 3-5.28, it would seem that we need to use only the first equation to find this derivative, until it is realized that $i_2(0+)$ is unknown. To determine the current $i_2(0+)$, we start with the second of Eqs. 3-5.28, and write this equation at $t = 0+$ as

$$100 = 20\,(0.952) + 23i_2(0+) + 0 + 4.76 \tag{3-5.37}$$

The current $i_2(0+)$ is found as

$$i_2(0+) = 3.31$$

This is substituted into the first of Eqs. 3-5.28 at $t = 0+$:

$$100 = 25\,(0.952) + 0.1\frac{di_1}{dt}(0+) + 20\,(3.31) \tag{3-5.38}$$

which yields

$$\frac{di_1}{dt}(0+) = 100 \tag{3-5.39}$$

The two constants in Eq. 3-5.36 are determined from these initial conditions, and the solution is

$$i_1 = 4 + \epsilon^{-255t}\,[-3.048 \cos 209t - 3.28 \sin 209t] \tag{3-5.40}$$

Example 3-5.3. The voltage e_s of Fig. 3-5.4(b) is applied to the circuit of (a), with the 10 volts applied long enough that the circuit is in steady-state before $t = 0$. The desired solution is the current i_1.

After $t = 0+$, the KVLE's are

$$10\epsilon^{-80t} = 50i_1 + 0.5\frac{di_1}{dt} + 20i_2 \qquad 10\epsilon^{-80t} = 20i_1 + 30i_2 + \frac{di_2}{dt} \tag{3-5.41}$$

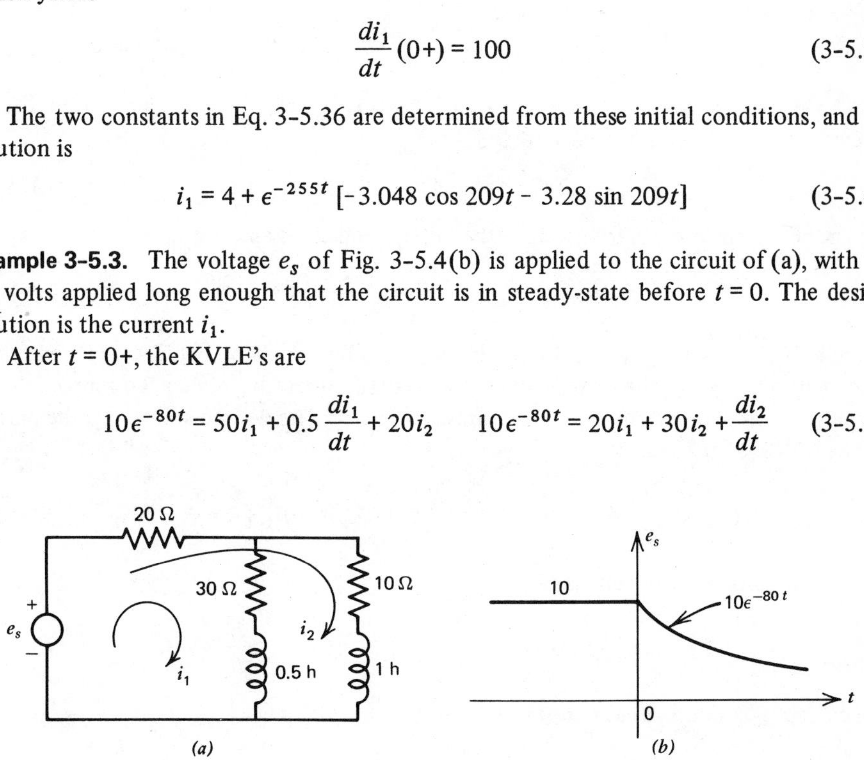

Fig. 3-5.4

The initial conditions can be found to be

$$i_1(0+) = 0.091 \qquad i_2(0+) = 0.273 \qquad \frac{di_1}{dt}(0+) = 0 \tag{3-5.42}$$

The particular component of each current is found from the equations

$$\begin{aligned} 10\epsilon^{-80t} &= 50i_{1p} + 0.5\frac{di_{1p}}{dt} + 20i_{2p} \\ 10\epsilon^{-80t} &= 20i_{1p} + 30i_{2p} + \frac{di_{2p}}{dt} \end{aligned} \tag{3-5.43}$$

We reason that the particular components have the forms

$$i_{1p} = K_1\,\epsilon^{-80t} \qquad i_{2p} = K_2\,\epsilon^{-80t} \tag{3-5.44}$$

These equations and their derivatives are substituted into Eq. 3-5.43, which at $t = 0+$ yields

$$10 = 10K_1 + 20K_2 \qquad 10 = 20K_1 - 50K_2 \tag{3-5.45}$$

These are solved as

$$K_1 = 0.778 \qquad K_2 = 0.111 \tag{3-5.46}$$

This circuit is identical to that of Ex. 3-5.1, except for the driving function. Hence, we can write the form of the current i_1 as

$$i_1 = 0.778\epsilon^{-80t} + A_1\,\epsilon^{-20t} + A_2\,\epsilon^{-110t} \tag{3-5.47}$$

The two constants are found from the initial conditions and the solution for i_1 is

$$i_1 = 0.778\epsilon^{-80t} - 0.1475\epsilon^{-20t} - 0.54\epsilon^{-110t} \tag{3-5.48}$$

Example 3-5.4. A Dependent Source Example. The circuit of Fig. 3-5.5(a) contains a dependent voltage source whose magnitude is 10 times the branch current j_2. Loop-currents are chosen and the circuit is redrawn as in (b). The desired solution is the current i_1. The KVLE's for $t > 0$ are

$$10 = 0.5\frac{di_1}{dt} + 50i_1 + 20i_2 \qquad 10 + 10i_1 = 20i_1 + 1\frac{di_2}{dt} + 30i_2 \tag{3-5.49}$$

These equations are rewritten as

$$10 = 0.5\frac{di_1}{dt} + 50i_1 + 20i_2 \qquad 10 = 10i_1 + 1\frac{di_2}{dt} + 30i_2 \tag{3-5.50}$$

The current i_2 is eliminated between these equations, yielding

$$0.025\frac{d^2i_1}{dt^2} + 3.25\frac{di_1}{dt} + 65i_1 = 5 \tag{3-5.51}$$

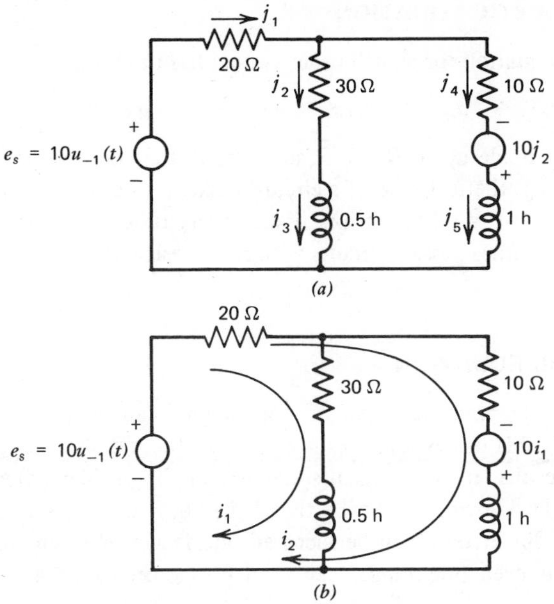

Fig. 3-5.5

The characteristic equation and its roots are

$$0.025s^2 + 3.25s + 65 = 0$$
$$s_1 = -24.66 \qquad s_2 = -105.34 \tag{3-5.52}$$

The form of the solution and the two initial conditions are

$$i_1 = 0.077 + A_1\,\epsilon^{-24.66t} + A_2\,\epsilon^{-105.34t}$$
$$i_1(0+) = 0 \qquad \frac{di_1}{dt}(0+) = 20 \tag{3-5.53}$$

The A's are evaluated, and the solution is

$$i_1 = 0.077 + 0.147\epsilon^{-24.66t} - 0.224\epsilon^{-105.34t} \tag{3-5.54}$$

3-6. THE SOLUTION TO THE *n*th-ORDER DIFFERENTIAL EQUATION

In Chapter 1, we started the discussion of the general properties of nth order differential equations. We proceeded to demonstrate these concepts, starting with simple examples in Chapter 2, and gradually adding complexities in this chapter. We return to the discussion of the nth order equation based upon the ideas that have been developed.

THE CHARACTERISTIC EQUATION

The characteristic equation for an nth order system has the form

$$a_n s^n + a_{n-1} s^{n-1} + \cdots + a_1 s + a_0 = 0 \tag{3-6.1}$$

For passive circuits made up of R's, L's, and C's, all the coefficients will be present and positive. For the hypothetical case of a circuit made up of only L's and C's, the equation will be either an even or an odd function; that is, every other coefficient will be zero. This general discussion assumes passive circuits with some resistance present.

FACTORING THE CHARACTERISTIC EQUATION

The only way an algebraic equation of high degree can be factored is by numerical methods. Although this discussion does not take up the details of these methods, we are interested in the results. In the original statement of the problem, the values of the R's, L's, and C's must be known, and hence the a's in Eq. 3-6.1 are known numerical numbers. If n is odd, a linear term can be factored out, leaving a polynomial of $n - 1$ degree, which is even. This even polynomial can then be factored into a set of quadratics, as indicated by

$$a_n(s + \alpha_1)(s^2 + b_1 s + c_1)(s^2 + b_2 s + c_2) \cdots (s^2 + b_j s + c_j) = 0 \tag{3-6.2}$$

where

$$j = \frac{n-1}{2}. \tag{3-6.3}$$

If n is even, the characteristic equation can be factored as

$$a_n(s^2 + b_1 s + c_1)(s^2 + b_2 s + c_2) \cdots (s^2 + b_j s + c_j) = 0 \tag{3-6.4}$$

where

$$j = \frac{n}{2}. \tag{3-6.5}$$

For either case, each of the coefficients in each of the quadratic terms will be a real positive number. The quadratic formula can be applied to each quadratic term and only one of three possibilities exist:

Case 1: The roots are negative, real, and distinct.

Case 2: The roots are negative and real, and the two roots are equal.

Case 3: The roots are a complex conjugate pair with a negative real part.

GEOMETRIC INTERPRETATION OF THE ROOTS

A brief geometric interpretation of these three possibilities will be helpful. The location of roots of this type can be shown in the complex s-plane. The s has a real part σ and an

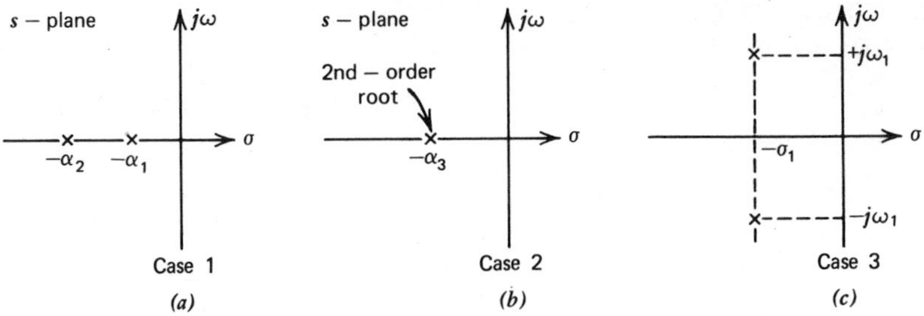

Fig. 3-6.1

imaginary part $j\omega$ as given by

$$s = \sigma + j\omega \tag{3-6.6}$$

The location of the roots for *Cases 1*, *2*, and *3* are shown by cross-marks (x) in (a), (b), and (c), respectively, of Fig. 3-6.1.

The contribution to the complementary component of the solution for each of these three cases is:

$$\begin{aligned} &\textit{Case } 1: \quad && A_1\,\epsilon^{-\alpha_1 t} + A_2\,\epsilon^{-\alpha_2 t} \\ &\textit{Case } 2: && \epsilon^{-\alpha_3 t}\,[B_1 + B_2 t] \\ &\textit{Case } 3: && \epsilon^{-\sigma_1 t}\,[C_1 \cos \omega_1 t + C_2 \sin \omega_1 t] \end{aligned} \tag{3-6.7}$$

THE COMPLETE COMPLEMENTARY COMPONENT OF THE SOLUTION

The complete complementary component for an *n*th-order system may be more complex than simply being the sum of the forms of *Cases 1*, *2,* and *3* just described. This is because several of the quadratic terms of Eq. 3-6.2 might contain the same root. If through these combinations a third-order real root appears, the form for the corresponding part of the solution would be

$$\epsilon^{-\alpha_1 t}\,[B_1 + B_2 t + B_3 t^2] \tag{3-6.8}$$

whereas, if a fourth-order real root occurs, the form would be

$$\epsilon^{-\alpha_1 t}\,[B_1 + B_2 t + B_3 t^2 + B_4 t^3] \tag{3-6.9}$$

Another possibility would be for two or more quadratic terms to have the same pair of complex conjugate roots. If the number is two, the corresponding part of the solution would be

$$\epsilon^{-\sigma_1 t}\,[(B_1 \cos \omega_1 t + B_2 \sin \omega_1 t)t + (B_3 \cos \omega_1 t + B_4 \sin \omega_1 t)] \tag{3-6.10}$$

Other factors similar to those just shown may occur, but these can be handled in a corresponding manner.

INTRODUCTION TO THE CONCEPT OF STABILITY

So far, this present discussion has been limited to passive R-L-C circuits. We now explore the more general situation.

Suppose some system has two roots that are to the right of the $j\omega$ axis as shown in Fig. 3-6.2. These roots are given as

$$s_1 \,\&\, s_2 = +\sigma_1 \pm j\omega_1 \qquad (3\text{-}6.11)$$

The factor appearing in the solution due to these roots is

$$\epsilon^{+\sigma_1 t}\,[B_1 \cos \omega_1 t + B_2 \sin \omega_1 t] \qquad (3\text{-}6.12)$$

The exponential factor now increases with time without bound.

A system of this sort is said to be unstable, and in most situations an unstable system is undesirable. No matter what input is applied to such a system, the output goes toward infinity.

We can reason that a passive R-L-C circuit will never behave this way. Suppose for example, we excite an R-L-C circuit with only a set of initial currents in inductors and initial voltages on capacitors. If oscillations are present, the energy is exchanged between the L's and the C's; but during each cycle, some of the energy will be dissipated in the R's. Hence, the system will gradually run down and approach a condition of rest. Therefore, all the roots of the characteristic equation must be in the left-half s-plane.

What is the explanation as to why a system can become unstable? We answer this within the context of circuit theory. Suppose we have a number of dependent sources each of which is related to some other quantity in the circuit. To be specific, suppose each of these sources is a voltage source whose magnitude depends on a current somewhere else in the circuit. Each of these dependent sources supplies energy to the circuit, but this energy is obtained from a power supply external to the circuit. Let us start this circuit with only a set of initial currents in the inductors and voltages on the capacitors. As the circuit oscillates, energy is lost in the resistors, but some of the energy may be resupplied from the power supply acting through the dependent sources. Now it is seen that if the dependent sources supply more energy than is dissipated, the oscillations can build up and the circuit is unstable.

In this book we will touch upon the concept of stability a number of times, but only as it affects the overall subject of systems. There are a number of stability criterions

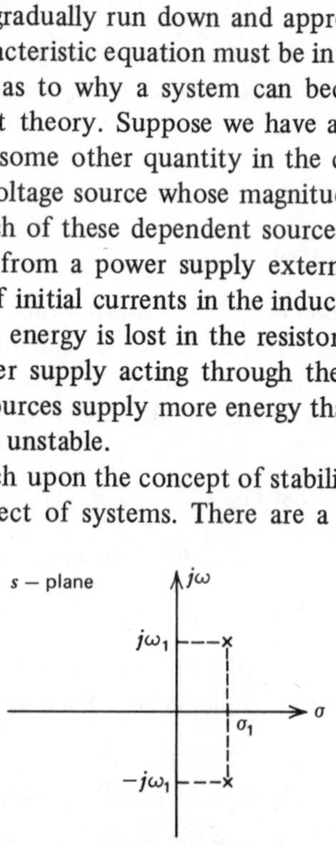

Fig. 3-6.2

accompanied by a set of tests that are more properly covered in a book on control system theory.[3]

STEPS IN OBTAINING THE SOLUTION TO THE nth-ORDER DIFFERENTIAL EQUATION

The solving of the nth-order differential equation by using classical methods is summarized in the following steps.

1 The circuit is examined from a topological point of view, and b, n_t, n, and l are determined. If $n < l$, the problem should be solved by using node-pair voltage equations. If $l < n$, the problem should be solved by using loop currents. If $n = l$, then the two methods are equally good.

2 After the choice is made in *Step 1*, the choice of variables is made so that these selected variables form an independent set.

Note: Another possibility is the writing of equations with mixed variables, which is discussed in Chapters 13 and 14, when state variables equations are studied.

3 The writing of the proper integro-differential equations is based on the variables chosen in Step 2.

Note: These first three steps are also necessary when Laplace transformation methods are used, as is discussed in Chapter 4.

4 The equations are differentiated to remove any integral terms.

5 If simultaneous equations are involved, they must be solved for the desired unknown.

6 If the equation obtained in *Step 5* is a non-homogeneous equation, the particular component must be found.

7 The characteristic equation is determined from the homogeneous equation.

8 The characteristic equation is factored. A suggested set of roots are:

$$\begin{aligned}&s_1 = -\alpha_1 \qquad s_2 = -\alpha_2 \qquad \cdots\\&s_h \;\&\; s_{h+1} = -\alpha_h \qquad s_{h+2} \;\&\; s_{h+3} = -\alpha_{h+2} \qquad \cdots\\&s_K \;\&\; s_{K+1} = -\sigma_1 \pm j\omega_1 \qquad s_{K+2} \;\&\; s_{K+3} = -\sigma_2 \pm j\omega_2 \qquad \cdots\end{aligned} \tag{3-6.13}$$

9 The solution is put in the desired form, as suggested by

$$\begin{aligned}i_1 = i_{1p} &+ A_1 \epsilon^{-\alpha_1 t} + A_2 \epsilon^{-\alpha_2 t} + \cdots\\&+ \epsilon^{-\alpha_h t}[B_1 + B_2 t] + \epsilon^{-\alpha_{h+2} t}[B_3 + B_4 t] + \cdots\\&+ \epsilon^{-\sigma_1 t}[C_1 \cos \omega_1 t + C_2 \sin \omega_1 t]\\&+ \epsilon^{-\sigma_2 t}[C_3 \cos \omega_2 t + C_4 \sin \omega_2 t] + \cdots\end{aligned} \tag{3-6.14}$$

where the number of constants adds up to n.

10 The original circuit had n independent storage elements. From the statement of the problem, we know the necessary voltages on the capacitors and currents in the induc-

[3] See *Control System Theory*, by Lago and Benningfield, The Ronald Press, Chapter 11.

tors at $t = 0$, and hence we know n initial conditions. These are not the set of initial conditions we need to solve for the n constants in Eq. 3-6.14. We need to know

$$i_1(0+)$$

$$\frac{di_1}{dt}(0+) \quad \cdots \quad \frac{d^{n-1}i}{dt^{n-1}}(0+)$$

To find this set, we go to the equations of *Step 3* and solve these simultaneously for the desired set.

11 The initial conditions of *Step 10* are imposed upon the form of the solution for i_1 given by Eq. 3-6.14. The result of this is a set of n equations containing n constants.

12 The set of n equations of *Step 11* are solved for the n constants.

13 The constants determined in *Step 12* are substituted into the form of the solution in *Step 9* to obtain the final solution.

The classical method involves a number of steps, and for that reason many attempts have been made to develop methods of solution that would eliminate some of these steps. This area of work is discussed in Chapter 4.

PROBLEMS

Note: For a discussion concerning hyperbolic functions, see Appendix B.

3-1 Show that (see above note):

(a) $\tanh j\beta t = j \tan \beta t$
(b) $\sinh(-bt) = -\sinh bt$
(c) $\cosh(-bt) = \cosh bt$
(d) $\sinh(b_1 t - b_2 t) = \sinh b_1 t \cosh b_2 t - \sinh b_2 t \cosh b_1 t$
(e) $\cosh(b_1 t - b_2 t) = \cosh b_1 t \cosh b_2 t - \sinh b_1 t \sinh b_2 t$
(f) $\cosh^2 bt - \sinh^2 bt = 1$

3-2 Sketch curves for each of the following (see above note):

(a) $f(t) = \sinh bt$
(b) $f(t) = \cosh bt$
(c) $f(t) = \tanh bt$
(d) $f(t) = \text{sech}\, bt$
(e) $f(t) = \text{csch}\, bt$
(f) $f(t) = \coth bt$

On each sketch use bt as the scale on the abscissa.

3-3 Find the solution that satisfies the differential equation and the initial conditions:

(a) $20\dfrac{d^2i}{dt^2} + 9\dfrac{di}{dt} + i = 0 \qquad i(0+) = 0 \qquad \dfrac{di}{dt}(0+) = 0.5$

(b) $\dfrac{d^2i}{dt^2} + 4\dfrac{di}{dt} + 3i = 0 \qquad i(0+) = 0 \qquad \dfrac{di(0+)}{dt} = 6$

(c) $\dfrac{5d^2e}{dt^2} + 6\dfrac{de}{dt} + e = 15 \qquad e(0+) = \dfrac{de(0+)}{dt} = 0$

(d) $3.5\dfrac{d^2e}{dt^2} + 15\dfrac{de}{dt} + 85e = 300 \qquad e(0+) = -100 \qquad \dfrac{de(0+)}{dt} = 1{,}000$

3-4 For the circuit of Example 3-3.1, start from

$$i = A_1\,\epsilon^{-71.9t} + A_2\,\epsilon^{-928.1t} \qquad q = A_1\,\epsilon^{-71.9t} + A_2\,\epsilon^{-928.1t}$$

and find both i and q. From these solutions put the answers in the hyperbolic forms.

3-5 For the circuit of Example 3-3.2 start from

$$i = A_1\,\epsilon^{(-35+j281)t} + A_2\,\epsilon^{(-35-j281t)}$$

and find A_1 and A_2; then put the answer in terms of circular functions.

3-6 Repeat Problem 3-5 for q.

3-7 Repeat Problem 3-5 for v_L.

3-8 Perform the integration suggested in Eq. 3-3.26.

3-9 Start with the form of the solution given by Eq. 3-3.38, and determine i.

3-10 Take the derivative of Eq. 3-3.39, and solve the resulting equation for i.

3-11 Write Eq. 3-3.60 in the form

$$i = i_p + A_1\,\epsilon^{(-35+j281)t} + A_2\,\epsilon^{(-35-j281)t}$$

and find A_1 and A_2, then put the answer in terms of circular functions.

3-12 Write Eq. 3-3.54 in terms of q, and solve this equation for q. Finally take the derivative of q, and check the solution for i with Eq. 3-3.65.

3-13 Write Eq. 3-3.66 in terms of q, and solve this equation for q. Finally take the derivative of this q, and check the solution for i with Eq. 3-3.78.

3-14 The circuit shown is in steady-state when S is opened at $t = 0$. Find the equation for i after $t = 0$.

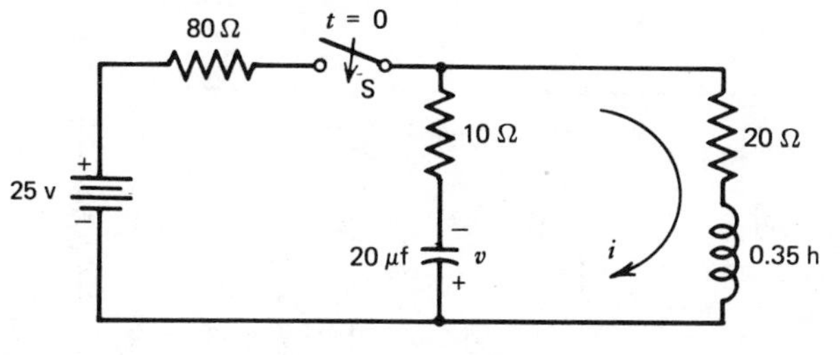

3-15 In Problem 3-14, find the equation for v.

3-16 The circuit shown is in steady-state when S is closed at $t = 0$. Find the equation for v.

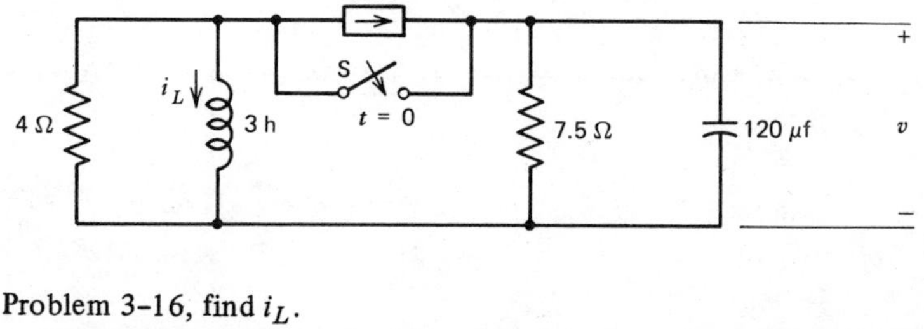

3-17 In Problem 3-16, find i_L.

3-18 The circuit shown is in steady-state when S is opened at $t = 0$. Find the equation for i_L. (The notation on e_s indicates that e_s is applied at the same instant S is opened.)

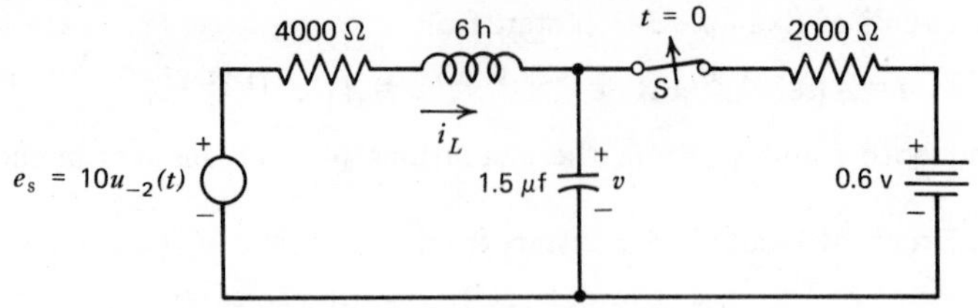

3-19 In Problem 3-18, find the voltage v on the capacitor.

3-20 The voltage e_s for (a) is shown in (b). The 60 volts is left long enough for the circuit to reach steady-state before $t = 0$. The switch S is opened at $t = 0$. Find the equation for v_C.

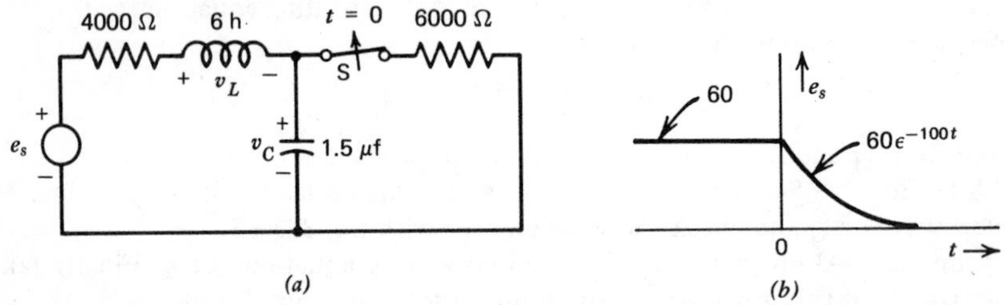

3-21 In Problem 3-20, find the equation for v_L.

3-22 The circuit shown is in steady-state and S is closed at $t = 0$, just as i_s is applied. Find the equation for v.

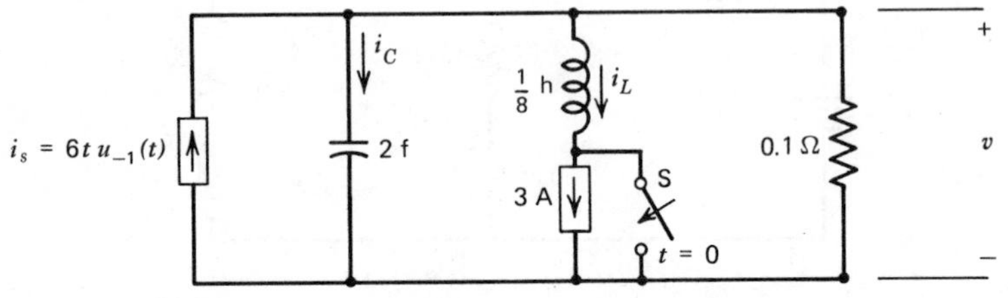

3-23 In Problem 3-22, find the current i_C.

3-24 The circuit shown is in steady-state when S is closed at $t = 0$. Find the equation for i_L.

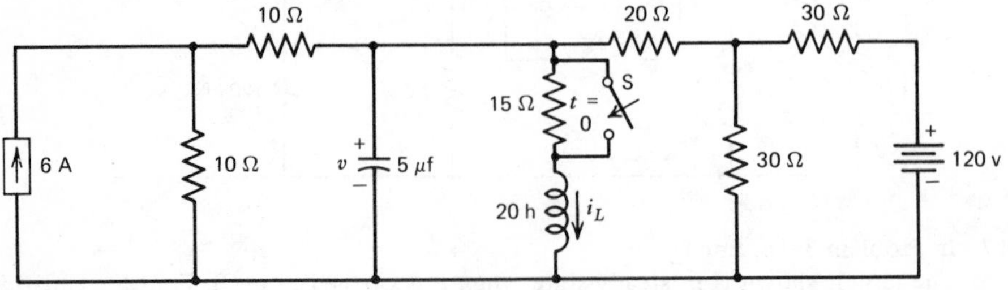

3-25 In Problem 3-24, find the voltage v.

3-26 The circuit shown is at rest before $t = 0$. Find i_R.

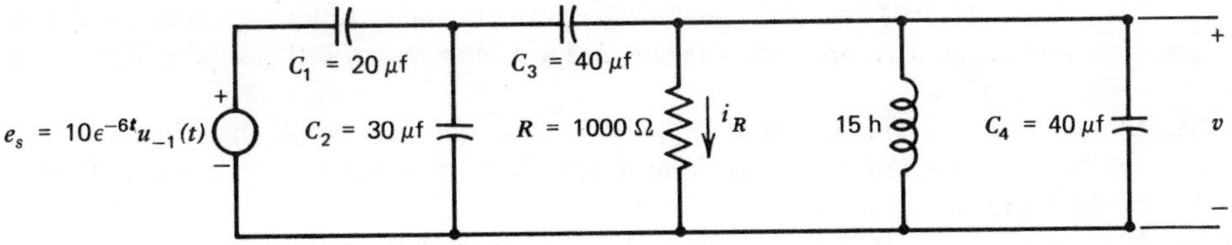

3-27 In Problem 3-26, find the voltage v on C_4.

3-28 The circuit shown is at rest before $t = 0$. Find i_R.

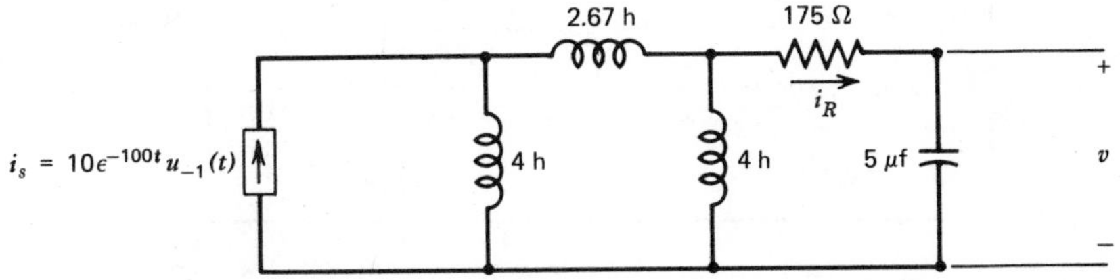

3-29 In Problem 3-28, find the voltage v.

3-30 The circuit shown is at rest before $t = 0$. Find the equation for i_{R3}.

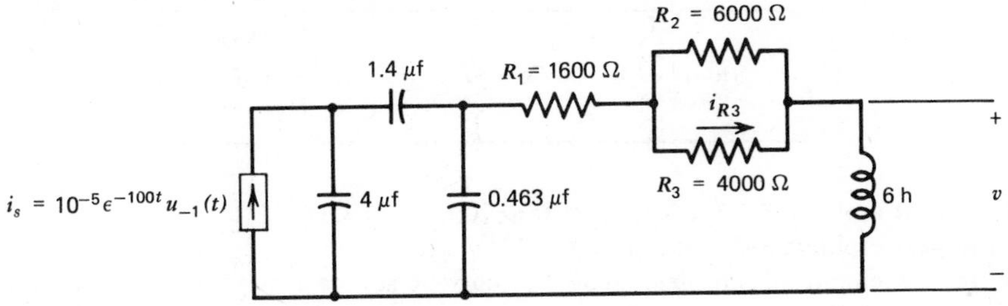

3-31 Repeat Problem 3-30, except find the equation for the voltage v.

3-32 In the circuit shown, the switch S is thrown to *Position 1* until steady-state is reached; and at $t = 0$, S is thrown to *Position 2*. The following values are given: $E = 20$ volts, $v(0) = 0$, and $\delta = 115°$. Find i.

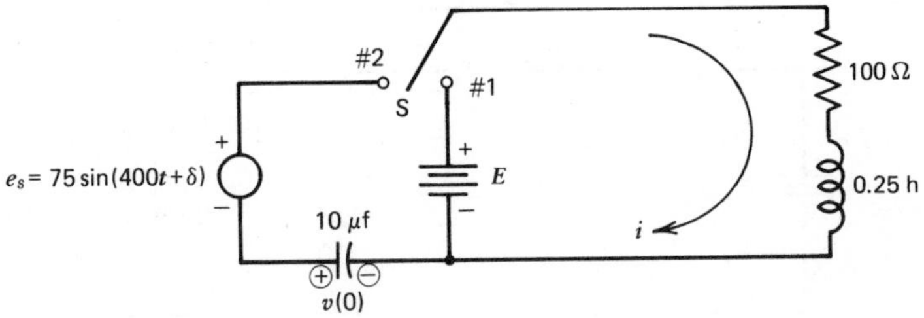

3-33 Repeat Problem 3-32, except use the following values: $E = 30$ volts, $v(0) = -50$ volts, and $\delta = 35°$.

3-34 Repeat Problem 3-32, except that $E = 20$ volts and $v(0)$ is a variable. Find such two values of $v(0)$ that it is possible to perform the switching to *Position 2* and have no transient component after $t = 0$. Find the value of δ that corresponds with each value of $v(0)$.

3-35 Repeat Problem 3-32, except find the largest such value of E that the switching can be done with no transient component after $t = 0$. Find the values of $v(0)$ and δ that correspond to this E.

3-36 The circuit shown is in steady-state when S is opened at $t = 0$. Find i.

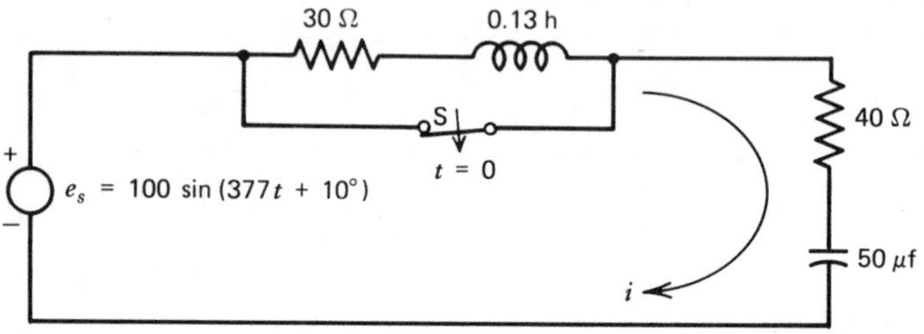

3-37 Given $i_s = 3u_{-1}(t)$. For the circuit shown, find such values of $i(0+)$ and $v(0+)$ (due to some previous switching operation) that the complementary component of the voltage v is zero.

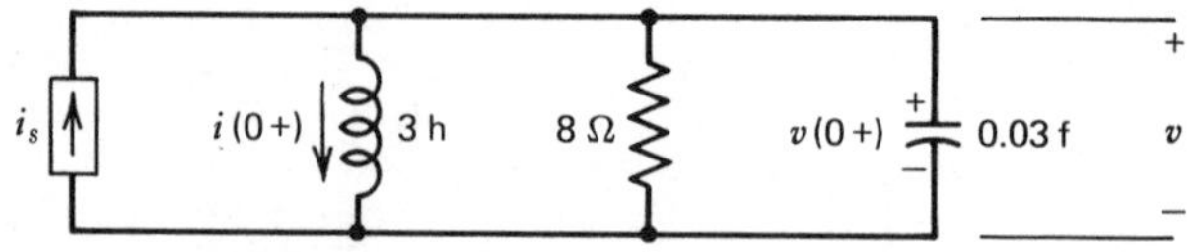

3-38 Repeat Problem 3-37, except $i_s = 3t\,u_{-1}(t)$.

3-39 Repeat Problem 3-37, except $i_s = 3\epsilon^{-5t}\,u_{-1}(t)$.

3-40 In the circuit of Problem 3-37 the current source $i_s = 3\sin(5t + \delta)\,u_{-1}(t)$. At $t = 0+$, the voltage $v(0+) = 0$, but the current $i(0+)$ can take on any desired value. Find such two values of δ that the transient component of the solution is zero, and find the appropriate value of $i(0+)$ for each value of δ.

3-41 The circuit shown is in steady-state when S is opened at $t = 0$. Find i_1.

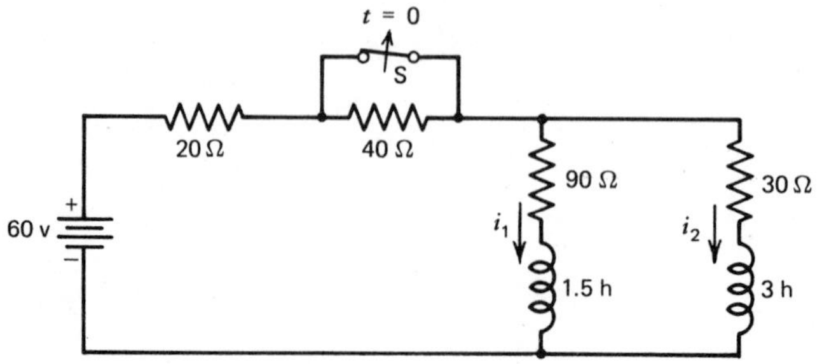

3-42 Repeat Problem 3-41, except find i_2.

3-43 In the circuit shown, the switch S is thrown to *Position 1* until steady-state is reached; and at $t = 0$, S is thrown to *Position 2*. Find the voltage from point A to point B.

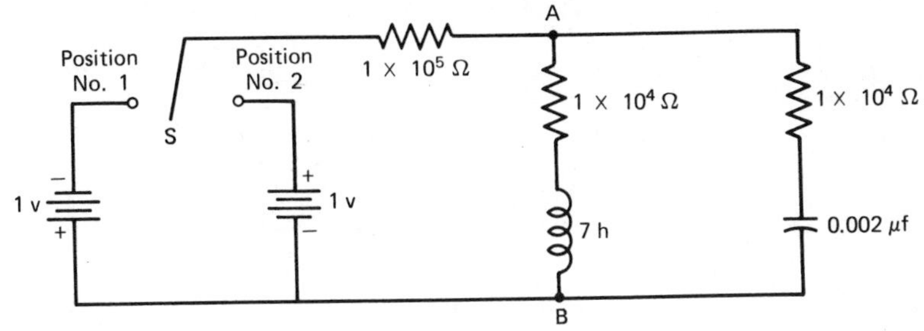

3-44 The circuit shown is in steady-state when S is opened at $t = 0$. Find i_1.

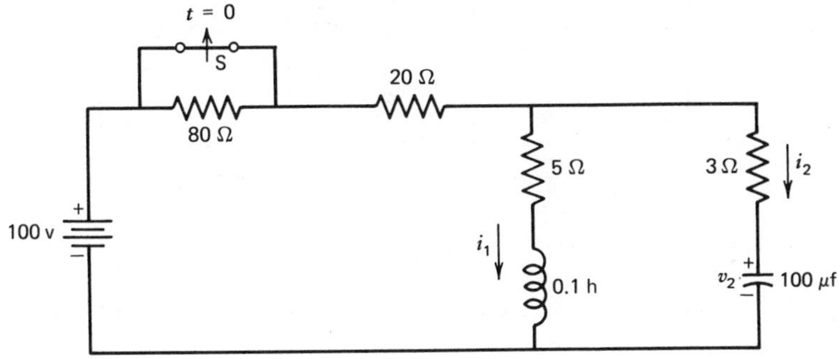

3-45 Repeat Problem 3-44, except find v_2.

3-46 The circuit shown is in steady-state when S is closed at $t = 0$. Find v_1.

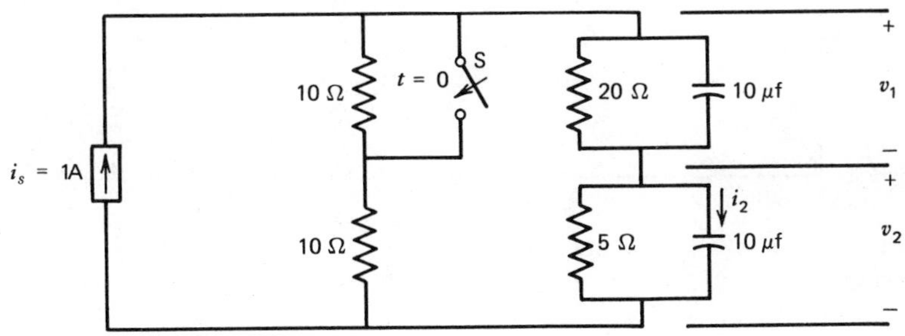

3-47 Repeat Problem 3-46, except find i_2.
3-48 The circuit shown is in steady-state when S is opened at $t = 0$. Find v_2.

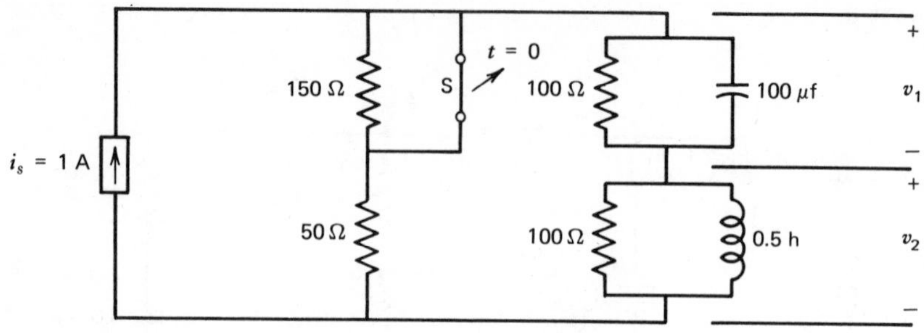

3-49 Repeat Problem 3-48, except find v_1.
3-50 The circuit shown is at rest before $t = 0$. Find i_1.

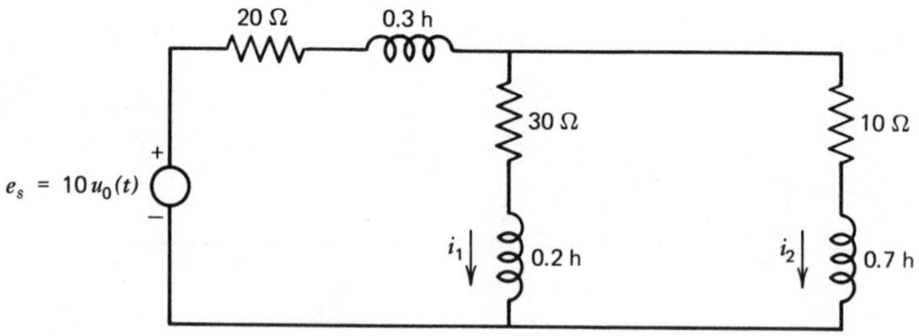

3-51 Repeat Problem 3-50, except $e_s = 10u_{-2}(t)$.
3-52 Repeat Problem 3-50, except $e_s = 10\epsilon^{-40t}\,u_{-1}(t)$.
3-53 The circuit shown is in steady-state when the switch S is closed at $t = 0$. Find i_1.

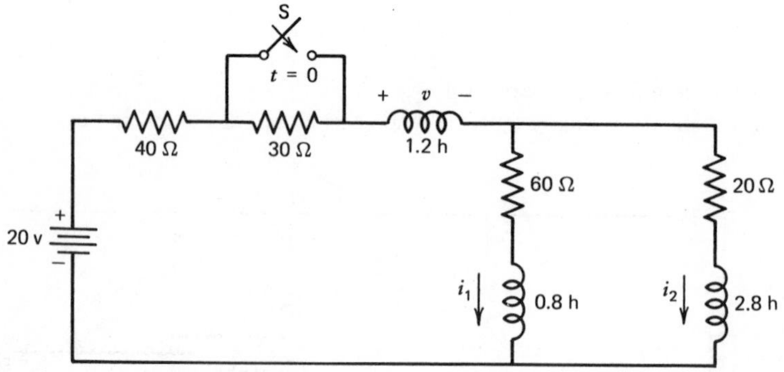

3-54 Repeat Problem 3-53, except find i_2.
3-55 Repeat Problem 3-54, except find v.

3-56 A system is described by a seventh-order differential equation, and the corresponding characteristic equation has the following roots:

$$-3 \quad -4 \quad -4 \quad -2 \pm j3 \quad -4 \pm j5$$

Write the form for the solution that is a current i.

3-57 Repeat Problem 3-56 for the ninth-order system whose roots are:

$$-2 \quad -2 \quad -2 \quad -2 \quad -3 \quad -2 \pm j4 \quad -2 \pm j4$$

3-58 Repeat Problem 3-56 for the eleventh-order system whose roots are:

$$-2 \quad -2 \quad -2 \quad -2 \quad -2 \quad -2 \quad -3 \quad -3 \quad -3 \quad -3 \pm j6$$

3-59 Repeat Problem 3-56 for the eighth-order system whose roots are:

$$-4 \pm 2 \quad -4 \pm 2 \quad -3 \pm j6 \quad -3 \pm j6$$

3-60 Repeat Problem 3-56 for the ninth-order system whose roots are:

$$-3 \quad -3 \quad -3 \quad -4 \pm 1 \quad 3 \pm j6 \quad 4 \pm j3$$

3-61 Repeat Problem 3-56 for the ninth-order system whose roots are:

$$-4 \quad -6 \quad -5 \pm 1 \quad -6 \quad 4 \pm j3 \quad 4 \pm j3$$

3-62 In Example 3-5.4, the dependent source $10j_2$ shown in Fig. 3-5.5(a) is replaced by a Kj_2, where K is a positive constant. Explore the nature of the roots of the characteristic equation, and draw a root-locus as K ranges from zero to infinity. Does the circuit ever become unstable?

3-63 Repeat Problem 3-62, except reverse the polarity of the dependent source.

3-64 Repeat Problem 3-62, except the dependent source is Kj_1.

3-65 Repeat Problem 3-64, except reverse the polarity of the dependent source.

Chapter **4**

One-Sided Laplace Transformation

4-1. INTRODUCTION

As discussed in Section 3-6, the classical method of solving differential equations involves a number of separate steps. The steps that are particularly bothersome are: (1) determining the n necessary initial conditions, and (2) solving the n simultaneous equations for the n arbitrary constants. Many attempts have been made to develop a method of solution involving the simplest possible set of initial conditions, and a method by which these conditions can be worked into a problem without using a group of auxiliary steps to evaluate the constants of integration.

One of these methods, widely used around the 1920's, is the Heaviside operational calculus. Although our purpose here is not that of an historian, it is interesting to note that Heaviside lived from 1850 to 1925. His methods were very controversial in that he did not prove the validity of his work, but more or less said you do this and this and you will get the answer.

Engineers and mathematicians spent much energy attempting to place a rigorous foundation under Heaviside's methods. In doing this, they turned to the work of an earlier mathematician, Laplace (1749-1827). Laplace had developed some integrals that were just the concept needed, and the result of this endeavor was the Laplace transformation methods.

4-2. THE DIRECT TRANSFORM INTEGRAL

The use of logarithms, in a sense, involves the idea of a transformation. By logarithm, the process of multiplication is transformed into addition, and the process of raising to a power is transformed into multiplication. In Chapters 2 and 3, the phasor transform is introduced and used.

The direct Laplace transformation, transforms a differential equation from the realm of time into the realm of a function of a complex variable. This may sound frightening until we say that for our purposes this is essentially the realm of algebra. The transform is manipulated as an algebraic function and the inverse Laplace transformation is performed to find the solution in the time domain.

By definition, the one-sided Laplace transform of a function of time $f(t)$ is obtained by multiplying $f(t)$ by ϵ^{-st}, and integrating the product between the limits of zero and infinity, as in

$$\mathcal{L}[f(t)] = \int_0^\infty f(t)\,\epsilon^{-st}\,dt = F(s) \tag{4-2.1}$$

The symbol $\mathcal{L}[f(t)]$ is read as the Laplace transform of $f(t)$. After integration, the time t is replaced with ∞ in the upper limit, and with 0 in the lower limit; therefore, t does not appear in the result, which is a function of s as indicated by the symbol $F(s)$. The $F(s)$ is said to be a one-sided Laplace transform, because the limits in the integral are zero and ∞. In Chapter 9 we take up the two-sided Laplace transform, where the limits on the integration are from $-\infty$ to $+\infty$. The s in $F(s)$ is a complex variable and has a real part σ, and an imaginary part $j\omega$, as given by

$$s = \sigma + j\omega \tag{4-2.2}$$

This s is sometimes thought of as a generalized frequency, and hence the s domain is called the frequency domain.

THE CONDITIONS FOR $F(s)$ TO EXIST

Since Eq. 4-2.1 is an improper integral, it exists only for a class of $f(t)$ that satisfy certain conditions. One way of stating these conditions is to say that $f(t)$ must be of exponential order and must be sectionally continuous over the range of $0 < t < \infty$. Fortunately, these conditions are usually satisfied by functions encountered in the physical sciences. We will justify these statements in later paragraphs.

A function $f(t)$ is sectionally continuous in a finite interval $t_1 \leq t \leq t_2$ if the $f(t)$ can be subdivided into a finite number of subintervals, in each of which $f(t)$ is continuous and has finite limits as t approaches either end point from the interior of the subinterval.

A function $f(t)$ is of exponential order if such a real positive constant σ exists that the product

$$|f(t)|\,\epsilon^{-\sigma t} \tag{4-2.3}$$

is bounded for all t greater than some finite number T and has the limit of zero as t approaches infinity. For each $f(t)$, this requirement places a limitation on the possible values of σ. The value of σ_0, termed the abscissa of convergence, is so defined that the limit of Expression 4-2.3 is zero if σ is greater than σ_0. For example, if $f(t) = \epsilon^{4t}$, then the limit of $\epsilon^{+4t}\,\epsilon^{-\sigma t}$ (as t approaches infinity) approaches zero only if σ is greater than 4. Therefore, for this $f(t)$, the value of σ_0 is 4.

The conditions just stated are the mathematical test to determine if an infinite integral converges. Since the Laplace transformation is defined as an infinite integral, these conditions apply here.

The condition that $f(t)$ is sectionally continuous is a sufficient condition and is not necessary; hence, it can be relaxed on occasion. A good example of this is the impulse function; which can be interpreted as having a Laplace transform, and yet the impulse is not sectionally continuous.

All the driving functions that have been examined thus far are of exponential order. Such functions as $E\,\epsilon^{-\sigma t}$, and $E_m \sin(\omega t + \sigma)$ are seen to be; the function t^n is more difficult. If t^n is of exponential order, then

$$\lim_{t\to\infty} t^n\,\epsilon^{-\sigma t} = 0 \tag{4-2.4}$$

The factor t^n goes to infinity, and $\epsilon^{-\sigma t}$ goes to zero, as t goes to infinity. Therefore in the limit, Eq. 4-2.4 becomes

$$\infty \cdot 0 \tag{4-2.5}$$

which is an undetermined form. To explore this, Eq. 4-2.4 is written

$$\lim_{t\to\infty} \frac{t^n}{\epsilon^{+\sigma t}} = \frac{\infty}{\infty} \tag{4-2.6}$$

which is evaluated by repeated use of L'Hopital's rule. After the numerator and the denominator are differentiated n times, the limit is zero as shown by

$$\lim_{t\to\infty} \frac{t^n}{\epsilon^{\sigma t}} = \lim_{t\to\infty} \frac{n!}{\sigma^n\,\epsilon^{\sigma t}} = 0 \tag{4-2.7}$$

as long as $\sigma > \sigma_0 = 0$.

As discussed in Chapter 3, the form for the solution to an nth-order differential equation is

$$i = i_p + A_1\,\epsilon^{-\alpha_1 t} + A_2\,\epsilon^{-\alpha_2 t} + \cdots + A_n\,\epsilon^{-\alpha_n t} \tag{4-2.8}$$

Each of the terms in the complementary component is of exponential order. Therefore, most of the functions encountered in the physical sciences fall within the restrictions necessary for the use of the Laplace transform.

AN $f(t)$ THAT DOES NOT HAVE AN $F(s)$

An example of an $f(t)$ that is not of exponential order is

$$f(t) = e^{+t^2}$$

If this $f(t)$ is multiplied by $\epsilon^{-\sigma t}$, as in

$$\epsilon^{+t^2}\epsilon^{-\sigma t} \tag{4-2.9}$$

no value σ exists, so the limit is zero as t goes to infinity.

COMMENTARY

The direct Laplace transform is defined by the integral, as in

$$\mathcal{L}[f(t)] = \int_0^\infty f(t)\,\epsilon^{-st}\,dt \tag{4-2.10}$$

This integral exists under certain conditions, but these conditions are satisfied by most of the time functions considered in this book. The direct transform is said to transform from the time domain to the frequency domain.

THE INVERSE LAPLACE TRANSFORM

The inverse Laplace transformation is indicated by

$$\mathcal{L}^{-1}[F(s)] = f(t) \tag{4-2.11}$$

The symbol $\mathcal{L}^{-1}[F(s)]$ is read as the inverse Laplace transform of $F(s)$. The inverse Laplace transform is said to transform from the frequency domain to the time domain. The inverse Laplace transform can also be expressed as an integral, which will be discussed briefly in this chapter. The mathematics necessary for handling this inverse integral depends upon a knowledge of the function of a complex variable, which is beyond the scope of this work. What we will do is develop a table of transform pairs, by starting with a variety of $f(t)$ and determining the corresponding $F(s)$. In any problem for which we want to find $\mathcal{L}^{-1}[F(s)]$, we manipulate the $F(s)$ into a form that is in the table of transform pairs and then from the table determine the corresponding $f(t)$.

4-3. PROPERTIES OF THE DIRECT TRANSFORM

The following properties are basic to the theory of the Laplace transformation.

Property 4-3.1. The transform of the product of (a constant) × (some function) is the product of the constant times the transform of the function, as in

$$\mathcal{L}[K\,f(t)] = \int_0^\infty K\,f(t)\,\epsilon^{-st}\,dt = K\int_0^\infty f(t)\,\epsilon^{-st}\,dt = K\,\mathcal{L}[f(t)] \tag{4-3.1}$$

Since the direct transform is based on an integral, the constant K can be moved outside the integral sign.

Property 4-3.2. The transform of a sum (or difference) of time functions is the sum (or difference) of the transforms of the individual time functions, as in

$$\mathcal{L}[f_1(t) + f_2(t) + \cdots + f_n(t)]$$

$$= \int_0^\infty [f_1(t) + f_2(t) + \cdots + f_n(t)]\, \epsilon^{-st}\, dt$$

$$= \int_0^\infty f_1(t)\, \epsilon^{-st}\, dt + \int_0^\infty f_2(t)\, \epsilon^{-st}\, dt + \cdots + \int_0^\infty f_n(t)\, \epsilon^{-st}\, dt$$

$$= \mathcal{L}[f_1(t)] + \mathcal{L}[f_2(t)] + \cdots + \mathcal{L}[f_n(t)] \qquad (4\text{-}3.2)$$

The direct transform is based on an integral and the integral of a sum is the sum of the integrals, provided that each of the integrals exists.

Property 4-3.3. The Laplace Transform of the Derivative of a Function. To determine the $\mathcal{L}[f'(t)]$, the defining integral is used, as in

$$\mathcal{L}[f'(t)] = \int_0^\infty f'(t)\, \epsilon^{-st}\, dt \qquad (4\text{-}3.3)$$

which can be evaluated using integration by parts

$$\int_0^\infty u\, dv = uv\Big|_0^\infty - \int_0^\infty v\, du \qquad (4\text{-}3.4)$$

with the following identifications:

$$\begin{aligned} dv &= f'(t)\, dt & u &= \epsilon^{-st} \\ v &= f(t) & du &= -s\, \epsilon^{-st}\, dt \end{aligned} \qquad (4\text{-}3.5)$$

Upon substitution, the integral becomes

$$\mathcal{L}[f'(t)] = [f(t)\, \epsilon^{-st}]\Big|_0^\infty - \int_0^\infty [-s\, \epsilon^{-st}]\, f(t)\, dt \qquad (4\text{-}3.6)$$

In the last term, the integration is with respect to t, therefore s is a constant and can be moved in front of the integral sign. This equation is rewritten, so as to examine the upper limit of the first term, as

$$\mathcal{L}[f'(t)] = \lim_{t \to \infty} [f(t)\, \epsilon^{-st}] - f(0)\, \epsilon^{-0} + s \int_0^\infty f(t)\, \epsilon^{-st}\, dt \qquad (4\text{-}3.7)$$

If $f(t)$ is of exponential order, and σ (the real part of $s = \sigma + j\omega$) is greater than σ_0, the upper limit term

$$\lim_{t \to \infty} [f(t)\, \epsilon^{-st}] \tag{4-3.8}$$

approaches zero. Therefore, Eq. 4-3.7 can be written

$$\mathcal{L}[f'(t)] = s\mathcal{L}[f(t)] - f(0) \tag{4-3.9}$$

where $f(0)$ is the value of $f(t)$ at $t = 0$.

This development shows that for Eq. 4-3.9 to be the Laplace transform of $f'(t)$, the $f(t)$ must be of exponential order. This fact adds no new restrictions to the work, because this is precisely what is needed for $\mathcal{L}[f(t)]$ to exist.

Property 4-3.4. The Laplace Transform of the Integral of a Function. To determine $\mathcal{L}[\int f(t)\, dt]$, we start from Eq. 4-3.9, repeated as

$$\mathcal{L}[f'(t)] = s\, \mathcal{L}[f(t)] - f(0) \tag{4-3.10}$$

and make the following replacements:

$$\begin{aligned} &\text{Replace } f(t) \text{ with } f^{-1}(t) \\ &\text{Replace } f(0) \text{ with } f^{-1}(0) \\ &\text{Replace } f'(t) \text{ with } f(t) \end{aligned} \tag{4-3.11}$$

What we are essentially doing is integrating Eq. 4-3.10 and writing the result as

$$\mathcal{L}[f(t)] = s\, \mathcal{L}[f^{-1}(t)] - f^{-1}(0) \tag{4-3.12}$$

which is solved for $\mathcal{L}[f^{-1}(t)]$ as

$$\mathcal{L}\left[\int f(t)\, dt\right] = \frac{\mathcal{L}[f(t)]}{s} + \frac{f^{-1}(0)}{s} \tag{4-3.13}$$

Equation 4-3.13 is the Laplace transform of the indefinite integral. Many times it is more convenient to write the original equation using the definite integral, and then taking the Laplace of the result. This is done as

$$\mathcal{L}\left[\int_0^t f(t)\, dt\right] = \frac{\mathcal{L}[f(t)]}{s} \tag{4-3.14}$$

An example is presented in the next section using both Eq. 4-3.13 and Eq. 4-3.14 to compare the two procedures.

Property 4-3.5. The Laplace Transform of ϵ^{-at} Times a Function. Often the transform of some $f(t)$ is known, and it is desired to determine the transform of this $f(t)$ multiplied by ϵ^{-at}, where a is an arbitrary constant.

We assume first that $f(t)$ has a Laplace transform, as indicated by

$$\int_0^\infty f(t)\,\epsilon^{-st}\,dt = F(s) \tag{4-3.15}$$

The transform of $\epsilon^{-at} f(t)$ is given formally by

$$\mathcal{L}[\epsilon^{-at} f(t)] = \int_0^\infty \epsilon^{-at} f(t)\,\epsilon^{-st}\,dt$$

$$= \int_0^\infty f(t)\,\epsilon^{-(s+a)t}\,dt = F(s+a) \tag{4-3.16}$$

Equations 4-3.15 and 16 can be interpreted in the following manner: if $f(t)$ has a transform $F(s)$, then $\epsilon^{-at} f(t)$ has a transform that can be obtained from $F(s)$ by replacing s by $s + a$.

Other properties are discussed later.

4-4. DEVELOPMENT OF TRANSFORM PAIRS

We are at a point analogous to an early stage in the discussion of calculus. Early in calculus, you learned to integrate (or differentiate) a number of functions to become familiar with the procedures. Later you depended more and more on tables.

We start by developing a short table of Laplace transform pairs. We see how we can expand this table almost indefinitely, and after this we use the table in Appendix E.

Example 4-4.1. The Transform of the Unit-Step Function. The first example is the unit-step function $u_{-1}(t)$. Since the one-sided Laplace transform is based on an integral from 0 to ∞, as far as the one-sided Laplace transform is concerned, $u_{-1}(t)$ is unity. The transform is obtained from

$$\mathcal{L}[1] = \int_0^\infty 1\,\epsilon^{-st}\,dt = -\frac{1}{s}\,\epsilon^{-st}\Big|_0^\infty$$

$$= -\frac{1}{s}\,[\lim_{t\to\infty} \epsilon^{-st} - \epsilon^{-s\cdot 0}] = \frac{1}{s} \tag{4-4.1}$$

In performing the integration, the term $\lim_{t\to\infty} \epsilon^{-st}$ is assumed to be zero. To examine this we start by remembering $s = \sigma + j\omega$, but we first replace s with its real part σ, getting

$$\lim_{t\to\infty} \epsilon^{-\sigma t} \tag{4-4.2}$$

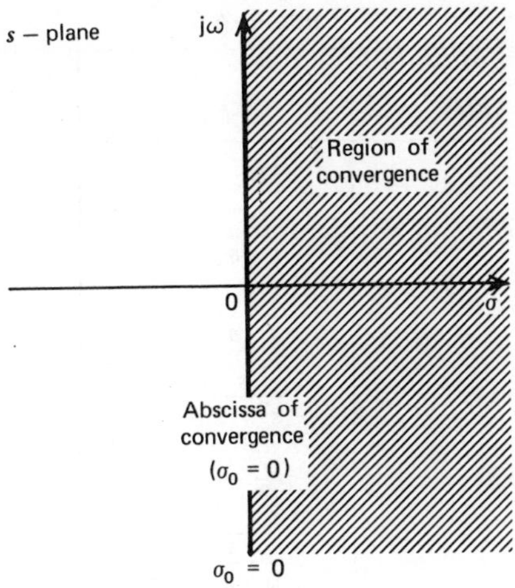

Fig. 4-4.1

This limit goes to zero as long as σ is greater than 0. Next we include the $j\omega$ part, and write

$$\lim_{t\to\infty} \epsilon^{-st} = \lim_{t\to\infty} \epsilon^{-\sigma t}\, \epsilon^{-j\omega t} \tag{4-4.3}$$

The factor $\epsilon^{-j\omega t}$ simply produces a rotation, and the factor $\epsilon^{-\sigma t}$ still goes to zero as $t \to \infty$, so long as σ is greater than 0. We summarize this by saying that σ_0 (the abscissa of convergence) is zero. Next, we show the s-plane as in Fig. 4-4.1, and draw a vertical line through $\sigma_0 = 0$. Finally we cross hatch the half of the s-plane to the right of σ_0, and we call this the region of convergence. The integral Eq. 4-4.1 converges as long as s is in the region of convergence.

We enter the result in Table 4-4.1.

Example 4-4.2. The Transform of ϵ^{-at}. The $f(t)$ to be used is $f(t) = \epsilon^{-at}$, and

$$\mathcal{L}[\epsilon^{-at}] = \int_0^\infty \epsilon^{-at}\, \epsilon^{-st}\, dt = \int_0^\infty \epsilon^{-(s+a)t}\, dt$$

$$= -\frac{1}{s+a}\, \epsilon^{-(s+a)t}\Big|_0^\infty = \frac{1}{s+a} \tag{4-4.4}$$

This result is entered into Table 4-4.1 as *Item 2*. In obtaining the result in Eq. 4-4.4, the upper limit $\lim_{t\to\infty} \epsilon^{-(s+a)t}$ is assumed to be zero, which implies an abscissa of convergence $\sigma_0 = -a$. Figure 4-4.2 indicates the vertical line through σ_0, with the half-plane to the right of σ_0 cross-hatched. The integral in Eq. 4-4.4 exists for all values of s in this half-plane.

Actually, the work in this example is not needed, because Property 4-3.5 and the

Table 4-4.1

Item no.	$f(t)$	$F(s)$	Abscissa of convergence σ_0
1	1	$1/s$	0
2	ϵ^{-at}	$\dfrac{1}{s+a}$	$-a$
3	ϵ^{+at}	$\dfrac{1}{s-a}$	$+a$
4	$\cosh bt$	$\dfrac{s}{s^2-b^2}$	$+b$
5	$\cos \beta t$	$\dfrac{s}{s^2+\beta^2}$	0
6	$\epsilon^{-at} \cosh bt$	$\dfrac{s+a}{(s+a)^2-b^2}$	$b-a$
7	$\epsilon^{-at} \cos \beta t$	$\dfrac{s+a}{(s+a)^2+\beta^2}$	$-a$
8	t^n	$\dfrac{n!}{s^{n+1}}$	0
9	$t^n \epsilon^{-at}$	$\dfrac{n!}{(s+a)^{n+1}}$	$-a$

first entry of Table 4-4.1 can be used to obtain the same results; that is, when $f(t)$ is 1, the corresponding $F(s)$ is $1/s$. If the original $f(t)$ is multiplied by ϵ^{-at}, the new $f(t)$ becomes $1\epsilon^{-at}$, and the corresponding $F(s)$ can be obtained from the original $F(s)$ by replacing s by $s + a$, which yields $1/(s + a)$.

Example 4-4.3. The Transform of ϵ^{+at}. Since a in *Item 2* of Table 4-4.1 is an arbitrary constant, this constant can be replaced by another constant, $-a$, leading to *Item 3*. *Item 2* and *Item 3* need not be listed separately, but are listed here to emphasize this point.

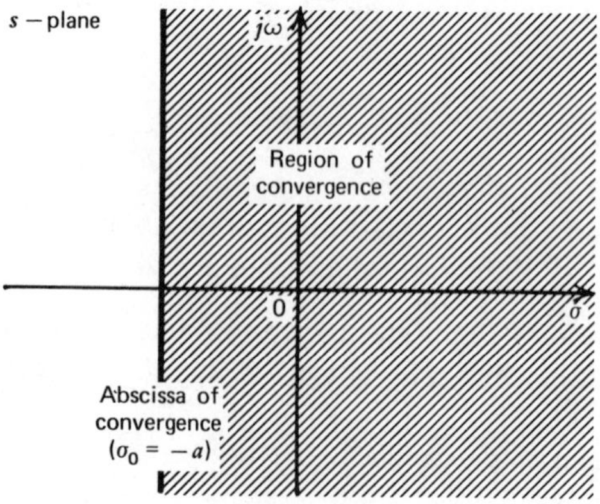

Fig. 4-4.2

Example 4-4.4. The Transform of cosh bt. We wish to find the transform of cosh bt. We could return to the integral as in

$$\mathcal{L}[\cosh bt] = \int_0^\infty \cosh bt\, \epsilon^{-st}\, dt = F(s) \tag{4-4.5}$$

Instead, we use the definition of cosh bt

$$\mathcal{L}[\cosh bt] = \mathcal{L}\left[\frac{\epsilon^{bt} + \epsilon^{-bt}}{2}\right] = \frac{1}{2}\{\mathcal{L}[\epsilon^{bt}] + \mathcal{L}[\epsilon^{-bt}]\}$$

$$= \frac{1}{2}\left[\frac{1}{(s-b)} + \frac{1}{(s+b)}\right] = \frac{1}{2}\left[\frac{s+b+s-b}{s^2-b^2}\right] = \frac{s}{s^2-b^2} \tag{4-4.6}$$

This result is entered into Table 4-4.1 as *Item 4*. Note that Properties 4-3.1 and 4-3.2 are used. Note also that the abscissa of convergence for $\mathcal{L}[\epsilon^{bt}]$ is $\sigma_0 = +b$, whereas the σ_0 for $\mathcal{L}[\epsilon^{-bt}]$ is $\sigma_0 = -b$. For the $\mathcal{L}[\cosh bt]$ we must take the one that is the more restrictive, hence for $\mathcal{L}[\cosh bt]$ the $\sigma_0 = +b$.

Example 4-4.5. The Transform of cos βt. The circular cosine and hyperbolic cosine are related, as

$$\cos \beta t = \cosh j\beta t \tag{4-4.7}$$

Therefore, we can find $\mathcal{L}[\cos \beta t]$ by using *Item 4* in the table, as

$$\mathcal{L}[\cos \beta t] = \mathcal{L}[\cosh j\beta t] = \frac{s}{s^2 - (j\beta)^2} = \frac{s}{s^2 + \beta^2} \tag{4-4.8}$$

where $\sigma_0 = 0$. This is entered as *Item 5* in the table.

Example 4-4.6. The Transform of ϵ^{-at} cosh bt. To find the $\mathcal{L}[\epsilon^{-at} \cosh bt]$, we return to *Item 4* and replace s with $s + a$, yielding

$$\mathcal{L}[\epsilon^{-at} \cosh bt] = \frac{s+a}{(s+a)^2 - b^2} \tag{4-4.9}$$

Here, $\sigma_0 = b - a$, and the result is entered as *Item 6* in the table.

Example 4-4.7. The Transform of ϵ^{-at} cos βt. Similarly to the last example

$$\mathcal{L}[\epsilon^{-at} \cos \beta t] = \frac{s+a}{(s+a)^2 + \beta^2} \tag{4-4.10}$$

with $\sigma_0 = -a$.

Example 4-4.8. The Transform of t. We could use the integral definition of $\mathcal{L}[t]$, but instead we start with the transform of the derivative

$$\mathcal{L}[f'(t)] = s\,\mathcal{L}[f(t)] - f(0) \tag{4-4.11}$$

If we let $f(t) = t$, then

$$f'(t) = 1$$

and we already know $\mathcal{L}[1] = 1/s$, and we reason that $f(0) = 0$. Upon substitution, Eq. 4-4.11 becomes

$$\mathcal{L}[1] = s\,\mathcal{L}[t] - 0$$

$$\mathcal{L}[t] = \frac{1}{s}\,\mathcal{L}[1] = \frac{1}{s^2} \tag{4-4.12}$$

Example 4-4.9. The Transform of t^2. We have a good thing going, so let us pursue it further by finding the $\mathcal{L}(t^2)$. We let

$$f(t) = t^2 \tag{4-4.13}$$

then

$$f'(t) = 2t \qquad f(0) = 0 \tag{4-4.14}$$

Substituting into Eq. 4-4.11 gives

$$\mathcal{L}[2t] = s\,\mathcal{L}[t^2] - 0$$

$$\mathcal{L}[t^2] = \frac{2\mathcal{L}[t]}{s} = \frac{2}{s^3} \tag{4-4.15}$$

Example 4-4.10. The Transform of t^n. The last two examples are generalized to

$$\mathcal{L}[t^n] = \frac{n!}{s^{n+1}} \tag{4-4.16}$$

Example 4-4.11. The Transform of $t^n\,\epsilon^{-at}$. Based on the last example and Property 4-3.5, we can write

$$\mathcal{L}[t^n\,\epsilon^{-at}] = \frac{n!}{(s+a)^{n+1}} \tag{4-4.17}$$

4-5. CIRCUIT EXAMPLES USING LAPLACE TRANSFORMS

Upon the first exposure to Laplace transforms, much of this material may not make sense until the beginner sees how all the parts fit together. This is done using a set of examples.

Example 4-5.1. The circuit shown in Fig. 4-5.1(a) is in steady-state before $t = 0$ with the switch S opened, and S is closed at $t = 0$. The desired solution is the current in the inductor after $t = 0$. Before $t = 0$ in steady-state, we observe that $i = 40/20 = 2$ amperes, therefore $i(0) = 2$; the circuit after $t = 0$ is shown in (b).

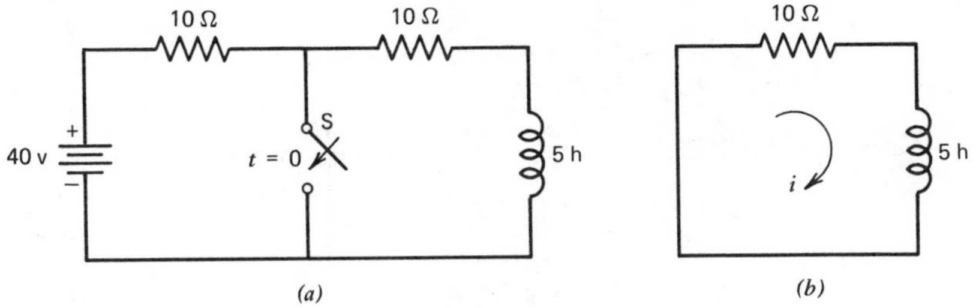

Fig. 4-5.1

Before we work the problem using Laplace methods, let us obtain the solution by classical methods. The solution has the form

$$i = A\,\epsilon^{-Rt/L} = A\,\epsilon^{-10t/5}$$
$$t = 0, i(0) = 2 = A \tag{4-5.1}$$

The solution is

$$i = 2\epsilon^{-2t} \tag{4-5.2}$$

To use Laplace methods, we write the equation

$$0 = 5\frac{di}{dt} + 10i \tag{4-5.3}$$

and the Laplace transformation is taken as

$$0 = 5\,[s\,I(s) - 2] + 10I(s) \tag{4-5.4}$$

The terms inside the brackets come from the Laplace of the derivative. This is the method whereby the initial condition is worked into the equation. Equation 4-5.4 is solved for $I(s)$ using the rules of algebra

$$[5s + 10]\,I(s) = 10 \tag{4-5.5}$$

$$I(s) = \frac{10}{(5s + 10)} \tag{4-5.6}$$

Next we consult Table 4-4.1 to see if there is an appropriate $F(s)$ similar in form to Eq. 4-5.6. If there is, we take the inverse transform by choosing the $f(t)$ that corresponds to the $F(s)$ selected. *Item 2* is almost what we want, except that the s term in the denominator is multiplied by 1 instead of by 5. Therefore, we modify Eq. 4-5.6 to get

$$I(s) = \frac{2}{(s + 2)} \tag{4-5.7}$$

The time function corresponding to this $F(s)$ is

$$\mathcal{L}^{-1}[I(s)] = i = 2\epsilon^{-2t} \tag{4-5.8}$$

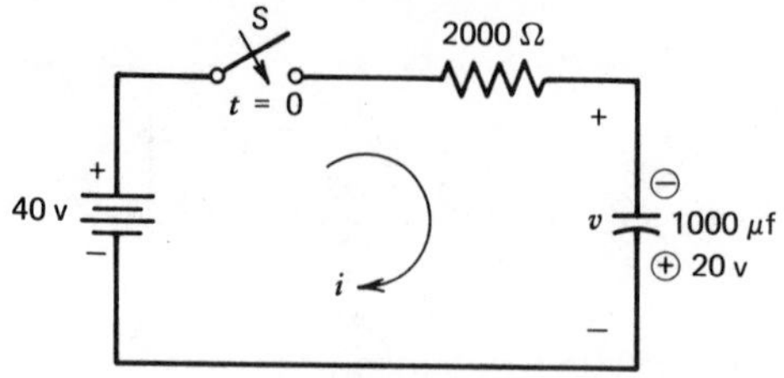

Fig. 4-5.2

Example 4-5.2. The circuit to be solved is shown in Fig. 4-5.2. The capacitor has 20 volts on it before $t = 0$, with polarity as shown by the ⊕, ⊖ markings. The direction for the current i is arbitrarily selected as shown. After $t = 0$, the KVLE becomes

$$40 = 2{,}000i + \frac{1}{1{,}000 \times 10^{-6}} \int_0^t i\,dt - 20 \tag{4-5.9}$$

The $\int_0^t i\,dt$ term has the polarity references as shown by the (+) and (−) marks, and since the 20 volts is on the capacitor in the opposite polarity, the 20 is minus.

Again, classical methods are used first in order to compare these with Laplace methods. At $t = 0+$, Eq. 4-5.9 yields

$$40 = 2{,}000i(0+) - 20 \tag{4-5.10}$$

$$i(0+) = \frac{60}{2{,}000} = 0.03 \tag{4-5.11}$$

The current i has the form

$$i = A\,\epsilon^{-t/RC} = A\,\epsilon^{-10^6 t/(2{,}000 \times 1{,}000)}$$

the A is found at $t = 0+$ as

$$A = 0.03 \tag{4-5.12}$$

and the solution is

$$i = 0.03\,\epsilon^{-0.5t} \tag{4-5.13}$$

Next the Laplace transform of Eq. 4-5.9 is taken, as in

$$\frac{40}{s} = 2{,}000\,I(s) + 1{,}000\frac{I(s)}{s} - \frac{20}{s} \tag{4-5.14}$$

Both the driving function of 40 volts and the initial voltage of −20 volts are treated as constants. Since Eq. 4-5.9 is written with the definite integral, the transform is taken as $I(s)/s$. We will rework this problem in just a moment using the indefinite integral to contrast the procedures.

Equation 4-5.14 is solved for $I(s)$, as

$$I(s) = \frac{60}{s[2{,}000 + 1000/s]} = \frac{60}{2{,}000s + 1{,}000} = \frac{0.03}{s + 0.5} \tag{4-5.15}$$

The inverse transform is found by using *Item 2* in Table 4-4.1, as

$$\mathcal{L}^{-1}[I(s)] = i = 0.03\,\epsilon^{-0.5t} \tag{4-5.16}$$

To demonstrate the use of the Laplace transform of the indefinite integral, Eq. 4-5.9 is written

$$40 = 2{,}000i + \frac{1}{1{,}000 \times 10^{-6}} \int i\,dt \tag{4-5.17}$$

To take the Laplace transform of this equation, Eq. 4-3.13 is used for the indefinite integral, which means $f^{-1}(0)$ needs to be determined. The $f(t)$ in this case is the current i; therefore $f^{-1}(0)$ is the charge $q(0)$. This is determined to be

$$q(0) = v(0) \times C = [-20] \times 10^{-3} = -0.02$$

The transform of Eq. 4-5.17 is

$$\frac{40}{s} = 2{,}000\,I(s) + \frac{1}{10^{-3}}\left[\frac{I(s)}{s} - \frac{0.02}{s}\right] \tag{4-5.18}$$

This equation is rewritten

$$\frac{40}{s} = 2{,}000\,I(s) + 1{,}000\,\frac{I(s)}{s} - \frac{20}{s} \tag{4-5.19}$$

This equation is identical to Eq. 4-5.14, so the remainder of the work is the same.

For most purposes, the definite integral is preferred and is used in this book.

Example 4-5.3. The circuit in Fig. 4-5.3 is the circuit used in Example 2-2.4. It might add to the comparison of the two methods to review this previous work. For $t > 0$, the

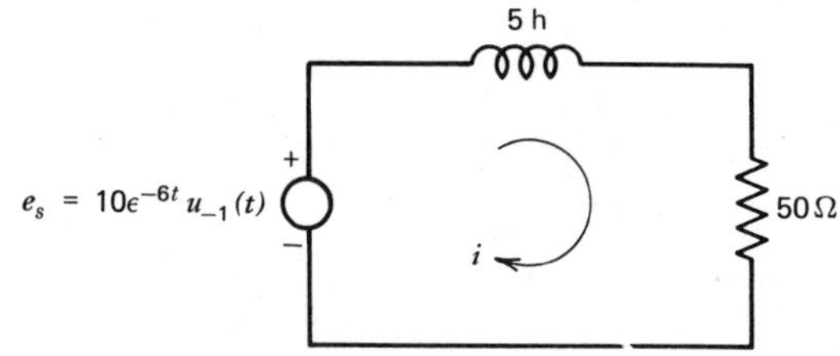

Fig. 4-5.3

KVLE is

$$10\epsilon^{-6t} = 5\frac{di}{dt} + 50i \tag{4-5.20}$$

The Laplace transform is taken and solved for $I(s)$, as

$$\frac{10}{s+6} = 5\,[s\,I(s) - 0] + 50I(s) \tag{4-5.21}$$

$$I(s) = \frac{2}{(s+6)(s+10)} \tag{4-5.22}$$

When Table 4-4.1 is examined, we find there is no appropriate entry for the inverse transform of Eq. 4-5.22. Therefore, the table must be extended to include an $F(s)$ of the form

$$F(s) = \frac{K}{(s+a)(s+b)} \tag{4-5.23}$$

What we want to do is *break up* this $F(s)$ into simpler terms already in the table. You may recall from algebra the idea of the partial fraction expansion. When used here, this $F(s)$ can be expanded as

$$\frac{K}{(s+a)(s+b)} = \frac{K_{-a}}{(s+a)} + \frac{K_{-b}}{(s+b)} \tag{4-5.24}$$

What we are doing is to demonstrate the need for the partial fraction expansion. In the next article we discuss this concept again and develop better methods for obtaining the expansion.

Both sides of Eq. 4-5.24 are multiplied by the denominator on the left side, giving

$$K = K_{-a}(s+b) + K_{-b}(s+a) \tag{4-5.25}$$

and this is rewritten as

$$K = (K_{-a} + K_{-b})\,s + (K_{-a}b + K_{-b}a) \tag{4-5.26}$$

The coefficients of corresponding powers of s on both sides of the equation are equated, and this leads to the two equations

$$K_{-a} + K_{-b} = 0 \qquad K_{-a}b + K_{-b}a = K \tag{4-5.27}$$

These are solved for K_{-a} and K_{-b} as

$$K_{-a} = \frac{K}{b-a} \qquad K_{-b} = \frac{K}{a-b} \tag{4-5.28}$$

and the desired expansion is

$$\frac{K}{(s+a)(s+b)} = \frac{K}{(b-a)(s+a)} + \frac{K}{(a-b)(s+b)} \tag{4-5.29}$$

where both of the terms on the right are in Table 4-4.1.

We return to Eq. 4-5.22, and use Eq. 4-5.29 to write

$$I(s) = \frac{2}{(s+6)(s+10)} = \frac{0.5}{s+6} - \frac{0.5}{s+10} \tag{4-5.30}$$

Item 2 in Table 4-4.1 is used on both terms, and the inverse transform of $I(s)$ is

$$\mathcal{L}^{-1}[I(s)] = i = 0.5\,(\epsilon^{-6t} - \epsilon^{-10t}) \tag{4-5.31}$$

Example 4-5.4. The circuit in Fig. 4-5.4(a) is in steady-state before $t = 0$ with the switch opened, and S is closed at $t = 0$. The desired solution is the current i in the inductor. Before $t = 0$ in steady-state, the current is $i = 40/20 = 2$ amperes; therefore $i(0) = 2$. After S is closed, Thevenin's Theorem is used to replace the circuit to the left of Points A-B and the resulting circuit is shown in (b).

The KVLE is

$$20 = 5\frac{di}{dt} + 15i \tag{4-5.32}$$

The Laplace is taken and solved for $I(s)$ as

$$\frac{20}{s} = 5\,[s\,I(s) - 2] + 15I(s)$$

$$I(s) = \frac{4}{s\,(s+3)} + \frac{2}{s+3} \tag{4-5.33}$$

The first term on the right can be expanded using Eq. 4-5.29, with the following identifications: $a = 0; b = 3; K = 4$

$$\frac{4}{s(s+3)} = \frac{4}{3s} - \frac{4}{3\,(s+3)} \tag{4-5.34}$$

Upon substitution and combining terms, Eq. 4-5.33 becomes

$$I(s) = \frac{1.333}{s} + \frac{0.666}{s+3} \tag{4-5.35}$$

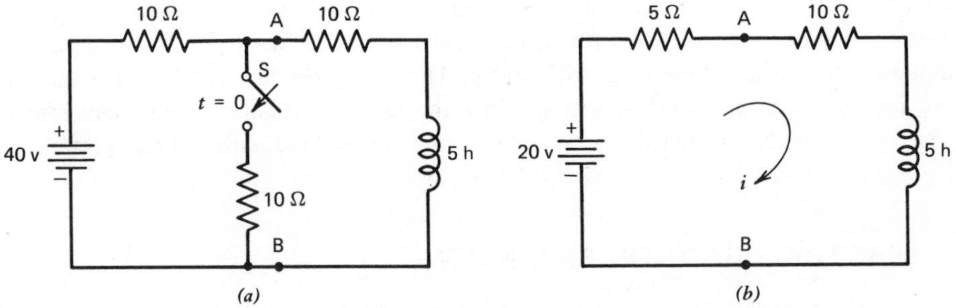

Fig. 4-5.4

The inverse transform is

$$\mathcal{L}^{-1}[I(s)] = i = 1.333 + 0.666\epsilon^{-3t} \tag{4-5.36}$$

COMMENTARY

These examples could be continued, but those presented give a feeling for how the Laplace transform is used. Other examples and comments are presented later.

The preliminary work (before the differential equation is written) is identical for the classical and Laplace methods. For example, the work needed to find the currents in the inductors and the voltages on the capacitors at $t = 0$ is the same. If a multi-loop-current circuit is reduced to a single-loop-current circuit by using Thevenin's theorem, the same work applies for both methods. The differences in methods result in how the problem is handled after the differential equation is written.

As soon as the differential equation is written, the Laplace transform is taken. The transformed equations contain all the information as to the initial currents in the inductors and the initial charges on the capacitors. The differential equation has been transformed into a function of s, which is manipulated by the rules of algebra to put it into a desired form that appears in a table, and the solution is obtained from the table. The solution has all the constants of integration (from the classical method) evaluated, and the particular and complementary components of the solution are found simultaneously with no distinction as to which is which.

The main advantage of the Laplace transform method is that it is a mechanistic procedure whereby information is plugged in, the crank is turned, and the solution comes out with a minimum of thinking on the part of the designer.

The main disadvantage of the Laplace transform method is that it is a mechanistic procedure whereby information is plugged in, the crank is turned and the solution comes out with a minimum of thinking on the part of the designer.

If you think you have just read the same statement twice, you are almost right. The point is that the very qualities that make the Laplace transforms so useful constitute its main disadvantage in that the work is so systematized that the user can almost stop thinking.

When it comes to manipulating equations, a person trained in pure mathematics should be better equipped to do this than the applied-scientist. If the designer adds any dimension to a problem, it is his physical interpretation of what goes on behind the mathematics. The classical method adds to this insight, but because of its systematized nature the Laplace method does not. If the Laplace methods are learned first, there is a temptation to use these methods exclusively. The best procedure is to use the method more appropriate in any given situation. Since the Laplace transform is just being introduced, its virtues are stressed at this time, but a more critical evaluation is made later.

4-6. PARTIAL FRACTION EXPANSION

In Ex. 4-5.3, the unknown coefficients in the partial fraction expansion were determined by multiplying the equation by the denominator of the function being expanded and equating the coefficients of like powers in s. This procedure leads to a set of simultaneous equations. Much of the drudgery of this method is eliminated by methods developed in the following examples.

Example 4-6.1. The first example is

$$F(s) = \frac{a_1 s + a_0}{(s+a)(s+b)} = \frac{K_{-a}}{s+a} + \frac{K_{-b}}{s+b} \tag{4-6.1}$$

This equation is examined with the thought of finding K_{-a}. To do this, both sides are multiplied by $s + a$, as

$$F(s)(s+a) = \frac{a_1 s + a_0}{(s+b)} = K_{-a} + \frac{K_{-b}(s+a)}{(s+b)} \tag{4-6.2}$$

Now examine this equation for a moment and ask yourself the question, what is the simplest way to find K_{-a}? The answer is to let $s = -a$, which eliminates the K_{-b} term and yields

$$K_{-a} = \frac{a_0 - a_1 a}{(b-a)} \tag{4-6.3}$$

In a similar manner, K_{-b} is found by multiplying Eq. 4-6.1 by $(s + b)$

$$F(s)(s+b) = \frac{a_1 s + a_0}{(s+a)} = \frac{K_{-a}(s+b)}{(s+a)} + K_{-b} \tag{4-6.4}$$

and letting $s = -b$ as

$$K_{-b} = \frac{a_0 - a_1 a}{(b-a)} \tag{4-6.5}$$

Upon substitution, Eq. 4-6.1 becomes

$$\frac{a_1 s + a_0}{(s+a)(s+b)} = \frac{a_0 - a_1 a}{(b-a)(s+a)} + \frac{a_0 - a_1 b}{(a-b)(s+b)} \tag{4-6.6}$$

and the inverse transform is

$$\mathcal{L}^{-1}\left[\frac{a_1 s + a_0}{(s+a)(s+b)}\right] = \left[\frac{a_0 - a_1 a}{(b-a)}\right]\epsilon^{-at} + \left[\frac{a_0 - a_1 b}{a-b}\right]\epsilon^{-bt} \tag{4-6.7}$$

This is *Item 14* in Appendix E. Many of the other items were determined in precisely this same manner.

THE COVER-UP TECHNIQUE

A rule of thumb that is convenient to remember can be stated thus: If a factor such as K_{-a} in Eq. 4-6.1 is to be evaluated, the $(s + a)$ factor in the denominator of $F(s)$ is *covered up*, and s throughout the rest of the $F(s)$ function is assigned the value that makes the covered-up term equal to zero, which in this case is $s = -a$. This technique is a simplified version of the multiplication of both sides of the equation by this factor. By use of this procedure, the partial fraction expansion can be written by inspection.

Example 4-6.2. The second example is

$$F(s) = \frac{a_2 s^2 + a_1 s + a_0}{s^2 (s + b)} = \frac{A}{s^2} + \frac{K_0}{s} + \frac{K_{-b}}{(s + b)} \tag{4-6.8}$$

A variation of the "cover-up" technique will work on two of the three terms in Eq. 4-6.8, but not on the third. This is best demonstrated by proceeding with the example.

If both sides of Eq. 4-6.8 are multiplied by s^2, the result is

$$F(s)\, s^2 = \frac{a_2 s^2 + a_1 s + a_0}{(s + b)} = A + K_0 s + \frac{K_{-b} s^2}{(s + b)} \tag{4-6.9}$$

The A can be found by letting $s = 0$, as

$$A = \frac{a_0}{b} \tag{4-6.10}$$

If both sides of Eq. 4-6.8 are multiplied by $(s + b)$, the result is

$$F(s)(s + b) = \frac{a_2 s^2 + a_1 s + a_0}{s^2} = \frac{A}{s^2}(s + b) + \frac{K_0}{s}(s + b) + K_{-b} \tag{4-6.11}$$

The K_{-b} can be found by letting $s = -b$, as

$$K_{-b} = \frac{a_2 b^2 - a_1 b + a_0}{b^2} \tag{4-6.12}$$

The steps just outlined for obtaining A and K_{-b} are precisely those of the cover-up technique. However, if we follow a parallel procedure to find K_0, the method does not work. In attempting to do this, Eq. 4-6.8 is multiplied by s, as

$$F(s)\, s = \frac{a_2 s^2 + a_1 s + a_0}{s (s + b)} = \frac{A}{s} + K_0 + \frac{K_{-b} s}{(s + b)} \tag{4-6.13}$$

If s is set equal to zero, the result for K_0 is

$$K_0 = +\infty - \infty \tag{4-6.14}$$

This indeterminant form can be evaluated by combining the two terms that produce the infinities before s takes on the value of zero.

The K_0 term can be found by a number of different methods, of which two are discussed.

Method 1. Since A and K_{-b} are known, Eq. 4-6.8 is rewritten with these values substituted, as

$$\frac{a_2 s^2 + a_1 s + a_0}{s^2 (s + b)} = \frac{a_0}{b s^2} + \frac{K_0}{s} + \frac{(a_2 b^2 - a_1 b + a_0)}{b^2 (s + b)} \tag{4-6.15}$$

The K_0 is the only unknown in this equation. Since this equation holds for all possible values of s, K_0 can be found by letting s take on any convenient non-zero value, such as $s = 1$, giving

$$\frac{a_2 + a_1 + a_0}{(1+b)} = \frac{a_0}{b} + K_0 + \frac{a_2 b^2 - a_1 b + a_0}{b^2(1+b)} \tag{4-6.16}$$

Upon solving, K_0 is

$$K_0 = \frac{a_1}{b} - \frac{a_0}{b^2} \tag{4-6.17}$$

Method 2. This method is more elegant mathematically and is the one usually presented in books, although it is not necessarily the easiest to apply. We return to Eq. 4-6.9, repeated as

$$F(s)\, s^2 = \frac{a_2 s^2 + a_1 s + a_0}{(s+b)} = A + K_0 s + \frac{K_{-b} s^2}{(s+b)} \tag{4-6.18}$$

and ask how can we find K_0 without first finding A or K_{-b}. The answer to this is to differentiate both sides of the equation with respect to s, as

$$\frac{d}{ds}\left[\frac{a_2 s^2 + a_1 s + a_0}{(s+b)}\right] = K_0 + K_{-b}\left[\frac{(s+b)\,2s - s^2(1)}{(s+b)^2}\right] \tag{4-6.19}$$

and let $s = 0$ in the result, giving

$$K_0 = \frac{a_1}{b} - \frac{a_0}{b^2} \tag{4-6.20}$$

No matter how the three constants of Eq. 4-6.8 are determined, the partial fraction expansion is obtained as

$$F(s) = \frac{a_2 s^2 + a_1 s + a_0}{s^2(s+b)} = \frac{a_0}{bs^2} + \left(\frac{a_1}{b} - \frac{a_0}{b^2}\right)\frac{1}{s} + \frac{a_2 b^2 - a_1 b + a_0}{b^2(s+b)} \tag{4-6.21}$$

and the $f(t)$ is found as

$$f(t) = \mathcal{L}^{-1}[F(s)] = \frac{a_0}{b}\, t + \left(\frac{a_1}{b} - \frac{a_0}{b^2}\right) + \frac{a_2 b^2 - a_1 b + a_0}{b^2}\, \epsilon^{-bt} \tag{4-6.22}$$

Example 4-6.3. The $F(s)$ for this example is

$$F(s) = \frac{a_3 s^3 + a_2 s^2 + a_1 s + a_0}{s^3(s+b)} = \frac{A}{s^3} + \frac{B}{s^2} + \frac{K_0}{s} + \frac{K_{-b}}{s+b} \tag{4-6.23}$$

The A and K_{-b} terms can each be found using the cover-up technique, as

$$A = \frac{a_0}{b} \tag{4-6.24}$$

$$K_{-b} = \frac{a_0 - a_1 b + a_2 b^2 - a_3 b^3}{-b^3} \tag{4-6.25}$$

We will find B and K_0 by using only *Method No. 2* of the last example. Eq. 4-6.23 is multiplied by s^3, as

$$\frac{a_3s^3 + a_2s^2 + a_1s + a_0}{(s+b)} = A + Bs + K_0s^2 + \frac{K_{-b}s^3}{(s+b)} \tag{4-6.26}$$

This equation is differentiated with respect to s, as

$$\frac{d}{ds}\left[\frac{a_3s^3 + a_2s^2 + a_1s + a_0}{(s+b)}\right] = B + 2K_0s + K_{-b}\frac{d}{ds}\left[\frac{s^3}{s+b}\right] \tag{4-6.27}$$

and s is set equal to zero to find B, giving

$$B = \left\{\frac{d}{ds}\left[\frac{a_3s^3 + a_2s^2 + a_1s + a_0}{(s+b)}\right]\right\}\Bigg|_{s=0} = \left(\frac{a_1}{b} - \frac{a_0}{b^2}\right) \tag{4-6.28}$$

Equation 4-6.27 is differentiated again, as

$$\frac{d^2}{ds^2}\left[\frac{a_3s^3 + a_2s^2 + a_1s + a_0}{(s+b)}\right] = 2K_0 + K_{-b}\frac{d^2}{ds^2}\left[\frac{s^3}{s+b}\right] \tag{4-6.29}$$

and s is set equal to zero to find K_0, giving

$$K_0 = \frac{1}{2}\left\{\frac{d^2}{ds^2}\left[\frac{a_3s^3 + a_2s^2 + a_1s + a_0}{(s+b)}\right]\right\}\Bigg|_{s=0} = \left(\frac{a_2}{b} - \frac{a_1}{b^2} + \frac{a_0}{b^3}\right) \tag{4-6.30}$$

These values for A, B, K_0 and K_{-b} can be substituted into Eq. 4-6.23, and the inverse transform taken as

$$\mathcal{L}^{-1}\left[\frac{a_3s^3 + a_2s^2 + a_1s + a_0}{s^3(s+b)}\right] = \frac{A}{2}t^2 + Bt + K_0 + K_{-b}\,\epsilon^{-bt} \tag{4-6.31}$$

where the constants have the values just determined.

4-7. HEAVISIDE EXPANSION THEOREM

The Heaviside expansion theorem is another method for obtaining the partial fraction expansion. The theorem is presented here for the case in which all the factors in the denominator are simple, that is, raised to the first power. The $F(s)$ is a ratio of two polynomials, as

$$F(s) = \frac{N(s)}{D(s)} \tag{4-7.1}$$

We assume that $(s + a)$ is one of the factors of $D(s)$ and that the product of all other factors in $D(s)$ is given by $G(s)$. In other words, $D(s)$ is

$$D(s) = (s+a)\,G(s) \tag{4-7.2}$$

The $F(s)$ can be written and the partial fraction expansion indicated as

$$F(s) = \frac{N(s)}{(s+a)\,G(s)} = \frac{K_{-a}}{(s+a)} + \cdots \tag{4-7.3}$$

The K_{-a} factor as found by using the cover-up technique is

$$K_{-a} = \frac{N(-a)}{G(-a)} \tag{4-7.4}$$

The theorem achieves this same result by an alternate procedure. The K_{-a} factor is found from

$$K_{-a} = \left.\frac{N(s)}{d/ds\,[D(s)]}\right|_{s=-a} \tag{4-7.5}$$

The K_{-a} from Eqs. 4-7.4 and 4-7.5 is the same as can be seen from:

$$\begin{aligned} K_{-a} &= \left.\frac{N(s)}{d/ds\,[D(s)]}\right|_{s=-a} = \left.\frac{N(s)}{d/ds\,[(s+a)\,G(s)]}\right|_{s=-a} \\ &= \left.\frac{N(s)}{(s+a)\,G'(s) + G(s)}\right|_{s=-a} = \frac{N(-a)}{G(-a)} \end{aligned} \tag{4-7.6}$$

The following numerical examples demonstrate the method.

Example 4-7.1. The $F(s)$ and its expansion already determined are

$$F(s) = \frac{(s+3)}{(s+1)(s+2)} = \frac{K_{-1}}{(s+1)} + \frac{K_{-2}}{(s+2)} = \frac{2}{(s+1)} + \frac{-1}{(s+2)} \tag{4-7.7}$$

The Heaviside expansion theorem is used to determine K_{-1} and K_{-2}, as

$$\frac{N(s)}{d/ds\,[D(s)]} = \frac{(s+3)}{d/ds\,[s^2+3s+2]} = \frac{(s+3)}{(2s+3)} \tag{4-7.8}$$

$$K_{-1} = \left.\frac{s+3}{2s+3}\right|_{s=-1} = 2 \qquad K_{-2} = \left.\frac{s+3}{2s+3}\right|_{s=-2} = -1 \tag{4-7.9}$$

These results check those of Eq. 4-7.7.

Example 4-7.2. The $F(s)$ and its expansion already determined are

$$\begin{aligned} F(s) = \frac{(s+3)(s+5)}{(s+1)(s+2)(s+4)} &= \frac{K_{-1}}{(s+1)} + \frac{K_{-2}}{(s+2)} + \frac{K_{-4}}{(s+4)} \\ &= \frac{8/3}{(s+1)} + \frac{-3/2}{(s+2)} + \frac{-1/6}{(s+4)} \end{aligned} \tag{4-7.10}$$

The Heaviside expansion theorem is used to determine the K's, as

$$\frac{N(s)}{d/ds\,[D(s)]} = \frac{(s+3)(s+5)}{d/ds\,[s^3+7s^2+14s+8]} = \frac{(s+3)(s+5)}{3s^2+14s+14} \tag{4-7.11}$$

$$K_{-1} = \left.\frac{(s+3)(s+5)}{3s^2+14s+14}\right|_{s=-1} = \frac{8}{3}$$

$$K_{-2} = \left.\frac{(s+3)(s+5)}{3s^2+14s+14}\right|_{s=-2} = -\frac{3}{2} \tag{4-7.12}$$

$$K_{-4} = \left.\frac{(s+3)(s+5)}{3s^2+14s+14}\right|_{s=-4} = -\frac{1}{6}$$

The results check those of Eq. 4-7.10.

4-8. DISCUSSION OF THE $F(s)$ FUNCTION

The direct Laplace transform changes an $f(t)$ into an $F(s)$ that is a function of a complex variable $s = \sigma + j\omega$. The field of study involving functions of a complex variable is well developed, and many of the concepts used in this book are applications of some of the simpler concepts of this theory.

Most $F(s)$ functions in circuit and system theory are single-valued functions of s. That is, for a specific value of the complex variable s, $F(s)$ is uniquely determined and also is complex, having a real and an imaginary part, as

$$F(s) = U(\sigma, \omega) + jV(\sigma, \omega)$$

where $U(\sigma, \omega)$ is the real part, and $jV(\sigma, \omega)$ is the imaginary part.

An $F(s)$ is said to be analytic at a point s, if $F(s)$ has a unique derivative at each point in a region including the point s. If the definition for the derivative of a real function of a single variable is extended to functions of a complex variable, the derivative is given by

$$\frac{d[F(s)]}{ds} = \lim_{\Delta s \to 0} \frac{\Delta F(s)}{\Delta s} = \lim_{\Delta s \to 0} \frac{F(s+\Delta s) - F(s)}{\Delta s} \tag{4-8.1}$$

where $\Delta F(s) = \Delta U + j\Delta V$, and the limit is independent of the choice of Δs. This equation is more complex than the corresponding case for real functions, in the sense that $\Delta s = \Delta\sigma + j\Delta\omega$ can approach zero along an infinite number of different paths. One such path is for $\Delta\omega = 0$ and $\Delta s = \Delta\sigma$. For this path, Eq. 4-8.1 becomes

$$\frac{d[F(s)]}{ds} = \lim_{\Delta\sigma \to 0} \left[\frac{\Delta U}{\Delta\sigma} + j\frac{\Delta V}{\Delta\sigma}\right] = \frac{\partial U}{\partial\sigma} + j\frac{\partial V}{\partial\sigma} \tag{4-8.2}$$

A second such path is for $\Delta\sigma = 0$ and $\Delta s = j\Delta\omega$. For this path, Eq. 4-8.1 becomes

$$\frac{d[F(s)]}{ds} = \lim_{j\Delta\omega \to 0} \left[\frac{\Delta U}{j\Delta\omega} + \frac{j\Delta V}{j\Delta\omega}\right] = -j\frac{\partial U}{\partial\omega} + \frac{\partial V}{\partial\omega} \tag{4-8.3}$$

For the derivative to be unique, Eqs. 4-8.2 and 4-8.3 must be equal, or

$$\frac{\partial U}{\partial \sigma} + j\frac{\partial V}{\partial \sigma} = \frac{\partial V}{\partial \omega} - j\frac{\partial U}{\partial \omega} \tag{4-8.4}$$

This leads to the relationships

$$\frac{\partial U}{\partial \sigma} = \frac{\partial V}{\partial \omega} \qquad \frac{\partial V}{\partial \sigma} = -\frac{\partial U}{\partial \omega} \tag{4-8.5}$$

These equations are the well-known Cauchy-Riemann conditions. It can be shown that if Δs can approach zero along the two paths just discussed, and the respective derivatives are equal, then Δs can approach zero along any path and the derivative is unique.

More generally, if both U and V and all four partial derivatives are continuous functions of s and satisfy the Cauchy-Riemann conditions at each point in a region including a specified point s, the function is analytic at the point s.

An example often given is

$$F(s) = s^2 = \sigma^2 - \omega^2 + j2\sigma\omega \tag{4-8.6}$$

In this case

$$\frac{\partial U}{\partial \sigma} = 2\sigma = \frac{\partial V}{\partial \omega} \qquad \frac{\partial V}{\partial \sigma} = 2\omega = -\frac{\partial U}{\partial \omega} \tag{4-8.7}$$

Therefore s^2 is analytic at every finite point in the s-plane and its derivative can be found to be $2s$ by use of either Eq. 4-8.2 or Eq. 4-8.3. In general, $F(s) = s^n$ also can be shown to be analytic at every finite point in the s-plane, and its derivative is ns^{n-1}.

Examples of functions of s that are not analytic are $|s|$, the magnitude of s, and $\bar{s}$, the conjugate of s.

The point (or points) at which a function $F(s)$ is not analytic is called a singular point. A typical $F(s)$ encountered in system theory, given in factored form, is

$$F(s) = \frac{K(s+z_1)(s+z_2)^2}{(s+p_1)(s+p_2)^2(s+p_3)^4} \tag{4-8.8}$$

This $F(s)$ has singular points at $s = -p_1$, $s = -p_2$, and $s = -p_3$. These are a specific type of singularities termed *poles*. A pole is defined in the following manner: If an $F(s)$ increases without bound as s approaches a value $-s_0$ in such a manner that the function

$$G(s) = (s + s_0)^n F(s) \tag{4-8.9}$$

has a finite non-zero value at $-s_0$ for some positive integer n, then $-s_0$ is termed a pole, of order n, of $F(s)$. When n is equal to unity, the pole is called a simple pole. For the $F(s)$ of Eq. 4-8.8, $s = -p_1$ is a simple pole; $s = -p_2$ is a second-order pole; and $s = -p_3$ is a fourth-order pole.

A value of s that makes $F(s)$ equal to zero is termed a zero of $F(s)$. For the $F(s)$ of Eq. 4-8.8, $s = -z_1$ is a simple zero and $s = -z_2$ is a second-order zero.

If the point at infinity is included, a function has as many zeros as it has poles, and vice versa. The example being considered has three finite zeros and seven finite poles.

Hence, there must be four zeros at infinity. The truth of this can be demonstrated by the fact that as $|s|$ approaches infinity, each of the z's and p's can eventually be ignored when compared with s. Therefore

$$\lim_{|s|\to\infty} F(s) = \lim_{|s|\to\infty} \frac{Ks^3}{s^7} = \lim_{|s|\to\infty} \frac{K}{s^4} \tag{4-8.10}$$

and the fourth-order zero at infinity is displayed.

The function

$$F(s) = \frac{K(s+z_1)(s+z_2)}{(s+p_1)(s+p_2)(s+p_3)} \tag{4-8.11}$$

is considered. It is helpful to locate the poles and zeros on a sketch of the complex s-plane as shown in Fig. 4-8.1. The locations of the zeros are shown by circles and those of the poles by x symbols. Except for the K factor of Eq. 4-8.11, the pole-zero configuration completely specifies the $F(s)$ function.

If the magnitude of the $F(s)$ function is considered as equal to the distance perpendicular to the s-plane of Fig. 4-8.1, then as s takes on all possible values, the plot of $|F(s)|$ forms a three-dimensional surface. For instance, as s approaches one of the zeros, say $-z_1$, $F(s)$ approaches the value zero. Likewise, if s approaches one of the poles, say $-p_1$, $|F(s)|$ increases in magnitude without bound. The geometry of $|F(s)|$ in the vicinity of a pole suggests the name pole.

A rather crude analogy to the shape of an $|F(s)|$ surface is that of a thin rubber membrane stretched taut over a large tabletop. If $F(s)$ has a zero at $s = \infty$, the outer portions of the membrane are clamped to the tabletop. At the locations of the finite zeros, the rubber membrane is held to the tabletop with tacks. At the poles, the membrane should be pushed upward to an infinite height by rods of zero diameter. The analogy is crude, but the shape of the rubber membrane represents, in a general way, the $F(s)$ surface.

Suppose that an $f(t)$ is given and the direct Laplace integral is used to find the corresponding $F(s)$. As a specific example, the $F(s)$ of Eq. 4-8.11 which has the pole-zero configuration shown in Fig. 4-8.1 is considered. As has been stressed, the direct Laplace integral is an improper integral and converges only if σ is larger than σ_0. In terms of the poles and zeros of an $F(s)$, σ_0 is the real part of the pole (or poles) located farthest to the right in the complex s-plane. For the configuration shown in Fig. 4-8.1, $\sigma_0 = -p_1$.

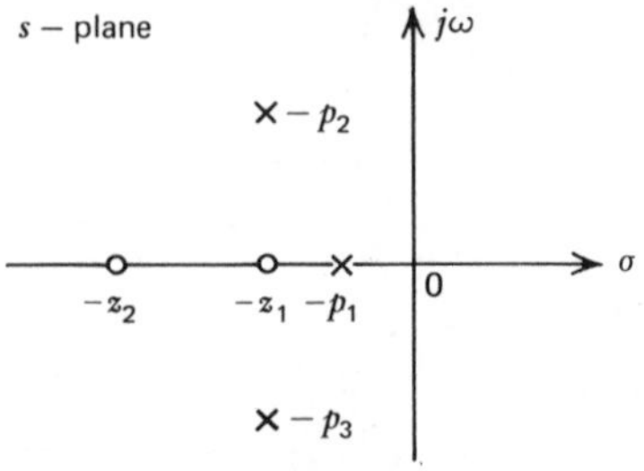

Fig. 4-8.1

For present purposes, the direct Laplace integral is written with s replaced by its real part σ, giving

$$F(\sigma) = \int_0^\infty f(t)\, \epsilon^{-\sigma t}\, dt \tag{4-8.12}$$

The integral converges along the portion of the real axis to the right of σ_0.

In the theory of functions of a complex variable, there is a concept known as analytic continuation. If a function is known to be analytic in some region and its values are known along some arc in this region, the function is uniquely determined throughout the region. This concept can be applied to the present discussion in the following manner: $F(s)$ can be found by using the direct Laplace transform as discussed earlier in this chapter; $F(\sigma)$ can be found by use of Eq. 4-8.12, and thus $F(\sigma)$ is certainly valid along an arc that is the real axis to the right of σ_0; $F(s)$ and $F(\sigma)$ are identical along the σ axis to the right of σ_0; therefore, $F(s)$ can be considered as the analytic continuation of $F(\sigma)$ throughout the entire s-plane, except at the points where $F(s)$ is not analytic.

Another term that comes from complex variable theory is residue. When all the poles of some $F(s)$ are simple, the coefficients of the partial fraction expansion are called residues. For example, return to Eq. 4-7.10, repeated as

$$F(s) = \frac{(s+3)(s+5)}{(s+1)(s+2)(s+4)} = \frac{K_{-1}}{(s+1)} + \frac{K_{-2}}{(s+2)} + \frac{K_{-4}}{(s+4)} \tag{4-8.13}$$

Each of the K terms is a residue. There doesn't seem to be any complete agreement in the literature as to how to say the following, but many books would say that K_{-1} is the residue of $F(s)$ in the pole at $s = -1$ (rather a fancy way of putting it).

If an $F(s)$ has a pole of higher order than the first, Eq. 4-6.23 is used as an example, repeated as

$$F(s) = \frac{a_3 s^3 + a_2 s^2 + a_1 s + a_0}{s^3 (s+b)} = \frac{A}{s^3} + \frac{B}{s^2} + \frac{K_0}{s} + \frac{K_{-b}}{(s+b)} \tag{4-8.14}$$

where the A and B coefficients are not residues, but the K_0 term is. In the partial fraction expansion of an $F(s)$ containing a pole of higher order, the only coefficient that is called residue is the coefficient over the s term (or in general, the $s + p$ term). If you look back over our work you will see we have saved the letter K for the residue terms and have used A, B, etc., for the terms that are not residues.

4-9. THE INVERSE LAPLACE TRANSFORM INTEGRAL

The Laplace transform table we have been using (Appendix E) was developed from the direct Laplace integral. To find the inverse Laplace transform the table is used in the reverse direction. However, we should at least comment that the inverse Laplace trans-

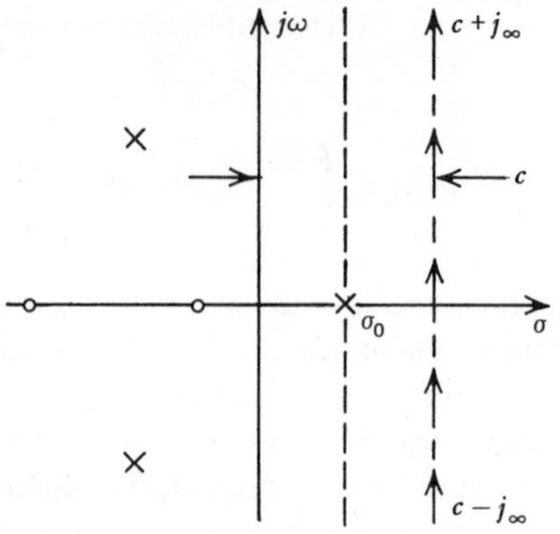

Fig. 4-9.1

form integral does exist, and examine it briefly. The inverse integral is given by

$$\mathcal{L}^{-1}[F(s)] = \frac{1}{2\pi j} \int_{c-j\infty}^{c+j\infty} F(s)\, \epsilon^{st}\, ds = f(t) \tag{4-9.1}$$

This integral is performed along a path in the complex s-plane and its use demands more knowledge of the theory of functions of a complex variable than seems appropriate in a book of this type. However, we can make the following comments.

The path of integration indicated by this equation is a path parallel to and c units away from the $j\omega$-axis. The c is determined in the following manner: An $F(s)$ has a certain pole-zero configuration as shown in Fig. 4-9.1. The abscissa of convergence σ_0 goes through the pole farthest to the right in the complex s-plane, and all of the plane to the right of this abscissa is the region of convergence for the direct Laplace integral. The quantity c in Eq. 4-9.1 must be so chosen that the path of integration in the complex plane is in the region of convergence. Stated more simply, this means that with the c chosen the path of integration is to the right of all the poles of $F(s)$ in the complex s-plane.

We will examine the inverse Laplace transform integral from time to time to remain familiar with its appearance. When the subject of the Fourier transform is discussed, we will find that the inverse Fourier transform integral is real and we will use this real integral at that time.

4-10. FINAL-VALUE THEOREM

A brief development of the final-value theorem shows how the theorem can be used and, of equal importance, it shows the situation in which the theorem cannot be used. The theorem is developed by using a specific $F(s)$ of the form

$$F(s) = \frac{K(s+z_1)(s+z_2)}{s(s+p_1)(s+p_2)} = \frac{K_0}{s} + \frac{K_{-p_1}}{s+p_1} + \frac{K_{-p_2}}{s+p_2} \tag{4-10.1}$$

The inverse transform of this equation is

$$\mathcal{L}^{-1}[F(s)] = K_0 + K_{-p_1}\epsilon^{-p_1 t} + K_{-p_2}\epsilon^{-p_2 t} \tag{4-10.2}$$

The final value of $f(t)$ is K_0, which can be obtained by letting $t \to \infty$ in Eq. 4-10.2. The coefficient K_0 is obtained from $F(s)$ by multiplying $F(s)$ by s and then letting s approach zero. Therefore, the final-value theorem can be stated as

$$\lim_{t \to \infty} [f(t)] = \lim_{s \to 0} s[F(s)] \tag{4-10.3}$$

where $F(s)$ is the transform of $f(t)$. The only time $\lim_{s \to 0} s[F(s)]$ can have a finite non-zero value is when the denominator of $F(s)$ has s as a factor.

The final-value theorem is valid if $\lim_{t \to \infty} f(t)$ exists. An examination of Eq. 4-10.2 indicates that if p_1 or p_2 or both are negative, the final value of $f(t)$ is infinite, whereas the theorem gives a finite value. Therefore, the theorem can be applied to an $F(s)$ that has all its poles in the left-half s-plane, except for a pole at the origin.

4-11. INITIAL-VALUE THEOREM

The initial-value theorem can be presented by using the equation for the Laplace transform of a derivative of a function in which differentiation is performed in the conventional manner, as

$$\int_0^\infty f'(t)\,\epsilon^{-st}\,dt = s\,F(s) - f(0) \tag{4-11.1}$$

The Laplace integral converges in a region in the s-plane to the right of $s = \sigma_0$. The limit as $s \to \infty$ can be taken on both sides of Eq. 4-11.1, giving

$$\lim_{s \to \infty} \int_0^\infty f'(t)\,\epsilon^{-st}\,dt = \lim_{s \to \infty} s\,F(s) - f(0) \tag{4-11.2}$$

The limiting process yields

$$0 = \lim_{s \to \infty} s\,F(s) - f(0) \tag{4-11.3}$$

which can be written as

$$\lim_{t \to 0} [f(t)] = \lim_{s \to \infty} s[F(s)] \tag{4-11.4}$$

In essence, this equation is a statement of the initial-value theorem.

EXAMPLES USING THE FINAL- AND INITIAL-VALUE THEOREMS

The initial-value and final-value theorems can be used as a partial check on the accuracy of the work as explained briefly here but discussed in more detail in later examples.

Example 4-11.1. The $F(s)$ from Ex. 4-7.1 and its inverse transform are

$$\mathcal{L}^{-1}[F(s)] = \mathcal{L}^{-1}\left[\frac{(s+3)}{(s+1)(s+2)}\right] = 2\epsilon^{-1t} - 1\epsilon^{-2t} \tag{4-11.5}$$

The initial-value theorem yields

$$\lim_{s \to \infty} [s\,F(s)] = \lim_{s \to \infty} \frac{s(s+3)}{(s+1)(s+2)} = 1 \tag{4-11.6}$$

and the final-value theorem gives

$$\lim_{s \to 0} [s\,F(s)] = \lim_{s \to 0} \frac{s(s+3)}{(s+1)(s+2)} = 0 \tag{4-11.7}$$

Both of these results are verified by the $f(t)$ in Eq. 4-11.5 and are a partial check on the accuracy of the work.

Example 4-11.2. The $F(s)$ from Ex. 4-7.2 and its inverse transform are

$$\mathcal{L}^{-1}\left[\frac{(s+3)(s+5)}{(s+1)(s+2)(s+4)}\right] = \frac{8}{3}\epsilon^{-1t} - \frac{3}{2}\epsilon^{-2t} - \frac{1}{6}\epsilon^{-4t} \tag{4-11.8}$$

The initial- and final-value theorems give

$$f(0) = \lim_{s \to \infty}\left[\frac{s(s+3)(s+5)}{(s+1)(s+2)(s+4)}\right] = 1 \tag{4-11.9}$$

$$f(\infty) = \lim_{s \to 0}\left[\frac{s(s+3)(s+5)}{(s+1)(s+2)(s+4)}\right] = 0 \tag{4-11.10}$$

which are a partial check on the accuracy of the inverse transform in Eq. 4-11.8.

An examination of the use of the initial value theorem shows that the $f(0)$ is of finite non-zero value only if the degree of s in the numerator is one less than the degree in the denominator.

4-12. OTHER PROPERTIES OF THE LAPLACE TRANSFORM

These are a continuation of the properties listed in Section 4-3.

Property 4-12.6. The Laplace Transform of $f''(t)$. The transform of the second derivative can be found from the integral

$$\mathcal{L}[f''(t)] = \int_0^\infty f''(t)\,\epsilon^{-st}\,dt \tag{4-12.1}$$

However, the following technique is easier to use.

The function $g(t)$ is defined as

$$g(t) = f'(t) \tag{4-12.2}$$

When $g(t)$ is differentiated, the result is

$$g'(t) = f''(t) \tag{4-12.3}$$

Therefore, we can write

$$\mathcal{L}[f''(t)] = \mathcal{L}[g'(t)] = s\,\mathcal{L}[g(t)] - g(0) \tag{4-12.4}$$

This last equation is written in terms of $f(t)$ as

$$\begin{aligned} \mathcal{L}[f''(t)] &= s\{\mathcal{L}[f'(t)]\} - f'(0) \\ &= s\{s\,\mathcal{L}[f(t)] - f(0)\} - f'(0) \\ &= s^2\,\mathcal{L}[f(t)] - s\,f(0) - f'(0) \end{aligned} \tag{4-12.5}$$

Property 4-12.7. The Laplace Transform of $f^n(t)$. By extending the technique just shown, the transform of $f'''(t)$ can be found as

$$\mathcal{L}[f'''(t)] = s^3\,\mathcal{L}[f(t)] - s^2 f(0) - s\,f'(0) - f''(0) \tag{4-12.6}$$

Finally, the transform of the nth derivative is

$$\mathcal{L}[f^n(t)] = s^n\,\mathcal{L}[f(t)] - s^{n-1} f(0) - s^{n-2} f'(0) - \cdots - f^{n-1}(0) \tag{4-12.7}$$

Property 4-12.8. The Laplace Transform of the nth Integral of a Function. The transform of a multiple integral can be found by techniques analogous to those just shown. The double integral

$$\int_0^t \int_0^t f(t)\,dt\,dt \tag{4-12.8}$$

can be written as

$$\int_0^t h(t)\,dt \tag{4-12.9}$$

where

$$h(t) = \int_0^t f(t)\,dt \tag{4-12.10}$$

The transform for the double integral is

$$\mathcal{L}\left[\int_0^t \int_0^t f(t)\,dt\,dt\right] = \mathcal{L}\left[\int_0^t h(t)\,dt\right] = \frac{\mathcal{L}[h(t)]}{s} = \frac{\mathcal{L}\left[\int_0^t f(t)\,dt\right]}{s} = \frac{\mathcal{L}[f(t)]}{s^2} \tag{4-12.11}$$

This process can be continued until the general term is obtained

$$\mathcal{L}\left[\int_0^t \int_0^t \cdots \int_0^t f(t)\, dt^n\right] = \frac{\mathcal{L}[f(t)]}{s^n} \tag{4-12.12}$$

Property 4-12.9. The Dead-Time Delay Operator. If $F(s)$ is multiplied by ϵ^{-Ts}, the time function $f_T(t)$ that has the transform $\epsilon^{-Ts} F(s)$ can be determined from the original $f(t)$ by translating the graph of $f(t)$ to the right through a distance of T units, with $f_T(t)$ equal to zero between $t = 0$ and $t = T$. This type of translation is sometimes referred to as a dead-time delay. The significance of this is shown in the following manner: The original $f(t)$ and $F(s)$ are related by the direct transform integral

$$F(s) = \int_0^\infty f(t)\, \epsilon^{-st}\, dt \tag{4-12.13}$$

This equation is multiplied by ϵ^{-Ts}

$$\epsilon^{-Ts} F(s) = \epsilon^{-Ts} \int_0^\infty f(t)\, \epsilon^{-st}\, dt \tag{4-12.14}$$

As far as Eq. 4-12.14 is concerned, ϵ^{-Ts} is a constant and can be moved inside the integral sign as

$$\epsilon^{-Ts} F(s) = \int_0^\infty f(t)\, \epsilon^{-s(t+T)}\, dt \tag{4-12.15}$$

This equation can be rewritten with a change of variable

$$\tau = t + T \tag{4-12.16}$$

The new limits of integration and the differential are

$$\begin{aligned} t = 0 &\qquad \tau = T \\ t = \infty &\qquad \tau = \infty \\ dt &= d\tau \end{aligned} \tag{4-12.17}$$

The equation with the change of variable is

$$\epsilon^{-Ts} F(s) = \int_0^T (0)\, d\tau + \int_T^\infty f(\tau - T)\, \epsilon^{-s\tau}\, d\tau \tag{4-12.18}$$

The new function $f_T(t)$ can be determined from this equation as

$$\begin{aligned} f_T(t) &= 0 \qquad 0 < t < T \\ f_T(t) &= f(t - T) \qquad T < t < \infty \end{aligned} \tag{4-12.19}$$

A number of examples are presented in a later section to demonstrate the use of this property.

Property 4-12.10. The Convolution Integral. The multiplication of two transforms such as

$$F_1(s)\,F_2(s) \tag{4-12.20}$$

gives the transform of a function which results from what is called the convolution of the two corresponding time functions. The $F_1(s)$ and $F_2(s)$ are the transforms of $f_1(t)$ and $f_2(t)$, respectively, as given by

$$\begin{aligned} \mathcal{L}[f_1(t)] &= F_1(s) \\ \mathcal{L}[f_2(t)] &= F_2(s) \end{aligned} \tag{4-12.21}$$

We want to determine the $f(t)$ in terms of $f_1(t)$ and $f_2(t)$ to give

$$f(t) = \mathcal{L}^{-1}[F_1(s)\,F_2(s)] \tag{4-12.22}$$

This is approached in the following manner:

$$F_1(s)\,F_2(s) = F_1(s)\int_0^\infty f_2(t)\,\epsilon^{-st}\,dt \tag{4-12.23}$$

Since the integration is with respect to t, $F_1(s)$ can be moved under the integral sign, as

$$F_1(s)\,F_2(s) = \int_0^\infty F_1(s)\,f_2(t)\,\epsilon^{-st}\,dt \tag{4-12.24}$$

In order to avoid confusion in the next few steps, t is replaced by τ. Also, the $\epsilon^{-s\tau}$ term is associated with $F_1(s)$, as

$$F_1(s)\,F_2(s) = \int_0^\infty [F_1(s)\,\epsilon^{-s\tau}]\,f_2(\tau)\,d\tau \tag{4-12.25}$$

By use of the dead-time delay operator (Property 4-12.9), the terms inside the brackets can be written as

$$\begin{aligned} F_1(s)\,\epsilon^{-s\tau} &= \mathcal{L}[f_1(t-\tau)\,u_{-1}(t-\tau)] \\ &= \int_0^\infty f_1(t-\tau)\,u_{-1}(t-\tau)\,\epsilon^{-st}\,dt \end{aligned}$$

Equation 4-12.25 is written as

$$F_1(s)\,F_2(s) = \int_0^\infty \left[\int_0^\infty f_1(t-\tau)\,u_{-1}(t-\tau)\,\epsilon^{-st}\,dt\right] f_2(\tau)\,d\tau \tag{4-12.26}$$

The $f_2(\tau)$ can be moved under the integral with respect to the t term, and the order of integration can be reversed, leading to

$$\int_0^\infty \left[\int_0^\infty f_1(t-\tau) f_2(\tau) u_{-1}(t-\tau)\, d\tau\right] \epsilon^{-st}\, dt \tag{4-12.27}$$

We are not discussing just any $f_1(t)$ or $f_2(t)$ but only the functions that have Laplace transforms as given by Eqs. 4-12.21. This means that both $f_1(t)$ and $f_2(t)$ are sectionally continuous and of exponential order and these conditions allow us to reverse the order of integration. An examination of the integral (with respect to τ) inside the brackets shows that if τ becomes larger than t, the $u_{-1}(t-\tau)$ term is zero, and when $\tau < t$ this term is unity. Hence, Eq. 4-12.27 can be written as

$$F_1(s) F_2(s) = \int_0^\infty \left[\int_0^t f_1(t-\tau) f_2(\tau)\, d\tau\right] \epsilon^{-st}\, dt \tag{4-12.28}$$

The expression inside the brackets is the $f(t)$ whose Laplace transform is $F_1(s)\,F_2(s)$, or

$$f(t) = \mathcal{L}^{-1}[F_1(s)\, F_2(s)] = \int_0^t f_1(t-\tau) f_2(\tau)\, d\tau = f_1(t) * f_2(t) \tag{4-12.29}$$

Since in this entire development, $f_1(t)$ and $f_2(t)$ could be exchanged, $f(t)$ could be written as

$$f(t) = \mathcal{L}^{-1}[F_2(s)\, F_1(s)] = \int_0^t f_2(t-\tau) f_1(\tau)\, d\tau = f_2(t) * f_1(t) \tag{4-12.30}$$

There is disagreement in the literature as to how to write these last two equations. Many books use the notation

$$f_1(t) * f_2(t) = \int_0^t f_1(t) f_2(t-\tau)\, d\tau$$

instead of that given in Eq. 4-12.29, and

$$f_2(t) * f_1(t) = \int_0^t f_2(t) f_1(t-\tau)\, d\tau$$

instead of that given in Eq. 4-12.30. Since

$$F_1(s)\, F_2(s) = F_2(s)\, F_1(s)$$

either form yields the same result. In this book, we follow the notation of Eqs. 4-12.29 and 4-12.30.

Examples of the use of the convolution integral and its graphical interpretation are presented in Chapter 6.

The preceding development and the resulting equations as given by Eqs. 4-12.29 and 4-12.30 are for time-functions defined for positive time. In Chapter 9, we will take up time functions defined for negative as well as positive time, and then the limits as given on these equations are not appropriate. The more general limits are given by

$$f(t) = \mathcal{L}^{-1}[F_1(s)\,F_2(s)] = \int_{-\infty}^{+\infty} f_1(t-\tau)\,f_2(\tau)\,d\tau \tag{4-12.31}$$

$$f(t) = \mathcal{L}^{-1}[F_2(s)\,F_1(s)] = \int_{-\infty}^{+\infty} f_2(t-\tau)\,f_1(\tau)\,d\tau \tag{4-12.32}$$

These equations can be used for both one-sided and two-sided situations.

Property 4-12.11. Convolution in the Complex Plane. If $f_1(t)$ and $f_2(t)$ both have Laplace transforms, as given by

$$\begin{aligned} F_1(s) &= \mathcal{L}[f_1(t)] \\ F_2(s) &= \mathcal{L}[f_2(t)] \end{aligned} \tag{4-12.33}$$

the Laplace transform of the product $f_1(t)\,f_2(t)$, is given by

$$\mathcal{L}[f_1(t)\,f_2(t)] = \frac{1}{2\pi j}\int_{c-j\infty}^{c+j\infty} F_1(s-p)\,F_2(p)\,dp \tag{4-12.34}$$

This integral is often denoted $F_1(s) * F_2(s)$, meaning that $F_1(s)$ is convolved with $F_2(s)$. Since $f_1(t)$ and $f_2(t)$ can be interchanged in the product $f_1(t)\,f_2(t)$, this equation can also be written as

$$\mathcal{L}[f_2(t)\,f_1(t)] = \frac{1}{2\pi j}\int_{c-j\infty}^{c+j\infty} F_2(s-p)\,F_1(p)dp \tag{4-12.35}$$

This integral is often denoted $F_2(s) * F_1(s)$, meaning that $F_2(s)$ is convolved with $F_1(s)$.

We are again back to the subject of integration in the complex plane, which is beyond the scope of this book. When we come to the subject of Fourier transforms, the integral that corresponds to this integral becomes a *real* integral, and we shall return this subject at that time.

Property 4-12.12. The Laplace Transform of a Change of Scale. An $f(t)$ is given that has a Laplace transform $F(s)$

$$\mathcal{L}[f(t)] = \int_0^{\infty} f(t)\,\epsilon^{-st}\,dt = F(s) \tag{4-12.36}$$

The t in $f(t)$ is replaced by a linear transformation (at) $(a > 0)$, and the effect upon $F(s)$ is to be determined.

$$\mathcal{L}[f(at)] = \int_0^\infty f(at)\,\epsilon^{-st}\,dt \tag{4-12.37}$$

with a change of variable

$$at = \tau \tag{4-12.38}$$

the equation becomes

$$\mathcal{L}[f(at)] = \frac{1}{a}\int_0^\infty f(\tau)\,\epsilon^{-s\tau/a}\,d\tau = \frac{1}{a}F\left(\frac{s}{a}\right) \tag{4-12.39}$$

This result is the fundamental theorem upon which scaling is based. The subject of scaling is discussed at several places in the book.

Property 4-12.13. The Laplace Transform of $f(t)$ Multiplied by $-t$. Differentiation of the transform of a function with respect to s corresponds to the multiplication of the time function by $-t$. The $f(t)$ and the corresponding $F(s)$ are given as

$$F(s) = \int_0^\infty \epsilon^{-st} f(t)\,dt \tag{4-12.40}$$

The derivative of $F(s)$ is given by

$$\frac{dF(s)}{ds} = \frac{\partial}{\partial s}\int_0^\infty [\epsilon^{-st} f(t)\,dt] \tag{4-12.41}$$

Since $f(t)$ is sectionally continuous and of exponential order, the differentiation can be performed under the integral sign, as

$$\frac{dF(s)}{ds} = \int_0^\infty \frac{\partial}{\partial s}[\epsilon^{-st} f(t)]\,dt = \int_0^\infty \epsilon^{-st}[-t\,f(t)]\,dt \tag{4-12.42}$$

which can be written as

$$\frac{dF(s)}{ds} = \mathcal{L}[-t\,f(t)] \tag{4-12.43}$$

This process can be generalized as

$$\frac{d^nF(s)}{ds^n} = \mathcal{L}[(-t)^n f(t)] \qquad (n = 1, 2, 3, \ldots) \tag{4-12.44}$$

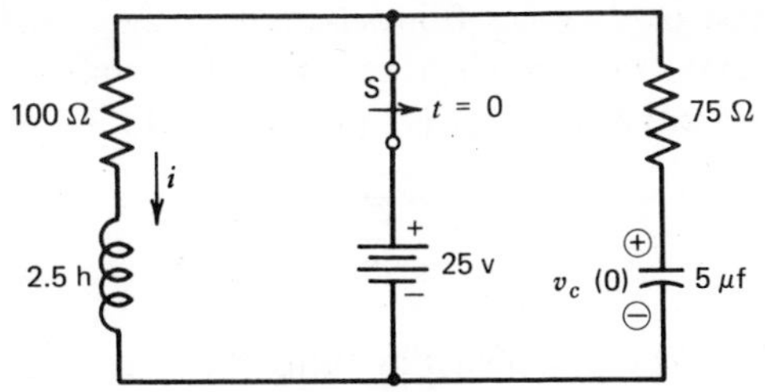

Fig. 4-13.1

4-13. ADDITIONAL CIRCUIT EXAMPLES

Example 4-13.1. The circuit shown in Fig. 4-13.1 is in steady-state when S is opened at $t = 0$. The desired solutions are the current i, and the q on the capacitor. This same circuit was solved in Example 3-3.2 by classical methods. Since all the work before $t = 0$ is the same for classical and Laplace methods, we restate $i(0)$ and $v_c(0)$ from Example 3-3.2 as $i(0) = 0.25$, and $v_c(0) = 25$ volts, with the ⊕ and ⊖ polarity marks as shown. The KVLE is

$$0 = 2.5\frac{di}{dt} + 175i + \frac{1}{5 \times 10^{-6}} \int_0^t i\,dt - 25 \tag{4-13.1}$$

The Laplace transformation is

$$0 = 2.5[s\,I(s) - 0.25] + 175I(s) + 2 \times 10^5 \frac{I(s)}{s} - \frac{25}{s} \tag{4-13.2}$$

This is solved for $I(s)$ as

$$I(s) = \frac{0.25\,(s + 40)}{s^2 + 70s + 80{,}000} \tag{4-13.3}$$

The denominator set equal to zero is the characteristic equation of the classical method

$$s^2 + 70s + 80{,}000 = 0 \tag{4-13.4}$$

The roots of this equation are

$$s_1 \;\&\; s_2 = -35 \pm j281 \tag{4-13.5}$$

Equation 4-13.4 in factored form is

$$(s + 35 + j281)(s + 35 - j281) = 0 \tag{4-13.6}$$

which can be written as

$$(s + 35)^2 + (281)^2 = 0 \tag{4-13.7}$$

The terminology just used comes from classical methods and is hardly appropriate here. Instead we talk about the characteristic polynomial and the zeros of the characteristic polynomial, which are the poles of $F(s)$. Equation 4-13.3 is written as

$$I(s) = \frac{0.25\,[s + 40]}{(s + 35)^2 + (281)^2} \tag{4-13.8}$$

The inverse transform could be taken by using the table in Appendix E. However, to keep the work within the scope of the material already presented, we use the identities

$$\begin{aligned} \mathcal{L}^{-1}\left[\frac{s+a}{(s+a)^2+\beta^2}\right] &= \epsilon^{-at}\cos\beta t \\ \mathcal{L}^{-1}\left[\frac{\beta}{(s+a)^2+\beta^2}\right] &= \epsilon^{-at}\sin\beta t \end{aligned} \tag{4-13.9}$$

$I(s)$ is put into these forms as

$$I(s) = 0.25\left\{\frac{(s+35)}{(s+35)^2+(281)^2} + \frac{5}{281}\left[\frac{281}{(s+35)^2+(281)^2}\right]\right\} \tag{4-13.10}$$

The inverse transform is

$$i = \epsilon^{-35t}\,[0.25 \cos 281t + 0.00445 \sin 281t] \tag{4-13.11}$$

To find q, we could start from the answer and find q as

$$q = \int_0^t i\,dt - 125 \times 10^{-6} \tag{4-13.12}$$

However, it is probably easier to use Laplace methods by taking the transform of this equation

$$Q(s) = \frac{I(s)}{s} - \frac{125 \times 10^{-6}}{s} \tag{4-13.13}$$

and substituting $I(s)$ from Eq. 4-13.8

$$Q(s) = \frac{0.25s + 10}{s\,[(s+35)^2 + (281)^2]} - \frac{125 \times 10^{-6}}{s} \tag{4-13.14}$$

An examination of the circuit reveals that $q_{ss} = 0$. If the final value theorem is used on the first part of this equation, a steady-state value results. The only way these two statements can both be true is for this steady-state value to cancel the second term. Therefore, it is easier to combine the two terms of Eq. 4-13.14 and to take the inverse transform of the result.

$$Q(s) = \frac{-125 \times 10^{-6}[s - 1{,}930]}{(s+35)^2 + (281)^2} \tag{4-13.15}$$

$$= -125 \times 10^{-6}\left\{\frac{(s+35)}{(s+35)^2 + (281)^2} + \frac{-1{,}930 - 35}{281}\left[\frac{281}{(s+35)^2 + (281)^2}\right]\right\} \tag{4-13.16}$$

$$q = \epsilon^{-35t}[-125 \times 10^{-6} \cos 281t + 875 \times 10^{-6} \sin 281t] \tag{4-13.17}$$

ANOTHER FORM FOR SOME TRANSFORMS

The form of some of the transform pairs in Appendix E differs from any used to this point. One such inverse transform is derived to show the method. *Item 13* in Appendix E is

$$\left[\frac{a_1 s + a_0}{(s+a)^2 + \omega^2}\right] \tag{4-13.18}$$

The inverse transform is first put in the form that has been used

$$\left[\frac{a_1 s + a_0}{(s+a)^2 + \omega^2}\right] = \frac{a_1(s+a)}{(s+a)^2 + \omega^2} + \frac{a_0 - a_1 a}{\omega} \frac{\omega}{[(s+a)^2 + \omega^2]}$$

$$\mathcal{L}^{-1}\left[\frac{a_1 s + a_0}{(s+a)^2 + \omega^2}\right] = \epsilon^{-at}\left[a_1 \cos \omega t + \frac{a_0 - a_1 a}{\omega} \sin \omega t\right] \tag{4-13.19}$$

The sine term can represented by a phasor that is drawn horizontally, and the cosine term by another phasor leading this sine phasor by 90 degrees, as shown in Fig. 4-13.2. The resultant phasor has a magnitude A and leads the sine term by an angle α, where

$$A\,\epsilon^{j\alpha} = \frac{a_0 - a_1 a}{\omega} + ja_1 \tag{4-13.20}$$

and Eq. 4-13.19 can be rewritten as

$$A\,\epsilon^{-at} \sin(\omega t + \alpha) \tag{4-13.21}$$

This checks the $f(t)$ in the table and is in more compact form than Eq. 4-13.19.

Many of the other items in the table can be obtained as an extension of these ideas. Therefore, we begin to use the Appendix E transform table.

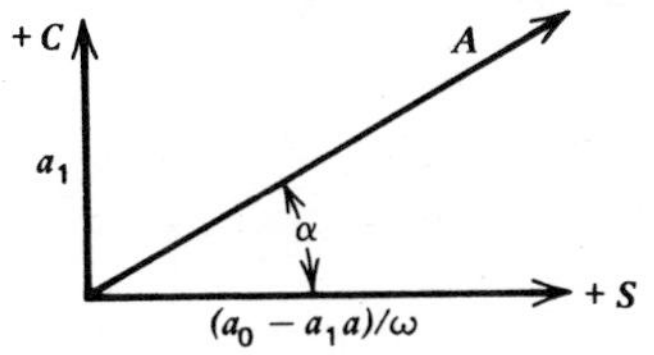

Fig. 4-13.2

Example 4-13.2. The same circuit of the last example is used again, except this time the inverse transform is taken from Appendix E. The KVLE of Eq. 4-13.1 is repeated as

$$0 = 2.5\frac{di}{dt} + 175i + \frac{1}{5 \times 10^{-6}}\int_0^t i\,dt - 25 \tag{4-13.22}$$

The $I(s)$ of Eq. 4-13.8 can be written as

$$I(s) = \frac{0.25s + 10}{(s+35)^2 + (281)^2} \tag{4-13.23}$$

To use the Appendix E table, we look down the first column until we find an $F(s)$ of the proper form to handle this $I(s)$. The $F(s)$ is listed as *Item 13* and has the form

$$F(s) = \frac{a_1 s + a_0}{(s+a)^2 + \omega^2} \tag{4-13.24}$$

Upon comparison of these two equations, the following identifications are made:

$$a_1 = 0.25 \qquad a_0 = 10 \qquad a = 35 \qquad \omega = 281 \tag{4-13.25}$$

We first evaluate $A\underline{/\alpha}$ as given in the $f(t)$ side of *Item 13* to be

$$A\underline{/\alpha} = \frac{10 - 0.25(35 - j281)}{281} = 0.25\underline{/89^\circ} \tag{4-13.26}$$

This result is finally substituted into the $f(t)$ to yield

$$i = 0.25\epsilon^{-35t}\sin(281t + 89^\circ) \tag{4-13.27}$$

Example 4-13.3. The circuit in Fig. 4-13.3 is at rest before $t = 0$, and the desired solution is the voltage v_C. This circuit was solved in Ex. 3-3.4. After $t = 0$, the KVLE is

$$10t = 0.1\frac{di}{dt} + 100i + \frac{1}{150 \times 10^{-6}}\int_0^t i\,dt \tag{4-13.28}$$

The transformed equation is

$$\frac{10}{s^2} = \left[0.1s + 100 + \frac{1}{150 \times 10^{-6}s}\right] I(s) \tag{4-13.29}$$

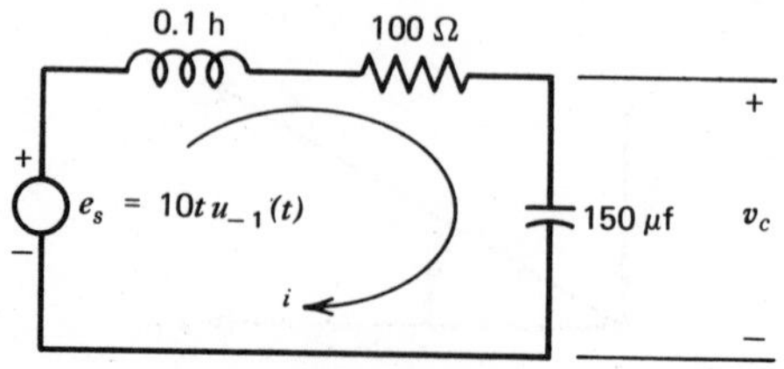

Fig. 4-13.3

$$I(s) = \frac{100}{s(s^2 + 1{,}000s + 6.67 \times 10^9)} = \frac{100}{s(s + 71.9)(s + 927.1)} \qquad (4\text{-}13.30)$$

The voltage v_c is related to the current as

$$v_c = \frac{1}{150 \times 10^{-6}} \int_0^t i\,dt \qquad (4\text{-}13.31)$$

The Laplace transform of this equation is

$$V_c(s) = \frac{I(s)}{150 \times 10^{-6} s} \qquad (4\text{-}13.32)$$

$I(s)$ from Eq. 4-13.30 is substituted as

$$V_c(s) = \frac{6.667 \times 10^5}{s^2 (s + 71.9)(s + 927.1)} \qquad (4\text{-}13.33)$$

The appropriate $F(s)$ in Appendix E is found to be *Item 21*

$$F(s) = \frac{a_3 s^3 + a_2 s^2 + a_1 s + a_0}{(s + a)^2 (s + b)(s + c)} \qquad (4\text{-}13.34)$$

The following identifications are made by comparing the last two equations

$$\begin{aligned} &a_3 = a_2 = a_1 = 0 \qquad a_0 = 6.667 \times 10^5 \\ &a = 0 \qquad b = 71.9 \qquad c = 927.1 \end{aligned} \qquad (4\text{-}13.35)$$

Upon substitution into the $f(t)$ of *Item 21*, the solution is

$$v_c = 10t - 0.15 + 0.151\epsilon^{-71.9t} - 0.001\epsilon^{-927.1t} \qquad (4\text{-}13.36)$$

Example 4-13.4. The circuit in Fig. 4-13.4 is at rest before $t = 0$, and the desired solution is the current i. This same circuit was solved in Ex. 3-3.6. The steps are summarized:

$$10\epsilon^{-100t} = 6\frac{di}{dt} + 4{,}000i + \frac{1}{1.5 \times 10^{-6}} \int_0^t i\,dt \qquad (4\text{-}13.37)$$

$$\frac{10}{s + 100} = \left[6s + 4000 + \frac{1}{1.5 \times 10^{-6} s}\right] I(s) \qquad (4\text{-}13.38)$$

$$I(s) = \frac{1.667s}{(s + 100)(s^2 + 666.7s + 1.11 \times 10^5)} \qquad (4\text{-}13.39)$$

$$I(s) = \frac{1.667s}{(s + 100)(s + 333.3)^2} \qquad (4\text{-}13.40)$$

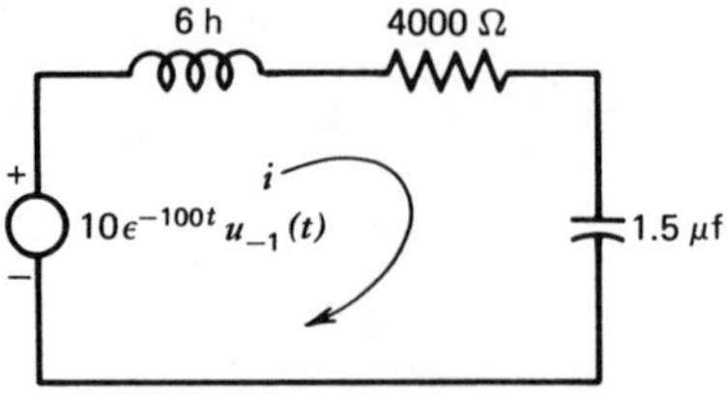

Fig. 4-13.4

Item 17 in the table is used with the following identifications:

$$a_2 = 0 \qquad a_1 = 1.667 \qquad a_0 = 0$$
$$a = 333.3 \qquad b = 100 \tag{4-13.41}$$

Upon substitution and simplification, the $f(t)$ becomes

$$i = -0.00306\epsilon^{-100t} + \epsilon^{-333.3t}\,[0.00306 + 2.38t] \tag{4-13.42}$$

As the last several examples show, the Laplace transform method is a routine crank turning procedure. When classical methods are used, we must stop at several points and decide how to proceed. If the roots are real and distinct, the complementary solution is put into the overdamped form, whereas if the roots are complex, the oscillatory form is used. Each new driving function requires a different form for the particular component of the solution. In Section 4-14 some of the disadvantages of the Laplace methods are pointed out, but one more example is presented to show additional advantages.

Example 4-13.5. The circuit of Fig. 4-13.5(a) is in steady-state when the switch S is closed at $t = 0$. This same circuit was used in Ex. 3-5.2. The currents i_1 and i_2 are shown in (b). At $t = 0$, the initial conditions are $i_1(0) = 0.952$, and $v_c(0) = 4.76$. The KVLE's are

$$100 = 25i_1 + 0.1\frac{di_1}{dt} + 20i_2$$
$$100 = 20i_1 + 23i_2 + 10^4\int_0^t i_2\,dt + 4.76 \tag{4-13.43}$$

These equations were solved in Ex. 3-5.2 by classical methods. The reader should look over that example to review the method by which these two equations were solved for $di_1(0)/dt$ and for

$$\frac{d^2i_1}{dt^2} + 511\frac{di_1}{dt} + 108{,}800i_1 = 435{,}000 \tag{4-13.44}$$

In using Laplace methods, the same steps could be used to find Eq. 4-13.44 and the Laplace taken as

$$\left[s^2 I_1(s) - i_1(0)\,s - \frac{di_1}{dt}(0)\right] + 511\,[s\,I_1(s) - i_1(0)] + 108{,}800\,I_1(s) = \frac{435{,}000}{s} \tag{4-13.45}$$

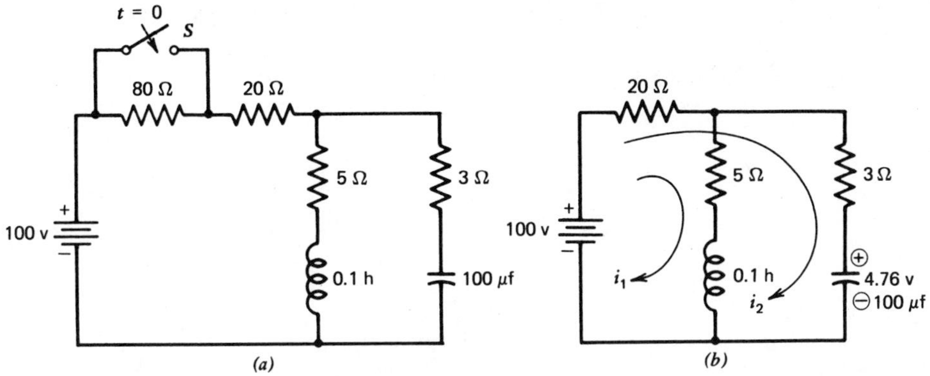

Fig. 4-13.5

However, some of the advantages of the Laplace method are lost. To show this, the transforms for Eqs. 4-13.43 are taken as

$$\frac{100}{s} = 25I_1(s) + 0.1\,[s\,I_1(s) - 0.952] + 20I_2(s)$$
$$\frac{100}{s} = 20I_1(s) + 23I_2(s) + \frac{10^4 I_2(s)}{s} + \frac{4.76}{s} \tag{4-13.46}$$

These are solved for $I_1(s)$, as

$$I_1(s) = \frac{\begin{vmatrix} 0.0952 + \dfrac{100}{s} & 20 \\ \dfrac{95.24}{s} & 23 + \dfrac{10^4}{s} \end{vmatrix}}{\begin{vmatrix} 0.1s + 25 & 20 \\ 20 & 23 + \dfrac{10^4}{s} \end{vmatrix}} \tag{4-13.47}$$

$$I_1(s) = \frac{0.952s^2 + 585s + 435{,}000}{s\,[s^2 + 511s + 108{,}800]} \tag{4-13.48}$$

Note that $di_1(0)/dt$ is not needed when the Laplace transformation is formed from Eqs. 4-13.43, whereas it is needed if Eq. 4-13.44 is used. The finding of $di_1(0)/dt$ represents a certain amount of effort and if the transforms are found from Eq. 4-13.43 this effort is saved. Also, the solving of Eqs. 4-13.46 for $I_1(s)$ uses only algebra and is much more straight forward than the method by which Eq. 4-13.44 was obtained from Eq. 4-13.43.

Taking the inverse transform of $I_1(s)$ represents a certain amount of drudgery and it seems sensible to try a few checks on this equation to see if any obvious mistakes have been made. Two easy checks are to use the initial and final value theorems. When applied to $I_1(s)$, these theorems yield

$$i_1(0) = 0.952 \qquad i_{1ss} = 4 \tag{4-13.49}$$

We know the $i_1(0)$ value is correct, and by looking at the circuit we see i_{1ss} is also correct. Thus we have a partial check on $I_1(s)$.

The zeros of the characteristic polynomial are

$$s_1 \;\&\; s_2 = -255 \pm j209 \tag{4-13.50}$$

$I_1(s)$ can be written as

$$I_1(s) = \frac{0.952s^2 + 585s + 435{,}000}{s\,[(s+255)^2 + (209)^2]} \tag{4-13.51}$$

Item 19 in Appendix E is used as

$$A\underline{/\alpha} = \frac{435{,}000 - 585\,(255 - j209) + 0.952\,(255 - j209)^2}{209\,(-255 + j209)} = -4.43\underline{/43.34^\circ} \tag{4-13.52}$$

The solution is

$$i_1 = 4 - 4.43\epsilon^{-255t}\sin(209t + 43.34^\circ) \tag{4-13.53}$$

4-14. ADVANTAGES AND DISADVANTAGES OF THE LAPLACE TRANSFORM METHOD

As mentioned earlier, the advantages and disadvantages of the Laplace method both stem from the fact that the method is a systematic crank turning process. If you encounter a new situation, a new type of driving function, the method works with a minimum of thought. However, if you have some previous knowledge of the system, the method may not be very adaptable.

Example 4-14.1 For example, consider the circuit of Fig. 4-14.1. The switch S is opened until the circuit reaches steady-state, and S is closed at $t = 0$. After $t = 0$, the cir-

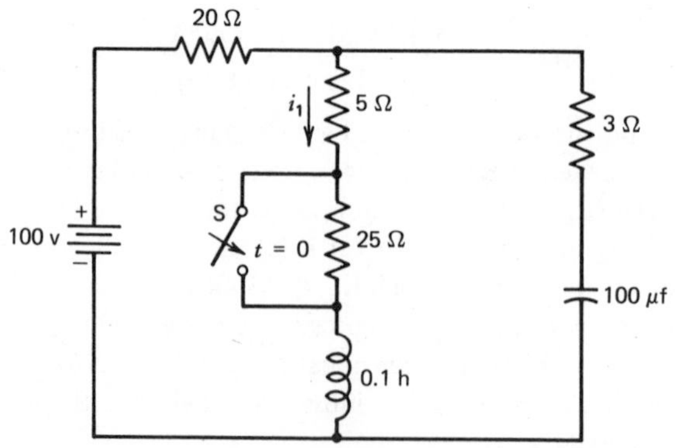

Fig. 4-14.1

cuit is identical with that of Ex. 4-13.5 except for the initial conditions. We can write the form of the solution based on previous knowledge as

$$i_1 = 4 + \epsilon^{-255t}\,[B_1 \cos 209t + B_2 \sin 209t] \tag{4-14.1}$$

and perhaps about $\frac{2}{3}$ of the work using the classical method is bypassed. We determine two initial conditions, and find B_1 and B_2, and we are through.

If Laplace methods are used, the same crank is turned and we go through almost the entire procedure needed to solve this same circuit the first time; the Laplace method does not adapt itself to the knowledge we already have about the circuit.

Example 4-14.2. Another example of this type is that of a circuit driven by a sinusoidal source. To discuss this, the circuit of Fig. 4-14.2 is used. This same circuit was used in Ex. 3-3.5. After $t = 0$, the KVLE is

$$10 \sin(200t + 30^\circ) = 2.5\frac{di}{dt} + 175i + \frac{1}{5 \times 10^{-6}}\int_0^t i\,dt \tag{4-14.2}$$

For convenience, the left side is expanded as

$$8.66 \sin 200t + 5 \cos 200t = 2.5\frac{di}{dt} + 175i + \frac{1}{5 \times 10^{-6}}\int_0^t i\,dt \tag{4-14.3}$$

The equation is transformed and solved for $I(s)$ as

$$\frac{8.66(200)}{s^2 + (200)^2} + \frac{5s}{s^2 + (200)^2} = 2.5s\,I(s) + 175I(s) + 2 \times 10^5\,\frac{I(s)}{s} \tag{4-14.4}$$

$$I(s) = \frac{2s^2 + 693s}{[s^2 + (200)^2]\,[s^2 + 70s + 8 \times 10^4]} \tag{4-14.5}$$

The characteristic polynomial is factored, and $I(s)$ is written as

$$I(s) = \frac{2s^2 + 693s}{[s^2 + (200)^2]\,[(s + 35)^2 + (281)^2]} \tag{4-14.6}$$

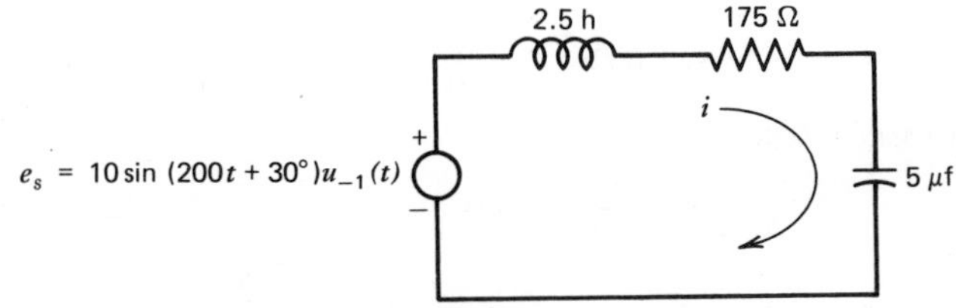

Fig. 4-14.2

Item 27 in Appendix E is used

$$A\underline{/\alpha} = \frac{-693\,(-j200) + 2\,(-j200)^2}{200\,[(35 + j200)^2 + (281)^2]} = 0.0189\underline{/100.75°} \qquad (4\text{-}14.7)$$

$$B\underline{/\beta} = \frac{-693\,(35 - j281) + 2\,(35 - j281)^2}{281\,[(-35 + j281)^2 + (200)^2]} = 0.0197\underline{/-68.3°} \qquad (4\text{-}14.8)$$

The solution is written as

$$i = 0.0189 \sin(200t + 100.7°) + 0.0197\,\epsilon^{-35t} \sin(281t - 68.3°) \qquad (4\text{-}14.9)$$

Note that the Laplace method took the transform of the sinusoid, the crank was turned, and out came the answer; a-c circuit methods were not used at all. The point is that a person with a background in a-c circuits already has some previous knowledge of part of this problem, and the Laplace method does not take advantage of this knowledge.

If you casually look over this example you will not appreciate the drudgery involved in finding $A\underline{/\alpha}$ and $B\underline{/\beta}$ until you work such a problem yourself. The reason for this is that the sinusoid adds an $(s^2 + \omega^2)$ factor to the denominator which automatically puts the $F(s)$ far down in the Laplace table. To find the inverse transform for the second-order system of this example, *Item 27* was used. Several problems in this chapter suggest you solve a given circuit by both classical and Laplace methods. The best thing to do is to work one or more of these problems and to decide for yourself. The point is that the classical method of solution takes advantage of all the time you have spent learning about a-c circuits, whereas the Laplace method ignores this knowledge.

4-15. USE OF THE DEAD-TIME DELAY OPERATOR

The dead-time delay operator was developed as Property 4-12.9, and its use makes it possible to obtain the transforms of many functions not yet considered.

Example 4-15.1. The rectangular pulse $f(t)$ of Fig. 4-15.1(a) is the first example. By inspection, we visualize the components necessary to build up this function. To match the rectangular pulse from 0 to T, we must have a step function with height K, as shown by $f_a(t)$ in Fig. 4-15.1(b). However, the step function is incorrect after $t = T$, so we must add a negative step of magnitude K that is delayed by T seconds, as shown by $f_b(t)$ in Fig. 4-15.1(a). In terms of time functions, we can write

$$f(t) = f_a(t) + f_b(t) \qquad (4\text{-}15.1)$$

The Laplace transform is taken as

$$F(s) = F_a(s) + F_b(s) \qquad (4\text{-}15.2)$$

$$F(s) = \frac{K}{s} - \frac{K}{s}\epsilon^{-Ts} = \frac{K(1 - \epsilon^{-Ts})}{s} \qquad (4\text{-}15.3)$$

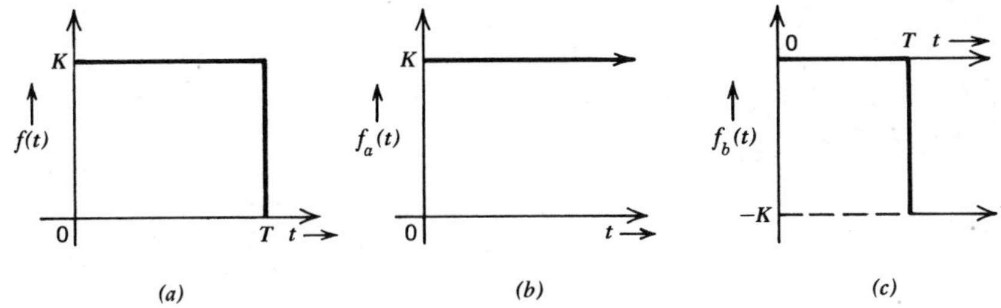

Fig. 4-15.1

Example 4-15.2. The triangular pulse shown in Fig. 4-15.2(a) is the second example. We first visualize the components necessary to build up the pulse. To match the triangular pulse from 0 to T, we must have a ramp function with a slope of K/T as shown by $f_a(t)$ in (b). This ramp function is incorrect after $t = T$, so we must add a ramp with a slope of $-2K/T$, delayed by T seconds as shown by $f_b(t)$ in (c). The sum of $f_a(t)$ and $f_b(t)$ has a slope of $-K/T$ for $t > T$, therefore the ramp $f_c(t)$ of (d) is necessary to kill this slope for $t > 2T$.

The $f(t)$ is made up of the three components

$$f(t) = f_a(t) + f_b(t) + f_c(t) \tag{4-15.4}$$

The transform of this equation is

$$F(s) = F_a(s) + F_b(s) + F_c(s) \tag{4-15.5}$$

$$F(s) = \frac{K}{Ts^2} - \frac{2K}{Ts^2}\epsilon^{-Ts} + \frac{K}{Ts^2}\epsilon^{-2Ts} = \frac{K(1 - \epsilon^{-Ts})^2}{Ts^2} \tag{4-15.6}$$

The next item is more of a property than it is an example. The last property presented was Property 4-12.13 so we pick up the numbering from here.

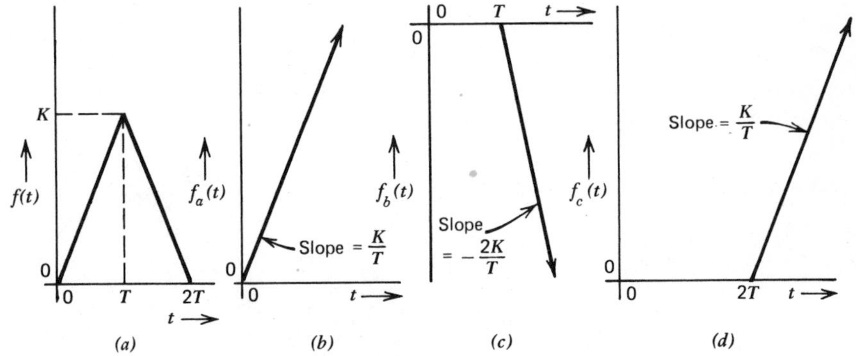

Fig. 4-15.2

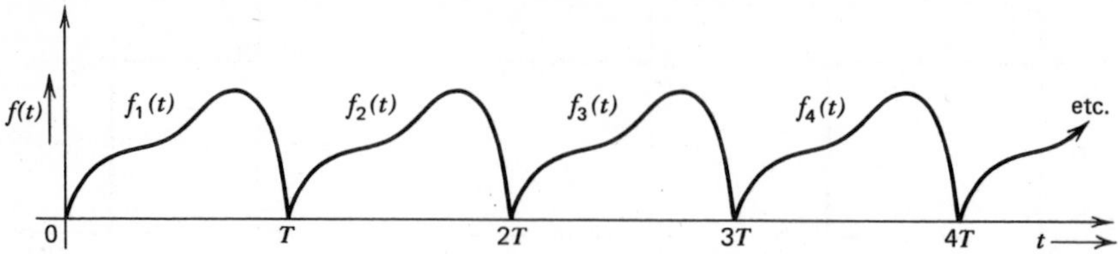

Fig. 4-15.3

Property 4-15.14. The Laplace Transform of Periodic Functions. Suppose we have a wave shape that starts at $t = 0$, and from there to $t = \infty$ is periodic with a period T. Figure 4-15.3 suggest this situation with $f_1(t)$ being the wave shape in the first period form 0 to T and the curve thereafter is a repeat of this shape.

We write for $f(t)$

$$f(t) = f_1(t) + f_2(t) + f_3(t) + \cdots \tag{4-15.7}$$

We designate the transform of $f_1(t)$ as $F_1(s)$. Since $f_2(t)$ is $f_1(t)$ delayed by T second, the transform $F_2(s) = F_1(s)\,\epsilon^{-Ts}$. Similarly, $F_3(s) = F_1(s)\,\epsilon^{-2Ts}$. Therefore, the transform of Eq. 4-15.7 is

$$F(s) = F_1(s) + F_1(s)\,\epsilon^{-Ts} + F_1(s)\,\epsilon^{-2Ts} + F_1(s)\,\epsilon^{-3Ts} + \cdots \tag{4-15.8}$$

$$F(s) = F_1(s)\,[1 + \epsilon^{-Ts} + \epsilon^{-2Ts} + \epsilon^{-3Ts} + \cdots] \tag{4-15.9}$$

The infinite series inside the brackets can be summed for $|\epsilon^{-Ts}| < 1$, yielding

$$F(s) = \frac{F_1(s)}{1 - \epsilon^{-Ts}} \quad \text{for} \quad |\epsilon^{-Ts}| < 1 \tag{4-15.10}$$

Since T is positive, the condition $|\epsilon^{-Ts}| < 1$ is satisfied for all values of s to the right of the $j\omega$ axis in the complex plane; hence $F(s)$ exists with an abscissa of convergence $\sigma_0 = 0$.

Example 4-15.3. Use is made of the Property 4-15.14 on the $f(t)$ shown in Fig. 4-15.4. We divide our work into 4 steps. The first step is to determine the function in the fundamental period $f_1(t)$ and this is shown in (b) of the figure. The second step is to break $f_1(t)$ into the functions necessary to build up $f_1(t)$. These functions are sketched as (c), (d), and (e). The third step is to determine $F_1(s)$ from these, as

$$F_1(s) = F_a(s) + F_b(s) + F_c(s) = \frac{K}{s} - \frac{2K}{s}\,\epsilon^{-Ts/2} + \frac{K}{s}\,\epsilon^{-Ts} = \frac{K(1 - \epsilon^{-Ts/2})^2}{s} \tag{4-15.11}$$

The fourth step is to find $F(s)$ from $F_1(s)$ as

$$F(s) = \frac{F_1(s)}{1 - \epsilon^{-Ts}} = \frac{K(1 - \epsilon^{-Ts/2})^2}{s(1 - \epsilon^{-Ts})} \tag{4-15.12}$$

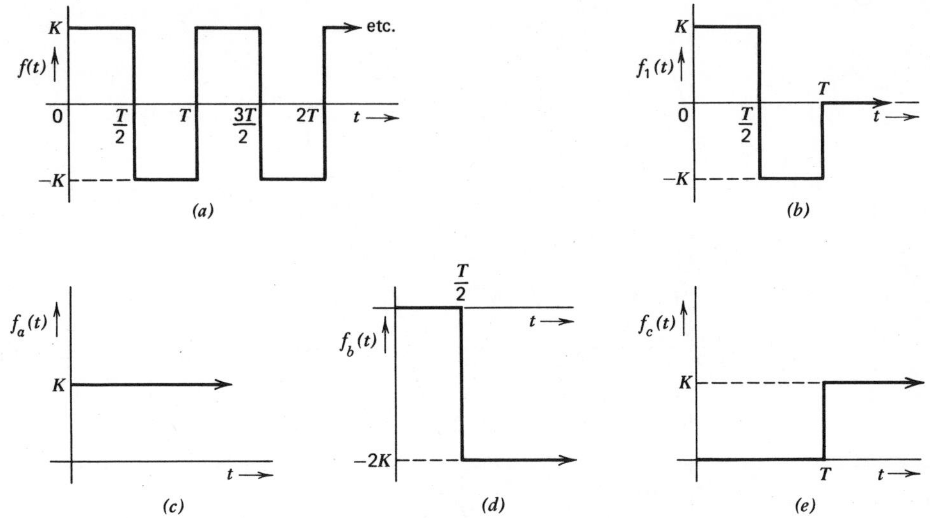

Fig. 4-15.4

This equation can be rewritten as

$$F(s) = \frac{K(1 - \epsilon^{-Ts/2})}{s(1 + \epsilon^{-Ts/2})} = \frac{K(\epsilon^{+Ts/4} - \epsilon^{-Ts/4})}{s(\epsilon^{+Ts/4} + \epsilon^{-Ts/4})} = \frac{K}{s} \tanh \frac{T}{4} s \tag{4-15.13}$$

Example 4-15.4. The transform for the periodic wave shape of Fig. 4-15.5(a) can be found by first visualizing the two components shown in (b) and (c) that make up the $f_1(t)$. The transform for the wave shape in the first period is

$$F_1(s) = \frac{\omega\,[1 + \epsilon^{-(\pi/\omega)s}]}{s^2 + \omega^2} \tag{4-15.14}$$

Therefore, the transform of the periodic wave is

$$F(s) = \frac{F_1(s)}{1 - \epsilon^{-(2\pi/\omega)s}} = \frac{\omega[1 + \epsilon^{-(\pi/\omega)s}]}{[s^2 + \omega^2]\,[1 - \epsilon^{-(2\pi/\omega)s}]} = \frac{\omega}{[s^2 + \omega^2]\,[1 - \epsilon^{-(\pi/\omega)s}]} \tag{4-15.15}$$

Example 4-15.5. The transform for the periodic wave of Fig. 4-15.6(a) can be expressed as the two periodic waves shown in (b) and (c). The transform for (b) is given in Eq. 4-15.15, and the transform for (c) is that of (b) delayed by π/ω seconds. The transform for the sum of the two parts is

$$F(s) = \frac{\omega[1 + \epsilon^{-(\pi/\omega)s}]}{[s^2 + \omega^2]\,[1 - \epsilon^{-(\pi/\omega)s}]} = \frac{\omega}{[s^2 + \omega^2]} \coth \frac{\pi}{2\omega} s \tag{4-15.16}$$

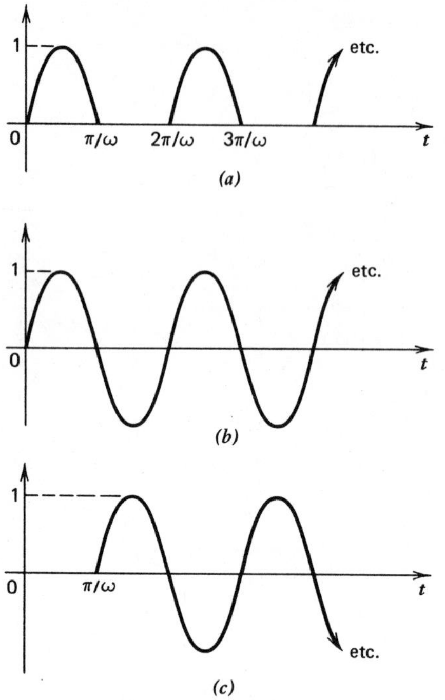

Fig. 4–15.5

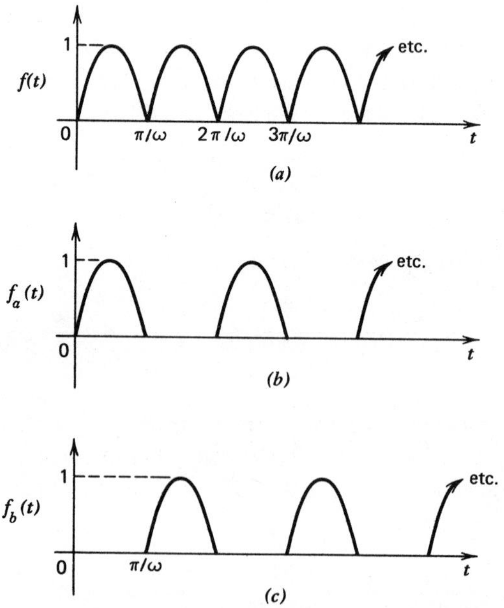

Fig. 4–15.6

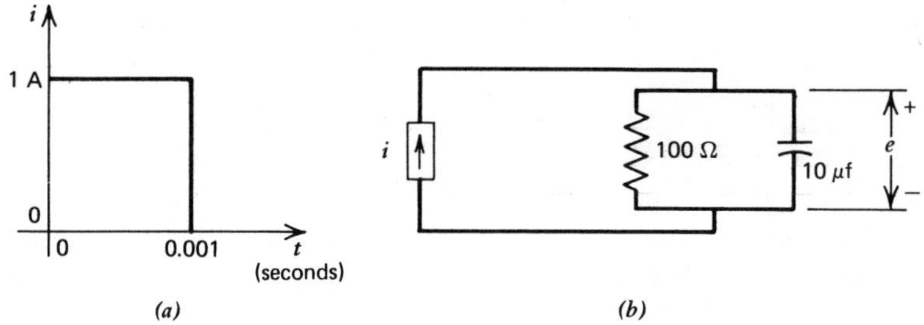

Fig. 4-15.7

Example 4-15.6. The pulse of current i of Fig. 4-15.7(a) is applied to the circuit of (b). Kirchhoff's current law can be written

$$i = i_R + i_C = \frac{e}{100} + 10 \times 10^{-6} \frac{de}{dt} \tag{4-15.17}$$

The transform is taken and solved for $E(s)$ as

$$\frac{1}{s} - \frac{1}{s} \epsilon^{-0.001s} = E(s)\left[\frac{1}{100} + 10 \times 10^{-6} s\right] \tag{4-15.18}$$

$$E(s) = \frac{10^5}{s\,[s + 10^3]} - \frac{10^5 \epsilon^{-0.001s}}{s\,[s + 10^3]} \tag{4-15.19}$$

$$E(s) = 100\left[\frac{1}{s} - \frac{1}{s + 10^3}\right] - 100\left[\frac{1}{s} - \frac{1}{s + 10^3}\right]\epsilon^{-0.001s} \tag{4-15.20}$$

The inverse transform of $E(s)$ is

$$e = 100\,[1 - \epsilon^{-10^3 t}] - 100\,[1 - \epsilon^{-10^3(t-0.001)}]\,u_{-1}(t - 0.001) \tag{4-15.21}$$

Example 4-15.7 Part 1. In Part 2 of this example, the wave shape of current of Fig. 4-15.8(a) is applied to the circuit of (b). Before considering the interaction of the circuit and the current source, an insight into this situation can be gained by examining the

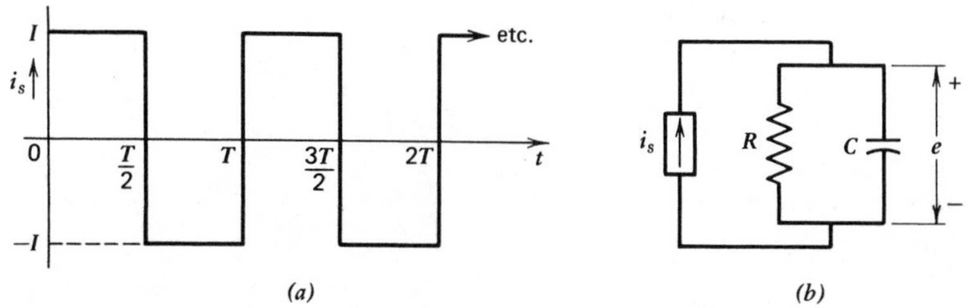

Fig. 4-15.8

inverse transform of $I_s(s)$. The transform for this wave shape of current is

$$I_s(s) = \frac{I}{s} \tanh \frac{T}{4} s = \frac{I}{s} \frac{\sinh (T/4)s}{\cosh (T/4)s} \tag{4-15.22}$$

At first glance it may seem that this $I_s(s)$ has a pole at $s = 0$, until we realize $\sinh (T/4)s$ also has a zero at $s = 0$; actually this $I_s(s)$ has the indeterminate form $\frac{0}{0}$ at $s = 0$, but when this form is evaluated it has a finite value, and hence there is no pole in $F(s)$ at $s = 0$. Therefore, the poles of $I_s(s)$ occur when

$$\cosh \frac{T}{4} s = 0 \tag{4-15.23}$$

The s is replaced by its complex representation $s = \sigma + j\omega$, and Eq. 4-15.23 is expanded as

$$\cosh \frac{T}{4} s = \cosh \frac{T}{4}(\sigma + j\omega) = \cosh \frac{T}{4} \sigma \cosh \frac{T}{4} j\omega + \sinh \frac{T}{4} \sigma \sinh \frac{T}{4} j\omega$$

$$= \cosh \frac{T}{4} \sigma \cos \frac{T}{4} \omega + j \sinh \frac{T}{4} \sigma \sin \frac{T}{4} \omega = 0 + j0 \tag{4-15.24}$$

For complex quantities to be equal, the real parts must be equal; as must be the imaginary parts. Hence, this equation leads to the two equations

$$\cosh \frac{T}{4} \sigma \cos \frac{T}{4} \omega = 0 \qquad \sinh \frac{T}{4} \sigma \sin \frac{T}{4} \omega = 0 \tag{4-15.25}$$

which are satisfied by

$$\sigma = 0 \qquad \frac{T}{4} \omega = \pm \frac{\pi}{2}(2n - 1) \qquad n = 1, 2, 3, \ldots \tag{4-15.26}$$

The poles of the $I_s(s)$ are located at

$$s = 0 \pm j \frac{2\pi}{T}(2n - 1) \qquad n = 1, 2, 3, \ldots \tag{4-15.27}$$

Figure 4-15.9 shows a part of the s-plane with the pole location indicated. The partial fraction expansion of the function can be written in the form

$$I_s(s) = \cdots + \frac{K_{-3}}{s + j(10\pi/T)} + \frac{K_{-2}}{s + j(6\pi/T)} + \frac{K_{-1}}{s + j(2\pi/T)} + \frac{K_{+1}}{s - j(2\pi/T)}$$

$$+ \frac{K_{+2}}{s - j(6\pi/T)} + \frac{K_{+3}}{s - j(10\pi/T)} + \cdots \tag{4-15.28}$$

Here is a partial fraction expansion in which the cover-up technique does not work. For example, if you wish to find K_{+1} in Eq. 4-15.28, what do you cover-up? However, the Heaviside expansion theorem will work because you can still take the derivative of the

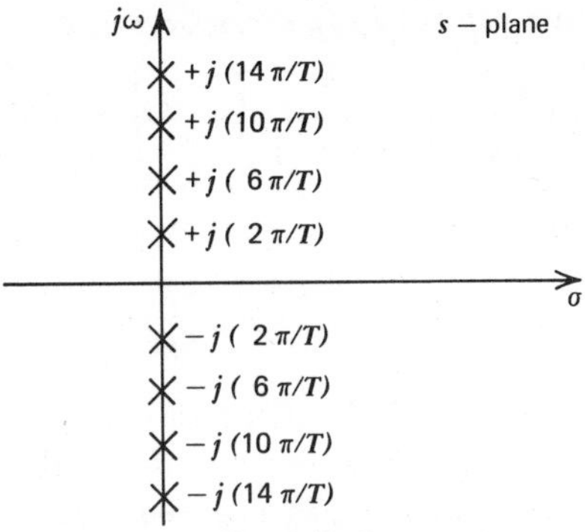

Fig. 4-15.9

denominator with respect to s, and then substitute $s = j(2\pi/T)$ into the result, as demonstrated by the first entry in Eq. 4-15.29 shown just below. In a certain sense, we are pulling a fast one on you, because we proved the legitimacy of the Heaviside expansion theorem in Section 4-7 in situations in which the cover-up technique would also work. The use of the theorem in this situation is valid however.

The residues of Eq. 4-15.28 are determined through use of the Heaviside expansion theorem as demonstrated on a few poles as given by

$$\begin{aligned} K_{+1} &= \left.\frac{I \sinh (T/4)s}{(d/ds)[s \cosh (T/4)s]}\right|_{s=+j(2\pi/T)} = -j\frac{2}{\pi}I \\ K_{+2} &= \left.\frac{I \sinh (T/4)s}{(d/ds)[s \cosh (T/4)s]}\right|_{s=+j(6\pi/T)} = -j\frac{2}{3\pi}I \\ K_{+3} &= \left.\frac{I \sinh (T/4)s}{(d/ds)[s \cosh (T/4)s]}\right|_{s=+j(10\pi/T)} = -j\frac{2}{5\pi}I \end{aligned} \tag{4-15.29}$$

The residues K_{-1}, K_{-2}, and K_{-3} are the conjugates of the residues K_{+1}, K_{+2}, and K_{+3} respectively; therefore, a few of the terms of $I_s(s)$ can be written

$$\begin{aligned} I_s(s) = \cdots \frac{j(2/5\pi)I}{s+j(10\pi/T)} + \frac{j(2/3\pi)I}{s+j(6\pi/T)} + \frac{j(2/\pi)I}{s+j(2\pi/T)} - \frac{j(2/\pi)I}{s-j(2\pi/T)} \\ - \frac{j(2/3\pi)I}{s-j(6\pi/T)} - \frac{j(2/5\pi)I}{s-j(10\pi/T)} + \cdots \end{aligned} \tag{4-15.30}$$

Corresponding terms of $I_s(s)$ can be combined in pairs, as

$$\frac{j(2/\pi)I}{s+j(2\pi/T)} - \frac{j(2/\pi)I}{s-j(2\pi/T)} = \frac{(8/T)I}{s^2+(2\pi/T)^2}$$

which can be written

$$\frac{4I}{\pi}\left[\frac{2\pi/T}{s^2+(2\pi/T)^2}\right] \tag{4-15.31}$$

When this is done for the other pairs of $I_s(s)$, it can be written

$$I_s(s) = \frac{4I}{\pi}\left[\frac{2\pi/T}{s^2 + (2\pi/T)^2}\right] + \frac{4I}{3\pi}\left[\frac{6\pi/T}{s^2 + (6\pi/T)^2}\right] + \frac{4I}{5\pi}\left[\frac{10\pi/T}{s^2 + (10\pi/T)^2}\right] + \cdots \quad (4\text{-}15.32)$$

The inverse transform for $I_s(s)$ is

$$i_s = \frac{4I}{\pi}\sin\frac{2\pi}{T}t + \frac{4I}{3\pi}\sin\frac{6\pi}{T}t + \frac{4I}{5\pi}\sin\frac{10\pi}{T}t + \cdots \quad (4\text{-}15.33)$$

The infinite series of Eq. 4-15.33 is the Fourier series development of the wave shape of Fig. 4-15.8(a). There is one big difference between the work here and the conventional Fourier series as found in a-c-circuit work. In a conventional Fourier series, the sinusoids exist for time t from $-\infty$ to $+\infty$, whereas here these sinusoids start at $t = 0$. This means that as Example 4-15.7 Part 2 is developed, the solution for the voltage e will contain both the complementary and particular components of the solution.

Example 4-15.7 Part 2. The wave shape of current of Fig. 4-15.8(a) is applied to the circuit of (b). The transform for the current is that already given in Eq. 4-15.22. The KCLE for this circuit is

$$i_s = i_R + i_C = \frac{e}{R} + C\frac{de}{dt} \quad (4\text{-}15.34)$$

and the transformed equation is

$$\frac{I}{s}\tanh\frac{T}{4}s = E(s)\left[\frac{1}{R} + Cs\right] \quad (4\text{-}15.35)$$

which, when solved for $E(s)$, yields

$$E(s) = \frac{I\tanh(T/4)s}{sC\,[s + 1/RC]} \quad (4\text{-}15.36)$$

To make the problem more specific, the following values are assumed:

$$I = 1 \text{ amp} \qquad R = 100 \text{ ohms} \qquad C = 100\ \mu\text{f} \qquad T = 0.02 \text{ sec} \quad (4\text{-}15.37)$$

With these values, $E(s)$ becomes

$$E(s) = \frac{10^4}{s(s+100)}\,\frac{\sinh 0.005s}{\cosh 0.005s} \quad (4\text{-}15.38)$$

A few of the poles of Eq. 4-15.28 are shown on the sketch of a portion of the s-plane of Fig. 4-15.10. Equation 4-15.38 can be expanded in partial fraction, as

$$E(s) = \cdots + \frac{K_{-j1,570}}{s + j1{,}570} + \frac{K_{-j942}}{s + j942} + \frac{K_{-j314}}{s + j314} + \frac{K_{+j314}}{s - j314} + \frac{K_{+j942}}{s - j942}$$

$$+ \frac{K_{+j1,570}}{s - j1{,}570} + \cdots + \frac{K_{-100}}{s + 100} \quad (4\text{-}15.39)$$

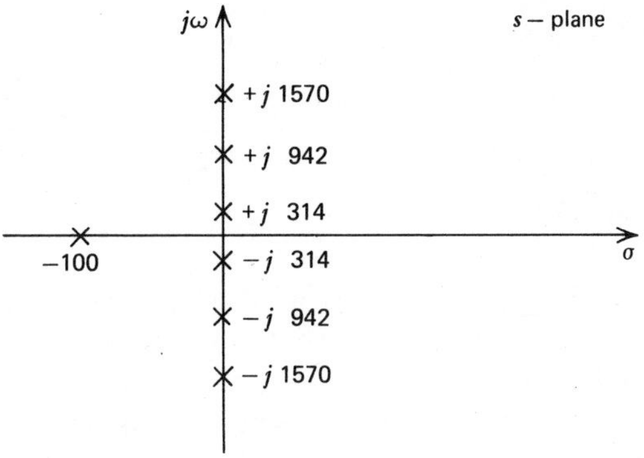

Fig. 4-15.10

The residue K_{-100} can be found by covering up the $(s+100)$ term in Eq. 4-15.38 and letting $s=-100$ in the result; thus

$$K_{-100} = \left.\frac{10^4 \sinh 0.005s}{s \cosh 0.005s}\right|_{s=-100} = 46.2 \tag{4-15.40}$$

For the other residues, the denominator of Eq. 4-15.38 is differentiated, as

$$\frac{d}{ds}\left[(s^2+100s)\cosh 0.005s\right] = (s^2+100s)(0.005)\sinh 0.005s + (2s+100)\cosh 0.005s \tag{4-15.41}$$

Since all the poles at which the other residues are to be evaluated are zeros of cosh 0.005s, the second term of Eq. 4-15.41 drops out in calculating these residues. The method of obtaining the remaining residues is demonstrated as

$$\begin{aligned} K_{+j314} &= \left.\frac{2\times 10^6}{s(s+100)}\right|_{s=j314} = 19.3\underline{/-162.3^\circ} \\ K_{+j942} &= \left.\frac{2\times 10^6}{s(s+100)}\right|_{s=j942} = 2.24\underline{/-173.9^\circ} \\ K_{+j1,570} &= \left.\frac{2\times 10^6}{s(s+100)}\right|_{s=j1,570} = 0.81\underline{/-176.3^\circ} \end{aligned} \tag{4-15.42}$$

Equation 4-15.39 can now be written as

$$\begin{aligned} E(s) = \cdots &+ \left\{\frac{19.3\underline{/162.3^\circ}}{s+j314} + \frac{19.3\underline{/-162.3^\circ}}{s-j314}\right\} + \left\{\frac{2.24\underline{/173.9^\circ}}{s+j942} + \frac{2.24\underline{/-173.9^\circ}}{s-j942}\right\} \\ &+ \left\{\frac{0.81\underline{/176.3^\circ}}{s+j1,570} + \frac{0.81\underline{/-176.3^\circ}}{s-j1,570}\right\} + \cdots + \frac{46.2}{s+100} \end{aligned} \tag{4-15.43}$$

The inverse transform of $E(s)$ is

$$e = 38.6 \sin(314t - 72.3°) + 4.48 \sin(942t - 83.9°) + 1.62 \sin(1570t - 86.3°) + \cdots + 46.2\epsilon^{-100t} \quad (4\text{-}15.44)$$

which is the desired response.

As stressed before, a person must use his judgment as to when to use Laplace methods. We don't propose that Laplace methods are necessarily the best way to solve such a problem. Classical methods could be used in a manner analogous to Example 2-3.2. To form an opinion as to which method to use, we suggest you work a problem both ways and decide for yourself.

4-16. THE IMPULSE FAMILY OF FUNCTIONS

In Appendix A, the impulse family of functions is discussed, and meaning is given to the symbols $u_0(t), u_{-1}(t), \ldots, u_{-n}(t)$. The symbols $u_{+1}(t), u_{+2}(t), \ldots$, are also introduced without giving any physical interpretation to these functions. We return to this family of functions to develop the corresponding Laplace transforms and to suggest a physical interpretation to the $u_{+n}(t)$ members of this family.

The Laplace transform for $u_{-1}(t)$ has already been determined as

$$\mathcal{L}[u_{-1}(t)] = \frac{1}{s} \quad (4\text{-}16.1)$$

The term $u_{-n}(t)$ can be expressed as shown in Eq. A-2.15, as

$$u_{-n}(t) = \frac{t^{n-1}}{(n-1)!} u_{-1}(t) \quad (4\text{-}16.2)$$

Item 8 in Table 4-4.1 is

$$\mathcal{L}[t^n] = \frac{n!}{s^{(n+1)}} \quad (4\text{-}16.3)$$

This last equation is written for $n - 1$ as

$$\mathcal{L}[t^{n-1}] = \frac{(n-1)!}{s^n} \quad (4\text{-}16.4)$$

and is combined with Eq. 4-16.2 to yield

$$\mathcal{L}[u_{-n}(t)] = \mathcal{L}\left[\frac{t^{n-1}}{(n-1)!}\right] = \frac{1}{(n-1)!} \times \frac{(n-1)!}{s^n} = \frac{1}{s^n} \quad (4\text{-}16.5)$$

THE $u_0(t)$ FUNCTION

The simplest way to find $\mathcal{L}[u_0(t)]$ is to write

$$\mathcal{L}[u_0(t)] = \int_{0-}^{\infty} u_0(t)\, \epsilon^{-st}\, dt = \int_{0-}^{0+} u_0(t)\, dt = 1 \tag{4-16.6}$$

The concept here is that the impulse occurs between $t = 0-$ and $t = 0+$, so that the defining integral for the Laplace transform must start at $t = 0-$ in order to include the impulse. Since the integral is zero for all time greater than $t = 0+$, the integral can be rewritten with the limits from $t = 0-$ to $t = 0+$, at which time $\epsilon^{-st} = 1$. Hence, the $\mathcal{L}[u_0(t)] = 1$.

Another way to obtain this same result is to use the rectangular pulse shown in Fig. 4-16.1 and to find its Laplace transform. In the limit as δ approaches zero, the $f(t)$ approaches the $u_0(t)$ and the limit of the corresponding $F(s)$ yields $\mathcal{L}[u_0(t)]$ as shown in the following development.

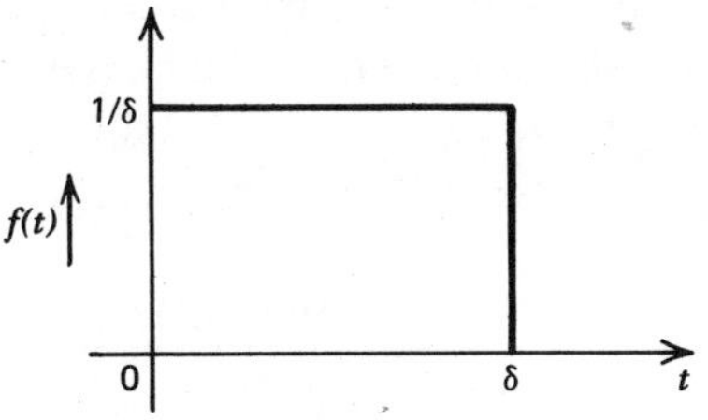

Fig. 4-16.1

The Laplace transform of the pulse shown in Fig. 4-16.1 is

$$\frac{1}{\delta s} - \frac{1\epsilon^{-\delta s}}{\delta s} = \frac{1 - \epsilon^{-\delta s}}{\delta s} \tag{4-16.7}$$

When the limit of Eq. 4-16.7 is taken as $\delta \to 0$, the result is the indeterminant form $\frac{0}{0}$, which can be evaluated by differentiating the numerator and the denominator with respect to δ:

$$\mathcal{L}[u_0(t)] = \lim_{\delta \to 0} \frac{1 - \epsilon^{-\delta s}}{\delta s} = \frac{0}{0} = \lim_{\delta \to 0} \frac{s\epsilon^{-\delta s}}{s} = 1 \tag{4-16.8}$$

which checks Eq. 4-16.6.

We should comment immediately that the rectangular pulse of Fig. 4-16.1 is not the only geometric form from which the unit-impulse functions can be derived. Actually, any wave shape could be used, subject to the limitation that the area under the curve be unity throughout the process of taking the limit as the width δ of the pulse approaches zero. Each of the three functions shown in Fig. 4-16.2 along with an infinite number of other functions can be used in obtaining the impulse function.

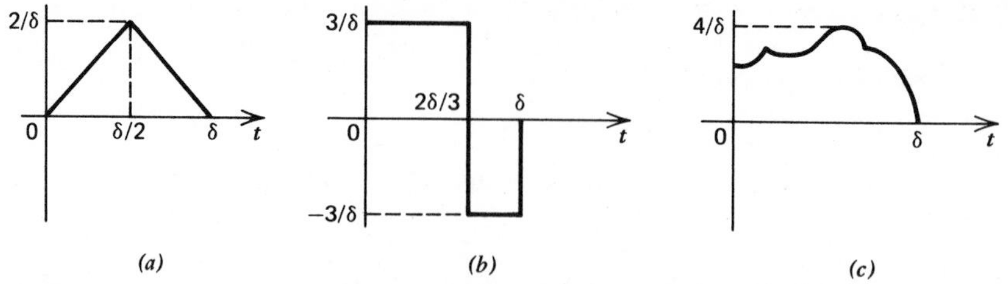

Fig. 4-16.2

To demonstrate this, we start with the wave shape shown in Fig. 4-16.2(a) and find the corresponding transform as

$$\frac{4}{\delta^2 s^2} - \frac{8\epsilon^{-\delta s/2}}{\delta^2 s^2} + \frac{4}{\delta^2 s^2}\epsilon^{-\delta s} = \frac{4(1 - \epsilon^{-\delta s/2})^2}{\delta^2 s^2} \tag{4-16.9}$$

The limit of this as δ approaches zero is again the indeterminate form, which can be evaluated by taking the derivative of the numerator and denominator with respect to δ the proper number of times and then passing to the limit, as in

$$\mathcal{L}[u_0(t)] = \lim_{\delta \to 0} \frac{4[1 - 2\epsilon^{-\delta s/2} + \epsilon^{-\delta s}]}{\delta^2 s^2} = \frac{0}{0} = \lim_{\delta \to 0} \frac{4[s\,\epsilon^{-\delta s/2} - s\,\epsilon^{-\delta s}]}{2\delta s^2} = \frac{0}{0}$$

$$= \lim_{\delta \to 0} \frac{4[-(s^2/2)\,\epsilon^{-\delta s/2} + s^2\,\epsilon^{-\delta s}]}{2s^2} = 1 \tag{4-16.10}$$

THE $u_{+1}(t)$ FUNCTION

The concept is that $u_{+1}(t)$ is the derivative of $u_0(t)$. Since the $u_0(t)$ can be determined by a limiting process of an infinite number of wave shapes, it seems reasonable that the same thing is also true of the $u_{+1}(t)$ function. With these thoughts in mind, we want to choose a wave shape to represent $u_0(t)$ which lends itself most conveniently to differentiation. The wave shape chosen here is shown in Fig. 4-16.3(a) and its derivative is shown in (b).

The curve of Fig. 4-16.3(b) consists of two spikes of equal area, one spike being positive and the other negative. For this reason the derivative of the unit impulse is

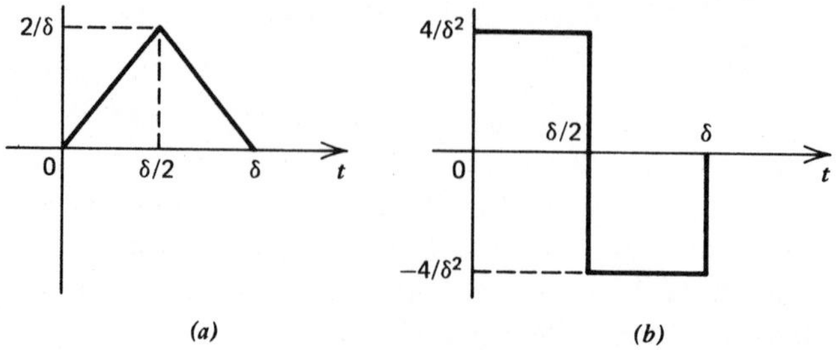

Fig. 4-16.3

referred to as a unit doublet or couple. The area under each spike is

$$\frac{4}{\delta^2} \times \frac{\delta}{2} = \frac{2}{\delta} \tag{4-16.11}$$

When δ approaches zero as a limit, the area under each spike increases without bound. The reason for commenting on this is to emphasize the fact that the $u_{+1}(t)$ function is not just two $u_0(t)$ functions of opposite signs spaced close together.

The wave shape of Fig. 4-16.3(b) can be visualized as being made up of three components and its transform is

$$\frac{4}{\delta^2 s} - \frac{8}{\delta^2 s} \epsilon^{-\delta s/2} + \frac{4}{\delta^2 s} \epsilon^{-\delta s} \tag{4-16.12}$$

As δ approaches zero, Eq. 4-16.12 approaches the transform of the unit doublet. If the limit is taken as the equation stands, the result is the indeterminate form $\frac{0}{0}$, which can be evaluated by differentiating the numerator and denominator with respect to δ the proper number of times and then taking the limit

$$\mathcal{L}[u_{+1}(t)] = \lim_{\delta \to 0} \frac{4 - 8\epsilon^{-\delta s/2} + 4\epsilon^{-\delta s}}{\delta^2 s} = \frac{0}{0} = \lim_{\delta \to 0} \frac{4s\,\epsilon^{-\delta s/2} - 4s\,\epsilon^{-\delta s}}{2\delta s} = \frac{0}{0}$$

$$= \lim_{\delta \to 0} \frac{-s\,\epsilon^{-\delta s/2} + 2s\,\epsilon^{-\delta s}}{1} = s \tag{4-16.13}$$

where $u_{+1}(t)$ stands for the unit doublet, and s is its transform.

THE $u_{+2}(t)$ FUNCTION

Based on what has been said, it seems reasonable to replace the wave shape of Fig. 4-16.3(b) with another wave shape that leads to the same limiting function as $\delta \to 0$, whose derivative is easier to handle. Such a function is shown in Fig. 4-16.4(a) and its derivative is shown in (b). The positive and negative lobes of (a) are chosen so that this wave shape has the same couple or turning action as the original wave of (b).

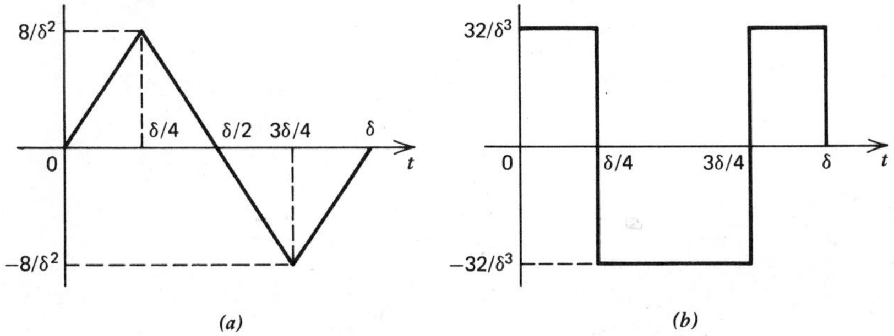

Fig. 4-16.4

We now turn our attention to the derivative curve shown in Fig. 4-16.4(b). First, let us look at the area under the first positive lobe, which is

$$\frac{32}{\delta^3} \times \frac{\delta}{4} = \frac{8}{\delta^2} \tag{4-16.14}$$

As $\delta \to 0$, this area goes off to infinity but as a different order of magnitude than did the corresponding area of $u_{+1}(t)$.

The wave shape is made up of four different components and its transform is

$$\frac{32}{\delta^3 s} - \frac{64}{\delta^3 s}\epsilon^{-\delta s/4} + \frac{64}{\delta^3 s}\epsilon^{-3\delta s/4} - \frac{32}{\delta^3 s}\epsilon^{-\delta s} \tag{4-16.15}$$

When the limit is taken as the equation stands the result is the indeterminant form $\frac{0}{0}$, which can be evaluated as

$$\mathcal{L}[u_{+2}(t)] = s^2 \tag{4-16.16}$$

The details of this are left as a problem.

THE $u_{+n}(t)$ FUNCTION

If you find satisfaction in doing so, you might extend the above idea to the $u_{+3}(t)$ and $u_{+4}(t)$ and other higher functions, and you will find this same pattern continues to develop. The $u_0(t)$ has one pulse up; the $u_{+1}(t)$ has one up and one down; the $u_{+2}(t)$ has one up, down, and up; the $u_{+3}(t)$ has one up, down, up, and down; etc. The areas under each pulse is some constant over δ^n and as $\delta \to 0$ each successive member of this family has an area under the pulse that goes off to infinity as a different order of magnitude than the preceding one.

No matter how diligently one pursues this geometric interpretation, after a while he has to give up and more or less accept the rest of this family of functions as a concept. With this thought in mind we finish the discussion by stating

$$\mathcal{L}[u_{+n}(t)] = s^n \tag{4-16.17}$$

TABLULAR SUMMARY OF THE TRANSFORMS OF THE IMPULSE FAMILY

This family of functions and the corresponding transforms are shown in Table 4-16.1.

4-17. EXAMPLES USING MEMBERS OF THE FAMILY OF FUNCTIONS

Up to the discussion of the impulse family of functions, all the $F(s)$ functions we have considered have had a numerator of lower degree in s than that of the denominator. If for example, you examine the $F(s)$ functions in the table in Appendix E you will find

Table 4-16.1

$f(t)$	$F(s)$
$u_{-n}(t)$	$1/s^n$
$u_{-2}(t)$	$1/s^2$
$u_{-1}(t)$	$1/s$
$u_0(t)$	1
$u_{+1}(t)$	s
$u_{+2}(t)$	s^2
$u_{+n}(t)$	s^n

that all the transforms are of this type. As soon as the driving functions are of the impulse type, the numerator of $F(s)$ may be of the same degree in s or of even higher degrees than the denominator. When this happens and the inverse transform is desired, all that needs to be done is to divide the denominator into the numerator a sufficient number of times until a remainder is reached that has a numerator of a lower degree than the denominator. An example will illustrate the method:

$$F(s) = \frac{s^3}{(s+5)} \tag{4-17.1}$$

$$\begin{array}{r|l} & s^2 - 5s + 25 \\ \hline s + 5 & s^3 \\ & \underline{s^3 + 5s^2} \\ & \quad -5s^2 \\ & \quad \underline{-5s^2 - 25s} \\ & \qquad\qquad 25s \\ & \qquad\qquad \underline{25s + 125} \\ & \qquad\qquad\quad -125 \end{array} \tag{4-17.2}$$

The $F(s)$ has been expanded, as

$$F(s) = \frac{s^3}{(s+5)} = s^2 - 5s + 25 - \frac{125}{(s+5)} \tag{4-17.3}$$

The terms other than the remainder represents the impulse family of functions and the remainder is the more traditional Laplace transform. Therefore, $f(t)$ is

$$f(t) = \mathcal{L}^{-1}[F(s)] = u_{+2}(t) - 5u_{+1}(t) + 25u_0(t) - 125\epsilon^{-5t} \tag{4-17.4}$$

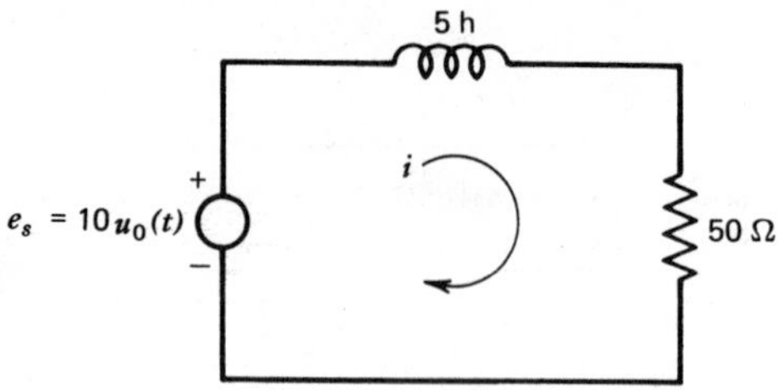

Fig. 4-17.1

Example 4-17.1. The circuit as shown in Fig. 4-17.1 is used in this example. This is the same circuit as used in Example 2-2.5. In Chapter 2, we had to do a lot of explaining when classical methods were used on the impulse because we were encountering a new situation. When Laplace transforms are used, the crank is turned and out comes the answer, as

$$10u_0(t) = 5\frac{di}{dt} + 50i \tag{4-17.5}$$

$$10 = 5[s\,I(s) - 0] + 50I(s) \tag{4-17.6}$$

$$I(s) = \frac{2}{(s+10)} \tag{4-17.7}$$

$$i = \mathcal{L}^{-1}[I(s)] = 2\epsilon^{-10t} \tag{4-17.8}$$

It is interesting to note that the initial condition $i(0) = 0$ was fed into Eq. 4-17.6 and yet Eq. 4-17.8 at $t = 0$ yields the value $i(0) = 2$.

The reason for this is that the current has a discontinuity at $t = 0$, and we are confusing $i(0-)$ and $i(0+)$. Before $t = 0$, the circuit is at rest and the current $i(0-)$ is zero. This value is placed in Eq. 4-17.6. The impulse hits at $t = 0$ and the current at $t = 0+$ is $i(0+) = 2$. We could compute the current at $t = 0+$ as $i(0+)$, and feed this condition into the transformed equation, but when we do, we have already calculated the effect of the impulse so we must ignore it as a driving function.

$$0 = 5\frac{di}{dt} + 50i \tag{4-17.9}$$

$$0 = 5[s\,I(s) - 2] + 50I(s) \tag{4-17.10}$$

$$I(s) = \frac{2}{(s+10)} \tag{4-17.11}$$

This checks Eq. 4-17.7.

Example 4-17.2. The circuit as shown in Fig. 4-17.2 is in steady-state before $t = 0$. At $t = 0-$, the value of $i(0-)$ is calculated as

$$i(0-) = \frac{150}{50} = 3 \tag{4-17.12}$$

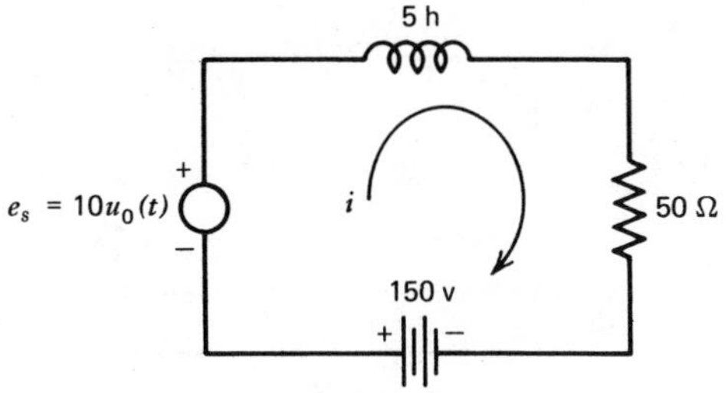

Fig. 4-17.2

The KVLE is

$$10u_0(t) + 150 = 5\frac{di}{dt} + 50i \qquad (4\text{-}17.13)$$

Since the impulse is included as part of the driving function, the initial condition to be fed into the Laplace transform must be the current that exists before the impulse is applied, which is $i(0-)$:

$$10 + \frac{150}{s} = 5[s\,I(s) - 3] + 50I(s) \qquad (4\text{-}17.14)$$

$$I(s) = \frac{5s + 30}{s\,(s + 10)} = \frac{3}{s} + \frac{2}{(s + 10)} \qquad (4\text{-}17.15)$$

$$i = \mathcal{L}^{-1}[I(s)] = 3 + 2\epsilon^{-10t} \qquad (4\text{-}17.16)$$

This example is worked again by computing the current at $t = 0+$ to be

$$i(0+) = 5 \qquad (4\text{-}17.17)$$

When the KVLE is written, the impulse is now omitted, since its effect is included in $i(0+)$:

$$150 = 5\frac{di}{dt} + 50i \qquad (4\text{-}17.18)$$

$$\frac{150}{s} = 5[s\,I(s) - 5] + 50I(s) \qquad (4\text{-}17.19)$$

$$I(s) = \frac{5s + 30}{s\,(s + 10)} \qquad (4\text{-}17.20)$$

which checks with Eq. 4-17.15.

In using classical methods in Chapter 1, 2, and 3, we talked about initial conditions at $t = 0+$, whereas in this chapter, the equation

$$\mathcal{L}[f'(t)] = s\,F(s) - f(0)$$

is vague about whether to use $f(0-)$ or $f(0+)$. The only time $f(0-)$ and $f(0+)$ differ is when a $u_0(t)$ (or higher-order impulse) is applied to the system and, as proposed in the

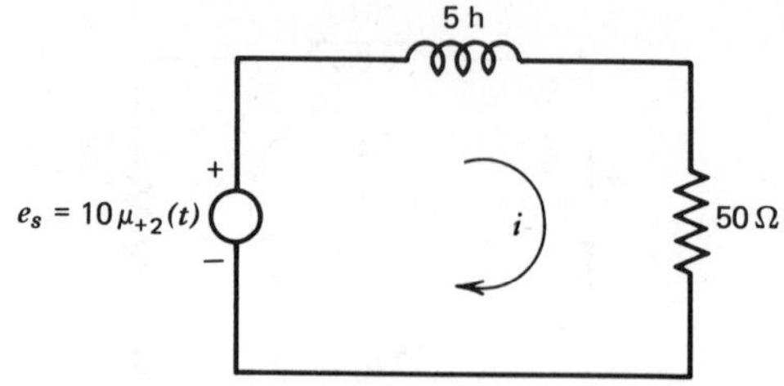

Fig. 4-17.3

preceding discussion, either $f(0-)$ or $f(0+)$ can be used assuming the impulse function is handled properly.

Example 4-17.3. The circuit shown in Fig. 4-17.3 is at rest before $t = 0$.

$$10u_{+2}(t) = 5\frac{di}{dt} + 50i \tag{4-17.21}$$

$$10s^2 = 5\,[s\,I(s) - 0] + 50I(s) \tag{4-17.22}$$

$$I(s) = \frac{2s^2}{s+10} = 2s - 20 + \frac{200}{s+10} \tag{4-17.23}$$

$$i = \mathcal{L}^{-1}[I(s)] = 2u_{+1}(t) - 20u_0(t) + 200\epsilon^{-10t} \tag{4-17.24}$$

The driving function used in this example is the second derivative of the driving function used in Example 4-17.1. Since the circuit is the same, we are able to obtain Eq. 4-17.24 as the second derivative of Eq. 4-17.8. We observe that i of Eq. 4-17.8 has a discontinuity of magnitude 2 at $t = 0$, therefore, the first derivative contains an impulse in addition to the usual derivative:

$$\text{1st derivative} = 2u_0(t) - 20\epsilon^{-10t} \tag{4-17.25}$$

The derivative of $2u_0(t)$ is $2u_{+1}(t)$, and we also observe the discontinuity of -20 in the second part of Eq. 4-17.25. Therefore, the derivative is

$$\text{2nd derivative} = 2u_{+1}(t) - 20u_0(t) + 200\epsilon^{-10t} \tag{4-17.26}$$

which checks Eq. 4-17.24.

4-18. ANOTHER USE FOR THE IMPULSE FUNCTIONS

With the addition of the impulse family, all functions that are zero for negative time can be differentiated and integrated as many times as desired. When a function has a discontinuity and is differentiated, a $K\,u_0(t)$ impulse results whose value K is equal to the magnitude of the discontinuity. If the resulting function is differentiated again, a $K\,u_{+1}(t)$ results, etc. Integration simply reverses the process. When the impulse family is added to the concept of differentiation, we refer to this as generalized differentiation as contrasted to the conventional differentiation.

In conventional mathematics, if a function is differentiated and the result in turn is integrated, the resulting function differs from the original by a constant of integration. However, if a function is defined for positive time and is set to zero for negative time, this function can be differentiated in the generalized sense, and then the result integrated and the original function recaptured. For example, take the function 2 for all time. The derivative is zero and the integral is a constant C to be determined. If however the original function is 2 for positive time and zero for negative time, then $f(t)$ becomes

$$f(t) = 2u_{-1}(t) \tag{4-18.1}$$

and the derivative is

$$f'(t) = 2u_0(t) \tag{4-18.2}$$

This result can be integrated as

$$f(t) = 2u_{-1}(t) \tag{4-18.3}$$

and the original function is recaptured.

Finally, if a function is defined only for positive time and is set equal to zero for negative time, the Laplace transform of the derivative of the function is s times the transform of the function, provided the derivative is performed in the generalized sense. The proof of this is left as a problem.

4-19. CLASSICAL AND LAPLACE TRANSFORM METHODS

Both the advantages and the disadvantages of the Laplace transform method stem from the fact that it is a systematized, crank-turning process. If the applied-scientist adds any dimension to system theory, it must be the physical interpretation of what goes on behind the mathematics. The classical methods add more to this physical interpretation than does the Laplace method. For example, the classical method emphasizes that the solution is made up of the particular and complementary components. The form of the complementary component remains the same unless the system is changed, whereas the nature of the particular component depends upon the driving function. Although the Laplace method yields the same results, its systematized nature does not stress this physical interpretation.

If the designer has some previous knowledge about the behavior of the system, this knowledge can usually be used with classical methods. If the roots of the characteristic equation of a system are known to be

$$s_1 \;\&\; s_2 = -300 \pm j600 \tag{4-19.1}$$

the form of the system solution can be written as

$$i = i_p + \epsilon^{-300t}\,[B_1 \cos 600t + B_2 \sin 600t] \tag{4-19.2}$$

and all that needs to be done is to find i_p and the constants B_1 and B_2. It should also be mentioned that if the designer knows the value of i and its derivative at some time other than $t = 0$, the constants B_1 and B_2 can still be found using classical methods.

If a system is described by a first-order differential equation, the solution can almost be written by inspection as

$$\begin{aligned} i &= i_p + [i(0+) - i_p(0+)]\, \epsilon^{-t/t_c} \\ v &= v_p + [v(0+) - v_p(0+)]\, \epsilon^{-t/t_c} \end{aligned} \tag{4-19.3}$$

The use of the Laplace transform is certainly not needed unless i_p or v_p are of a different form than has been encountered before.

As far as this discussion is concerned, the knowledge of a-c circuit theory is classified as previous knowledge. The theory of a-c circuits can be used with classical methods to find the particular component of the solution, whereas this theory cannot be used with Laplace methods.

ADVANTAGES OF CLASSICAL METHODS

Some of the advantages of classical methods are:

1 Adds to the physical interpretation.

2 Takes advantage of previous knowledge.

3 The constants of integration can be found from conditions other than those of $t = 0$.

ADVANTAGES OF LAPLACE METHODS

One of the main advantages of the Laplace methods is that it is so systematized. Example 4-17.3 represents a situation not discussed before in this book, and yet the solution was determined without any difficulty. The equation solved in this example was

$$10u_{+2}(t) = 5\frac{di}{dt} + 50i \tag{4-19.4}$$

We suggest you solve this same equation by using classical methods. This is one of the problems at the end of this chapter. When you do, we think you will decide that the Laplace method is much simpler.

The steps that are particularly bothersome when solving an nth-order equation using classical methods are: determining the n necessary initial conditions, and solving n simultaneous equations for n arbitrary constants. The Laplace method essentially eliminates these difficulties. For example, consider a circuit with l-loops and n independent storage elements. When KVLE's are written around these loops, the only type of terms encountered are

$$L\frac{di}{dt} \qquad Ri \qquad \frac{1}{C}\int i\,dt \tag{4-19.5}$$

If the Laplace is taken of these equations as they stand, the only initial conditions needed are the currents in the inductors and the voltages on the capacitors and these conditions are already known from the statement of the problem.

Another advantage of the Laplace method is that of notational convenience. We shall continue the study of transformed methods in Chapter 5, at which time block diagrams are introduced. These block diagrams are also used in Chapters 7, 8, 9, and 10. The signal-flow diagram is introduced in Chapter 12. As you shall learn, these diagrams are most conveniently manipulated using transformed quantities.

Some of the advantages of Laplace methods are:

1 A systematized method.

2 Simplest possible set of initial conditions used.

3 No separate steps needed to evaluate the constants of integration.

4 Notational convenience.

We say again that the designer should not stubbornly use only the classical method or the Laplace method, but rather he should be able to think in terms of both and to use the one that is more conveneient in any given situation.

PROBLEMS

4-1 Extend Table 4-4.1 by finding the $F(s)$ and σ_0 for each of the following:

(a) $\sinh bt$ (b) $\sin \beta t$ (c) $\epsilon^{-at} \sinh bt$ (d) $\epsilon^{-at} \sin \beta t$

(e) $t \cosh bt$ (f) $t \cos \beta t$ (g) $t \sinh bt$ (h) $t \sin \beta t$

(i) $t^2 \cos \beta t$ (j) $t^2 \sin \beta t$ (k) $t^2 \sinh bt$ (l) $t^2 \cosh bt$

4-2 In a manner similar to that of Example 4-4.8, start with the $\mathcal{L}[f_1(t)]$ in Table 4-4.1 and find $\mathcal{L}[f_2(t)]$. The $f_1(t)$ and $f_2(t)$ for each part are given below:

(a) $f_1(t) = \cosh bt \quad f_2(t) = \sinh bt$

(b) $f_1(t) = \epsilon^{-at} \cosh bt \quad f_2(t) = \epsilon^{-at} \sinh bt$

(c) $f_1(t) = \epsilon^{-at} \cos \beta t \quad f_2(t) = \epsilon^{-at} \sin \beta t$

(d) $f_1(t) = t\,\epsilon^{-at} \quad f_2(t) = t^2\,\epsilon^{-at}$

4-3 Find the inverse transform $f(t)$ corresponding to each of the following $F(s)$ (do not use the Appendix E table).

(a) $\dfrac{(s+3)}{(s+2)(s+4)}$

(b) $\dfrac{(s+2)}{(s+1)(s+6)}$

(c) $\dfrac{(s+4)}{(s+2)(s+3)}$

(d) $\dfrac{(s+6)}{(s+1)(s+3)(s+7)}$

(e) $\dfrac{(s+1)(s+6)}{(s+2)(s+3)(s+4)}$

(f) $\dfrac{(s+6)(s+10)}{(s+2)(s+4)(s+8)}$

(g) $\dfrac{(s+4)}{(s+2)^2+4^2}$

(h) $\dfrac{s}{(s+2)^2+4^2}$

(i) $\dfrac{(s+4)}{(s+2)^2-4^2}$

(j) $\dfrac{(s+6)}{(s+1)(s+3)^2}$

(h) $\dfrac{(s+2)(s+3)}{(s+1)^2(s+4)}$

(l) $\dfrac{(s+2)^2}{(s+1)(s+3)^2}$

4-4 Find the inverse transform $f(t)$ corresponding to each of the first six $F(s)$ given in Problem 4-3 except use the Heaviside expansion theorem.

Work the following problems using Laplace transforms, without using the Appendix E table.

4-5 Find v in Problem 2-1.
4-6 Find v_c in Problem 2-2.
4-7 Find v in Problem 2-4.
4-8 Find v in Problem 2-6.
4-9 Find v in Problem 2-8.
4-10 Find v in Problem 2-10.
4-11 Find v in Problem 2-11.
4-12 Find v in Problem 2-12.
4-13 Find v in Problem 2-13.
4-14 Find v in Problem 2-14.
4-15 Find v in Problem 2-15.
4-16 Find v in Problem 2-16.
4-17 Find i in Problem 2-17.
4-18 Find i in Problem 2-18.
4-19 Find i in Problem 2-19.
4-20 Find i in Problem 2-20.
4-21 Find i in Problem 2-21.
4-22 Find i in Problem 2-43.
4-23 Find i in Problem 2-44.
4-24 Find i in Problem 2-45.
4-25 Problem 3-14.
4-26 Problem 3-16.
4-27 Problem 3-18.
4-28 Problem 3-20.

Verify the following items in Appendix E.

4-29 *Item 14.*
4-30 *Item 15.*
4-31 *Item 16.*
4-32 *Item 17.*
4-33 *Item 18.*
4-34 *Item 20.*
4-35 *Item 21.*
4-36 *Item 24.*

Given the $f(t)$'s as listed in the following problems, find the $\int_0^t f(t)\,dt$ using Appendix E.

4-37 $f(t) = t\,\epsilon^{-2t}$.
4-38 $f(t) = \epsilon^{-2t} \sin 4t$.
4-39 $f(t) = t \sin 4t$.
4-40 $f(t) = \epsilon^{-1t} \cos 1t$.
4-41 $f(t) = t \cos 4t$.
4-42 $f(t) = \epsilon^{-2t} \cos 4t$.

Work the following problems using Laplace transforms and Appendix E.

4-43 Problem 3-14.
4-44 Problem 3-16.
4-45 Problem 3-18.
4-46 Problem 3-20.
4-47 Problem 3-22.
4-48 Problem 3-24.
4-49 Problem 3-41.
4-50 Problem 3-43.
4-51 Problem 3-44.
4-52 Problem 3-46.
4-53 Problem 3-48.
4-54 Problem 3-50.

Work the following problems using Laplace transforms and Appendix E. Compare the work required with that using classical methods.

4-55 Problem 2-26.
4-56 Problem 2-27.

4-57 Problem 2–28.
4-58 Problem 2–35.
4-59 Problem 2–36.
4-60 Problem 2–37.
4-61 Problem 3–32.
4-62 Problem 3–33.
4-63 Problem 3–34.
4-64 Problem 3–35.
4-65 Problem 3–36.
4-66 Problem 3–40.

4-67 Determine the transforms for the wave shapes shown.

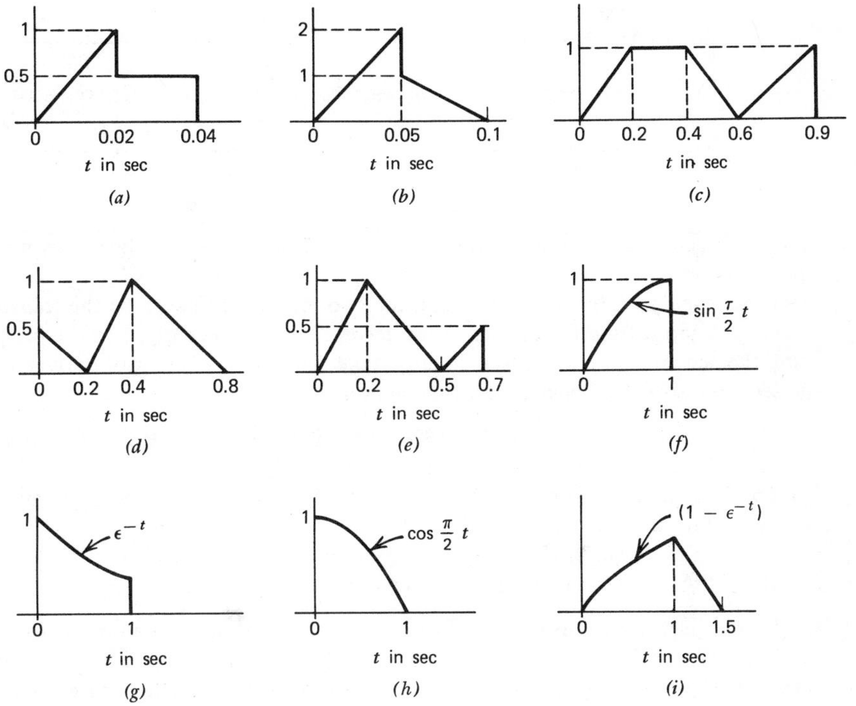

4-68 The wave shapes to be used are those shown in P4–67, except these wave shapes are periodic for $t > 0$ and repeat on to infinity with the period T as given. Find:

(a) Wave shape of (a), $T = 0.08$ sec.
(b) Wave shape of (b), $T = 0.3$ sec.
(c) Wave shape of (e), $T = 2.0$ sec.
(d) Wave shape of (f), $T = 4$ sec.

4-69 Start with the transform of Problem 4–68(d), and perform the inverse transform, thus obtaining the Fourier series. Carry the solution out to three terms.

4-70 The wave shape shown in P4–67(f) describes a voltage source applied to a 1-henry inductor in series with a 1-Ω resistor. Find the current i in the circuit, using Laplace transforms.

4-71 Work out each part of the problem, (a) to (d) as shown, using classical methods first, then rework using Laplace transforms.

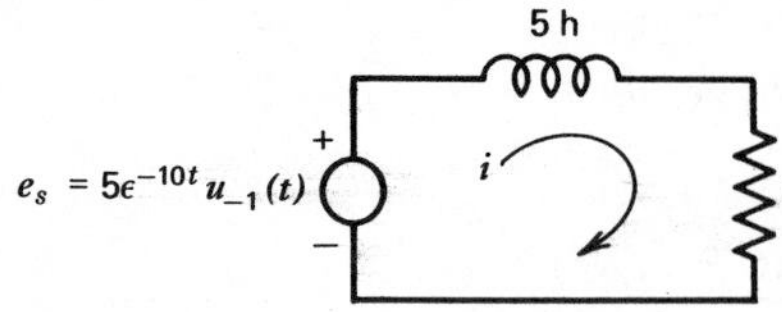

(a) Find the current i.

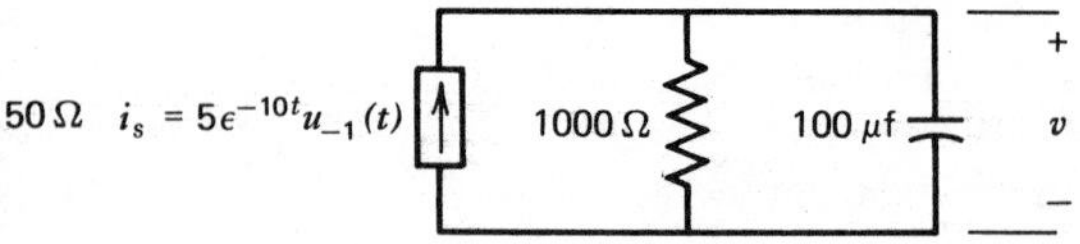

(b) Find the voltage v.

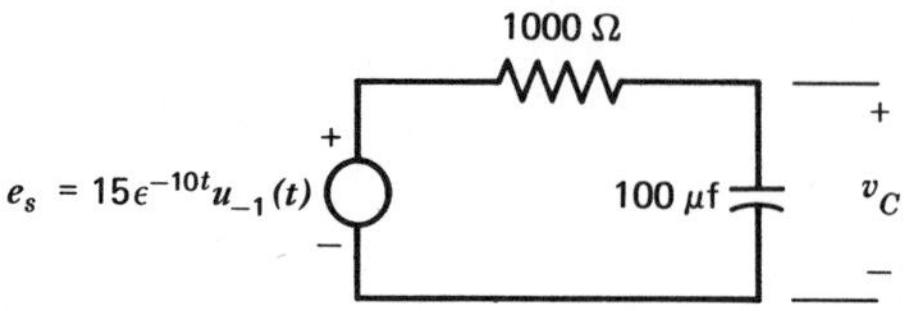

(c) Find the voltage v_C.

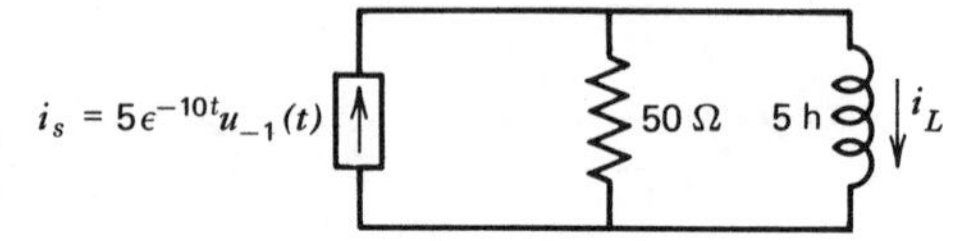

(d) Find the current i_L.

4-72 Start with the transform as given in Eq. 4-16.15, and let $\delta \to 0$, thus showing that $\mathcal{L}[u_{+2}(t)] = s^2$.

4-73 Work out each part, (a) to (d) below, using two methods: First, take the transform of the first input function and solve for desired output. Second, solve the problem using the second input function. Finally, take the proper number of derivatives of the second answer and check with first answer.

(a) Use the circuit shown in P4-71(a): the first input is $e_s = 5u_{+1}(t)$, and the second input is $e_s = 5u_{-1}(t)$. Find i.

(b) Use the circuit shown in P4-71(b): the first input is $i_s = 5u_{+2}(t)$ and the second input is $i_s = 5u_{-2}(t)$. Find v.

(c) Repeat (a) above for the circuit shown in P4-71(c). Find v_C.

(d) Repeat (b) above for the circuit shown in P4-71(d). Find i_L.

4-74 Work each part using two methods: First, include the impulse as an input and take the Laplace transform for $t > 0-$. Second, exclude the impulse and take the transform for $t > 0+$. Compare answers. (Each circuit is in steady-state before $t = 0$.)

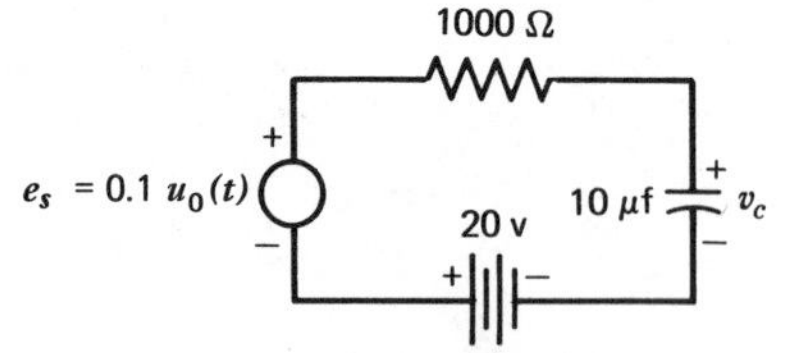

(a) Find the voltage v_c.

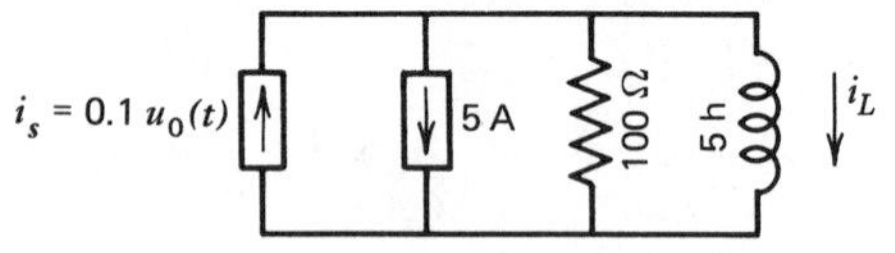

(b) Find the current i_L.

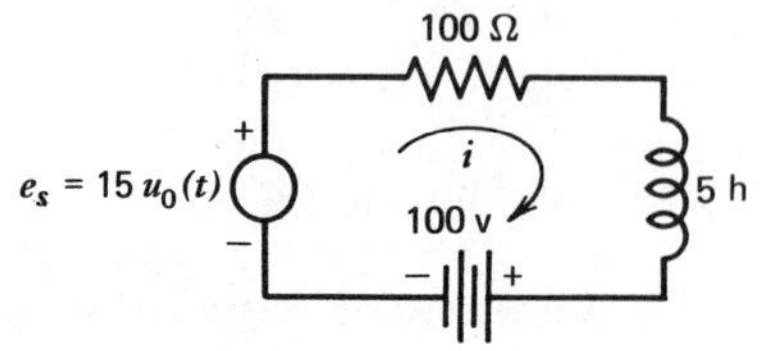

(c) Find the current i.

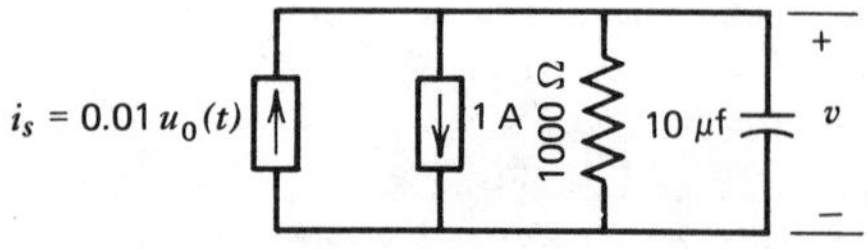

(d) Find the voltage v.

4-75 Solve the equation

$$10u_{+2}(t) = 5\frac{di}{dt} + 50i$$

for i, using classical methods.

4-76 Prove the statement in the last paragraph of Section 4-18.

Chapter 5

Operational Methods

5-1. INTRODUCTION

The a-c circuit theory has developed a number of techniques, such as combining impedances in series and in parallel. These techniques can be incorporated in transform methods, and we refer to this work as operational methods. We start with the special case of no energy stored in the circuit at $t = 0$, and define operational impedance. Later we add initial conditions and show how these can be moved into equivalent sources.

5-2. OPERATIONAL IMPEDANCE

In a-c circuit theory, phasor impedance is defined as the phasor voltage divided by the phasor current. Operational impedance is defined in an analogous manner as the transformed voltage divided by the transformed current with all initial conditions neglected.

$$Z(s) = \frac{V(s)}{I(s)} \tag{5-2.1}$$

OPERATIONAL ADMITTANCE

Operational admittance is defined as

$$Y(s) = \frac{1}{Z(s)} = \frac{I(s)}{V(s)} \tag{5-2.2}$$

The operational impedance and admittance for the L, C, and R elements are:

Inductance Element:

$$v = L\frac{di}{dt} \qquad V(s) = sL\, I(s)$$

$$Z(s) = \frac{V(s)}{I(s)} = sL \quad Y(s) = \frac{I(s)}{V(s)} = \frac{1}{sL} \tag{5-2.3}$$

Capacitive Element:

$$\begin{aligned} v &= \frac{1}{C}\int i\,dt \qquad & Z(s) &= \frac{1}{sC} \\ V(s) &= \frac{I(s)}{sC} & Y(s) &= sC \end{aligned} \tag{5-2.4}$$

Resistive Element:

$$\begin{aligned} v &= Ri \qquad & Z(s) &= R \\ V(s) &= R\,I(s) & Y(s) &= G \end{aligned} \tag{5-2.5}$$

IMPEDANCES CONNECTED IN SERIES

Figure 5-2.1 is used to show that the impedance of a number of elements connected in series is the sum of the individual impedances. The KVL and its transform are

$$v = v_1 + v_2 + \cdots + v_n \tag{5-2.6}$$

$$V(s) = V_1(s) + V_2(s) + \cdots + V_n(s) \tag{5-2.7}$$

Equation 5-2.7 is divided by $I(s)$ and rewritten

$$\frac{V(s)}{I(s)} = \frac{V_1(s)}{I(s)} + \frac{V_2(s)}{I(s)} + \cdots + \frac{V_n(s)}{I(s)} \tag{5-2.8}$$

$$Z(s) = Z_1(s) + Z_2(s) + \cdots + Z_n(s) \tag{5-2.9}$$

The $Z(s)$ is the operational impedance of the entire combination. As a specific example, the impedance of a series *R-L-C* circuit is

$$Z(s) = sL + R + \frac{1}{sC} \tag{5-2.10}$$

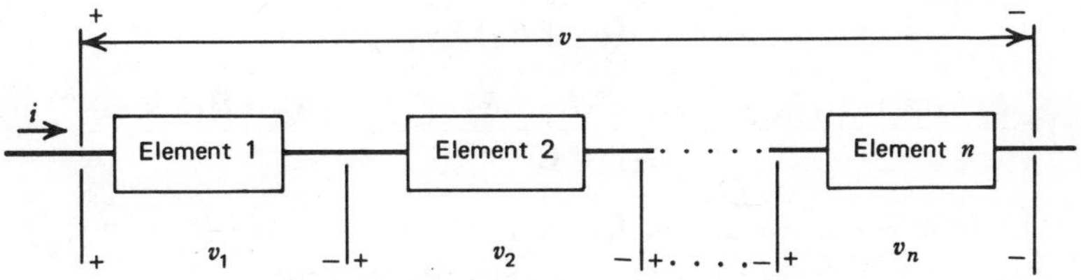

Fig. 5-2.1

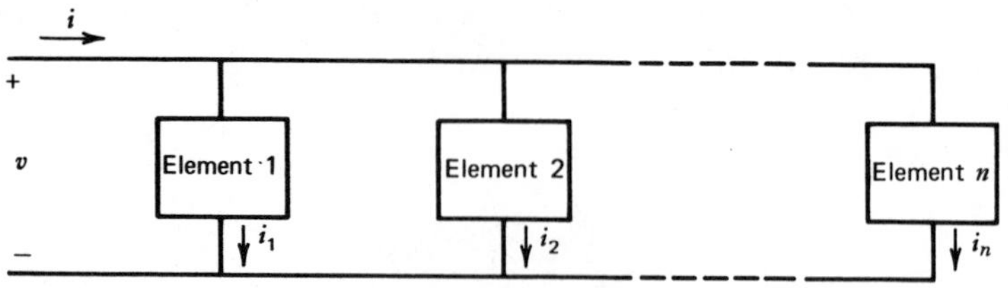

Fig. 5-2.2

ADMITTANCES CONNECTED IN PARALLEL

Figure 5-2.2 is used to show that the admittance of a number of elements connected in parallel is the sum of the individual admittances. The KCLE and its transform are

$$i = i_1 + i_2 + \cdots + i_n \tag{5-2.11}$$

$$I(s) = I_1(s) + I_2(s) + \cdots + I_n(s) \tag{5-2.12}$$

Equation 5-2.12 is divided by $V(s)$ and rewritten

$$\frac{I(s)}{V(s)} = \frac{I_1(s)}{V(s)} + \frac{I_2(s)}{V(s)} + \cdots + \frac{I_n(s)}{V(s)} \tag{5-2.13}$$

$$Y(s) = Y_1(s) + Y_2(s) + \cdots + Y_n(s) \tag{5-2.14}$$

The $Y(s)$ is the operational admittance of the entire combination. As a specific example, the admittance of a parallel R-L-C circuit is

$$Y(s) = \frac{1}{sL} + \frac{1}{R} + sC \tag{5-2.15}$$

Most other a-c circuit techniques have counterparts in the operational approach. For examples: voltage division across impedances in series, and current division among impedances in parallel, are identical in form. No effort is made to prove these items, but they are used as needed.

Example 5-2.1. The circuit is shown in Fig. 5-2.3(a) and the transformed circuit in (b). The impedance of the circuit from the voltage source is

$$Z_{in}(s) = Z(s) + \frac{Z_1(s)\,Z_2(s)}{Z_1(s) + Z_2(s)} = R + \frac{(R_1 + sL_1)(R_2 + sL_2)}{(R_1 + R_2) + s\,(L_1 + L_2)} \tag{5-2.16}$$

$$Z_{in}(s) = \frac{L_1 L_2 s^2 + (R_1 L_2 + R_2 L_1 + RL_1 + RL_2)\,s + (R_1 R_2 + R_1 R + R_2 R)}{(L_1 + L_2)\,s + (R_1 + R_2)} \tag{5-2.17}$$

To complete the example, we use the following values: $R = 20\ \Omega$, $R_1 = 30\ \Omega$, $R_2 = 10\ \Omega$; $L_1 = 0.5$ henry, $L_2 = 1$ henry. The $I(s)$ and its inverse transform are

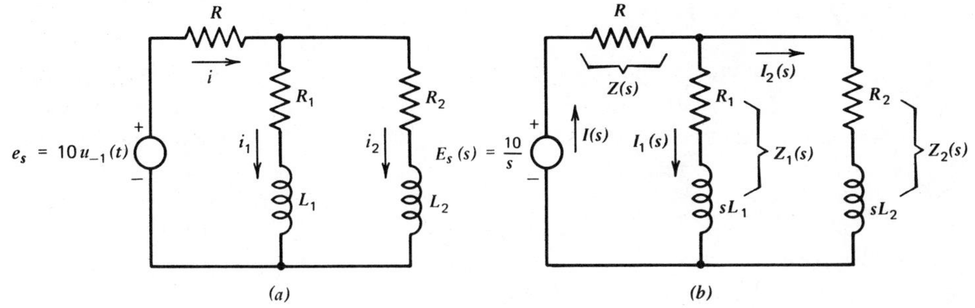

Fig. 5-2.3

$$I(s) = \frac{E(s)}{Z_{in}(s)} = \frac{30s + 800}{s\,[s^2 + 130s + 2{,}200]} \tag{5-2.18}$$

$$i = 0.364 - 0.111\epsilon^{-20t} - 0.253\epsilon^{-110t} \tag{5-2.19}$$

The currents $I_1(s)$ and $I_2(s)$ and their inverse transforms are

$$I_1(s) = \frac{Z_2(s)}{Z_1(s) + Z_2(s)}\,I(s) = \frac{20s + 200}{s\,[s^2 + 130s + 2{,}200]} \tag{5-2.20}$$

$$I_2(s) = \frac{Z_1(s)}{Z_1(s) + Z_2(s)}\,I(s) = \frac{10s + 600}{s\,[s^2 + 130s + 2{,}200]} \tag{5-2.21}$$

$$i_1 = 0.091 + 0.111\epsilon^{-20t} - 0.202\epsilon^{-110t} \tag{5-2.22}$$

$$i_2 = 0.273 - 0.222\epsilon^{-20t} - 0.051\epsilon^{-110t} \tag{5-2.23}$$

The voltage across the L_1 inductor is

$$V_{L1}(s) = sL_1\,I_1(s) = \frac{10s + 100}{(s^2 + 130s + 2{,}200)} \tag{5-2.24}$$

$$v_{L1} = -1.11\epsilon^{-20t} + 11.11\epsilon^{-110t} \tag{5-2.25}$$

Note that i, i_1, i_2, and v_{L1}, have been found and yet not a single differential equation has been written. This work is parallel to a-c circuit method. When phasor transformed circuits are used, you almost forget that you are solving some differential equation for the particular component. Here we not only solve for the particular component, but for the complete solution $i = i_p + i_c$.

The methods of finding the inverse transforms are discussed in Chapter 4 and add nothing to the present discussion; therefore, for the most part we stop finding the time solutions.

5-3. THE $Z(s)$ FUNCTION

Sometimes the characteristic polynomial for the circuit appears in the numerator, and sometimes in the denominator of $Z(s)$. This seems like the "shell game" until we understand the situation. Assume we have a $Z(s)$ that is made up of a numerator polynomial over a denominator polynomial as

$$Z(s) = \frac{N(s)}{D(s)} \tag{5-3.1}$$

If this circuit is driven by a voltage source, the response is a current

$$I(s) = \frac{E(s)}{Z(s)} = \frac{E(s)\,D(s)}{N(s)} \tag{5-3.2}$$

Since the characteristic polynomial always appears in the denominator of the function whose inverse transform is to be taken, $N(s)$ is the characteristic polynomial. However, if this circuit is now driving by a current source, the response is a voltage

$$E(s) = I(s)\,Z(s) = \frac{I(s)\,N(s)}{D(s)} \tag{5-3.3}$$

By the argument just used, $D(s)$ is now the characteristic polynomial.

At several places, we have stated that for a given circuit, all the voltages and currents have the same form, which means they all have the same characteristic polynomial. If this statement is compared with those in the last two paragraphs, it seems that something is wrong. This apparent confusion is understood when it is realized that the circuit has been changed.

To clarify this situation, we use the circuit of Fig. 5-2.3 with the voltage source removed and a number of terminals added as shown in Fig. 5-3.1. The circuit is first opened at *Point 1* and $Z_1(s)$ is determined. This is Eq. 5-2.17 repeated as

$$Z_1(s) = \frac{L_1L_2s^2 + (R_1L_2 + R_2L_1 + RL_1 + RL_2)s + (R_1R_2 + R_1R + R_2R)}{(L_1 + L_2)s + (R_1 + R_2)} \tag{5-3.4}$$

The impedance of a voltage source is zero; therefore, when the circuit of Fig. 5-3.1 is driven at *Point 1* by a voltage source, the resulting circuit is shown in Fig. 5-3.2(a). The

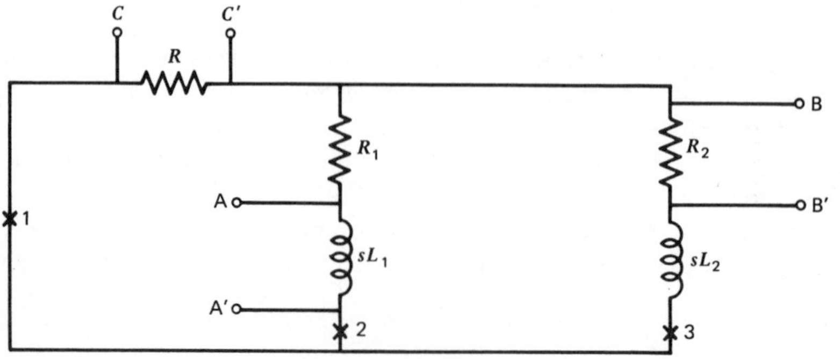

Fig. 5-3.1

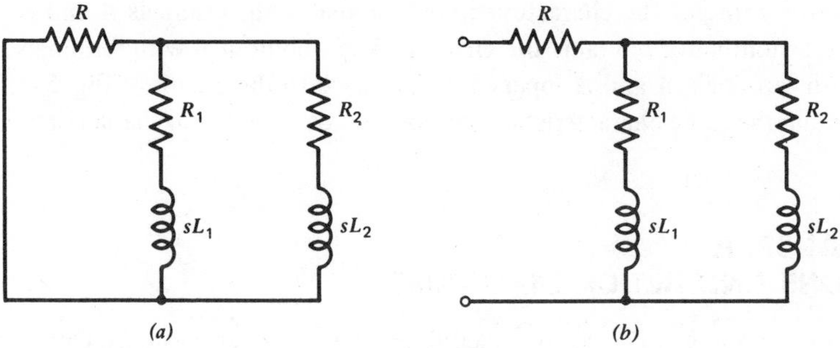

Fig. 5-3.2

impedance of a current source is infinite; therefore, when the circuit of Fig. 5-3.1 is driven at *Point 1* by a current source, the resulting circuit is shown in Fig. 5-3.2(b). Since the circuits are different, the characteristic polynomials are different.

When a $Z(s)$ is obtained at two terminals, the numerator is the characteristic polynomial for the resulting circuit when these two terminals are shorted and the denominator is the characteristic polynomial for the resulting circuit when these two terminals are opened. For the example just considered.

$$L_1L_2s^2 + (R_1L_2 + R_2L_1 + RL_1 + RL_2)s + (R_1R_2 + R_1R + R_2R) \qquad (5\text{-}3.5)$$

is the characteristic polynomial for the circuit of Fig. 5-3.2(a), whereas

$$(L_1 + L_2)s + (R_1 + R_2) \qquad (5\text{-}3.6)$$

is the characteristic polynomial for the circuit of Fig. 5-3.2(b).

To continue, the circuit of Fig. 5-3.1 is opened at *Point 2* and $Z_2(s)$ is determined

$$Z_2(s) = sL_1 + R_1 + \frac{R(sL_2 + R_2)}{sL_2 + R + R_2}$$

$$= \frac{L_1L_2s^2 + (R_2L_1 + R_1L_2 + RL_1 + RL_2)s + R_1R_2 + R_1R + R_2R}{L_2s + (R + R_2)} \qquad (5\text{-}3.7)$$

The circuit with the terminals at *Point 2* shorted is the same as the circuit of Fig. 5-3.2(a), and hence the numerator is the same as in Eq. 5-3.4. However, the circuit with the terminals at *Point 2* opened is a different circuit, and hence Eq. 5-3.7 has a different denominator.

Finally, the impedance $Z_{AA'}(s)$ across terminals A and A' is

$$\frac{1}{Z_{AA'}(s)} = \frac{1}{L_1s} + \cfrac{1}{R_1 + \cfrac{R(L_2s + R_2)}{L_2s + (R_2 + R)}} \qquad (5\text{-}3.8)$$

$$Z_{AA'}(s) = \frac{L_1s\,[(R_1L_2 + RL_2)s + (R_1R_2 + R_1R + R_2R)]}{L_1L_2s^2 + (R_2L_1 + R_1L_2 + RL_1 + RL_2)s + (R_1R_2 + R_1R + R_2R)}$$

The numerator contains the characteristic polynomial with terminals A and A shorted, whereas the denominator contains the characteristic polynomial with terminals A and A' opened. With terminals A and A' opened we are back to the circuit of Fig. 5-3.2(a), and we again obtain the same characteristic equation except now it is in the denominator.

5-4. TRANSFER FUNCTIONS AND BLOCK DIAGRAMS

The operational approach can be extended to all linear physical systems. One method for doing this is through use of transfer functions. The overall system is displayed in block diagram form, with the transfer function for various parts of the system shown in the corresponding block. These diagrams can be manipulated to obtain the operational expression for the entire system. The present discussion is an introduction to this subject.

The transfer function is defined as the ratio of the transformed output to the transformed input of the system or portion of a system being considered, assuming that there is no energy stored before $t = 0$. One symbol used for the transfer function is $G(s)$, where $G(s)$ is defined as

$$G(s) = \frac{O(s)}{I(s)} \tag{5-4.1}$$

$G(s)$ is the transfer function.

$O(s)$ is the Laplace transform of the output.

$I(s)$ is the Laplace transform of the input.

The information contained in Eq. 5-4.1 is displayed in block diagram form as shown in Fig. 5-4.1.

The $I(s)$ is the input to the system and can be thought of as a signal. The transfer function is displayed inside the block and $O(s)$ is the output. The arrows indicate that the signal flows into the block and is modified by the $G(s)$ to produce the $O(s)$. Equation 5-4.1 can be solved for $O(s)$ as

$$O(s) = G(s)\, I(s) \tag{5-4.2}$$

To explore the $G(s)$, suppose the input as a time function is $u_0(t)$. The corresponding $I(s)$ is

$$I(s) = \mathcal{L}[u_0(t)] = 1 \tag{5-4.3}$$

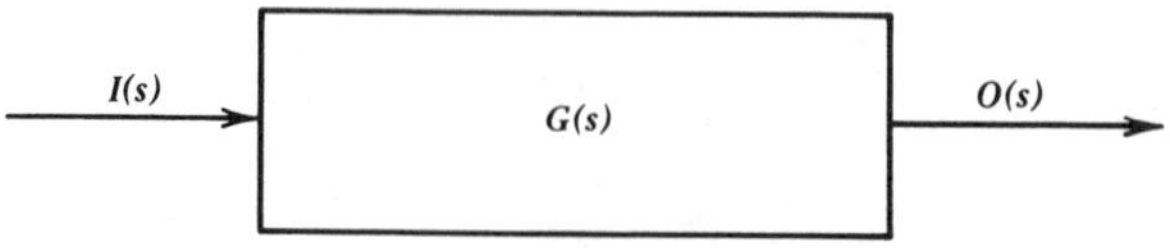

Fig. 5-4.1

With this input, Eq. 5-4.2 becomes

$$O(s) = G(s) \cdot 1 = G(s) \tag{5-4.4}$$

Under these conditions, the transfer function equals the output of the system and $G(s)$ can be given a physical interpretation as being the transform of the impulse response of the system and $g(t)$ is the corresponding time function. In Chapter 2 we used block diagrams in this same manner and placed $g(t)$ inside the block. Other notations are used for $G(s)$ in the literature and several of these are used later in this book.

The definition of transfer function is completely general in that the input and output can be any combination of voltage, current, displacement, velocity, pressure, temperature torque or other physical quantity.

Example 5-4.1. The circuit of Ex. 5-2.1 is shown again in Fig. 5-4.2(a) and the transformed circuit in (b). The first transfer function to be found is $I(s)/E_s(s)$, which can be found from Eq. 5-2.18 as

$$G(s) = \frac{I(s)}{E_s(s)} = \frac{(3s + 80)}{(s^2 + 130s + 2{,}200)} \tag{5-4.5}$$

Since the voltage and the current are at the same pair of terminals, this $G(s)$ is also called a driving point admittance. This transfer function is displayed in the block diagram of Fig. 5-4.3.

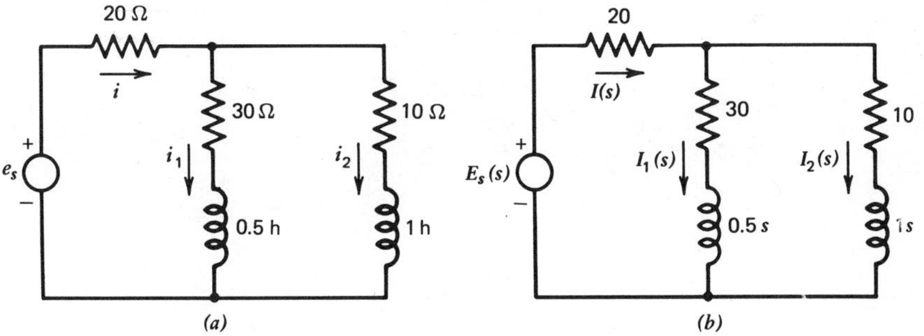

Fig. 5-4.2

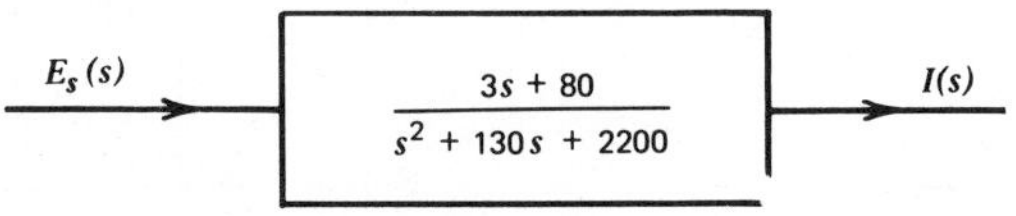

Fig. 5-4.3

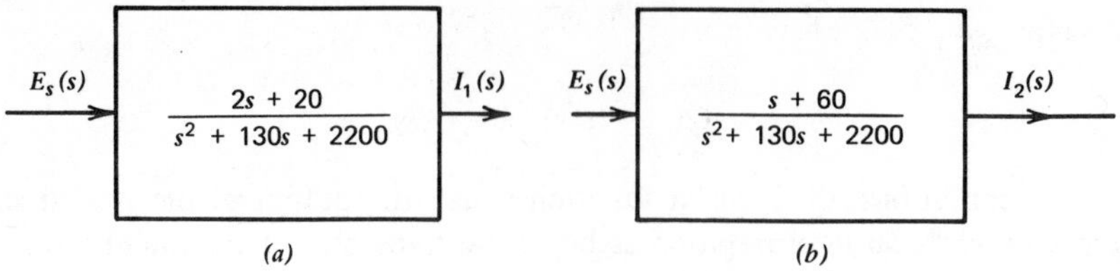

Fig. 5-4.4

The second and third transfer functions are found from Eq. 5-2.20 and 5-2.21, respectively, as

$$G(s) = \frac{I_1(s)}{E_s(s)} = \frac{(2s + 20)}{(s^2 + 130s + 2{,}200)} \tag{5-4.6}$$

$$G(s) = \frac{I_2(s)}{E_s(s)} = \frac{(s + 60)}{(s^2 + 130s + 2{,}200)} \tag{5-4.7}$$

Since the voltage is applied in one part of the circuit and the current is found in another part, this $G(s)$ is also called a transfer admittance. The block diagrams are shown in Fig. 5-4.4(a) and (b).

Example 5-4.2. The circuit is shown in Fig. 5-4.5(a) and the transformed circuit in (b). The transfer function and the block diagram (see also Fig. 5-4.6) are found as

$$I_2(s) = \frac{1/2s}{(1/2s) + (1s/2) + 1} I_s(s) = \frac{I_s(s)}{(s + 1)^2} \tag{5-4.8}$$

$$V_0(s) = I_2(s) \times 1 = \frac{I_s(s)}{(s + 1)^2} \tag{5-4.9}$$

$$\frac{V_0(s)}{I_s(s)} = \frac{1}{(s + 1)^2} \tag{5-4.10}$$

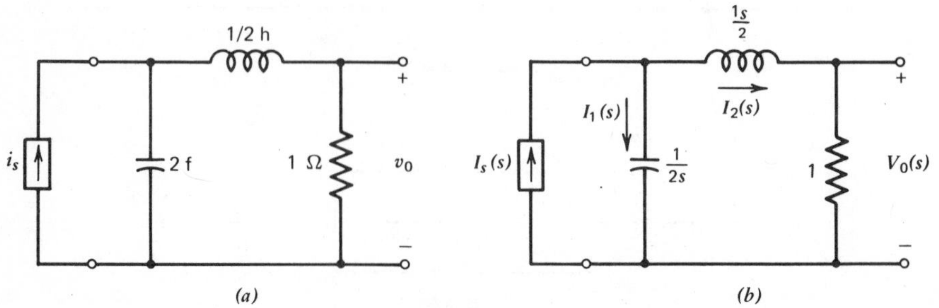

Fig. 5-4.5

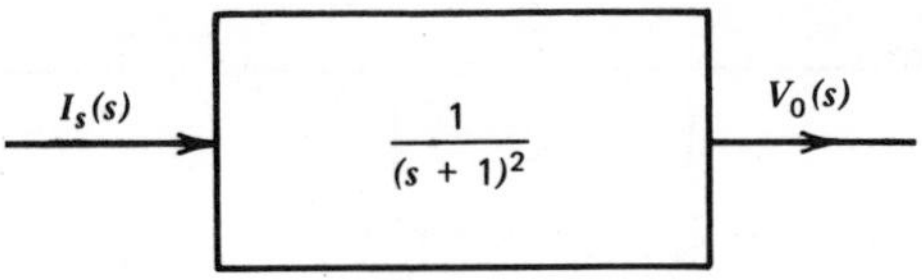

Fig. 5-4.6

Example 5-4.3. The circuit is shown in Fig. 5-4.7(a) and the transformed circuit in (b). The transfer function and the block diagram (see also Fig. 5-4.8) are found as

$$V_0(s) = \frac{Z_2(s)}{Z_1(s) + Z_2(s)} E_s(s) = \frac{1/s}{1 + 1/s} E_s(s) \tag{5-4.11}$$

$$G(s) = \frac{V_0(s)}{E_s(s)} = \frac{1/s}{1 + 1/s} = \frac{1}{(s+1)} \tag{5-4.12}$$

This $G(s)$ is called a voltage ratio.

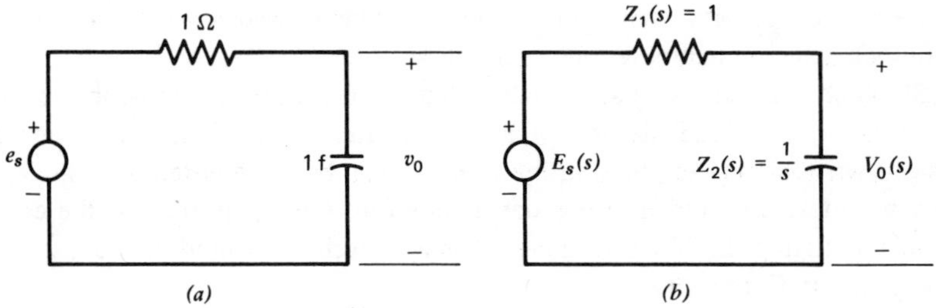

Fig. 5-4.7

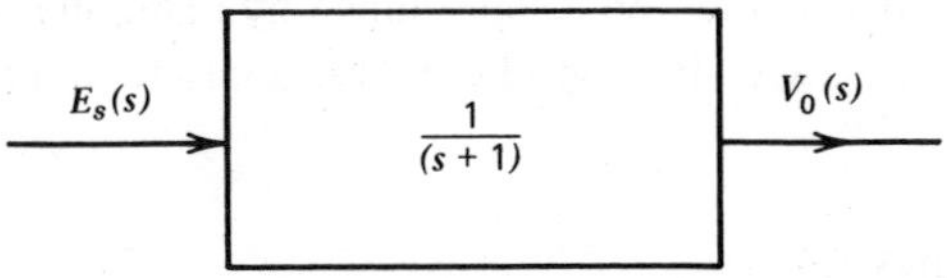

Fig. 5-4.8

Example 5-4.4. The circuit is shown in Fig. 5-4.9(a) and the transformed circuit in (b). The transfer function is found in

$$V_0(s) = \frac{1}{1+s} E_s(s) \tag{5-4.13}$$

$$\frac{V_0(s)}{E_s(s)} = \frac{1}{(s+1)} \tag{5-4.14}$$

The block diagram is again the one shown in Fig. 5-4.8 and the conclusion is that the systems of Fig. 5-4.7 and 5-4.9 systems behave in the same manner.

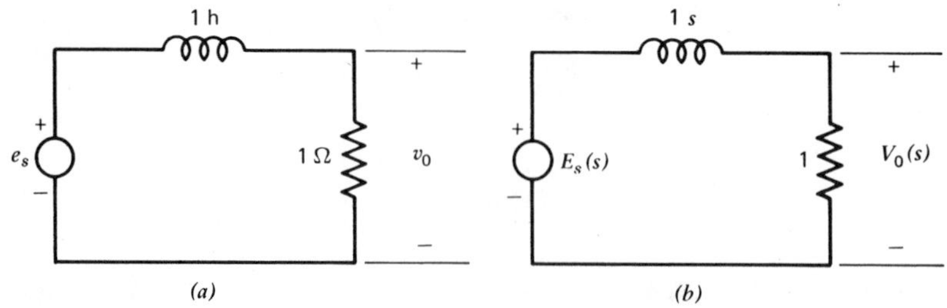

Fig. 5-4.9

5-5. BLOCK DIAGRAM MANIPULATIONS

In Chapter 12, we study the signal-flow graph in detail. Block diagrams and signal-flow graphs are closely related and accomplish the same purpose. The signal-flow graph has the advantage of a more uniform notation. The signal-flow graph is more convenient to draw and easier to visualize. Elegant methods have been developed for reducing the signal-flow graph. Although these methods can be applied to block diagrams, we only take them up during the discussion of the signal-flow graph. For these reasons we do not pursue the subject of block diagram manipulations in any detail.

A single block may represent an overall system or only a single component of the system depending on the detail desired. A set of equations describing the behavior of a system can be written and each term in the equation can be represented by a block. These blocks can be interconnected in a manner dictated by the equations and the equations solved by manipulating the block diagram. This approach is covered in the discussion of signal-flow graphs in Chapter 12.

Another approach is to divide the system into subsystems and to develop the transfer function for each subsystem. The block diagrams for the subsystems are interconnected and the overall system transfer function is determined by manipulating the block diagram. Circuit theory examples are used although these ideas apply to all linear physical systems.

TRANSFER FUNCTIONS OF NETWORKS CONNECTED IN TANDEM

Two networks are connected in tandem if the output of one network is the input to the next. If the second network is connected to the first without "loading" the first network, the transfer function of the combined network is the product of the transfer function of the individual networks.

Example 5-5.1. Two *R-C* networks of Fig. 5-4.7(a) are connected in tandem, as shown in Fig. 5-5.1. The box labeled *1* represents an isolation stage so that the second stage does not load the first stage, yet at the same time the input v_a to the isolation stage appears at its output and applies v_a to the input of the second stage. The transfer function for each

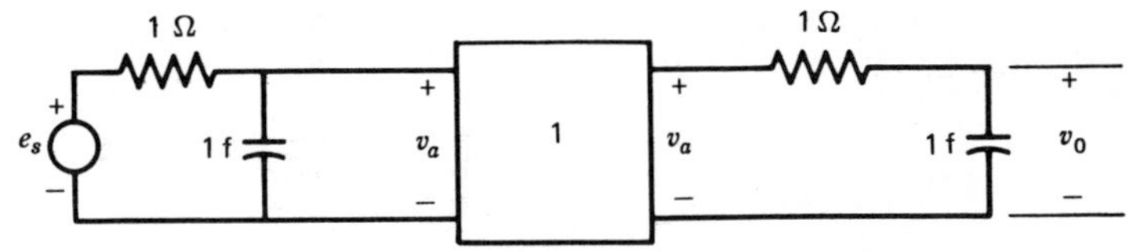

Fig. 5-5.1

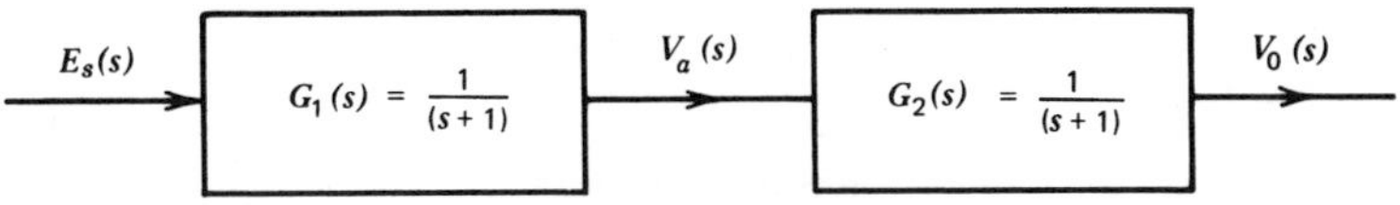

Fig. 5-5.2

stage was found in Eq. 5-4.12 as

$$G(s) = \frac{1}{(s+1)} \tag{5-5.1}$$

Because there is no loading between stages, the two transfer functions can be placed in tandem as shown in Fig. 5-5.2.

Under the conditions just described, the ratio of $V_0(s)$ to $E_s(s)$ can be written as

$$\frac{V_0(s)}{E_s(s)} = \frac{V_a(s)}{E_s(s)} \frac{V_0(s)}{V_a(s)} \tag{5-5.2}$$

We define $G(s)$ as the transfer function of the two stages in tandem and can write

$$G(s) = G_1(s)\, G_2(s) = \frac{1}{(s+1)^2} = \frac{1}{(s^2 + 2s + 1)} \tag{5-5.3}$$

If n networks are connected in tandem, the overall transfer function would be

$$G(s) = G_1(s)\, G_2(s) \cdots G_n(s) \tag{5-5.4}$$

Example 5-5.2. The isolation stage of Fig. 5-5.1 is removed, leaving the circuit shown in Fig. 5-5.3. The output of the first R-C combination is now affected by the presence of the second R-C combination, and the block diagram of Fig. 5-5.2 does not apply. When

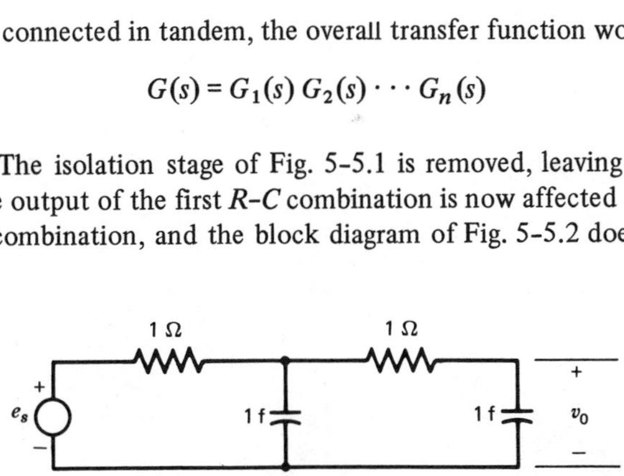

Fig. 5-5.3

the entire circuit is transformed, the proper equations can be written, and the overall transfer function can be found as

$$G(s) = \frac{V_0(s)}{E_s(s)} = \frac{1}{(s^2 + 3s + 1)} \qquad (5\text{-}5.5)$$

Note that the $G(s)$ of Eq. 5-5.5 differs from the $G(s)$ of Eq. 5-5.3.

THE SUMMING POINT

A summing point is a point in a system where two or more signals combine. Figure 5-5.4(a) indicates that $F_1(s)$ and $F_2(s)$ combine to yield $F_3(s)$

$$F_1(s) - F_2(s) = F_3(s) \qquad (5\text{-}5.6)$$

The plus sign on $F_1(s)$ indicates that $F_1(s)$ goes through the summing point with a plus sign whereas the minus sign on $F_2(s)$ means $F_2(s)$ changes signs in going through the summing point. Figure 5-5.4(b) shows that both signals go through the summing point with plus signs as

$$F_1(s) + F_2(s) = F_3(s) \qquad (5\text{-}5.7)$$

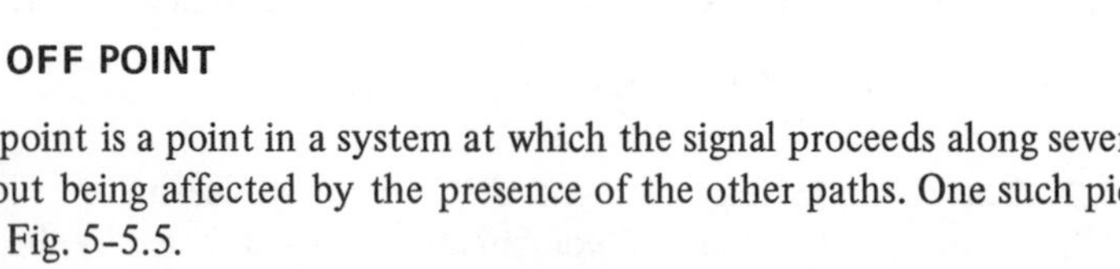

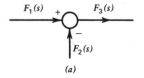

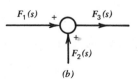

Fig. 5-5.4

THE PICK OFF POINT

A pick-off point is a point in a system at which the signal proceeds along several different paths without being affected by the presence of the other paths. One such pick-off point is shown in Fig. 5-5.5.

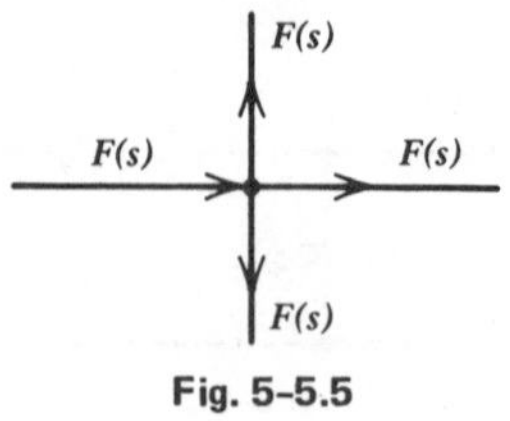

Fig. 5-5.5

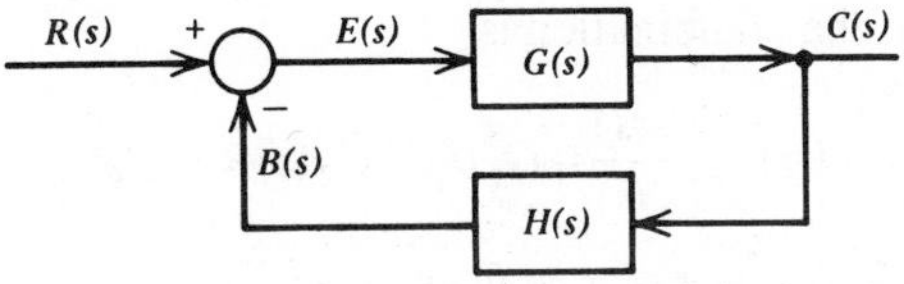

Fig. 5-5.6

A CLOSED-LOOP SYSTEM

The block-diagram for the basic closed-loop system is shown in Fig. 5-5.6. The input to the system is labeled $R(s)$. $G(s)$ is the transfer function of the forward part of the system, and $H(s)$ is the transfer function of the feedback portion. The signal $C(s)$ is the output, but this is also fed back to $H(s)$. The following three equations describe the system.

$$E(s) = R(s) - B(s)$$

$$C(s) = G(s)\,E(s) \tag{5-5.8}$$

$$B(s) = H(s)\,C(s)$$

These three equations can be solved to yield the transfer function for the closed loop system as

$$\frac{C(s)}{R(s)} = \frac{G(s)}{1 + H(s)\,G(s)} \tag{5-5.9}$$

TRANSFER FUNCTIONS CONNECTED IN PARALLEL

Figure 5-5.7(a) indicates three transfer functions connected in parallel. The output of the summing point is

$$F_5(s) = F_2(s) - F_3(s) + F_4(s) \tag{5-5.10}$$

which can be written as

$$F_5(s) = F_1(s)\,G_1(s) - F_1(s)\,G_2(s) + F_1(s)\,G_3(s) \tag{5-5.11}$$

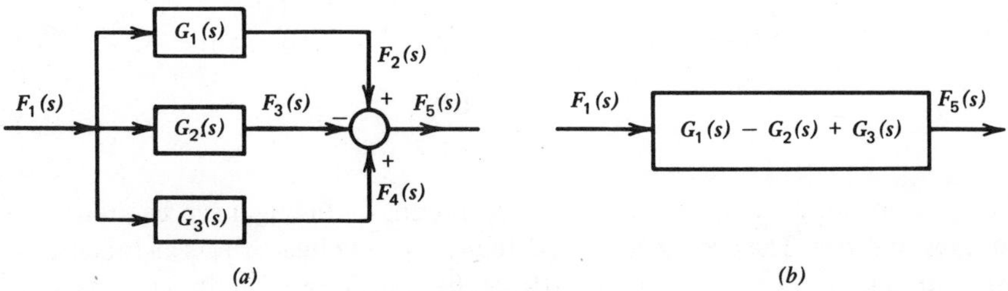

Fig. 5-5.7

The transfer function for the combination is

$$G(s) = \frac{F_5(s)}{F_1(s)} = G_1(s) - G_2(s) + G_3(s) \tag{5-5.12}$$

The resulting diagram is shown in Fig. 5-5.7(b).

If a group of transfer functions are connected in parallel, the transfer function for the combination is the sum of the individual transfer functions with the proper signs.

COMMENTARY

The basic closed-loop system of Fig. 5-5.6 can be studied in much detail. One of the items of interest is that of the stability of the closed-loop system. A complete discussion of this is beyond the scope of this work and is left to such sources as control system theory.[1]

5-6. ADDING INITIAL CONDITIONS TO THE OPERATIONAL METHODS

The operational methods developed in Section 5-2 assumed no energy was stored in the circuit before $t = 0$. The initial conditions can be added as shown in the following.

ADDING INITIAL CONDITIONS TO THE INDUCTOR

An inductor with an initial current is shown in Fig. 5-6.1(a) with the two terminals labeled A and B. The volt-ampere relationship can be expressed in either of two forms:

$$v = L\frac{di}{dt} \qquad i = \frac{1}{L}\int_0^t v\,dt + i(0) \tag{5-6.1}$$

The transforms of these two equations are

$$V(s) = L\,[s\,I(s) - i(0)] = sL\,I(s) - L\,i(0) \tag{5-6.2}$$

$$I(s) = \frac{V(s)}{sL} + \frac{i(0)}{s} \tag{5-6.3}$$

The units on Eq. 5-6.2 are volts, and the two terms on the right can be thought of as two voltages in series. The first term, $sLI(s)$ represents a voltage across an inductor with no initial current, and the second term $Li(0)$ is the transform of an impulse. An equiva-

[1] For example see Lago and Benningfield, *op. cit.*

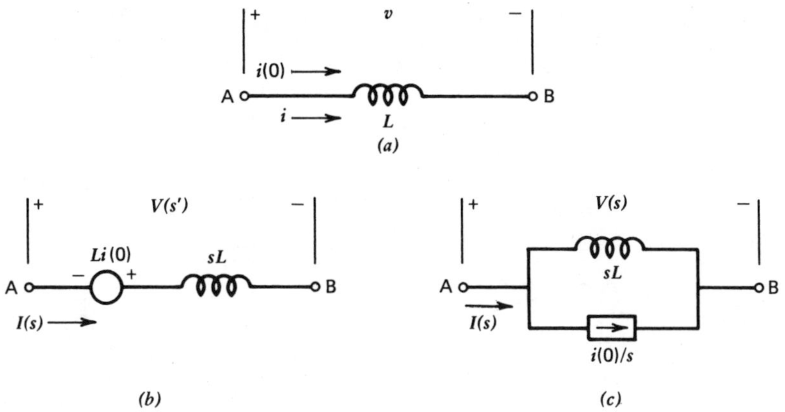

Fig. 5-6.1

lent transformed circuit that has the same A-B terminal characteristic is shown in Fig. 5-6.1(b).

The units of Eq. 5-6.3 are amperes, and these two terms can be thought of as two currents in parallel. The first term $V(s)/sL$ represents the current in an inductor with no initial conditions, and the second term is the transform of a constant current source. An equivalent transformed circuit that has the same A-B terminal characteristics is shown in Fig. 5-6.1(c).

ADDING INITIAL CONDITIONS TO THE CAPACITOR

A capacitor with an initial voltage is shown in Fig. 5-6.2(a) with the two terminals labeled A and B. The volt-ampere relationship can be expressed in either of two forms:

$$v = \frac{1}{C}\int_0^t i\,dt + v(0) \qquad i = \frac{C\,dv}{dt} \tag{5-6.4}$$

The transforms of these two equations are

$$V(s) = \frac{I(s)}{sC} + \frac{v(0)}{s} \tag{5-6.5}$$

$$I(s) = C\,[s\,V(s) - v(0)] = s\,C\,V(s) - C\,v(0) \tag{5-6.6}$$

The units on Eq. 5-6.5 are volts, and the two terms can be thought of as two voltages in series. The first term $I(s)/sC$ represents a voltage on a capacitor with no initial conditions, and the second term $v(0)/s$ is the transform of a constant voltage source. An equivalent transformed circuit that has the same A-B terminal characteristics is shown in Fig. 5-6.2(b).

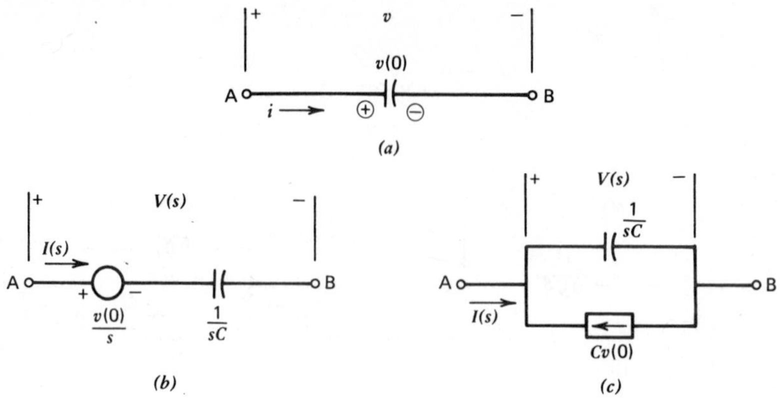

Fig. 5-6.2

The units on Eq. 5-6.6 are amperes, and these two terms can be thought of as two currents in parallel. The first term $sC\,V(s)$ represents the current in a capacitor with no initial charge and the second term $C\,v(0)$ is the transform of an impulse. An equivalent transformed circuit that has the same A-B terminal characteristics is shown in Fig. 5-6.2(c).

EQUIVALENT CIRCUITS

It should be emphasized that the equivalent circuit can replace the original circuit only so far as conditions external to the A-B terminals are concerned after $t = 0$. In Fig. 5-6.1(c) and Fig. 5-6.2(c), the currents into the parallel combinations are the currents in the original circuits. In Fig. 5-6.1(b), the impulse of voltage at $t = 0$ throws the proper $i(0)$ into the inductor, and for $t > 0$, the two circuits cannot be distinguished. In Fig. 5-6.2(c), the impulse of current at $t = 0$ throws the proper charge on the capacitor and for $t > 0$, the equivalent circuit and the original circuit are identical.

For both the inductor and the capacitor, we have two choices. The initial conditions can be brought out in series with the element as a voltage source or in parallel as a current source. Although other methods can be followed, it is convenient to move the initial conditions into voltage sources when loop-currents are used, and to move them into current sources when node-pair voltages are used.

Example 5-6.1. The circuit shown in Fig. 5-6.3 is analyzed using operational methods. From some previous switching operation, the inductor contains a current and the capacitor a voltage at $t = 0$, as shown. The desired solution is the current i_L.

We first look the situation over from a topological point of view.[2] The topology is shown in Fig. 5-6.4(a), from which we see that

$$b = 6 \qquad n_t = 4 \qquad n = 3 \qquad l = 3$$

[2] See Appendix C for a more detailed discussion about writing circuit equations.

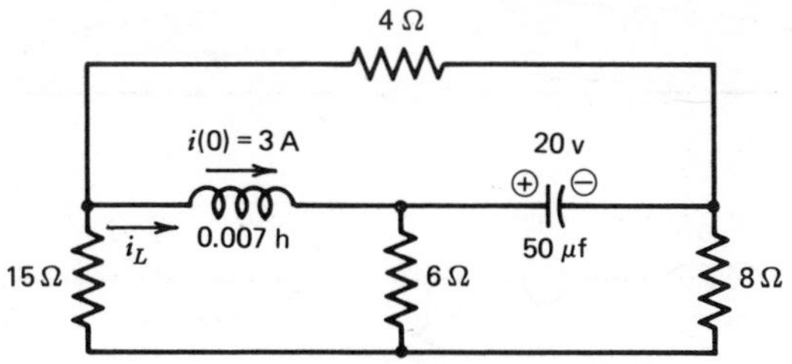

Fig. 5-6.3

If loop-currents are used, $l = 3$ equations are needed. If node-pair voltages are used, $n = 3$ equations are needed. Therefore, we should be able to work the problems either way with equal ease. We use loop-currents first.

If we use window currents, the current i_L would be the difference between two window currents, and this method represents more work than if we choose a tree and make certain the inductor is one of the links. Although this leaves several choices, we use the tree shown in Fig. 5-6.4(b). We drop the links in place one at a time and determine the paths of the respective loop-currents. This result is shown in (c).

Since we are using loop currents, we move the initial conditions into voltage sources. The transformed circuit is shown in Fig. 5-6.5.

The KVLE's are given by

$$
\begin{aligned}
0.021 &= (6 + 15 + 0.007s)\, I_1(s) - 6I_2 + 15I_3(s) \\
\frac{-20}{s} &= -6I_1(s) + \left(6 + 8 + \frac{2 \times 10^4}{s}\right) I_2(s) + 8I_3(s) \\
0 &= 15I_1(s) + 8I_2(s) + (4 + 8 + 15)\, I_3(s)
\end{aligned}
\tag{5-6.7}
$$

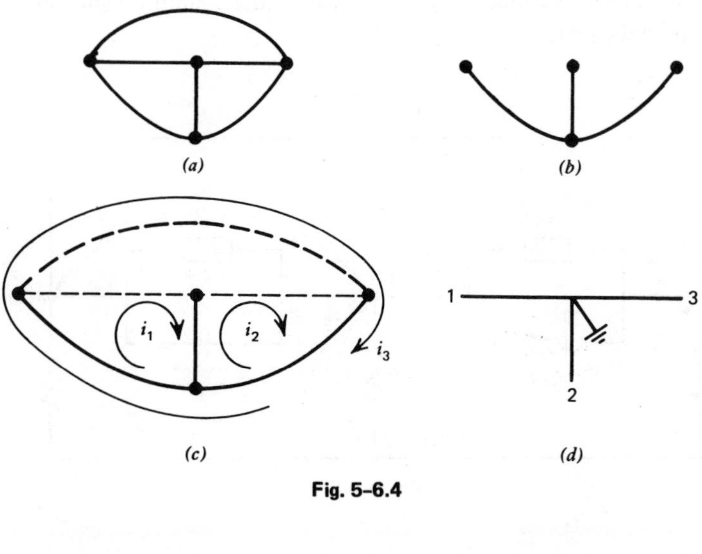

Fig. 5-6.4

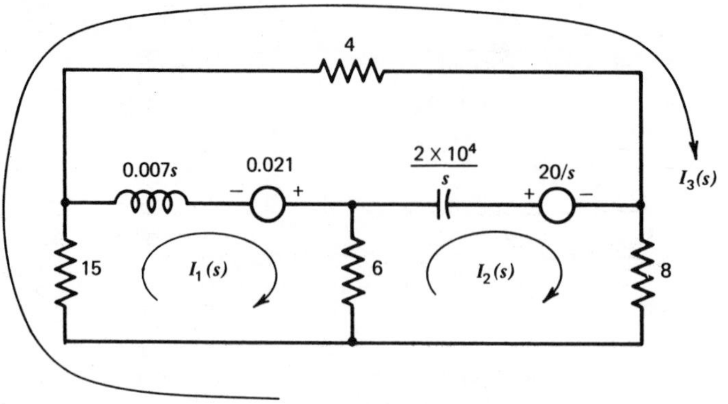

Fig. 5-6.5

Determinants are used to find $I_1(s)$

$$I_L(s) = I_1(s) = \frac{\begin{vmatrix} 0.021 & -6 & 15 \\ \dfrac{-20}{s} & \left(14 + \dfrac{2 \times 10^4}{s}\right) & 8 \\ 0 & 8 & 27 \end{vmatrix}}{\begin{vmatrix} 0.007s + 21 & -6 & 15 \\ -6 & \left(14 + \dfrac{2 \times 10^4}{s}\right) & 8 \\ +15 & 8 & 27 \end{vmatrix}} = \frac{3s + 2594}{s^2 + 2{,}189s + 3.11 \times 10^6} \qquad (5\text{-}6.8)$$

The actual inverse transform is not taken.

Again we have solved a problem, this time with initial conditions and yet we have not written a differential equation.

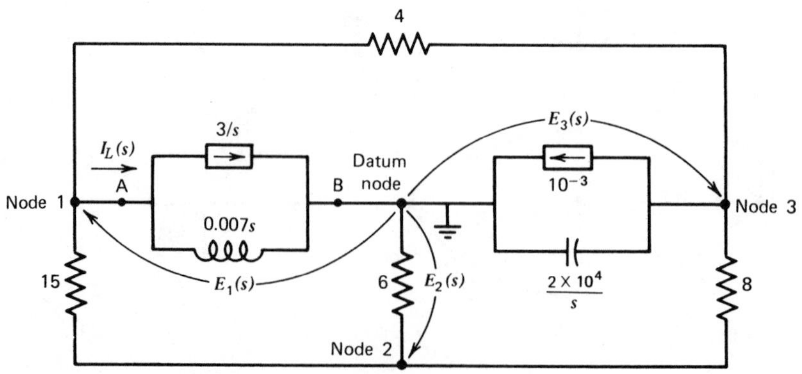

Fig. 5-6.6

Example 5-6.2. The example is solved again except node-pair voltage equations are used. We solve for the voltage across the inductor, and from this we then find $I_L(s)$. If we use the tree of Fig. 5-6.4(c), the voltage across the inductor is the difference between two tree-branch voltages. Therefore, we choose a different tree in which the inductor is one of the tree branches. The tree chosen is shown in Fig. 5-6.4(d). For our present purposes the node to datum method is sufficient. We use the datum node shown, with the other three nodes numbered.

The initial conditions are moved into current sources, and the transformed circuit is shown in Fig. 5-6.6. The three KCLE's are

$$\frac{-3}{s} = \left(\frac{1}{4} + \frac{1}{15} + \frac{1}{0.007s}\right) E_1(s) - \frac{1}{15} E_2(s) - \frac{1}{4} E_3(s)$$

$$0 = -\frac{1}{15} E_1(s) + \left(\frac{1}{15} + \frac{1}{6} + \frac{1}{8}\right) E_2(s) - \frac{1}{8} E_3(s) \tag{5-6.9}$$

$$-10^{-3} = -\frac{1}{4} E_1(s) - \frac{1}{8} E_2(s) + \left(\frac{1}{4} + \frac{1}{8} + 50 \times 10^{-6} s\right) E_3(s)$$

The $E_1(s)$ voltage is

$$E_1(s) = \frac{\begin{vmatrix} \dfrac{-3}{s} & -0.06667 & -0.25 \\ 0 & 0.3583 & -0.125 \\ -10^{-3} & -0.125 & (0.375 + 50 \times 10^{-6} s) \end{vmatrix}}{\begin{vmatrix} \left(\dfrac{0.31667s + 142.86}{s}\right) & -0.06667 & -0.25 \\ -0.06667 & 0.3583 & -0.125 \\ -0.25 & -0.125 & (0.375 + 50 \times 10^{-6} s) \end{vmatrix}}$$

$$E_1(s) = \frac{(-27.82s - 6.535 \times 10^4)}{(s^2 + 2{,}189s + 3.11 \times 10^6)} \tag{5-6.10}$$

The $I_L(s)$ is found from

$$I_L(s) = \frac{E_1(s)}{0.007s} + \frac{3}{s} = \frac{3\,s + 2{,}594}{s^2 + 2{,}189s + 3.11 \times 10^6} \tag{5-6.11}$$

5-7. THE GENERAL SOLUTION

We want to visualize a complicated circuit and generalize our thinking. This circuit contains b passive branches made up of R, L, and C elements. Each storage element can contain initial conditions. In addition to the passive elements, there can be any number of voltage and current sources. Each source can have any possible wave-shape. The circuit has n_t nodes. Throughout this discussion we use the general notation that

$$n = n_t - 1 \tag{5-7.1}$$

Where n is the number of node-to-datum voltage equations needed, and

$$l = b - n \tag{5-7.2}$$

Where l is the number of loop-current equations needed.

LOOP-CURRENT EQUATIONS

Since window-currents do not work on a non-flat network, and we want to discuss a general method, we choose loop-currents. In using loop-current equations, it is convenient to have only voltage sources. Therefore, as a preliminary step, all current sources are transformed into voltage sources and all initial conditions are brought out as voltage sources.

Next we choose a tree and drop the links in place one at a time to establish the loop-currents. The resulting equations have the following form.

$$\begin{aligned}
E_{11}(s) &= z_{11}(s)\,I_1(s) + z_{12}(s)\,I_2(s) + \cdots + z_{1\ell}(s)\,I_\ell(s)\\
E_{22}(s) &= z_{21}(s)\,I_1(s) + z_{22}(s)\,I_2(s) + \cdots + z_{2\ell}(s)\,I_\ell(s)\\
&\vdots \quad \cdots\cdots\cdots\cdots\cdots\cdots\cdots\cdots\cdots\cdots\\
E_{\ell\ell}(s) &= z_{\ell 1}(s)\,I_1(s) + z_{\ell 2}(s)\,I_2(s) + \cdots + z_{\ell\ell}(s)\,I_\ell(s)
\end{aligned} \tag{5-7.3}$$

These equations can be written in matrix form[3] as

$$E(s)]_{\ell 1} = [z(s)]_{\ell\ell}\, I(s)]_{\ell 1} \tag{5-7.4}$$

The subscripts are added to show the number of rows and columns.
The inverse matrix is found as

$$[z(s)]_{\ell\ell}^{-1} = [Y(s)]_{\ell\ell} \tag{5-7.5}$$

Equations 5-7.4 are premultiplied by $[Y(s)]_{\ell\ell}$ as

$$[Y(s)]_{\ell\ell}\, E(s)]_{\ell 1} = [Y(s)]_{\ell\ell}\, [z(s)]_{\ell\ell}\, I(s)]_{\ell 1} \tag{5-7.6}$$

which can be written as

$$I(s)]_{\ell 1} = [Y(s)]_{\ell\ell}\, E(s)]_{\ell 1} \tag{5-7.7}$$

To observe these solutions, we write out the corresponding equations as

$$\begin{aligned}
I_1(s) &= Y_{11}(s)\,E_{11}(s) + Y_{12}(s)\,E_{22}(s) + \cdots + Y_{1\ell}(s)\,E_{\ell\ell}(s)\\
I_2(s) &= Y_{21}(s)\,E_{11}(s) + Y_{22}(s)\,E_{22}(s) + \cdots + Y_{2\ell}(s)\,E_{\ell\ell}(s)\\
&\vdots \quad \cdots\cdots\cdots\cdots\cdots\cdots\cdots\cdots\cdots\cdots\\
I_\ell(s) &= Y_{\ell 1}(s)\,E_{11}(s) + Y_{\ell 2}(s)\,E_{22}(s) + \cdots + Y_{\ell\ell}(s)\,E_{\ell\ell}(s)
\end{aligned} \tag{5-7.8}$$

[3] The reader not familiar with matrix theory should consult Appendix D.

In Eqs. 5-7.3, the $E(s)$ and the $z(s)$ are known, but the $I(s)$ are unknown. Equations 5-7.8 are the solutions for the unknown $I(s)$.

NODE-TO-DATUM VOLTAGE EQUATIONS

Although tree-branch voltage equations are more general, node-to-datum voltage equations can handle non-flat networks and are sufficient for our present purposes. In using nodal equations, it is convenient to have only current sources. Therefore as a preliminary step, all voltage sources are transformed into current sources and all initial conditions are brought out as current sources.

Next, we choose a datum node, and number the remaining nodes from 1 through n. The resulting equations have the following form.

$$\begin{aligned} I_{11}(s) &= y_{11}(s)E_1(s) + y_{12}(s)E_2(s) + \cdots + y_{1n}(s)E_n(s) \\ I_{22}(s) &= y_{21}(s)E_1(s) + y_{22}(s)E_2(s) + \cdots + y_{2n}(s)E_n(s) \\ &\vdots \quad \cdots\cdots\cdots\cdots\cdots\cdots\cdots\cdots \\ I_{nn}(s) &= y_{n1}(s)E_1(s) + y_{n2}(s)E_2(s) + \cdots + y_{nn}(s)E_n(s) \end{aligned} \tag{5-7.9}$$

These equations can be written in matrix form as

$$I(s)]_{n1} = [y(s)]_{nn}\, E(s)]_{n1} \tag{5-7.10}$$

The inverse matrix is found as

$$[y(s)]_{nn}^{-1} = [Z(s)]_{nn} \tag{5-7.11}$$

Equations 5-7.10 are premultiplied by $[Z(s)]_{nn}$ as

$$[Z(s)]_{nn}\, I(s)]_{n1} = [Z(s)]_{nn}\, [y(s)]_{nn}\, E(s)]_{n1} \tag{5-7.12}$$

which can be written as

$$E(s)]_{n1} = [Z(s)]_{nn}\, I(s)]_{n1} \tag{5-7.13}$$

To observe these solutions, we write out the corresponding equations as

$$\begin{aligned} E_1(s) &= Z_{11}(s)I_{11}(s) + Z_{12}(s)I_{22}(s) + \cdots + Z_{1n}(s)I_{nn}(s) \\ E_2(s) &= Z_{21}(s)I_{11}(s) + Z_{22}(s)I_{22}(s) + \cdots + Z_{2n}(s)I_{nn}(s) \\ &\vdots \quad \cdots\cdots\cdots\cdots\cdots\cdots\cdots\cdots \\ E_n(s) &= Z_{n1}(s)I_{11}(s) + Z_{n2}(s)I_{22}(s) + \cdots + Z_{nn}(s)I_{nn}(s) \end{aligned} \tag{5-7.14}$$

In Eqs. 5-7.9, the $I(s)$ and the $y(s)$ are known but the $E(s)$ are unknown. Equations 5-7.14 are the solution for the unknown $E(s)$.

THE $z(s)$ AND $Y(s)$ FUNCTIONS (LOOP-CURRENT EQUATIONS)

The $z(s)$ terms on the principal diagonal are the sum of all the impedances in a circuit-set made up of tree branches and one link branch. All the other link currents are set equal to zero. To review the terms off the principal diagonal we examine $z_{jk}(s)$ as a general term. The $z_{jk}(s)$ is the intersection of the $z_{jj}(s)$ set and the $z_{kk}(s)$ set. The $z_{jk}(s)$ carries its own sign by comparing the directions of $I_j(s)$ and $I_k(s)$ in elements of the $z_{jk}(s)$ set. The situation is more complex when mutual inductance is present.

To understand the $Y(s)$ terms, we turn our attention to Eqs. 5-7.8. We pick $Y_{jj}(s)$ as being representative of the terms on the principal diagonal. We set all the $E(s)$ terms equal to zero, except $E_{jj}(s)$, and solve the resulting equations for $I_j(s)$ and for $Y_{jj}(s)$ as

$$I_j(s) = Y_{jj}(s)\, E_{jj}(s) \qquad Y_{jj}(s) = \frac{I_j(s)}{E_{jj}(s)} \tag{5-7.15}$$

The $E_{jj}(s)$ is a voltage source in series with the jth link and $I_j(s)$ is the current in the jth link. Therefore, $Y_{jj}(s)$ is the driving point admittance seen by opening the circuit in series with the jth link.

We pick $Y_{jk}(s)$ as being representative of the terms off the principal diagonal. We set all the $E(s)$ terms equal to zero, except $E_{kk}(s)$, and solve the resulting equations for $I_j(s)$ and for $Y_{jk}(s)$, as

$$I_j(s) = Y_{jk}(s)\, E_{kk}(s) \qquad Y_{jk}(s) = \frac{I_j(s)}{E_{kk}(s)} \tag{5-7.16}$$

The $E_{kk}(s)$ is a voltage source in series with the kth link, and $I_j(s)$ is the current in the jth link. Therefore, $Y_{jk}(s)$ is a transfer admittance, and the double subscript describes its properties. The first subscript designates the output, and the second subscript the input.

THE $y(s)$ AND $Z(s)$ FUNCTIONS (NODAL EQUATIONS)

The $y(s)$ terms on the principal diagonal are the sum of all the admittance of a cut-set made up by shorting all the nodes to the datum except one. To review the terms off the principal diagonal, we examine $y_{jk}(s)$ as a general term. The $y_{jk}(s)$ is the intersection of the $y_{jj}(s)$ set and the $y_{kk}(s)$ set. The $y_{jk}(s)$ carries its sign, which for the node-to-datum equations is negative. The situation is more complex when mutual inductance is present.

To understand the $Z(s)$ terms, we turn our attention to Eqs. 5-7.14. We pick $Z_{jj}(s)$ as being representative of the terms on the principal diagonal. We set all the $I(s)$ terms equal to zero, except $I_{jj}(s)$, and solve the resulting equations for $E_j(s)$ and for $Z_{jj}(s)$, as

$$E_j(s) = Z_{jj}(s)\, I_{jj}(s) \qquad Z_{jj}(s) = \frac{E_j(s)}{I_{ii}(s)} \tag{5-7.17}$$

The $I_{jj}(s)$ is a current source from datum to the jth node, and $E_j(s)$ is the voltage from datum to the jth node. Therefore, $Z_{jj}(s)$ is a driving point impedance seen between the jth node and the datum node.

We pick $Z_{jk}(s)$ as being representative of the terms off the principal diagonal. We set all the $I(s)$ terms equal to zero, except $I_{kk}(s)$, and solve the resulting equations for $E_j(s)$

and for $Z_{jk}(s)$, as

$$E_j(s) = Z_{jk}(s)\, I_{kk}(s) \qquad Z_{jk}(s) = \frac{E_j(s)}{I_{kk}(s)} \tag{5-7.18}$$

The $I_{kk}(s)$ is a current source from datum to the kth node, and $E_j(s)$ is the voltage from datum to the jth node. Therefore, $Z_{jk}(s)$ is a transfer impedance, and the double subscript describes its properties. The first subscript designates the output and the second subscript the input.

Example 5-7.1. The circuit of Fig. 5-7.1(a) is in steady-state when S is closed at $t = 0$. In this example, we use loop currents. Our purpose is not so much to find a solution as it is to explore the forms of $[z(s)]$ and $[Y(s)]$, etc. Figure 5-7.1(a) shows the initial current in the inductor and the initial voltage in the capacitor; (b) shows the transformed circuit.

The two resistors are chosen as the tree branches, the inductor as *Link No. 1* and the capacitor as *Link No. 2.* The resulting loop currents are shown as (b).

The following matrices are determined.

$$\begin{bmatrix} E_{11}(s) \\ E_{22}(s) \end{bmatrix} = \begin{bmatrix} \dfrac{60}{s} + 4 \\ \dfrac{60}{s} - \dfrac{20}{s} \end{bmatrix} \tag{5-7.19}$$

$$[z(s)] = \begin{bmatrix} (2s + 15) & 5 \\ 5 & (5 + 1/4s) \end{bmatrix} \tag{5-7.20}$$

$$[Y(s)] = [z(s)]^{-1} = \frac{1}{10s + 50.5 + 3.75/s} \begin{bmatrix} (5 + 1/4s) & -5 \\ -5 & (2s + 15) \end{bmatrix} \tag{5-7.21}$$

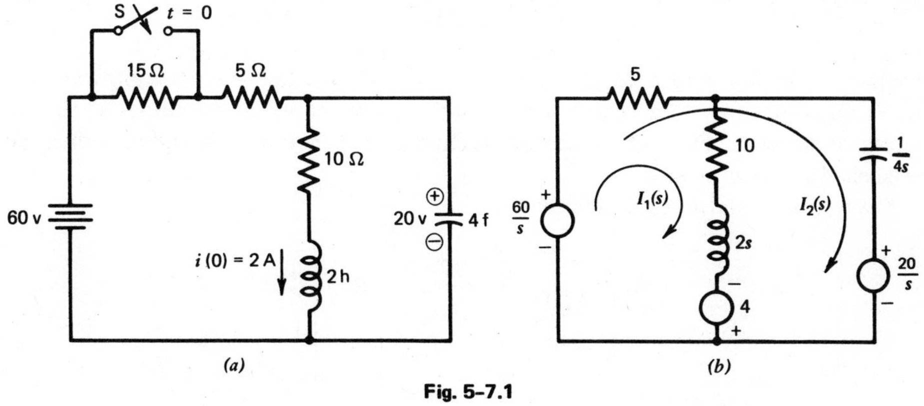

Fig. 5-7.1

After multiplying top and bottom by s, the terms of $[Y(s)]$ are

$$Y_{11}(s) = \frac{5s + 0.25}{10s^2 + 50.5s + 3.75}$$

$$Y_{12}(s) = Y_{21}(s) = \frac{-5s}{10s^2 + 50.5s + 3.75} \qquad (5\text{-}7.22)$$

$$Y_{22}(s) = \frac{2s^2 + 15s}{10s^2 + 50.5s + 3.75}$$

To continue, we find $Y_{11}(s)$ from the physical interpretation as discussed above. To do this we first remove all the voltage sources in Fig. 5-7.1(b) and open the circuit in series with *Link No. 1*, and write

$$Y_{11}(s) = \frac{1}{2s + 10 + [(5 \times 1/4s)/(5 + 1/4s)]} = \frac{5s + 0.25}{10s^2 + 50.5s + 3.75} \qquad (5\text{-}7.23)$$

Next we desire to find $Y_{21}(s)$. The first subscript indicates the output and the second the input. Therefore, we place a unit impulse of voltage in series with *Link No. 1* and solve for the current in *Link No. 2*. First we solve for $I_1^*(s)$.

$$I_1^*(s) = 1Y_{11}(s) = \frac{5s + 0.25}{10s^2 + 50.5s + 3.75} \qquad (5\text{-}7.24)$$

This current divides and $I_2^*(s)$ is found as

$$I_2^*(s) = Y_{21}(s) = -I_1^* \left[\frac{5}{5 + 1/4s}\right] = \frac{-5s}{10s^2 + 50.5s + 3.75} \qquad (5\text{-}7.25)$$

The $Y_{11}(s)$ and $Y_{21}(s)$ just found check with Eqs. 5-7.22. The above $I_1^*(s)$ and $I_2^*(s)$ are not the $I_1(s)$ and $I_2(s)$ of the circuit of Fig. 5-7.1(b), but are those needed to find $Y_{11}(s)$ and $Y_{21}(s)$. To complete the example, we find $I_1(s)$ in the circuit of Fig. 5-7.1(b) as

$$I_1(s) = Y_{11}(s)E_{11}(s) + Y_{12}(s)E_{22}(s) = \frac{20s^2 + 101s + 15}{s\,[10s^2 + 50.5s + 3.75]} \qquad (5\text{-}7.26)$$

Example 5-7.2. We return to the circuit of Fig. 5-7.1(a) but make a transformation from a voltage source to a current source and remove the initial conditions as current sources, as shown in Fig. 5-7.2. One node is chosen as the datum node and the other two are labelled as shown.

The following matrices are determined

$$\begin{bmatrix} I_{11}(s) \\ I_{22}(s) \end{bmatrix} = \begin{bmatrix} \left(-\dfrac{12}{s} + \dfrac{2}{s} - 80\right) \\ \left(-\dfrac{2}{s}\right) \end{bmatrix} \qquad (5\text{-}7.27)$$

$$[y(s)] = \begin{bmatrix} 1/5 + 1/2s + 4s & -\ 1/2s \\ -\ 1/2s & 1/10 + 1/2s \end{bmatrix} \tag{5-7.28}$$

$$[Z(s)] = [y(s)]^{-1} = \frac{1}{0.4s + 2.02 + 0.15/s} \begin{bmatrix} \left(0.1 + \dfrac{0.5}{s}\right) & \left(\dfrac{0.5}{s}\right) \\ \left(\dfrac{0.5}{s}\right) & \left(4s + 0.2 + \dfrac{0.5}{s}\right) \end{bmatrix} \tag{5-7.29}$$

After multiplying the top and bottom by $25s$, the terms of $[Z(s)]$ are

$$Z_{11}(s) = \frac{2.5s + 12.5}{10s^2 + 50.5s + 3.75}$$

$$Z_{12}(s) = Z_{21}(s) = \frac{12.5}{10s^2 + 50.5s + 3.75} \tag{5-7.30}$$

$$Z_{22}(s) = \frac{100s^2 + 5s + 12.5}{10s^2 + 50.5s + 3.75}$$

To continue the example, we find $Z_{22}(s)$ from the physical interpretation described above. To do this, we first remove all the current sources in Fig. 5-7.2 and write an expression for the impedance between the *No. 2* node and datum as

$$Z_{22}(s) = \frac{10[2s + (5 \times 1/4s)/(5 + 1/4s)]}{10 + 2s + (5 \times 1/4s)/(5 + 1/4s)} = \frac{100s^2 + 5s + 12.5}{10s^2 + 50.5s + 3.75} \tag{5-7.31}$$

Next we desire to find $Z_{12}(s)$. The first subscript indicates the output, and the second subscript indicates the input. Therefore, we place a unit impulse of current from the datum node to the *No. 2* node, and we solve for the voltage from datum to the *No. 1* node. First we solve for $E_2^*(s)$ as

$$E_2^*(s) = 1Z_{22}(s) = \frac{100s^2 + 5s + 12.5}{10s^2 + 5.05s + 3.75} \tag{5-7.32}$$

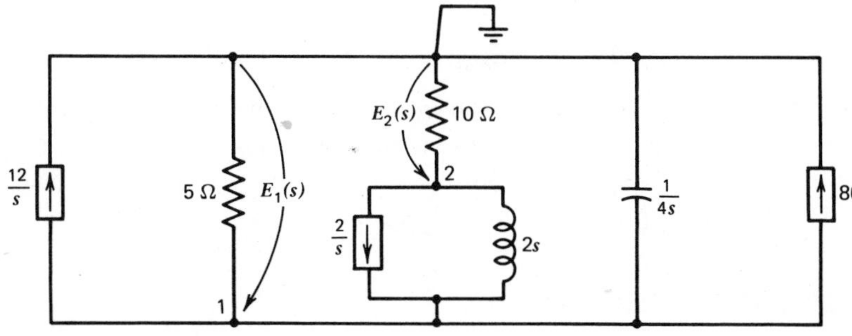

Fig. 5-7.2

This voltage divides across two impedances in series, and $E_1^*(s)$ is found as

$$E_1^*(s) = Z_{12}(s) = \frac{E_2^*(s)(5 \times 1/4s)/(5 + 1/4s)}{2s + (5 \times 1/4s)/(5 + 1/4s)} = \frac{12.5}{10s^2 + 50.5s + 3.75} \qquad (5\text{-}7.33)$$

The $Z_{22}(s)$ and $Z_{12}(s)$ just found check with Eqs. 5-7.30. The above $E_1^*(s)$ and $E_2^*(s)$ are not the $E_1(s)$ and $E_2(s)$ of the circuit of Fig. 5-7.2, but are the voltages needed to find the $Z(s)$ terms from a physical interpretation of the problem.

To complete the example, we find $E_2(s)$ of Fig. 5-7.2 as

$$E_2(s) = Z_{21}(s)\,I_{11}(s) + Z_{22}(s)\,I_{22}(s) = \frac{-200s^2 - 1{,}010s - 150}{s\,[10s^2 + 50.5s + 3.75]} \qquad (5\text{-}7.34)$$

SOME PROPERTIES OF THE GENERAL SOLUTIONS

Using loop-current equations, we start with

$$E(s)] = [z(s)]\ I(s)] \qquad (5\text{-}7.35)$$

and obtained the solutions written out as

$$\begin{aligned} I_1(s) &= Y_{11}(s)\,E_{11}(s) + Y_{12}(s)\,E_{22}(s) + \cdots + Y_{1\ell}(s)\,E_{\ell\ell}(s) \\ I_2(s) &= Y_{21}(s)\,E_{11}(s) + Y_{22}(s)\,E_{22}(s) + \cdots + Y_{2\ell}(s)\,E_{\ell\ell}(s) \\ &\vdots \quad \cdots\cdots\cdots\cdots\cdots\cdots\cdots\cdots \\ I_\ell(s) &= Y_{\ell 1}(s)\,E_{11}(s) + Y_{\ell 2}(s)\,E_{22}(s) + \cdots + Y_{\ell\ell}(s)\,E_{\ell\ell}(s) \end{aligned} \qquad (5\text{-}7.36)$$

Using node-to-datum voltage equations, we start with

$$I(s)] = [y(s)]\ E(s)] \qquad (5\text{-}7.37)$$

and obtain the solutions written out as

$$\begin{aligned} E_1(s) &= Z_{11}(s)\,I_{11}(s) + Z_{12}(s)\,I_{22}(s) + \cdots + Z_{1n}(s)\,I_{nn}(s) \\ E_2(s) &= Z_{21}(s)\,I_{11}(s) + Z_{22}(s)\,I_{22}(s) + \cdots + Z_{2n}(s)\,I_{nn}(s) \\ &\vdots \quad \cdots\cdots\cdots\cdots\cdots\cdots\cdots\cdots \\ E_n(s) &= Z_{n1}(s)\,I_{11}(s) + Z_{n2}(s)\,I_{22}(s) + \cdots + Z_{nn}(s)\,I_{nn}(s) \end{aligned} \qquad (5\text{-}7.38)$$

Suppose you were the first person in the world to work out these two sets of solution equations. With this thought in mind, examine these equations and see what properties you see. One of the first things that is obvious is that both solutions sets are made up of the addition of terms. This property is given the name *superposition theorem.*

THE SUPERPOSITION THEOREM

To discuss this theorem, we first look at Eqs. 5-7.36. Suppose that all the $E(s)$ terms except $E_{11}(s)$ are set equal to zero. The solution equations would be the first column. Next suppose that all the $E(s)$ terms except $E_{22}(s)$ are set equal to zero. The solution equations would be the second column. We can continue this line of reasoning for all the $E(s)$ terms. Finally suppose all the $E(s)$ terms act simultaneously; then all the solutions found separately can be added together for the complete solution. Thus the form of Eqs. 5-7.36 is the proof of superposition theorem.

Exactly the same line of reasoning can be applied to Eqs. 5-7.38.

This proof is completely general in that some of the voltage terms in Eqs. 5-7.36 can come from initial conditions, and the actual sources can have any wave shapes. These equations include both the particular and complementary components of the solutions.

The sources need not be added one at a time for the superposition to hold. For example, suppose $n = 20$ in Eqs. 5-7.38. Fourteen of the current sources could be applied at one time giving a set of components for the $E(s)$, and then the other six sources could be applied. The solutions when all 20 sources are applied simultaneously is the sum of the two sets of components.

We again turn our attention to Eqs. 5-7.35, and 5-7.36. Because we are considering passive circuits (no gyrators) and have written our equations in a systematic manner, the $[z(s)]$ matrix has symmetry off the principal diagonal; that is

$$z_{jk}(s) = z_{kj}(s) \tag{5-7.39}$$

There is a well-known property of matrices (this is left as an exercise), that the inverse matrix of a symmetric matrix is also symmetric. In the present case, this means that

$$Y_{jk}(s) = Y_{kj}(s) \tag{5-7.40}$$

Again, suppose you were the first person in the world to notice the $[Y(s)]$ matrix of Eqs. 5-7.36 is symmetric. What can you do with this property?

THE RECIPROCITY THEOREM

The reciprocity theorem is nothing more than giving a physical interpretation to the fact that the solution matrix $[Y(s)]$ is symmetric.

We return to the physical interpretation of the $Y(s)$ functions and examine $Y_{jk}(s)$. Since the transfer function is an admittance, we know the output is a current and the input is a voltage. The double-subscript notation tells us the output is a current in the jth link, and the voltage source is placed in series with the kth link, as

$$Y_{jk}(s) = \frac{I_j(s)}{E_{kk}(s)} \tag{5-7.41}$$

By the same line of reasoning

$$Y_{kj}(s) = \frac{I_k(s)}{E_{jj}(s)} \tag{5-7.42}$$

Equation 5-7.40 yields

$$\frac{I_j(s)}{E_{kk}(s)} = Y_{jk}(s) = Y_{kj}(s) = \frac{I_k(s)}{E_{jj}(s)} \tag{5-7.43}$$

This is one form of the reciprocity theorem.

Suppose we now examine Eqs. 5-7.37 and 5-7.38. Because we are considering passive circuits (no gyrators) and have written our equations in a systematic way, the $[y(s)]$ matrix has symmetry off the principal diagonal, and $[y(s)]^{-1} = [Z(s)]$ is also symmetric. In the present case, this means that

$$Z_{jk}(s) = Z_{kj}(s) \tag{5-7.44}$$

The physical interpretation of these terms yields

$$\frac{E_j(s)}{I_{kk}(s)} = Z_{jk}(s) = Z_{kj}(s) = \frac{E_k(s)}{I_{jj}(s)} \tag{5-7.45}$$

This is the other form of the reciprocity theorem. Even though these specific equations were developed in a node-to-datum set of voltages, all the steps are equally valid for a tree-branch set of voltages.

Example 5-7.3. The transformed circuit of Fig. 5-7.3(a) has a voltage $E_i(s)$ applied, and the response $I_0(s)$ is found as

$$I_0(s) = \frac{(s+60)\,E_i(s)}{s^2 + 130s + 2{,}200} \tag{5-7.46}$$

In (b) the location of the input voltage and output current response are exchanged. When this circuit is solved for $I_0(s)$, the result is

$$I_0(s) = \frac{(s+60)\,E_i(s)}{s^2 + 130s + 2{,}200} \tag{5-7.47}$$

which is the same as Eq. 5-7.46.

Example 5-7.4. The transformed circuit of Fig. 5-7.4(a) has a current $I_i(s)$ applied, and the response $V_0(s)$ is found as

$$V_0(s) = \frac{2 \times 10^4 s\, I_i(s)}{2.3s^2 + 1{,}175s + 25 \times 10^4} \tag{5-7.48}$$

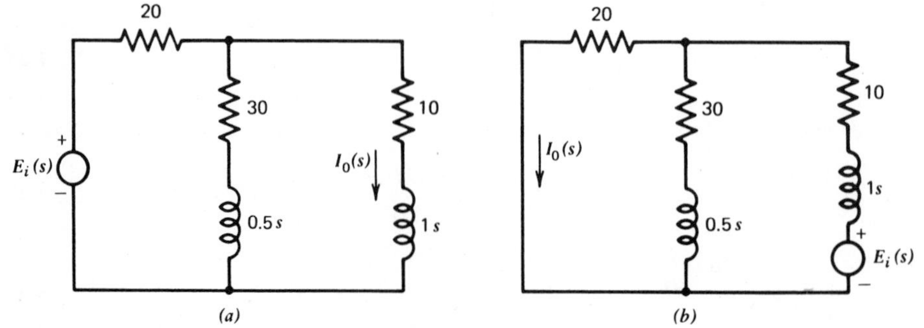

Fig. 5-7.3

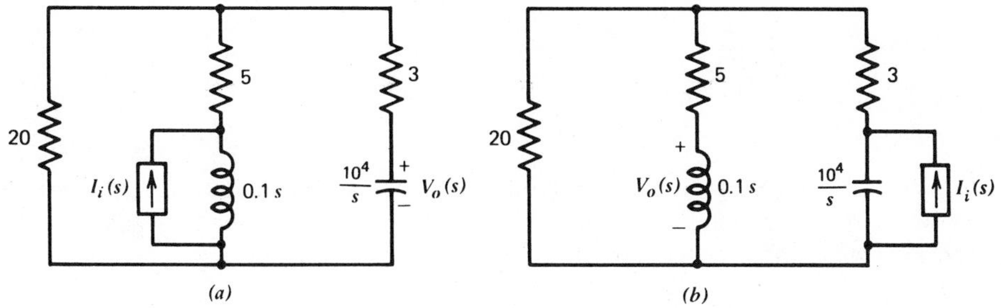

Fig. 5-7.4

In (b), the location of the input current source and the output voltage response are exchanged. When this circuit is solved for $V_0(s)$, the result is

$$V_0(s) = \frac{2 \times 10^4 s\, I_i(s)}{2.3s^2 + 1{,}175s + 25 \times 10^4} \tag{5-7.49}$$

which is the same as Eq. 5-7.48.

EQUATIONS IN THE MIXED FORM

When loop-current equations are written, all the unknowns are currents. When tree-branch voltage or node-to-datum equations are written, all unknowns are voltages. It is possible to write a third type of equation in which a mixture of voltages and currents are used as the unknowns. In recent years, equations of this type have received a good deal of attention because of the interest in the subject of state variables. In Chapter 13, methods of setting up equations of this type are discussed.

5-8. MUTUAL INDUCTANCE

We know from Faraday's law that a voltage v is induced in a coil of N turns as given by the equation

$$v = N \frac{d\phi}{dt} \tag{5-8.1}$$

This law states nothing about where the flux originates. If the flux is produced by a current existing in the coil itself, the resulting voltage is due to the effect known as self-inductance, and Eq. 5-8.1 becomes

$$v_2 = N \frac{d\phi}{dt} = L \frac{di}{dt} \tag{5-8.2}$$

The polarity reference for v is always fixed in relation to the direction of the current i, as shown in Fig. 5-8.1.

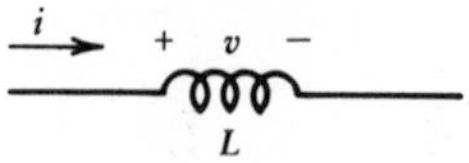

Fig. 5-8.1

It is possible that some of the flux produced in one coil links the turns of a second coil. Thus, a voltage is induced in the second coil by current in the first coil, and this effect is known as mutual inductance. For this case, Eq. 5-8.1 becomes

$$v_2 = N\frac{d\phi}{dt} = M_{21}\frac{di_1}{dt} \tag{5-8.3}$$

A new factor to consider is the question as to the polarity of the voltage v_2. For example, if the two coils L_1 and L_2 of Fig. 5-8.2 are wound in a certain manner, the polarity of v_2 will be in one direction; but if the direction of the winding of one of the coils is reversed, the polarity of v_2 will be in the opposite direction.

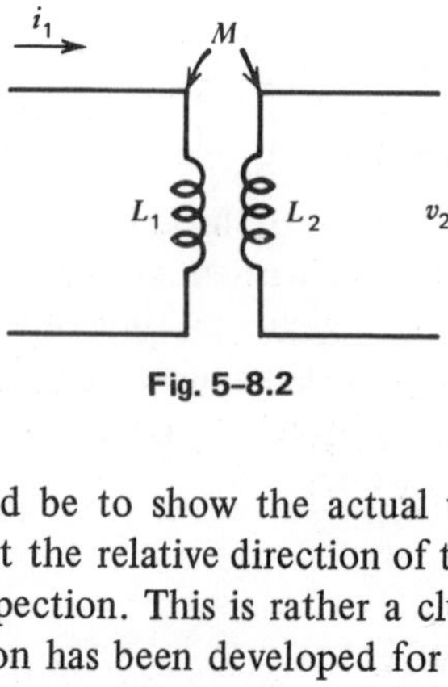

Fig. 5-8.2

One way to proceed would be to show the actual winding directions and physical orientation of the coils, so that the relative direction of the self-induced flux and mutual flux can be determined by inspection. This is rather a clumsy method of conveying this information, and a *dot* notation has been developed for the two-winding situation. The location of the first dot is arbitrary. The location of the second dot is determined by the physical arrangement of the two windings. The second dot is so placed that if both i_1 and i_2 are directed into (or out of) their respective dots, the mutual flux aids the flux of self-induction. A transformer with the dots already determined is shown in Fig. 5-8.3. The current i_1 is sent into one dot, and the voltage v_1 due to the self-inductance has the polarity markings as shown. The polarity of the voltage v_2 due to mutual inductance is determined in the following manner. Since the (+) for v_1 is at the dotted end of L_1, the (+) for v_2 is at the dotted end of L_2.

When many coils are mutually coupled, the extension of the dot notation becomes very clumsy and can be replaced by a matrix notation, but that is beyond the scope of this book.

It can be shown for air-core transformers that $M_{12} = M_{21}$. Therefore, we use M to represent either one.

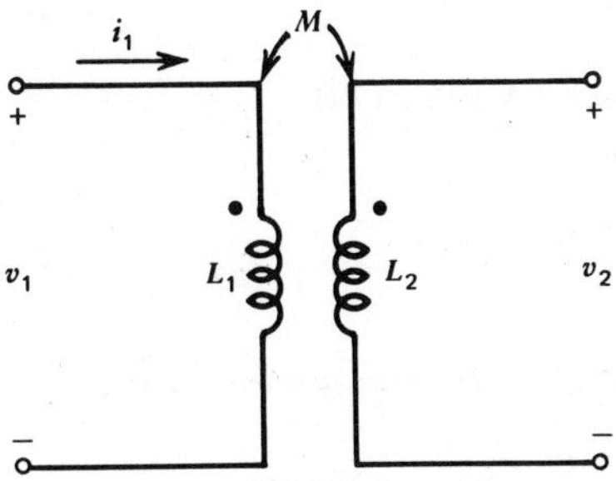

Fig. 5-8.3

Example 5-8.1. The circuit shown in Fig. 5-8.4 is in steady-state when S is closed at $t = 0$. The desired solution is $I_1(s)$. The currents at $t = 0-$ are

$$i_1(0-) = 1.54 \qquad i_2(0-) = 1.15 \tag{5-8.4}$$

After $t = 0$, the KVLE's are

$$\begin{aligned} 10 &= 3i_1 + 0.03\frac{di_1}{dt} + 0.02\frac{di_2}{dt} \\ 10 &= 0.02\frac{di_1}{dt} + 4i_2 + 0.025\frac{di_2}{dt} \end{aligned} \tag{5-8.5}$$

Since both i_1 and i_2 are into their respective dots, the mutual terms are written with the same sign as the self-inductance terms.

The transformed equations are

$$\begin{aligned} \frac{10}{s} &= 3I_1(s) + 0.03\,[s\,I_1(s) - 1.54] + 0.02\,[s\,I_2(s) - 1.15] \\ \frac{10}{s} &= 0.02\,[s\,I_1(s) - 1.54] + 4I_2(s) + 0.025\,[s\,I_2(s) - 1.15] \end{aligned} \tag{5-8.6}$$

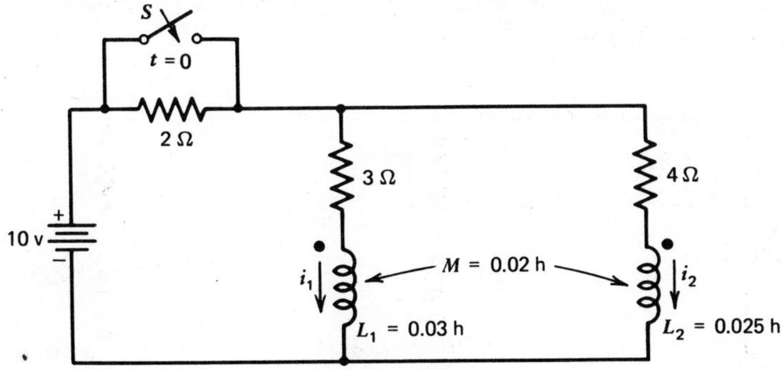

Fig. 5-8.4

These can be rewritten as

$$\frac{(0.03 \times 1.54 + 0.02 \times 1.15)\,s + 10}{s} = [0.03s + 3]\ I_1(s) + [0.02s]\ I_2(s)$$

$$\frac{(0.02 \times 1.54 + 0.025 \times 1.15)\,s + 10}{s} = [0.02s]\ I_1(s) + [0.025s + 4]\ I_2(s) \tag{5-8.7}$$

It should be noted in passing that $z_{12}(s)$ is not the intersection of $z_{11}(s)$ and $z_{22}(s)$ due to the mutual inductance.

These equations can be solved for $I_1(s)$ as

$$I_1(s) = \frac{1.54s^2 + 936s + 114{,}200}{s\,(s + 70.6)\,(s + 486.5)} \tag{5-8.8}$$

This problem can be solved by using operational methods. Care must be used in pulling the initial conditions out as voltage sources, because of the interaction of the two coils through the mutual inductance. An inspection of Eqs. 5-8.6 shows the procedure to be used. The voltage source to go in series with the 0.03-henry inductor is made up of two terms. One term is the product of $L_1\,i_1(0)$ and the other term is the product of $M\,i_2(0)$. Therefore, this voltage is

$$0.03 \times 1.54 + 0.02 \times 1.15 = 0.0693 \tag{5-8.9}$$

In a similar manner, the voltage source to go in series with the L_2 inductor is

$$0.02 \times 1.54 + 0.025 \times 1.15 = 0.0596 \tag{5-8.10}$$

The transformed circuit is shown in Fig. 5-8.5. The $I_1(s)$ solved from this circuit checks with the $I_1(s)$ of Eq. 5-8.8.

Example 5-8.2. In the circuit shown in Fig. 5-8.6(a), the desired solution is to find the operational impedance seen looking into terminals A-B.

There is a tendency to transform the circuit and to start combining things in series and parallel. This can be done in certain cases by first going to an equivalent circuit. Here we use a more general approach.

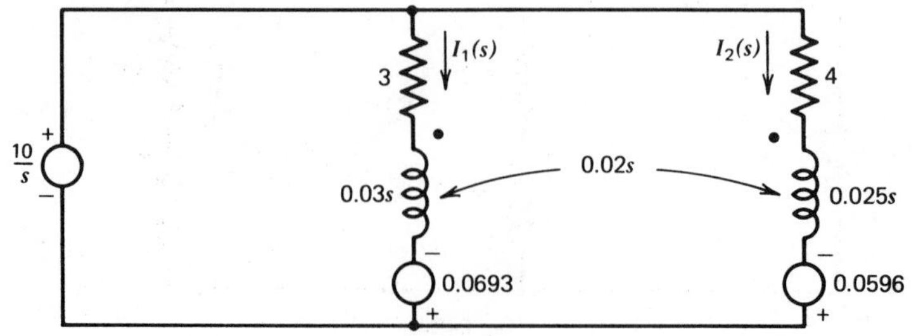

Fig. 5-8.5

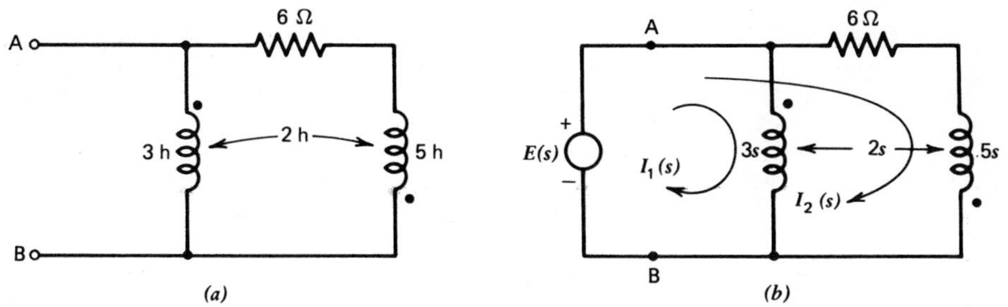

Fig. 5-8.6

The definition of $Z(s)$ is

$$Z(s) = \frac{E(s)}{I(s)} \tag{5-8.11}$$

We can find $Z(s)$ by assuming an $E(s)$ is applied to terminals A-B, solving for the current $I(s)$, and finally forming the ratio of Eq. 5-8.11. In Fig. 5-8.6(b), the transformed circuit and choice of current directions are shown. The KVLE's are written and solved for $Z(s)$ as

$$\begin{aligned} E(s) &= 3s\, I_1(s) - 2s\, I_2(s) \\ E(s) &= -2s\, I_1(s) + (5s + 6)\, I_2(s) \end{aligned} \tag{5-8.12}$$

$$I_1(s) = \frac{E(s)\,(7s + 6)}{(11s^2 + 18s)} \qquad I_2(s) = \frac{E(s)\,(5s)}{(11s^2 + 18s)} \tag{5-8.13}$$

$$Z(s) = \frac{E(s)}{I_1(s) + I_2(s)} = \frac{11s^2 + 18s}{12s + 6} \tag{5-8.14}$$

5-9. THEVENIN'S AND NORTON'S THEOREMS

In the present discussion, Thevenin's theorem is featured, since Norton's theorem is essentially a source transformation on the Thevenin equivalent circuit. Some of the ramifications of Thevenin's theorem are discussed later, with their proof. This present discussion is more in keeping with the traditional presentation, except it is extended to use the operational approach; which incidentally allows an initial current in the load element.

Example 5-9.1. This example is the same as Ex. 4-13.5. The original circuit is shown in Fig. 4-13.5(a) where a switching arrangement puts initial conditions into the circuit. The operational equivalent of this circuit is shown in Fig. 5-9.1(a). The desired solution is the current $I_1(s)$.

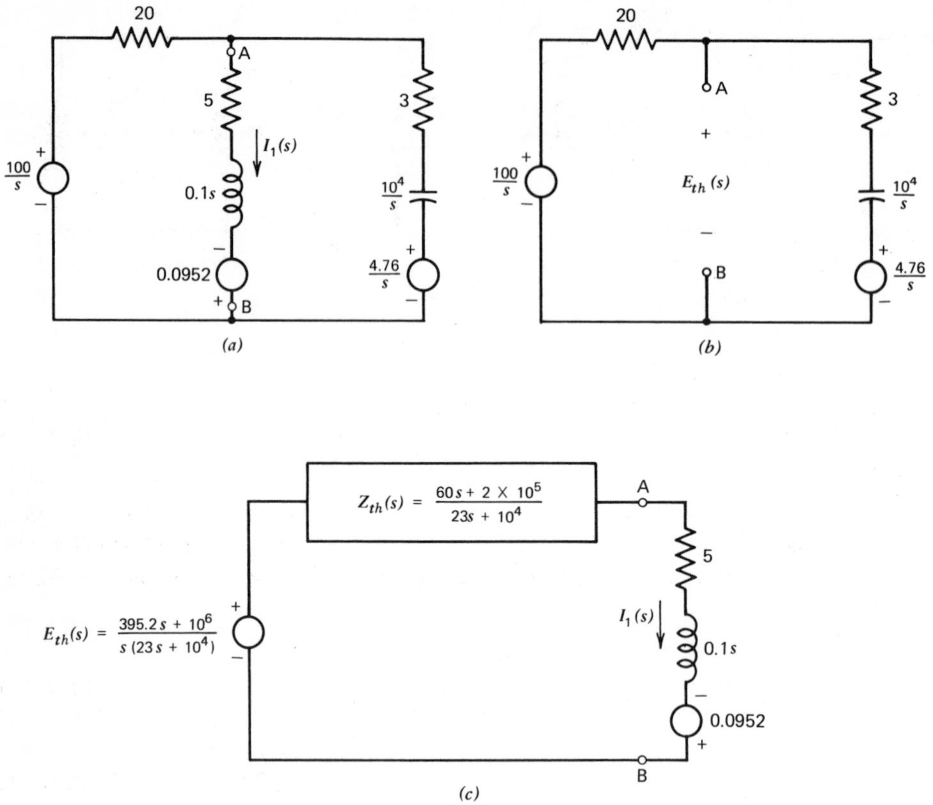

Fig. 5-9.1

The load impedance along with its associated voltage source is removed, and the voltage $E_{th}(s)$ is found across the terminals A-B in (b) as

$$E_{th}(s) = \frac{395.2s + 10^6}{s\,(23s + 10^4)}. \tag{5-9.1}$$

The impedance looking into terminals A-B is

$$Z_{th}(s) = \frac{60s + 2 \times 10^5}{23s + 10^4} \tag{5-9.2}$$

The equivalent circuit is shown in (c), from which $I_1(s)$ is found to be

$$I_1(s) = \frac{E_{th}(s) + 0.0952}{Z_{th}(s) + 0.1s + 5} = \frac{0.952s^2 + 585s + 435{,}000}{s\,(s^2 + 511s + 108{,}800)} \tag{5-9.3}$$

This checks Eq. 4-13.48.

AN HISTORICAL COMMENT ABOUT THEVENIN'S THEOREM

Although our purpose is not that of an historian, it is of interest to comment about Thevenin's original paper[4] written in 1883. The paper is about $1\frac{1}{2}$ pages long, and in the publication of the French Academy of Science was classified as a note. The paper contains no figures and essentially only one equation. The paper assumed d-c voltage sources and resistive elements. Yet the proof is so general that it holds under conditions unknown in 1883. In the proof that follows, no attempt is made to literally follow Thevenin's original proof; however this presentation is similar to his.

A GENERAL PROOF OF THEVENIN'S THEOREM

The proof to follow is presented in such a way that mutual inductance can be present throughout the circuit. We direct our attention to one specific current somewhere in the circuit. The element (or group of elements) in which this current exists is called the load impedance. To make the proof as general as possible, we assume that there is mutual inductance coupling between the load impedance and the rest of the circuit.

The load impedance is separated from the rest of the circuit, as shown in Fig. 5-9.2(a). The circuit can contain voltage and current sources of any wave shape. The inductors and capacitors can have initial conditions that are moved into sources. Therefore, when we speak about sources we include all initial conditions.

In (b), we open the circuit in series with the load impedance, and we refer to the voltage that appears across these open terminals as e_{th}, the Thevenin-theorem voltage. We could connect an oscilloscope across these open terminals and observe the wave shape of this voltage. Since there is an open circuit in series with the load impedance, the current in the load is zero.

In (c), we conceptually connect a voltage source that generates e_{th} across these open terminals. As far as the current in the load impedance is concerned, the circuit of (c) is no different from that of (b); therefore,

$$i_A = 0 \tag{5-9.4}$$

We now turn to the use of the superposition theorem that was proved in Section 5-7. We think of the current i_A of (c) as being made up of two components. For the first component, we remove e_{th} but leave the other sources, and for the second component we remove all the rest of the sources and leave only e_{th}. The first component is i, as shown in (d), and the second component is i_{th} as shown in (e). By superposition, the current i_A is

$$i_A = 0 = i - i_{th} \tag{5-9.5}$$

This equation can be solved for

$$i_{th} = i \tag{5-9.6}$$

[4] Academie des Sciences, Comptes Rendus 1883, Vol. 97, pp. 159–161.

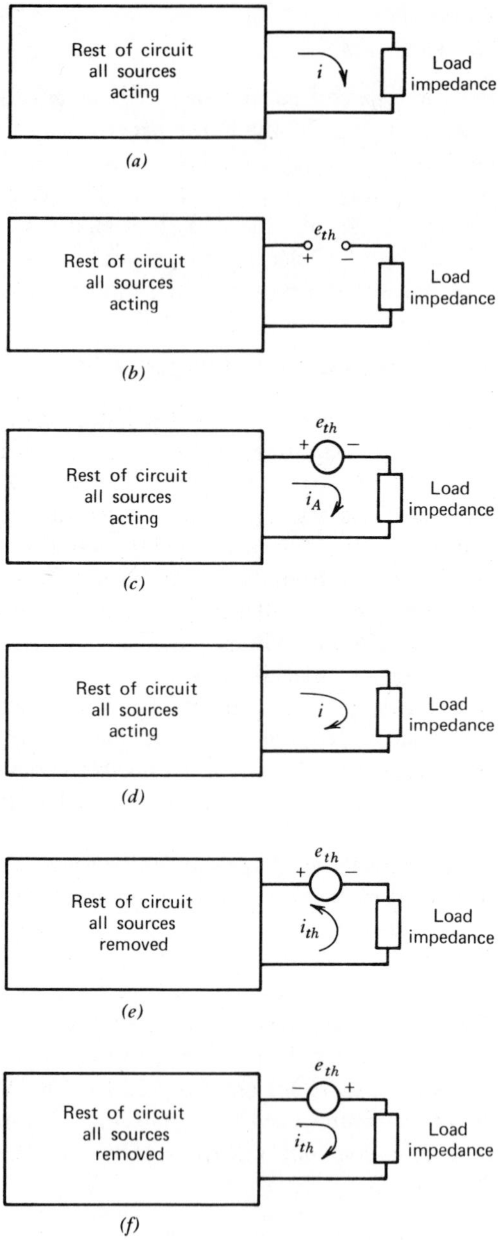
Rest of circuit
all sources
acting
i
Load
impedance
(a)
Rest of circuit
all sources
acting
e_{th}
+
−
Load
impedance
(b)
Rest of circuit
all sources
acting
e_{th}
+
−
i_A
Load
impedance
(c)
Rest of circuit
all sources
acting
i
Load
impedance
(d)
Rest of circuit
all sources
removed
e_{th}
+
−
i_{th}
Load
impedance
(e)
Rest of circuit
all sources
removed
e_{th}
−
+
i_{th}
Load
impedance
(f)

Fig. 5-9.2

Finally we note that the circuit of (d) is identical to the original circuit of (a). This completes the proof.

In (e), the current i_{th} is reversed from the original direction of i in (a). Although there is no need to do so, the direction of e_{th} in (e) can be reversed as shown in (f), which makes the direction of i_{th} the same as the original i.

This proof is completely general and holds for the complete solution, that is the complementary component plus the particular. If all the sources are sinusoidal, we can wait until the complementary component has gone to zero and we have proved that Thevenin's theorem is valid for a-c circuit theory. Similarly, the proof holds for d-c circuits. We continue the discussion through examples.

Example 5-9.2. The circuit shown in Fig. 5-9.3(a) is in steady-state, and the switch S is closed at $t = 0$. The desired solution is the current $I_2(s)$ obtained by the use of Thevenin's theorem. We find $I_2(s)$ first by more conventional methods to check the results.

The transformed circuit is shown in (b), with the initial conditions brought out as voltage sources, as in Ex. 5-8.1. The KVLE's are written and solved for $I_2(s)$ as

$$\begin{aligned} \frac{10}{s} + 3.85 &= (2s+9)\,I_1(s) + (s-5)\,I_2(s) \\ 3.85 &= (s-5)\,I_1(s) + (3s+5)\,I_2(s) \end{aligned} \tag{5-9.7}$$

$$I_2(s) = \frac{3.85s^2 + 43.9s + 50}{s\,(5s^2 + 47s + 20)} \tag{5-9.8}$$

The circuit of (c) is used to find $E_{th}(s)$ in the following steps:

$$\begin{aligned} I_A(s) &= \frac{3.85s + 10}{s\,(2s+9)} \\ E_1(s) &= 5I_A(s) = \frac{19.25s + 50}{s\,(2s+9)} \\ E_2(s) &= s\,I_A(s) - 3.85 = \frac{-3.85s^2 - 24.65s}{s\,(2s+9)} \\ E_{th}(s) &= E_1(s) - E_2(s) = \frac{3.85s^2 + 43.9s + 50}{s\,(2s+9)} \end{aligned} \tag{5-9.9}$$

The voltage $E_{th}(s)$ is applied to the circuit of (d) with the other sources removed. The KVLE's are written and solved for $I_2'(s)$ as

$$\begin{aligned} 0 &= (2s+9)\,I_1'(s) + (s-5)\,I_2'(s) \\ E_{th}(s) &= (s-5)\,I_1'(s) + (3s+5)\,I_2'(s) \end{aligned} \tag{5-9.10}$$

$$I_2'(s) = \frac{3.85s^2 + 43.9s + 50}{s\,(5s^2 + 47s + 20)} \tag{5-9.11}$$

This $I_2'(s)$ is equal to the $I_2(s)$ of Eq. 5-9.8.

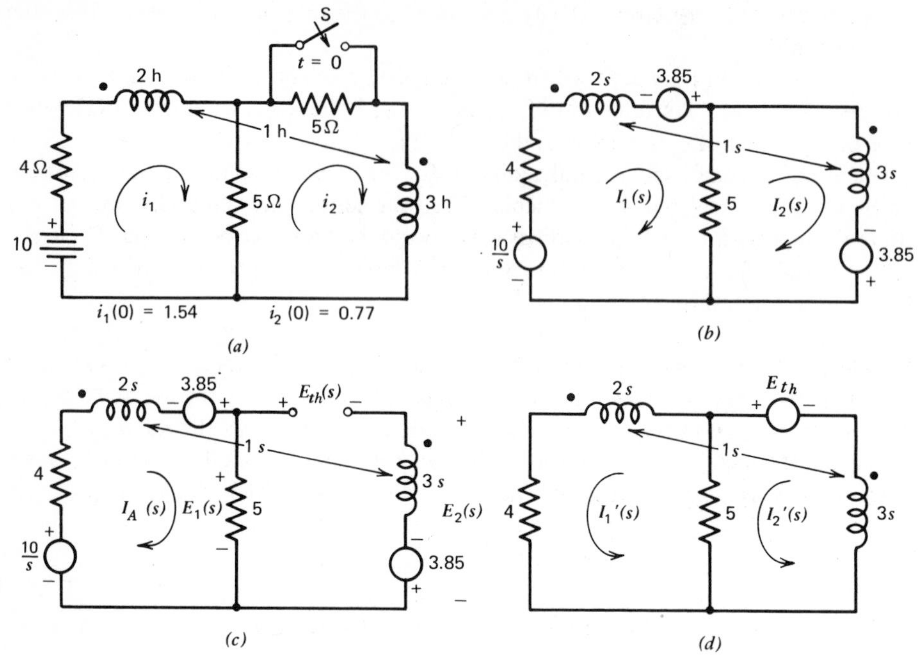

Fig. 5-9.3

THE USE OF THEVENIN'S THEOREM

If by removing an element or a group of elements the rest of the circuit contains elements of a single type, the Thevenin-theorem equivalent of the rest of the circuit will be a voltage source in series with a single element. In many situations, this can lead to rapid solution to the problem (we did this in Chapters 2 and 3). In a-c circuit theory, the rest of the circuit reduces to a single phasor impedance. However, when the complete solution is desired, and the load impedance is coupled into the rest of the circuit with mutual inductance, the use of Thevenin's theorem may not be practical. In Ex. 5-9.2, the Thevenin-circuit of (d) of Fig. 5-9.3 is just as complex as the original circuit of (a), but requires several extra steps before it can be obtained. Thevenin's theorem does hold, however, and if the designer can keep this in mind he may use it to advantage.

5-10. DEPENDENT SOURCES

Examples of circuits with dependent sources are included in Chapters 2 and 3, and this situation is again discussed. The circuit of Fig. 3-5.5(b) is redrawn again for Fig. 5-10.1. Equations 3-5.49 are written again as

$$10 = 0.5\frac{di_1}{dt} + 50i_1 + 20i_2$$

$$10 + 10i_1 = 20i_1 + 1\frac{di_2}{dt} + 30i_2 \qquad (5\text{-}10.1)$$

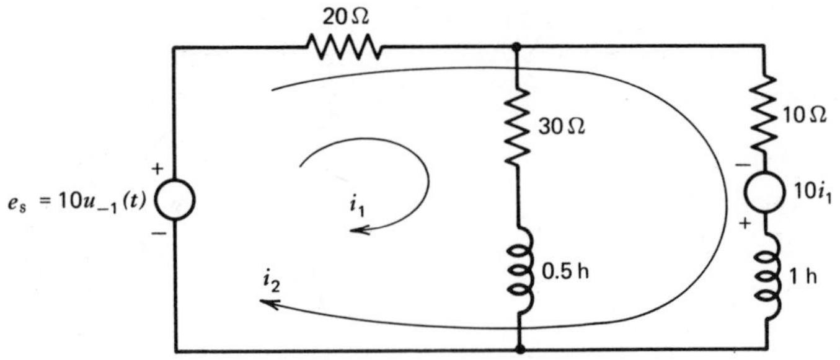

Fig. 5-10.1

These are transformed and written as

$$\begin{bmatrix} \dfrac{10}{s} \\ \dfrac{10}{s} \end{bmatrix} = \begin{bmatrix} (0.5s + 50) & 20 \\ (20 - 10) & (s + 30) \end{bmatrix} \begin{bmatrix} I_1(s) \\ I_2(s) \end{bmatrix} \tag{5-10.2}$$

The -10 in the second-row, first column element of the square matrix on the right is a result of the dependent source. If the dependent source were zero, the $[z(s)]$ in the equations

$$[E(s)] = [z(s)]\ I(s)] \tag{5-10.3}$$

would be symmetric, and the

$$[Y(s)] = [z(s)]^{-1} \tag{5-10.4}$$

would also be symmetric. The fact that $[Y(s)]$ is symmetric is the necessary ingredient in the proof of the reciprocity theorem.

Since the $[z(s)]$ in Eq. 5-10.2 is not symmetric, the $[z(s)]^{-1}$ also is not symmetric; and this result shows that the reciprocity theorem, as a result of the presence of the dependent source, does not hold for the original circuit.

If a person were to sit at his desk and dream up examples, he can add dependent sources in matched pairs and the reciprocity theorem will still hold. In the circuit of Fig. 5-10.1, another dependent voltage source could be set equal to $10i_2$. The resulting circuit is shown in Fig. 5-10.2. The new set of equations corresponding to Eqs. 5-10.2 are

$$\begin{bmatrix} \dfrac{10}{s} \\ \dfrac{10}{s} \end{bmatrix} = \begin{bmatrix} (0.5s + 50) & (20 - 10) \\ (20 - 10) & (s + 30) \end{bmatrix} \begin{bmatrix} I_1(s) \\ I_2(s) \end{bmatrix} \tag{5-10.5}$$

and the necessary symmetry for the reciprocity theorem to hold has been restored.

A person dreaming up circuit examples is not the usual case; rather he is presented with a circuit to be analyzed. If the circuit contains a solid-state device, the model for this

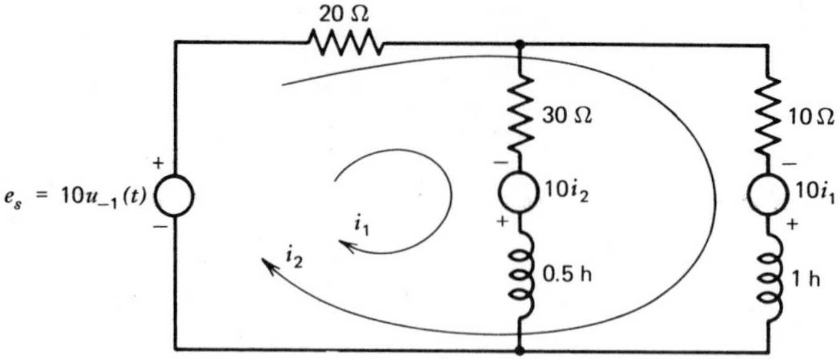

Fig. 5-10.2

device will more than likely contain some dependent sources. The probability that these sources appear in matched pairs is very low.

The voltage induced in a coil due to the action of mutual inductance can be thought of as a dependent source. The voltage in one coil depends on the derivative of the current in the other coil. For air-core transformers, $M_{21} = M_{12}$, and the two dependent sources act as a matched pair and do not destroy the reciprocity property of the circuit.

As mentioned briefly before, the symmetry of the $[z(s)]$ when reciprocity holds is the result of following a systematic procedure in writing the equations. This systematic procedure is the result of using the same tree in determining the loop-currents and in choosing the paths around which the KVLE's are written. The discussion just presented assumes that this systematic procedure is used.

THE GYRATOR

A number of times we are close to a subject which cannot be pursued in any detail in a book of this sort but should at least receive a comment in passing. The gyrator was proposed a number of years ago by B. D. H. Tellegen to add to the theoretical set of tools in circuit analysis and synthesis. Since that time, actual devices that closely approximate the gyrator have been placed on the market. The gyrator has two pairs of terminals and is shown by the symbol in Fig. 5-10.3. The main point to be made in this discussion is that by definition, the volt-ampere relationships for this device are given by

$$\begin{bmatrix} v_1 \\ v_2 \end{bmatrix} = \begin{bmatrix} 0 & -r \\ +r & 0 \end{bmatrix} \begin{bmatrix} i_1 \\ i_2 \end{bmatrix} \tag{5-10.6}$$

When this device is inserted into a network, the symmetric property just discussed is destroyed. Based on the definition of the gyrator, it can be shown to be passive, and this

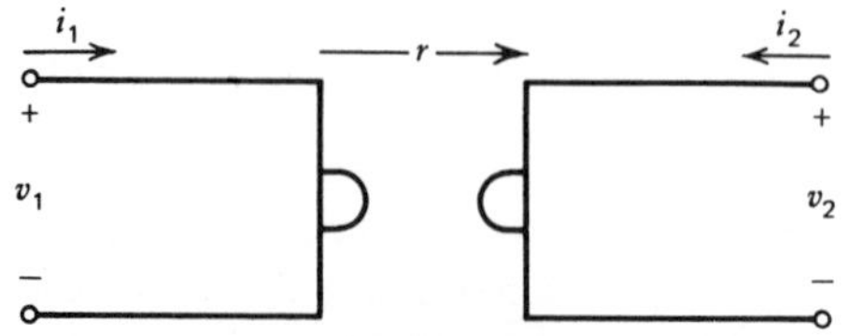

Fig. 5-10.3

leads to an example of a passive, linear network in which the reciprocity principal does not apply.

5-11. SCALING

The subject of scaling can be approached from several points of view. Certain aspects are seen from Property 4-12.12.

$$\mathcal{L}[f(t)] = F(s) \tag{5-11.1}$$

$$\mathcal{L}[f(at)] = \frac{1}{a}F\left(\frac{s}{a}\right) \tag{5-11.2}$$

Without going into details, Eq. 5-11.2 shows that a multiplies t but divides s. The result is an inverse relationship between the time and the frequency domains. Suppose an $f(t)$ is plotted vs. time, and the poles and zeros of the corresponding $F(s)$ are plotted in the s-plane. If the plot of the time function is stretched in the direction of time, the pole-zero plot shrinks, and vice versa.

MAGNITUDE SCALING

Suppose for some reason we want to increase the impedance level of a circuit by a factor of K. We use an intuitive approach to this and look at the operational impedances for the separate elements as

$$R \qquad Ls \qquad \frac{1}{Cs} \tag{5-11.3}$$

To increase the impedance level by K, each impedance is multiplied by K.

$$KR \qquad KLs \qquad \frac{K}{Cs} \tag{5-11.4}$$

Stated in words, the value of R and L are multiplied by K but C is divided by K.

FREQUENCY SCALING

Under magnitude scaling, the poles and zeros of a $Z(s)$ are not changed but only the magnitude. Under frequency scaling, the magnitude is unchanged, but the location of the poles and zeros are expanded or contracted. The operational impedances of the elements in Eq. 5-11.3 are again observed. If the value of R is left unchanged, but the values of L and C are both divided by K, the corresponding pole-zero plot expands by a factor of K.

MAGNITUDE AND FREQUENCY SCALING COMBINED

On a number of occasions in this book a value of a capacitance, such as 2 farads, has been used. Any person with practical experience knows that this is a very large capacitance.

For most practical circuits, such an element value can be rescaled. If the magnitude of the impedance is scaled up by a factor of 1,000, the capacitor now has a value of 2×10^{-3} farads. If in addition the frequencies are scaled up by a factor of 1,000, the capacitor now has a value of 2×10^{-6} farads = 2 μf, which is of a practical size.

PROBLEMS

These problems are to be worked by methods developed in Chapter 5.

5-1 In the circuit of Problem 2-73, find $I_{C1}(s)$, and by current division find $I_R(s)$. Finally, find i_R.

5-2 Repeat Problem 5-1, except find $I_{L4}(s)$ and i_{L4}.

5-3 In the circuit of Problem 2-75, use current division to find $I_R(s)$. Finally, find i_R.

5-4 Repeat Problem 5-3, except find $I_{C4}(s)$ and i_{C4}.

5-5 The circuit shown is at rest when $e_{s1} = 10u_{-1}(t)$ and $e_{s2} = 0$ are applied. Find $I_1(s)$, $I_2(s)$, and $I_3(s)$.

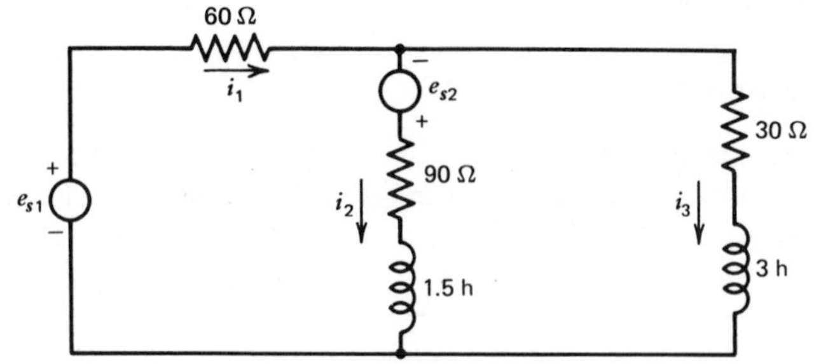

5-6 Repeat Problem 5-5, except $e_{s1} = 0$ and $e_{s2} = 10u_{-1}(t)$.

5-7 The circuit shown is at rest when $i_{s1} = 1u_{-1}(t)$ and $i_{s2} = 0$ are applied. Find $V_1(s)$, $V_2(s)$, and $V_3(s)$.

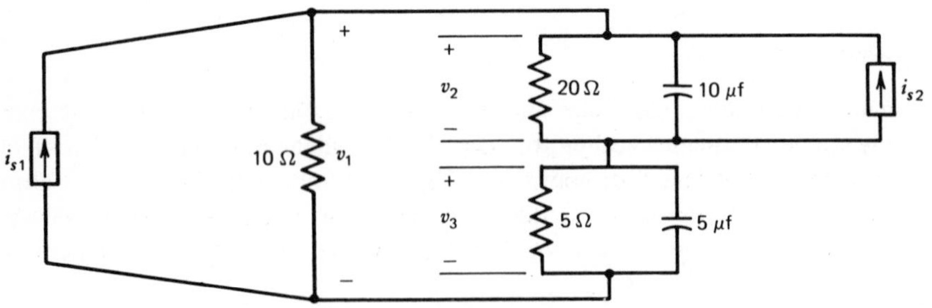

5-8 Repeat Problem 5-7, except $i_{s1} = 0$ and $i_{s2} = 1u_{-1}(t)$.

5-9 Repeat Problem 5-7, except replace the 5-μf capacitor with a 0.05-henry inductor.

5-10 Repeat Problem 5-9, except $i_{s1} = 0$ and $i_{s2} = 1u_{-1}(t)$.

5-11 Open the circuit shown at *Point 2*, and find the impedance $Z_2(s)$ at these terminals. From this $Z_2(s)$, find the characteristic polynomial for the circuit.

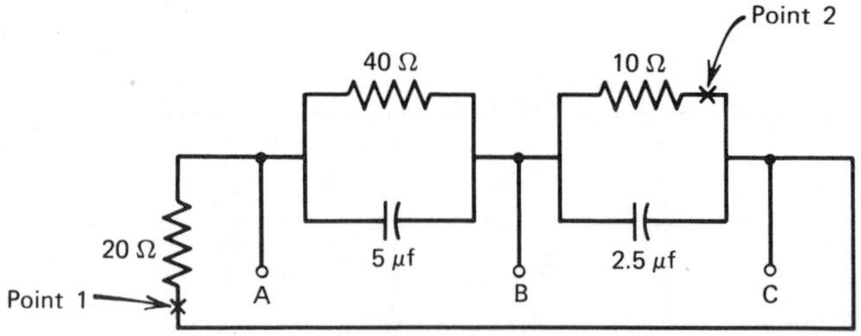

5-12 Repeat Problem 5-11, except look across terminals B-C and find $Z_{BC}(s)$; and from this find the characteristic polynomial for the circuit as shown.

5-13 Repeat Problem 5-11, except open the circuit at *Point 1*.

5-14 Repeat Problem 5-12, except look across terminals A-B.

5-15 Repeat Problem 5-11, except replace the 5-μf capacitor with a 0.05-henry inductor.

5-16 Repeat Problem 5-12, except replace the 5-μf capacitor with a 0.05-henry inductor.

5-17 Repeat Problem 5-13, except replace the 5-μf capacitor with a 0.05-henry inductor.

5-18 Repeat Problem 5-14, except replace the 5-μf capacitor with a 0.05-henry inductor.

5-19 The circuit of Problem 5-5 is at rest when $e_{s1} = 10u_{-1}(t)$ and $e_{s2} = 0$ are applied. Find the transfer functions $G_1(s) = I_1(s)/E_{s1}(s)$, $G_2(s) = I_2(s)/E_{s1}(s)$, and $G_3(s) = I_3(s)/E_{s1}(s)$, and show the results in block-diagram form.

5-20 The circuit of Problem 5-5 is at rest when $e_{s1} = 0$ and $e_{s2} = 10u_{-1}(t)$ are applied. Find the transfer functions $G_1(s) = I_1(s)/E_{s2}(s)$, $G_2(s) = I_2(s)/E_{s2}(s)$ and $G_3(s) = I_3(s)/E_{s2}(s)$, and show the results in block-diagram form.

5-21 The circuit of Problem 5-7 is at rest when $i_{s1} = 1u_{-1}(t)$ and $i_{s2} = 0$ are applied. Find the transfer functions $G_1(s) = V_1(s)/I_{s1}(s)$, $G_2(s) = V_2(s)/I_{s1}(s)$, and $G_3(s) = V_3(s)/I_{s1}(s)$, and show the results in block-diagram form.

5-22 The circuit of Problem 5-7 is at rest when $i_{s1} = 0$ and $i_{s2} = 1u_{-1}(t)$ are applied. Find the transfer functions $G_1(s) = V_1(s)/I_{s2}(s)$, $G_2(s) = V_2(s)/I_{s2}(s)$, and $G_3(s) = V_3(s)/I_{s2}(s)$, and show the results in block-diagram form.

5-23 The voltages $e_{s1} = 2u_{-1}(t)$ and $e_{s2} = 0$ are applied to the resting circuit shown. Find the transfer functions $G_1(s) = I_1(s)/E_{s1}(s)$ and $G_2(s) = I_2(s)/E_{s1}(s)$ and show the results in block diagram form.

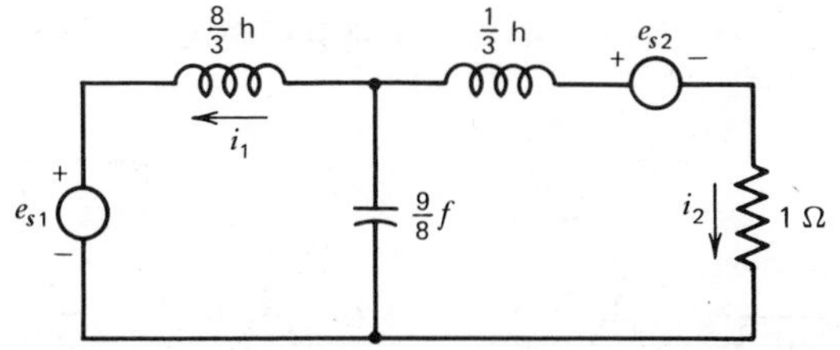

5-24 In the circuit of Problem 5-23, $e_{s1} = 0$ and $e_{s2} = 2u_{-3}(t)$. Find $G_1(s) = I_1(s)/E_{s2}(s)$ and $G_2(s) = I_2(s)/E_{s2}(s)$, and show the result in block-diagram form.

5-25 The currents $i_{s1} = 2u_{-2}(t)$ and $i_{s2} = 0$ are applied to the resting circuit shown. Find the $G_1(s) = V_1(s)/I_{s1}(s)$ and $G_2(s) = V_2(s)/I_{s1}(s)$, and show the results in block-diagram form.

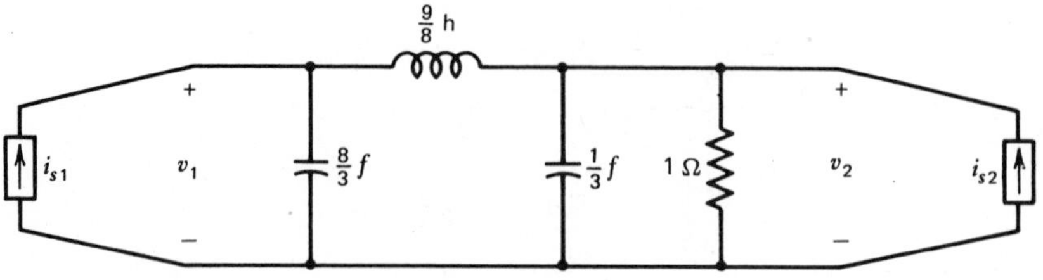

5-26 In the circuit of Problem 5-25, $i_{s1} = 0$ and $i_{s2} = 2u_{-3}(t)$. Find $G_1(s) = V_1(s)/I_{s2}(s)$ and $G_2(s) = V_2(s)/I_{s2}(s)$, and show the results in block-diagram form.

5-27 Demonstrate that the block diagrams shown yield the same result, so far as $F_1(s)$, $F_2(s)$, and $F_4(s)$ are concerned.

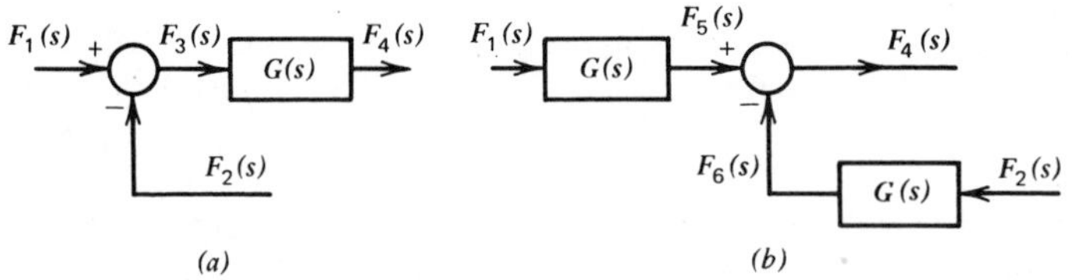

5-28 Demonstrate that the block diagrams shown yield the same results so far as $F_1(s)$ and $F_2(s)$ are concerned.

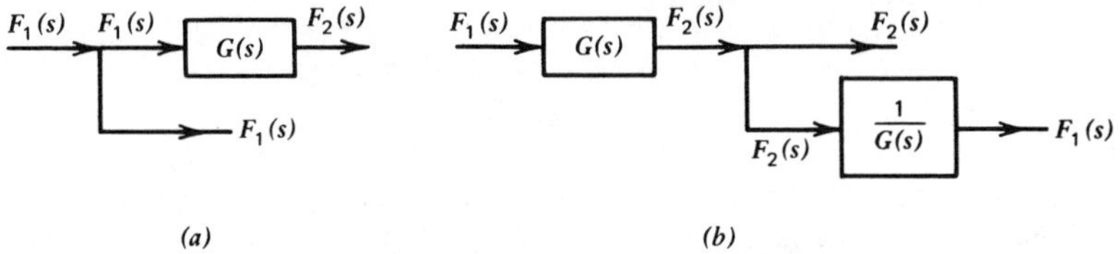

5-29 Demonstrate that the block diagrams yield the same results so far as $F_1(s)$, $F_3(s)$, and $F_4(s)$ are concerned.

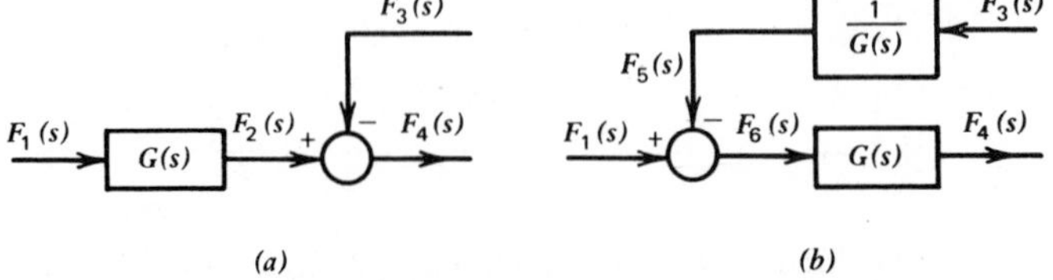

5-30 Reduce the block diagram shown to a single block with one transfer function.

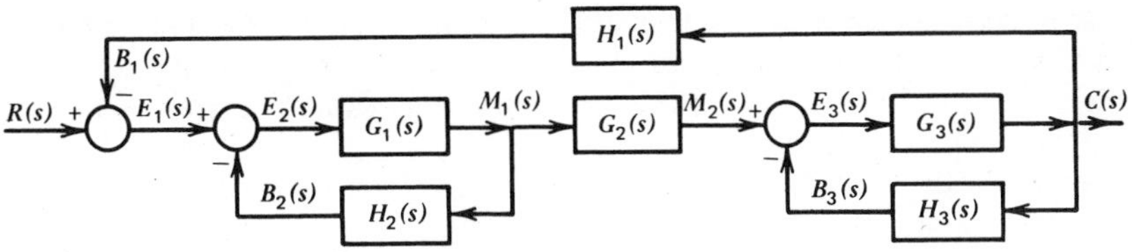

5-31 The circuit shown has the initial conditions as shown. Using the window currents as given, solve the resulting equations for $I_C(s)$.

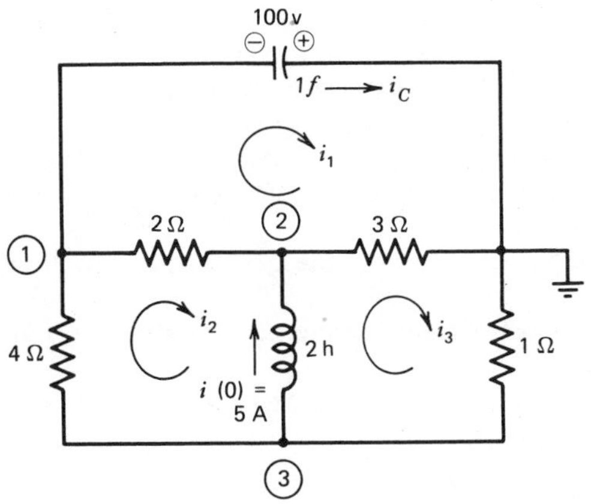

5-32 Repeat Problem 5-31, except use node-to-datum voltages with node numbering and datum node as shown. Solve first for the proper node-to-datum voltage, and from this find $I_C(s)$.

5-33 Repeat Problem 5-31 for the circuit shown.

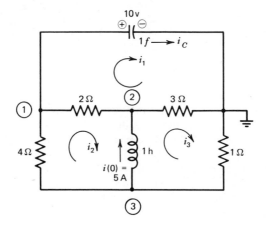

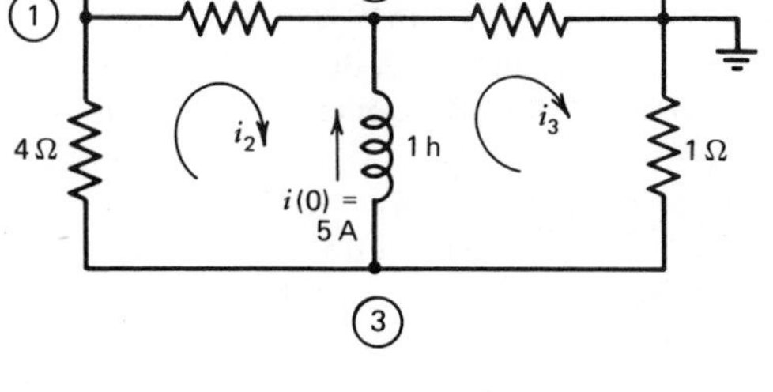

5-34 Repeat Problem 5-32, except use the circuit of Problem 5-33.

5-35 Use the circuit of Problem 5-5 with the 1.5-henry inductor as *link branch 1* and the 3-henry inductor as *link branch 2*. Write the $[z(s)]$ matrix, and find its inverse, which is the $[Y(s)]$ matrix. Use physical interpretation, as is done in Ex. 5-7.1, to determine $Y_{11}(s)$ and $Y_{21}(s)$. Check these results with the corresponding terms in $[Y(s)]$.

5-36 Repeat Problem 5-35, except use physical interpretation to determine $Y_{22}(s)$ and $Y_{12}(s)$.

5-37 Repeat Problem 5-35, except use the circuit of Problem 5-23 with the $^9/_8$-f capacitor as *link branch 1* and the 1-Ω resistor as *link branch 2*.

5-38 Repeat Problem 5-37, except use physical interpretation to determine $Y_{22}(s)$ and $Y_{12}(s)$.

5-39 Use the circuit shown in Problem 5-7, using node-to-datum voltages with the negative end of v_1 being used as datum, the (+) end of v_1 as *node 1* and the (+) end of v_3 as *node 2*. Write the $[y(s)]$ matrix, and find its inverse, which is the $[Z(s)]$ matrix. Use physical interpretation, as is in Ex. 5-7.2, to determine $Z_{11}(s)$ and $Z_{21}(s)$. Check these results with the corresponding terms in $[Z(s)]$.

5-40 Repeat Problem 5-39, except use physical interpretation to determine $Z_{22}(s)$ and $Z_{12}(s)$.

5-41 Repeat Problem 5-39, except use the circuit shown in Problem 5-25. Let $e_1 = v_1$ and $e_2 = v_2$.

5-42 Repeat Problem 5-41, except use physical interpretation to determine $Z_{22}(s)$ and $Z_{12}(s)$.

5-43 The circuit shown is in steady-state when S is closed at $t = 0$. Write the differential equations, take the Laplace transforms, and solve for $I_1(s)$.

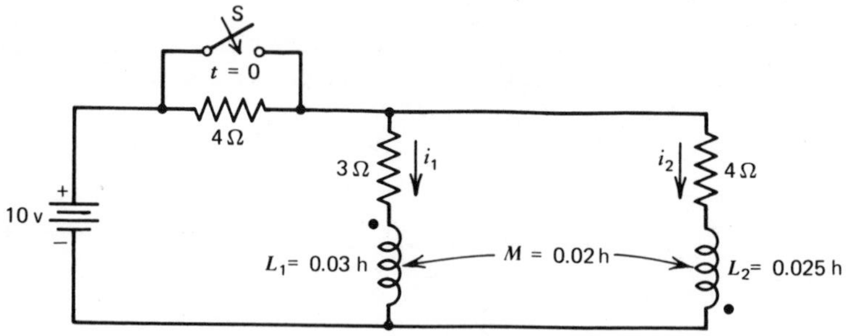

5-44 Repeat Problem 5-43, except solve for $I_1(s)$ using operational methods.

5-45 Repeat Problem 5-43, except move the dot on L_2 to the other end of the coil.

5-46 Solve for $I_1(s)$ in Problem 5-45, using operational methods.

5-47 Look into the circuit of Problem 5-43 from the voltage source with S opened, and find the expression for the operational impedance $Z(s)$.

5-48 Repeat Problem 5-47, except move the dot on L_2 to the other end of the coil.

5-49 The circuit shown is in steady-state when S is closed at $t = 0$. Write the differential equations, and solve for $I_2(s)$ using Laplace transforms.

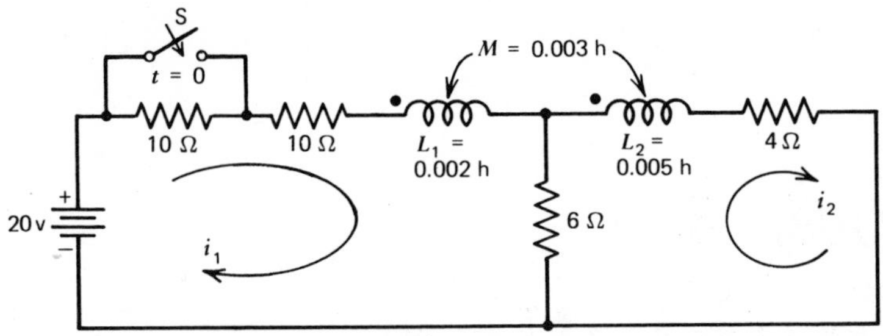

5-50 Repeat Problem 5-49, except solve for $I_2(s)$ using operational methods.

5-51 Repeat Problem 5-49, except move the dot on L_2 to the other end of the coil.

5-52 Solve for $I_2(s)$ in Problem 5-51 using operational methods.

5-53 Look into the circuit of Problem 5-49 from the voltage source with S opened and find the expression for the operational impedance $Z(s)$.

5-54 Repeat Problem 5-47, except move the dot on L_2 to the other end of the coil.

5-55 Solve for $I_2(s)$ in the circuit of Problem 5-5 using Thevenin's theorem.

5-56 Solve for $I_3(s)$ in the circuit of Problem 5-5 using Thevenin's theorem.

5-57 Solve for $I_1(s)$ in the circuit of Problem 5-6 using Thevenin's theorem.

5-58 Solve for $I_3(s)$ in the circuit of Problem 5-6 using Thevenin's theorem.

5-59 Solve for $I_1(s)$ in the circuit of Problem 5-43 using Thevenin's theorem.

5-60 Solve for $I_1(s)$ in the circuit of Problem 5-45 using Thevenin's theorem.

5-61 Solve for $I_2(s)$ in the circuit of Problem 5-49 using Thevenin's theorem.

5-62 Solve for $I_2(s)$ in the circuit of Problem 5-51 using Thevenin's theorem.

5-63 Develop a general proof of Norton's theorem that is analogous to the proof of Thevenin's theorem shown in Fig. 5-9.2.

5-64 Show that the inverse of a symmetric matrix is also symmetric.

5-65 The circuit of Problem 5-23 is designed as a low-pass filter with the cut-off angular frequency of $\omega_c = 1$. Scale the element values so that ω_c is increased to $\omega_c = 2{,}000$ and the impedance level is increased by a factor of 1,000.

Chapter 6

Convolution

6-1. INTRODUCTION

The convolution integral was developed as Property 4-12.10 of the Laplace transform in Chapter 4. In one sense it is unfortunate to introduce convolution in this manner, because this implies that convolution is only one facet of the larger subject of the Laplace transform. The convolution integral can be developed and used without mentioning, or for that matter knowing, anything about the Laplace transform. From this point of view, the convolution integral and the Laplace transform are two separate subjects that are related through the Property 4-12.10. In a number of the examples that follow, the solution is obtained by use of both the convolution integral and the Laplace transform in order to show how both methods achieve the same result.

6-2. THE CONVOLUTION INTEGRAL

We want to apply the convolution integral to a system whose block diagram is shown in Fig. 6-2.1. The input to the system is the time function $f_1(t)$ and its Laplace transform

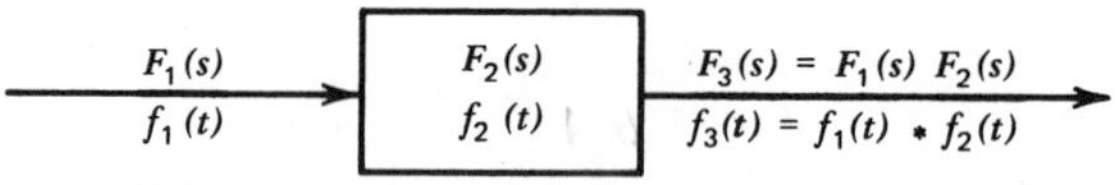

Fig. 6-2.1

is $F_1(s)$.

$$\mathcal{L}[f_1(t)] = F_1(s) \tag{6-2.1}$$

The impulse response of the system is $f_2(t)$ and its Laplace transform is $F_2(s)$.

$$\mathcal{L}[f_2(t)] = F_2(s) \tag{6-2.2}$$

The output of the system if $f_3(t)$ and its Laplace transform is $F_3(s)$.

$$\mathcal{L}[f_3(t)] = F_3(s) \tag{6-2.3}$$

Through block diagram manipulation, $F_3(s)$ is equal to

$$F_3(s) = F_1(s)F_2(s) \tag{6-2.4}$$

The time function $f_3(t)$ can be found as the inverse transform of $F_3(s)$,

$$f_3(t) = \mathcal{L}^{-1}[F_3(s)] = \mathcal{L}^{-1}[F_1(s)F_2(s)] \tag{6-2.5}$$

or it can be found as the convolution of $f_1(t)$ and $f_2(t)$. The convolution integral can be expressed in either of the two forms

$$f_3(t) = \int_{-\infty}^{+\infty} f_1(t-\tau)f_2(\tau)\,d\tau = f_1(t) * f_2(t) \tag{6-2.6}$$

$$f_3(t) = \int_{-\infty}^{+\infty} f_2(t-\tau)f_1(\tau)\,d\tau = f_2(t) * f_1(t) \tag{6-2.7}$$

Since the product of $F_1(s)F_2(s) = F_2(s)F_1(s)$, there has been no need in the literature to make a great distinction between these two forms. Instead of Eq. 6-2.6, many authors write

$$f_1(t) * f_2(t) = \int_{-\infty}^{+\infty} f_1(\tau)f_2(t-\tau)\,d\tau$$

and instead of Eq. (6-2.7), they write

$$f_2(t) * f_1(t) = \int_{-\infty}^{+\infty} f_2(\tau)f_1(t-\tau)\,d\tau$$

In Chapter 2, we stressed that the input and output of a linear system are tied together by the system itself. If a new input is applied that is the integral of the old input, the new output will be the integral of the old output. The convolution integral in a certain sense is a generalization of this idea in that if the impulse response of a system is known, the output of this system to any input can be found by use of the convolution integral. If you stop and think about this you will find it a very remarkable idea.

The convolution integral does not imply the restrictions that $f_1(t)$ and $f_2(t)$ must be zero for negative time. In Chapter 9, the two-sided Laplace transform is introduced, and functions defined for all time are applied to the system. However, as long as we think of $f_2(t)$ as the impulse response of the system, our work assumes that $f_2(t)$ is zero for negative time. When convolution in the frequency domain is used in Chapter 10, the more general situation of convolution will be encountered. In this chapter, however, both $f_1(t)$ and $f_2(t)$ are zero for negative time.

A GRAPHICAL INTERPRETATION OF CONVOLUTION

A graphical interpretation of these integrals is helpful. To examine this, we must take the form of Eq. 6-2.6, repeated as

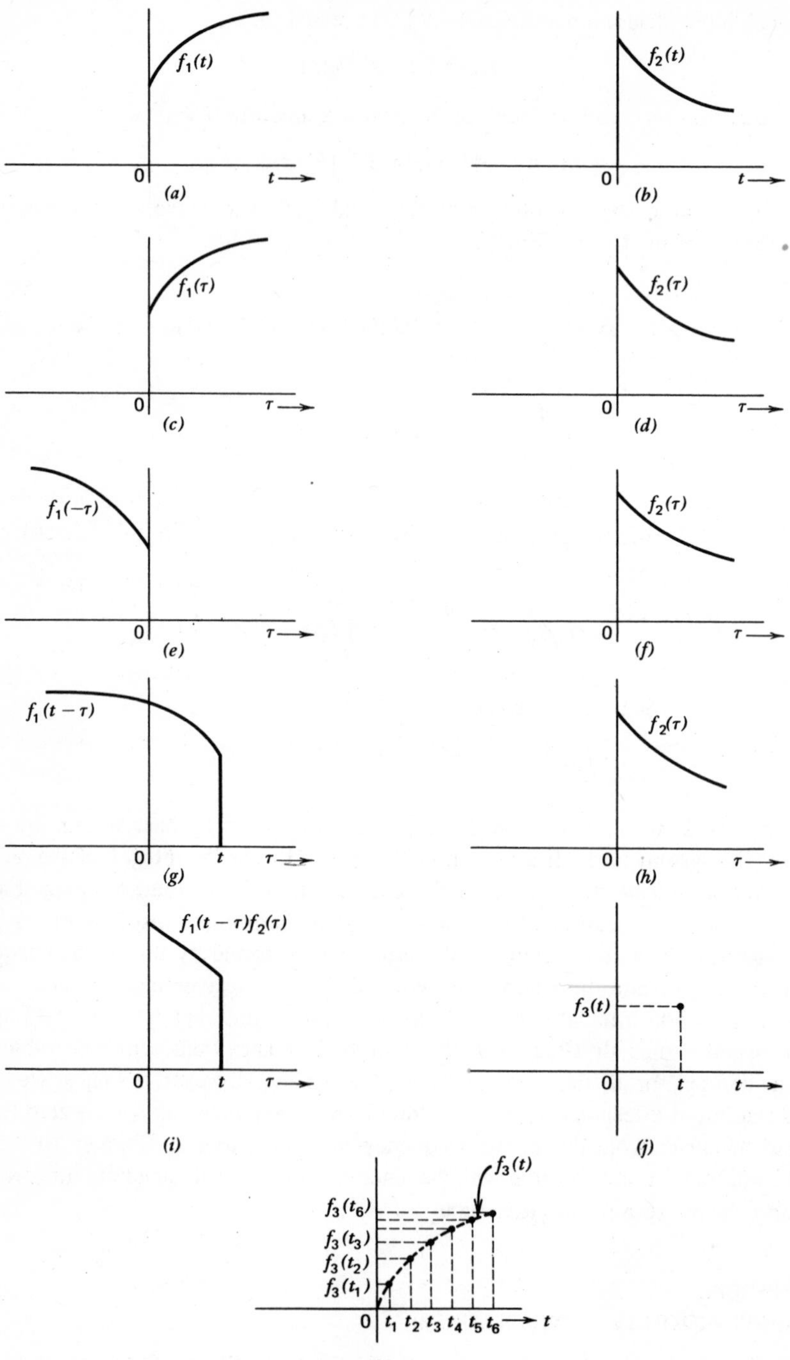

$f_1(t)$
0
t
(a)
$f_2(t)$
0
t
(b)
$f_1(\tau)$
0
τ
(c)
$f_2(\tau)$
0
τ
(d)
$f_1(-\tau)$
0
τ
(e)
$f_2(\tau)$
0
τ
(f)
$f_1(t-\tau)$
0
t
τ
(g)
$f_2(\tau)$
0
τ
(h)
$f_1(t-\tau)f_2(\tau)$
0
t
τ
(i)
$f_3(t)$
0
t
t
(j)
$f_3(t)$
$f_3(t_6)$
$f_3(t_3)$
$f_3(t_2)$
$f_3(t_1)$
0
t_1 t_2 t_3 t_4 t_5 t_6
t
(k)

Fig. 6–2.2

$$f_3(t) = \int_{-\infty}^{+\infty} f_1(t - \tau) f_2(\tau)\, d\tau \tag{6-2.8}$$

To start, we assume the two functions $f_1(t)$ and $f_2(t)$, as suggested, in Fig. 6-2.2(a) and (b), respectively. Since the integration in Eq. 6-2.8 is with respect to τ, we make a change of variables, and $f_1(t)$ and $f_2(t)$ become $f_1(\tau)$ and $f_2(\tau)$ as shown in (c) and (d), respectively. In the next step, $f_2(\tau)$ is left alone, as shown in (f), but the τ in $f_1(\tau)$ is replaced by $(-\tau)$, as shown in (e). This is commonly referred to by saying $f_1(\tau)$ is *folded over* to produce the $f_1(-\tau)$. Since the integration is with respect to τ, t is a constant. We proceed by letting $f_1(-\tau)$ slide to the right, by a value of t, as shown in (g). The $f_2(\tau)$ remains unchanged, as shown in (h). Next, the product curve of $f_1(t - \tau) f_2(\tau)$ is sketched as shown in (i). This product curve is the integrand of Eq. 6-2.8 for a specific value of t. The area under this product curve becomes the value of $f_3(t)$ for a specific value of t, as suggested in (j).

The entire $f_3(t)$ curve is the result of a repetition of this process over and over. For this example, when the $f_1(\tau)$ curve is folded over and t is given a value of zero, the area under the product curve is zero as shown by a dot at $t = 0$ on the $f_3(t)$ curve of (k). As t is given values greater than zero, the $f_1(t - \tau)$ curve slides across the $f_2(\tau)$ curve and the area under the product curve $f_1(t - \tau) f_2(\tau)$ increases as t increases.[1] Since the value of $f_3(t)$ is equal to the area under the product curve, the value of $f_3(t)$ also increases. The time t is given the sequence of values t_1, t_2, t_3, etc., and the corresponding values of $f_3(t)$ are suggested in curve of (k). If enough values of t are taken, the shape of the $f_3(t)$ curve is plotted out.

It should be pointed out that the general limits from $-\infty$ to $+\infty$ on Eq. 6-2.8 reduce to the more specific limits appropriate for this case. When $t < 0$, there is no area under the product curve and $f_3(t) = 0$. When $0 < t$, the product curve is non-zero only for τ between 0 and t. Hence, $f_3(t)$ can be written as

$$f_3(t) = \begin{cases} 0 & \text{for } 0 > t \\ \displaystyle\int_0^t f_1(t - \tau) f_2(\tau)\, d\tau & \text{for } 0 < t \end{cases} \tag{6-2.9}$$

The entire discussion can be repeated for the other form of the convolution integral, namely

$$f_3(t) = \int_{-\infty}^{+\infty} f_2(t - \tau) f_1(\tau)\, d\tau \tag{6-2.10}$$

This is left as an exercise.

Example 6-2.1. The block diagram for this system is shown in Fig. 6-2.3.

[1]This statement is true for the specific $f_1(t)$ and $f_2(t)$ curves of this example. It is possible for the area to decrease as time increases, given other types of $f_1(t)$ and $f_2(t)$.

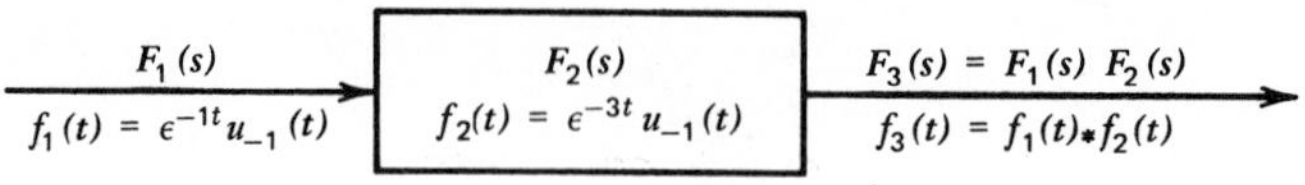

Fig. 6-2.3

The $f_1(t)$ and $f_2(t)$ are given as

$$f_1(t) = \epsilon^{-1t} u_{-1}(t) \qquad f_2(t) = \epsilon^{-3t} u_{-1}(t) \tag{6-2.11}$$

and the desired solution is the function $f_3(t)$. Before we finish, we shall work this problem by three methods.

Method 1. This method uses Laplace transforms and is not an example of convolution. The $F_1(s)$ and $F_2(s)$ are found as

$$F_1(s) = \frac{1}{(s+1)} \qquad F_2(s) = \frac{1}{(s+3)} \tag{6-2.12}$$

The $F_3(s)$ is

$$F_3(s) = F_1(s) F_2(s) = \frac{1}{(s+1)(s+3)} = \frac{1/2}{(s+1)} - \frac{1/2}{(s+3)} \tag{6-2.13}$$

The $f_3(t)$ is the inverse transform of this, or

$$f_3(t) = \tfrac{1}{2} \epsilon^{-1t} - \tfrac{1}{2} \epsilon^{-3t} \tag{6-2.14}$$

The result is sketched in Fig. 6-2.4.

Method 2. The $f_3(t)$ is found as the convolution of $f_1(t)$ and $f_2(t)$ using the form of Eq. 6-2.9, which for $t > 0$ becomes

$$f_3(t) = \int_0^t \epsilon^{-(t-\tau)} \epsilon^{-3\tau} \, d\tau \qquad \text{for } 0 < t \tag{6-2.15}$$

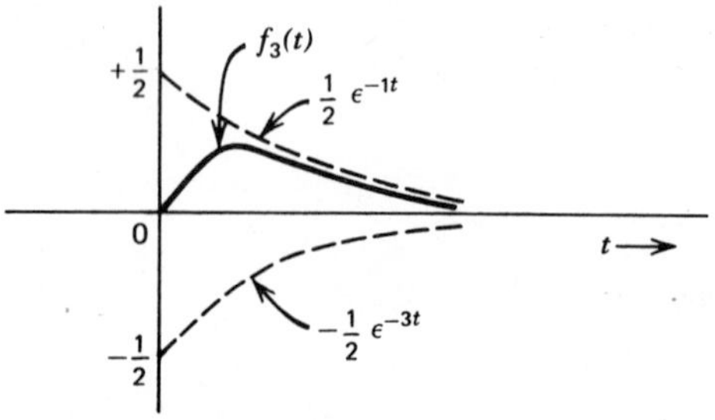

Fig. 6-2.4

Since the integration is with respect to τ, the ϵ^{-t} term can be taken outside the integral as

$$f_3(t) = \epsilon^{-t} \int_0^t \epsilon^{-2\tau}\, d\tau = \epsilon^{-t}\,[-\tfrac{1}{2}\epsilon^{-2\tau}|_0^t] = \epsilon^{-t}[-\tfrac{1}{2}\epsilon^{-2t} + \tfrac{1}{2}] = \tfrac{1}{2}\epsilon^{-t} - \tfrac{1}{2}\epsilon^{-3t} \quad (6\text{-}2.16)$$

Method 3. The $f_3(t)$ is found using the other form of convolution, as

$$f_3(t) = \int_{-\infty}^{+\infty} f_2(t-\tau)\, f_1(\tau)\, d\tau \quad (6\text{-}2.17)$$

which for $t > 0$ becomes

$$f_3(t) = \int_0^t \epsilon^{-3(t-\tau)}\, \epsilon^{-\tau}\, d\tau \quad (6\text{-}2.18)$$

$$= \epsilon^{-3t} \int_0^t \epsilon^{2\tau}\, d\tau = \epsilon^{-3t}\,[\tfrac{1}{2}\epsilon^{2\tau}|_0^t] = \epsilon^{-3t}\,[\tfrac{1}{2}\epsilon^{+2t} - \tfrac{1}{2}] = \tfrac{1}{2}\epsilon^{-t} - \tfrac{1}{2}\epsilon^{-3t} \quad (6\text{-}2.19)$$

Example 6-2.2. The $f_1(t)$ and $f_2(t)$ are given as

$$f_1(t) = \epsilon^{-2t}\, u_{-1}(t) \qquad f_2(t) = \epsilon^{-2t}\, u_{-1}(t) \quad (6\text{-}2.20)$$

Working this example by using Laplace transforms is left as an exercise. Since $f_1(t) = f_2(t)$, the two forms of the convolution integral are identical. The $f_3(t)$ for $t > 0$ is

$$f_3(t) = \int_0^t \epsilon^{-2(t-\tau)}\, \epsilon^{-2\tau}\, d\tau \quad (6\text{-}2.21)$$

$$= \epsilon^{-2t} \int_0^t 1\, d\tau = \epsilon^{-2t}[\tau|_0^t] = t\,\epsilon^{-2t} \quad (6\text{-}2.22)$$

Example 6-2.3. The $f_1(t)$ and $f_2(t)$ for $t > 0$ are given as

$$f_1(t) = t \qquad f_2(t) = \epsilon^{-2t} \quad (6\text{-}2.23)$$

Working this example using Laplace transforms is left as an exercise. Convolution is performed using

$$f_3(t) = \int_0^t \epsilon^{-2(t-\tau)}\tau\, d\tau \qquad t > 0 \quad (6\text{-}2.24)$$

$$= \epsilon^{-2t} \int_0^t \tau\, \epsilon^{2\tau}\, d\tau = \epsilon^{-2t}\left[\frac{\epsilon^{2\tau}}{4}(2\tau - 1)\Big|_0^t\right]$$

$$= \epsilon^{-2t}\left[\frac{\epsilon^{2t}}{4}(2t-1) - (1/4)(-1)\right] = (1/2)t - (1/4) + (1/4)\epsilon^{-2t} \quad (6\text{-}2.25)$$

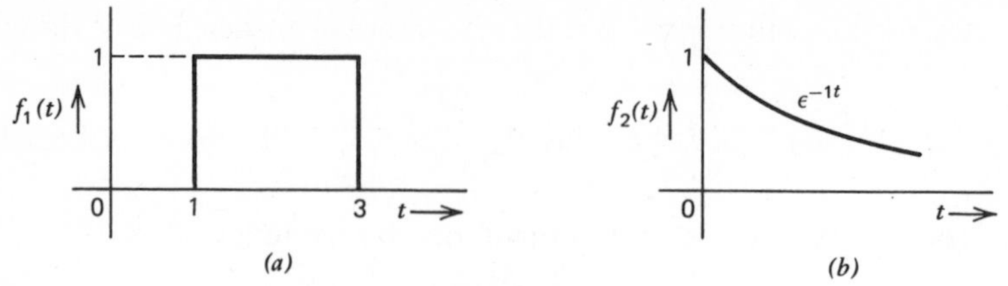

Fig. 6-2.5

Example 6-2.4. The input to the system is the $f_1(t)$ shown graphically in Fig. 6-2.5(a), and the impulse response of the system is the $f_2(t)$ shown in (b). We work the problem first by using Laplace transform methods.

Method 1. The $F_1(s)$ and $F_2(s)$ are found as:

$$F_1(s) = \frac{1}{s}\epsilon^{-1s} - \frac{1}{s}\epsilon^{-3s} \qquad F_2(s) = \frac{1}{(s+1)} \tag{6-2.26}$$

$$F_3(s) = F_1(s)\,F_2(s) = \frac{1\epsilon^{-1s}}{s(s+1)} - \frac{1\epsilon^{-3s}}{s(s+1)} \tag{6-2.27}$$

$$F_3(s) = \left[\frac{1}{s} - \frac{1}{(s+1)}\right]\epsilon^{-1s} - \left[\frac{1}{s} - \frac{1}{(s+1)}\right]\epsilon^{-3s} \tag{6-2.28}$$

The inverse transform is

$$f_3(t) = [1 - \epsilon^{-(t-1)}]\,u_{-1}(t-1) - [1 - \epsilon^{-(t-3)}]\,u_{-1}(t-3) \tag{6-2.29}$$

This function is sketched in Fig. 6-2.6.

Method 2. We fold $f_2(\tau)$ for this discussion. The $f_2(t-\tau)$ and $f_1(\tau)$ are shown in Fig. 6-2.7, with $f_2(t-\tau)$ sketched for a negative t. As t increases, you can visualize the $f_2(t-\tau)$ curve sliding to the right. There is no overlap between the two functions until $t = +1$; therefore,

$$f_3(t) = 0 \qquad t < 1 \tag{6-2.30}$$

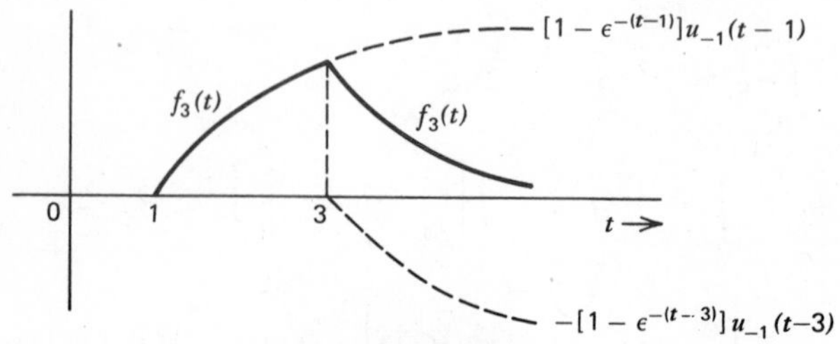

Fig. 6-2.6

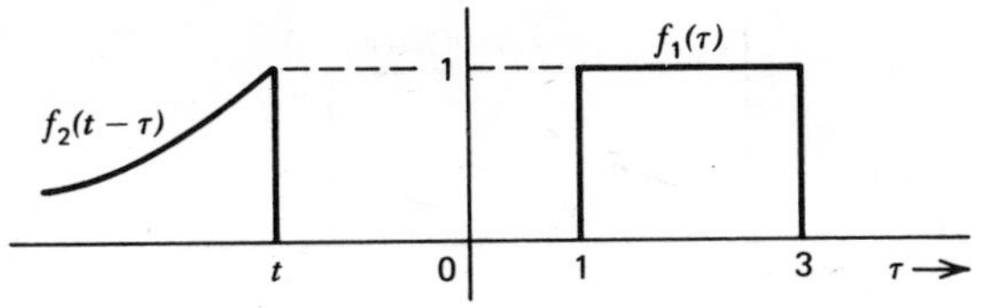

Fig. 6-2.7

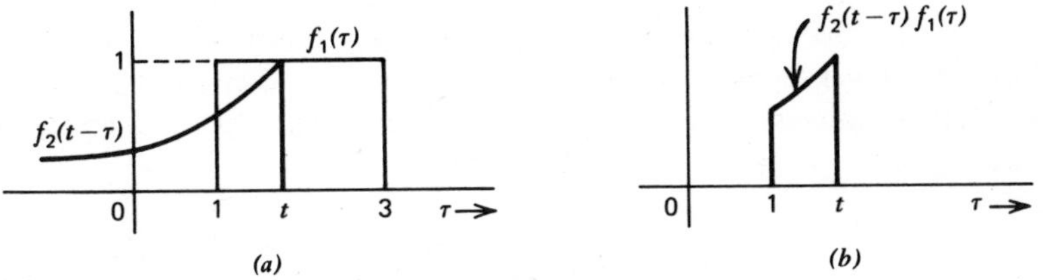

Fig. 6-2.8

We redraw this figure in Fig. 6-2.8(a) to suggest the $f_2(t-\tau)$ function with $1 < t < 3$. The product function exists from 1 to t as shown in (b). Therefore, in this range of t, the $f_3(t)$ function is given by

$$f_3(t) = \int_1^t \epsilon^{-(t-\tau)}\,(1)\,d\tau = \epsilon^{-t} \int_1^t \epsilon^{+\tau}\,d\tau = \epsilon^{-t}\,[\epsilon^{+\tau}|_1^t]$$

$$= \epsilon^{-t}[\epsilon^t - \epsilon^1] = 1 - \epsilon^{-(t-1)} \qquad 1 < t < 3 \qquad (6\text{-}2.31)$$

We redraw the figure in Fig. 6-2.9(a) to suggest the $f_2(t-\tau)$ function when $3 < t$. The product function exists from 1 to 3 as shown in (b). Therefore, in this range to t, the $f_3(t)$ function is given by

$$f_3(t) = \int_1^3 \epsilon^{-(t-\tau)}\,(1)\,d\tau = \epsilon^{-t} \int_1^3 \epsilon^{+\tau}\,d\tau = \epsilon^{-t}[\epsilon^{+\tau}|_1^3]$$

$$= \epsilon^{-t}[\epsilon^3 - \epsilon^1] = \epsilon^{-(t-3)} - \epsilon^{-(t-1)} \qquad 3 < t \qquad (6\text{-}2.32)$$

It is left to the reader to convince himself that Eqs. 6-2.30, 6-2.31 and 6-2.32 yield the same result as Eq. 6-2.29.

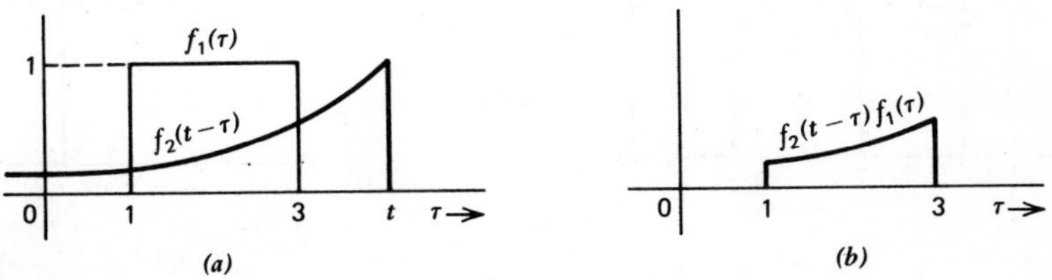

Fig. 6-2.9

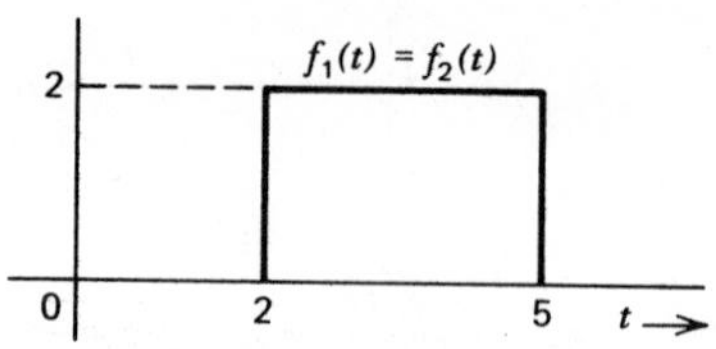

Fig. 6-2.10

Example 6-2.5. In this example, $f_1(t) = f_2(t)$ and are shown in Fig. 6-2.10. Figure 6-2.11 shows the $f_1(\tau)$ function folded and slid across the other curve by a time t. An inspection of these curves shows the product curve is zero until t is greater than 4.

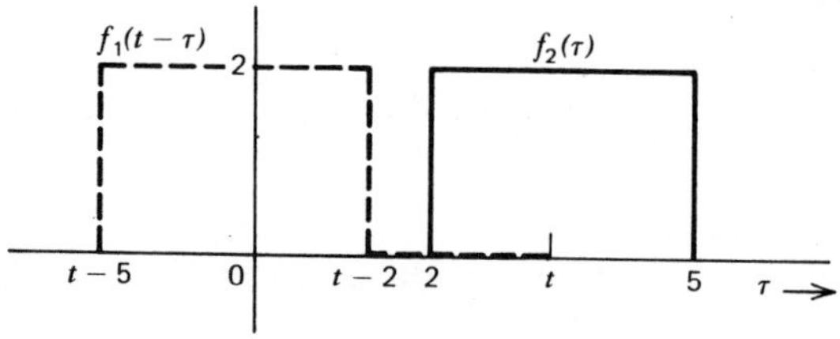

Fig. 6-2.11

Therefore

$$f_3(t) = 0 \qquad -\infty < t < 4 \tag{6-2.33}$$

Figure 6-2.12(a) suggests the $f_1(t-\tau)$ curve when t is in the interval $4 < t < 7$, and (b) shows the corresponding product curve. The $f_3(t)$ is obtained from

$$f_3(t) = \int_2^{t-2} (2)(2)\,d\tau = 4\tau\Big|_2^{t-2} = 4(t-2) - 4(2) = 4t - 16 \qquad 4 < t < 7 \tag{6-2.34}$$

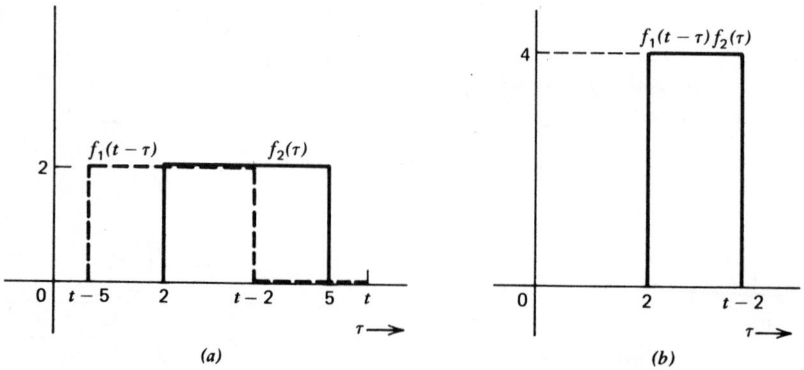

Fig. 6-2.12

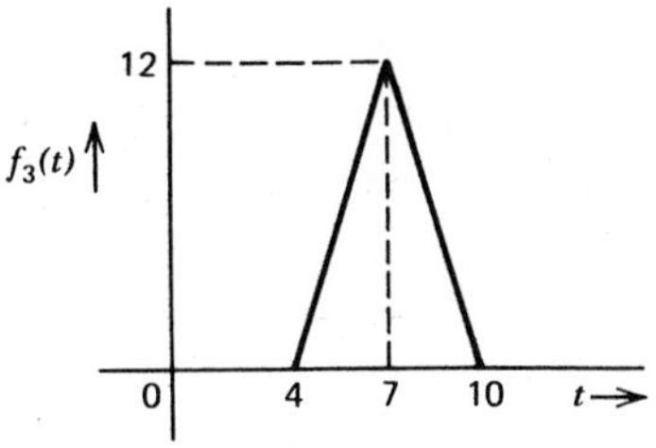

Fig. 6-2.13

An inspection of Fig. 6-2.12(a) shows that when t is in the interval $7 < t < 10$, the integral becomes

$$f_3(t) = \int_{t-5}^{5} (2)(2)\, d\tau = 4\tau\Big|_{t-5}^{5} = -4t + 40 \qquad 7 < t < 10 \tag{6-2.35}$$

Finally, when $10 < t$, there is no overlap between $f_1(t - \tau)$ and $f_2(\tau)$, and

$$f_3(t) = 0 \qquad 10 < t \tag{6-2.36}$$

The four equations for $f_3(t)$ are sketched in Fig. 6-2.13.

6-3. USING THE GRAPHICAL INTERPRETATION OF THE CONVOLUTION

The solution to certain problems can be determined from the graphical interpretation of the convolution integral. Example 6-2.5 is an excellent illustration of this. However, we introduce this idea in simpler terms before returning to that example.

Example 6-3.1. For this example, both $f_1(t)$ and $f_2(t)$ are the unit-step function. That is

$$f_1(t) = f_2(t) = u_{-1}(t) \tag{6-3.1}$$

The $f_3(t)$ by use of Laplace transforms can be easily found as

$$f_3(t) = \mathcal{L}^{-1}[F_1(s)F_2(s)] = \mathcal{L}^{-1}[1/s^2] = t\,u_{-1}(t) \tag{6-3.2}$$

We wish to find $f_3(t)$ using convolution without actually setting up the integrals. In Fig. 6-3.1(a) $f_1(t - \tau)$ and $f_2(\tau)$ are sketched. It is easy to visualize from the product

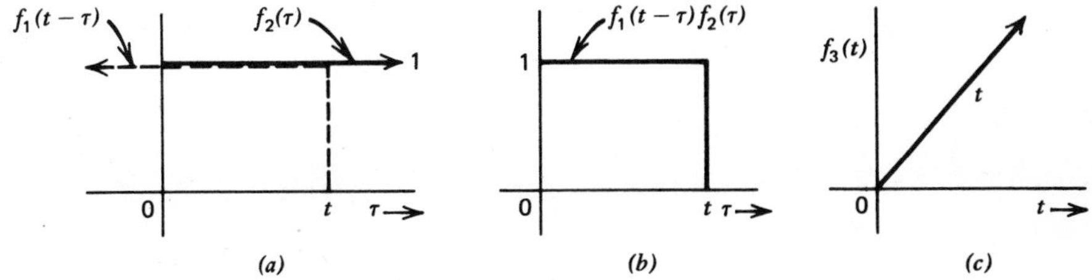

Fig. 6-3.1

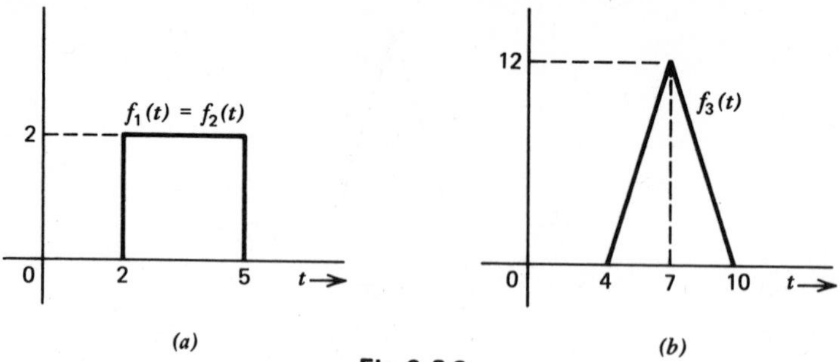

Fig. 6-3.2

curve in (b) that $f_3(t)$ is zero if t is negative, and is equal to t if t is positive. Hence, the result is sketched in (c).

Example 6-3.2. We return to Ex. 6-2.5, which was worked rather laboriously in the last section. In this example, $f_1(t) = f_2(t)$, and these are shown again in Fig. 6-3.2(a). We visualize folding $f_1(\tau)$ and sliding it along. The product curve is zero until $t = 4$. The area under the product curve increases linearly until the two curves coincide, at which times the product curve is a rectangle that has a height of 4, a base of 3, and an area of 12. Hence, $f_3(t)$ equals 12 at $t = 7$, and we draw in a straight line as the curve of (b) from the zero point at $t = 4$ and the 12 point at $t = 7$.

Next we observe that the area under the product curve decreases linearly from 12 at $t = 7$ to zero at $t = 10$, and is zero after $t = 10$. We draw the appropriate straight lines in (b) and our work is finished.

6-4. PROPERTIES OF THE CONVOLUTION INTEGRAL

Several of these properties have been used in the preceding discussion. The proofs of some of these properties are suggested using Laplace methods.

Property 6-4.1. The Commutative Property.

$$f_3(t) = f_1(t) * f_2(t) = f_2(t) * f_1(t) \tag{6-4.1}$$

This can be seen to hold from

$$F_3(s) = F_1(s)\,F_2(s) = F_2(s)\,F_1(s) \tag{6-4.2}$$

Property 6-4.2. The Distributive Property. Suppose that $f_1(t)$ is made up of a sum of other functions, such as

$$f_1(t) = f_1^A(t) + f_1^B(t) + \cdots + f_1^K(t) \tag{6-4.3}$$

then

$$f_1(t)*f_2(t) = [f_1^A(t) + f_1^B(t) + \cdots + f_1^K(t)] * f_2(t)$$
$$= f_1^A(t)*f_2(t) + f_1^B(t)*f_2(t) + \cdots + f_1^K(t)*f_2(t) \qquad (6\text{-}4.4)$$

In Chapter 4, a number of time functions were built up of a group of simpler functions, which leads to

$$F_1(s)\,F_2(s) = [F_1^A(s) + F_1^B(s) + \cdots + F_1^K(s)]\,F_2(s)$$
$$= F_1^A(s)\,F_2(s) + F_1^B(s)\,F_2(s) + \cdots + F_1^K(s)\,F_2(s) \qquad (6\text{-}4.5)$$

Property 6-4.3. The Associative Property.

$$[f_1(t)*f_2(t)]*f_3(t) = f_1(t)*[f_2(t)*f_3(t)] \qquad (6\text{-}4.6)$$

The transfer function of a group of transfer functions connected in tandem is

$$G(s) = G_1(s)\,G_2(s)\,G_3(s) \qquad (6\text{-}4.7)$$

The final $G(s)$ is independent of the manner in which the product is formed.

Property 6-4.4. The Derivative of $f_3(t)$.

$$f_3'(t) = f_1'(t)*f_2(t) = f_1(t)*f_2'(t) \qquad (6\text{-}4.8)$$

Given $F_3(s) = F_1(s)\,F_2(s)$, both sides can be multiplied by s and written as

$$[sF_3(s)] = [sF_1(s)]\,[F_2(s)] \qquad (6\text{-}4.9)$$

or as

$$[sF_3(s)] = [F_1(s)]\,[sF_2(s)] \qquad (6\text{-}4.10)$$

The multiplication of an $F(s)$ by s corresponds to generalized differentiation of $f(t)$.

Property 6-4.5. The Integral of $f_3(t)$.

$$f_3^{-1}(t) = f_1^{-1}(t)*f_2(t) = f_1(t)*f_2^{-1}(t) \qquad (6\text{-}4.11)$$

The suggested proof is similar to that for Property 6-4.4.

Property 6-4.6. Convolving the Derivative of One Function with the Integral of the Other.

$$f_3(t) = f_1'(t)*f_2^{-1}(t) = f_1^{-1}(t)*f_2'(t) \qquad (6\text{-}4.12)$$

Given $F_3(s) = F_1(s)\,F_2(s)$. One of these functions on the right can be multiplied by s, and the other divided by s without changing $F_3(s)$, as

$$F_3(s) = [sF_1(s)]\left[\frac{F_2(s)}{s}\right] \qquad (6\text{-}4.13)$$

or

$$F_3(s) = \left[\frac{F_1(s)}{s}\right] [sF_2(s)] \qquad (6\text{–}4.14)$$

Equation 6–4.12 has some philosophical implication. To this point, the impulse response of a system has been featured. The $f_2^{-1}(t)$ is the integral of the impulse response, which means that it is the step-function response of the system. Equation 6–4.12 implies that the step-function response is just as fundamental, but it must be convolved with the derivative of the input function. Again, differentiation must be done in the generalized manner.

Equation 6–4.12 can be generalized as

$$f_3(t) = f_1^n(t) * f_2^{-n}(t) = f_1^{-n}(t) * f_2^n(t) \qquad (6\text{–}4.15)$$

Property 6–4.7. Convolving with a Unit Impulse. The convolution of a unit impulse with some $f(t)$ results in the same $f(t)$.

$$u_0(t) * f(t) = f(t) \qquad (6\text{–}4.16)$$

Given $F_3(s) = F_1(s) F_2(s)$, where $f_1(t) = u_0(t)$

$$F_3(s) = 1\, F_2(s) = F_2(s) \qquad (6\text{–}4.17)$$

The interpretation of this concept in the time domain is helpful. Figure 6–4.1(a) indicates a unit-impulse that is to be convolved with an $f(t)$ in (b). As the impulse is folded and slides across the $f(\tau)$, the product curve is zero everywhere except at the point at which the impulse is located, and here it is an impulse of value $f(\tau)$. Therefore, the $f(t)$ curve simply duplicates the $f(\tau)$ curve, as shown in (c).

Equation 6–4.16 can be generalized as

$$u_{+n}(t) * f(t) = f^{+n}(t) \qquad (6\text{–}4.18)$$

$$u_{-n}(t) * f(t) = f^{-n}(t) \qquad (6\text{–}4.19)$$

Again each derivative must be done in the generalized manner.

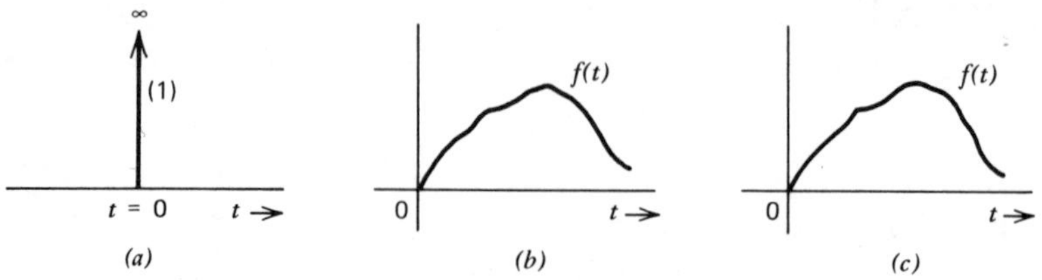

Fig. 6–4.1

6-5. ADDITIONAL CONVOLUTION EXAMPLES

Example 6-5.1. The $f_1(t)$ in Fig. 6-5.1(a) is to be convolved with the $f_2(t)$ of (b). We do this by building up the $f_1(t)$ with the two step-functions as given by

$$f_1(t) = f_1^A(t) + f_1^B(t) = 1u_{-1}(t-1) - 1u_{-1}(t-7) \tag{6-5.1}$$

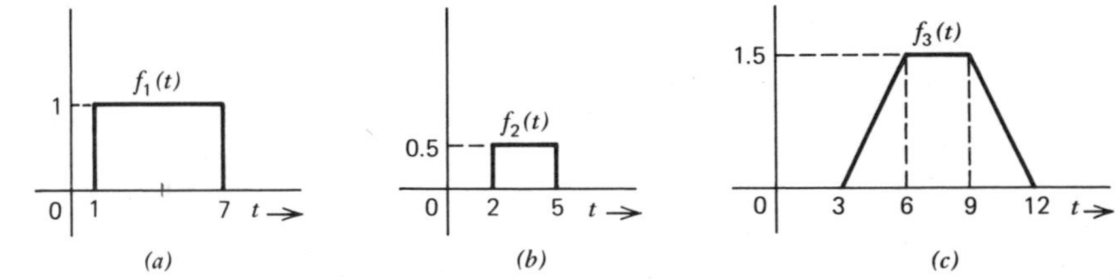

Fig. 6-5.1

The $f_1^A(t)$ is convolved with $f_2(t)$ to yield $f_3^A(t)$, and $f_1^B(t)$ is convolved with $f_2(t)$ to yield $f_3^B(t)$. Finally, $f_3(t) = f_3^A(t) + f_3^B(t)$, as shown in (c). You should carry out the details to check the result.

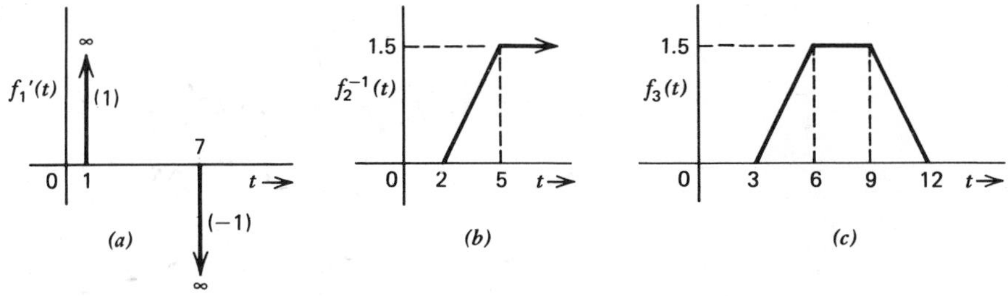

Fig. 6-5.2

Example 6-5.2. The $f_1(t)$ and $f_2(t)$ of Ex. 6-5.1 are used again. This time the $f_1(t)$ is differentiated, and $f_2(t)$ integrated, as shown in Fig. 6-5.2(a) and (b), respectively. The $f_1'(t)$ is convolved with $f_2^{-1}(t)$ and the result is shown in (c). The impulses can be convolved separately yielding

$$f_3(t) = f_3^A(t) + f_3^B(t) \tag{6-5.2}$$

Example 6-5.3. The $f_1(t)$ in Fig. 6-5.3(a) is to be convolved with the $f_2(t)$ of (b). Since the derivative of $f_1(t)$ contains only impulses, the $f_1'(t)$ shown in (c) is convolved with $f_2^{-1}(t)$ shown in (d), and the result, $f_3(t)$, is shown in (e).

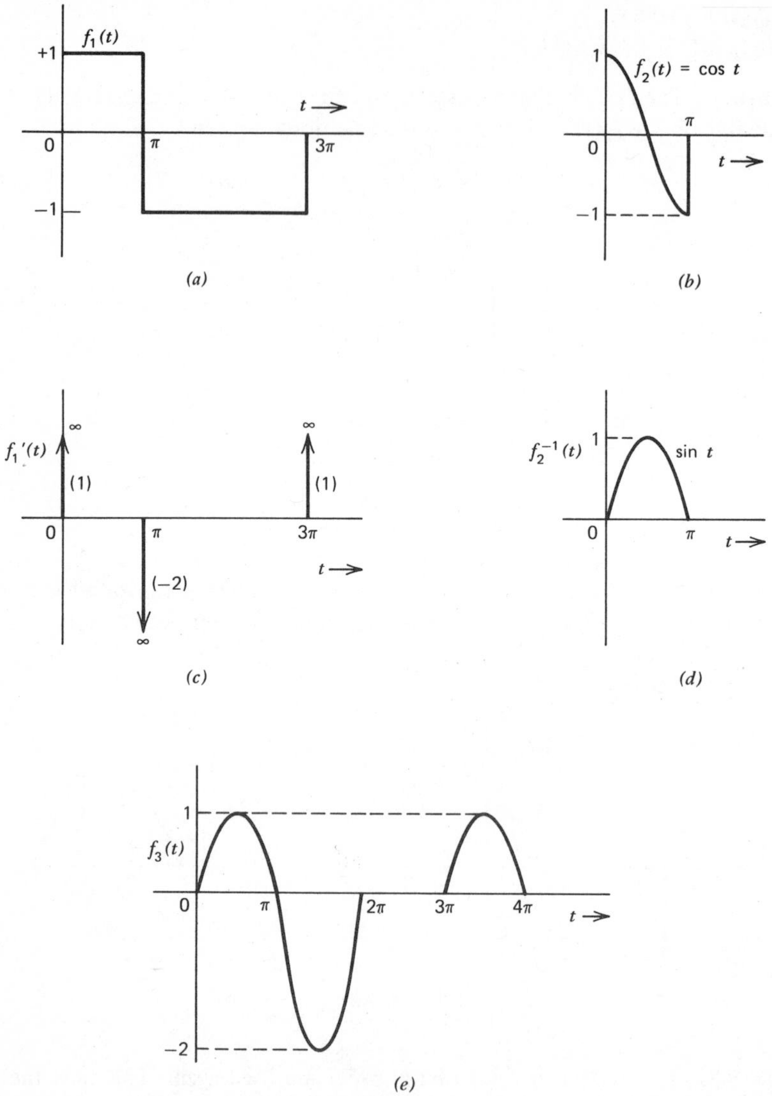

Fig. 6-5.3

Example 6-5.4. For this example, $f_1(t)$ and $f_2(t)$ are shown in Fig. 6-5.4(a). The derivative of $f_1(t)$ and the integral of $f_2(t)$ are shown in (b) and (c), respectively. You should draw several of the $f_3^A(t)$, $f_3^B(t)$, $f_3^C(t)$, etc., curves resulting from the individual impulse terms in $f_1'(t)$ and add the first few terms together to verify that the $f_3(t)$ curve shown in (d) is correct.

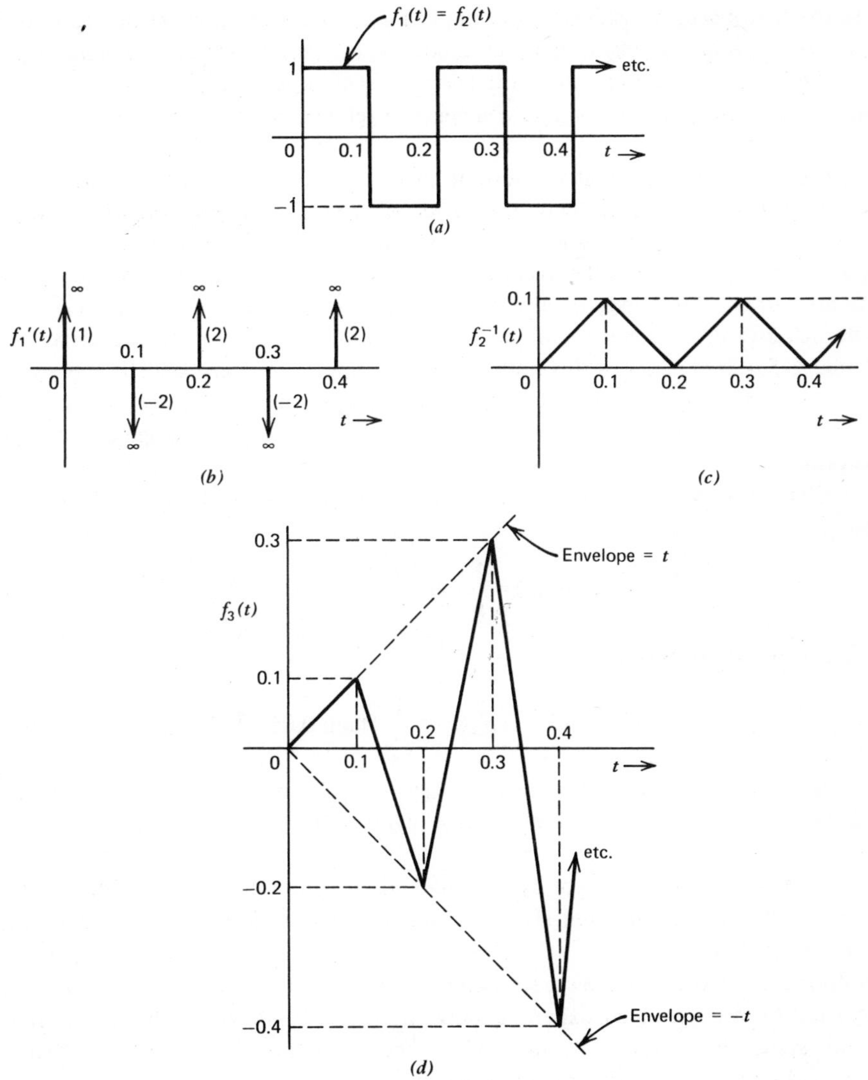

Fig. 6-5.4

6-6. PRACTICAL ASPECTS OF CONVOLUTION

One of the questions usually asked by students when they study a subject for the first time is "Yes, but what good is it?" This obviously is a question that always deserves an answer. In this section, we will attempt to answer this question when asked about convolution by approaching it from several points of view.

In the first place, the philosophical implications of convolution are tremendous. The concept of convolution tells us that just as soon as the impulse response of a linear physical system is known, the response of this system to any input can be uniquely determined. If you think about that for a while, you begin to get a feeling for what a remarkable idea this really is.

The second point, one that we hope to make many times, is this: the more different ways you have of doing something, the better off you are. There are occasions when the impulse response of a system and the input to the system are known analytically and the output can be determined either using Laplace transforms or by the convolution integral. There are certainly occasions when it is easier to determine the output by using Laplace transforms. However, there are other occasions when it is easier to use convolution. Example 6–5.4 demonstrates this point. In this example, we convolved the square wave of Fig. 6–5.4(a) with itself and determined the output as shown in (d) without much labor. For contrast, let us see how we would find this same output through use of Laplace transforms.

In Chapter 4, we found the $F(s)$ corresponding to the $f_1(t)$ and $f_2(t)$ of this example as

$$F_1(s) = F_2(s) = \frac{1}{s} \tanh \frac{T}{4} s = \frac{1}{s} \tanh 0.05s \qquad (6\text{–}6.1)$$

The transform of the output is

$$F_3(s) = F_1(s) F_2(s) = \frac{1}{s^2} [\tanh 0.05s]^2 \qquad (6\text{–}6.2)$$

If you recall what a jolly time we had in Chapter 4 taking the inverse transform of Eq. 6–6.1, you can imagine the amount of work involved in taking the inverse transform of Eq. 6–6.2.

If you try to solve a problem by one method and you run into difficulties, drop this method and try another method. There are many situations in which this other method might be convolution.

Another point is that there are situations in which the impulse response of an existing system has been determined experimentally and hence is known graphically. It may be that this system has never been subjected to a certain type of input and it is desired to explore its behavior before the input is actually applied.

Let us examine this situation with respect to Laplace transforms. How do we find the $F_2(s)$ corresponding to this $f_2(t)$ when $f_2(t)$ is known graphically? There are two ways of proceeding that are outside the material covered in this book; however, we can make brief comments. Both of these methods are known as approximation procedures.

One way involves *time-domain* methods, which consist of approximating the graphical curve with a time function. As soon as this is done, $F_2(s)$ can be found from the time function and you can proceed to find the output.

The other involves *frequency-domain* approximation. The direct Fourier transform can be taken on $f_2(t)$ using numerical methods to convert the data to frequency plot. We

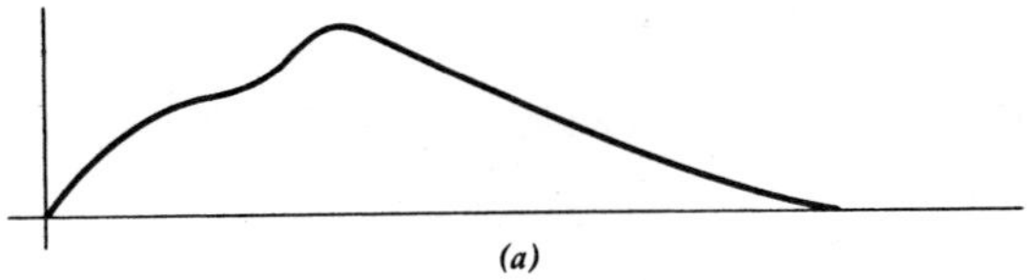

(a)

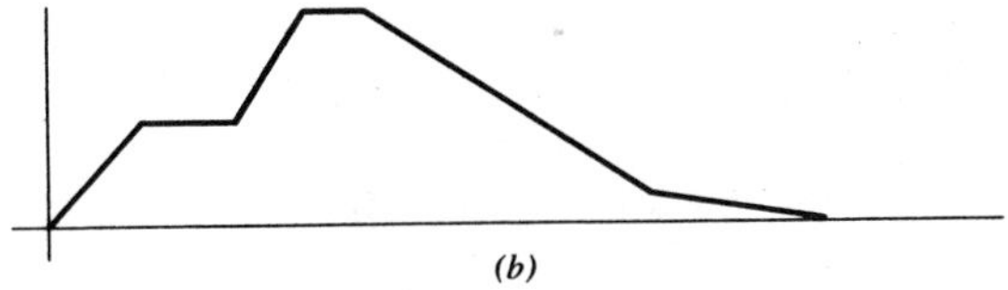

(b)

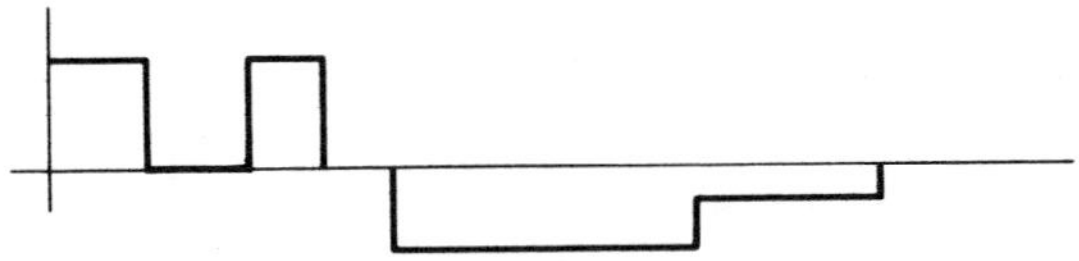

(c)

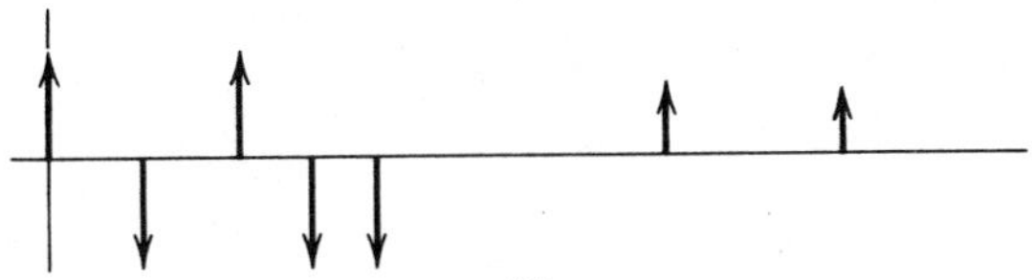

(d)

Fig. 6-6.1

study the Fourier transform in Chapter 10. The frequency plot thus obtained, is approximated in such a way that $F_2(s)$ is found. The output is again determined from this and the known input.

When the convolution integral is used, the output can be determined directly from the input and the graphical impulse response, using some form of numerical methods. Within the context of what we have been discussing, the following procedure is suggested. Figure 6-6.1(a) shows an impulse response that has been determined experimentally; (b) suggests an approximation of the curve of (a) by a set of straight lines; (c) shows the derivative of the curve of (b); and (d) shows the derivative of the curve of (c). In all likelihood, the input to be applied to the system is known analytically. The input function can be integrated two times and the result convolved with the impulse functions of (d) to obtain a fairly decent approximation of the output of the system.

The final point to be made is that there are many occasions when an exact answer is not required but only the general nature of the response of a system to some input is desired. This could occur in the early stages of some design procedure. It is possible to visualize folding the input curve and to observe the general nature of $f_3(t)$ from the product curve as the input curve is moved across the $f_2(t)$. This visualization may be all the information the designer needs to make some basic decisions.

PROBLEMS

6-1 The $f_3(t)$ in each part is to be found using methods 1, 2, and 3 as in Example 6-2.1.

(a) $f_1(t) = 2\epsilon^{-3t}\, u_{-1}(t)$ $\quad f_2(t) = 4\epsilon^{-6t}\, u_{-1}(t)$
(b) $f_1(t) = 3\epsilon^{-1t}\, u_{-1}(t)$ $\quad f_2(t) = 6\epsilon^{-4t}\, u_{-1}(t)$
(c) $f_1(t) = 6\epsilon^{-3t}\, u_{-1}(t)$ $\quad f_2(t) = 5\epsilon^{-4t}\, u_{-1}(t)$
(d) $f_1(t) = 4\epsilon^{-4t}\, u_{-1}(t)$ $\quad f_2(t) = 5\epsilon^{-5t}\, u_{-1}(t)$
(e) $f_1(t) = 2t\, \epsilon^{-3t}\, u_{-1}(t)$ $\quad f_2(t) = 4\epsilon^{-6t}\, u_{-1}(t)$
(f) $f_1(t) = 3t\, \epsilon^{-1t}\, u_{-1}(t)$ $\quad f_2(t) = 6\epsilon^{-4t}\, u_{-1}(t)$
(g) $f_1(t) = 6t\, \epsilon^{-3t}\, u_{-1}(t)$ $\quad f_2(t) = 5\epsilon^{-4t}\, u_{-1}(t)$
(h) $f_1(t) = 4t\, \epsilon^{-4t}\, u_{-1}(t)$ $\quad f_2(t) = 5\epsilon^{-5t}\, u_{-1}(t)$

6-2 Work Examples 6-2.2 and 6-2.3 using Laplace methods.

6-3 Rework Example 6-2.4 by folding $f_1(t)$.

6-4 Perform the convolution of the two time functions in a manner similar to that of Example 6-2.4; that is, find the equations for $f_3(t)$ in the appropriate intervals, and sketch $f_3(t)$. Fold the first of the two time functions. (See the sketches for items a to h.)

(a) $f_3(t) = f_a(t) * f_b(t)$
(b) $f_3(t) = f_b(t) * f_a(t)$
(c) $f_3(t) = f_c(t) * f_d(t)$
(d) $f_3(t) = f_d(t) * f_c(t)$
(e) $f_3(t) = f_d(t) * f_e(t)$
(f) $f_3(t) = f_e(t) * f_d(t)$
(g) $f_3(t) = f_f(t) * f_g(t)$
(h) $f_3(t) = f_g(t) * f_f(t)$

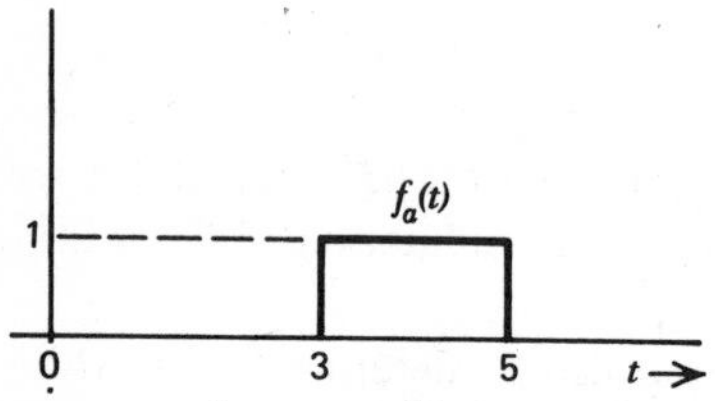

(*a*)

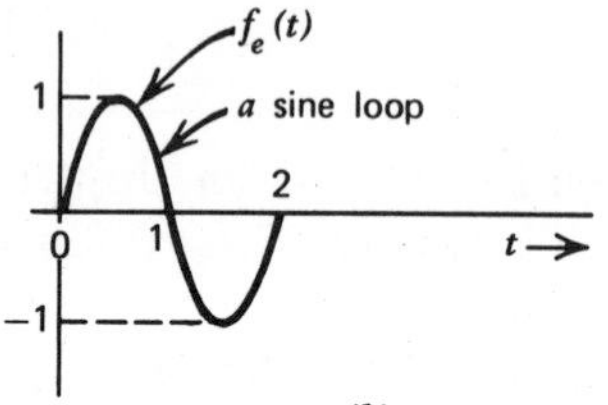

(*b*)

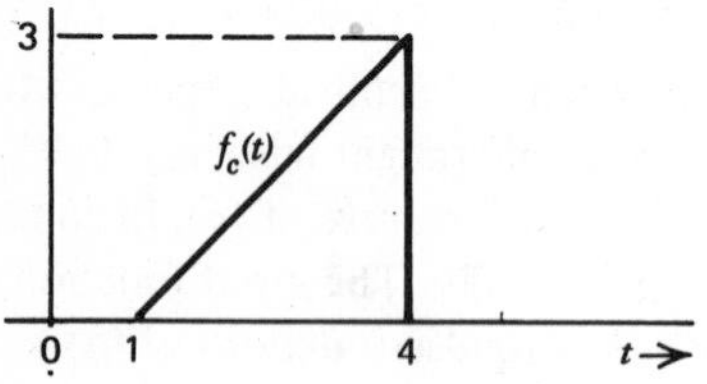

(*c*)

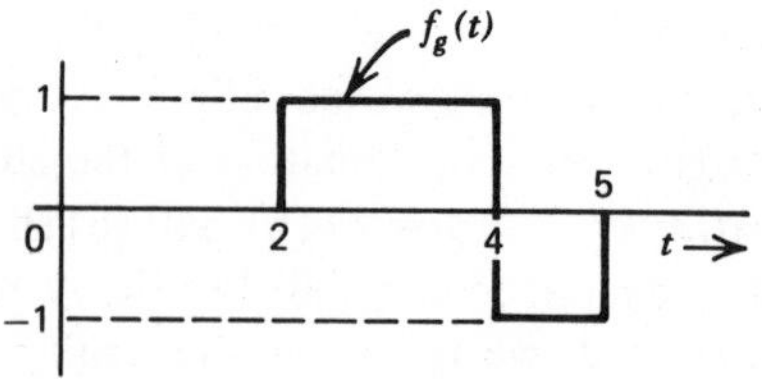

(*d*)

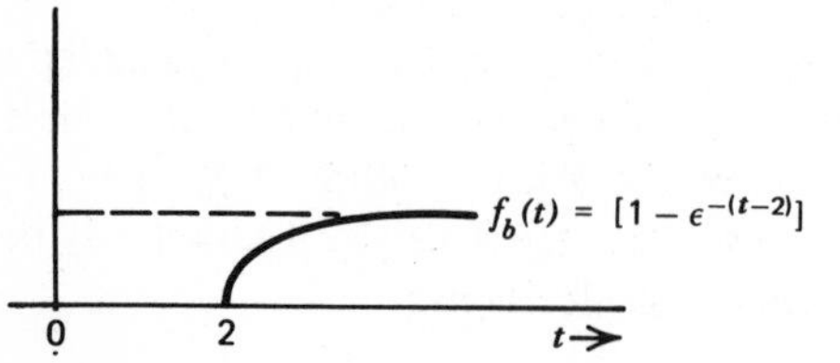

(*e*)

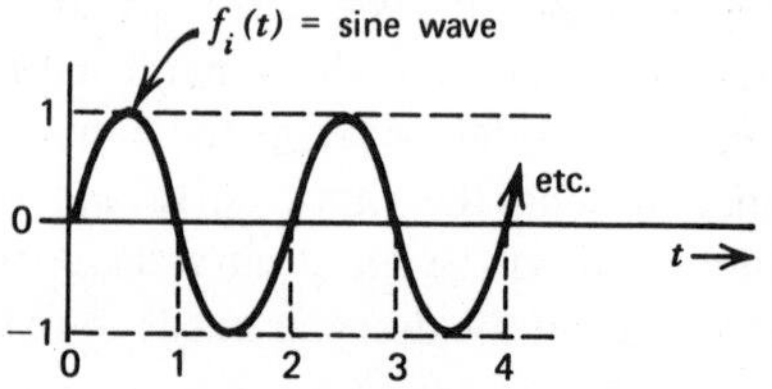

(*f*)

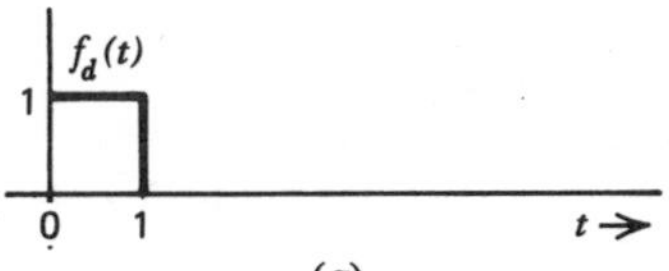

(g)

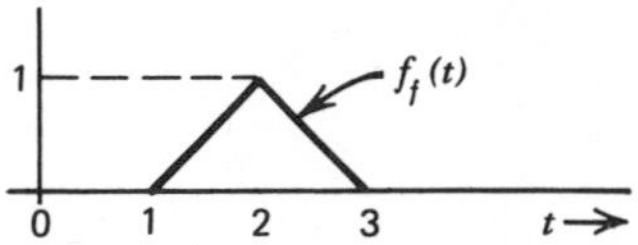

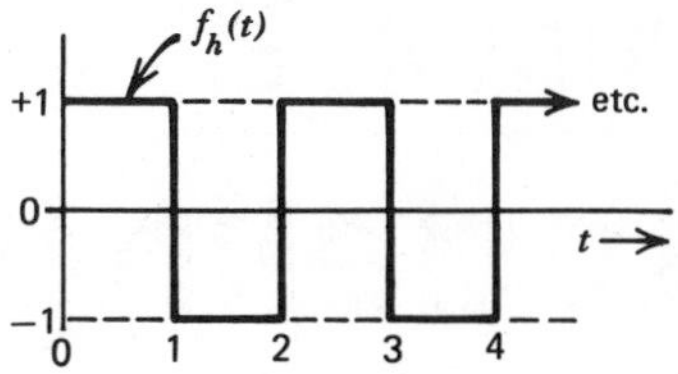

(h)

6-5 Carry out the details of Example 6–5.1, showing $f_3^A(t)$, $f_3^B(t)$, and $f_3(t) = f_3^A(t) + f_3^B(t)$.

6-6 Similar to Problem 6–5, except use Example 6–5.2.

6-7 Similar to Problem 6–5, except use Example 6–5.4.

6-8 Use the time functions of Problem 6–4. The first of the time functions given is to be built up with a set of step functions. Pattern your work after Example 6–5.1 to find $f_3(t)$.

(a) $f_3(t) = f_a(t) * f_b(t)$
(b) $f_3(t) = f_a(t) * f_e(t)$
(c) $f_3(t) = f_d(t) * f_c(t)$
(d) $f_3(t) = f_g(t) * f_f(t)$

6-9 The $f_3(t)$ in each part is to be found by the convolution of the derivative of one function with the integral of the other. The $f_1(t)$ and $f_2(t)$ in (a) refer to the functions in (a) of Problem 6–1 and (b) to (b) of Problem 6.1, etc.

(a) $f_3(t) = f_1^1(t) * f_2^{-1}(t)$
(b) $f_3(t) = f_2^1(t) * f_1^{-1}(t)$
(c) $f_3(t) = f_1^{-1}(t) * f_2^1(t)$
(d) $f_3(t) = f_1^{-1}(t) * f_2^1(t)$
(e) $f_3(t) = f_1^1(t) * f_2^{-1}(t)$
(f) $f_3(t) = f_1^1(t) * f_2^{-1}(t)$
(g) $f_3(t) = f_1^1(t) * f_2^{-1}(t)$
(h) $f_3(t) = f_1^1(t) * f_2^{-1}(t)$

6-10 This is a continuation of Problem 6–9, except the time functions referred to are those of Problem 6–4.

(a) $f_3(t) = f_a^1(t) * f_b^{-1}(t)$
(b) $f_3(t) = f_d^1(t) * f_c^{-1}(t)$
(c) $f_3(t) = f_d^1(t) * f_e^{-1}(t)$
(d) $f_3(t) = f_g^1(t) * f_f^{-1}(t)$
(e) $f_3(t) = f_d^1(t) * f_h^{-1}(t)$
(f) $f_3(t) = f_d^1(t) * f_i^{-1}(t)$
(g) $f_3(t) = f_a^1(t) * f_h^{-1}(t)$
(h) $f_3(t) = f_a^1(t) * f_i^{-1}(t)$
(i) $f_3(t) = f_g^1(t) * f_h^{-1}(t)$
(j) $f_3(t) = f_g^1(t) * f_i^{-1}(t)$
(k) $f_3(t) = f_h^1(t) * f_i^{-1}(t)$
(l) $f_3(t) = f_h^1(t) * f_e^{-1}(t)$

6-11 Develop sketches similar to those of Figure 6–2.2 except for $f_3(t) = \int_{-\infty}^{+\infty} f_2(t-\tau) f_1(\tau)\, d\tau$.

Chapter 7

Modeling of Mechanical and Electro-Mechanical Systems

7-1. INTRODUCTION

Underlying much of present-day so-called system analysis and design are certain fundamental methods of attack, certain common mathematical techniques, and certain common physical laws involving *conservation* or *preserving of the totality* concepts. Because of these commonality features, it has become more or less standard practice to build on these features in teaching systems analysis. This leads to greater efficiency in utilizing the student's and the teacher's time and in the amount of *memory space* required. It also allows an insight in one field to be applied in other fields. Certain aspects of the commonality mentioned above are introduced in this chapter as simplified models of nonelectrical systems are introduced. In particular, the similarity in the form of the differential equation models is of importance.

One of the more important simplifying assumptions made in this development is that of *lumping* effects or actions so that *lumped parameters* may be used. In effect, this amounts to the assumption that mechanical effects, such as mass, occur at points; that the mass of a spring can be lumped at one location and an otherwise massless spring considered; that electrical resistance of a piece of material can be considered as a single lumped effect and not distributed throughout the bulk of the material. Such assumptions allow us to develop mathematical models for our systems in the form of ordinary differential equations rather than partial differential equations. The space-distribution effect or variable is not present. This does not mean that space or distance does not enter the differential equations as a dependent variable describing the location of a point mass, because we most certainly will have such situations.

The similarity of form of the differential equations that model various systems leads to the possibility of utilizing various analogies or analogs. For some students the use of such analogs greatly simplifies the extension of previous knowledge to new systems. There is no question whether such analogies are valid, provided there is the required similarity of form of the defining equations, and conditions for unique solutions exist. Some attention is devoted to the development of analogies at various points in the following, but it is not made a major issue. If the reader finds it easier to develop the material using analogies, he should by all means pursue this route.

7-2. THROUGH AND ACROSS VARIABLES

It is presumed that the reader is familiar, from basic physics, with the fact that various physical phenomena are described by use of quantities, such as voltage, current, force, velocity, and temperature. In particular, such quantities are paired in some form of relationship in order to describe the behavior of various devices. For example, Ohm's Law relates voltage, current, and resistance for a lumped resistor. Thus, voltage and current are paired in the defining relationship, $e = Ri$ for a linear resistor. Here we have an example of the *lumping* effect and the representation of it with a lumped parameter resistance R. The pairing of voltage and current, however, leads to another concept. This is that our lumped element has two terminals associated with it, with the voltage measured *across* the element or terminals and the current measured as a quantity flowing *through* the element. Voltmeters are placed across or in parallel with the element for the desired *across* voltage, and ammeters are placed in series with the element to measure the *through* variable.

A similar development can be made for all of our other parameters and systems so that the *through* and *across* variable concepts can be made an underlying common feature. In general, each two-terminal element of the type mentioned in the resistor example above is defined by a single *through* variable and a single *across* variable related by the mathematical model for the two-terminal element. If only single lumped effects are considered between the two terminals, then the relationship is generally a simple algebraic, differential, or integral one. Attention will be devoted to the *through* and *across* variable concepts as we proceed.

7-3. CONSERVATION LAWS

It is somewhat natural to think of the *conservation* of matter when considering basic physical laws. In some sense this philosophy can be extended to conservation of quantity of material in developing the models that follow. We will need to account for all fluid, all charge, all quantities of a chemical, all quantities of heat, etc. Such a *conservation-of-*

quantity concept is particularly appropriate when thinking about or setting up equations in terms of the *through* quantities of the previous section. From this conservation concept we have Kirchhoff's Current Law, which states that the sum of all currents leaving a junction (or closed surface) must be zero. In other words, charge is not created or destroyed. A similar relationship or law can be written for fluids, although it may be necessary to add a term to account for compressibility effects. As various models are developed, attention is directed to the use of the conservation-of-quantity concept mentioned above.

Kirchhoff's Voltage Law, which states that the sum of voltages around a closed path is zero, is in essence a statement of another form of conservation concept. This *conservation-of-potential* concept can be extended to the *across* variables of the previous section. A directly analogous situation occurs in hydraulic systems when dealing with pressures. Use of the two conservation concepts described above is of major importance in the work that follows.

7-4. MECHANICAL SYSTEMS—IDEAL SPRINGS

Many physical systems involve mechanical elements in the form of masses, springs, and damper elements. The analysis of such systems is generally based on some form of lumped modeling of these components of the system. Simple versions of such systems utilize linear models, low velocities relative to the speed of light, and single-direction-of-motion constraints. Translation and rotation are generally treated separately, although it is certainly possible to have combined translation and rotation.

The linear spring is considered as the first mechanical element. In particular this is an idealized mechanical component, with no mass and with a deflection proportional to the force or torque transmitted through the element. If it is necessary to consider some mass, the spring is divided into segments and the mass lumped at appropriate points in the spring. Figure 7-4.1(a) and (b) show common symbols for the ideal spring, for translational and rotational systems, respectively.

In order to mathematically describe the ideal spring, it is necessary to define the *through* and *across* variables for this component. In particular, spring deflection and force are needed, and positive reference directions for these quantities must be specified. Specifying positive reference directions simply means stating the direction for force and displacement which will be considered numerically positive. This process is completely arbitrary. However, once these directions are specified, then the physics of the device determines the signs used in the various models.

For the positive reference directions in Fig. 7-4.1, with the displacement having its *origin at the end of the unstretched spring*, the mathematical models for the ideal spring with one end fixed are

$$F = K_s x \qquad T = K_s \theta \tag{7-4.1}$$

The linear relationships between force (torque) and displacement are shown in Fig. 7-4.1. It would be a simple matter to consider a non-linear relationship, especially in graphical

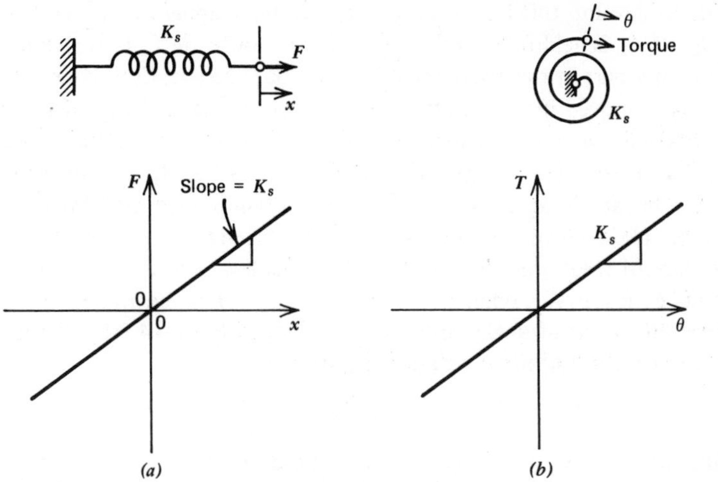

Fig. 7-4.1

form. However, we would not be able to use the simple spring constant, K_s, as describing the non-linear spring. It should be noted that the spring constant, K_s, is the slope of the force-displacement curve for the ideal spring. Such a fact can be used in experimentally determining the model for the ideal spring.

The association of *through* and *across* variables with the physical quantities force and displacement in this case are not quite as evident as in the cases mentioned in Section 7-2. It would seem most appropriate to associate the *across* variable with *displacement*, and the *through* variable with *force*. This follows from the fact that force is transmitted unaltered *through* the spring from one end to the other, whereas the displacement is measured between the two ends of the spring in an *across* fashion. Such an association is used here.

If both ends of the spring move with respect to some other fixed coordinate system, it is a simple matter to modify the mathematical models of Eqs. 7-4.1 to accommodate this change. The relative elongation or compression of the spring is the important factor. For positive reference directions, as shown in Fig. 7-4.2, the translation relation becomes

$$F = K_s\,(x_1 - x_2) \qquad (7\text{-}4.2)$$

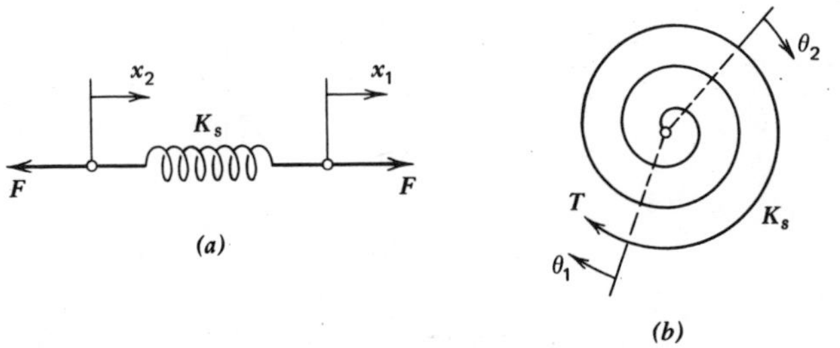

Fig. 7-4.2

It is possible to develop this relationship in a two-step manner as follows. If the displacement coordinate x_2 is held fixed at $x_2 = 0$, and the coordinate x_1 is moved in the positive direction, the first part of the above relationship is obtained as, $K_s x_1$. If x_1 is now held at $x_1 = 0$, and x_2 is moved in the positive x_2 direction, the second part of the relationship, $-K_s x_2$, is obtained. Since this element is linear, it is possible to combine these two parts in a superposition of effects to give the original Eq. 7–4.2. Such a superposition is not valid if a non-linear spring is used, as the relationship may well involve $(x_1 - x_2)^2$, $(x_1 - x_2)^3$, etc. These terms are drastically different from $(x_1^2 - x_2^2)$, etc.

For the case of rotational motion, the spring displacement is an angular displacement and the spring constant, K_s, relates an angle and torque. For the arbitrarily assigned positive reference directions of angles and torque shown in Fig. 4–4.2(b), the mathematical model of the linear ideal spring in rotational systems is

$$T = K_s (\theta_1 - \theta_2) \tag{7–4.3}$$

Superposition of motions of the ends of the spring could again be used as described above for the translational spring. It should be noted that the spring constant, K_s, for rotational systems has dimensions different from those for translational systems. For example, it would be possible to have rotational spring constants as newton-meters/radian while translational constants could be newtons/meter.

7-5. MECHANICAL SYSTEMS—FRICTION

A second lumped mechanical system component is that of viscous friction. This is a phenomena that relates the relative velocity of the two terminals of the element to the force transmitted through it. Such behavior is an idealization of various fluid piston dampers that force fluid through an orifice or depend upon the shearing force of a viscous fluid. It is at best an idealization that is much less realistic than the spring of the previous paragraphs. Figure 7–5.1 shows the symbol commonly used for the viscous friction element and also a graph indicating the physical relationship between velocity and force.

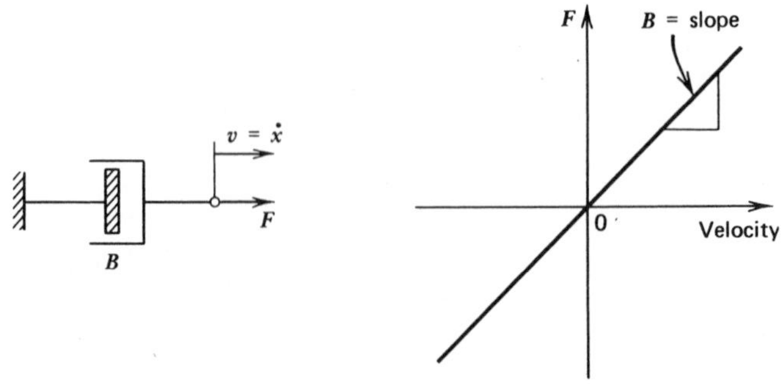

Fig. 7–5.1

In this case the mathematical model becomes

$$F = Bv = B\,dx/dt \tag{7-5.1}$$

for the positive assumed reference directions of velocity and force. These positive reference directions are arbitrary, however, and some other choice could have been used with a corresponding change in Eq. 7-5.1.

An extension of the *through* and *across* variable associations used for the spring is possible for viscous friction. Thus, the velocity becomes the *across* variable and force becomes the *through* variable. Force is transmitted unaltered through the viscous friction element. A force applied at one terminal of the viscous friction element is instantly transmitted to the other terminal. This idealization assumes no mass for the friction element.

A rotational viscous friction element is also useful with its mathematical model relating angular velocity and torque by using a viscous friction coefficient, B, as above. For both terminals of the element in motion relative to some coordinate system, the model for the rotational viscous friction element is

$$T = B(\dot{\theta}_1 - \dot{\theta}_2) \tag{7-5.2}$$

It might be noted that the origin, or zero location, for the displacement coordinates used in the model for viscous friction is immaterial. Only the velocity or derivative of position is involved. A constant can be added to the position coordinates without effecting the velocity.

Two other commonly encountered forms of friction in mechanical systems are coulomb friction and stiction. *Coulomb friction* is encountered when most bodies slide relative to each other and is a constant force opposing motion in any direction, as shown in Fig. 7-5.2(a). It can be represented mathematically as

$$F = f_c v/|v| = f_c \operatorname{sign} v \tag{7-5.3}$$

In general, the force coefficient f_c is proportional to the force normal to the sliding surface. *Stiction* is a frictional force phenomena that exists only for zero velocity and opposes the initiation of motion. This effect is shown in Fig. 7-5.2(b).

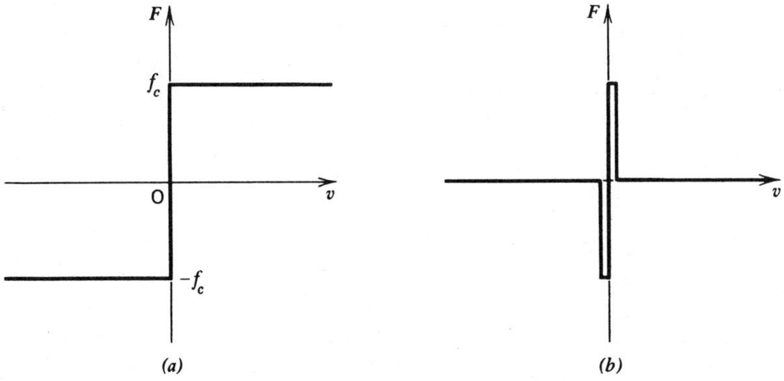

Fig. 7-5.2

Generally the total frictional effect is a combination of the various types mentioned above together with other parts that depend upon higher powers of the velocity. However, it is common to consider only viscous friction in a first analysis, because it is the only linear type of friction and therefore lends itself to simplified analysis.

7-6. MECHANICAL SYSTEMS—MASS AND INERTIA

The final mechanical system element to be introduced before considering overall mechanical systems is that of mass for translational systems, or mass moment of inertia for rotational systems. This element is basically a parameter involved in relating acceleration to force(torque) or in defining and relating momentum to force(torque). The relationship involved in the mathematical model is obtained from Newton's Second Law as the time rate of change of momentum equals the algebraic sum of the external forces or

$$F = d(Mv)/dt = Ma = M\, d^2x/dt^2 \tag{7-6.1}$$

where mass M can only be taken outside the derivative if M is constant.

It is also possible to consider that any mass M requires an accelerating force $d(Mv)/dt$ to establish an acceleration dv/dt. Therefore this equivalent force can be used in a summation of force relationship in developing overall system equations. The difference between this *reaction* force and the F force term in Eq. 7-6.1 should be noted. The F term in Eq. 7-6.1 is an algebraic sum of all *external* forces acting on the mass and not the internal inertial or reaction force due to an acceleration of the mass.

Again attention must be given to positive reference directions for displacement or acceleration and force. Equation 7-6.1 is written for the positive reference directions shown in Fig. 7-6.1(a) and indicates that a numerically positive force, F, tends to cause positive accelerations or motion in the positive displacement direction.

For rotational systems the mathematical model involves the polar mass moment of inertia, J, from mechanics and leads to

$$T = J\, d^2\theta/dt^2 \tag{7-6.2}$$

for the positive reference directions in Fig. 7-6.1(b).

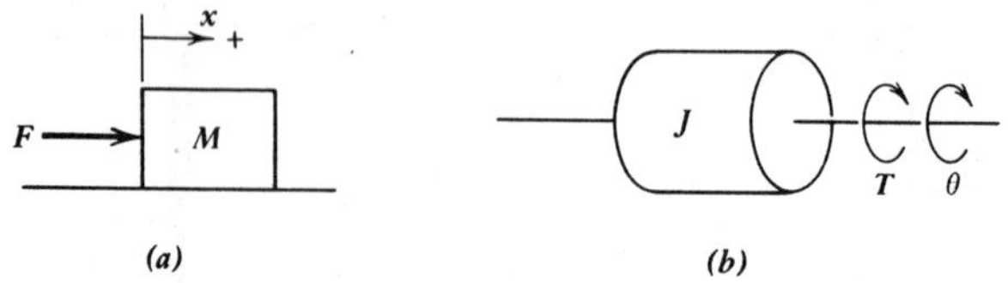

Fig. 7-6.1

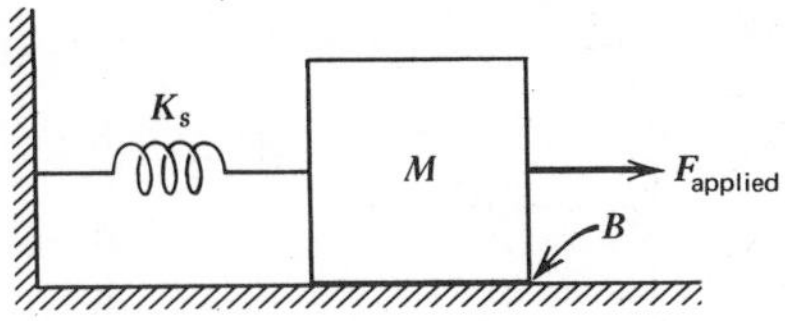

Fig. 7-7.1

7-7. MECHANICAL SYSTEMS–GENERAL

Realistic mechanical systems involve combinations of the mechanical system parameters described previously. The most classical of such systems consists of the spring-mass-viscous friction system as depicted in Fig. 7-7.1.

In general, it is desired to determine the dynamic behavior of such a system, or to determine the position, velocity, and acceleration of the mass when subjected to some form of applied forcing function and/or combination of initial conditions. Because the system is linear, it is possible to consider forcing functions and initial conditions separately and then superimpose the results to obtain a total response for both forms of excitation.

As a first step in developing the mathematical model for the whole mechanical system, it is necessary to determine the number of variables needed to define the motion of the system and to choose the positive reference directions for the *through* and *across* variables. For the example being considered, only one position variable is needed to determine the displacement of the spring, the relative velocity of the terminals of the viscous friction element, and the position of the mass, presuming that motion parallel to only one coordinate direction is allowed. Vertical motion is not possible, because of the rigid base upon which sliding occurs and the assumption that the mass will not rise off of this sliding surface.

Figure 7-7.2(a) shows a choice of the positive reference direction for the one position coordinate or *across* variable needed. The origin for x is assumed to be at the position of the mass with the spring unstretched. Figure 7-7.2(b) shows the various individual mechanical components isolated with a corresponding choice for the positive reference directions for the forces or *through* variables transmitted by the spring and viscous friction element and the reaction forces on the mass.

Figure 7-7.2(b) for the mass separately is a *free-body* diagram showing all of the external forces acting on the mass. It might be noted that the spring and viscous friction forces are transmitted through these elements from the wall, or fixed structure, to the mass.

Several approaches can now be used to combine the various relationships into a single differential equation whose solution gives the required dynamical behavior. The method used first is that of applying Newton's Second Law as described previously. This involves equating the algebraic sum of external forces acting on the mass to the time rate of change of momentum. Also, as mentioned before, forces tending to cause motion in the assumed positive x reference direction are taken as positive in the force summation. For this ap-

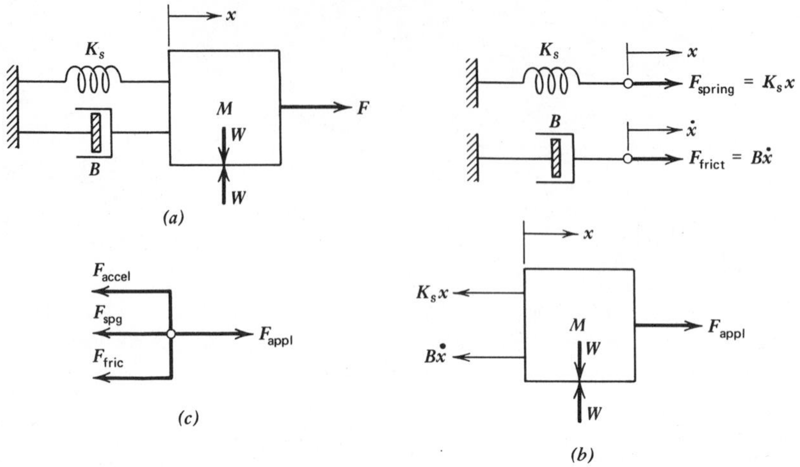

Fig. 7-7.2

proach we have

$$\begin{pmatrix}\text{Rate of change}\\ \text{of momentum}\end{pmatrix} = \begin{pmatrix}\text{Applied}\\ \text{force}\end{pmatrix} - \begin{pmatrix}\text{Spring}\\ \text{force}\end{pmatrix} - \begin{pmatrix}\text{Viscous friction}\\ \text{force}\end{pmatrix} \tag{7-7.1}$$

$$M\, d^2x/dt^2 = F(t) - K_s x - B\, dx/dt \tag{7-7.2}$$

A simple rearrangement of terms in Eq. 7-7.2 gives the differential equation to be solved to obtain the position of the mass as a function of the independent variable time.

$$M\ddot{x} + B\dot{x} + K_s x = F(t) \tag{7-7.3}$$

It might be noted that the rate of change of momentum is really an accelerating force and could be considered as such in an alternate method of formulating the differential equation. This method is that of summing all *through* variables at a position coordinate and setting them equal to zero in a manner analogous to Kirchhoff's Current Law in circuits. By treating the accelerating force, $M\, d^2x/dt^2$, as opposing motion, we find that all forces except the applied force oppose motion and the summation of force equation is

$$F_{\text{appl.}} - F_{\text{accel.}} - F_{\text{spr.}} - F_{\text{vis. fric.}} = 0 \tag{7-7.4}$$

Substitution of the individual *through–across* variable relationships for each element also leads to the result of Eq. 7-7.3.

Example 7-7.1. It is desired to study in detail the mechanical system previously described, with the mechanical parameters having the values

$$M = W/g = 1 \text{ kilogram} \qquad B = 1 \text{ newton/(m/sec)}$$

$$K_s = 1 \text{ newton/meter} \qquad F_{\text{appl.}} = \begin{cases} 0.2 \text{ newtons}, & t \geqslant 0 \\ 0 \text{ newtons}, & t < 0 \end{cases}$$

$$x(0) = -0.1 \text{ meter} \qquad dx/dt|_{t=0} = 0.1 \text{ meter/sec}$$

For the above numerical values, Eq. 7-7.3 becomes

$$d^2x/dt^2 + dx/dt + x = 0.2 \qquad t \geqslant 0$$
$$x(0) = -0.1 \qquad \dot{x}(0) = 0.1 \tag{7-7.5}$$

This equation can now be Laplace transformed and $X(s)$ solved for, as in

$$[s^2 X(s) - s\,x(0) - \dot{x}(0)] + [s\,X(s) - x(0)] + X(s) = 0.2/s \tag{7-7.6}$$

$$X(s) = \frac{0.2/s}{s^2 + s + 1} + \frac{(s+1)\,x(0)}{s^2 + s + 1} + \frac{\dot{x}(0)}{s^2 + s + 1} \tag{7-7.7}$$

The poles of the system response due to the mechanical system itself are found by factoring $(s^2 + s + 1)$ to determine the values $s = -1/2 \pm j\sqrt{3}/2$ with the complex nature of these poles indicating a decaying oscillatory response. The arrangement of Eq. 7-7.7 indicates that the various initial condition and forcing function parts can be separated and studied individually.

Equations 7-7.8, 9, 10, and 11 are the results of inverting each part of the $X(s)$ expression separately, with $x(0)$ and $\dot{x}(0)$ values substituted, and then combining all of the terms to give the total response for $t > 0$.

$$x(t)|_{x(0)} = (-0.2/\sqrt{3})\,\epsilon^{-t/2} \sin(\sqrt{3}\,t/2 + 60°) \tag{7-7.8}$$

$$x(t)|_{\dot{x}(0)} = (0.2/\sqrt{3})\,\epsilon^{-t/2} \sin(\sqrt{3}\,t/2) \tag{7-7.9}$$

$$x(t)_{\text{forced}} = 0.2 - (0.4/\sqrt{3})\,\epsilon^{-t/2} \sin(\sqrt{3}\,t/2 + 60°) \tag{7-7.10}$$

$$x(t) = 0.2 - (0.1)(\sqrt{28/3})\epsilon^{-t/2} \sin(\sqrt{3}\,t/2 + \tan^{-1} 3\sqrt{3}) \tag{7-7.11}$$

Figures 7-7.3 and 4 illustrate the dynamic responses corresponding to the four cases of Eqs. 7-7.8, 9, 10, and 11 above. It is of interest to consider a few features shown in the figures. The initial values of x and $\dot{x}$ can easily be checked at $t = 0$. The final values can

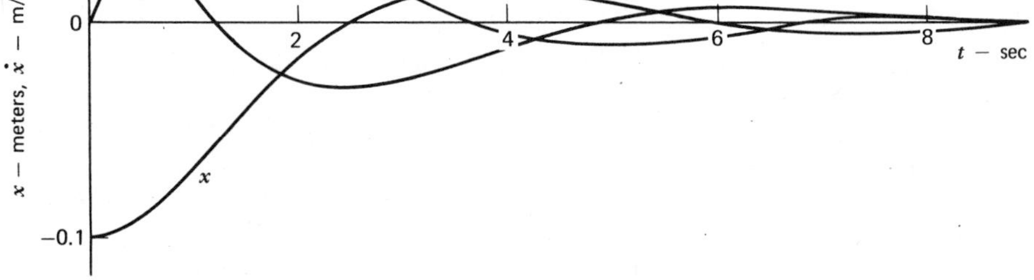

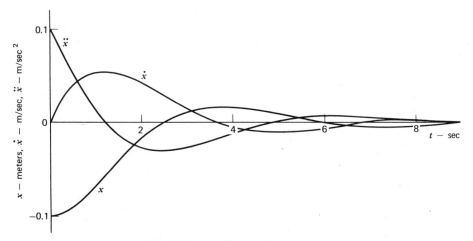

Fig. 7-7.3(a). Initial position response.

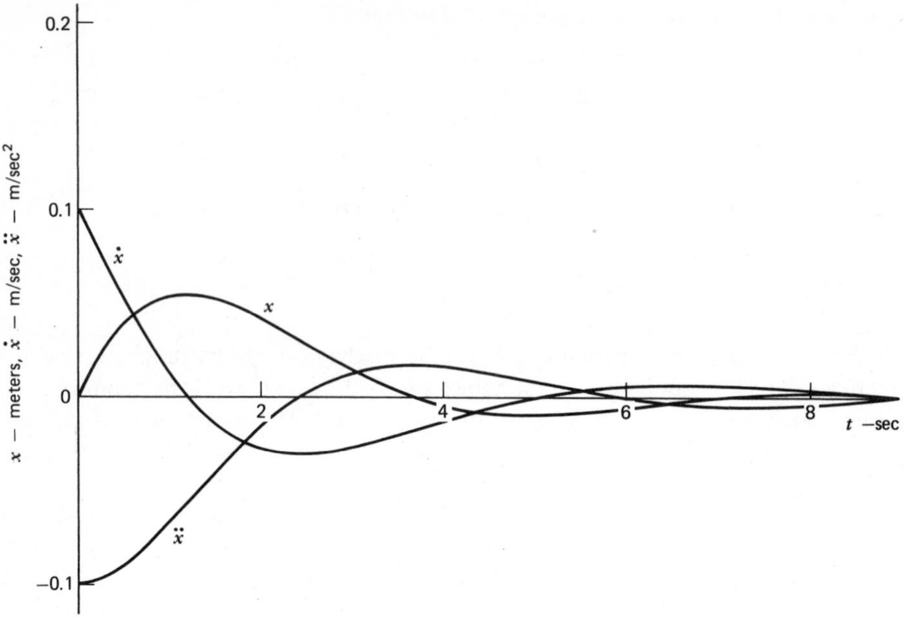

Fig. 7-7.3(b). Initial velocity response.

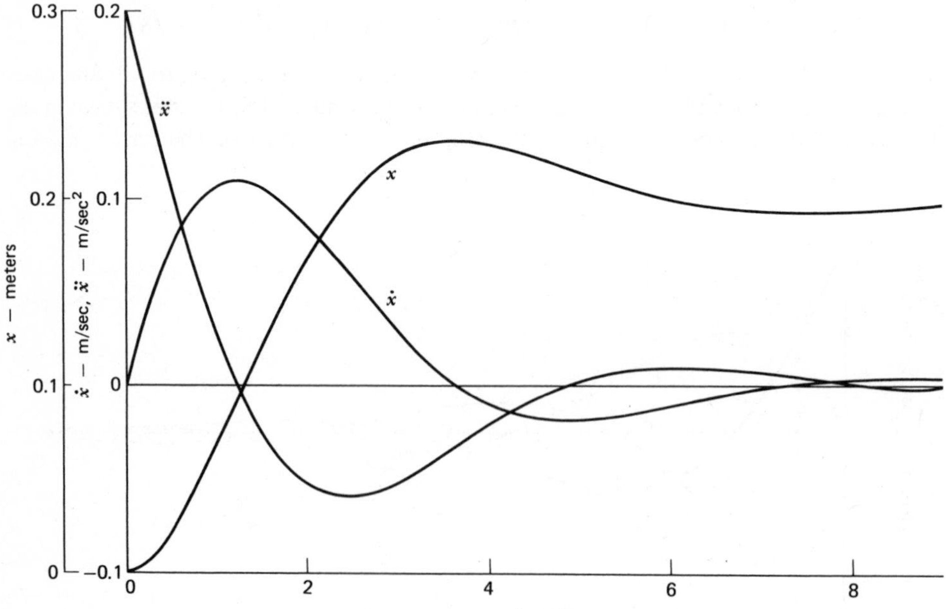

Fig. 7-7.3(c). Forced response.

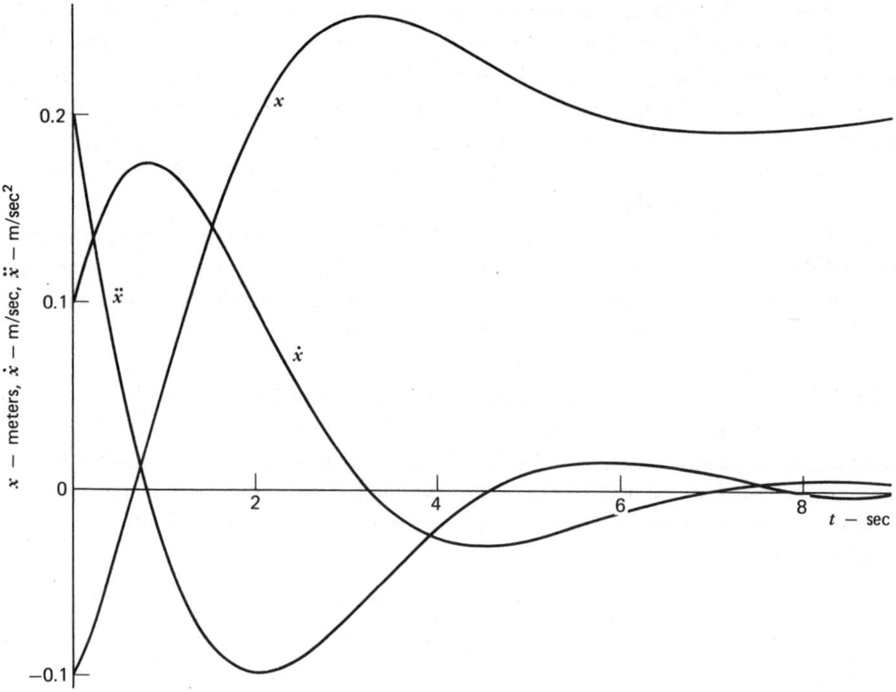

Fig. 7-7.4. Total response.

also be checked for large time as the system settles into a new equilibrium condition with all rates of change equal to zero. The final total position value of 0.2 meters can also be checked by using either the final value theorem of Laplace transforms and Eq. 7-7.7, or by setting all derivative terms equal to zero in Eq. 7-7.5. The period of the oscillation can be checked by determining an average period in any of the figures. From the fact that the imaginary part of the system poles is $\sqrt{3}/2$ rad/sec, the period should be $4\pi/\sqrt{3} = 7.25520$ seconds.

Since the spring force is $K_s x$, and $K_s = 1$, the graph of $x(t)$ is also a graph of the spring force for the assumed reference direction. When this curve is greater than zero, the spring is stretched and pulls back on the mass. This occurs for all positive x values. When the $x(t)$ curve is negative, the spring is compressed and pushes the mass in the positive x direction.

The viscous friction element transmits a force $B\dot{x}$ or simply dx/dt in this case. Thus, the graph of velocity is also a graph of the force transmitted by the viscous friction element. It should be noted that this force is positive when the velocity is positive, and pulls back on the mass for such positive velocity. When the velocity becomes negative, and the mass is moving in the negative x direction (its position may still be positive), the friction force is also negative and acts physically to oppose this motion. Thus, the viscous friction force is always in a direction whereby it opposes the motion of the mass.

The rate of change of momentum of a mass is non-zero only when there is acceleration. For $M = 1$, the plot of the acceleration thus gives a plot of the inertial force. The part of the constant 0.2-newton applied force that goes into acceleration is given by the plot of the acceleration curve. As acceleration increases, position and velocity become non-zero and part of the applied force goes into overcoming spring and friction forces. In the final steady state, an equilibrium is reached with no acceleration or velocity; the total applied force has matched the spring force, and the spring on its part has stretched to offer a force equal to the applied force.

Detailed consideration of the above problem for different choices of positive reference directions is instructive in gaining a real understanding of the mechanism of arbitrary choice of reference directions. The shape of the solutions remain the same with only possible reversals of signs. Such a study is left as problem work.

Example 7-7.2. Figure 7-7.5 illustrates a second mechanical system to be studied. Here torque is applied to a flywheel that is coupled to a fixed structure through a viscous clutch. Figure 7-7.5(b) illustrates a choice for the one angle necessary to define the motion of the system and also a choice of positive reference directions for torques.

Newton's Second Law can again be applied to obtain the overall mathematical model for the system, as

$$J\, d^2\theta/dt^2 = T_{\text{appl.}} - B\, d\theta/dt \tag{7-7.12}$$

For the numerical values of J, B and $T_{\text{appl.}}$, Eq. 7-7.12 can be solved by Laplace transform methods as

$$0.003\ddot{\theta} + 0.3\dot{\theta} = 0.3 \quad (t \geqslant 0) \qquad \theta(0) = 0 = \dot{\theta}(0) \tag{7-7.13}$$

$$0.003s^2\,\Theta(s) + 0.3s\,\Theta(s) = 0.3/s \tag{7-7.14}$$

$$\Theta(s) = \frac{0.3}{s\,(0.003s^2 + 0.3s)} = \frac{100}{s^2\,(s + 100)} \tag{7-7.15}$$

$$\theta(t) = t - 0.01 + 0.01\epsilon^{-100t} \tag{7-7.16}$$

This solution indicates that the position of the flywheel increases as a ramp function with a 0.01-radian lag. The constant applied torque initially all goes into angular acceleration of the flywheel. As the flywheel begins to achieve a non-zero velocity, part of the

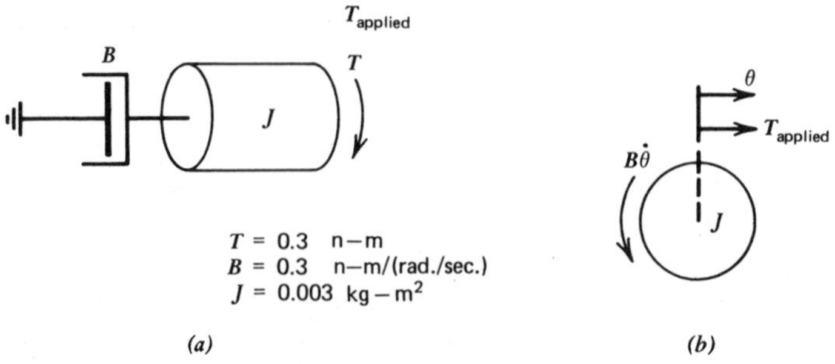

Fig. 7-7.5

applied torque goes to overcome the torque required to drive the viscous damper. A new equilibrium is reached, with the flywheel turning at a constant speed of 1 rad/sec, all of the applied torque going to supply the torque transmitted by the viscous damper and none left for acceleration.

Example 7-7.3. As a final example of a purely mechanical system, Fig. 7-7.6 is considered. In this system, an inner cylinder is rigidly coupled to the main shaft, and an outer-sleeve cylinder can turn relative to the inner cylinder but is coupled to it by viscous friction and a spring. It is desired to formulate the differential equations required to solve for the dynamic behavior of this system.

To specify the behavior of the system under consideration, two angular position coordinates are needed, one for the inner cylinder and another for the outer cylinder. Figure 7-7.7 illustrates a choice of positive reference directions for two such angles, each cylinder is isolated as a free body showing the torques acting on it. For this case Newton's Second Law results in the two rate-change of momentum equations

$$J_1\ddot{\theta}_1 = T_{\text{appl.}} - B(\dot{\theta}_1 - \dot{\theta}_2) - K_s(\theta_1 - \theta_2) \qquad (7\text{-}7.17)$$

$$J_2\ddot{\theta}_2 = B(\dot{\theta}_1 - \dot{\theta}_2) + K_s(\theta_1 - \theta_2) \qquad (7\text{-}7.18)$$

It is possible to transform these two equations and solve for both $\Theta_1(s)$ and $\Theta_2(s)$; this is left as an exercise to be pursued by the reader.

The various torques acting on the two cylinders in this system could have been developed one at a time, first holding one cylinder fixed at its zero position and moving the other cylinder in its positive angular direction, and then reversing the process with the other cylinder clamped. The torque components from the two motions are then added. This superposition of motion process is useful when in doubt about forces or torques in a linear system.

Use of the basic equations for the mechanical parameters K_s, B, M, and J, together with Newton's Second Law or the algebraic summation of *through* variables, allows one to

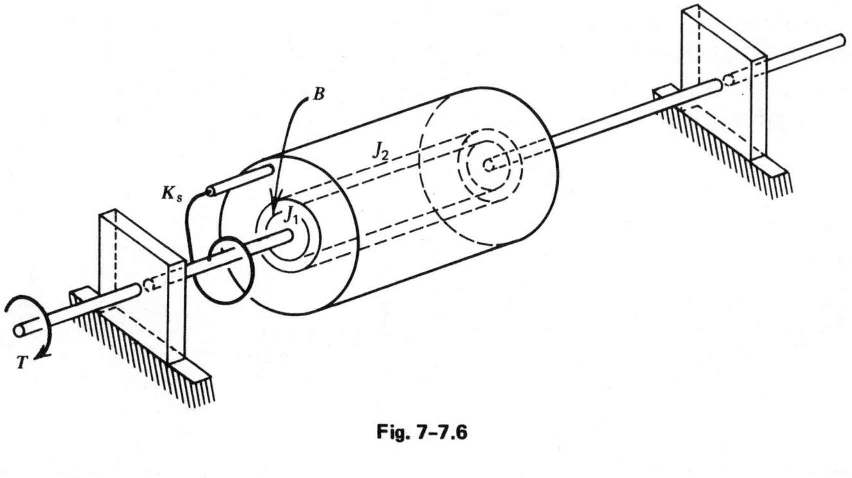

Fig. 7-7.6

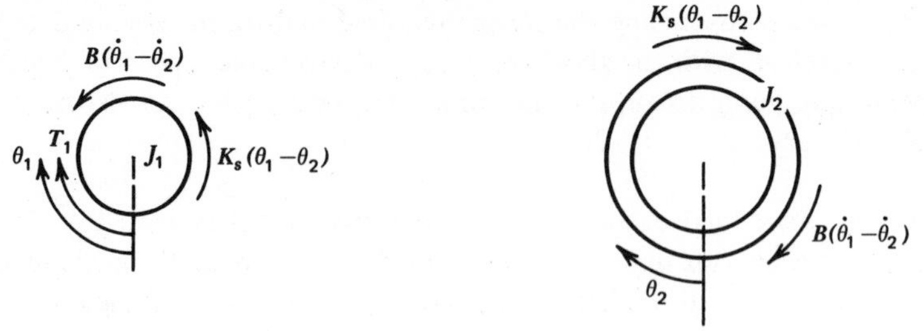

Fig. 7-7.7

solve many mechanical-system problems. As illustrated in the preceding examples, the mathematical models of overall systems of linear, lumped-parameter, mechanical components are generally ordinary differential equations.

7-8. GEAR TRAINS AND LEVERS

Devices for transforming torque or force are essential in many mechanical systems. The simplest of these devices is the lever, as depicted in Fig. 7-8.1(a). For small displacements, the device can be considered linear and the geometry of the device is pictured in (b). This ignores the fact that each lever arm rotates at a constant length from the fulcrum.

For this case the relationships between the forces and displacements are

$$F_1 L_1 = F_2 L_2 \tag{7-8.1}$$

$$\frac{x_1}{x_2} = \frac{L_1}{L_2} = \frac{F_2}{F_1} \tag{7-8.2}$$

where Eq. 7-8.1 expresses a balance of torques and Eq. 7-8.2 simply extends this using similar-triangle relationships.

Rotational systems utilize gear trains for transforming torque and angular displacement or speed. Perhaps the most commonly encountered need for gear trains is for speed reduction when using motors. In any event, Fig. 7-8.2 illustrates a typical stage in an

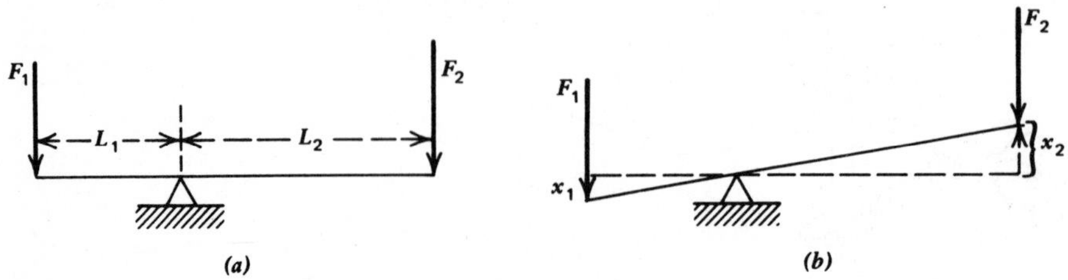

Fig. 7-8.1

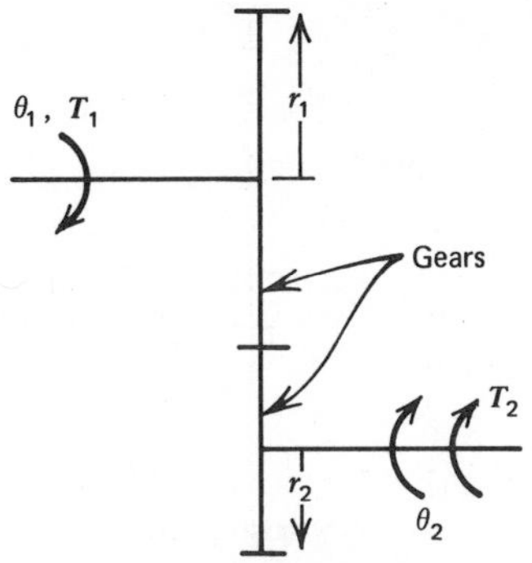

Fig. 7-8.2

idealized gear train that has no inertia or friction. By using the fact that there is no slippage, the following geometrical relationship can be written.

$$r_1\theta_1 = r_2\theta_2 \tag{7-8.3}$$

At the point of contact of gears a common force acts to result in a torque r_1F for *gear 1*, and r_2F for *gear 2*. From equilibrium considerations these torques have to equal the torques at the shafts of the gear train, as no inertia or friction was assumed for the gears. Thus,

$$T_1 = r_1F \qquad T_2 = r_2F \tag{7-8.4}$$

$$\frac{T_1}{T_2} = \frac{r_1}{r_2} = \frac{1}{n} \tag{7-8.5}$$

where n is a gear ratio.

From Eq. 7-8.3 it is also possible to bring in the angle or speed ratio as in

$$\frac{T_1}{T_2} = \frac{r_1}{r_2} = \frac{\theta_2}{\theta_1} = \frac{\dot{\theta}_2}{\dot{\theta}_1} = \frac{1}{n} \tag{7-8.6}$$

Several facts indicated by the above ratios should be investigated. If the gear ratio n is greater than one, then $\dot{\theta}_2$ is less than $\dot{\theta}_1$ and a speed reduction results. This also requires that $T_2 = nT_1$, with T_2 greater than T_1, for a torque multiplication. Consideration of the conservation of energy or equality of work at the two shafts indicates that this must be the case. The smaller angular change θ_2 *times* a larger torque T_2 is equal to the larger angular change θ_1 *times* a smaller torque T_1.

The torques on the two gear-train shafts can be from many different sources. In particular, let us suppose that T_2 is a torque being transmitted through a viscous clutch as shown in Fig. 7-8.3 and is $B_2\dot{\theta}_2$. On the *number-1* side, this represents a torque

$$T_1 = T_2/n = B_2\dot{\theta}_2/n \tag{7-8.7}$$

It is helpful to express this torque in terms of the angular velocity $\dot{\theta}_1$, as

$$T_1 = B_2\dot{\theta}_1/n^2 = (B_2/n^2)\dot{\theta}_1 \tag{7-8.8}$$

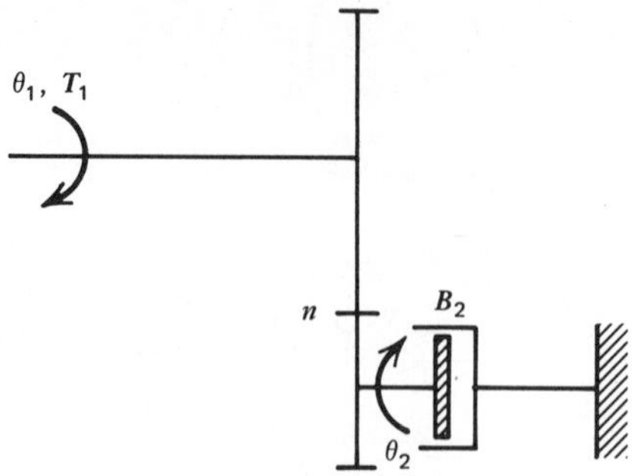

Fig. 7-8.3

The net effect of the coupling of this viscous friction through the gear train is to cause it to behave at the *number-1 shaft* as a new equivalent viscous clutch, with viscous friction coefficient B_2/n^2. We can thus transfer such coefficients through gear trains by using the square of the gear ratio in an appropriate manner. A similar relationship can be developed for spring constants and inertias.

Example 7-8.1. Figure 7-8.4 depicts a system whose dynamic behavior is desired. The problem is approached by transferring the load inertia and viscous friction to the driving *number-1 shaft*. The inertia transfers as J_2/n^2, while the viscous friction transfers as B_2/n^2. These transferred elements result in torques that are added to those caused by J_1 and B_1. The net result is a total equivalent inertia of $J_1 + J_2/n^2$, and an equivalent viscous friction coefficient of $B_1 + B_2/n^2$. The applied torque, $T_{\text{appl.}}$, can now be considered as driving an equivalent load made up of the total equivalent inertia and friction.

$$(J_1 + J_2/n^2)\,\ddot{\theta}_1 + (B_1 + B_2/n^2)\,\dot{\theta}_1 = T_{\text{appl.}} \tag{7-8.9}$$

Again, this equation can be Laplace transformed and solved for $\theta_1(t)$ or $\dot{\theta}_1(t)$. From this, the load angle θ_2 or velocity $\dot{\theta}_2$ can be found by dividing by the gear ratio n.

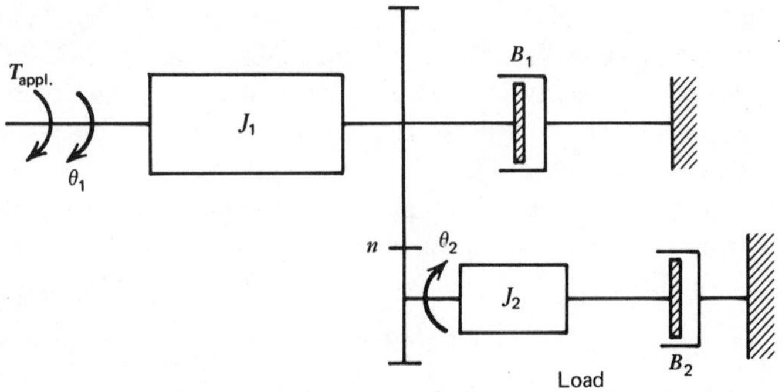

Fig. 7-8.4

7-9. ELECTRO-MECHANICAL SYSTEMS

The coupling of electrical and mechanical forms of system components is an essential technique in solving many systems engineering problems. Two basic mechanisms are involved in the coupling. The first of these is the generation of a voltage in a circuit when the magnetic flux linkages of this circuit change due to either a conductor moving in a magnetic field, or a time varying magnetic field linking a circuit. The second basic mechanism linking electrical and mechanical systems is that of a force acting on an electrical conductor carrying current in a magnetic field. While these two phenomena depend upon a number of parameters, simple mathematical models can be developed by considering only the essential variables and lumping all other factors into one overall constant.

Consideration of the generated voltage phenomena leads to the conclusion that the value of the voltage depends on the strength of the magnetic field and on the rate at which the magnetic flux linkages change. Certainly other factors, involving the number of conductors linked, the angle of the conductors relative to the direction of the magnetic field, the material in which the circuit is embedded, etc., affect the value of the generated voltage. It frequently happens, however, that all of these items are constant for a given situation and that what we are really interested in is the generated voltage, as related to the speed with which the conductors move through a magnetic field together with the density of the magnetic field. This is the situation in both d-c generators and d-c motors. For such situations, we can write

$$e_{\text{gen.}} = K_g \text{ (Speed) (Magnetic field density)} \tag{7-9.1}$$

where K_g lumps together numerous effects, as noted above.

Fig. 7-9.1 shows a commonly encountered representation of a d-c generator in which the magnetic field is established by current in a generator field circuit, the rotor or armature is driven by an external prime mover, and the generated voltage is connected to an external circuit at the armature terminals. If the speed is maintained constant, it also can be lumped with K_g in Eq. 7-9.1. Additionally, the density of the magnetic field linking the armature conductors is directly proportional to the magnitude of the generator field current. By lumping this proportionality factor with all of the other constants, Eq.

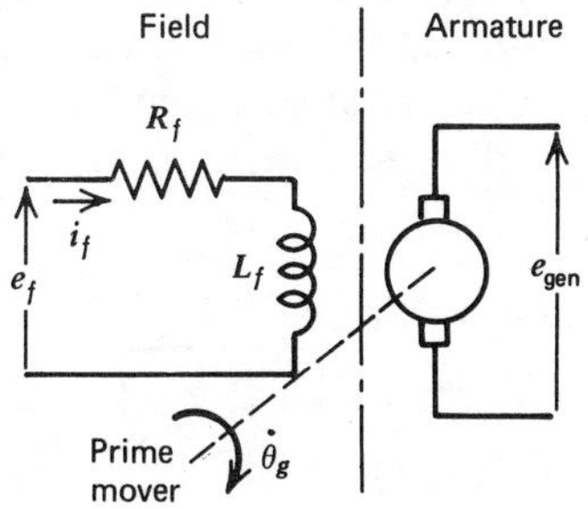

Fig. 7-9.1

7-9.2 can be used as a linear approximation, relating generator field current as input, to generated voltage as output.

$$e_{gen.} = K_{gt}\, i_f \tag{7-9.2}$$

A number of qualifications concerning this relationship must be stated. Because of the magnetic material used in the generator, a linear relationship between field current and magnetic field density is a tenuous one. Saturation and hysteresis effects certainly exist. The strength of the magnetic field is also affected by the amount of current supplied to the external circuit connected to the armature circuit. This effect has been totally ignored. Nevertheless, it exists and may tend to oppose the magnetizing effect of the field unless the machine is perfectly *compensated*.

It is also a rather gross approximation to consider the speed of the armature as constant if varying currents are drawn from the armature circuit. Such currents flow in conductors in a magnetic field and cause a torque to exist, thereby mechanically loading the prime mover. This variable loading may very well affect the prime-mover speed. If such is the case, then the nonlinear product relationship

$$e_{gen.} = K_{gn}\, \dot{\theta}_g\, i_f \tag{7-9.3}$$

is necessary relating the three variables, $e_{gen.}$, i_f, and $\dot{\theta}_g$. This complication indicates the desirability of using Eq. 7-9.2, if at all reasonable.

A final point to be noted is that the field of the generator is frequently driven by a voltage source, so the relationship between this voltage and the field current i_f must also be considered as a part of the model of the idealized d-c generator. For the simple circuit shown in Fig. 7-9.1, a single loop circuit equation, as follows, suffices.

$$L_f\, di_f/dt + R_f\, i_f = e_f \tag{7-9.4}$$

This equation, together with Eq. 7-9.2, can be Laplace transformed, and a block diagram with transfer functions developed as shown in Fig. 7-9.2. This idealized system is characterized by a single pole, $s = -R_f/L_f$, on the negative real axis of the s-plane, indicating a decaying, real exponential transient response mode due to the generator.

If the arrangement of field and armature excitation of the generator described previously is modified, it is possible to obtain d-c motor action. In this configuration, as shown in Fig. 7-9.3, both the field and the armature circuits are electrically excited. Use is made of the torque produced when the armature current flows in conductors linked by the magnetic field established by the field current of the device. This torque is then used to drive a mechanical load.

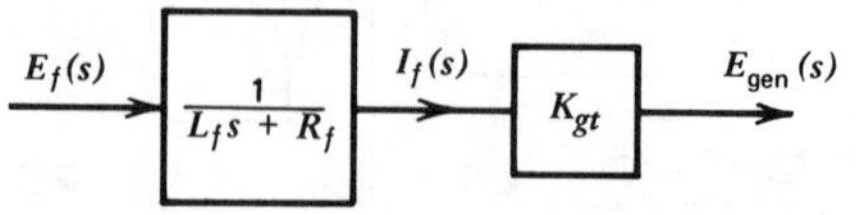

Fig. 7-9.2

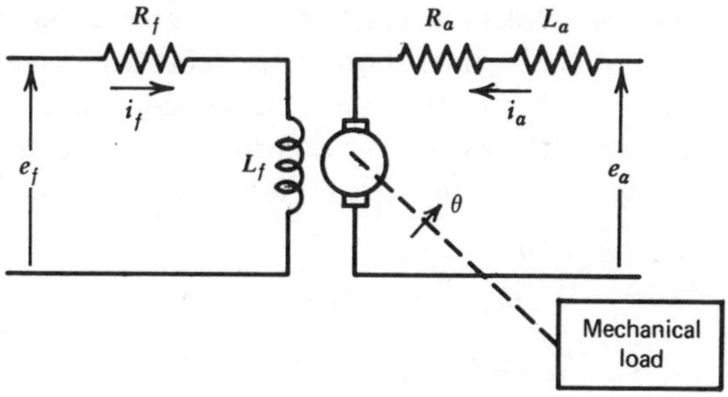

Fig. 7-9.3

The key factors in attacking the modeling of the d-c motor are:

1 the magnetic field strength linking the armature is proportional to the field current of the motor.

2 the torque produced on the armature is proportional to the magnetic field strength and the current flowing in the armature conductors;

3 As the armature conductors move, a generated voltage is produced in them just as in the d-c generator described above.

These factors can be combined into the mathematical relations

$$e_{\text{gen.}} = K_{gm}\,\dot{\theta}_m\, i_f \text{ volts} \tag{7-9.5}$$

$$T_{\text{dev.}} = K_{Tm}\, i_a\, i_f \text{ newton-meters} \tag{7-9.6}$$

Both equations are non-linear if $\dot{\theta}_m$, i_a, and i_f are all allowed to vary. It appears desirable to fix i_f, since this will result in linear relationships for both the torque and the generated voltage. However, it indicates that motor speed will be a variable, as will developed torque, generated voltage, and armature current.

Using the restriction described above indicates that it is possible to control the d-c motor and obtain variable torque to drive a mechanical load while retaining linear relationships, by using armature current as a control variable or input. To proceed with the modeling of this configuration, the only additional relationship needed is the mechanical torque equation

$$J_m\ddot{\theta}_m + B_m\dot{\theta}_m + K_s\theta_m = T_{\text{dev.}} = (K_{Tm}\, i_f)\, i_a = K_T\, i_a \tag{7-9.7}$$

If i_a, rather than armature voltage e_a, is used as the input, then Eq. 7-9.7 can be transformed and a transfer function obtained directly. However, it is also common practice to drive the armature circuit with a voltage, e_a, which then requires additional effort in modeling the system.

When using armature voltage, e_a, as the control variable, it is necessary to utilize the electrical loop equation for the armature circuit. A possible example of this configuration is the d'Arsonval d-c meter movement in which a voltage is measured by applying it to the armature terminals and noting the final steady-state deflection. The mechanical load of this device must contain a spring so that the constant d-c voltage applied results in a constant final deflection angle. Ordinary d-c control motors are also examples of this configuration. Considering an armature circuit resistance and inductance, R_a and L_a, respectively, leads to

$$L_a \, di_a/dt + R_a \, i_a + e_{\text{gen.}} = e_a \tag{7-9.8}$$

$$L_a \, di_a/dt + R_a i_a + K_{ga} \dot{\theta}_m = e_a \tag{7-9.9}$$

Equation 7-9.7 must now be solved simultaneously with Eq. 7-9.9 for the transfer function model for this armature-controlled motor. The appearance of armature current in the mechanical system equation, and of motor speed in the electrical equation, couple these domains and complicate the analysis. Figure 7-9.4 shows the resulting transfer function.

Another possibility for controlling the d-c motor while retaining linear behavior is that of using field voltage or current as the control input while driving the armature with a constant fixed current, I_{ao}. While it is not immediately apparent that this leads to a linear model, this fact can be obtained as follows. The mechanical system equation is still basically Eq. 7-9.7, with i_a constant instead of i_f. This is rewritten as

$$J_m \ddot{\theta}_m + B_m \dot{\theta}_m + K_s \theta_m = (K_{Tm} I_{ao}) \, i_f \tag{7-9.10}$$

In this case the armature circuit equation is not needed, as the constant current I_{ao} is established by an infinite impedance source and the various component voltages in the armature circuit don't affect this current. Thus, the effect of a varying generated voltage, $e_{\text{gen.}}$, will not affect the armature current. Equation 7-9.10, together with the motor field circuit Eq. 7-9.4, defines this device. The transfer function of the field-controlled motor is shown in Fig. 7-9.5.

Several facts about the two d-c motor configurations above should be noted. In the field-controlled case, the mechanical and electrical contributions to the motor dynamics are separated. The electrical parameters R_f and L_f determine one system pole on the

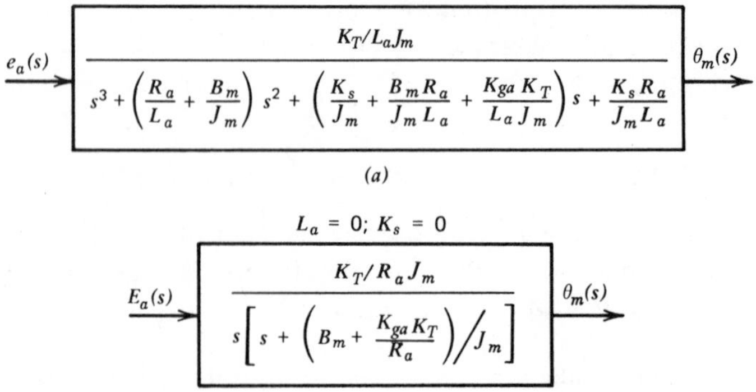
$e_a(s)$
$\dfrac{K_T/L_a J_m}{s^3 + \left(\dfrac{R_a}{L_a} + \dfrac{B_m}{J_m}\right) s^2 + \left(\dfrac{K_s}{J_m} + \dfrac{B_m R_a}{J_m L_a} + \dfrac{K_{ga} K_T}{L_a J_m}\right) s + \dfrac{K_s R_a}{J_m L_a}}$
$\theta_m(s)$
(a)
$L_a = 0; \; K_s = 0$
$E_a(s)$
$\dfrac{K_T/R_a J_m}{s\left[s + \left(B_m + \dfrac{K_{ga} K_T}{R_a}\right) \Big/ J_m\right]}$
$\theta_m(s)$

(b)

Fig. 7-9.4

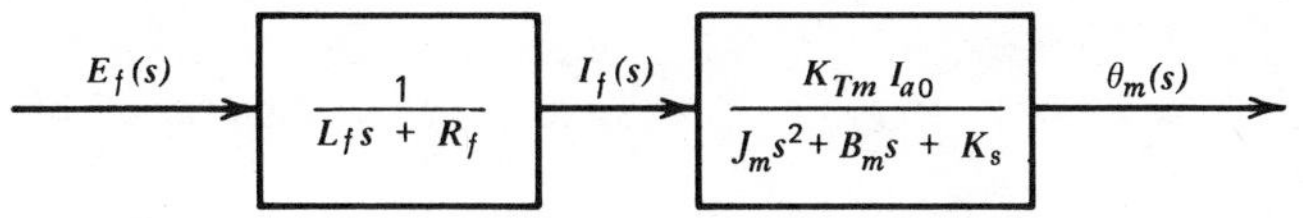

Fig. 7-9.5

negative real s-plane axis, while the mechanical parameters J_m, B_m, and K_s determine the remaining system poles. In contrast, the armature-controlled case illustrated in Fig. 7-9.4(a) has the mechanical and electrical parts of the system truly coupled together. The system poles are determined by a complex mixture of mechanical and electrical parameters. If this case is simplified by assuming that there is no spring loading, $K_s = 0$, and that the armature inductance, L_a, is negligible, the simplified transfer function of Fig. 7-9.4(b) results. For this case, the coupling amounts to having the term $K_T K_{ga}/R_a$ add to the viscous friction coefficient, B_m. The armature-controlled motor model now has a pole at the origin of the s-plane, and at $s = -(B_m + K_T K_{ga}/R_a)/J_m$ on the negative real s-plane axis. The presence of the pole at the origin leads to an *integration*, in the sense that a constant applied armature voltage will lead to a constant motor *speed* rather than a constant motor-armature-shaft *angle*.

Example 7-9.1. The case of an armature-controlled motor with the parameters below and a 120-volt-step function armature input is to be studied. The equations for motor speed and armature position

$$(B_m + K_T K_{ga}/R_a)/J_m = 100 \text{ sec}^{-1} \qquad K_T/R_a J_m = 400 \text{ rad/volt-sec}^2$$

are to be obtained and plots made for these quantities. Direct utilization of the parameter values allow us to write

$$\Theta_m(s) = \frac{400 E_a(s)}{s(s+100)} = \frac{48{,}000}{s^2(s+100)} \tag{7-9.11}$$

$$s\,\Theta_m(s) = \frac{48{,}000}{s(s+100)} = \omega(s) \tag{7-9.12}$$

These equations can now be inverted to yield the desired angle and speed. Plots of the angle and speed are shown in Fig. 7-9.6.

$$\theta_m(t) = 480t + 4.8\epsilon^{-100t} - 4.8 \text{ rad} \tag{7-9.13}$$

$$\omega_m(t) = 480(1 - \epsilon^{-100t}) \text{ rad/sec} \tag{7-9.14}$$

Since the input to the system, e_a, is a constant, the fact that the steady-state response, $\theta_m(t)$, is a ramp function should be noted. This represents the *integration* property mentioned above. It should also be noted that *only two* overall motor parameters had to be specified to completely determine the transfer function.

One last aspect of the models for d-c motors is considered. Equations 7-9.5 and 6 introduce two constants, K_{gm} and K_{Tm}. These two constants can be related by use of

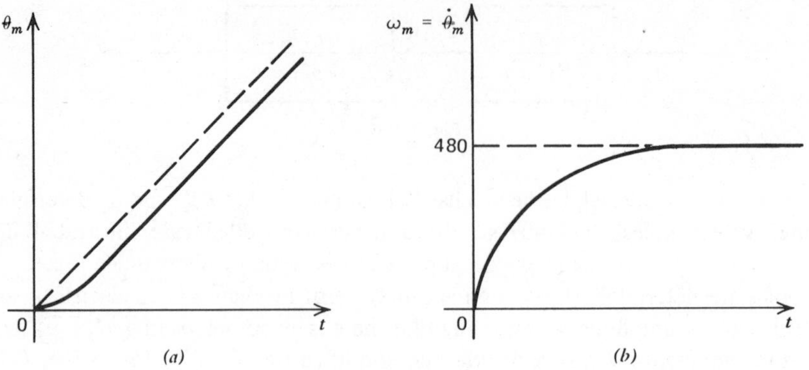

Fig. 7-9.6

the equality of electrical and mechanical forms of power. In electrical units the power converted to mechanical form is found from Eq. 7-9.5 by multiplying it by the armature current, i_a, as shown in

$$e_{\text{gen.}}\, i_a = K_{gm}\, \dot{\theta}_m\, i_f\, i_a \text{ watts} \tag{7-9.15}$$

The mechanical power relationship is found by multiplying the mechanical torque (Eq. 7-9.6) by angular velocity. This follows by considering that work in a rotational system is angular displacement *times* torque, and that power is the time rate of change of work or energy.

$$T_{\text{dev.}}\dot{\theta}_m = K_{Tm}\, i_a\, i_f\, \dot{\theta}_m \text{ (e.g., N-m/sec)} \tag{7-9.16}$$

The two forms of power in Eqs. 7-9.15 and 16 must be the same when converted to the same dimension.

The preceding development of linearized models for simple electromechanical systems extends our modeling concepts to a realm that includes many useful physical systems. While it in no way covers all such cases, it nevertheless admits first approximations to many systems.

PROBLEMS

7-1 In some mechanical systems several springs may be used in series or parallel arrangements, as shown. Determine an equivalent spring constant for each of these arrangements. Is there an analogous situation in electrical circuits?

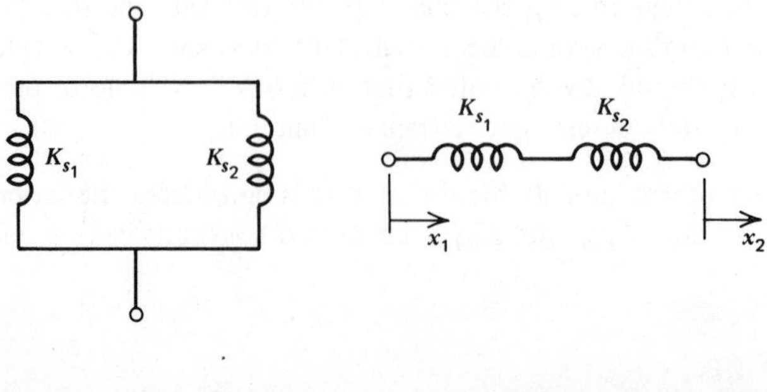

7-2 In order to determine the value of K_s experimentally for the system shown, a constant force of 50 newtons is applied as F, and it is found that the spring is elongated 0.1 meter due to the force. The spring is then disconnected and the same force applied with a resulting velocity in steady-state of 2 m/sec. From the above data, determine K_s and B.

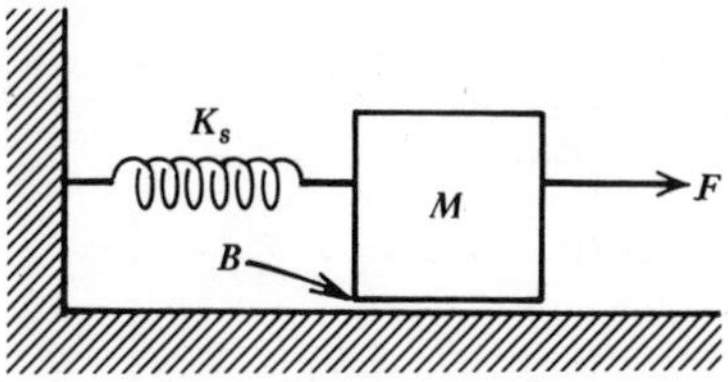

7-3 A form of accelerometer is pictured. A steady-state acceleration results in what form of output from the electrical leads from the device?

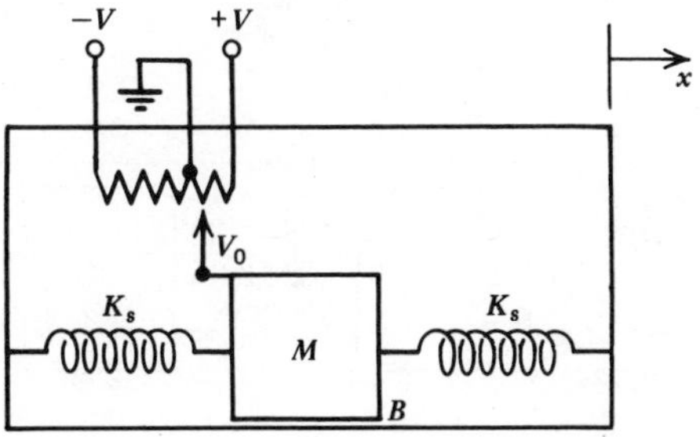

7-4 Write the differential equation(s) describing the motion of the mechanical system depicted. The bottom of the spring and damper are given a sinusoidal displacement of the form $x_1(t) = \sin 3t$ cm. For $K_s = 45$ n/m, $B = 4$ n/(m/sec), and $M = 5$ kg, determine the amplitude of the motion of the mass M with this input displacement.

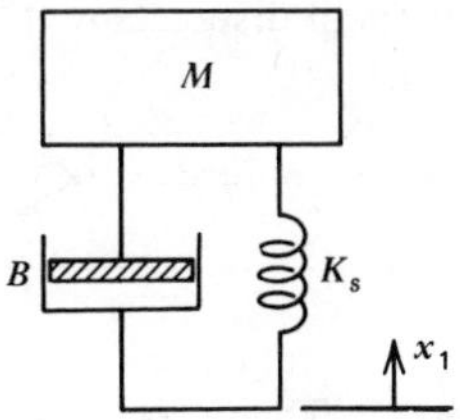

7-5 Write the differential equations describing motion for the system depicted.

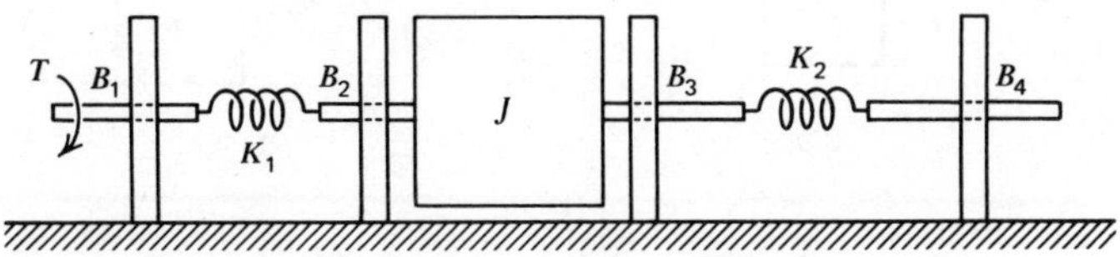

7-6 A cylinder, as shown, has a piston with mass M coupled to the end, with a spring with constant K_s and with viscous friction between the mass and the cylinder wall with coefficient B. For displacement of the piston designated as positive upward, write the differential equation describing motion. Use an applied force of $1\ \epsilon^{-t}$ newtons. For $M = 2$ kg, $K_s = 40$ n/m, and $B =$ n/(m/sec), determine the equation of $x(t)$, $v(t)$, and $\ddot{x}(t)$. Repeat for the positive direction of $x(t)$ reversed.

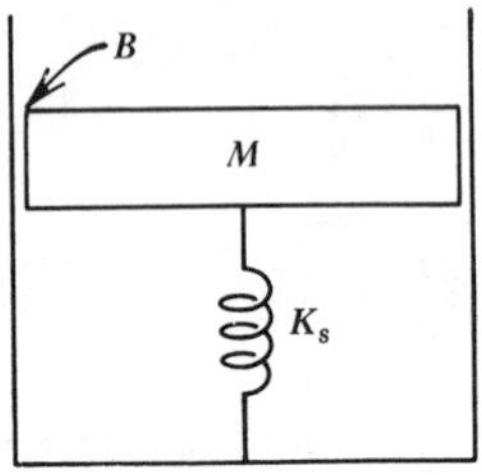

7-7 Write the differential equations describing motion for the system shown. Is this a linear system?

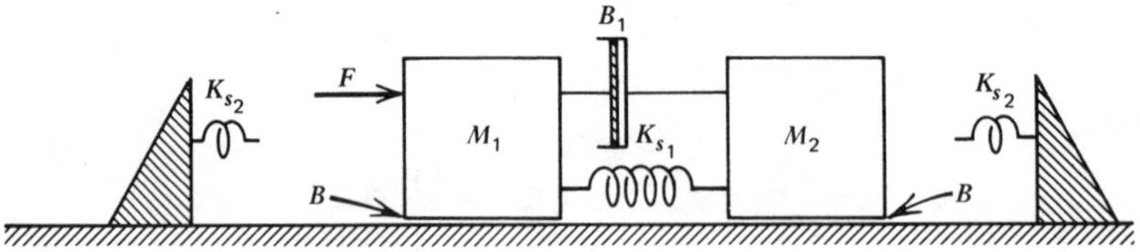

7-8 Write the system of differential equations describing motion for the mechanical configuration shown. Assume displacements of the ends of the lever are small compared to the lever arm lengths, and that the lever does not bend. Also assume that the system is balanced in equilibrium in a horizontal position before mass M_1 is given a forced displacement of 2 cm as a step function. For the indicated parameter values, compute the equations of motion of mass M_2 and of the left end of the lever. Is the assumption about small displacements justified?

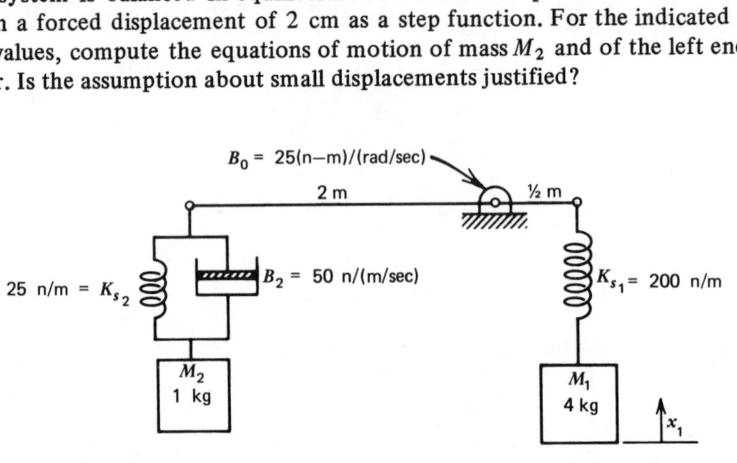

7-9 Write the differential equations of motion for the mechanical system shown, and describe any limitations on the validity of the equations.

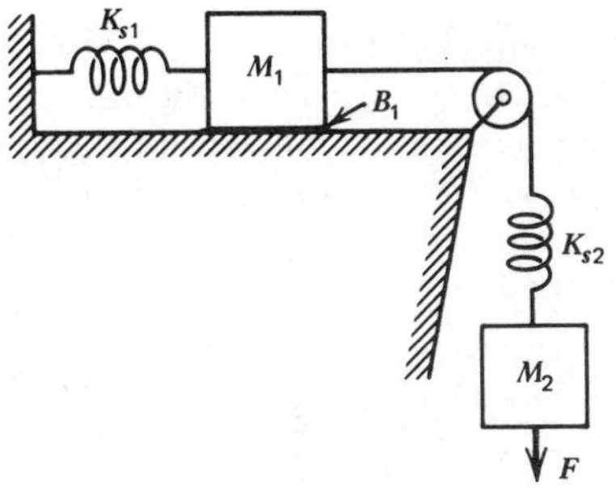

7-10 The system shown is to be analyzed for its dynamic behavior. Write the differential equation(s) required, and solve these with the indicated parameter values when the applied torque is given by $10(1 - \epsilon^{-10t})$ n-m. Also determine the equations for torque and velocity at the output shaft.

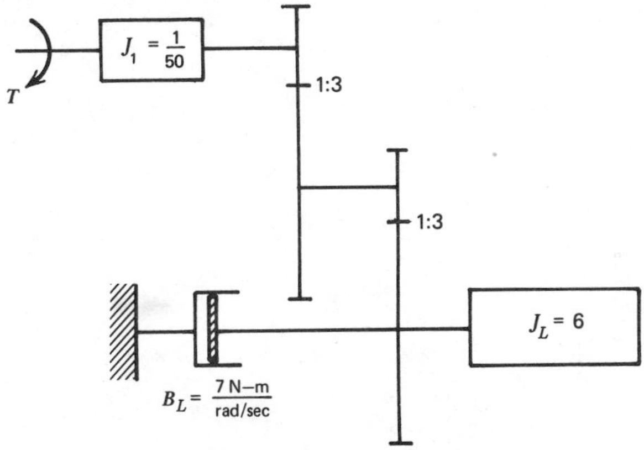

7-11 A d-c generator like the one shown in Fig. 7-9.1 has $R_f = 100$ ohms, $L_f = 5$ henry, $E_f = 50$ volts d-c and the resulting $E_g = 150$ volts at a speed of $20\ \pi$ rad/sec. Determine I_f and the lumped generator constant, K_{gt}. If the speed drops to $(50\ \pi/3)$ rad/sec, what is the new generated voltage?

7-12 Construct a block diagram for the generator of Fig. 7-9.1, but allow speed to be an input variable as well as field current. Also assume that the effect of load current on the generator is to counteract the flux due to field current with a constant of K_{fL} relating load current, i_L, to the equivalent current effect on field flux. Show this effect with a feedback path from i_L when a resistance R_L loads the generator.

7-13 An armature-controlled d-c motor, with transfer function as given in Fig. 7-9.4 for $L_a = 0 = K_s$, is to be considered. An armature voltage of 48 volts d-c results in a speed of 30 π rad/sec. Determine the transient pole location, if the response following switch closure on the 48-volt source has a time constant of 0.5 sec. What is the complex parameter (K_T/R_aT_m) or the numerator constant in the transfer function under consideration?

7-14 A field-controlled d-c motor as shown in Fig. 7-9.5 has a field time constant of 0.02 sec, with a field resistance of 50 ohms. The rest of the transfer function of the motor has a complex pole pair in the s-plane with $K_s/J_m = 2{,}500$, and $B_m/J_m = 60$. A constant field voltage of 50 volts results in an angular deflection of 120° in the steady-state. Determine the complete transfer function for the motor, and describe the response due to a step function input.

Chapter **8**

Modeling of Fluidic, Thermal, Chemical, and Socio-Economic Systems

8-1 INTRODUCTION

Extension of the areas of physical science for which we develop simple models leads naturally to the fields of hydraulic, thermal, chemical, and socio-economic systems. While we can model social and economic systems without having to involve the other forms of physical systems under discussion, it frequently happens that several types of systems are all coupled together in a given engineering problem. The extension of the conservation of quantity concept introduced in Chapter 7, together with the similarity of form of the differential equations involved, provides a common base from which to proceed.

Many of the systems considered in this chapter are inherently non-linear, so there is a need for some form of linearizing approximation if we are to apply our linear analysis and design tools. Thus, the concept of incremental variables is introduced and related to a Taylor series expansion of the functions describing non-linear systems.

8-2. FLUIDIC SYSTEMS

There are complex systems and components of a great many types that could be considered under the category of fluidic systems. We consider, however, only fluid handling systems of tanks, valves, and pipes, together with hydraulic power pistons and valves.

Consideration of the basic variables in such systems indicates that the primary ones are pressure and volume-flow rate. It is also necessary to introduce various geometrical parameters, but these are normally lumped into overall constants much as we have done for electro-mechanical systems. The basic equation relating flow rate and pressure drop across various restrictions is non-linear, with the equation for incompressible flow through an orifice being

$$p_2 - p_1 = K_o q\,|q| \tag{8-2.1}$$

Here, K_o is a constant involving area of the orifice, density of the fluid, and a discharge coefficient; while $p_2 - p_1$ is the pressure drop across the orifice, and q is the volume flow rate through the orifice. Further discussion of the assumptions made is given as various models are developed.

The category of fluid handling systems mentioned above is considered first. Figure 8-2.1 illustrates such a system consisting of a tank, an inflow of fluid, and a restriction in an outflow line. One basic *conservation of quantity* relation must be satisfied in this system. This relationship is

$$\begin{pmatrix}\text{Rate of accumulation}\\ \text{of fluid in the tank}\end{pmatrix} = \begin{pmatrix}\text{Rate of fluid}\\ \text{inflow}\end{pmatrix} - \begin{pmatrix}\text{Rate of fluid}\\ \text{outflow}\end{pmatrix} \tag{8-2.2}$$

It is now necessary to relate the height or *head* of the fluid in the tank to the rate of accumulation of fluid and also to relate the rate of outflow to the pressure differential across the restriction in the outflow line.

The easiest of the above steps is that of introducing the head, H, into the accumulation rate term. This simply amounts to expressing the volume of the tank in terms of its

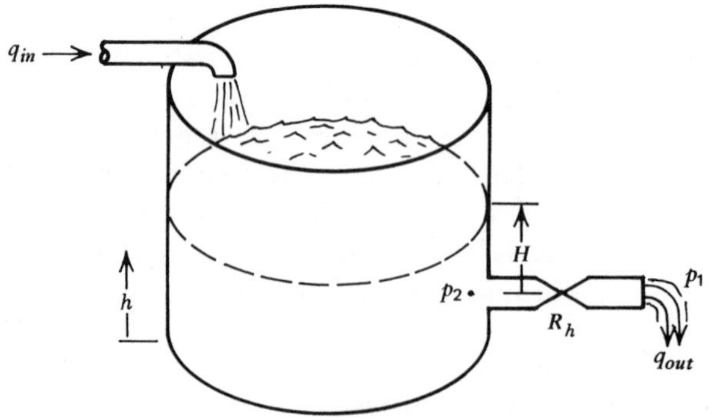

Fig. 8-2.1

height. The result is

$$(\text{Volume of fluid in tank}) = \int_0^H A\,dh \tag{8-2.3}$$

where A is the cross-sectional area of the tank. If A is not a function of h, the above expression simplifies to AH = (Volume), and the rate of change of this volume is given by

$$(\text{Rate of accumulation of fluid in tank}) = A\,dH/dt \tag{8-2.4}$$

The rate of fluid outflow will undoubtedly depend on the height of fluid in the tank which causes a pressure, p_2, to exist at the input to the outflow restriction. Figure 8-2.2 shows a typical plot of Eq. 8-2.1 for the pressure-drop-vs.-flow-rate relationship for incompressible flow through an orifice. As one method of approximating this curve with a linear relationship, a straight line could be used to approximate the parabolic curve over the region of normal operation. This leads to the concept of a hydraulic resistance and

$$p_2 - p_1 \simeq R_h q \tag{8-2.5}$$

when an analogy to electrical circuits is considered, with volume flow rate analogous to current, and pressure drop to voltage drop. The form of Eq. 8-2.5 can be used with many of the non-linear flow-vs.-pressure relationships to obtain gross linear models for such flow.

A somewhat more realistic approach is that of using an incremental variable approach as described below. Total variables are equated to a constant part plus a variable part. For example, the total volume flow rate and pressure are expressed as in

$$\begin{aligned} q(t) = q_o + \delta q(t) \qquad p_1(t) &= p_{1o} + \delta p_1(t) \\ p_2(t) &= p_{2o} + \delta p_2(t) \end{aligned} \tag{8-2.6}$$

In these relations q_o, p_{1o}, and p_{2o} are normally constant operating values of these variables. The δq, δp_1 and δp_2 values are the amounts by which the total quantities vary

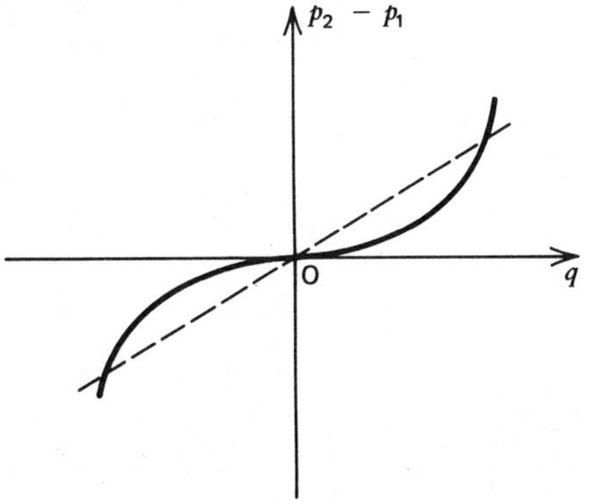

Fig. 8-2.2

from the operating values. These expressions can be substituted into the appropriate flow-rate-vs.-pressure relationship, and only linear terms in the incremental parts retained.

For example, using the incremental approach, Eq. 8-2.1 can be modified to

$$(p_{2o} + \delta p_2) - (p_{1o} + \delta p_1) = K_o(q_o + \delta q)\,|(q_o + \delta q)| \tag{8-2.7}$$

In this equation the operating values chosen for p_{2o}, p_{1o}, and q_o must be such that $p_{2o} - p_{1o} = K_o q_o\,|q_o|$. For total $q(t)$ positive, Eq. 8-2.7 can be rewritten

$$(p_{2o} + \delta p_2) - (p_{1o} + \delta p_1) = K_o(q_o + \delta q)^2 \qquad q(t) \geqslant 0 \tag{8-2.8}$$

and this rearranged as

$$(p_{2o} - p_{1o}) + (\delta p_2 - \delta p_1) = K_o q_o^2 + 2K_o q_o(\delta q) + K_o(\delta q)^2 \tag{8-2.9}$$

By canceling the parts involving only the operating values on each side of Eq. 8-2.9 as being equal, the incremental part is obtained as

$$(\delta p_2 - \delta p_1) = 2K_o q_o(\delta q) + K_o(\delta q)^2 \tag{8-2.10}$$

This relationship is still non-linear, because of the squared term. If the variational part, $K_o(\delta q)^2$, is small relative to the operating values, the squared term can be dropped and a linear approximation obtained as

$$(\delta p_2 - \delta p_1) \simeq 2K_o q_o(\delta q) \qquad |\delta q| \ll q_o \qquad q(t) \geqslant 0 \tag{8-2.11}$$

Figure 8-2.3 illustrates the above manipulations.

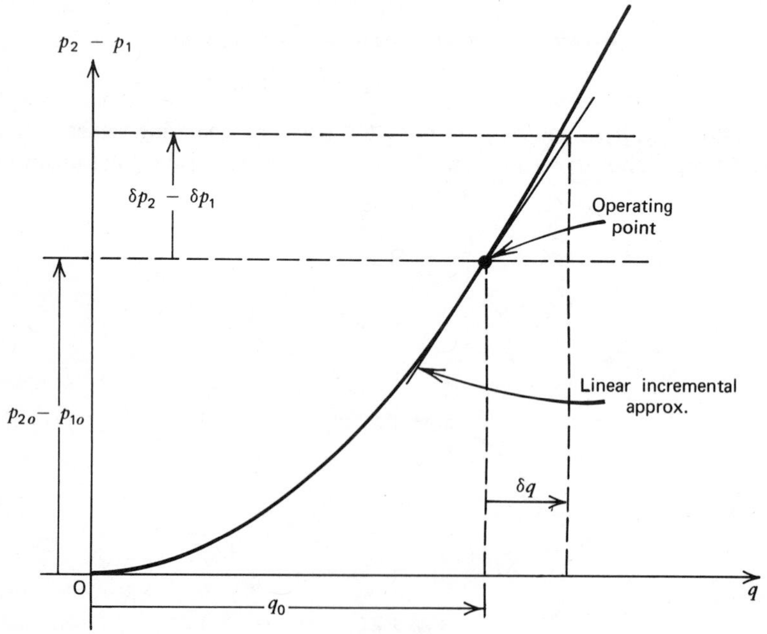

Fig. 8-2.3

A more general method of obtaining the linearized relationship of Eq. 8-2.11 is by the use of a Taylor series expansion about the operating point of the pressure-vs.-flow-rate relationship. Equations 8-2.12 through 15 illustrate this technique for the case being considered (substitution of $(\delta q = q - q_o)$ and of $(p_2 = p_{2o} + \delta p_2)$, $(p_1 = p_{1o} + \delta p_1)$ is made to obtain Eq. 8-2.14 from Eq. 8-2.13).

$$p_2 - p_1 = K_o q^2 = f(q) \qquad q \geqslant 0 \tag{8-2.12}$$

$$= f(q_o) + f'(q_o)(q - q_o) + f''(q_o)/2!\,(q - q_o)^2 + \cdots \tag{8-2.13}$$

$$(p_{2o} + \delta p_2) - (p_{1o} + \delta p_1) = f(q_o) + f'(q_o)(\delta q) + f''(q_o)/2!\,(\delta q)^2 + \cdots \tag{8-2.14}$$

$$(p_{2o} + \delta p_2) - (p_{1o} + \delta p_1) = K_o q_o^2 + 2K_o q_o(\delta q) + K_o(\delta q)^2 + \cdots \tag{8-2.15}$$

Considering that the operating values were so chosen that the original equation is satisfied for all incremental values equal to zero again allows cancelation of $(p_{2o} - p_{1o} = K_o q_o^2)$. This yields the relationship of Eq. 8-2.11 when higher order terms in δq are dropped as being small. This Taylor series approach to the development of the incremental equations is more generally useful than the direct substitution method used previously.

It is now possible to return to the original problem of studying the fluid handling system of Fig. 8-2.1 and Eq. 8-2.2. Use can be made of Eqs. 8-2.4 and 5, or of a version of Eq. 8-2.4, using incremental variables, together with Eq. 8-2.11. Total variable equations are used first. Equation 8-2.2 can now be written as

$$A\, dH/dt = q_{\text{in}} - q_{\text{out}} = q_{\text{in}} - (p_2 - p_1)/R_h \tag{8-2.16}$$

The main problem with this equation is that two forms of pressure are involved, H for head and p for pressure. The head of fluid in the tank is also related to the pressure difference between the top and the bottom of the tank. For the case of atmospheric pressure exerted on the top of the fluid and at the output of the outflow line, the following relations apply.

$$\gamma H = p_2 - p_1 \qquad \gamma = (\text{specific weight of fluid}) \tag{8-2.17}$$

It is now possible to convert all pressure terms to actual pressure, p, or to head, H. For this case, head H is retained to facilitate relating the solution to possible overflow of the fluid in the tank. Equation 8-2.16, by substitution of Eq. 8-2.17, can now be converted to

$$A\, dH/dt = q_{\text{in}} - \gamma H/R_h \tag{8-2.18}$$

It should be noted that use of gauge pressures will cause p_1 to be zero and simplify the equations. The equation

$$dH/dt + \gamma H/R_h A = q_{\text{in}}/A \tag{8-2.19}$$

is a final form of a linearized model of the tank system of Fig. 8-2.1. It is a first-order differential equation with q_{in} as the forcing function. The rather gross approximation made in using a total hydraulic resistance, R_h, for modeling the flow through the restriction should be recalled. Additional factors that need attention are the range of validity of the variables in the model. If H becomes too large, the tank overflows and the equation for rate of accumulation is no longer valid. If H becomes negative, this implies less than zero fluid in the tank, and again the model is invalid. Additionally, the flow through the restriction would need to be reversed—not likely in this case. As in all mathematical models, care must be used in their application.

It is also possible to develop a model for the system based upon incremental variables. Equation 8-2.11 is already in suitable form, but Eq. 8-2.4 needs modification by introducing the operating and incremental parts of H as $H = H_o + \delta H$. Since H_o is a constant, the rate of accumulation as expressed in Eq. 8-2.4 can be modified to

$$A\, dH/dt = A\, d(H_o + \delta H)/dt = A\, d(\delta H)/dt \qquad (8\text{-}2.20)$$

This relationship can now be substituted into Eq. 8-2.2 to give

$$A\, d(\delta H)/dt = q_{\text{in}} - (q_{\text{out},o} + \delta q_{\text{out}}) \qquad (8\text{-}2.21)$$

It is apparent that q_{in} and q_{out} must be equal for equilibrium to exist. Conversion of all pressure variables to the dimensions of head together with the use of Eq. 8-2.11 yields

$$A\, d(\delta H)/dt = (q_{\text{in},o} + \delta q_{\text{in}}) - [q_{\text{out},o} - (\delta p_2 - \delta p_1)/2K_o q_{\text{out},o}] \qquad (8\text{-}2.22)$$

$$= \delta q_{\text{in}} - \gamma\,(\delta H)/2K_o q_{\text{out},o} \qquad (8\text{-}2.23)$$

$$d(\delta H)/dt + \gamma(\delta H)/(2K_o q_{\text{out},o} A) = \delta q_{\text{in}}/A \qquad (8\text{-}2.24)$$

A first-order equation also results for this mathematical model relating incremental variables. It should be noted that this approach breaks down if the operating value of q_{out} is zero. This follows from the fact that division by $q_{\text{out},o}$ is necessary or that the linear incremental approximation for the flow-vs.-pressure-drop relationship for the orifice predicts zero incremental pressure variation. Use of the total variable approach developed previously is necessary in this case. A note of caution in the use of Eq. 8-2.24 is offered. In order to discard the higher-order terms in the flow-vs.-pressure-drop relationship, it was assumed that the incremental parts were sufficiently small relative to the operating values. Just how small is a question of judgment, and of the use to be made of the results. If 80% accuracy is sufficient, then the incremental parts can be larger than if 95% accuracy is required.

The above development has introduced hydraulic quantities that are analogous to electrical resistance and capacitance. By associating fluid flow rate with the *through* variable, it is easily possible to see the analogy to electrical current. Similarly, pressure can be considered as an *across* variable analogous to electrical voltage. With these associations it is apparent that the hydraulic resistance, R_h, of Eq. 8-2.5 is directly analogous to electrical resistance. Equation 8-2.5 defines a *total* hydraulic resistance, whereas the $2K_o q_o$ coefficient in Eq. 8-2.11 is an incremental hydraulic resistance. It is also possible to define a hydraulic resistance in terms of fluid head vs. flow rate, with this latter resistance being related to the ones above by $R_{h\,\text{head}} = R_{h\,press}/\gamma$.

Considering the electrical relationship involving capacitance, it can be seen, in view of the analogies mentioned above, that an analogous hydraulic relationship would be that given by

$$q = C_h\, dp/dt \qquad (8\text{-}2.25)$$

This is related to Eq. 8-2.4 with a change from pressure, p, to head. Introducing this relationship results in

$$q_{\text{accum.}} = C_h \gamma\, dH/dt \tag{8-2.26}$$

and the ability to identify hydraulic capacitance in terms of tank area and fluid specific weight as given in

$$C_h = A/\gamma \tag{8-2.27}$$

It is also possible to define hydraulic capacitance in terms of the derivative of head instead of pressure.

Use of the definitions of hydraulic resistance and capacitance allow the final system model equations to be written as

$$dH/dt + H/R_h C_h = q_{\text{in}}/A \tag{8-2.28}$$

instead of Eq. 8-2.19 for the total quantities, or in terms of

$$d(\delta H)/dt + (\delta H)/R_{\delta h} C_{\delta h} = \delta q_{\text{in}}/A \tag{8-2.29}$$

instead of Eq. 8-2.24.

Example 8-2.1. It is desired to study the fluid handling system of Fig. 8-2.1 with the parameters given below and to compare the *total* and *incremental* approaches with the *exact* solution.

Cylindrical tank area = 80 m^2
Tank height = 20 m.
Operating head = 9 m.
Operating input and output flow rates = 75 m^3/min

The system is to be considered for step inputs, δq_{in}, equal in amplitude to 20% of the operating value. The *exact* solution is obtained by direct integration of the non-linear first-order differential equation for the system, or by numerical integration of it, which is not considered in detail here.

The *exact* equation is found by using Eq. 8-2.2 and the equation for flow through the orifice repeated as

$$\gamma H = K_o q^2 \qquad q \geqslant 0 \tag{8-2.30}$$

$$A\, dH/dt + \sqrt{\gamma/K_o}\,\sqrt{H} = q_{\text{in}} \qquad H \geqslant 0 \tag{8-2.31}$$

For equilibrium conditions to exist before any input disturbances are applied, it is necessary that Eq. 8-2.30 be satisfied so that K_o/γ can be found as $H_o/q_o^2 = 9/(75)^2$. This allows Eq. 8-2.31 to be rewritten as

$$dH/dt + 75/(3 \times 80)\,\sqrt{H} = q_{\text{in}}/80 \qquad H \geqslant 0 \tag{8-2.32}$$

$$dH/dt + 0.3125\,\sqrt{H} = q_{\text{in}}/80 \qquad H(0) = 9 \tag{8-2.33}$$

which has to be solved to give the *exact* solution for $H(t)$. While this is not a really difficult task, it is messy and is not given here. The solution for the step change in q_{in} is plotted in Fig. 8-2.5.

In order to use Eq. 8-2.28 to obtain approximate solutions involving linearized total variables, it is necessary to determine $R_h C_h$. From the previous discussion of hydraulic resistance and capacitance, it can be seen that

$$R_h/\gamma = H_o/q_o = 9/75 \text{ min/m}^2 \qquad (8\text{-}2.34)$$

$$\gamma C_h = A = 80 \text{ m}^2 \qquad (8\text{-}2.35)$$

$$R_h C_h = (9)\,(80)/75 = 9.6 \text{ min} \qquad (8\text{-}2.36)$$

Equation 8-2.28 can now be written as

$$dH/dt + H/9.6 = q_{\text{in}}/80 \qquad H(0) = 9 \qquad (8\text{-}2.37)$$

and its solution as

$$H(t) = 0.12q_{\text{in}} + (9 - 0.12q_{\text{in}})\,\epsilon^{-t/9.6} \text{ meters} \qquad (8\text{-}2.38)$$

The solution for the q_{in} value is also plotted in Fig. 8-2.5.

For the incremental case of Eq. 8-2.29 it is also possible to express $R_{\delta h}$ and $C_{\delta h}$ in terms of the given parameters as

$$R_{\delta h}/\gamma = \delta H/\delta q_{\text{out}} = 2K_o q_{\text{out},o}/\gamma = 2(9)\,(75)/(75)^2 \qquad (8\text{-}2.39)$$

$$\gamma C_{\delta h} = A = 80 \qquad (8\text{-}2.40)$$

$$R_{\delta h} C_{\delta h} = 2\,(9)\,(80)/(75) = 19.2 \text{ min} \qquad (8\text{-}2.41)$$

Equation 8-2.29 can now be written as

$$d(\delta H)/dt + (\delta H)/19.2 = \delta q_{\text{in}}/80 \qquad \delta H(0) = 0 \qquad (8\text{-}2.42)$$

$$\delta H(t) = 0.24\,(\delta q_{\text{in}})\,(1 - \epsilon^{-t/19.2}) \qquad (8\text{-}2.43)$$

and, with the total H variable, as

$$H(t) = H_o + \delta H(t) = 9 + 0.24\,(\delta q_{\text{in}})\,(1 - \epsilon^{-t/19.2}) \qquad (8\text{-}2.44)$$

The total solution is also plotted in Fig. 8-2.5.

It is helpful to consider the various approximations used above by examining the curve relating drop in head across the output valve vs. the volume flow rate through it. This curve is given in Fig. 8-2.4. The exact curve is parabolic in shape.

From the differential equations above, it is possible to determine the final or steady-state value of the head that corresponds to equality of the input and output flow rates. These values are marked on Fig. 8-2.4. It should be pointed out that the choice of the hydraulic resistance value to use for the linearized total variable case was based on a straight line drawn through the origin and the operating point at $H_o = 9$ meters, $q_{\text{in},o} = q_{\text{out},o} = 75$ m^3/min. This is obviously a poor approximation, as can be seen by inspecting Fig. 8-2.4 and comparing the *exact* parabolic curve and the straight line approximation for the total variable linear approximation case.

A straight line approximation is also the result of the incremental variable linearized approach, but this straight line is tangent to the *exact* curve at the operating point. It is apparently a better approximation so long as the variational parts are small. The reader

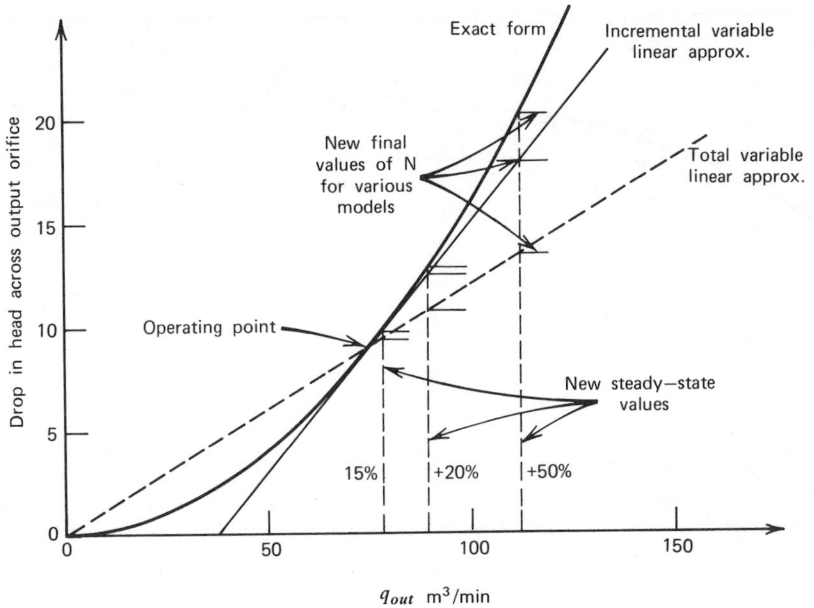

Fig. 8-2.4

should consider the fact that the solutions given do not include the output flow rate, $q_{\text{out}}(t)$, and that this can be obtained directly from $H(t)$.

In some situations it is important to consider the acceleration, caused by pressure or head differentials, of the fluid in a pipe. This case can be modeled by considering Newton's Second Law involving the rate of change of momentum. The accelerating force is found from the pressure differential as

$$A_p\,(p_2 - p_1) = (\text{Force accelerating fluid}) \tag{8-2.45}$$

while the mass of fluid in the pipe between the points at which p_2 and p_1 exist, presuming incompressible fluid, is

$$(\text{Mass of fluid}) = \rho A_p L \tag{8-2.46}$$

In the above equation for mass, L is the length of the pipe between pressures p_2 and p_1, A_p is pipe cross-sectional area, and ρ is the mass density of the fluid. It is now possible, using Newton's Second Law, to put these relationships together as

$$d(\rho A_p L v)/dt = A_p\,(p_2 - p_1) \tag{8-2.47}$$

The volume flow rate can be introduced, and common terms eliminated, to give Eq. 8-2.48.

$$(\rho L/A_p)\,dq/dt = (p_2 - p_1) \tag{8-2.48}$$

$$(L/gA_p)\,dq/dt = (H_2 - H_1) \tag{8-2.49}$$

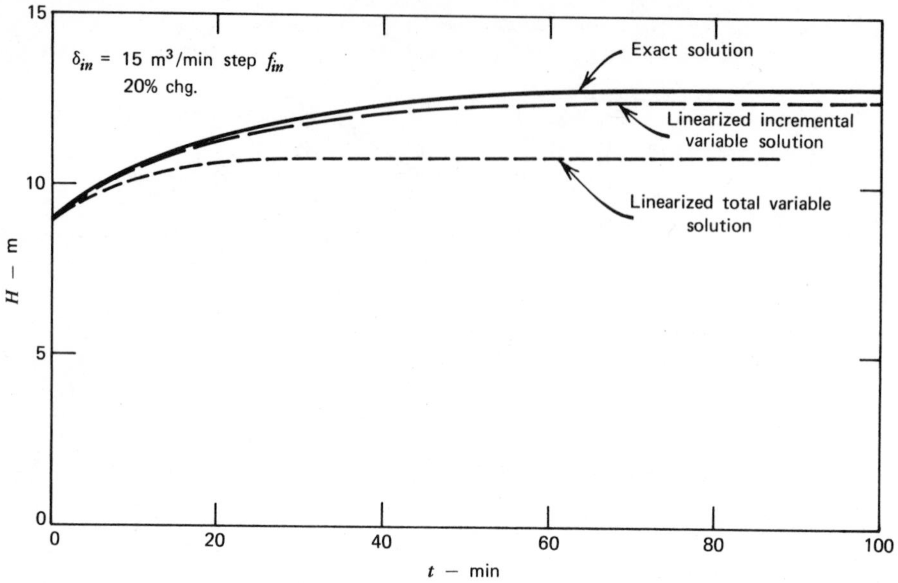

Fig. 8-2.5

Use has to be made of the fact that $q = A_p v$, and that mass density, ρ, and specific weight, γ, are related by $\rho g = \gamma$. The above relationships are particularly useful when appreciable masses of fluid are being accelerated and the steady-state flow relations, such as used in the example above, are of questionable validity.

8-3. HYDRAULIC ACTUATORS

Hydraulic valves and actuators are considered as a final category of fluidic systems. Figure 8-3.1 illustrates a basic actuator piston. The behavior of this device can involve very complex physical phenomena, if the fine details are considered. It is possible, however, to formulate a fairly simple model of it that is adequate for our purposes. The fundamental consideration is that when fluid flows into the actuator, the fluid must go

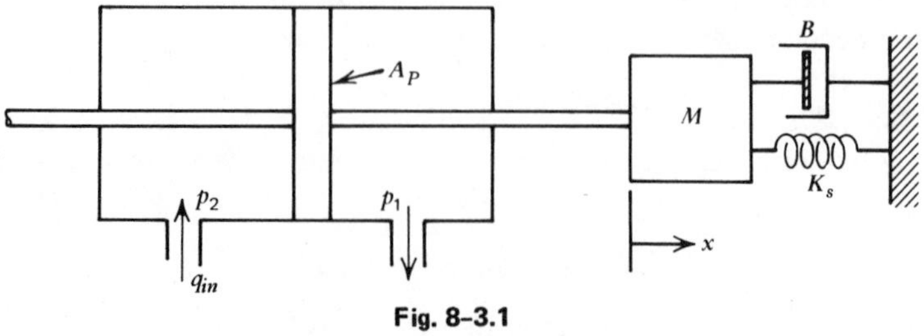

Fig. 8-3.1

somewhere in view of our *conservation of quantity* law. It can go to move the piston, leak around it, or compress the fluid already there increasing its density and the pressure. We will consider these effects one at a time.

The quantity of fluid going to move the piston is easily found from the volume displaced as the piston moves, and is given by

$$\text{(Volume flow rate due to piston motion)} = A_p \text{ (Piston velocity)} \tag{8-3.1}$$

$$q_v = A_p\, dx/dt \tag{8-3.2}$$

The volume flow rate of fluid leaking around the piston is probably small and difficult to predict theoretically, but most likely it depends mainly on the pressure difference across the piston and other physical and geometrical factors, such as seal condition and piston clearance. It can be approximated by using a proportionality constant, as in Eq. 8-3.3.

$$\text{(Leakage vol. flow rate)} = q_L = K_L\,(p_2 - p_1) = K_L p_L \tag{8-3.3}$$

The proportionality constant can be determined experimentally, using a blocked piston test, and measuring the fluid that leaks through in a given time interval.

In considering the volume flow rate that can go into so-called compression flow, it is necessary to relate the incremental change in pressure resulting from the forcing of an incremental volume of fluid into a given volume of fluid. This relationship can be expressed as

$$\begin{matrix}\text{Incremental volume of fluid}\\ \text{forced into piston}\end{matrix} = \frac{\text{(Total volume)(Pressure increment)}}{\beta} \tag{8-3.4}$$

where β is a parameter of the fluid called the bulk modulus. Then

$$\frac{\text{(Volume forced in)}}{\text{(Time increment)}} = q_c = (V/\beta)\, dp/dt \tag{8-3.5}$$

Equation 8-3.5 gives the final result of dividing by a time increment and going to differentials. The result is a compression flow rate term that can be used to account for the part of the input flow rate going into compression.

Example 8-3.1. To illustrate the basic behavior of an actuator such as pictured in Fig. 8-3.1, it is presumed that total input flow is an independent quantity and all of this flow goes into movement of the piston. The transfer function from this input flow rate to output position is to be determined.

For this case Eq. 8-3.2 provides the only required model, with the q_v component as q_{in}, and piston position x indicating the output.

$$q_{\text{in}} = A_p\, dx/dt \tag{8-3.6}$$

When this is transformed, assuming zero initial conditions, the result is given by

$$Q_{\text{in}}(s) = A_p s\, X(s) \qquad X(s)/Q_{\text{in}}(s) = 1/A_p s \tag{8-3.7}$$

This result indicates an s-plane pole at the origin, which is an integration property. A constant input flow rate leads to a constant rate of change of the output position.

The example above is very much simplified in that the actual mechanical load on the piston and the pressures in the cylinder are not parts of the transfer function. Surely these factors affect a realistic system of this type. Additionally, it is most likely that the input flow rate will not be an independent variable but will be controlled by some other quantity, such as a pilot valve opening as pictured in Fig. 8-3.2. This brings us back to the matter of flow through orifices or openings. In general, the pilot valve spindle would be moved by some mechanical or electromechanical means, such as a solenoid, thereby creating openings for the fluid to flow into and out of the actuator. Effectively, such openings are orifices, with the area of the orifice dependent on the distance the pilot valve spindle moves. The flow through these openings will depend on the area of the opening and on the pressure drops across the openings. Such a relationship can be expressed as

$$q_{\text{in}} = f(x_1, p_L) \qquad p_L = p_2 - p_1 \tag{8-3.8}$$

The supply pressure, p_s, and sump pressure, p_o, are presumed to be constant. This can represent a general nonlinear (or linear) relationship. Since it will generally be highly non-linear, it is appropriate to consider the incremental variable approach described earlier, in the paragraphs on general fluid handling systems. Using a Taylor series in two variables, and retaining only the first order terms, gives

$$q_{\text{in}}(x_1, p_L) \simeq q_{\text{in}}(x_{1o}, p_{Lo}) + (\partial q_{\text{in}}/\partial x_1)_o\,(x_1 - x_{1o}) + (\partial q_{\text{in}}/\partial p_L)_o\,(p_L - p_{Lo}) \tag{8-3.9}$$

$$\delta q_{\text{in}} \simeq K_1\,\delta x_1 + K_2\,\delta p_L \tag{8-3.10}$$

where the o subscripts indicate operating point values. The partial derivatives are, thus, treated as constants, since they are numerical values determined at the operating point of the device. In particular, K_1 is the incremental change in input flow due to incremental

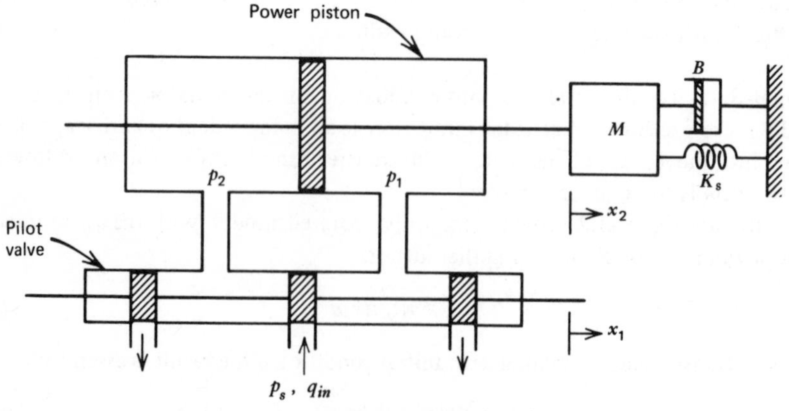

Fig. 8-3.2

variations in the position of the pilot valve spindle at constant load pressure. Similarly, K_2 is the incremental change in input flow rate due to incremental changes in load pressure at constant position of the pilot valve spindle.

Consideration of the normal operating point for pilot valve operated actuators of the type being studied is in order. The common equilibrium condition is that of zero valve opening and zero load pressure unless the mechanical load includes a spring. For this operating point, $x_{1o} = 0 = p_{Lo}$, the incremental variables are also total variables, and the relationship of Eq. 8-3.10 can be used in either incremental or linearized total variable equations.

One additional relationship is needed before models of the pilot-valve controlled actuator can be developed. This is the relationship between the hydraulic and the mechanical parts of the system. What is needed is the force relationship of

$$A_p p_L = M\, d^2x_2/dt^2 + B\, dx_2/dt + K_s x_2 \tag{8-3.11}$$

It is now possible to put the various relationships developed above together into a realistic linearized model of the system of Fig. 8-3.2. The first step is to formulate the conservation-of-fluid relationship establishing that the input flow goes into the components indicated in Eqs. 8-3.2, 3, and 5, and that q_{in} is also given by Eq. 8-3.10. The second step is to utilize Eq. 8-3.11 as found by applying Newton's Second Law. It is then possible to transform the equations and obtain a desired transfer relationship as illustrated in the following example.

Example 8-3.2. It is desired to develop the transfer function models relating pilot valve position x_1 as input to load position x_2, and actuator pressure drop p_L as outputs for a system of the type shown in Fig. 8-3.2. The mechanical load is to consist of only mass and viscous friction with no springs.

The operating point for the whole system will be presumed to be at zero for all variables so that incremental and total variables are identical in the linearized model. For this case the input flow rate is given as

$$q_{in}(t) = K_1 x_1 + K_2 p_L = A_p\, dx_2/dt + K_L p_L + (V/\beta)\, dp_L/dt \tag{8-3.12}$$

and the mechanical force equation as

$$A_p p_L = M\, d^2x_2/dt^2 + B\, dx_2/dt \tag{8-3.13}$$

The input flow rate is presumed to be made up of displacement, leakage and compression flow as described previously.

In order to obtain the transfer function relationship between input x_1 and output x_2, it is necessary to eliminate the load pressure variable, p_L. This is most easily done by first Laplace transforming Eqs. 8-3.12 and 13, and then eliminating $P_L(s)$. The result of transforming and rearranging the above equations is given by

$$K_1\, X_1(s) = (K_L - K_2)\, P_L(s) + (V/\beta)\, s\, P_L(s) + A_p s\, X_2(s) \tag{8-3.14}$$

$$A_p\, P_L(s) = (Ms + B)\, s\, X_2(s) \tag{8-3.15}$$

The transfer function for $X_2(s)$ is

$$\frac{X_2(s)}{X_1(s)} = \frac{K_1 A_p}{\{(VM/\beta)s^2 + [VB/\beta + M(K_L - K_2)]s + [B(K_L - K_2) + A_p^2]\}s} \tag{8-3.16}$$

and the transfer function for $P_L(s)$ is

$$\frac{P_L(s)}{X_1(s)} = \frac{K_1(Ms + B)}{(VM/\beta)s^2 + \{VB/\beta + M(K_L - K_2)\}s + [B(K_L - K_2) + A_p^2]} \tag{8-3.17}$$

The first of these includes an integrating pole at the origin of the s-plane indicating that a constant input, x_1, would lead to a constant velocity of the output actuator. It is apparent that the complexity of the model is greater than that of some of the previous models. Nevertheless, this model is linear and can be used to study the effects of various parameters. If desired, the compressibility flow can be omitted and a more simplified model obtained.

Several more or less commonly encountered fluidic systems have been considered in the last two sections. While no attempt at complete in-depth studies has been made, at least several reasonable approaches to the development of hydraulic system models have been presented. The reader is encouraged to consider models developed in other literature.

8-4. THERMAL SYSTEMS

The modeling of thermal systems follows the same basic procedures used in the systems described previously. The system *through* and *across* variables need to be determined, and the relationships between them formulated. System parameters then enter the relationships between variables. For thermal systems the main concern is flow of heat, by the mechanisms of conduction, convection, and radiation, together with the associated temperatures of materials. Our main concern is the development of lumped models for simplified versions of systems involving the control of temperatures or heat flow. As contrasted to previously considered systems, the *through* variable, which is heat flow in this case, is an energy variable itself; before, *through* variables were not energy or work. The *across* variables in heat flow relationships are temperatures. As a final point, it should be mentioned that thermal problems generally lead to partial differential equations involving space and time as independent variables. However, by considering that the temperatures of certain masses can be maintained uniform by stirring, or some such mechanism, we can lump the space effect and assume that a quantity of heat is instantly distributed throughout the mass, thereby resulting in a uniform temperature.

Perhaps the most familiar heat flow relationship relates the temperature change of a mass to the quantity of heat input to it, presuming no change in state (solid, liquid, or gas) of the material. This relationship is expressed

$$\delta Q = Mc_p(\theta_2 - \theta_1) \tag{8-4.1}$$

where M is mass, c_p is the specific heat of the mass at constant pressure, δQ is the quantity of heat transferred, θ_1 is the initial temperature, and θ_2 is the final temperature. Dividing both sides of this equation by a time increment, and letting the increment in heat approach zero together with the time increment

$$q_h = dQ/dt = Mc_p\, d\theta/dt \tag{8-4.2}$$

In view of the previous definitions of *through* and *across* variables, it can be seen that Eq. 8-4.2 is analogous to that for an electrical capacitance, and as such the quantity Mc_p can be defined as a thermal capacitance, C_t.

A second basic relationship is that for the transfer of heat by conduction between two points at different temperatures. This relationship is Fourier's Law stated as

$$q = (\sigma A/L)(\theta_2 - \theta_1) \tag{8-4.3}$$

where σ is a parameter called the thermal conductivity, A is the surface area normal to the direction of heat flow, and L is the distance between the surfaces at which θ_1 and θ_2 are measured. This indicates a proportionality between the rate of heat flow, q, and the temperature difference between two surfaces. By analogy to electrical circuits, it leads to the definition of a thermal resistance as $R_t = L/\sigma A$. This concept of thermal resistance can be extended to certain situations involving convection by simply defining thermal resistance as the ratio of temperature difference to heat flow rate.

The remaining heat flow relationship is associated with radiation and is the Stefan–Boltzmann Law expressed by

$$q = C_r(\theta_2^4 - \theta_1^4) \tag{8-4.4}$$

where C_r is a proportionality constant. It is obvious that this equation is non-linear, because of the fourth-power terms. It can be linearized for small variations in temperatures, but this is left as a problem assignment.

Example 8-4.1. It is desired to consider a lumped, linear approximation model for the temperature of an oven as pictured in Fig. 8-4.1. In this situation the internal temperature of the oven and its contents is assumed to be uniform. The input heat rate, q_{in}, is an independent variable. There is heat loss through the walls of the oven due to the temperature differential between the oven temperature and the outside air temperature.

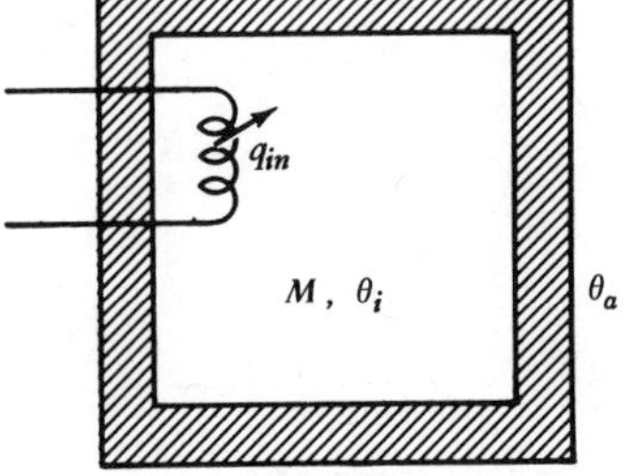

Fig. 8-4.1

The system is modeled by considering the inside of the oven and its contents as thermally equivalent to a lumped mass, M, at a temperature θ_i, with a thermal resistance representing the relationship due to heat loss through the walls. Resorting to our conservation-of-quantity relationship provides the key to combining the necessary versions of Eqs. 8-4.2 and 3 as

$$\begin{pmatrix}\text{Heat rate}\\ \text{input}\end{pmatrix} - \begin{pmatrix}\text{Heat rate}\\ \text{output}\end{pmatrix} = \begin{pmatrix}\text{Rate of accumulation}\\ \text{of heat in the oven}\end{pmatrix} \tag{8-4.5}$$

Then

$$q_{\text{in}} - (\theta_i - \theta_a)/R_t = C_t\, d\theta_i/dt \tag{8-4.6}$$

expresses the result after substitution of the thermal resistance and capacitance relationships from above. One new aspect enters this equation. That is the presence of a second independent quantity, which is the outside air temperature, θ_a. Something must be given to define this quantity. If it is assumed that the outside air mass is so large that the heat flow out of the oven will not affect the outside air temperature, then we merely have to specify the constant temperature of the outside air. For this case, θ_a will be assumed to be constant at θ_{ao} °C.

It is now possible to transform and rearrange Eq. 8-4.6 to give Eq. 8-4.7.

$$(C_t s + 1/R_t)\Theta_i(s) = C_t\,\theta_i(0) + \Theta_a(s)/R_t + q_{\text{in}}(s) \tag{8-4.7}$$

$$\Theta_i(s) = \frac{\theta_{ao}}{(s + 1/R_t C_t)} + \frac{\theta_{ao}}{R_t C_t\, s\,(s + 1/R_t C_t)} + \frac{q_{\text{in}}(s)}{C_t\,(s + 1/R_t C_t)} \tag{8-4.8}$$

$$\Theta_i(s) = \frac{R_t C_t\,\theta_{ao}}{s} + \frac{q_{\text{in}}(s)}{C_t\,(s + 1/R_t C_t)} \tag{8-4.9}$$

The oven temperature, $\theta_i(0)$, at $t = 0$ is also assumed to be θ_{ao}, as would be the case for the equilibrium condition with no heat input. A block diagram can now be drawn showing the relationships between the variables as in Fig. 8-4.2. In this case, the input due to the outside air temperature has been combined with the initial value of oven temperature to provide the single input, θ_{ao}.

Applying the conservation of heat relationship of Eq. 8-4.5 to each of the masses, it is not difficult to develop models for thermal systems in which more than one thermal capacitance and thermal resistance exist. Additional examples in the next article illustrate

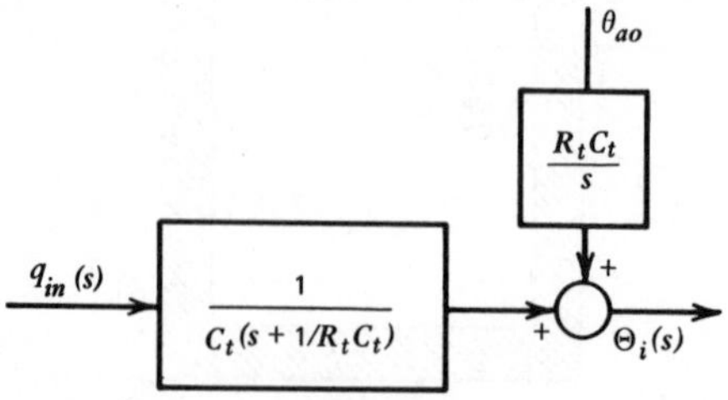

Fig. 8-4.2

applications of the thermal system equations for systems involving fluid flow and chemical reactions of an exothermic or endothermic nature.

8-5. CHEMICAL SYSTEMS[1]

In most cases, modeling the dynamics of chemical reactions is a difficult problem, because of the lack of knowledge concerning the way parameters enter the required relationships and the general complexity of the phenomena. It is possible, however, to consider a few relatively simple cases, as illustrated in the following paragraphs.

A simple chemical reaction can be illustrated by

$$A \xrightarrow{k_1} B \tag{8-5.1}$$

which indicates that chemical A reacts to form chemical B, with a rate coefficient k_1. If this reaction is a so-called first-order reaction, the previous statement can be expressed as

$$dN_A/dt = -k_1 N_A \qquad N_B = N_o - N_A \tag{8-5.2}$$

where N_A and N_B are the concentrations of the chemicals A and B, and N_o is the concentration of A at $t = 0$. In this relationship it can be seen that the rate coefficient k_1 expresses the fractional change of chemical A per unit time, as found by rearranging Eq. 8-5.2 in the form of

$$k_1 = -(dN_A/N_A)/dt \tag{8-5.3}$$

It is also possible to have second order reactions in which the non-linear differential equation

$$dN_A/dt = -k_2 N_A^2 \tag{8-5.4}$$

results. It generally happens that the reaction rate coefficients are functions of other parameters, such as temperature. A linear dependence, such as $k_1 = k_{1o} + \gamma T$, requires development of a model to predict the temperature of the material as well as the concentrations. Later examples illustrate this interaction and that it generally leads to a non-linear system.

Example 8-5.1. It is desired to model mathematically the dynamics of the chemical reaction indicated by

$$A \xrightarrow{k_1} B \xrightarrow{k_2} C \tag{8-5.5}$$

It may be assumed that the reactions are first-order, with rate coefficients k_1 and k_2.

The various relationships can be formed by consideration of the conservation of quantity relationship we have been using. The required differential equations are ex--

[1]See "The Small Analog Computer in the Teaching of Chemical Kinetics at the Advanced Freshman Level," by Myron L. Corrin, *Analog/Hybrid Computer Educational User's Group Applications*, Vol. 1., No. 4, June, 1966.

pressed by

$$\begin{pmatrix}\text{Rate of change of}\\ \text{concentration of } A\\ \text{going into } B\end{pmatrix} = dN_A/dt = -k_1 N_A \tag{8-5.6}$$

$$\begin{pmatrix}\text{Rate of change of}\\ \text{concentration of } B\end{pmatrix} = \begin{pmatrix}\text{Rate } A \text{ is con-}\\ \text{verted to } B\end{pmatrix} - \begin{pmatrix}\text{Rate } B \text{ is con-}\\ \text{verted to } C\end{pmatrix} \tag{8-5.7}$$

$$dN_B/dt = k_1 N_A - k_2 N_B \tag{8-5.8}$$

$$N_C = N_o - N_A - N_B \tag{8-5.9}$$

While it is easily possible to reduce these equations by elimination, this is not pursued here. It might be pointed out that this model is in the form of two first-order differential equations rather than a single higher-order DE. Such a situation will be encountered frequently in the so-called *state-variable* method of analysis, which is developed in Chapters 13 and 14.

Example 8-5.2.[2] The continuous stirred tank reactor is a chemical plant unit in which chemical reactions progress under either heating or cooling conditions. It is desired to develop a mathematical model for such a device, as pictured in Fig. 8-5.1. The variables and parameters used are defined below.

T_i = Temperature of the inlet stream
T_o = Temperature of the outlet stream and mixed batch
T_c = Temperature of the cooling (or heating) fluid
N_{ci} = Inlet reactant concentration (kg-mole/m^3)
N_{co} = Outlet reactant concentration & concentration in tank
F = Inlet and outlet volume flow rate
V = Reactor volume
Q = Heat given off during reaction (exothermic) (Joule/kg-mole)
ρ = specific mass (kg/m^3)
c_p = Specific heat of reactant
k = Reaction rate coefficient
h = Wall heat transfer coefficient
A = Wall heat transfer area

The reaction taking place can be represented as a first-order reaction with the rate coefficient linearly dependent on the reactant temperature, as shown in

$$A \xrightarrow{k} B \qquad k = k_o + \gamma T_o \tag{8-5.10}$$

The differential equations defining the system can be obtained from the material and heat balance or conservation of quantity relationships as follows.

[2] "Start-Up and Heat Transfer in a Stirred-Tank Reactor" by L. Burkhart, *Analog/Hybrid Educational User's Group Application*, Note No. 3, 1969.

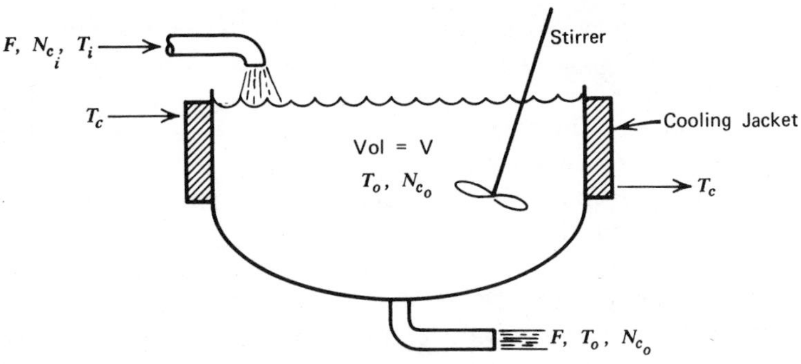

Fig. 8-5.1

Material balance for the reactor:

$$\begin{pmatrix}\text{Accumulated}\\ \text{material } A/\\ \text{unit time}\end{pmatrix} = \begin{pmatrix}\text{Material } A\\ \text{input rate}\end{pmatrix} - \begin{pmatrix}\text{Material } A\\ \text{output rate}\end{pmatrix} - \begin{pmatrix}\text{Material } A \text{ converted to}\\ \text{product } B/\text{unit time}\end{pmatrix} \quad (8\text{-}5.11)$$

$$V\, dN_{co}/dt = F\, N_{ci} - F\, N_{co} - k\, VN_{co} \quad (8\text{-}5.12)$$

Heat balance:

$$\begin{pmatrix}\text{Heat}\\ \text{accumulated/unit time}\end{pmatrix} = \begin{pmatrix}\text{Heat}\\ \text{input}\\ \text{rate}\end{pmatrix} - \begin{pmatrix}\text{Heat}\\ \text{output}\\ \text{rate}\end{pmatrix} - \begin{pmatrix}\text{Heat loss rate}\\ \text{through walls}\end{pmatrix} + \begin{pmatrix}\text{Heat generated by}\\ \text{reaction/unit time}\end{pmatrix}$$

(8-5.13)

$$V\rho c_p\, dT_o/dt = F\rho c_p T_i - F\rho c_p T_o - hA\,(T_o - T_c) + QVkN_{co} \quad (8\text{-}5.14)$$

In general, the temperature of the outlet reactant, T_o, and its concentration, N_{co}, are the dependent variables that are to be solved for by use of Eqs. 8-5.10, 12, and 14. All other quantities must be specified as parameters or independent variables. When the temperature dependence of the reaction rate coefficient, k, expressed in Eq. 8-5.10, is substituted into Eqs. 8-5.12 and 14, a non-linear product involving the dependent variables T_o and N_{co} is obtained. Thus, although the model equations are first-order, they are complicated by being non-linear. The most suitable method of solution is to use either analog or digital computer methods of obtaining numerical results. Incremental linearization is possible, as described in Section 8-2, but this is not pursued here.

8-6. SOCIO-ECONOMIC SYSTEMS

The mathematical modeling of phenomena in the social and business sections of our world is still in an early developmental stage relative to the state of much of physical science. While certain efforts at such modeling of economic systems are several decades old, most of social system modeling has been done in the last two decades. That quantitative understanding of such phenomena is still evolving, and that there are enormous numbers of variables involved, have been major deterrents to model development. The present emphasis on social and environmental problems will, no doubt, accelerate the development of the state-of-the-art in these areas. It should be pointed out that many of the models developed in the areas of socio-economic systems are discrete rather than continuous models. This means that quantities are defined only at discrete time instants, or the variables have values only in the form of sequences. Perhaps business conditions are assessed weekly or monthly rather than at every instant in time; for example, sales orders are filed daily, and pay checks issued weekly. In some cases, however, it is possible to model such situations with continuous variables.

Before considering specific models of socio-economic systems, some general consideration should be given to the types of variables and parameters encountered. When dealing with economic systems, the quantities of concern include such as product and material inventories, production rates, investment, sales rates, demand for products, profit and loss, income, tax rates, and regulatory constraints. It is not difficult to appreciate the fact that the interrelationships among such factors will be most complex, if any appreciable segment of the economy is to be included in the model. Apparently, it is essential that we consider segments of the economic system or of individual businesses small enough to allow comprehensible models to be built. However, the question always remains as to the validity of such segment models when run independently of the rest of the economy. In reality no such segment is free of ties to the economy of the whole world. Understanding and judgment, however, allow useful information to be obtained from such studies of segments of the economy or businesses. Perhaps the most extensive economic models are those for the U.S. economy; however, it appears that even such large models are not doing well enough in providing information for control of the economy to enable us to avoid inflation or unemployment.

Selection of variables and parameters to quantitatively model social systems is an even more tenuous problem. The models that have been used involve such quantities as amount of communication, social activity, hostility, leadership, social status, and desire to maintain social status. It appears that the assignment of quantitative values to such variables is not an easy task, if results are to be of any real value. In the development of such models, it is desirable to consider the segment of society to be modeled, and the purpose of the model, and then to list the obvious variables and reasonable relationships or functional dependences that are known to exist. Other quantities and secondary relationships will undoubtedly emerge in the process, necessitating revision and modification of the system model. After a first-trial model is devised, it is then desirable to test it with certain trial cases and to examine its performance with respect to data from the actual social system being modeled. Such iterative steps are really the same as needed in modeling any *real world* system.

Example 8-6.1. It is desired to develop a simplified mathematical model, for investigation of the variation of inventory of a single product, when subjected to sales as an independent input and when decoupled from such other items as raw material supply, and

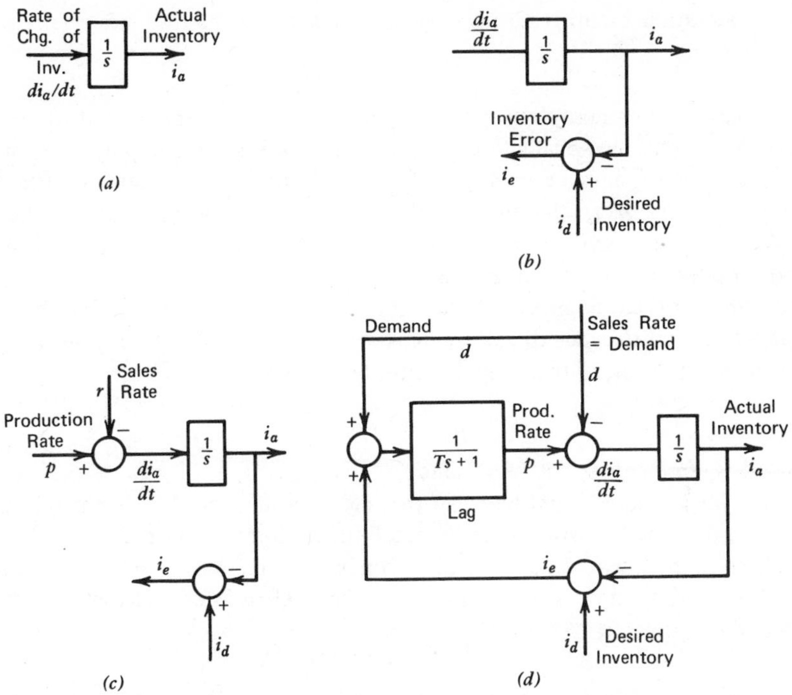

Fig. 8-6.1

shortage of plant capacity. Since it is frequently easier to develop such models via some form of block diagram indicating functional relationships in either mathematical or narrative form, this approach is used. If desired, mathematical equations can be written from the block diagram.

Fig. 8-6.1 illustrates the steps in the development of a suitable model for this example. It starts in Fig. 8-6.1(a) with actual inventory level as output, and indicates the integral relationship between rate of change of inventory and inventory. The difference between desired inventory level and actual inventory is added in the (b). The relationship of sales rate and production rate to inventory rate are included next, in (c), as a simple difference relationship. It is now necessary to consider how the production rate should be related to the error in inventory level and the demand for the product. To simplify matters in this example, a simple proportional relationship with a time lag is presumed to exist and is added in (d) to complete the model. It is apparent that ties to management decisions, availability of raw materials, lag between demand and sales, etc., can be added if desired.

It is now possible to write

$$di_a/dt = p - d \qquad T\,dp/dt + p = i_e + d \qquad i_e = i_d - i_a \tag{8-6.1}$$

from the block diagram. In this case, sales and desired inventory level are required independent inputs to the system model. The model is linear, so normal Laplace transform methods can be used to solve for $i_a(t)$ for particular sales and desired inventory levels.

For consideration of other forms of economic systems the reader is referred to the literature.[3]

Example 8-6.2. A commonly encountered term in social systems is that of *group dynamics*. H. A. Simon[4] has characterized the behavior of a social group by four variables: $I(t)$, the intensity of interaction among the members; $F(t)$, the level of friendliness among the members; $A(t)$, the amount of activity carried on by members within the group; and $E(t)$, the amount of activity imposed on the group by external factors. It is desired to consider Simon's model for such a social system.

In beginning the development of this model, it is reasonable to assume that the intensity of interaction depends directly upon the level of friendliness and the amount of internal activity. A linear, additive relationship, such as

$$I(t) = c_1 F(t) + c_2 A(t) \tag{8-6.2}$$

is postulated as representing this dependency.

A second relationship is that between the rate at which friendliness changes and the level of interaction and existing friendliness. With no interaction it is reasonable to assume that friendliness would decrease to zero while interaction would lead to a rate of change of friendliness. Simon's representation of these relationships is expressed by

$$dF(t)/dt = c_3 I(t) - c_4 F(t) \tag{8-6.3}$$

The use of the $-c_4 F(t)$ term implies that the decay of friendship with no interaction is of an exponential nature.

The final equation relates the level of internal activity to the level of friendliness, external demands, and existing level of internal activity. In the absence of friendliness and external demand for activity, it is postulated that the internal activity would decay exponentially to zero. It is also presumed that the rate of change of internal activity would depend directly upon the level of friendliness existing and upon the external demand for activity. These factors are expressed by

$$dA(t)/dt = c_5 F(t) + c_6 E(t) - c_7 A(t) \tag{8-6.4}$$

It is now possible to consider the form of behavior predicted by the model consisting of Eqs. 8-6.2, 3, and 4. Since the model postulated is linear, it is no problem to Laplace transform the above equations and determine a block diagram relating the various variables. Figure 8-6.2 shows the result of this operation together with the inclusion of initial conditions. The characteristic equation for this system can be found to be

$$s^2 + (c_4 + c_7 - c_1 c_3) s + (c_4 c_7 - c_1 c_3 c_7 - c_2 c_3 c_5) = 0 \tag{8-6.5}$$

[3] "Cybernetics of Economic Systems," by H. M. Runyan, *IEEE Trans. on Systems, Man, and Cybernetics*, Vol. SMC-1, No. 1, Jan., 1971, pp. 8–18.

[4] *Models of Man, Social and Rational*, by Herbert A. Simon, John Wiley & Sons, Inc., 1957, pp. 100-7.

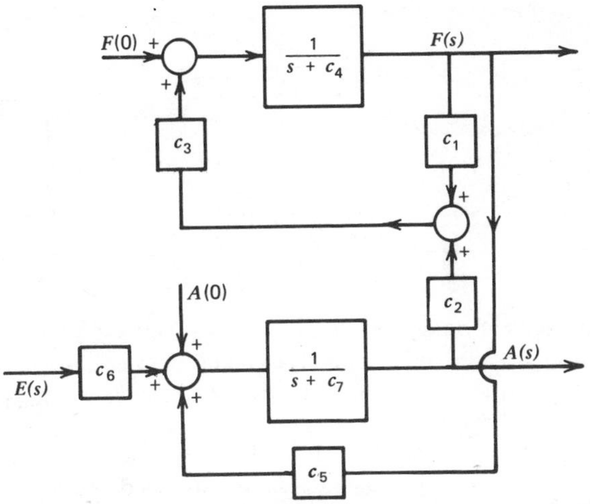

Fig. 8-6.2

It can be seen that the quantity $(c_4 + c_7 - c_1c_3)$ determines whether the roots of the characteristic equation lie in the left-hand-half s-plane. It is possible for this system to be unstable in the sense that the variables become infinitely large. It is also easily possible to determine the equilibrium values for the variables in this social system by setting the rates of change in Eqs. 8-6.3 and 4 equal to zero and solving the resulting algebraic equations (presuming that an equilibrium exists). This results in

$$F_e = c_2c_3c_6\, E_e/(c_4c_7 - c_1c_3c_7 - c_2c_3c_5)$$

$$A_e = c_6(c_4 - c_1c_3)\,E_e/(c_4c_7 - c_1c_3c_7 - c_2c_3c_5) \quad (8\text{-}6.6)$$

Hopefully, the above example illustrates the nature of many mathematical models of social systems. It would certainly seem appropriate and interesting to pursue the development of such systems in spite of the complexity encountered.

PROBLEMS

8-1 Find the transfer function $\delta H(s)/\delta Q(s)$ for incremental changes in the tank system of Fig. 8-2.1.

8-2 The simple tank model of Fig. 8-2.1 is to be used to experimentally determine incremental parameters for the model of Eq. 8-2.29. The tank being used has a uniform cross-sectional area of 1 m^2. It is found that a step-function reduction in input flow rate results in a decrease in the level of fluid in the tank from an original steady-state level of 5 m to a new steady state level of 4 m, with 63.2% of this change occurring in 10 min. What is the incremental transfer function $\delta H(s)/\delta Q_{in}(s)$ with numerical values for the model parameters?

8-3 An incremental model is to be developed for the two-tank system shown.

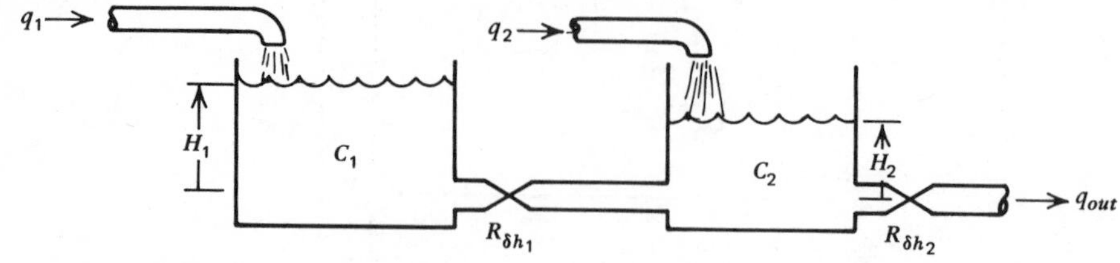

Assume that steady-state values exist for q_1 and q_2, so that $q_{1o} + q_{2o} = q_{oo}$, and H_{1o} and H_{2o} are the steady operating values. Develop the necessary differential equation(s) for finding the effects of incremental changes in q_1 and/or q_2. What are the system's s-plane poles?

8-4 A soft-drink firm desires to study the sales of a new returnable bottle by a consumer field test. The generalized model shown is considered adequate for the test.[5] The following assumptions are made: (a) sales rate is proportional to market-inventory level $h(t)$; (b) container-loss rate is proportional to the number of new containers purchased by consumers, thereby decreasing the number of empty new bottles returned to the bottling plant; (c) consumers are presumed to simply hold the bottles for τ days while consuming the soft drink, before returning empty containers, less some fractional loss component; (d) delays and losses in the bottling plant and market place are negligible as compared to those due to consumers.

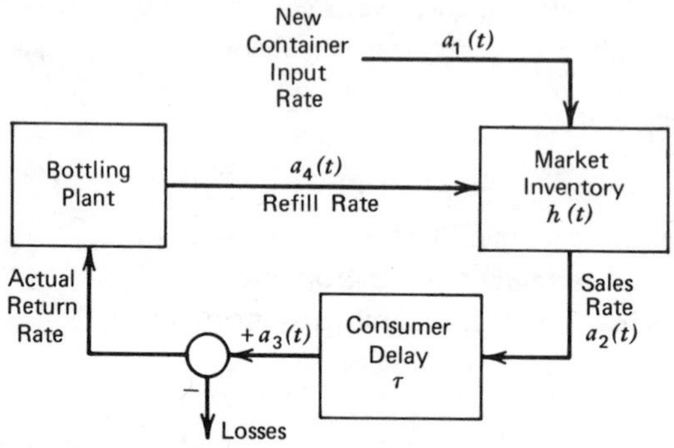

Develop the differential equation for market–inventory level $h(t)$ as a function of $a_1(t)$. If the consumer delay is approximated by the lag function $A_3(s)/A_2(s) = 1/(\tau s + 1)$, determine the transfer function $H(s)/A_1(s)$. For a step function input of 200 new containers/day, determine $h(t)$.

8-5 The hydraulic piston model of Eq. 8-3.16 is to be converted to a numerical model by experimental means. What is the minimum number of parameters that must be

[5] "Analog Computer Demonstration Problem," by David B. Greenberg, *Simulation*, April, 1968, pp. 157-162. The Society for Computer Simulation (Simulation Councils, Inc.).

determined? If one of the tests involves a step change in $x_1(t)$, what happens to $x_2(t)$?

8–6 Develop an incremental linear model for the piston system of Fig. 8–3.2 when the spring load is retained.

8–7 Develop an incremental linear model for the piston system of Fig. 8–3.2 when compressibility flow is ignored and there is no spring load.

8–8 Develop a linear incremental model for the radiation heat flow equation, Eq. 8–4.4.

8–9 An agitated heating tank simplified model of a fluid heating system is shown.[6] It is assumed that the temperature of the fluid in the tank is uniform throughout, due to stirring, and that there is negligible heat loss through tank walls. It is also assumed that there is no accumulation of fluid in the tank, or $F_{in} = F_{out} =$ constant = 45 Kg/min. Derive the mathematical model for this system, with T_i and q as independent inputs. If equilibrium initially exists with $T_i = 10°\text{C}$ and $T_o = 40°\text{C}$, what is the initial heat input? If T_i is increased by a step function to 25°C, while $q(t)$ is decreased by a step function to 2.6×10^6 Joules/min, find the time variation of the outlet temperature, $T_o(t)$.

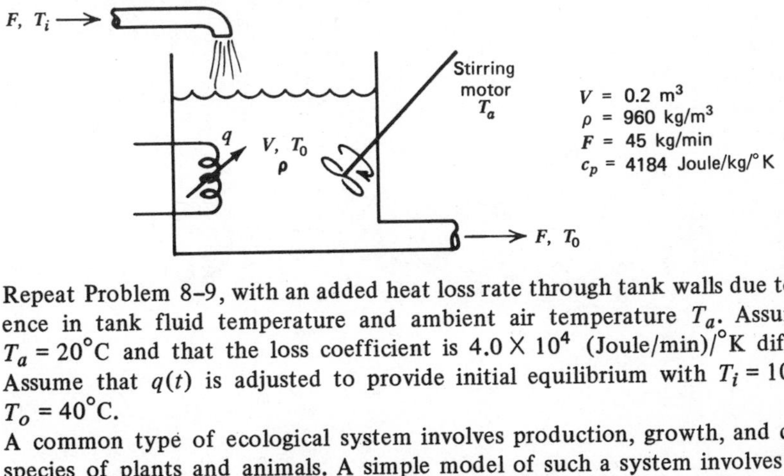

8–10 Repeat Problem 8–9, with an added heat loss rate through tank walls due to difference in tank fluid temperature and ambient air temperature T_a. Assume that $T_a = 20°\text{C}$ and that the loss coefficient is 4.0×10^4 (Joule/min)/°K difference. Assume that $q(t)$ is adjusted to provide initial equilibrium with $T_i = 10°\text{C}$ and $T_o = 40°\text{C}$.

8–11 A common type of ecological system involves production, growth, and death of species of plants and animals. A simple model of such a system involves a single plant species with population, PLNT, and a single herbivore with population, HB.[7] The following conditions are given to assist in developing a mathematical model:

(a) The time rate of increase of PLNT is proportional to PLNT and also $(\text{PLNT})^2$. It also is reduced by a term proportional to the product (HB)(PLNT).

(b) The time rate of increase of HB is proportional to the (HB)(PLNT) product. The death rate for herbivores is proportional to HB and also $(\text{HB})^2$.

Write two first-order differential equations describing the system using the above assumptions. Is this model linear? If not, develop a linear incremental model for this system.

[6] "The Analog Computer is Dead," by Donald C. Martin, *Simulation*, Sept. 1970, pp. 127–130. The Society for Computer Simulation (Simulation Councils, Inc.).

8-12 A certain system is approximated by the mathematical model[7]

$$\frac{dx_1}{dt} = x_1 - 10^{-5}x_1^2 - 10^{-3}x_1x_2 \qquad x_1(0) = 5 \times 10^4$$

$$\frac{dx_2}{dt} = 10^{-3}\alpha x_1x_2 - x_2 - 10^{-3}x_2^2 \qquad x_2(0) = 750$$

Develop a linear increment model from these equations about $x_1(0)$ and $x_2(0)$. What critical value of α causes the incremental model to have poles on the imaginary axis of the Laplace transform s-plane if either s_1 or s_2 are displaced from the initial operating values?

[7] "A Simulation Study of the Effect on Simple Ecological Systems of Making Rate of Increase of Population Density-Dependent," by David Garfinkel, *Simulation*, May, 1967, pp. 275–280. The Society for Computer Simulation (Simulation Councils, Inc.).

Chapter 9

The Two-Sided Laplace Transforms

9-1. INTRODUCTION

There are many situations in which it is of interest to analyze a system if a signal is applied before $t = 0$. Since the defining integral for the one-sided Laplace transform

$$F(s) = \int_0^{\infty} f(t)\, \epsilon^{-st}\, dt \tag{9-1.1}$$

assumes the $f(t)$ is zero for negative time, this integral may place an unwanted restriction on the transform method.

For a very simple example, consider the $f(t)$ of Fig. 9-1.1(a). This $f(t)$ has a one-sided transform $F(s)$. If $F(s)$ is multiplied by ϵ^{-Ts}, the new time function corresponding to the new transform is the original time function shifted to the right by T seconds, as shown in (b).

It seems intuitively evident that T should be able to take on negative values, and the resulting shift of $f(t)$ would be to the left as suggested in (c). However, even an idea as simple as this cannot be applied in the case of the one-sided Laplace transform, because the integral of Eq. 9-1.1 demands that $f(t)$ be zero for negative time.

It would seem that the lower limit on the integral of Eq. 9-1.1 places a restriction on the potential use of this method that can be alleviated if the integral is defined for all time. When this is done, Eq. 9-1.1 becomes

$$F_2(s) = \int_{-\infty}^{+\infty} f_2(t)\, \epsilon^{-st}\, dt \tag{9-1.2}$$

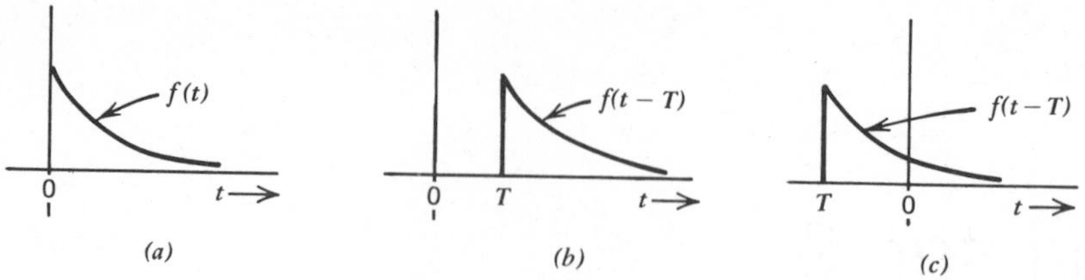

Fig. 9-1.1

This integral is called the two-sided Laplace transform, to distinguish it from Eq. 9-1.1. The subscript in $f_2(t)$ and $F_2(s)$ is to indicate that these functions are two-sided. If the $f_2(t)$ in Eq. 9-1.2 is zero for negative time, Eq. 9-1.2 reduces to Eq. 9-1.1, which indicates the one-sided Laplace transform is a special case of the two-sided Laplace transform.

There is a large body of literature available about methods for handling functions defined for all values of time known as the Fourier transform. The two-sided Laplace transformation methods and the Fourier transformation methods are very closely related, and there is no one way of presenting this relationship. Some authors prefer to start the entire discussion from the Fourier series, then move to the Fourier transform, and finally introduce the Laplace transform. We have obviously started with the Laplace transforms, and then introduced the two-sided Laplace transforms. This book will move on to the Fourier transform, and finally go from the Fourier transform to the Fourier series.

In certain quarters, much energy has been expended debating whether Fourier transforms are superior to Laplace transforms or it is the Laplace transforms that are superior. To us, this debate seems to be a waste of time. A person should not commit himself exclusively to either method, but should use whichever is the more appropriate in a given situation.

9-2. A BRIEF REVIEW OF THE ONE-SIDED LAPLACE TRANSFORM

Before starting on the subject of the two-sided Laplace transform, we review briefly certain aspects of the one-sided Laplace transform.

For $f(t)$ to have a one-sided Laplace transform, certain conditions must be satisfied. The $f(t)$ must be of exponential order, and must be sectionally continuous. If these conditions are satisfied, $F(s)$ exists. If the system is made up of lumped elements, the $F(s)$ is a ratio of polynomials in s.

$$F(s) = \frac{a(s)}{b(s)} \tag{9-2.1}$$

The pole-zero configuration for such an $F(s)$ is suggested in Fig. 9-2.1. The abscissa of convergence $[\sigma_0]$ goes through the pole that is the farthest to the right in the complex s-plane. The region of convergence is to the right of the abscissa of convergence and is

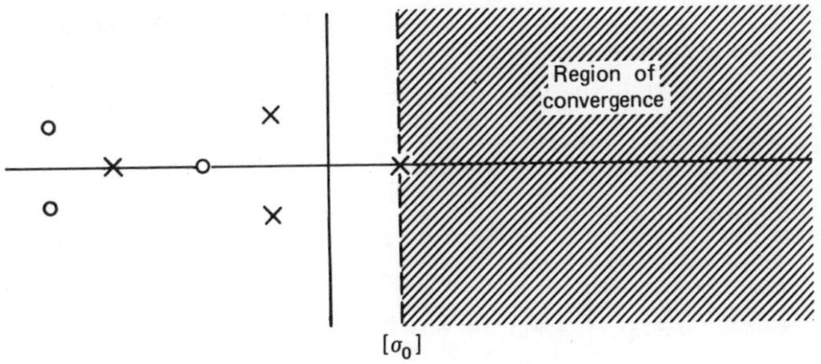

Fig. 9-2.1

shown by the shaded area in the figure. The integral of Eq. 9-1.1 exists within this region of convergence, but does not exist outside or on the boundary of the region of convergence.

Very little has been done with the inverse Laplace transform integral, but let us write it again in order to remain familiar with its significance.

$$f(t) = \frac{1}{2\pi j} \int_{c-j\infty}^{c+j\infty} F(s)\, \epsilon^{st}\, ds \tag{9-2.2}$$

The integration is in the complex plane along a line c units distant from the $j\omega$ axis, from $(c - j\infty)$ to $(c + j\infty)$, as shown in Fig. 9-2.2. The c factor is chosen so that the path of integration is in the region of convergence.

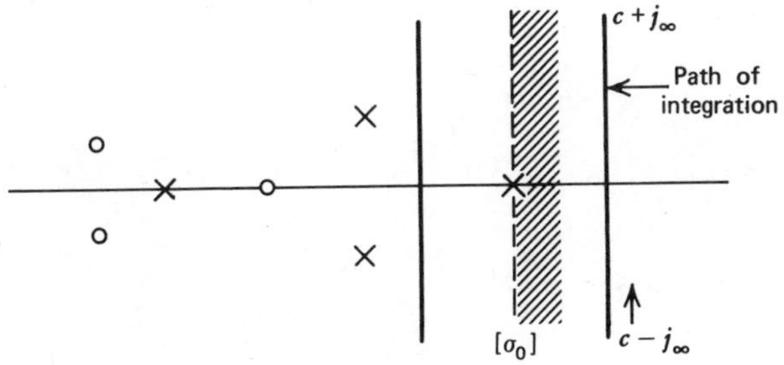

Fig. 9-2.2

9-3. THE DIRECT TWO-SIDED LAPLACE TRANSFORM

The direct two-sided Laplace transform is given by

$$F_2(s) = \int_{-\infty}^{+\infty} f_2(t)\, \epsilon^{-st}\, dt \tag{9-3.1}$$

where in general, the $f_2(t)$ is defined for all values of time. For convenience, $f_2(t)$ is divided into the two functions defined in their respective intervals as

$$f_2(t) = \begin{cases} f_a(t) & \text{for } t < 0 \\ f_b(t) & \text{for } t > 0 \end{cases} \tag{9-3.2}$$

Equation 9-3.1 can be written as

$$F_2(s) = \int_{-\infty}^{0} f_a(t)\,\epsilon^{-st}\,dt + \int_{0}^{\infty} f_b(t)\,\epsilon^{-st}\,dt \tag{9-3.3}$$

The discussion of the convergence of these integrals is postponed until the mechanics of finding $F_2(s)$ are covered.

For notational convenience, $F_a(s)$ and $F_b(s)$ are defined as

$$F_a(s) = \int_{-\infty}^{0} f_a(t)\,\epsilon^{-st}\,dt \tag{9-3.4}$$

$$F_b(s) = \int_{0}^{\infty} f_b(t)\,\epsilon^{-st}\,dt \tag{9-3.5}$$

If $F_2(s)$ exists, it can be written as

$$F_2(s) = F_a(s) + F_b(s) \tag{9-3.6}$$

The $F_b(s)$ of Eq. 9-3.5 is precisely the direct one-sided Laplace transform of Eq. 9-1.1, and all the concepts that have been learned about the one-sided Laplace transform apply to this integral. However, $F_a(s)$ is new and requires some attention

$$F_a(s) = \int_{-\infty}^{0} f_a(t)\,\epsilon^{-st}\,dt \tag{9-3.7}$$

A change in variables is made

$$t = -\tau \tag{9-3.8}$$

when

$$\begin{array}{ll} t = -\infty & \tau = +\infty \\ t = 0 & \tau = 0 \end{array} \tag{9-3.9}$$

The differential becomes

$$dt = -d\tau \tag{9-3.10}$$

and $F_a(s)$ can be written as

$$F_a(s) = -\int_{\infty}^{0} f_a(-\tau)\,\epsilon^{-(-s)\tau}\,d\tau \tag{9-3.11}$$

The limits are exchanged as

$$F_a(s) = \int_0^{\infty} f_a(-\tau)\, \epsilon^{-(-s)\tau}\, d\tau \tag{9-3.12}$$

To be completely explicit, $-s$ is replaced by p, for

$$F_a(p) = \int_0^{\infty} f_a(-\tau)\, \epsilon^{-p\tau}\, d\tau \tag{9-3.13}$$

This integral is precisely the same as the usual one-sided Laplace transform except the result is a function of p. The procedure is that the $f_a(t)$ can be manipulated in such a way as to obtain $F_a(s)$ by using previous knowledge about the one-sided Laplace transform. This method is itemized in the following four steps:

Step 1. The $f_2(t)$ is examined, and $f_a(t)$ is determined.

Step 2. The $f_a(t)$ is folded around the $t = 0$ axis; that is, $f_a(-\tau)$ is determined from $f_a(t)$ by replacing t with $-\tau$.

Step 3. The Laplace transform of $f_a(-\tau)$ is taken in the usual one-sided manner; however, this is now a function of p.

Step 4. The p in $\mathcal{L}[f_a(-\tau)]$ is replaced by $-s$ to yield the desired $F_a(s)$.

These ideas are demonstrated through the use of examples.

Example 9-3.1. The $f_2(t)$ for this example is given by

$$f_2(t) = \begin{cases} \epsilon^{+2t} & t < 0 \\ \epsilon^{-3t} & t > 0 \end{cases} \tag{9-3.14}$$

and a sketch of $f_2(t)$ is shown in Fig. 9-3.1. The $F_a(s)$ is found from

Step 1. $$f_a(t) = \epsilon^{+2t} \tag{9-3.15}$$

Step 2. $$f_a(-\tau) = \epsilon^{-2\tau} \tag{9-3.16}$$

Step 3. $$\mathcal{L}[f_a(-\tau)] = \frac{1}{p+2} \tag{9-3.17}$$

Step 4. $$F_a(s) = \frac{1}{-s+2} = -\frac{1}{s-2} \tag{9-3.18}$$

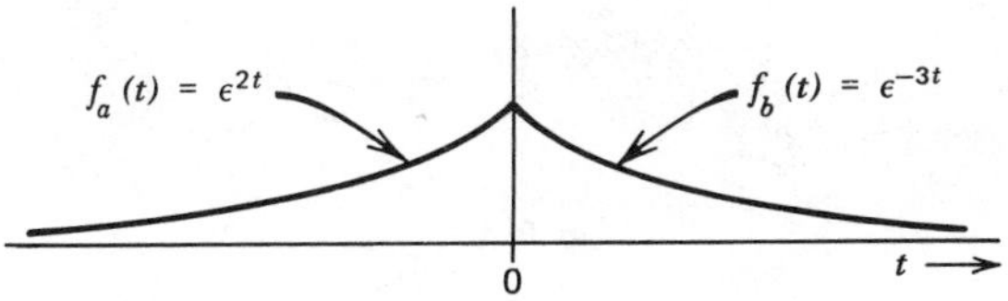

Fig. 9-3.1

The $F_b(s)$ is found from

$$f_b(t) = \epsilon^{-3t}$$

using conventional one-sided Laplace methods, as

$$F_b(s) = \frac{1}{s+3} \tag{9-3.19}$$

The step of replacing p by $-s$, as used in finding $F_a(s)$, has an important implication. This step essentially folds the s-plane around the $j\omega$-axis. As a result, the abscissa of convergence for $F_a(s)$ goes through the pole farthest to the left in the s-plane, and the region of convergence is to the left of the abscissa of convergence.

For this example, there is only one pole at $s = +2$ (as shown in Fig. 9-3.2(a)), and the abscissa of convergence $[\sigma_0]_a$ goes through this pole and the region of convergence is to the left of $[\sigma_0]_a$.

The location of the pole of $F_b(s)$, the abscissa of convergence $[\sigma_0]_b$, and the region of convergence of $F_b(s)$ are shown in Fig. 9-3.2(b).

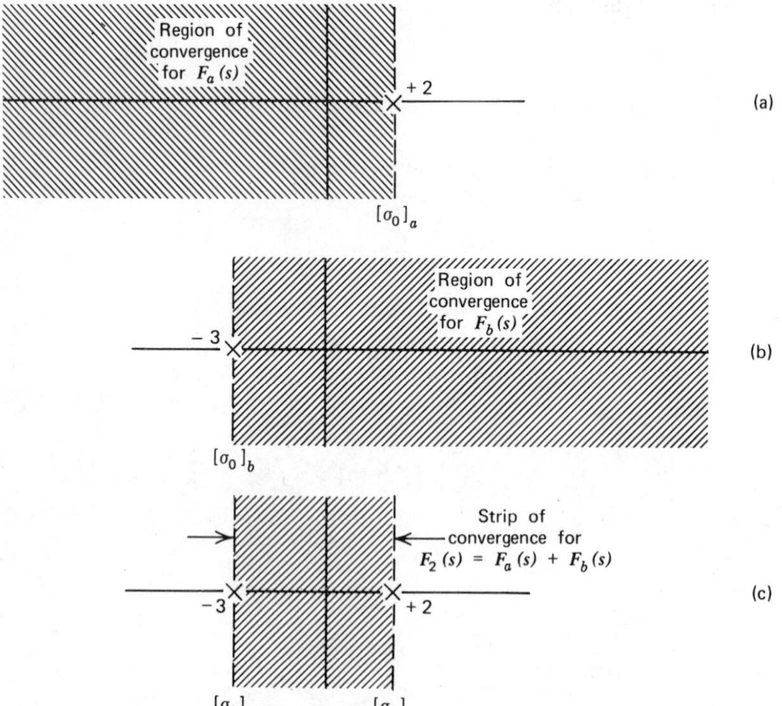

Fig. 9-3.2

The $F_a(s)$ cannot be added to the $F_b(s)$ to obtain $F_2(s)$ without making certain that this is a legitimate step. To explore this, the integral is examined.

$$\int_{-\infty}^{+\infty} f_2(t)\,\epsilon^{-st}\,dt = \int_{-\infty}^{0} f_a(t)\,\epsilon^{-st}\,dt + \int_{0}^{\infty} f_b(t)\,\epsilon^{-st}\,dt \tag{9-3.20}$$

The question is: Does the integral of $f_2(t)$ exist? The integral of $f_a(t)$ exists to the left of $s = +2$. The integral of $f_b(t)$ exists to the right of $s = -3$. The integral of $f_2(t)$ can exist only if the regions of convergence for $F_a(s)$ and $F_b(s)$ have a common portion, so that both regions are satisfied simultaneously. The common portion of these two regions is called *the strip of convergence*, and for this example exists between $\sigma = -3$ and $\sigma = +2$, as shown in Fig. 9-3.2(c). Therefore, $F_2(s)$ exists in this strip and is the sum of $F_a(s)$ and $F_b(s)$

$$F_2(s) = F_a(s) + F_b(s) \tag{9-3.21}$$

For this specific example:

$$F_2(s) = \frac{-1}{s-2} + \frac{1}{s+3} = \frac{-5}{(s-2)(s+3)} \tag{9-3.22}$$

with the strip of convergence as given by

$$-3 < \sigma < +2 \tag{9-3.23}$$

For a two-sided Laplace transform to be completely specified, $F_2(s)$ and the strip of convergence must be given.

Example 9-3.2. The $f_2(t)$ for this example is given by

$$f_2(t) = \begin{cases} \epsilon^{+4t} & t < 0 \\ \epsilon^{+3t} & t > 0 \end{cases} \tag{9-3.24}$$

and a sketch of $f_2(t)$ is shown in Fig. 9-3.3. The $F_a(s)$ is found from—

Step 1.

$$f_a(t) = \epsilon^{+4t} \tag{9-3.25}$$

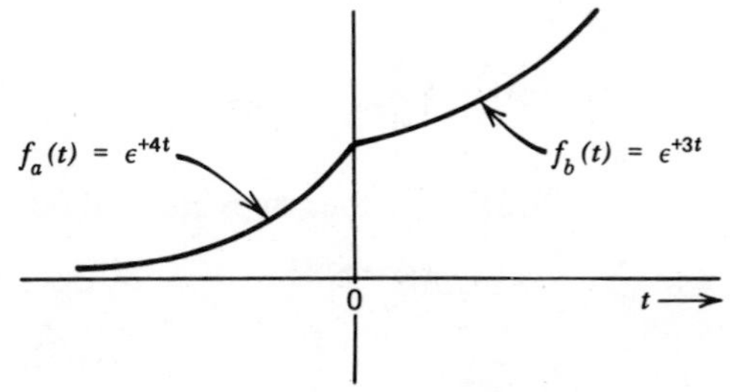

Fig. 9-3.3

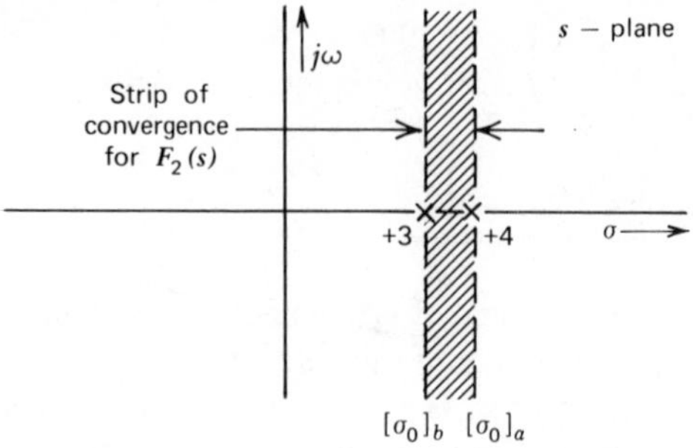

Fig. 9-3.4

Step 2. $$f_a(-\tau) = \epsilon^{-4\tau} \tag{9-3.26}$$

Step 3. $$\mathcal{L}[f_a(-\tau)] = \frac{1}{p+4} \tag{9-3.27}$$

Step 4. $$F_a(s) = \frac{1}{-s+4} = -\frac{1}{s-4} \tag{9-3.28}$$

The $F_b(s)$ is found as

$$f_b(t) = \epsilon^{+3t} \tag{9-3.29}$$

$$F_b(s) = \frac{1}{s-3} \tag{9-3.30}$$

The abscissas of convergence for $F_a(s)$ and $F_b(s)$, and the strip of convergence, are shown in Fig. 9-3.4. The $F_2(s)$ exists, and is

$$F_2(s) = F_a(s) + F_b(s) = \frac{-1}{s-4} + \frac{1}{s-3} = \frac{-1}{(s-4)(s-3)} \tag{9-3.31}$$

With a strip of convergence

$$3 < \sigma < 4 \tag{9-3.32}$$

Example 9-3.3. The $f_2(t)$ for this example is given by

$$f_2(t) = \begin{cases} \epsilon^{+3t} & t < 0 \\ \epsilon^{+4t} & t > 0 \end{cases} \tag{9-3.33}$$

A sketch of $f_2(t)$ is shown in Fig. 9-3.5. The four steps are used on

$$f_a(t) = \epsilon^{+3t} \tag{9-3.34}$$

to find

$$F_a(s) = \frac{1}{-s+3} = -\frac{1}{s-3} \tag{9-3.35}$$

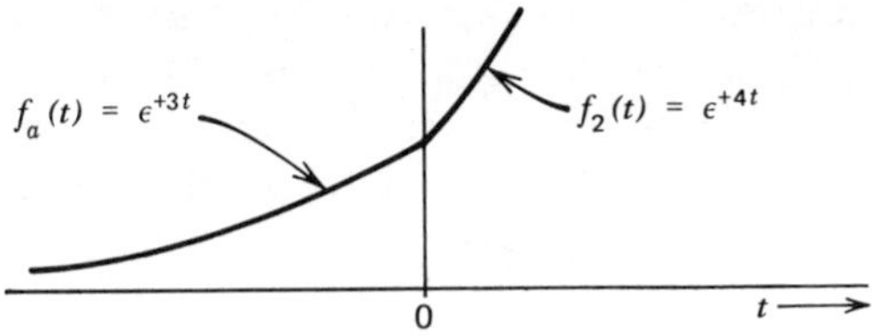

Fig. 9-3.5

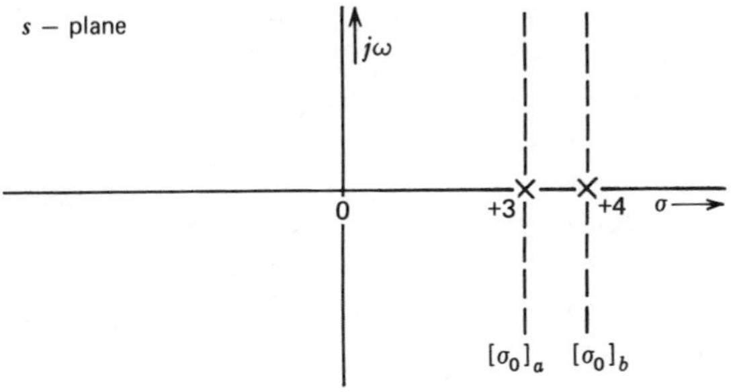

Fig. 9-3.6

and the $F_b(s)$ is found as

$$F_b(s) = \frac{1}{s - 4} \tag{9-3.36}$$

The abscissas of convergence for $F_a(s)$ and $F_b(s)$ are shown in Fig. 9-3.6. The $F_a(s)$ integral converges to the left of $[\sigma_0]_a$ and the $F_b(s)$ integral converges to the right of $[\sigma_0]_b$. These two regions of convergence have no common portion and the $F_2(s)$ integral does not converge anywhere. Therefore, this $f_2(t)$ does not have a two-sided Laplace transform.

Example 9-3.4. The $f_2(t)$ for this example is given by

$$f_2(t) = \epsilon^{-5t} \qquad \text{for all } t \tag{9-3.37}$$

The four steps are used on

$$f_a(t) = \epsilon^{-5t} \tag{9-3.38}$$

to find

$$F_a(s) = \frac{1}{-s - 5} = -\frac{1}{s + 5} \tag{9-3.39}$$

The $F_b(s)$ is found as

$$F_b(s) = \frac{1}{s + 5} \tag{9-3.40}$$

The abscissas of convergence for both $F_a(s)$ and $F_b(s)$ are at $\sigma = -5$. The region of convergence for $F_a(s)$ is to the left of -5, and the region of convergence for $F_b(s)$ is to the right of -5. Therefore, there is no strip of convergence, and strictly speaking, the $F_2(s)$ does not exist. However, we are able to give meaning to this particular situation through a limiting process, as will be taken up in Section 9-6.

COMMENTARY

When all possible problems are considered, it is seen that the one-sided Laplace transform handles a certain class of these problems. This class is made up of functions defined only for positive time. The $f(t)$ must be of expotential order and sectionally continuous. As discussed several times, most problems of interest to the physical scientist automatically satisfy these last two restrictions.

Problems involving functions defined for all values of time can be handled with relatively simple extensions to the one-sided Laplace transform theory. For convenience the $f_2(t)$ is divided into an $f_a(t)$ and an $f_b(t)$. The $f_a(t)$ and the $f_b(t)$ must individually be of exponential order and sectionally continuous before $F_a(s)$ and $F_b(s)$ can exist separately.

The $F_a(s)$ is obtained from $f_a(t)$ by use of the four steps as discussed. The resulting $F_a(s)$ has a set of poles and zeros, as suggested in Fig. 9-3.7. The abscissa of convergence $[\sigma_0]_a$ goes through the pole of $F_a(s)$ farthest to the left in the s-plane.

The $F_b(s)$ is obtained from $f_b(t)$ in the same manner as any one-sided Laplace transform is obtained. The resulting $F_b(s)$ has a set of poles and zeros as suggested in Fig. 9-3.7. The abscissa of convergence $[\sigma_0]_b$ goes through the pole of $F_b(s)$ farthest to the right in the s-plane.

If $[\sigma_0]_a$ is to the right of $[\sigma_0]_b$, a strip of convergence exists, and $F_2(s)$ is the sum of $F_a(s)$ and $F_b(s)$.

$$[\sigma_0]_b < \sigma < [\sigma_0]_a \tag{9-3.41}$$

$$F_2(s) = F_a(s) + F_b(s) \tag{9-3.42}$$

If $[\sigma_0]_a$ is to the left of $[\sigma_0]_b$, there is no region in the s-plane where both $F_a(s)$ and $F_b(s)$ exist simultaneously, and hence $F_2(s)$ does not exist anywhere. The corresponding $f_2(t)$ does not have a two-sided Laplace transform.

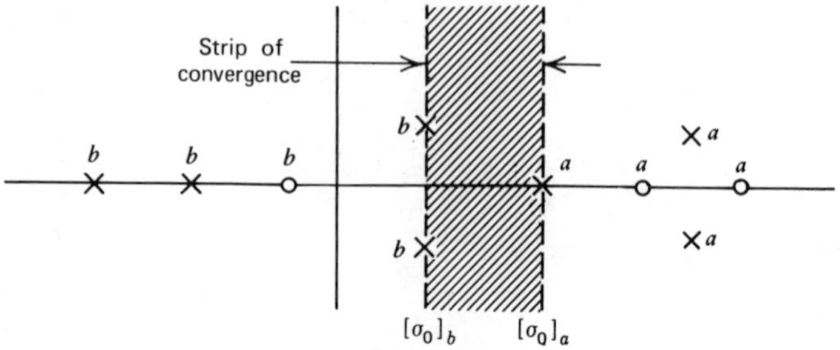

Fig. 9-3.7

If $[\sigma_0]_a = [\sigma_0]_b$, the $f_2(t)$ does not have a two-sided Laplace transform in the strict sense of the term. However, through the use of a limiting process an interpretation can be given to this situation. This is done later.

9-4. THE INVERSE TWO-SIDED LAPLACE TRANSFORM

The inverse two-sided Laplace transform can be found through use of the inverse transform integral

$$f_2(t) = \frac{1}{2\pi j} \int_{c-j\infty}^{c+j\infty} F_2(s)\, \epsilon^{st}\, ds \tag{9-4.1}$$

For the $F_2(s)$ to exist, there must be a strip of convergence in the s-plane, as suggested in Fig. 9-4.1. A constant c must be chosen that puts the path of integration within this strip of convergence. This integration is in the complex plane along this line from $(c - j\infty)$ to $(c + j\infty)$.

Again, in order to avoid using complex variable theory, the inverse transform integral is not used. The $f_2(t)$ is found from $F_2(s)$ by reversing the above steps, as demonstrated in the next few examples.

Example 9-4.1. The problem is to find the $f_2(t)$, given

$$F_2(s) = \frac{-2}{(s-4)(s-6)} \qquad 4 < \sigma < 6 \tag{9-4.2}$$

The $F_2(s)$ is expanded as

$$F_2(s) = \frac{-2}{(s-4)(s-6)} = \frac{-1}{s-6} + \frac{1}{s-4} \tag{9-4.2a}$$

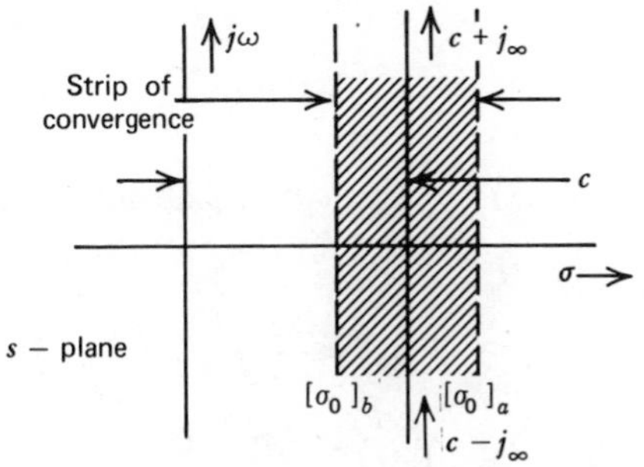

Fig. 9-4.1

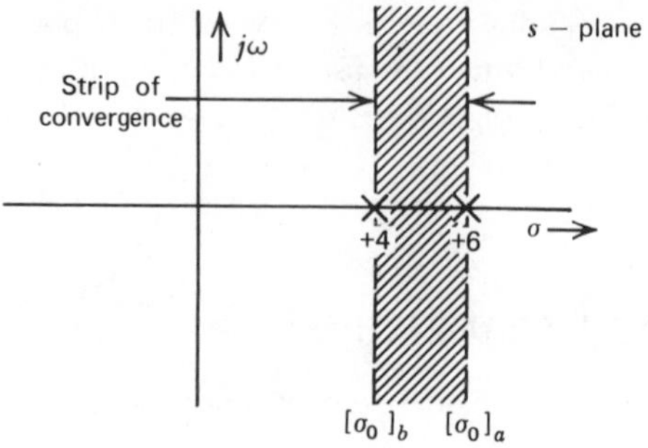

Fig. 9-4.2

The poles are located in the complex s-plane, and the strip of convergence is labeled as shown in Fig. 9–4.2.

The abscissa of convergence contains the information as to how the poles of $F_2(s)$ are to be divided between $F_a(s)$ and $F_b(s)$. The following statements are stated in general terms that can be applied to any problem.

All the poles of $F_2(s)$ that are either on $[\sigma_0]_a$ or to the right belong to the $F_a(s)$. All of the poles of $F_2(s)$ that are either on $[\sigma_0]_b$ or to the left belong to $F_b(s)$. These statements must be true, because this is how the strip of convergence is determined.

When these rules are applied to this specific example, $F_a(s)$ and $F_b(s)$ are determined to be

$$F_a(s) = -\frac{1}{s-6} \qquad F_b(s) = \frac{1}{s-4} \tag{9–4.3}$$

The $f_a(t)$ is found from $F_a(s)$ by using the four steps procedure in the reverse direction:

Step 4. $$F_a(s) = -\frac{1}{s-6}$$

Step 3. $$\mathcal{L}[f_a(-\tau)] = -\frac{1}{-p-6} = \frac{1}{p+6} \tag{9–4.4}$$

Step 2. $$f_a(-\tau) = \epsilon^{-6\tau}$$

Step 1. $$f_a(t) = \epsilon^{+6t}$$

The $f_b(t)$ is found from $F_b(s)$ by using the usual one-sided inverse transformation methods, as

$$F_b(s) = \frac{1}{s-4}$$
$$f_b(t) = \epsilon^{+4t} \tag{9–4.5}$$

The desired $f_2(t)$ has been determined as

$$f_2(t) = \begin{cases} \epsilon^{+6t} & t<0 \\ \epsilon^{+4t} & t>0 \end{cases} \tag{9-4.6}$$

Example 9-4.2. The problem is to find the $f_2(t)$, given

$$F_2(s) = \frac{-2s^2 - 4s - 25}{(s^2 + 5^2)(s+4)} \qquad -4 < \sigma < 0 \tag{9-4.7}$$

The $F_2(s)$ is expanded as

$$F_2(s) = \frac{-2s^2 - 4s - 25}{(s^2 + 5^2)(s+4)} = \frac{-2s^2 - 4s - 25}{(s+j5)(s-j5)(s+4)} = \frac{-\frac{1}{2}}{s+j5} + \frac{-\frac{1}{2}}{s-j5} + \frac{-1}{s+4} \tag{9-4.8}$$

The poles are located on the complex s-plane and the given strip of convergence is marked and labeled as shown in Fig. 9-4.3.

The poles on $[\sigma_0]_a$ are assigned to $F_a(s)$, as

$$F_a(s) = \frac{-\frac{1}{2}}{s+j5} + \frac{-\frac{1}{2}}{s-j5} \tag{9-4.9}$$

and the pole on $[\sigma_0]_b$ is assigned to $F_b(s)$, as

$$F_b(s) = -\frac{1}{s+4} \tag{9-4.10}$$

The $f_a(t)$ is found from $F_a(s)$ by using the four steps in reverse:

Step 4. $$F_a(s) = \frac{-\frac{1}{2}}{s+j5} + \frac{-\frac{1}{2}}{s-j5}$$

Step 3. $$\mathcal{L}[f_a(-\tau)] = \frac{\frac{1}{2}}{p-j5} + \frac{\frac{1}{2}}{p+j5} \tag{9-4.11}$$

Step 2. $$f_a(-\tau) = \tfrac{1}{2}\epsilon^{+j5\tau} + \tfrac{1}{2}\epsilon^{-j5\tau}$$

Step 1. $$f_a(t) = \tfrac{1}{2}(\epsilon^{-j5t} + \epsilon^{+j5t}) = \cos 5t$$

The $f_b(t)$ is found as:

$$f_b(t) = -\epsilon^{-4t} \tag{9-4.12}$$

The desired $f_2(t)$ has been determined as:

$$f_2(t) = \begin{cases} \cos 5t & t<0 \\ -\epsilon^{-4t} & t>0 \end{cases} \tag{9-4.13}$$

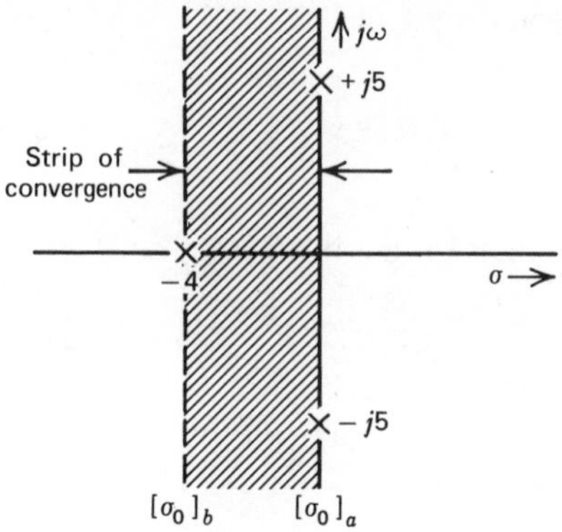

Fig. 9-4.3

9-5. THE STRIP OF CONVERGENCE MUST BE SPECIFIED

In Section 9-4, reference was made several times to the need for some knowledge of the strip of convergence. To dramatize this situation, suppose an $F_2(s)$ is given but nothing is said about the strip of convergence. As an example

$$F_2(s) = \frac{(2s+1)(s+4)}{(s-2)(s-1)(s+1)(s+2)} \tag{9-5.1}$$

This $F_2(s)$ can be expanded as

$$F_2(s) = \frac{\frac{5}{2}}{s-2} - \frac{\frac{5}{2}}{s-1} - \frac{\frac{1}{2}}{s+1} + \frac{\frac{1}{2}}{s+2} \tag{9-5.2}$$

The problem is now encountered as to how the system poles are to be divided between $F_a(s)$ and $F_b(s)$.

To begin the example, the poles of $F_2(s)$ are divided in an arbitrary manner as in Fig. 9-5.1. The poles labeled with a's are to be assigned to $F_a(s)$, and those with b's to $F_b(s)$. The abscissa of convergence $[\sigma_0]_a$ goes through the pole of $F_a(s)$ farthest to the left in the complex s-plane, whereas $[\sigma_0]_b$ goes through the pole of $F_b(s)$ farthest to the right. The $[\sigma_0]_a$ and $[\sigma_0]_b$ are added to the figure as shown.

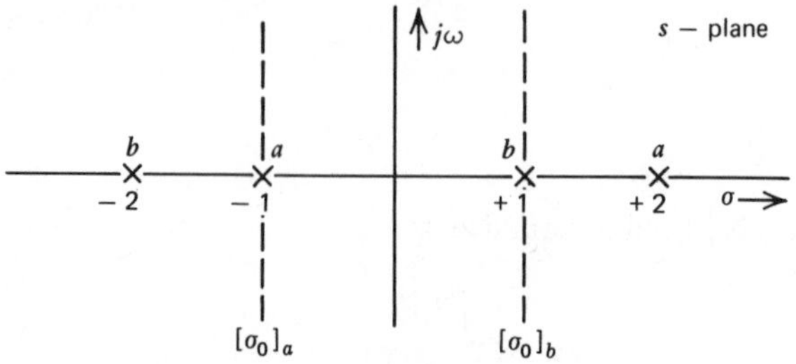

Fig. 9-5.1

Table 9-5.1

Strips of convergence	Possible answers
$-\infty < \sigma < -2$	Case 1
$-2 < \sigma < -1$	Case 2
$-1 < \sigma < +1$	Case 3
$+1 < \sigma < +2$	Case 4
$+2 < \sigma < +\infty$	Case 5

An examination of Fig. 9-5.1 shows that a contradiction has been reached. If we had started with an $f_2(t)$ that yielded the $[\sigma_0]_a$ and $[\sigma_0]_b$ as shown, we would know that no strip of convergence exists and therefore that $F_2(s)$ does not exist. Yet the statement of the problem says that $F_2(s)$ does exist. Therefore, the $F_2(s)$ has been divided into $F_a(s)$ and $F_b(s)$ in the wrong manner.

For a strip of convergence to exist, all the poles of $F_a(s)$ must be to the right and all the poles of $F_b(s)$ must be to the left of the strip. For the $F_2(s)$ of this example, there are five possible solutions, as shown in Table 9-5.1.

The inverse transform is determined for two of these cases, and the other three cases are left as problems.

Case I. For Case I, the strip of convergence is $-\infty < \sigma < -2$, and all four poles of $F_2(s)$ are assigned to $F_a(s)$, as

$$F_a(s) = \frac{\frac{5}{2}}{s-2} - \frac{\frac{5}{2}}{s-1} - \frac{\frac{1}{2}}{s+1} + \frac{\frac{1}{2}}{s+2} \tag{9-5.3}$$

with

$$F_b(s) = 0 \tag{9-5.4}$$

The $f_a(t)$, $f_b(t)$, and $f_2(t)$ are found in the now familiar way.

$$f_2(t) = \begin{cases} -\frac{5}{2}\epsilon^{+2t} + \frac{5}{2}\epsilon^{+1t} + \frac{1}{2}\epsilon^{-t} - \frac{1}{2}\epsilon^{-2t} & t < 0 \\ 0 & t > 0 \end{cases} \tag{9-5.5}$$

Case IV. For Case IV, the strip of convergence is $+1 < \sigma < +2$, and the four poles of $F_2(s)$ are divided between $F_a(s)$ and $F_b(s)$, as

$$F_a(s) = \frac{\frac{5}{2}}{s-2} \tag{9-5.6}$$

and

$$F_b(s) = \frac{-\frac{5}{2}}{s-1} + \frac{-\frac{1}{2}}{s+1} + \frac{\frac{1}{2}}{s+2} \tag{9-5.7}$$

and the $f_2(t)$ is found to be

$$f_2(t) = \begin{cases} -\frac{5}{2}\epsilon^{+2t} & t<0 \\ -\frac{5}{2}\epsilon^{+1t} - \frac{1}{2}\epsilon^{-1t} + \frac{1}{2}\epsilon^{-2t} & t>0 \end{cases} \tag{9-5.8}$$

The conclusion to be reached from this is that the strip of convergence must be specified in order to determine a unique inverse transform.

9-6. TWO-SIDED LAPLACE TRANSFORMS THAT EXIST IN THE LIMIT

There are a number of situations in which the two-sided Laplace transform does not exist in the literal sense of the term. However, the original function can be modified to give the modified function a two-sided Laplace transform. Then a limiting process is performed on the modified function in such a way that it approaches the original function in the limit. Next, the transform of the modified function is observed during this limiting process, and if a limit for this modified transform exists, it is said to be the two-sided Laplace transform in the limit of the original function.

This concept can best be illustrated with several examples.

Example 9-6.1. In Example 9-3.4, the function

$$f_2(t) = \epsilon^{-5t} \qquad \text{for all } t \tag{9-6.1}$$

was explored, and we said that this function did not have a two-sided Laplace transform in the literal meaning of that term.

The $F_a(s)$ and $F_b(s)$ were found to be

$$F_a(s) = -\frac{1}{s+5} \qquad F_b(s) = \frac{1}{s+5} \tag{9-6.2}$$

The $F_a(s)$ function converges to the left of a line $s = -5$, and the $F_b(s)$ function converges to the right of a line $s = -5$. Since $s = -5$ is the boundary of both regions, it does not exist in either region and hence there is no strip of convergence.

The $f_2(t)$ is modified to an $f_{2m}(t)$ in such a way that the modified function has a strip of convergence in the s-plane.

$$f_{2m}(t) = \begin{cases} \epsilon^{(-5+\delta)t} & t<0 \\ \epsilon^{(-5-\delta)t} & t>0 \end{cases} \tag{9-6.3}$$

The transforms of these functions are:

$$F_{am}(s) = -\frac{1}{s+5-\delta} \tag{9-6.4}$$

and

$$F_{bm}(s) = \frac{1}{s+5+\delta} \tag{9-6.5}$$

The strip of convergence is shown in Fig. 9-6.1.

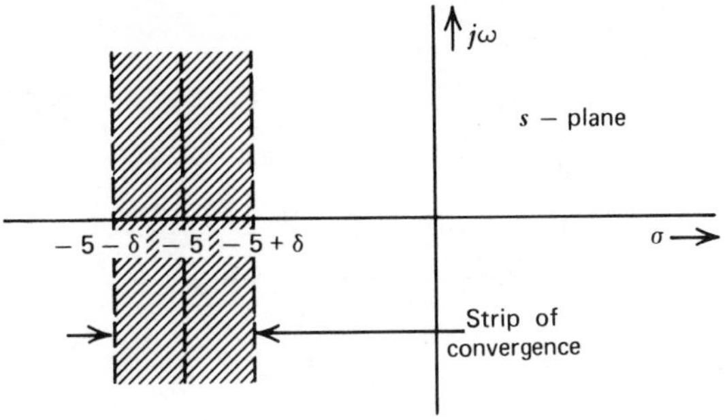

Fig. 9-6.1

As long as $\delta > 0$, $F_{2m}(s)$ exists, and is given by

$$F_{2m}(s) = -\frac{1}{s+5-\delta} + \frac{1}{s+5+\delta} \tag{9-6.6}$$

Next we want to explore the $f_{2m}(t)$ and $F_{2m}(s)$ functions as δ approaches zero in the limit. The $f_{2m}(t)$ was chosen in such a way that

$$f_{2m}(t)|_{\delta\to 0} \longrightarrow f_2(t) \tag{9-6.7}$$

The real problem comes about in determining the limit of $F_{2m}(s)$ as δ approaches zero.

A hasty examination of the $F_{2m}(s)$ of Eq. 9-6.6 indicates that as δ approaches zero, $F_{2m}(s)$ takes on the value of zero. The fact that this is not so is actually very disturbing. The thing we are doing is taking the limit on δ too soon. We should "take it easy" and get a feel for the situation before jumping to any conclusions.

The s in $F_{2m}(s)$ is replaced by $s = \sigma + j\omega$, and $F_{2m}(s)$ becomes

$$F_{2m}(\sigma + j\omega) = -\frac{1}{\sigma + j\omega + 5 - \delta} + \frac{1}{\sigma + j\omega + 5 + \delta} \tag{9-6.8}$$

The two parts of this function are rationalized as

$$F_{2m}(\sigma + j\omega) = \frac{-[(\sigma+5-\delta) - j\omega]}{[(\sigma+5-\delta)+j\omega]\,[(\sigma+5-\delta)-j\omega]} + \frac{[(\sigma+5+\delta) - j\omega]}{[(\sigma+5+\delta)+j\omega]\,[(\sigma+5+\delta)-j\omega]} \tag{9-6.9}$$

$$F_{2m}(\sigma + j\omega) = \frac{-(\sigma+5-\delta)}{(\sigma+5-\delta)^2+\omega^2} + \frac{j\omega}{(\sigma+5-\delta)^2+\omega^2} + \frac{\sigma+5+\delta}{(\sigma+5+\delta)^2+\omega^2} - \frac{j\omega}{(\sigma+5+\delta)^2+\omega^2} \tag{9-6.10}$$

This $F_{2m}(\sigma + j\omega)$ function is explored along the line of $\sigma = -5$.

$$F_{2m}(-5 + j\omega) = \frac{\delta}{\delta^2+\omega^2} + \frac{j\omega}{\delta^2+\omega^2} + \frac{\delta}{\delta^2+\omega^2} - \frac{j\omega}{\delta^2+\omega^2}$$

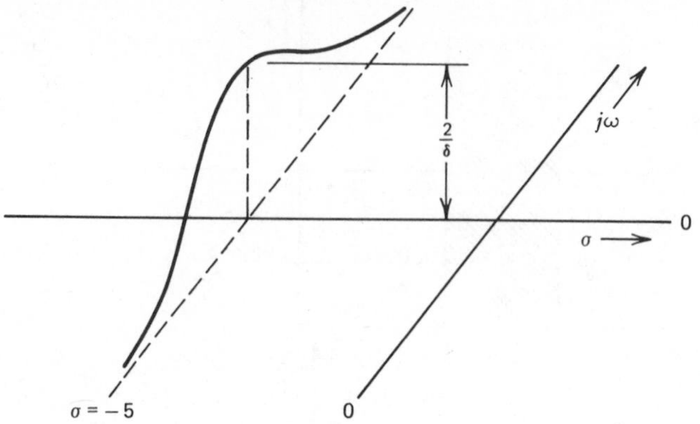

Fig. 9-6.2

The j-parts cancel, leaving

$$F_{2m}(-5 + j\omega) = \frac{2\delta}{\delta^2 + \omega^2} \tag{9-6.11}$$

This equation defines a curve that is sketched in Fig. 9-6.2. The area under the curve is shown to be independent of δ by setting up the integral

$$\int_{-\infty}^{+\infty} \frac{2\delta}{\delta^2 + \omega^2}\, d\omega = \int_{-\infty}^{+\infty} \frac{2\, d(\omega/\delta)}{1 + (\omega/\delta)^2} = 2 \tan^{-1} \frac{\omega}{\delta} \bigg|_{-\infty}^{+\infty} = 2\pi \tag{9-6.12}$$

As δ becomes smaller, the height becomes greater and the width becomes less, but the area under the curve remains constant. In the limit, as δ approaches zero, the height is infinite and the width is zero, but the area remains as 2π. The function being described is an impulse of area 2π located at $s = -5 + j0$.

In summary

$$f_2(t) = \epsilon^{-5t} \qquad \text{for all } t \tag{9-6.13}$$

and the $f_2(t)$ is said to have the two-sided Laplace transform in the limit

$$F_2(s) = 2\pi u_0(s + 5) \tag{9-6.14}$$

The strip of convergence has shrunk to a line, and we say that $F_2(s)$ converges on the line

$$\sigma = -5 \tag{9-6.15}$$

Fig. 9-6.3 summarizes what has been done in pictorial form.

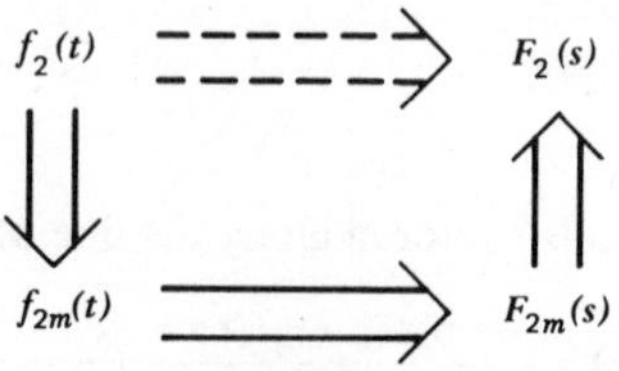

Fig. 9-6.3

It is desired to go from $f_2(t)$ to an $F_2(s)$, as shown by the dashed lines, but this cannot be done in the strict sense because the $F_2(s)$ does not exist. Therefore, a roundabout approach is used, going to an $f_{2m}(t)$, to an $F_{2m}(s)$, and then to an $F_2(s)$, through a limiting process.

There is an expression, "arm waving," used in many quarters. You probably won't find a definition for this in any dictionary, but the term does apply in certain situations. If a man is lecturing at the board, and knows that what he is saying is not exactly rigorous mathematically, he may feel that if he waves his arms around enough, he may be able to distract attention from his lack of rigor and put across his point without anyone's noticing.

There is a certain amount of "arm waving" in the preceding development. As long as δ is small but non-zero, everything is all right; and it is certainly more convenient to identify the results with an impulse than to write out such expressions as Eq. 9-6.10 every time.

GRAPHICAL REPRESENTATION OF IMPULSE IN THE FREQUENCY DOMAIN

In the *time domain*, the notation

$$f(t) = u_0(t - \tau) \qquad (9\text{-}6.16)$$

indicates an impulse that occurs at $t = \tau$. Since $f(t)$ can be plotted as the ordinant vs. t as the abscissa, the graphical interpretation of the impulse is relatively simple, and this has been used throughout this book.

The representation of the impulse in the *frequency domain* is more difficult since the impulse is a function of the variable s, which is

$$s = \sigma + j\omega \qquad (9\text{-}6.17)$$

The graphical representation used in this book is to show the impulse perpendicular to the s-plane, and the location of the impulse in the s-plane is indicated by the location of the base of the impulse. For example, the $F_2(s)$

$$F_2(s) = 2\pi\, u_0(s + 5) \qquad (9\text{-}6.18)$$

is shown in Fig. 9-6.4.

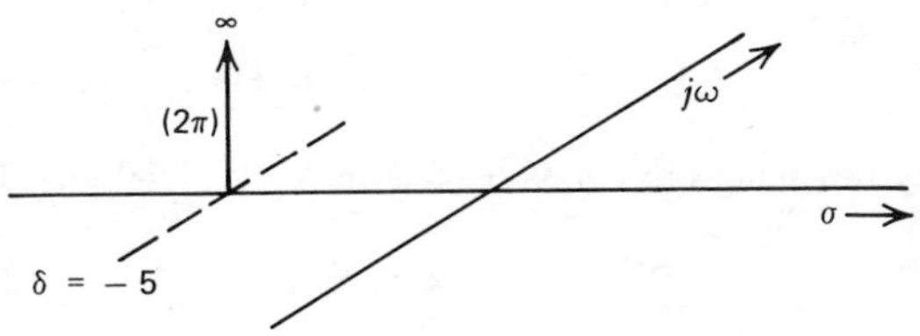

Fig. 9-6.4

Example 9-6.2. For this example

$$f_2(t) = 1 \qquad -\infty < t < +\infty \tag{9-6.19}$$

Many of the properties of one-sided Laplace transform carry over to the two-sided situation. This idea is discussed in more detail later; however, one of these properties is helpful at this time. The property in question restated in terms of the two-sided case is the following: suppose an $f_2(t)$ has an $F_2(s)$, as indicated by

$$f_2(t) \Longrightarrow F_2(s) \tag{9-6.20}$$

If the $f_2(t)$ is multiplied by an ϵ^{-at}, the two-sided Laplace transform of the product is

$$f_2(t)\,\epsilon^{-at} \Longrightarrow F_2(s+a) \tag{9-6.21}$$

With this property, the $F_2(s)$ for Eq. 9-6.19 can be found from the $F_2(s)$ of the last example. The $f_2(t)$ and $F_2(s)$ are repeated as

$$f_2(t) = \epsilon^{-5t} \qquad F_2(s) = 2\pi\, u_0(s+5) \tag{9-6.22}$$

The new $f_2(t)$ is the old $f_2(t)$ multiplied by ϵ^{+5t}

$$f_2(t) \text{ new} = f_2(t) \text{ old } [\epsilon^{+5t}] = \epsilon^{-5t}\,[\epsilon^{+5t}] = 1 \tag{9-6.23}$$

The new $F_2(s)$ is found from the old $F_2(s)$ by replacing s with $(s - 5)$

$$F_2(s) \text{ new} = 2\pi\, u_0(s - 5 + 5) = 2\pi\, u_0(s) \tag{9-6.24}$$

The result converges on the line $\sigma = 0$, and is shown in Fig. 9-6.5.

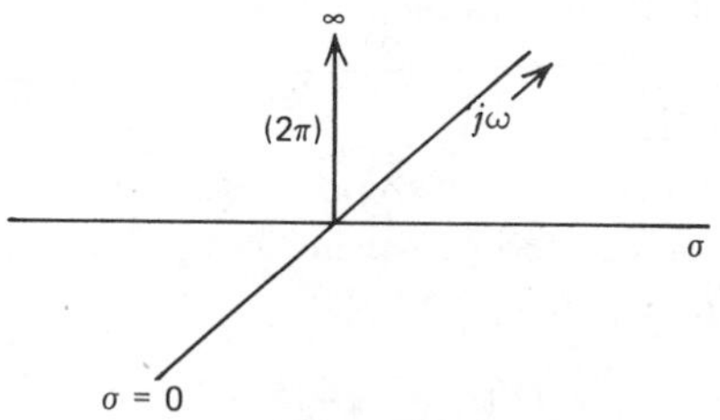

Fig. 9-6.5

Example 9-6.3. For this example

$$f_2(t) = \epsilon^{+4t} \qquad -\infty < t < +\infty \tag{9-6.25}$$

The $F_2(s)$ is found by replacing s in Eq. 9-6.24 with $(s - 4)$, yielding

$$F_2(s) = 2\pi\, u_0(s-4) \tag{9-6.26}$$

The result converges on the line $\sigma = 4$, and is shown in Fig. 9-6.6.

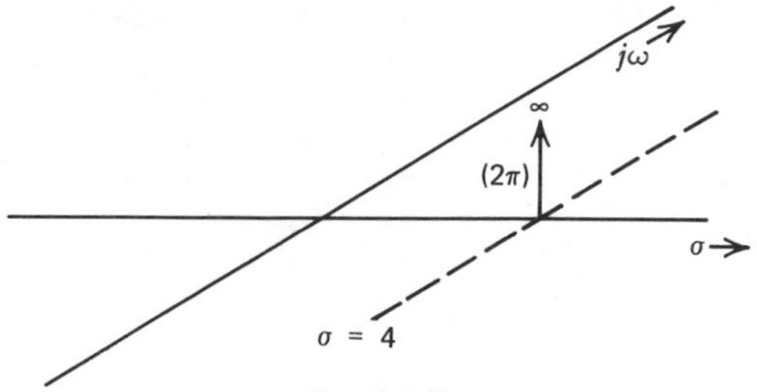

Fig. 9-6.6

A PHYSICAL INTERPRETATION OF THE LAST THREE EXAMPLES

It is of interest to observe the last three $f_2(t)$ functions and the corresponding $F_2(s)$ functions. When $f_2(t)$ decreases exponentially, the impulse of $F_2(s)$ is located on the negative real axis. When $f_2(t)$ is a constant the impulse of $F_2(s)$ is at the origin, and when $f_2(t)$ increases exponentially the impulse is on the positive real axis. In your mind's eye, you can imagine moving this impulse back and forth and observe the change in the $f_2(t)$.

Example 9-6.4. For this example

$$f_2(t) = \cosh 6t \qquad -\infty < t < +\infty \tag{9-6.27}$$

The cosh $6t$ can be expanded as

$$\cosh 6t = \frac{\epsilon^{+6t} + \epsilon^{-6t}}{2} \tag{9-6.28}$$

If the transforms of the two exponential terms are taken separately, they are

$$\mathcal{L}_2\left[\frac{\epsilon^{+6t}}{2}\right] = \pi\, u_0(s - 6) \tag{9-6.29}$$

and

$$\mathcal{L}_2\left[\frac{\epsilon^{-6t}}{2}\right] = \pi\, u_0(s + 6) \tag{9-6.30}$$

These are shown in Fig. 9-6.7. The $\pi\, u_0(s + 6)$ converges on the line $\sigma = -6$ and the $\pi\, u_0(s - 6)$ converges on the line $\sigma = +6$. Since these two lines of convergence are not the

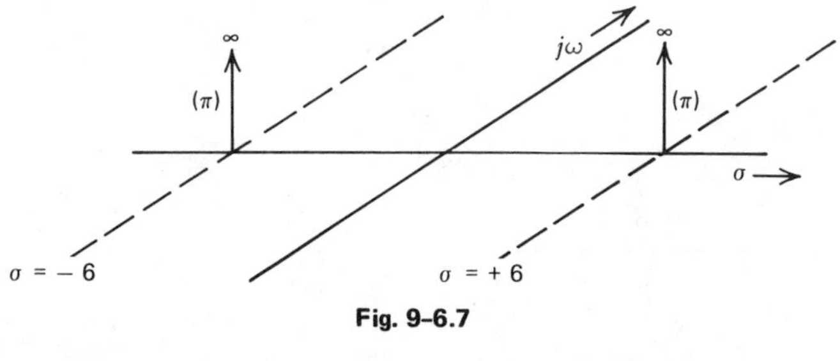

Fig. 9-6.7

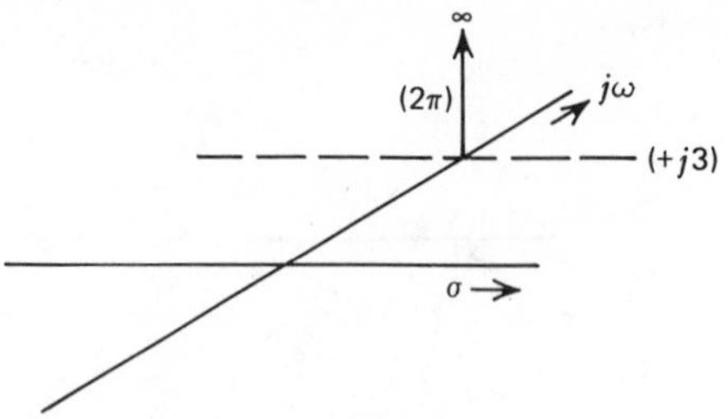

Fig. 9-6.8

same, the two transforms cannot be added. Therefore, cosh $6t$, defined for all t, does not have a two-sided Laplace transform.

Example 9-6.5. We have been looking at $f_2(t)$ functions having $F_2(s)$ functions that are impulses on the real axis in the s-plane. Suppose for the moment we move the impulse in $F_2(s)$ along the $j\omega$-axis to $s = +j3$ as shown in Fig. 9-6.8, and find the corresponding $f_2(t)$. The time functions examined thus far have had transforms that are impulses on the real axis in the s-plane. *We ask the question*, What kind of time function will have a transform that is an impulse that is not on the real axis? To begin with, we observe the $F_2(s)$ of Fig. 9-6.8 that is an impulse located at $s = +j3$. The corresponding time function is

$$f_2(t) = \epsilon^{+j3t} \qquad -\infty < t < \infty \tag{9-6.31}$$

This $f_2(t)$ is complex, as can be seen by

$$\epsilon^{+j3t} = \cos 3t + j \sin 3t \tag{9-6.32}$$

and since $f_2(t)$ must be real, placing one impulse on the $j\omega$-axis does not lead to "physically realizable" results.

What we need to do is to divide the impulse of weight 2π into two impulses placed symmetrically along the $j\omega$-axis as in Fig. 9-6.9. The impulse at $s = +j3$ corresponds to

$$\tfrac{1}{2}\epsilon^{+j3t} \tag{9-6.33}$$

and the one at $s = -j3$ to

$$\tfrac{1}{2}\epsilon^{-j3t} \tag{9-6.34}$$

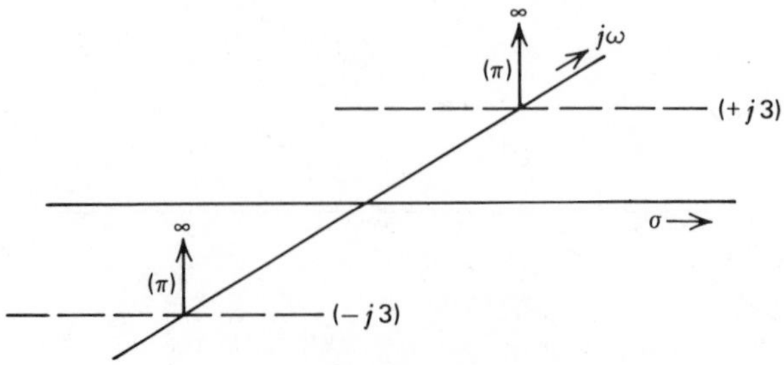

Fig. 9-6.9

The two together correspond to

$$f_2(t) = \frac{\epsilon^{+j3t} + \epsilon^{-j3t}}{2} = \cos 3t \qquad -\infty < t < +\infty \tag{9-6.35}$$

Both $\frac{1}{2}\epsilon^{+j3t}$ and $\frac{1}{2}\epsilon^{-j3t}$ converge on the line

$$\sigma = 0 \tag{9-6.36}$$

The two components can be added, and $\cos 3t$ also converges on this same line. The resulting $F_2(s)$ is

$$F_2(s) = \pi\, u_0(s - j3) + \pi\, u_0(s + j3) \tag{9-6.36a}$$

Example 9-6.6. For this example

$$f_2(t) = \epsilon^{-4t} \cos 3t \qquad -\infty < t < +\infty \tag{9-6.37}$$

To be completely graphic about the situation, we can take the impulse of weight 2π and send it along the real axis to $s = -4$. Then we saw the impulse in half, and send one half along the $s = -4$ line to $s = -4 + j3$, and the other half along the same line in the other direction to $s = -4 - j3$. The resulting $F_2(s)$ is shown in Fig. 9-6.10, and is given by

$$F_2(s) = \pi\, [u_0(s + 4 + j3) + u_0(s + 4 - j3)] \tag{9-6.38}$$

and converges on the line

$$\sigma = -4 \tag{9-6.39}$$

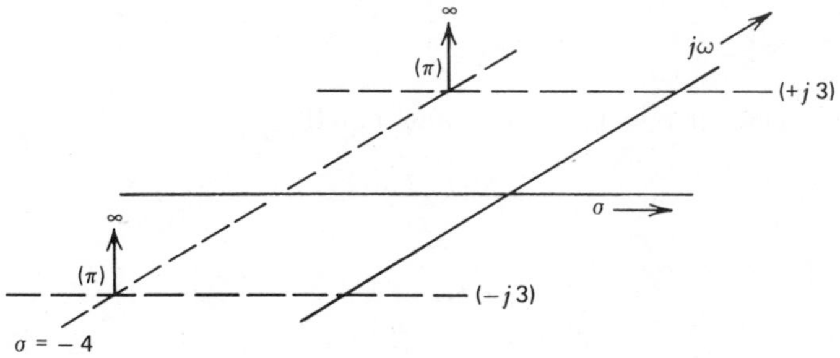

Fig. 9-6.10

A RETURN TO THE PHYSICAL INTERPRETATION

After the first three examples, a few comments were made associating the impulse in the s-plane on the negative real axis with the corresponding time functions. These comments can now be extended to include the entire s-plane.

Figure 9-6.9 indicates a pair of impulses placed symmetrically along the $j\omega$-axis. The corresponding time function is $\cos 3t$. If this pair of impulses is moved farther out on the

$j\omega$-axis, the $\cos \omega t$ corresponds a higher frequency; but if they are moved closer to the origin, the time function is of a lower frequency. If these two impulses move together and combine at the origin, the corresponding time function is a d-c signal. In a certain sense, a d-c signal is a cosine term of zero frequency.

Figure 9-6.10 is referred to in order to complete this discussion. If the ω is fixed, but this pair of impulses moves into the left-half plane, the time function is an exponential decaying type of sinusoid. If this pair of impulses moves into the right-half plane, the sinusoid increases exponentially.

OTHER TWO-SIDED LAPLACE TRANSFORMS THAT EXIST IN THE LIMIT

There are many other time functions having two-sided Laplace transforms that exist in the limit, but we don't push the subject any further at this time. The subject is brought up again during the discussion of Fourier transforms, and at that time comments are made concerning the two-sided Laplace case.

9-7. TWO-SIDED TIME SIGNALS APPLIED TO LINEAR SYSTEMS

The block diagram for the system being discussed is shown in Fig. 9-7.1.

The $R_2(s)$ is the Laplace transform of the two-sided input $r_2(t)$. The $G(s)$ is the transform of $g(t)$, and $C_2(s)$ is the Laplace transform of the two-sided output $c_2(t)$ and is given by

$$C_2(s) = R_2(s)\, G(s) \tag{9-7.1}$$

The nature of $G(s)$ is explored in the following step. If

$$r_2(t) = u_0(t) \tag{9-7.2}$$

then

$$R_2(s) = 1 \tag{9-7.3}$$

$$C_2(s) = G(s) \tag{9-7.4}$$

Therefore $g(t)$ is the impulse response of the system and has precisely the same meaning as it did for the one-sided case. In the two-sided case the meaning of $g(t)$ does not change, but since the input $r_2(t)$ is a two-sided function, the output $c_2(t)$ is also two-sided.

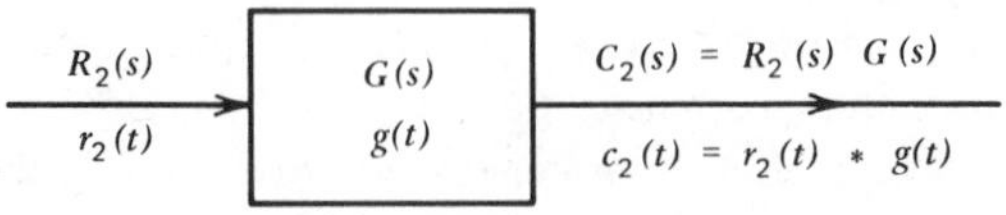

Fig. 9-7.1

The output $c_2(t)$ can also be found as the convolution of $r_2(t)$ with $g(t)$, as

$$c_2(t) = r_2(t) * g(t) \tag{9-7.5}$$

The subject of convolution is taken up later in this chapter.

Suppose that $r_2(t)$ is such that $R_2(s)$ exists and has a strip of convergence in the complex s-plane, as suggested in Fig. 9-7.2(a). Next, suppose $G(s)$ has the abscissa of convergence and the region of convergence as shown in (b). Since $C_2(s)$ is the product $R_2(s)\,G(s)$, the strip of convergence of $C_2(s)$ is the common portion of the regions for convergence between $R_2(s)$ and $G(s)$, shown in (c). This situation can be summarized in three cases.

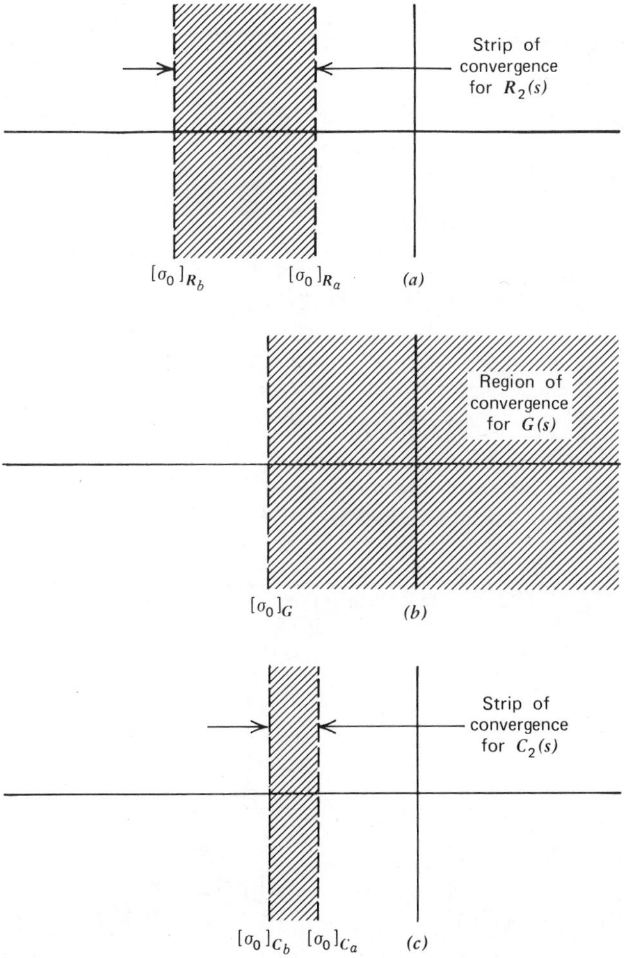

Fig. 9-7.2

Case I. If $[\sigma_0]_G$ is to the left of $[\sigma_0]_{Rb}$, then the strip of convergence for $C_2(s)$ is the same as that for $R_2(s)$.
Case II. If $[\sigma_0]_G$ is to the right of $[\sigma_0]_{Rb}$ but to the left of $[\sigma_0]_{Ra}$, then $C_2(s)$ has a strip of convergence that is bounded on the left by $[\sigma_0]_G$ and on the right by $[\sigma_0]_{Ra}$.
Case III. If $[\sigma_0]_G$ is to the right of $[\sigma_0]_{Ra}$, then no strip of convergence exists for $C_2(s)$, and hence $C_2(s)$ does not exist.

These ideas and many others are demonstrated through the following set of examples.

Example 9-7.1. The system of Fig. 9-7.1 has $g(t)$ and $r_2(t)$ as given by

$$g(t) = 1\epsilon^{-10t} \qquad t>0 \tag{9-7.6}$$

$$r_2(t) = \begin{cases} \epsilon^{2t} & t<0 \\ 0 & t>0 \end{cases} \tag{9-7.7}$$

The desired solution is $c_2(t)$.
The $R_a(s)$ is found from $r_a(t)$ using the four steps, as

$$R_a(s) = \frac{-1}{s-2} \tag{9-7.8}$$

The $G(s)$ is found from $g(t)$, as

$$G(s) = \frac{1}{s+10} \tag{9-7.9}$$

The $C_2(s)$ is the product

$$C_2(s) = R_2(s)\,G(s) = \frac{-1}{(s-2)(s+10)} \tag{9-7.10}$$

and has the strip of convergence shown in Fig. 9-7.3.
The $C_2(s)$ is expanded in partial fractions as

$$C_2(s) = \frac{-1}{(s-2)(s+10)} = \frac{-\frac{1}{12}}{s-2} + \frac{\frac{1}{12}}{s+10} \tag{9-7.11}$$

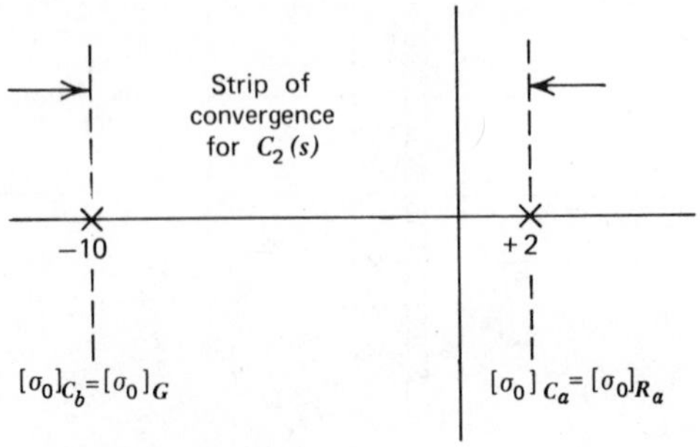

Fig. 9-7.3

The poles on, and to the right of, $[\sigma_0]_{Ca}$ are assigned to $C_a(s)$

$$C_a(s) = \frac{-\frac{1}{12}}{s-2} \tag{9-7.12}$$

and the poles on, and to the left of, $[\sigma_0]_{Cb}$ are assigned to $C_b(s)$

$$C_b(s) = \frac{\frac{1}{12}}{s+10} \tag{9-7.13}$$

The $c_a(t)$ and $c_b(t)$ are found as the inverse transform of $C_a(s)$ and $C_b(s)$, respectively, and $c_2(t)$ is written as

$$c_2(t) = \begin{cases} \frac{1}{12}\epsilon^{+2t} & t<0 \\ \frac{1}{12}\epsilon^{-10t} & t>0 \end{cases} \tag{9-7.14}$$

The $c_2(t)$ is sketched in Fig. 9-7.4.

In the terminology of classical differential equations, a solution is made up of the particular component and the complementary component. The particular component takes on the nature of the driving function, and the complementary component the nature of the system itself.

In the example the driving function is $r_a(t) = \epsilon^{+2t}$; one component of the solution is $c_a(t) = \frac{1}{12}\epsilon^{+2t}$, and hence this is the particular component of the solution. The system's natural response is $g(t) = \epsilon^{-10t}$; and one component of the solution is $c_b(t) = \frac{1}{12}\epsilon^{-10t}$, and hence this is the complementary component of the solution. Therefore, the total response of this system is divided between the particular component and the complementary component. The amazing thing is that the particular component exists before $t = 0$, and the complementary component exists after $t = 0$.

A physical interpretation of this result will help us understand some of the situations we will encounter in later examples. The system by itself wants to follow an exponential decay, as given by $g(t) = \epsilon^{-10t}$. Before $t = 0$, the input is $r_a(t) = \epsilon^{+2t}$ and forces the system to respond in a way that goes counter to its own natural desires. Hence, the $K\,\epsilon^{+2t}$ is forced onto the system. After $t = 0$, the input is zero and the system dies away from its $t = 0$ value according to the natural behavior of the system $K\,\epsilon^{-10t}$.

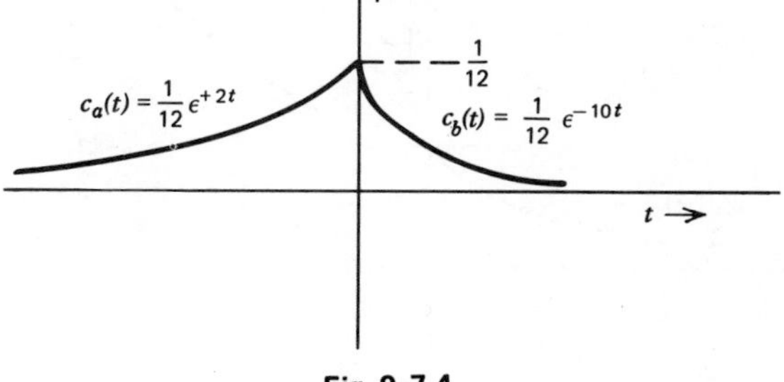

Fig. 9-7.4

Example 9-7.2. The impulse response of the system is

$$g(t) = 1\epsilon^{+1t} \qquad t > 0 \tag{9-7.15}$$

and the same $r_2(t)$ of Eq. 9-7.7 is used again, which means that

$$R_a(s) = \frac{-1}{s-2} \tag{9-7.16}$$

The $G(s)$ is found from $g(t)$, as

$$G(s) = \frac{1}{s-1} \tag{9-7.17}$$

The $C_2(s)$ is the product

$$C_2(s) = R_2(s)\,G(s) = \frac{-1}{(s-1)(s-2)} \tag{9-7.18}$$

and has the strip of convergence $1 < \sigma < 2$.

The $C_2(s)$ is expanded in partial fractions

$$C_2(s) = \frac{-1}{(s-1)(s-2)} = \frac{1}{s-1} - \frac{1}{s-2} \tag{9-7.19}$$

The strip of convergence divides $C_2(s)$ into $C_a(s)$ and $C_b(s)$, as

$$C_a(s) = \frac{-1}{s-2} \tag{9-7.20}$$

$$C_b(s) = \frac{1}{s-1} \tag{9-7.21}$$

The $c_a(t)$ and $c_b(t)$ are found, and $c_2(t)$ is written as

$$c_2(t) = \begin{cases} \epsilon^{+2t} & t < 0 \\ \epsilon^{+1t} & t > 0 \end{cases} \tag{9-7.22}$$

The resulting $c_2(t)$ is sketched in Fig. 9-7.5.

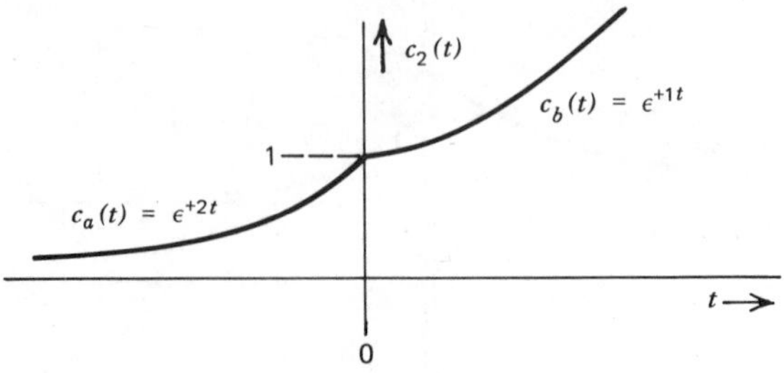

Fig. 9-7.5

A physical explanation is again given to these results. The system itself is unstable and wants to go off to infinity as $K\epsilon^{+1t}$. Before $t = 0$, the input is going off to infinity as ϵ^{+2t}, which is faster than the system wants to respond, so the input maintains control of the output up to $t = 0$. After $t = 0$, the system continues off to infinity according to its own natural desires.

Example 9-7.3. The impulse response of the system is

$$g(t) = 1\epsilon^{+4t} \qquad t > 0 \tag{9-7.23}$$

and the same $r_2(t)$ of Eq. 9-7.7 is used, which means that

$$R_a(s) = \frac{-1}{s-2} \tag{9-7.24}$$

The $G(s)$ is found from $g(t)$, as

$$G(s) = \frac{1}{s-4} \tag{9-7.25}$$

The region of convergence for $R_a(s)$ is to the left of $\sigma = +2$, whereas the region of convergence for $G(s)$ is to the right of $\sigma = +4$. Therefore, no strip of convergence exists and $C_2(s)$ does not exist.

A physical explanation is again given to these results. The system itself is unstable and wants to go off to infinity as $K\epsilon^{+4t}$. Before $t = 0$, the input goes off to infinity as ϵ^{+2t}. Once some small pertabation gets the system started, the system output outruns the input and the input has no control over the system.

COMMENTARY

In Examples 9-7.1 to 3, in essence the new thing that is added by the two-sided Laplace transform is the ability to study what happens to a system before $t = 0$. This is why the last three examples had an $f_b(t)$ equal to zero in order to concentrate on the $f_a(t)$ function.

If you are like most people, you probably haven't even considered functions like the ones just presented. The interesting thing to note is how easily the two-sided Laplace transform handled these problems.

The remaining examples have $f_b(t)$ functions not equal to zero.

Example 9-7.4. The impulse response of the system is

$$g(t) = \epsilon^{-3t} \qquad t > 0 \tag{9-7.26}$$

The following input is applied to the system

$$r_2(t) = \begin{cases} \epsilon^{-2t} & t < 0 \\ \epsilon^{-4t} & t > 0 \end{cases} \tag{9-7.27}$$

The $R_a(s)$ and $R_b(s)$ are found as

$$R_a(s) = \frac{-1}{s+2} \qquad R_b(s) = \frac{1}{s+4} \tag{9-7.28}$$

A strip of convergence for $R_2(s)$ exists

$$-4 < \sigma < -2 \tag{9-7.29}$$

and $R_2(s)$ is given by

$$R_2(s) = R_a(s) + R_b(s) = \frac{-1}{s+2} + \frac{1}{s+4} \tag{9-7.30}$$

The $G(s)$ is found as

$$G(s) = \frac{1}{s+3} \tag{9-7.31}$$

which converges to the right of $\sigma = -3$.

The product $R_2(s)\,G(s)$ converges in the strip common to $R_2(s)$ and $G(s)$, which is

$$-3 < \sigma < -2 \tag{9-7.32}$$

and $C_2(s)$ is given by

$$C_2(s) = \frac{-1}{(s+2)(s+3)} + \frac{1}{(s+3)(s+4)} \tag{9-7.33}$$

The $C_2(s)$ is expanded as

$$C_2(s) = \frac{-1}{s+2} + \frac{2}{s+3} - \frac{1}{s+4} \tag{9-7.34}$$

The strip of convergence divides $C_2(s)$ into $C_a(s)$ and $C_b(s)$ as

$$C_a(s) = \frac{-1}{s+2} \tag{9-7.35}$$

$$C_b(s) = \frac{2}{s+3} - \frac{1}{s+4} \tag{9-7.36}$$

The $c_a(t)$ is found from $C_a(s)$, and $c_b(t)$ from $C_b(s)$, to yield

$$c_2(t) = \begin{cases} \epsilon^{-2t} & t < 0 \\ 2\epsilon^{-3t} - 1\epsilon^{-4t} & t > 0 \end{cases} \tag{9-7.37}$$

Example 9-7.5. The impulse response of the system is

$$g(t) = \epsilon^{-1t} \qquad t > 0 \tag{9-7.38}$$

The following input is applied to the system

$$r_2(t) = \begin{cases} \epsilon^{-2t} & t < 0 \\ \epsilon^{-4t} & t > 0 \end{cases} \tag{9-7.39}$$

The input is the same as the last example; hence, the $R_2(s)$ is

$$R_2(s) = \frac{-1}{s+2} + \frac{1}{s+4} \tag{9-7.40}$$

with a strip of convergence

$$-4 < \sigma < -2 \tag{9-7.41}$$

The $G(s)$ is found as

$$G(s) = \frac{1}{s+1} \tag{9-7.42}$$

and this converges to the right of $s = -1$. There is no common area between the strip of convergence of $R_2(s)$ and the region of convergence of $G(s)$, and hence the $C_2(s)$ does not exist.

Example 9-7.6. The impulse response of the system is

$$g(t) = \epsilon^{-7t} \qquad t > 0 \tag{9-7.43}$$

The following input is applied to the system

$$r_2(t) = \epsilon^{-5t} \qquad -\infty < t < +\infty \tag{9-7.44}$$

The transform for $R_2(s)$ exists in the limit and was found in Section 9-6 as

$$R_2(s) = 2\pi\, u_0(s+5) \tag{9-7.45}$$

and converges on the line

$$\sigma = -5 \tag{9-7.46}$$

The $G(s)$ is found

$$G(s) = \frac{1}{s+7} \tag{9-7.47}$$

and converges to the right of $\sigma = -7$. The line of convergence of $R_2(s)$ is in the region of convergence of $G(s)$; therefore $C_2(s)$ exists

$$C_2(s) = R_2(s)\, G(s) = \frac{2\pi}{s+7}\, u_0(s+5) \tag{9-7.48}$$

and also exists in the limit and converges on the line

$$\sigma = -5 \tag{9-7.49}$$

Since the $u_0(s+5)$ function is zero everywhere except at $s = -5$, the $C_2(s)$ can be written as

$$C_2(s) = \frac{2\pi}{-5+7}\, u_0(s+5) = \pi\, u_0(s+5) \tag{9-7.50}$$

The inverse transform of $C_2(s)$ is

$$c_2(t) = \tfrac{1}{2}\epsilon^{-5t} \qquad -\infty < t < +\infty \tag{9-7.51}$$

Example 9-7.7. The impulse response of the system is

$$g(t) = \epsilon^{-3t} \qquad t > 0 \tag{9-7.52}$$

The following input is applied to the system

$$r_2(t) = \epsilon^{-5t} \qquad -\infty < t < +\infty \tag{9-7.53}$$

The transform for $R_2(s)$ exists in the limit as

$$R_2(s) = 2\pi\, u_0(s+5) \tag{9-7.54}$$

and converges on the line

$$\sigma = -5 \tag{9-7.55}$$

The $G(s)$ is found as

$$G(s) = \frac{1}{s+3} \tag{9-7.56}$$

and converges to the right of $\sigma = -3$. Since the line of convergence of $R_2(s)$ is not in the region of convergence of $G(s)$, the $C_2(s)$ does not exist.

Example 9-7.8. The impulse response of the system is

$$g(t) = \epsilon^{-7t} \qquad t > 0 \tag{9-7.57}$$

The following input is applied to the system

$$r_2(t) = \cos 4t \qquad -\infty < t < +\infty \tag{9-7.58}$$

The transform for $R_2(s)$ exists in the limit and was found in Section 9-6 as

$$R_2(s) = \pi\, u_0(s - j4) + \pi\, u_0(s + j4) \tag{9-7.59}$$

and converges on the line

$$\sigma = 0 \tag{9-7.60}$$

The $G(s)$ is found as

$$G(s) = \frac{1}{s+7} \tag{9-7.61}$$

and converges to the right of $\sigma = -7$. Since the line of convergence of $R_2(s)$ is in the region of convergence of $G(s)$, $C_2(s)$ exists

$$C_2(s) = \frac{\pi}{s+7} u_0(s - j4) + \frac{\pi}{s+7} u_0(s + j4) \tag{9-7.62}$$

and also exists in the limit and converges on the line

$$\sigma = 0 \tag{9-7.63}$$

Since the first term of $C_2(s)$ is zero everywhere except at $s = +j4$, and the second term is zero everywhere except at $s = -j4$, this equation can be rewritten as

$$C_2(s) = \frac{\pi}{j4+7} u_0(s - j4) + \frac{\pi}{-j4+7} u_0(s + j4) \tag{9-7.64}$$

The inverse transform is

$$c_2(t) = \frac{1}{2}\left[\frac{\epsilon^{+j4t}}{7+j4} + \frac{\epsilon^{-j4t}}{7-j4}\right] \tag{9-7.65}$$

and can be written as

$$c_2(t) = 0.124 \cos(4t - 29.72°) \tag{9-7.66}$$

and the answer holds for all t.

Example 9-7.9. The impulse response of the system is

$$g(t) = \epsilon^{-7t} \qquad t > 0 \tag{9-7.67}$$

and the following input is applied to the system

$$r_2(t) = \epsilon^{-5t} \cos 4t \qquad -\infty < t < +\infty \tag{9-7.68}$$

The transform for $R_2(s)$ exists in the limit and is

$$R_2(s) = \pi\, u_0(s + 5 - j4) + \pi\, u_0(s + 5 + j4) \tag{9-7.69}$$

and converges on the line

$$\sigma = -5 \tag{9-7.70}$$

The $G(s)$ is found as

$$G(s) = \frac{1}{s+7} \tag{9-7.71}$$

and converges to the right of $\sigma = -7$. Since the line of convergence of $R_2(s)$ is in the region for convergence for $G(s)$, $C_2(s)$ exists

$$C_2(s) = \frac{\pi}{s+7} u_0(s + 5 - j4) + \frac{\pi}{s+7} u_0(s + 5 + j4) \tag{9-7.72}$$

and also exists in the limit and converges on the line

$$\sigma = -5 \tag{9-7.73}$$

Since the first term of $C_2(s)$ is zero everywhere except at $s = -5 + j4$, and the second term is zero everywhere except at $s = -5 - j4$, the equation can be rewritten as

$$C_2(s) = \frac{\pi}{-5 + j4 + 7} u_0(s + 5 - j4) + \frac{\pi}{-5 - j4 + 7} u_0(s + 5 + j4) \tag{9-7.74}$$

The inverse transform is

$$c_2(t) = \frac{1}{2}\left[\frac{\epsilon^{(-5+j4)t}}{2+j4} + \frac{\epsilon^{(-5-j4)t}}{2-j4}\right] \tag{9-7.75}$$

and can be written as

$$c_2(t) = 0.224\epsilon^{-5t} \cos(4t - 63.4°) \tag{9-7.76}$$

and the answer holds for all t.

COMMENTARY

In Example 9-7.8, the input

$$r_2(t) = \cos 4t \qquad \text{for all } t \tag{9-7.77}$$

was applied to a system, and the output was found to be

$$c_2(t) = 0.124 \cos(4t - 29.72°) \qquad \text{for all } t \tag{9-7.78}$$

If you look this example over very carefully, you will see that this development can be made to be a very elegant justification of all a-c circuit theory. Not only that, but the extension of this example quickly shows the situations under which a-c circuit theory cannot be used. For example, if $g(t)$ is such that $G(s)$ has a pole, or poles, to the right of the $j\omega$-axis, then a-c circuit methods cannot be used.

In Example 9-7.9, the input

$$r_2(t) = \epsilon^{-5t} \cos 4t \qquad \text{for all } t \tag{9-7.79}$$

was applied to a system, and the output was found to be

$$c_2(t) = 0.224\epsilon^{-5t} \cos(4t - 63.4°) \qquad \text{for all } t \tag{9-7.80}$$

This development can be made to extend a-c circuit techniques to driving functions that are of the type of exponential decay functions. Again this theory will yield the information as to when these methods cannot be used. Example 9-7.6 shows a situation in which these methods can be used, and Example 9-7.7 a situation in which they cannot be used.

9-8. CONVOLUTION AGAIN

The discussion of convolution in this book is divided between Chapter 6 and this chapter. As far as the convolution integral is concerned, this division is not necessary. This division is the result of the division of Laplace transform theory into one-sided and two-sided cases. Chapter 6 was concerned with the convolution of time functions that could be handled by the one-sided Laplace transform, whereas this discussion does the same thing for the two-sided Laplace transform. The convolution integral handles both types of function in the same manner.

9-9. EXAMPLES USING CONVOLUTION

The discussion proceeds by solving several of the examples of Section 9-7 by use of the convolution integral.

Example 9-9.1. This example is the same as Example 9-7.1, and the impulse response of the system is

$$g(t) = \epsilon^{-10t} \qquad t > 0 \tag{9-9.1}$$

The input to the system is

$$r_2(t) = \begin{cases} \epsilon^{2t} & t < 0 \\ 0 & t > 0 \end{cases} \tag{9-9.2}$$

Since $r_2(t)$ is identical with $r_a(t)$, here the input is referred to as $r_a(t)$.

This example is worked using both forms of the convolution integral, which are

$$c_2(t) = \int_{-\infty}^{+\infty} r_a(t - \tau)\, g(\tau)\, d\tau \tag{9-9.3}$$

$$c_2(t) = \int_{-\infty}^{+\infty} g(t - \tau)\, r_a(\tau)\, d\tau \tag{9-9.4}$$

On both equations the limits reduce to simpler limits, depending on the example at hand. Perhaps the best way to understand this is through sketches similar to those used in Chapter 6.

In using Eq. 9-9.3, the t in $r_a(t)$ and $g(t)$ is replaced by τ. The resulting $r_a(\tau)$ and $g(\tau)$ are shown in Fig. 9-9.1(a) and (b), respectively. The $r_a(\tau)$ is then folded, and is drawn in (c) with t given a negative value. The product curve $r_a(t - \tau)\, g(\tau)$ is suggested in (d), and from this the limits on the integral of Eq. 9-9.3 (for negative t) are seen to be

$$c_a(t) = \int_0^{\infty} r_a(t - \tau)\, g(\tau)\, d\tau$$

$$= \int_0^{\infty} \epsilon^{2(t-\tau)}\, \epsilon^{-10\tau}\, d\tau = \epsilon^{2t} \int_0^{\infty} \epsilon^{-12\tau}\, d\tau$$

$$= \epsilon^{2t}\, [-\tfrac{1}{12}\epsilon^{-12\tau}|_0^{\infty}] = \tfrac{1}{12}\epsilon^{+2t} \tag{9-9.5}$$

The subscript of $c_a(t)$ is used to emphasize that this is the equation for the output when t is less than zero.

When t is positive, the curve of $r_a(t - \tau)$ becomes that shown in Fig. 9-9.1(e), and the product curve that of (f). From (f), the limits on the integral of Eq. 9-9.3 (for positive time) are seen to be

$$c_b(t) = \int_t^{\infty} r_a(t - \tau)\, g(\tau)\, d\tau$$

$$= \int_t^{\infty} \epsilon^{2(t-\tau)}\, \epsilon^{-10\tau}\, d\tau = \epsilon^{2t} \int_t^{\infty} \epsilon^{-12\tau}\, d\tau$$

$$= \epsilon^{2t}\, [-\tfrac{1}{12}\epsilon^{-12\tau}|_t^{\infty}] = \tfrac{1}{12}\epsilon^{-10t} \tag{9-9.6}$$

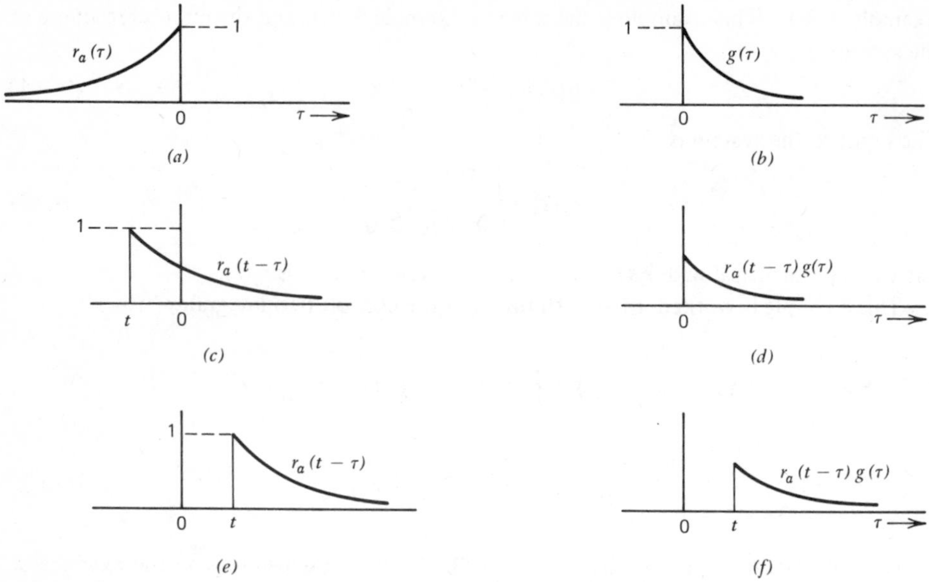

Fig. 9-9.1

The complete $c_2(t)$ is

$$c_2(t) = \begin{cases} \frac{1}{12}\epsilon^{2t} & t<0 \\ \frac{1}{12}\epsilon^{-10t} & t>0 \end{cases} \tag{9-9.7}$$

This answer checks Eq. 9-7.14.

This example is worked a second time by folding $g(t)$. The $g(t-\tau)$ for negative t is shown in Fig. 9-9.2(a), and for positive t in (b). The product curve for negative t is suggested in (c), and for positive t in (d).

The limits on the integrals for $c_a(t)$ and $c_b(t)$ are determined from Fig. 9-9.2(c) and (d), respectively, as

$$c_a(t) = \int_{-\infty}^{t} g(t-\tau)\, r_a(\tau)\, d\tau$$

$$= \int_{-\infty}^{t} \epsilon^{-10(t-\tau)}\, \epsilon^{2\tau}\, d\tau = \tfrac{1}{12}\epsilon^{2t} \tag{9-9.8}$$

$$c_b(t) = \int_{-\infty}^{0} g(t-\tau)\, r_a(\tau)\, d\tau$$

$$= \int_{-\infty}^{0} \epsilon^{-10(t-\tau)}\, \epsilon^{2\tau}\, d\tau = \tfrac{1}{12}\epsilon^{-10t} \tag{9-9.9}$$

Thus, the $c_2(t)$ of Eq. 9-9.7 is again determined.

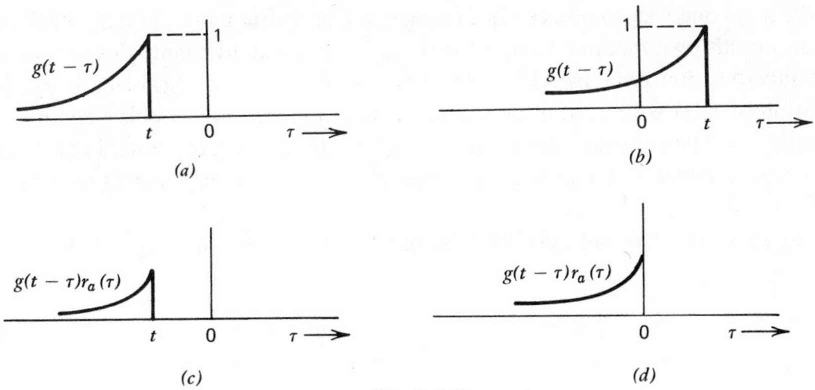

Fig. 9-9.2

Example 9-9.2. This example is the same as Example 9-7.3, and the system impulse response and the input to the system are

$$g(t) = \epsilon^{+4t} \qquad t>0$$

$$r_2(t) = \begin{cases} \epsilon^{+2t} & t<0 \\ 0 & t>0 \end{cases} \tag{9-9.10}$$

The work using two-sided Laplace transforms revealed that the solution to this problem does not exist. The point of interest here is to see how convolution displays this same information.

This time the $r_a(t)$ is folded, and Fig. 9-9.1(d) can be consulted for the limits on the integral for negative t.

$$c_a(t) = \int_0^\infty r_a(t-\tau)\, g(\tau)\, d\tau$$

$$= \int_0^\infty \epsilon^{2(t-\tau)}\, \epsilon^{+4\tau}\, d\tau = \epsilon^{2t}\left[\int_0^\infty \epsilon^{2\tau}\, d\tau\right]$$

$$= \epsilon^{2t}\,[\tfrac{1}{2}\epsilon^{+2\tau}|_0^\infty] \tag{9-9.11}$$

This integral does not exist, because upon substitution of the upper limit an infinite value is obtained.

Example 9-9.3. This example is the same as Example 9-7.4, and the system impulse response and the input to the system are

$$g(t) = \epsilon^{-3t} \qquad t>0$$

$$r_2(t) = \begin{cases} \epsilon^{-2t} & t<0 \\ \epsilon^{-4t} & t>0 \end{cases} \tag{9-9.12}$$

There is no one way to proceed in finding $c_2(t)$, and after a little practice each person can work out his own method. Conceptually, the easiest way to handle this situation is to use the superposition principal. The $r_2(t)$ can be divided into an $r_a(t)$ and an $r_b(t)$. The convolution of $r_a(t)$ with $g(t)$, discussed in the last two examples, yields a component of $c_a(t)$ and $c_b(t)$. The convolution of $r_b(t)$ with $g(t)$ is the subject discussed in Section 6-2. A superscript is added to the notation to identify which function produces which component of the response.

The $r_a(t)$ is convolved with $g(t)$ by folding $r_a(t)$. For $t < 0$, the integral is

$$c_a^a(t) = \int_0^{\infty} \epsilon^{-2(t-\tau)} \epsilon^{-3\tau} \, d\tau = \epsilon^{-2t} \tag{9-9.13}$$

For $t > 0$, the integral becomes

$$c_b^a(t) = \int_t^{\infty} \epsilon^{-2(t-\tau)} \epsilon^{-3\tau} \, d\tau = \epsilon^{-3t} \tag{9-9.14}$$

Next, $r_b(t)$ is convolved with $g(t)$ by folding $r_b(t)$. For $t < 0$, the result is zero, as

$$c_a^b(t) = 0 \tag{9-9.15}$$

For $t > 0$, the integral is

$$c_b^b(t) = \int_0^{t} \epsilon^{-4(t-\tau)} \epsilon^{-3\tau} \, d\tau = \epsilon^{-3t} - \epsilon^{-4t} \tag{9-9.16}$$

The resulting $c_2(t)$ is written as

$$c_2(t) = \begin{cases} c_a^a(t) + c_a^b(t) = \epsilon^{-2t} & t < 0 \\ c_b^a(t) + c_b^b(t) = 2\epsilon^{-3t} - \epsilon^{-4t} & t > 0 \end{cases} \tag{9-9.17}$$

which checks with Eq. 9-7.37.

Example 9-9.4. This example is the same as Example 9-7.6, and the system impulse response and the input to the system are

$$\begin{aligned} g(t) &= \epsilon^{-7t} \qquad && t > 0 \\ r_2(t) &= \epsilon^{-5t} && \text{for all } t \end{aligned} \tag{9-9.18}$$

Since one equation describes $r_2(t)$ for all time, separate steps are not needed for $r_a(t)$ and $r_b(t)$. Figure 9-9.3(a) suggests the $r_2(t)$ folded and drawn for $t < 0$, whereas (b) shows the $g(\tau)$. The product curve exists from 0 to ∞, and the resulting $c_2(t)$, good for all t, is

$$c_2(t) = \int_0^{\infty} \epsilon^{-5(t-\tau)} \epsilon^{-7\tau} \, d\tau = \tfrac{1}{2}\epsilon^{-5t} \qquad \text{for all } t \tag{9-9.19}$$

This checks with Eq. 9-7.51.

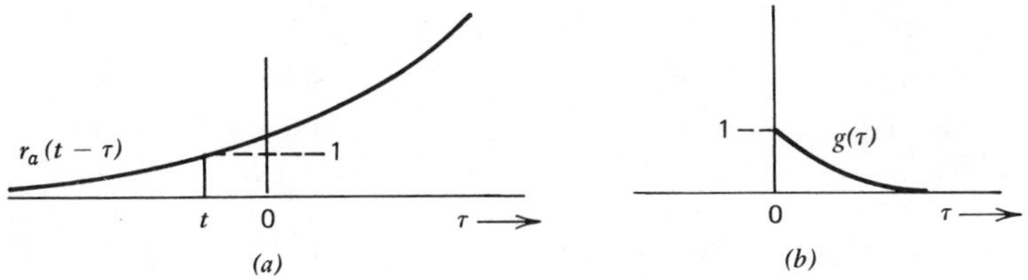

Fig. 9-9.3

Example 9-9.5. This example is the same as Example 9-7.8, and the system impulse response and the input to the system are

$$g(t) = \epsilon^{-7t} \qquad t > 0$$
$$r_2(t) = \cos 4t \qquad \text{for all } t \tag{9-9.20}$$

The $g(t)$ is folded, and the resulting equation for $c_2(t)$, good for all t, is

$$c_2(t) = \int_{-\infty}^{t} \epsilon^{-7(t-\tau)} \cos 4\tau \, d\tau$$

$$= \epsilon^{-7t} \int_{-\infty}^{t} \epsilon^{7\tau} \cos 4\tau \, d\tau$$

$$= \frac{7}{7^2 + 4^2} \cos 4t + \frac{4}{7^2 + 4^2} \sin 4t$$

$$= 0.124 \cos (4t - 29.75^\circ) \qquad \text{for all } t \tag{9-9.21}$$

This checks with Eq. 9-7.66.

PROBLEMS

9-1 Find the one-sided Laplace transform for the following functions, plot the poles, and show the region of convergence on the complex s-plane.

(a) $f(t) = 4\epsilon^{-6t} + 6\epsilon^{-4t} + t^2 \epsilon^{-2t} \qquad t > 0$
(b) $f(t) = 5 \sin 4t + 6 \cos 7t + t^2 \epsilon^{+2t} \qquad t > 0$
(c) $f(t) = 3 \cosh 5t + 4\epsilon^{+2t} + 2 \sin 3t \qquad t > 0$
(d) $f(t) = t \sin 4t + \epsilon^{-2t} \sin 4t + t\, \epsilon^{-3t} \qquad t > 0$
(e) $f(t) = 3 \cosh 7t + 4\epsilon^{+7t} + 6\epsilon^{-3t} \qquad t > 0$
(f) $f(t) = 7 \sin 5t + 7\epsilon^{-3t} \cosh 3t + t^2 \epsilon^{-1t} \qquad t > 0$

9-2 If $f_2(t)$ has a two-sided Laplace transform, find $F_2(s)$ and the strip of convergence. If $F_2(s)$ does not exist, explain why.

(a) $f_2(t) = \begin{cases} \epsilon^{+5t} & t < 0 \\ \epsilon^{+4t} & t > 0 \end{cases}$ (b) $f_2(t) = \begin{cases} \epsilon^{+4t} & t < 0 \\ \epsilon^{+5t} & t > 0 \end{cases}$

(c) $f_2(t) = \begin{cases} \sinh 3t & t<0 \\ \epsilon^{-2t} & t>0 \end{cases}$

(d) $f_2(t) = \begin{cases} \sinh 2t & t<0 \\ \epsilon^{-3t} & t>0 \end{cases}$

(e) $f_2(t) = \begin{cases} \sin 2t & t<0 \\ \epsilon^{-1t} & t>0 \end{cases}$

(f) $f_2(t) = \begin{cases} \epsilon^{-1t} & t<0 \\ \sin 2t & t>0 \end{cases}$

(g) $f_2(t) = \begin{cases} \cos 4t & t<0 \\ \epsilon^{+2t} & t>0 \end{cases}$

(h) $f_2(t) = \begin{cases} \epsilon^{+2t} & t<0 \\ \cos 4t & t>0 \end{cases}$

9-3 Given the $F_2(s)$ and the associated regions of convergence. Find the $f_2(t)$.

(a) $F_2(s) = \dfrac{-2}{s^2 - 12s + 35}$ $(+5 < \sigma < +7)$

(b) $F_2(s) = \dfrac{2s - 8}{s^2 - 8s + 15}$ $(+3 < \sigma < +5)$

(c) $F_2(s) = \dfrac{-9}{s^2 + 3s - 18}$ $(-6 < \sigma < +3)$

(d) $F_2(s) = \dfrac{-3}{s^2 + 9s + 18}$ $(-6 < \sigma < -3)$

(e) $F_2(s) = \dfrac{2s + 4}{s^2 + 4s - 32}$ $(-8 < \sigma < +4)$

(f) $F_2(s) = \dfrac{-4}{s^2 - 12s + 32}$ $(+4 < \sigma < +8)$

(g) $F_2(s) = \dfrac{-6}{s^2 + 2s - 8}$ (see note below)

(h) $F_2(s) = \dfrac{-4}{s^2 - 4}$ (see note below)

Note: For (g) and (h) it is known that the $j\omega$-axis is in the strip of convergence.

9-4 In Section 9-5, complete the example for:

(a) Case No. 2 (b) Case No. 3 (c) Case No. 5

9-5 For each part of Problem 9-3 as assigned, assume that the $F_2(s)$ exists, but that the strip of convergence is not known. Determine each possible $f_2(t)$ that is a solution, and sketch.

9-6 Repeat Problem 9-5 for the following $F_2(s)$:

(a) $F_2(s) = \dfrac{s^2 - 2s - 1}{s\,[s^2 - 1]}$ (b) $F_2(s) = \dfrac{2s^2 - s + 2}{s\,[s^2 + s - 2]}$ (c) $F_2(s) = \dfrac{-5s - 2}{s\,[s^2 - s - 2]}$

9-7 Someone noticed while obtaining an $F_a(s)$ from an $f_a(t)$ that he could obtain the same result another way. He treats $f_a(t)$ as if it were defined for positive time, and takes the usual one-sided Laplace transform, and then he multiplies the result by -1 to obtain $F_a(s)$. Explore his method and see what you think of it.

9-8 Initial and final value theorems exist for the one-sided Laplace transforms. See if you can develop similar theorems for the two-sided Laplace transforms.

9-9 The following $f_2(t)$'s are defined for all time. Find the corresponding $F_2(s)$'s, and sketch.

(a) $f_2(t) = \epsilon^{-4t}$	(e) $f_2(t) = \epsilon^{-10t}$	(i) $f_2(t) = \epsilon^{-1t}$
(b) $f_2(t) = \epsilon^{-4t} \cos 3t$	(f) $f_2(t) = \epsilon^{-10t} \cos 4t$	(j) $f_2(t) = \epsilon^{-1t} \cos 1t$
(c) $f_2(t) = \epsilon^{+4t}$	(g) $f_2(t) = \epsilon^{+10t}$	(k) $f_2(t) = \epsilon^{+1t}$
(d) $f_2(t) = \epsilon^{+4t} \cos 3t$	(h) $f_2(t) = \epsilon^{+10t} \cos 4t$	(l) $f_2(t) = \epsilon^{+1t} \cos 1t$

9-10 A few terms of the Fourier series for periodic wave shapes, defined for all t, are given. Find the corresponding $F_2(s)$, and sketch.

(a) $f_2(t) = 50 + 63.7 \cos 2t - 21.2 \cos 6t + 12.7 \cos 10t + \cdots$

(b) $f_2(t) = \cos 3t + \dfrac{1}{3^2} \cos 9t + \dfrac{1}{5^2} \cos 15t + \cdots$

(c) $f_2(t) = 0.318 + 0.5 \cos 2t + 0.212 \cos 4t - 0.0424 \cos 8t + \cdots$

9-11 The system of Fig. 9-7.1 is used. The $r_2(t)$ is the same for all four parts. For each part find $R_2(s)$, $G(s)$, $C_2(s)$, and the corresponding strips of convergence. If $C_2(s)$ exists, find the $c_2(t)$, and sketch. If $C_2(s)$ does not exist, explain why.

$$r_2(t) = \begin{cases} \epsilon^{+2t} & t < 0 \\ \epsilon^{-2t} & t > 0 \end{cases}$$

(a) $g(t) = 2\epsilon^{-4t} \quad t > 0$	(c) $g(t) = 2\epsilon^{+1t} \quad t > 0$
(b) $g(t) = 2\epsilon^{-1t} \quad t > 0$	(d) $g(t) = 2\epsilon^{+4t} \quad t > 0$

9-12 Repeat Problem 9-11 for the following functions.

$$r_2(t) = \begin{cases} \epsilon^{+4t} & t < 0 \\ \epsilon^{-1t} & t > 0 \end{cases}$$

(a) $g(t) = 2\epsilon^{+6t} \quad t > 0$	(c) $g(t) = 2\epsilon^{+1t} \quad t > 0$
(b) $g(t) = 2\epsilon^{+3t} \quad t > 0$	(d) $g(t) = 2\epsilon^{-4t} \quad t > 0$

9-13 Repeat Problem 9-11 for the following functions.

$$r_2(t) = \begin{cases} \epsilon^{+2t} & t < 0 \\ \epsilon^{-6t} & t > 0 \end{cases}$$

(a) $g(t) = 4\epsilon^{-8t} \quad t > 0$	(c) $g(t) = 4\epsilon^{+1t} \quad t > 0$
(b) $g(t) = 4\epsilon^{-4t} \quad t > 0$	(d) $g(t) = 4\epsilon^{+3t} \quad t > 0$

9-14 This problem is similar to Problem 9-11, except that $r_b(t)$ and $g(t)$ remain fixed and $r_a(t)$ is varied in each part of the problem.

$$r_b(t) = \epsilon^{-6t} \quad t > 0 \qquad\qquad g(t) = \epsilon^{-3t} \quad t > 0$$

(a) $r_a(t) = \epsilon^{+3t} \quad t < 0$	(c) $r_a(t) = \epsilon^{-4t} \quad t < 0$
(b) $r_a(t) = \epsilon^{-2t} \quad t < 0$	(d) $r_a(t) = \epsilon^{-6t} \quad t < 0$

9-15 Repeat Problem 9-14 for the following functions.

$$r_b(t) = \epsilon^{+4t} \quad t > 0 \qquad\qquad g(t) = \epsilon^{+6t} \quad t > 0$$

(a) $r_a(t) = \epsilon^{+10t} \quad t < 0$	(c) $r_a(t) = \epsilon^{+5t} \quad t < 0$
(b) $r_a(t) = \epsilon^{+8t} \quad t < 0$	(d) $r_a(t) = \epsilon^{+3t} \quad t < 0$

9–16 Repeat Problem 9–14 for the following functions.

$$r_b(t) = \epsilon^{-4t} \quad t > 0 \qquad g(t) = \epsilon^{-2t} \quad t > 0$$

(a) $r_a(t) = \epsilon^{+1t} \quad t < 0$
(b) $r_a(t) = \epsilon^{-1t} \quad t < 0$
(c) $r_a(t) = \epsilon^{-3t} \quad t < 0$
(d) $r_a(t) = \epsilon^{-5t} \quad t < 0$

9–17 The impulse response of a system is

$$g(t) = \epsilon^{+2t} \quad t > 0$$

and the input to the system is

$$r_2(t) = \begin{cases} \epsilon^{+2t} & t < 0 \\ \epsilon^{-2t} & t > 0 \end{cases}$$

The line $s = 2 + j\omega$ is on the boundary of both $G(s)$ and $R_2(s)$. Try to develop a limiting process whereby $c_2(t)$ can be found using the two-sided Laplace transform methods.

9–18 This problem is similar to Problem 9–11, except $r_2(t)$ is defined by one function for all t.

$$r_2(t) = \epsilon^{+4t} \quad \text{for all } t$$

(a) $g(t) = \epsilon^{+6t} \quad t > 0$
(b) $g(t) = \epsilon^{+5t} \quad t > 0$
(c) $g(t) = \epsilon^{-2t} \quad t > 0$
(d) $g(t) = \epsilon^{-1t} \quad t > 0$

9–19 Repeat Problem 9–18 for the following systems.

$$r_2(t) = \epsilon^{+8t} \quad \text{for all } t$$

(a) $g(t) = \epsilon^{-4t} \quad t > 0$
(b) $g(t) = \epsilon^{-6t} \quad t > 0$
(c) $g(t) = \epsilon^{+10t} \quad t > 0$
(d) $g(t) = \epsilon^{+12t} \quad t > 0$

9–20 Repeat Problem 9–18 for the following systems.

$$r_2(t) = \cos 6t \quad \text{for all } t$$

(a) $g(t) = \epsilon^{+4t} \quad t > 0$
(b) $g(t) = \epsilon^{+3t} \quad t > 0$
(c) $g(t) = \epsilon^{+1t} \quad t > 0$
(d) $g(t) = \epsilon^{-1t} \quad t > 0$

9–21 Repeat Problem 9–18 for the following systems.

$$r_2(t) = \epsilon^{+3t} \cos 6t \quad \text{for all } t$$

(a) $g(t) = \epsilon^{-1t} \quad t > 0$
(b) $g(t) = \epsilon^{-4t} \quad t > 0$
(c) $g(t) = \epsilon^{+5t} \quad t > 0$
(d) $g(t) = \epsilon^{+6t} \quad t > 0$

9–22 Each part of this problem refers to systems described in earlier problems, except this time the $c_2(t)$ function is to be found using convolution. If any of the $c_2(t)$ functions does not exist, explain why in terms of the convolution integral.

(a) Problem 9–11: fold $r_2(t)$.
(b) Problem 9–11: fold $g(t)$.
(c) Problem 9–12: fold $r_2(t)$.
(d) Problem 9–12: fold $g(t)$.
(e) Problem 9–13: fold $r_2(t)$.
(f) Problem 9–13: fold $g(t)$.
(g) Problem 9–14: fold $g(t)$.
(h) Problem 9–15: fold $r_2(t)$.
(i) Problem 9–16: fold $r_2(t)$.

9-23 See if the system suggested in Problem 9-17 can be solved using convolution.

9-24 Repeat Problem 9-22 for the following situations.

(a) Problem 9-18: fold $r_2(t)$.
(b) Problem 9-19: fold $g(t)$.
(c) Problem 9-20: fold $r_2(t)$.
(d) Problem 9-21: fold $g(t)$.

9-25 In each part of the problem, the input to the system and its impulse response are given. Attempt to solve this problem using two-sided Laplace transforms. Next find the solution using convolution. Explain the results.

(a) $r_2(t) = \begin{cases} \epsilon^{+3t} & t<0 \\ \epsilon^{+6t} & t>0 \end{cases} \qquad g(t) = \epsilon^{-2t} \quad t>0$

(b) $r_2(t) = \begin{cases} \epsilon^{+4t} & t<0 \\ \epsilon^{+7t} & t>0 \end{cases} \qquad g(t) = \epsilon^{-1t} \quad t>0$

(c) $r_2(t) = \begin{cases} \epsilon^{+3t} & t<0 \\ \epsilon^{+7t} & t>0 \end{cases} \qquad g(t) = \epsilon^{-3t} \quad t>0$

Chapter **10**

The Fourier Transform

10-1. THE DIRECT AND THE INVERSE FOURIER TRANSFORMS

Let us review the direct and the inverse two-sided Laplace transforms. The direct two-sided Laplace transform is found from the integral

$$F_2(s) = \int_{-\infty}^{+\infty} f_2(t)\, \epsilon^{-st}\, dt \tag{10-1.1}$$

provided that the integral exists. We approached the existence of this integral by dividing the $f_2(t)$ into two parts.

$$f_2(t) = \begin{cases} f_a(t) & t<0 \\ f_b(t) & t>0 \end{cases} \tag{10-1.2}$$

For the $F_a(s)$ and $F_b(s)$ to exist separately, both the $f_a(t)$ and the $f_b(t)$ must be of exponential order and sectionally continuous. In addition if the regions of convergence of $F_a(s)$ and $F_b(s)$ have a common strip, $F_2(s)$ exists and is the sum of $F_a(s)$ and $F_b(s)$. One possible region of convergence is shown in Fig. 10-1.1(a).

The inverse two-sided Laplace transform is found from the integral

$$f_2(t) = \frac{1}{2\pi j} \int_{c-j\infty}^{c+j\infty} F_2(s)\, \epsilon^{+st}\, ds \tag{10-1.3}$$

where the integration is performed along a line c-units away from the $j\omega$-axis, as indicated in Fig. 10-1.1(b). A constant c is chosen that puts this line within the strip of convergence.

One situation occurs so often and is so important that it deserves special attention. This is when the $j\omega$-axis is within the strip of convergence. As a result, the integral of

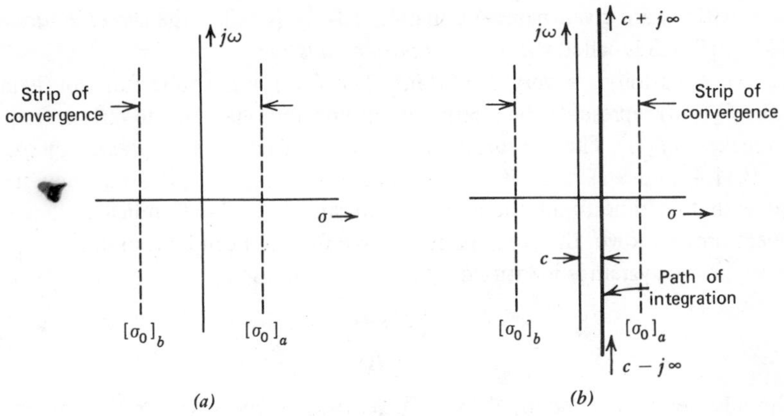

Fig. 10-1.1

Eq. 10-1.1 converges on the line $s = j\omega$, and hence s can be replaced by $j\omega$, yielding

$$F_2(j\omega) = \int_{-\infty}^{+\infty} f_2(t)\, \epsilon^{-j\omega t}\, dt \tag{10-1.4}$$

The strip of convergence for $F_2(j\omega)$ need not be stated since the $j\omega$-axis conveys the necessary information. To recover the $f_2(t)$, the poles to the right of the $j\omega$-axis are assigned to $F_a(s)$ and those to the left to $F_b(s)$.

For the inverse transform, the $F_2(s)$ is replaced by $F_2(j\omega)$, the c is chosen as zero since the $j\omega$-axis is in the strip of convergence, and Eq. 10-1.3 becomes

$$f_2(t) = \frac{1}{2\pi j} \int_{-j\infty}^{+j\infty} F_2(j\omega)\, \epsilon^{+j\omega t}\, d(j\omega) \tag{10-1.5}$$

Rather than integrate with respect to $j\omega$, we integrate with respect to ω. The limits are changed as

$$\begin{aligned} j\omega &= -j\infty \qquad \omega = -\infty \\ j\omega &= +j\infty \qquad \omega = +\infty \end{aligned} \tag{10-1.6}$$

The j inside the differential is brought out, as

$$f_2(t) = \frac{j}{2\pi j} \int_{-\infty}^{+\infty} F_2(j\omega)\, \epsilon^{+j\omega t}\, d\omega \tag{10-1.7}$$

The j factors cancel, and the equation is written as

$$f_2(t) = \frac{1}{2\pi} \int_{-\infty}^{+\infty} F_2(j\omega)\, \epsilon^{+j\omega t}\, d\omega \tag{10-1.8}$$

These equations are given names. Equation 10–1.4 is called the *direct Fourier transform*, and Eq. 10–1.8 is called the *inverse Fourier transform.*

These two equations are very important. The $f_2(t)$ is the time function being considered. The $F_2(j\omega)$ represents the relative amplitude and phase of the various frequencies that go to make up $f_2(t)$. The $F_2(j\omega)$ is commonly referred to as the *frequency spectrum.* Equation 10–1.4 indicates that if a time function is known, the frequency spectrum associated with this time function can be found, and Eq. 10–1.8 indicates that if a frequency spectrum is known, the corresponding time function can be found.

Next, suppose a system is identified by a transfer function

$$G(s) = \frac{O(s)}{I(s)} \tag{10–1.9}$$

and the impulse response is denoted by $g(t)$. We assume the system's Fourier transform exists; then

$$G(j\omega) = \frac{O(j\omega)}{I(j\omega)} \tag{10–1.10}$$

This equation is essentially of the phasor transform type, and if the input phasor and $G(j\omega)$ are known, the output phasor can be determined. With these thoughts in mind, Eq. 10–1.4 tells us that if the impulse response of a system is known, the frequency response of the system can be uniquely determined at all frequencies. Equation 10–1.8 tells us that if the frequency response of a system is known at all frequencies, the impulse response of the system can be uniquely determined.

This interrelationship between transient response and frequency response opens up a whole new understanding of the behavior of systems. A person can go into the laboratory and run sinusoidal steady-state frequency curves on a linear system, and it is theoretically possible to go from these curves to the impulse response of the system. Since the data for the frequency response are in graphical form, the integration must be carried out by using numerical methods, and the impulse response will also be presented in graphical form.

The notation used in the literature concerning the Fourier transforms differs widely. Sometimes $g(\omega)$ and $h(t)$ are used. Sometimes the $\frac{1}{2\pi}$ factor appears in front of the direct transform integral instead of the inverse transform integral. Some authors divide up (so to speak) the $\frac{1}{2\pi}$ factor by placing a $1/\sqrt{2\pi}$ in front of both the direct and inverse transform integrals. Instead of integrating with respect to $\omega = 2\pi f$, some authors integrate with respect to f, and the $\frac{1}{2\pi}$ factor does not appear in either the direct or the inverse integral. What these systems of notation have in common is that the exponent in the direct Fourier transform contains the $-j$ factor and the inverse transform contains the $+j$ factor. Some authors refer to these as the $-j$ and the $+j$ integrals.

It would seem from this presentation that the Fourier transform is just a special case of the two-sided Laplace transform. Although this argument can be defended from a mathematical point of view, the physical interpretation of the Fourier transform opens up a new world of understanding and interpretation. Also the integral for the inverse Fourier transform is real, and we are finally in a position to use the inverse transform integral.

The notation we shall use for the direct Fourier transform is

$$\mathcal{F}[f_2(t)] = F_2(j\omega) \tag{10-1.11}$$

The symbol is read as the direct Fourier transform of $f_2(t)$. The notation for the inverse fourier transform is

$$\mathcal{F}^{-1}[F_2(j\omega)] = f_2(t) \tag{10-1.12}$$

The symbol is read as the inverse Fourier transform of $F_2(j\omega)$.

10-2. A GEOMETRIC INTERPRETATION OF THE FOURIER TRANSFORM

Th two-sided Laplace transform $F_2(s)$ is a function of the complex variable $s = \sigma + j\omega$, and is also complex.

$$F_2(s) = F_2(\sigma + j\omega) = U(\sigma, \omega) + j\, V(\sigma, \omega) \tag{10-2.1}$$

One possible geometric interpretation of $F_2(s)$ would be to display two different three-dimensional surfaces. For example, the real part of $F_2(s)$, Re $[F_2(s)] = U(\sigma, \omega)$ could be displayed as the third-dimension perpendicular to the s-plane, as indicated in Fig. 10-2.1(a). The other surface could be the imaginary part of $F_2(s)$, Im $[F_2(s)] = V(\sigma, \omega)$, which is not shown.

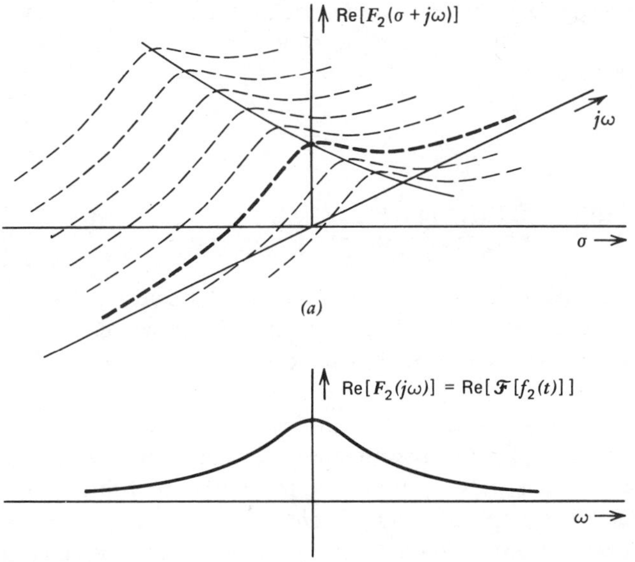

Fig. 10-2.1

When we talk about $F_2(s)$, in a certain sense we sit back and observe both these surfaces relative to the entire s-plane. However, when we talk about the Fourier transform we are only interested in one part of these three-dimensional surfaces. To be completely explicit, we could take the three-dimensional surface of Fig. 10-2.1(a) and run it through a buzz saw that cuts along the $j\omega$-axis. Then we turn the resulting surface so that we are looking perpendicular to the cut and observe only the shape of the cut surface. Because of the canceling of the j terms in Eq. 10-1.7, we label the abscissa with an ω, as shown in Fig. 10-2.1(b). The same argument can be used on the imaginary part of $F_2(s)$.

10-3. THE DIRICHLET CONDITIONS

The approach we have taken to the existence of the direct Fourier transform is through the use of the two-sided Laplace transform. If the $f_2(t)$ has a two-sided Laplace transform $F_2(s)$, and the $j\omega$-axis is in the strip of convergence, then the Fourier transform of $f_2(t)$ exists and can be obtained from $F_2(s)$ by replacing s with $j\omega$.

An alternate set of conditions for the existence of the Fourier transform are known as the *Dirichlet conditions*. These are stated as:

1 The $f_2(t)$ must be absolutely integrable, that is $\int_{-\infty}^{+\infty} |f_2(t)|\, dt$ must be finite.

2 The $f_2(t)$ must have a finite number of maxima and minima in any finite interval.

3 The $f_2(t)$ must have a finite number of finite discontinuities in any finite interval.

In a manner analogous to the conditions for the existence of the Laplace transform, these conditions can be relaxed in a number of situations. For example, the impulse function violates *Condition 3*, and yet through a limiting process can be interpreted as having a Fourier transform.

10-4. SIGNALS CAN BE REPRESENTED EITHER IN THE TIME DOMAIN OR IN THE FREQUENCY DOMAIN

The direct Laplace transform changes a time function $f_2(t)$ into an $F_2(s)$. For lumped systems, $F_2(s)$ is made up of a set of poles and zeros. With a certain amount of experience, the location of the poles and zeros gives the designer a great deal of information about $f_2(t)$ but in an abstract sort of a way.

Although the Fourier transform can also be regarded as a mathematical relationship, it can also be given a physical interpretation. For example, a signal can be described either as a time function $f_2(t)$ or by $F_2(j\omega)$, the Fourier transform of $f_2(t)$. The $F_2(j\omega)$ represents the relative amplitude and the phase of the various frequency components that go to make up $f_2(t)$. Both concepts are equally familiar to the designer. For example, most people are accustomed to referring to the specification for an amplifier by saying it passes frequencies for 20 to 20,000 hertz.

We continue by presenting several examples.

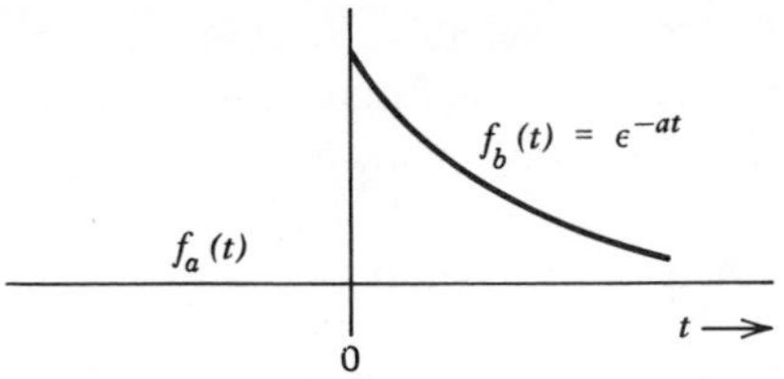

Fig. 10-4.1

Example 10-4.1. The time function to be examined is

$$f_2(t) = \begin{cases} 0 & t < 0 \\ \epsilon^{-at} & t > 0 \end{cases} \tag{10-4.1}$$

and the function is shown in Fig. 10-4.1. In order to satisfy the Dirichlet condition, the a in Eq. 10-4.1 must be greater than zero. If $a \leqslant 0$, then the integral does not converge and $F_2(j\omega)$ does not exist. The $F_2(j\omega)$ is found by using the direct Fourier transform integral, as

$$F_2(j\omega) = \int_{-\infty}^{+\infty} f_2(t)\, \epsilon^{-j\omega t}\, dt = \int_0^{\infty} \epsilon^{-at}\, \epsilon^{-j\omega t}\, dt$$

$$= \frac{-1}{a + j\omega}\, \epsilon^{-(+a+j\omega)t} \Big|_0^{\infty} = \frac{1}{a + j\omega} \tag{10-4.2}$$

If we did not already have a Laplace transform table, we would stop and develop a Fourier transform table. However, since a Laplace transform table is available, we will make use of it. We again obtain the Fourier transform of the $f_2(t)$ given in Eq. 10-4.1, in the following way.

$$\begin{aligned} f_a(t) &= 0 \qquad F_a(s) = 0 \\ f_b(t) &= \epsilon^{-at} \qquad F_b(s) = \frac{1}{s + a} \end{aligned} \tag{10-4.3}$$

A region of convergence exists, and therefore $F_2(s)$ exists, as

$$F_2(s) = F_a(s) + F_b(s) = \frac{1}{s + a} \tag{10-4.4}$$

So long as $a > 0$, the strip of convergence contains the $j\omega$-axis, and $s = j\omega$ is a legitimate substitution. Therefore, the Fourier transform is

$$F_2(j\omega) = \frac{1}{a + j\omega} \tag{10-4.5}$$

DISPLAYING THE FOURIER TRANSFORM

In general, the $F_2(j\omega)$ is complex and requires more effort to be displayed graphically than does the original time function. For a fixed value of a, $F_2(j\omega)$ of Eq. 10-4.5 can be shown on a polar plot with ω as the variable. However, it is more customary to use two plots. One such pair of plots would be magnitude and phase vs. ω, and another would be the real and imaginary parts of $F_2(j\omega)$ vs. ω. Both sets of plots are developed for this example.

The $F_2(j\omega)$ for this example can be written as

$$F_2(j\omega) = \frac{1}{a + j\omega} = \frac{1}{\sqrt{a^2 + \omega^2}} \underline{/-\tan^{-1} \omega/a} \tag{10-4.6}$$

The magnitude spectrum $|F_2(j\omega)|$ is

$$|F_2(j\omega)| = \frac{1}{\sqrt{a^2 + \omega^2}} \tag{10-4.7}$$

and the phase spectrum is

$$\theta(\omega) = -\tan^{-1} \frac{\omega}{a} \tag{10-4.8}$$

The plots of magnitude and phase are shown in Fig. 10-4.2(a) and (b), respectively.

This $F_2(j\omega)$ can be rationalized as

$$F_2(j\omega) = \frac{1}{a + j\omega} \frac{a - j\omega}{a - j\omega} = \frac{a}{a^2 + \omega^2} - \frac{j\omega}{a^2 + \omega^2} \tag{10-4.9}$$

The $F_2(j\omega)$ is made up of a real part, which is an even function, and an imaginary part, which is an odd function. A plot of the even part, and j times the odd part, are shown in Fig. 10-4.3(a) and (b), respectively.

Example 10-4.2. The time function to be examined is

$$f_2(t) = \begin{cases} \epsilon^{+at} & t < 0 \\ 0 & t > 0 \end{cases} \tag{10-4.10}$$

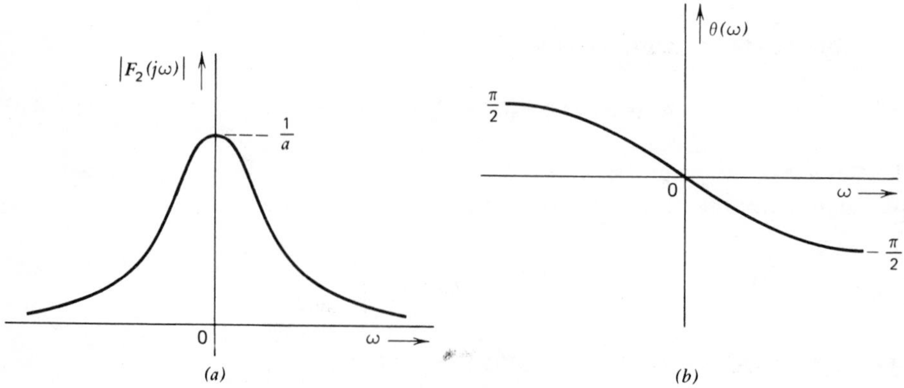

Fig. 10-4.2

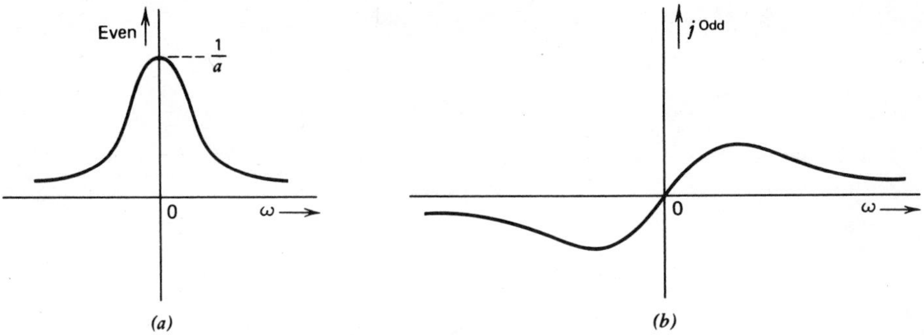

Fig. 10-4.3

We proceed using the two-sided Laplace transform approach.

$$f_a(t) = \epsilon^{+at} \Longrightarrow F_a(s) = \frac{-1}{s-a}$$
$$f_b(t) = 0 \Longrightarrow F_b(s) = 0 \tag{10-4.11}$$

The two-sided Laplace transform exists and is

$$F_2(s) = F_a(s) + F_b(s) = \frac{1}{a-s} \tag{10-4.12}$$

So long as $a > 0$, the $j\omega$-axis is in the strip of convergence, and $s = j\omega$ is a legitimate substitution.

$$F_2(j\omega) = \frac{1}{a - j\omega} = \frac{1}{\sqrt{a^2 + \omega^2}} \angle -\tan^{-1}\left(\frac{-\omega}{a}\right) \tag{10-4.13}$$

The magnitude and phase of $F_2(j\omega)$ are shown in Fig. 10-4.4(a) and (b), respectively.

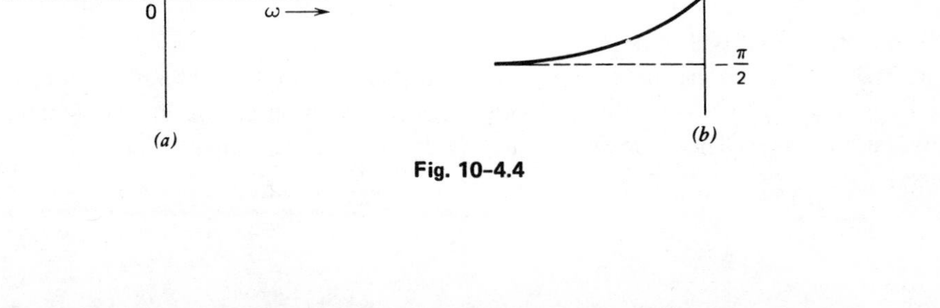

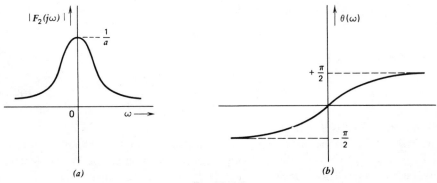

Fig. 10-4.4

Example 10-4.3. The time function to be examined is

$$f_2(t) = \begin{cases} \epsilon^{+at} & t<0 \\ \epsilon^{-at} & t>0 \end{cases} \tag{10-4.14}$$

This time function is the sum of the time functions in Examples 10-4.1 and 10-4.2. Therefore, the $F_2(s)$ is the sum of the transforms found in the previous two examples, assuming $a > 0$.

$$F_2(s) = \frac{1}{a-s} + \frac{1}{a+s} \tag{10-4.15}$$

The Fourier transform can be found from $F_2(s)$ by replacing s with $j\omega$.

$$F_2(j\omega) = \frac{1}{a-j\omega} + \frac{1}{a+j\omega} = \frac{2a}{a^2+\omega^2} \tag{10-4.16}$$

Since the odd part of $F_2(j\omega)$ is zero, only one curve is necessary to display the function.

Example 10-4.4. The time function to be examined is

$$f_2(t) = \begin{cases} \epsilon^{+3t} & t<0 \\ \epsilon^{+2t} & t>0 \end{cases} \tag{10-4.17}$$

An examination of this function shows that the $\int_{-\infty}^{+\infty} |f_2(t)|\,dt$ is infinite; therefore, this function violates the Dirichlet conditions and the Fourier transform does not exist. Let us follow the two-sided Laplace transform approach and see how this function is rejected by these methods. The $F_2(s)$ and its strip of convergence are given by

$$F_2(s) = \frac{-1}{s-3} + \frac{1}{s-2} \tag{10-4.18}$$

$$2 < \sigma < 3$$

Since the strip of convergence does not include the $j\omega$-axis, the substitution $s = j\omega$ cannot be made and the Fourier transform does not exist.

A little thought reveals that if the $j\omega$-axis is in the strip of convergence, the $\int_{-\infty}^{+\infty} |f_2(t)|\,dt$ is finite.

Example 10-4.5. The time function to be examined is

$$f_2(t) = \begin{cases} 1 & -T/2 < t < +T/2 \\ 0 & \text{elsewhere} \end{cases} \tag{10-4.19}$$

and is shown in Fig. 10-4.5(a).

This time function can be built up with two step-functions in a manner analogous to that done in Chapter 4, but this procedure brings up questions we are unprepared to answer at this time. Therefore, we find $F_2(j\omega)$ by using the direct Fourier transform integral.

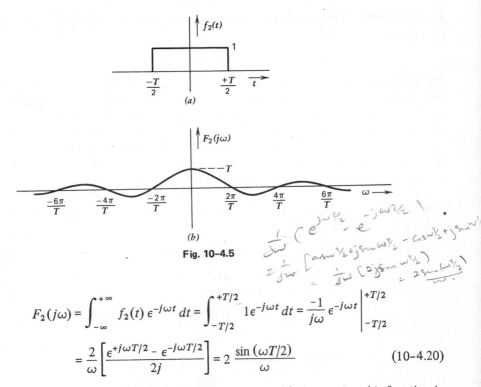

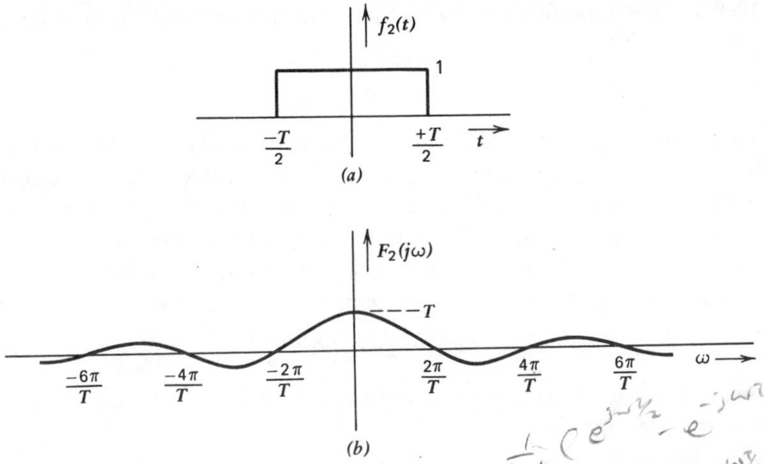

Fig. 10-4.5

$$F_2(j\omega) = \int_{-\infty}^{+\infty} f_2(t)\,\epsilon^{-j\omega t}\,dt = \int_{-T/2}^{+T/2} 1\epsilon^{-j\omega t}\,dt = \frac{-1}{j\omega}\,\epsilon^{-j\omega t}\Bigg|_{-T/2}^{+T/2}$$

$$= \frac{2}{\omega}\left[\frac{\epsilon^{+j\omega T/2} - \epsilon^{-j\omega T/2}}{2j}\right] = 2\,\frac{\sin(\omega T/2)}{\omega} \qquad (10\text{–}4.20)$$

We take advantage of the well-known properties of $(\sin x)/x$ by putting this function in the form:

$$F_2(j\omega) = T\left[\frac{\sin(\omega T/2)}{\omega T/2}\right] \qquad (10\text{–}4.21)$$

This function is shown in Fig. 10–4.5(b).

The first crossing of the ω-axis occurs when

$$\sin\frac{\omega T}{2} = 0 \quad \text{or} \quad \frac{\omega T}{2} = \pi \quad \text{or} \quad \omega = \frac{2\pi}{T} \qquad (10\text{–}4.22)$$

The other crossings are found in a similar manner. The $f_2(t)$ in this example is an even function of t, and the $F_2(j\omega)$ is an even function of ω.

10-5. DIRECT FOURIER TRANSFORMS THAT EXIST IN THE LIMIT

In Section 9–6, two-sided Laplace transforms that exist in the limit are discussed. A similar situation occurs in connection with the Fourier transform. This situation is explored through a set of examples.

Example 10-5.1. The time function to be examined is a d-c signal equal to unity, or an $f_2(t)$ given by

$$f_2(t) = 1 \qquad -\infty < t < +\infty \tag{10-5.1}$$

The area under the $|f_2(t)|$ curve is infinite, and hence the Dirichlet conditions are not met. In the strict sense of the term, this $f_2(t)$ does not have a Fourier transform. This situation is similar to the two-sided Laplace transforms that do not exist in the strict sense of the word but can be said to exist in the limit, as discussed in Section 9-6. As a matter of fact, Example 9-6.2 is precisely the same time function as Eq. 10-5.1, and the two-sided Laplace transform in the limit was found in Eq. 9-6.24, as

$$F_2(s) = 2\pi\, u_0(s) \tag{10-5.2}$$

This function converged on the line $\sigma = 0$, and since $\sigma = 0$ is the $j\omega$-axis, s can be replaced by $j\omega$, and the Fourier transform results.

$$F_2(j\omega) = 2\pi\, u_0(j\omega) \tag{10-5.3}$$

We wish to obtain this same result from an entirely different limiting process. In the Example 10-4.5, we started with the time function

$$f_2(t) = \begin{cases} 1 & -T/2 < t < +T/2 \\ 0 & \text{elsewhere} \end{cases} \tag{10-5.4}$$

and the Fourier transform was found to be

$$F_2(j\omega) = T\left[\frac{\sin(T\omega/2)}{T\omega/2}\right] \tag{10-5.5}$$

For the following steps, we suggest sketches be made to observe both the $f_2(t)$ and $F_2(j\omega)$ curves as the discussion proceeds. First, T is given a relatively small value, say T_1. Next, we replace T_1 with T_2, where $T_2 = 2T_1$. The time function obviously spreads over twice as much time. An examination of Eq. 10-5.5 and Fig. 10-4.5(b) indicates that the magnitude of the new $F_2(j\omega)$ curve is 2 times the original curve, but the zero crossings occur at one-half the original ω. We continue by letting T take on the value of $T_3 = 2T_2 = 4T_1$. The magnitude of the $F_2(j\omega)$ curve again doubles, and zero crossings again occur at one-half the ω. We obviously can keep this process up indefinitely by letting T take on values of $T_4 = 2T_3 = 8T_1; T_5 = 2T_4 = 16T_1; \ldots$. In each successive step, the new $F_2(j\omega)$ has a height twice that of the previous function, and the zero crossings occur at $\omega/2$.

Next we explore the area under the $F_2(j\omega)$ curve by setting up the integral

$$\text{Area} = \int_{-\infty}^{+\infty} T\left[\frac{\sin(\omega T/2)}{\omega T/2}\right] d\omega = 2\pi \tag{10-5.6}$$

We observe that this area is independent of T. That is, as $T \to \infty$, the height of this $F_2(j\omega)$ curve goes to infinity and the width approaches zero but the area throughout this limiting process remains at 2π.

Everything about this discussion is legitimate until T approaches infinity as a limit, and then the Dirichlet conditions are not met. For example, the $f_2(t)$ pulse could start 10^6 years ago and continue into the future for another 10^6 years, and all the steps are legitimate. Certainly within our lifetime we cannot tell the difference between this finite duration pulse and a true d-c wave shape. In terms of the mathematical ease of talking about the situation, it is convenient to let $T \to \infty$, and the $F_2(j\omega)$ then becomes an impulse of area 2π, or

$$F_2(j\omega) = 2\pi\, u_0(j\omega) \tag{10-5.7}$$

We then say that $f_2(t)$ has a Fourier transform in the limit, because a meaning is given to the function through a limiting process.

One other comment is in order. The $f_2(t)$ we are talking about is a d-c signal. The $F_2(j\omega)$ is an impulse function occuring at $\omega = 0$. This result fits what we already known by intuition, which is that there is only one frequency present in a d-c signal and this is a frequency of zero.

Example 10-5.2. The $f_2(t)$ in the last example was a d-c signal and represents the ultimate as far as flatness is concerned. This example goes to the other extreme and finds the Fourier transform of an impulse.

$$f_2(t) = u_0(t) \tag{10-5.8}$$

The two-sided Laplace transform is

$$F_2(s) = 1$$

This function converges in the entire s-plane, and hence $s = j\omega$ is a legitimate substitution; therefore, the Fourier transform is

$$F_2(j\omega) = 1 \qquad \text{for all } \omega \tag{10-5.9}$$

COMMENTARY

Examples 10-5.1 and 10-5.2 are extreme cases of a general property of Fourier transforms. The more spread out a time function is, the more peaked is the frequency spectrum. The d-c time function of the first example has an $F_2(j\omega)$ that is an impulse. At the other extreme, the more peaked the time function is, the more spread out the frequency spectrum. The impulse time function has a flat frequency spectrum, as indicated by the second example. In a certain sense all the other cases lie somewhere between these two extremes.

Example 10-5.3. The next time function is the unit-step function.

$$f_2(t) = u_{-1}(t) \tag{10-5.10}$$

Again the area under the $|f_2(t)|$ curve is infinite, and the only way a Fourier transform can exist is through some limiting process. For variety, a different approach to finding the limit is used. The Laplace transform of the unit-step function is

$$F_2(s) = \frac{1}{s} \tag{10-5.11}$$

Since this function has a pole at the origin, the region of convergence is to the right of the $j\omega$-axis. Hence, the substitution $s = j\omega$ cannot be made. We use a slightly altered version of the idea developed in Section 9-6, by modifying the original function. The modified $f_2(t)$ function is chosen to be

$$f_{2m}(t) = \epsilon^{-at} u_{-1}(t) \tag{10-5.12}$$

where $a > 0$, and the transform is

$$F_{2m}(s) = \frac{1}{s+a} \tag{10-5.13}$$

Since the $j\omega$-axis is in the region of convergence, $s = j\omega$ is a legitimate substitution. Therefore, the modified function does have a Fourier transform, as given by

$$F_{2m}(j\omega) = \frac{1}{a + j\omega} \tag{10-5.14}$$

The equation is rationalized as

$$F_{2m}(j\omega) = \frac{1}{a + j\omega}\,\frac{a - j\omega}{a - j\omega} = \frac{a}{a^2 + \omega^2} - j\frac{\omega}{a^2 + \omega^2} \tag{10-5.15}$$

The real and the imaginary parts are examined separately.

A sketch of the real part is shown in Fig. 10-5.1. The area under the curve is

$$\int_{-\infty}^{+\infty} \frac{a}{a^2 + \omega^2}\,d\omega = \int_{-\infty}^{+\infty} \frac{d(\omega/a)}{1 + (\omega/a)^2} = \tan^{-1}\left(\frac{\omega}{a}\right)\Bigg|_{-\infty}^{+\infty} = \pi \tag{10-5.16}$$

As $a \to 0$, the height goes to infinity, and the width to zero, but the area under the curve is constant throughout the limiting process. In the limit this function can be represented by an impulse of value π.

As $a \to 0$, the imaginary part of Eq. 10-5.15 becomes

$$\text{Im}\,[F_2(j\omega)]\,|_{a\to 0} = \frac{1}{j\omega} \tag{10-5.17}$$

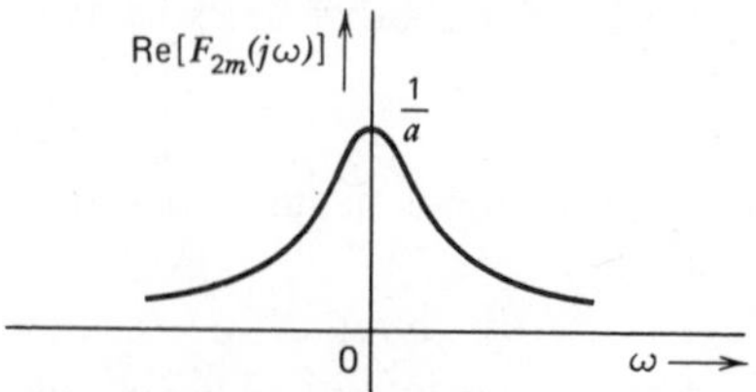

Fig. 10-5.1

These two parts are added to yield the Fourier transform that exists in the limit for the unit-step function

$$\mathfrak{F}\,[u_{-1}(t)] = \pi\, u_0(j\omega) + \frac{1}{j\omega} \qquad (10\text{-}5.18)$$

If a person objects to letting a take on the limiting value of zero, he can let it take on an arbitrarily small but non-zero value and the Dirichlet conditions are not violated. For example, a could be set so that the modified time function departs from the $u_{-1}(t)$ by 1% after 10^6 years have passed. None of us could tell the difference between these two functions in our lifetime. The real part of the Fourier transform will not be an impulse, but it would certainly be convenient to approximate it by one.

Example 10-5.4. The signum function abbreviated as $Sgn(t)$ is defined as

$$Sgn(t) = \begin{cases} -1 & t<0 \\ +1 & t>0 \end{cases} \qquad (10\text{-}5.19)$$

Two properties of Fourier transforms discussed in Chapter 11 are needed. One property is

$$\mathfrak{F}[f_A(t) + f_B(t)] = \mathfrak{F}[f_A(t)] + \mathfrak{F}[f_B(t)] \qquad (10\text{-}5.20)$$

The other is called the reversal property, and is given by:

$$\text{If} \quad \mathfrak{F}[f_2(t)] = F_2(j\omega) \quad \text{then} \quad \mathfrak{F}[f_2(-t)] = F_2(-j\omega) \qquad (10\text{-}5.21)$$

We write

$$\mathfrak{F}[Sgn(t)] = \mathfrak{F}[-u_{-1}(-t)] + \mathfrak{F}[u_{-1}(t)] = -\pi\, u_0(j\omega) + \frac{1}{j\omega} + \pi\, u_0(j\omega) + \frac{1}{j\omega} = \frac{2}{j\omega}$$

$$(10\text{-}5.22)$$

Example 10-5.5. We wish to find the Fourier transform of a unit ramp function

$$f_2(t) = u_{-2}(t) = t\, u_{-1}(t) \qquad (10\text{-}5.23)$$

Another property of the Fourier transform discussed in Chapter 11 is needed. We state the property: Given, an $f_2(t)$ which has a Fourier transform.

$$\mathfrak{F}[f_2(t)] = F_2(j\omega) \qquad (10\text{-}5.24)$$

If the $f_2(t)$ is multiplied by t^n, the corresponding Fourier transform of the product is given by

$$\mathfrak{F}[t^n f_2(t)] = j^n \frac{d^n}{d\omega^n} F_2(j\omega) \qquad (10\text{-}5.25)$$

We observe that the Fourier transform of $u_{-1}(t)$ is known from Example 10-5.3, and that we can obtain the transform of $t\, u_{-1}(t)$ by using Eq. 10-5.25, with $n = 1$, as

$$\mathfrak{F}[t\, u_{-1}(t)] = j\left[\pi\, u_{+1}(j\omega) - \frac{1}{j\omega^2}\right] = j\pi\, u_{+1}(j\omega) - \frac{1}{\omega^2} = j\pi\, u_{+1}(j\omega) + \frac{1}{(j\omega)^2} \qquad (10\text{-}5.26)$$

Table 10-5.1

1st Family of Functions

$f_2(t)$	$F_2(j\omega)$	$F_2(s)$	region of convergence for $F_2(s)$
$u_{-1}(t)$	$\pi u_0(j\omega) + \frac{1}{j\omega}$	$\frac{1}{s}$	$0 < \sigma < \infty$
$u_{-2}(t)$	$j\pi u_{+1}(j\omega) + \frac{1}{(j\omega)^2}$	$\frac{1}{s^2}$	$0 < \sigma < \infty$
$u_{-3}(t)$	$\frac{j^2}{2}\pi u_{+2}(j\omega) + \frac{1}{(j\omega)^3}$	$\frac{1}{s^3}$	$0 < \sigma < \infty$
$u_{-n}(t)$	$\frac{(j)^{n-1}}{(n-1)!}\pi u_{+(n-1)}(j\omega) + \frac{1}{(j\omega)^n}$	$\frac{1}{s^n}$	$0 < \sigma < \infty$

2nd Family of Functions

$f_2(t)$	$F_2(j\omega)$	$F_2(s)$	region of convergence for $F_2(s)$
1, for all t	$2\pi u_0(j\omega)$	$2\pi u_0(s)$	$\sigma = 0$
t, for all t	$j2\pi u_{+1}(j\omega)$	$-2\pi u_{+1}(s)$	$\sigma = 0$
$\frac{t^2}{2}$, for all t	$\frac{(j)^2}{2} 2\pi u_{+2}(j\omega)$	$(-1)^2 \frac{2\pi u_{+2}(s)}{2}$	$\sigma = 0$
$\frac{t^n}{n!}$, for all t	$\frac{(j)^n}{n!} 2\pi u_{+n}(j\omega)$	$(-1)^n \frac{2\pi u_{+n}(s)}{n!}$	$\sigma = 0$

3rd Family of Functions

$f_2(t)$	$F_2(j\omega)$	$F_2(s)$	region of convergence for $F_2(s)$
$sgn(t)$	$\frac{2}{(j\omega)}$	$\frac{2}{s}$	$\sigma = 0$
$t\, sgn(t)$	$\frac{2}{(j\omega)^2}$	$\frac{2}{s^2}$	$\sigma = 0$
$\frac{t^2}{2} sgn(t)$	$\frac{2}{(j\omega)^3}$	$\frac{2}{s^3}$	$\sigma = 0$
$\frac{t^n}{n!} sgn(t)$	$\frac{2}{(j\omega)^{n+1}}$	$\frac{2}{s^{n+1}}$	$\sigma = 0$

Example 10-5.6. We continue by finding the Fourier transform of

$$f_2(t) = |t| \qquad \text{for all } t \tag{10-5.27}$$

We observe that if we multiply the Signum function by t, we have the present $f_2(t)$ function. Therefore we obtain

$$\mathfrak{F}[|t|] = j\frac{d}{d\omega}\left(\frac{2}{j\omega}\right) = \frac{-2}{\omega^2} = \frac{2}{(j\omega)^2} \tag{10-5.28}$$

Example 10-5.7. Another similar example is

$$f_2(t) = t \qquad \text{for all } t \tag{10-5.29}$$

We start from the result of Example 10-5.1, given as

$$\mathfrak{F}[1] = 2\pi\, u_0(j\omega) \tag{10-5.30}$$

and multiply this $f_2(t)$ by t, and use Eq. 10-5.25, with $n = 1$, as

$$\mathfrak{F}[(t)] = j\, 2\pi\, u_{+1}(j\omega) \tag{10-5.31}$$

COMMENTARY

We have obviously started on an infinite family of functions in the last few examples, or actually on three families. Table 10-5.1 summarizes the results for the Fourier transform and extends the work to include the general terms. The table also includes the two-sided Laplace transform for these functions.

The sum of a member of the 2nd Family to the corresponding member of the 3rd Family divided by 2 yields the corresponding member of the 1st Family for the Fourier transform part of the table. Additional comments will be made concerning this table in Chapter 11.

10-6. THE INVERSE FOURIER TRANSFORM

We turn our attention to the inverse transform through a set of examples.

Example 10-6.1. To introduce this situation, we return to the $f_2(t)$ of Example 10-4.5, which is repeated as

$$f_2(t) = \begin{cases} 1 & -T/2 < t < T/2 \\ 0 & \text{elsewhere} \end{cases} \tag{10-6.1}$$

The corresponding $F_2(j\omega)$ was found in Eq. 10-4.21 to be

$$F_2(j\omega) = T\left[\frac{\sin(T\omega/2)}{T\omega/2}\right] \tag{10-6.2}$$

In this example, we start with the $F_2(j\omega)$ and wish to find the corresponding $f_2(t)$. By methods analogous to the way we have used the Laplace transform, the $f_2(t)$ of Eq. 10-6.1 and the $F_2(j\omega)$ of Eq. 10-6.2 could be entered into a Fourier transform table as a transform pair. When we are presented with the $F_2(j\omega)$ of Eq. 10-6.2, we could consult the table for the inverse transform and determine the $f_2(t)$ of Eq. 10-6.1.

However, we do not wish to work this example in that manner. We have stressed several times that the inverse Fourier transform integral is real, and this integral is used to find the $f_2(t)$.

$$f_2(t) = \frac{1}{2\pi}\int_{-\infty}^{+\infty} F_2(j\omega)\,\epsilon^{+j\omega t}\,d\omega = \frac{1}{2\pi}\int_{-\infty}^{+\infty}\left[T\frac{\sin(T\omega/2)}{T\omega/2}\right]\epsilon^{+j\omega t}\,d\omega$$

$$= \frac{1}{\pi}\left[\int_{-\infty}^{+\infty}\frac{\sin(T\omega/2)\cos\omega t}{\omega}\,d\omega + j\int_{-\infty}^{+\infty}\frac{\sin(T\omega/2)\sin\omega t}{\omega}\,d\omega\right] \qquad (10\text{-}6.3)$$

The term multiplied by the j has an integrand that is an odd function[1] of ω, and hence the integral from $-\infty$ to 0 is equal to, but opposite in sign from, the integral from 0 to ∞. Thus, the entire term is zero. The first term has an integrand that is an even function[1] of ω, and hence the integration need be only from 0 to ∞, with the result multiplied by 2. Therefore, the $f_2(t)$ can be found from

$$f_2(t) = \frac{2}{\pi}\int_0^{\infty}\frac{\sin(T\omega/2)\cos\omega t}{\omega}\,d\omega \qquad (10\text{-}6.4)$$

The following improper integral is needed

$$\int_0^{\infty}\frac{\sin au}{u}\,du = \begin{cases} +\pi/2 & a>0 \\ -\pi/2 & a<0 \end{cases} \qquad (10\text{-}6.5)$$

The following identity will put Eq. 10-6.4 in the desired form.

$$\sin A\cos B = \frac{\sin(A+B) - \sin(B-A)}{2} \qquad (10\text{-}6.6)$$

$$f_2(t) = \frac{1}{\pi}\left[\int_0^{\infty}\frac{\sin\omega(t+T/2)\,d\omega}{\omega} - \int_0^{\infty}\frac{\sin\omega(t-T/2)\,d\omega}{\omega}\right] \qquad (10\text{-}6.7)$$

The first term, found from Eq. 10-6.5, is equal to $-\frac{1}{2}$ when $t<-T/2$, and is equal to $+\frac{1}{2}$ when $t>-T/2$. The second term is equal to $+\frac{1}{2}$ when $t<T/2$ and is equal to $-\frac{1}{2}$ when $t>+T/2$. When these two terms are added, the $f_2(t)$ of Eq. 10-6.1 is recaptured.

Example 10-6.2. In the last example, the inverse Fourier transform integral was used. A similar possibility exists in finding the inverse Laplace transform by using the inverse

[1] The subject of even and odd functions is discussed in more detail in Section 11-7.

transform integral, but we avoided this because of the need to use complex variable theory. Instead, a Laplace transform table was developed by using the direct Laplace transform, and this table was used in the reverse direction to find the inverse transform.

This same thing can be done with the Fourier transform. A transform table can be built up using the direct Fourier transform, and this table can be used in the reverse direction. We modify this procedure in order to use the Laplace table.

The following $F_2(j\omega)$ is given.

$$F_2(j\omega) = \frac{10}{\omega^2 + 24 + j2\omega} \tag{10-6.8}$$

and we desire to find the corresponding $f_2(t)$. Just for experience, you might solve the inverse Fourier transform integral as given by

$$f_2(t) = \frac{1}{2\pi} \int_{-\infty}^{+\infty} \frac{10}{\omega^2 + 24 + j2\omega} \epsilon^{+j\omega t} \, d\omega \tag{10-6.9}$$

We solve the problem by reversing the steps that were followed in finding the direct Fourier transform through use of the two-sided Laplace transform. If the Fourier transform exists, the two-sided Laplace transform exists and has a strip of convergence that includes the $j\omega$-axis. The s in $F_2(s)$ is replaced with $j\omega$ to find $F_2(j\omega)$.

To reverse these steps, Eq. 10-6.8 is rewritten as a function of $j\omega$, as

$$F_2(j\omega) = \frac{10}{-(j\omega)^2 + 24 + j2\omega} \tag{10-6.10}$$

The two-sided Laplace transform is found by replacing $j\omega$ with s.

$$F_2(s) = \frac{10}{-s^2 + 2s + 24} = \frac{-10}{(s-6)(s+4)} = \frac{-1}{s-6} + \frac{1}{s+4} \tag{10-6.11}$$

The pole to the right of the $j\omega$-axis goes into $F_a(s)$, and the one to the left into $F_b(s)$. The inverse transform is

$$\begin{aligned} F_a(s) &= \frac{-1}{s-6} \qquad & F_b(s) &= \frac{1}{s+4} \\ f_a(t) &= \epsilon^{+6t} \qquad & f_b(t) &= \epsilon^{-4t} \end{aligned} \tag{10-6.12}$$

The resulting time function is

$$f_2(t) = \begin{cases} \epsilon^{+6t} & t < 0 \\ \epsilon^{-4t} & t > 0 \end{cases} \tag{10-6.13}$$

Example 10-6.3. In certain situations it is convenient to use the inverse transform integral. The following is an example:

$$F_2(j\omega) = \begin{cases} 1 & -\Omega/2 < \omega < +\Omega/2 \\ 0 & \text{elsewhere} \end{cases} \tag{10-6.14}$$

$$f_2(t) = \frac{1}{2\pi}\int_{-\infty}^{+\infty} F_2(j\omega)\,\epsilon^{+j\omega t}\,d\omega = \frac{1}{2\pi}\int_{-\Omega/2}^{+\Omega/2} 1\epsilon^{+j\omega t}\,d\omega = \frac{\Omega}{2\pi}\left[\frac{\sin(\Omega/2)\,t}{(\Omega/2)\,t}\right] \tag{10-6.15}$$

When this example is compared with Example 10-4.5, you can observe how remarkably similar the two examples are. This subject is discussed in more detail later.

Example 10-6.4. The inverse transform integral is used again in following steps.

$$F_2(j\omega) = \begin{cases} +j & 0<\omega<\Omega \\ -j & -\Omega<\omega<0 \end{cases} \tag{10-6.16}$$

The inverse integral is set up as

$$f_2(t) = \frac{1}{2\pi}\left[\int_{-\Omega}^{0} (-j)\,\epsilon^{+j\omega t}\,d\omega + \int_{0}^{\Omega} j\,\epsilon^{+j\omega t}\,d\omega\right] \tag{10-6.17}$$

and the result can be shown to be:

$$f_2(t) = \frac{\Omega}{\pi}\left[\frac{\sin(\Omega/2)\,t}{(\Omega/2)\,t}\right]\sin\Omega t \tag{10-6.18}$$

Example 10-6.5. As an example of a different type, suppose we are given a Fourier transform such as

$$F_2(j\omega) = \frac{2}{j\omega} \tag{10-6.19}$$

and are asked to find the corresponding time function. If we consult Table 10-5.1, we find there is only one such entry, and the corresponding time function is

$$f_2(t) = Sgn(t) \tag{10-6.20}$$

This is certainly true, but we wish to use a reasoning process that will help handle other similar type functions and not just this specific problem.

As we look at the $F_2(j\omega)$ of Eq. 10-6.19, we note there is a pole on the $j\omega$-axis in the s-plane. We also know the only way an $F_2(j\omega)$ can have a pole on the $j\omega$-axis is the result from some limiting operation. If we assume that this pole moves to the $j\omega$-axis from a position in the left-half plane, the inverse transform would exist for only positive time. However, the Fourier transform for a unit-step function is

$$u_0(j\omega) + \frac{1}{j\omega} \tag{10-6.21}$$

Similarly, we reason that the pole cannot move from a position in the right-half plane. The only other possibility is to divide the $F_2(j\omega)$, as

$$F_2(j\omega) = \frac{1}{j\omega} + \frac{1}{j\omega} \tag{10-6.22}$$

and to think of one term as representing a pole moving in from the left-half plane, and the other term as representing a pole moving in from the right-half plane. We now can switch to the two-sided Laplace transform, and write

$$F_2(s) = \frac{1}{s} + \frac{1}{s} \tag{10-6.23}$$

One of the $1/s$ terms is assigned to $F_a(s)$ and the other to $F_b(s)$. The inverse transform follows immediately, as

$$f_2(t) = Sgn(t) \tag{10-6.24}$$

COMMENTARY

One of the advantages usually stressed for the inverse Fourier transform is that the integral is real. The fact that the integral is real doesn't necessarily mean that the integration is easy to perform. In a course in complex variable theory, one of the points made is that integration in the complex plane is often used as a method of solving real integrals.

Quite often if a real integral is difficult to solve in closed form, numerical methods are used. The use of numerical methods and the digital computer is one of the reasons the inverse Fourier transform integral is so attractive.

10-7. INVERSE FOURIER TRANSFORMS THAT EXIST IN THE LIMIT

Before the direct Fourier transform can exist, the following condition must be satisfied:

$$\int_{-\infty}^{+\infty} |f_2(t)|\, dt < \infty \tag{10-7.1}$$

A similar condition must be satisfied for the inverse Fourier transform to exist:

$$\int_{-\infty}^{+\infty} |F_2(j\omega)|\, d\omega < \infty \tag{10-7.2}$$

An $F_2(j\omega)$ may violate this condition but still have an inverse transform in the limit.

Example 10-7.1. The $F_2(j\omega)$ to be examined is

$$F_2(j\omega) = 1 \qquad \text{for all } \omega \tag{10-7.3}$$

Table 10-7.1.

$F_2(j\omega)$ That Are Even and Real	
$F_2(j\omega)$	$f_2(t)$
1 for all ω	$u_0(t)$
ω^2 for all ω	$-u_{+2}(t)$
ω^4 for all ω	$u_{+4}(t)$
$\omega\, Sgn(\omega)$	$-1/\pi t^2$
$\omega^3\, Sgn(\omega)$	$3!/\pi t^4$
$F_2(j\omega)$ That Are Odd and Imaginary	
$j[F_2(j\omega)]$	$f_2(t)$
$+\omega$ for all ω	$-u_{+1}(t)$
$+\omega^3$ for all ω	$u_{+3}(t)$
$Sgn(\omega)$	$+1/\pi t$
$\omega^2\, Sgn(\omega)$	$-2/\pi t^3$

In a manner exactly parallel to that of Example 10-5.1, we start with the pulse of finite width of Example 10-6.3, and let $\Omega \to \infty$. The result is

$$F_2(t) = \frac{2\pi}{2\pi} u_0(t) = u_0(t) \tag{10-7.4}$$

This same transform pair was determined in Example 10-5.2, except the direct transform was used.

Table 10-7.1 lists other inverse Fourier transforms and the corresponding time functions that exist in the limit.

10-8. APPLYING SIGNALS TO LINEAR SYSTEMS USING FOURIER TRANSFORMS

The block diagram for applying a signal to a linear system is shown in Fig. 10-8.1. The mathematics of using Fourier transforms for this situation can be presented in steps that are almost identical to those used for the two-sided Laplace, particularly if we depend upon the Laplace transform table for finding the inverse transform. However, the physical interpretation of the situation is quite different.

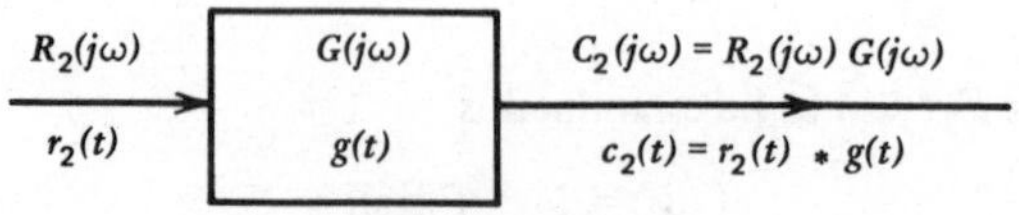

Fig. 10-8.1

The time signal $r_2(t)$ is applied to the system in Fig. 10-8.1. The $R_2(j\omega)$ is the Fourier transform for the input and displays the spectrum of the frequencies that go to make up $r_2(t)$. The $g(t)$ is the impulse response of the system, and $G(j\omega)$ displays the spectrum of the frequencies that go to make up this time signal. The $c_2(t)$ can be found as the convolution of $r_2(t)$ and $g(t)$, which is identical to what we have done before, or it can be found as the inverse Fourier transform of the product of $R_2(j\omega)\,G(j\omega)$. The physical interpretation to be stressed here is that the spectrum $R_2(j\omega)$ of the input signal is applied to the system, and as the signal passes through the system, the spectrum is modified due to the action of the system into a new spectrum $C_2(j\omega) = R_2(j\omega)\,G(j\omega)$. In other words, the system acts essentially as a filter, and this filtering action changes the spectrum of the input signal into that of the output. The inverse Fourier transform of $C_2(j\omega)$ yields an output that is the result of this filtering action. This concept should be kept in mind as one goes through the following examples.

Example 10-8.1. In the first example, the direct Fourier transform is used to find $C_2(j\omega)$, and the inverse Fourier transform integral is used to find the $c_2(t)$. This is somewhat analogous to the first example in Section 10-6. The circuit is shown in Fig. 10-8.2, and the current i is the desired solution.

By using methods developed in Chapter 2, the desired solution is

$$i = 0.2\,(1 - \epsilon^{-10t})\,u_{-1}(t) \tag{10-8.1}$$

Next, Fourier transforms are used. The Fourier transform of the input is

$$R_2(j\omega) = \mathcal{F}[10u_{-1}(t)] = 10\pi\,u_0(j\omega) + \frac{10}{j\omega} \tag{10-8.2}$$

The transfer function of the system is

$$G(s) = \frac{0.2}{s+10} \tag{10-8.3}$$

Since the $j\omega$-axis is in the region of convergence, $s = j\omega$ is a legitimate substitution, and $G(j\omega)$ is

$$G(j\omega) = \frac{0.2}{j\omega + 10} \tag{10-8.4}$$

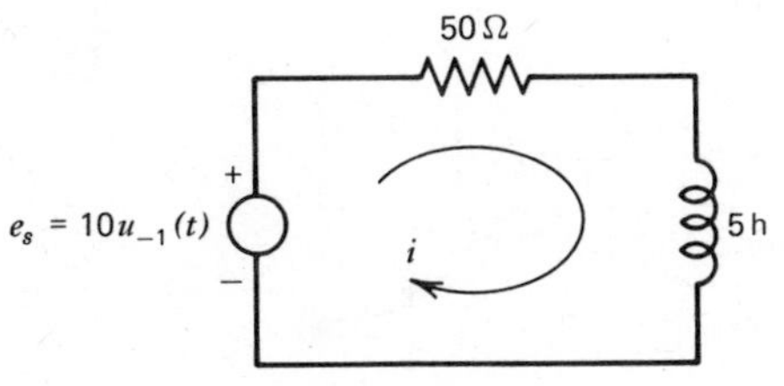

Fig. 10-8.2

The transformed output is

$$C_2(j\omega) = R_2(j\omega)\, G(j\omega) = \frac{2\pi\, u_0(j\omega)}{j\omega + 10} + \frac{2}{j\omega\,(j\omega + 10)} \tag{10-8.5}$$

The two terms of Eq. 10-8.5 are solved separately. The inverse transform of the first term is labeled $c_2'(t)$, and is found from

$$c_2'(t) = \frac{1}{2\pi} \int_{-\infty}^{+\infty} \frac{2\pi\, u_0(j\omega)}{j\omega + 10}\, \epsilon^{+j\omega t}\, d\omega \tag{10-8.6}$$

The impulse term is zero everywhere except at $\omega = 0$; hence

$$c_2'(t) = 0.1 \int_{-\infty}^{+\infty} u_0(j\omega)\, d\omega = 0.1 \tag{10-8.7}$$

The inverse transform of the second term of Eq. 10-8.5 is labeled $c_2''(t)$, and is found from

$$c_2''(t) = \frac{1}{2\pi} \int_{-\infty}^{+\infty} \frac{2}{j\omega\,(j\omega + 10)}\, \epsilon^{+j\omega t}\, d\omega$$

$$c_2''(t) = \frac{1}{\pi} \int_{-\infty}^{+\infty} \frac{1}{j\omega}\left[\frac{10}{\omega^2 + 10^2} - \frac{j\omega}{\omega^2 + 10^2}\right] \epsilon^{+j\omega t}\, d\omega$$

$$c_2''(t) = \frac{1}{\pi} \int_{-\infty}^{+\infty} \left[\frac{(10 \cos \omega t + \omega \sin \omega t) + j(10 \sin \omega t - \omega \cos \omega t)}{j\omega\,(\omega^2 + 10^2)}\right] d\omega \tag{10-8.8}$$

The first portion of the integrand is an odd function[2] of ω and integrates to zero, whereas the second portion is an even function[2] and can be written as

$$c_2''(t) = \frac{2}{\pi} \int_0^{\infty} \left[\frac{10 \sin \omega t}{\omega\,(\omega^2 + 10^2)} - \frac{\cos \omega t}{(\omega^2 + 10^2)}\right] d\omega \tag{10-8.9}$$

The two improper integrals that are useful in evaluating this equation are

$$\int_0^{\infty} \frac{\sin a u}{u\,(u^2 + b^2)}\, du = \begin{cases} (+\pi/2b^2)(1 - \epsilon^{-ab}) & a > 0 \\ (-\pi/2b^2)(1 - \epsilon^{+ab}) & a < 0 \end{cases}$$

$$\int_0^{\infty} \frac{\cos a u}{u^2 + b^2}\, du = \begin{cases} (\pi/2b)\, \epsilon^{-ab} & a > 0 \\ (-\pi/2b)\, \epsilon^{+ab} & a < 0 \end{cases} \tag{10-8.10}$$

These integrals are applied to Eq. 10-8.9 for the condition $t > 0$.

[2] The subject of even and odd functions is discussed in more detail in Section 11-7.

$$c_2''(t) = \frac{2}{\pi}\left[\frac{10\pi}{2(10)^2}(1 - \epsilon^{-10t}) - \frac{\pi}{2(10)}\epsilon^{-10t}\right] = 0.1 - 0.2\epsilon^{-10t} \qquad (10\text{-}8.11)$$

Next these integrals are applied for the condition $t < 0$.

$$c_2''(t) = \frac{2}{\pi}\left[\frac{-10\pi}{2(10)^2}(1 - \epsilon^{+10t}) - \frac{\pi}{2(10)}\epsilon^{+10t}\right] = -0.1 \qquad (10\text{-}8.12)$$

Finally the two parts are added to obtain $c_2(t)$.

$$c_2(t) = c_2'(t) + c_2''(t) = \begin{cases} 0 & t<0 \\ 0.2\,(1 - \epsilon^{-10t}) & t>0 \end{cases} \qquad (10\text{-}8.13)$$

This example is a simple problem, and yet you can see the amount of work necessary to use the inverse Fourier transform integral. For these reasons, the remaining examples are worked using the Laplace transform table in the opposite direction.

Example 10-8.2. The input voltage

$$r_2(t) = e_s(t) = \begin{cases} \epsilon^{+2t} & t<0 \\ \epsilon^{-2t} & t>0 \end{cases} \qquad (10\text{-}8.14)$$

is applied to the circuit of Fig. 10-8.3 and the desired output $c_2(t)$ is the voltage across the capacitor, $c_2(t) = v_c(t)$. The Fourier transform of the input signal is

$$R_2(j\omega) = \frac{-1}{j\omega - 2} + \frac{1}{j\omega + 2} = \frac{4}{\omega^2 + 4} \qquad (10\text{-}8.15)$$

The impulse response of the system and the Fourier transform are

$$g(t) = 1\epsilon^{-1t} \qquad (10\text{-}8.16)$$

$$G(j\omega) = \frac{1}{j\omega + 1} \qquad (10\text{-}8.17)$$

The Fourier transform of the output is

$$V_c(j\omega) = E_s(j\omega)\,G(j\omega) = \frac{4}{(\omega^2 + 4)(j\omega + 1)} \qquad (10\text{-}8.18)$$

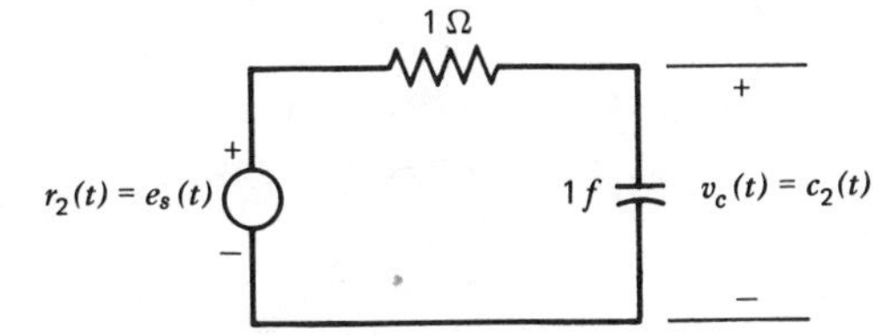

Fig. 10-8.3

Again, since we have not developed an extensive table of Fourier transforms, we return to the two-sided Laplace methods to determine the corresponding time function.

$$V_c(s) = \frac{-4}{(s^2 - 4)(s + 1)} = \frac{-4}{(s + 2)(s - 2)(s + 1)} = \frac{-1}{s + 2} + \frac{-\frac{1}{3}}{s - 2} + \frac{\frac{4}{3}}{s + 1} \quad (10\text{-}8.19)$$

The $j\omega$-axis divides this into the two functions

$$[V_c(s)]_a = \frac{-\frac{1}{3}}{s - 2}$$
$$[V_c(s)]_b = \frac{-1}{s + 2} + \frac{\frac{4}{3}}{s + 1} \quad (10\text{-}8.20)$$

and the corresponding time function is

$$v_c(t) = \begin{cases} \frac{1}{3}\epsilon^{+2t} & t < 0 \\ -1\epsilon^{-2t} + \frac{4}{3}\epsilon^{-1t} & t > 0 \end{cases} \quad (10\text{-}8.21)$$

Example 10-8.3. The following input

$$r_2(t) = \begin{cases} \epsilon^{-2t} & t < 0 \\ \epsilon^{-4t} & t > 0 \end{cases} \quad (10\text{-}8.22)$$

is applied to a system whose impulse response is

$$g(t) = \epsilon^{-3t} \qquad t > 0 \quad (10\text{-}8.23)$$

The $r_2(t)$ does not satisfy the Dirichlet conditions, and $R_2(j\omega)$ does not exist. Therefore, this example can not be worked using Fourier transforms.

Example 10-8.4. The following input

$$r_2(t) = \begin{cases} \epsilon^{+3t} & t < 0 \\ \epsilon^{-2t} & t > 0 \end{cases} \quad (10\text{-}8.24)$$

is applied to a system whose impulse response is

$$g(t) = \epsilon^{+1t} \quad (10\text{-}8.25)$$

The $R_2(j\omega)$ exists, but the region of convergence for $G(s)$ is to the right of $\sigma = +1$; and since the $j\omega$-axis is not in the region of convergence, the $G(j\omega)$ does not exist. Therefore, this example cannot be worked using Fourier transforms.

Example 10-8.5. The impulse response of a system is

$$g(t) = \epsilon^{-1t} \qquad t > 0 \quad (10\text{-}8.26)$$

and the following input is applied to the system

$$r_2(t) = Sgn(t) \quad (10\text{-}8.27)$$

The $G(j\omega)$ exists, and is

$$G(j\omega) = \frac{1}{j\omega + 1} \tag{10-8.28}$$

The $R_2(j\omega)$ exists in the limit, and is

$$R_2(j\omega) = \frac{2}{j\omega} \tag{10-8.29}$$

The $C_2(j\omega)$ also exists in the limit.

$$C_2(j\omega) = \frac{2}{j\omega\,(j\omega + 1)} \tag{10-8.30}$$

The argument used in Example 10-6.5 can be applied to finish this example.

$$C_2(s) = \frac{2}{s\,(s + 1)} = \frac{1}{s} + \frac{1}{s} - \frac{2}{s + 1} \tag{10-8.31}$$

The $C_2(s)$ is divided as

$$C_a(s) = \frac{1}{s} \qquad C_b(s) = \frac{1}{s} - \frac{2}{s + 1} \tag{10-8.32}$$

and the solution is

$$c_2(t) = \begin{cases} -1 & t < 0 \\ 1 - 2\epsilon^{-1t} & t > 0 \end{cases} \tag{10-8.33}$$

10-9. FREQUENCY RESPONSE (BODE) PLOTS

As indicated in the previous section, the Fourier transform of the output of a system can be found from the product of the transform of the input signal and the transfer function of the system. The transfer function, $G(j\omega)$, provides the key to the effect the system has on various frequency components in any input signal. In order to quickly observe the effect the system has on the total spectrum of frequencies, it is convenient to use some form of a plot of $G(j\omega)$ vs. frequency ω. Such a plot is the frequency response of the system. It should be noted that $G(j\omega)$ is a complex quantity with real and imaginary parts or magnitude and phase parts.

One of the more common methods of presenting frequency response information is the so-called *Bode plot*, which shows the logarithm of the gain or magnitude of $G(j\omega)$ vs. the logarithm of frequency. In fact, it is common to introduce an extra scaling factor of 20 and use the gain in terms of a rather loose interpretation of decibels (db) defined as

$$(\text{Number of db}) = 20 \log_{10} |G(j\omega)| \tag{10-9.1}$$

An advantage of using logarithmic scales is the ability to present much larger ranges of frequency and gain on a given plot size than would be possible with simple linear scales. A second and most important advantage is the simplification afforded in combining gain terms and in graphically approximating terms.

In addition to the gain or magnitude part of $G(j\omega)$, there is also an angle or phase part. The display of this information is generally done by plotting the angle in degrees vs. the logarithm of frequency.

It is possible to use log–log coordinate graph paper in plotting the frequency response. However, it has become somewhat more common practice to use semi-log coordinate paper in order to allow more versatility in selecting gain scales and in plotting the phase on a linear scale on the same plot as the gain curve. More information on the details of Bode plots and the practical considerations involved is presented in the following sections.

10–10. SOME LOGARITHMIC SCALE CONSIDERATIONS

While knowledge of various coordinate systems is presumed in this text, it is probably desirable to present some significant points that are frequently new to many students. Let us consider the relationship

$$(\text{db gain}) = 20 \log_{10} \omega \tag{10–10.1}$$

$$y = 20x \tag{10–10.2}$$

and the alternate form of it using the variables $x = \log_{10} \omega$, and $y =$ db gain. On a plot using linear x and y scales, Eq. 10–10.2 is obviously a straight line with a slope of 20 units. If it is desired to omit the actual calculation of $\log_{10} \omega$, one can use semi-log graph paper with the ω-axis along the log scale. Figure 10–10.1 illustrates the plot of Eq. 10–10.1 and 2, together with various coordinate systems. Two additional gain axes have been included to show some possible simplifications in the numbers involved.

A second relationship, also shown in Fig. 10–10.1, is given by

$$(\text{db gain}) = 20 \log_{10} (1/\omega) = -20 \log_{10} (\omega) \tag{10–10.3}$$

$$y = -20x \tag{10–10.4}$$

It is of some interest to consider the relationship between increments in the variable x and corresponding ω values, as in

$$x_2 - x_1 = \log_{10} (\omega_2) - \log_{10} (\omega_1) = \log_{10} (\omega_2/\omega_1) \tag{10–10.5}$$

It is apparent that an increment in x corresponds to the ratio of the corresponding frequencies. It is common practice to speak of an *octave* up or down in frequency which corresponds to a ratio of 2 or $\frac{1}{2}$, respectively, for the end frequencies which are an octave apart. Thus, $\omega = 3$ is an octave below $\omega = 6$, while $\omega = 12$ is an octave above $\omega = 6$. Similarily, a *decade* of frequency is a ratio of either 10 or $\frac{1}{10}$ between frequencies.

The slopes of the lines in Fig. 10–10.1 can now be expressed in a number of different

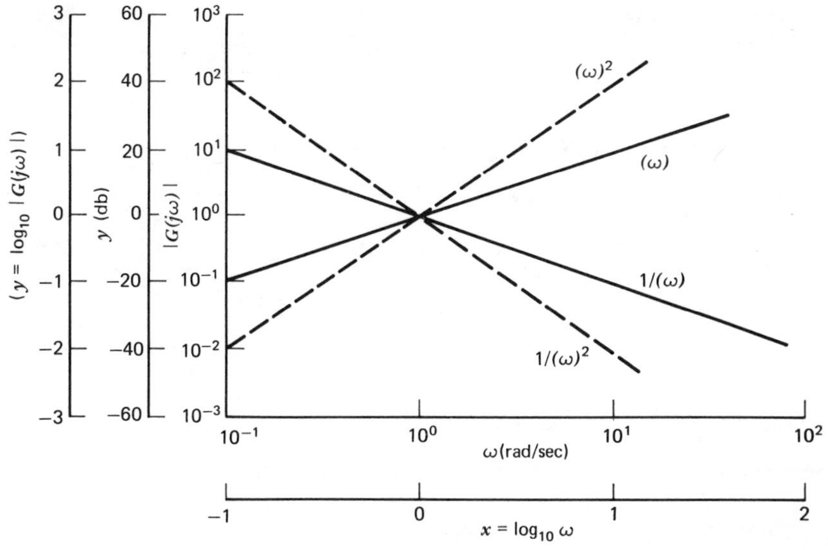

Fig. 10-10.1

ways. Perhaps the forms most commonly encountered are in db/octave and db/decade. To determine the numerical values for these forms, we simply use two end frequencies that are separated by an octave and a decade and substitute into Eq. 10-10.5. The result for an octave is given in

$$y_2 - y_1 = 20 \log_{10} (2\omega_1) - \log_{10} (\omega_1) = 20 \log_{10} (2) \approx 6 \tag{10-10.6}$$

and that for a decade is given in

$$y_2 - y_1 = 20 \log_{10} (10\omega_1) - \log_{10} (\omega_1) = 20 \log_{10} (10) = 20 \tag{10-10.7}$$

Thus, common slopes for simple ω terms, when plotted on Bode gain plots, are 6 db/octave or 20 db/decade. For $(1/\omega)$ terms, the slopes are simply −6 db/octave and −20 db/decade.

It is also possible to express the slopes of the lines of Fig. 10-10.1 in terms of the coordinates x and v, where $v = \log_{10} |G(j\omega)|$. For these variables, the slopes are simply +1 for $G(j\omega) = j\omega$, and −1 for $G(j\omega) = (1/j\omega)$. These are particularly simple slopes and would indicate that it might be desirable to use such slopes in lieu of the db/octave and db/decade terms. However, the latter expressions have been well established in the literature.

As a further extension of the above discussion, let us consider the transfer function given in Eq. 10-10.8.

$$G(j\omega) = (j\omega)^n \tag{10-10.8}$$

Using the x and v coordinates of Fig. 10-10.1, it can be seen that we again have a straight line, but this time with slope of n. It is also of some interest to consider the relationships between magnitudes of $G(j\omega)$ and corresponding frequencies for the transfer function of Eq. 10-10.6. In this case the relationships required at two frequencies, ω_1 and ω_2, are

$$G(j\omega_2) = (j\omega_2)^n \qquad G(j\omega_1) = (j\omega_1)^n \tag{10-10.9}$$

$$v_2 - v_1 = \log_{10} |G_2| - \log_{10} |G_1| = \log_{10} (\omega_2)^n - \log_{10} (\omega_1)^n \tag{10-10.10}$$

$$|G_2/G_1| = (\omega_2/\omega_1)^n \tag{10-10.11}$$

It is apparent that the ratio of gain magnitudes depends upon the nth power of the ratio of frequencies. Since there has been no restriction on the n's being positive or negative, the relationship of Eq. 10-10.11 is applicable for both signs of n.

10-11. BODE MAGNITUDE PLOTS

Let us now consider some simple examples where $G(j\omega)$ involves several terms.

Example 10-11.1. As a first example, consider

$$G(j\omega) = K/[(j\omega)(1 + j\omega T)] \tag{10-11.1}$$

The use of the decibel relationship of Eq. 10-1.1 converts Eq. 10-11.1 into

$$(\text{db}) = 20 \log_{10} |G(j\omega)| = 20 \log_{10} K/|[(j\omega)(1 + j\omega T)]| \tag{10-11.2}$$

with further reductions leading to

$$(\text{db}) = 20 \log_{10} K - 20 \log_{10} (\omega) - 20 \log_{10} |(1 + j\omega T)| \tag{10-11.3}$$

It can be seen that the process of taking logarithms has converted the products and ratios of $G(j\omega)$ into sums and differences. This allows separate construction of the gain curves for each term, such as the fixed gain K, the term $(1/j\omega)$, and also $1/(1 + j\omega T)$. This sum and difference feature is a key to the simplicity of the Bode plots.

For the fixed gain term K in Eq. 10-11.3 the plot of Fig. 10-11.1 simply reflects a constant amplitude independent of frequency. The $(1/j\omega)$ term is discussed in the previous section, where it was shown that the resulting plot was a straight line with a slope of -6 db/octave, -20 db/decade or simply (-1) in x, v coordinate terms.

The remaining term, $1/(1 + j\omega T)$, in Eq. 10-11.3 leads us to another key concept in making Bode plots. When the magnitude of this term is taken, the result is

$$20 \log_{10} |(1 + j\omega T)| = 20 \log_{10} [1 + (\omega T)^2]^{1/2} \tag{10-11.4}$$

$$= 10 \log_{10} [1 + (\omega T)^2] \tag{10-11.5}$$

Attention is now devoted to a key approximation. For frequencies that cause the term $(\omega T)^2$ to be much smaller than unity, it is permissible to approximate Eq. 10-11.5 as

$$20 \log_{10} |(1 + j\omega T)| \approx 10 \log_{10} (1) = 0 \qquad \omega T \ll 1 \tag{10-11.6}$$

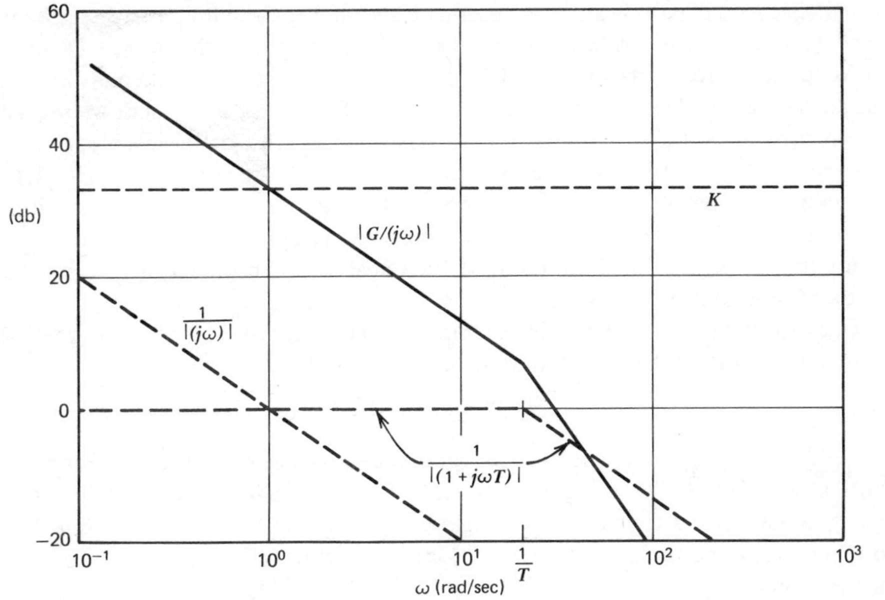

Fig. 10-11.1

This results in zero db gain for all such frequencies. For frequencies that cause $(\omega T)^2$ to be much larger than unity, it is possible to approximate Eq. 10-11.5 as

$$20 \log_{10} |(1 + j\omega T)| \approx 20 \log_{10} (\omega T) \qquad \omega T \gg 1 \qquad (10\text{-}11.7)$$

The result of this approximation is a term identical in form to the $\log_{10} (\omega)$ term considered in Section 10-10, except that (ωT) appears in place of simply (ω). The curve that results for this term is a straight line that goes through zero decibels when $\omega = (1/T)$. This particular frequency is termed a break frequency for the original transfer function, since the two straight-line approximations developed above intersect at this frequency and the total approximating curve breaks at $\omega = (1/T)$. Figure 10-11.1 shows the various straight lines developed so far for the Bode plot of Eq. 10-11.1.

In order to improve the accuracy of the approximation to the term $-20 \log_{10} |(1 + j\omega T)|$ for the frequency range near $(\omega T) = 1$, a simple computation at the break frequency indicates an actual value of -3 db. Similarily, computations at frequencies an octave below the break frequency, or $\omega = (1/2T)$, lead to approximately -1 db, while at an octave above the break the result is approximately -7 db. These three correction points allow a smooth transition curve to be drawn between the straight-line approximations where they intersect.

Terms of the type of $(1 + j\omega T)$ and $1/(1 + j\omega T)$ are commonly encountered in transfer functions. When the logarithm of $|G(j\omega)|$ is taken, these terms lead respectively to positive and negative terms that can be approximated by two straight-line segments. In more general transfer functions, there may be several terms, such as those being considered. Each such term is approximated as above and the resulting curves algebraically added to give the composite curve. Figure 10–11.1 also shows the result of adding the curves for the three parts of Eq. 10–11.1 to give the composite Bode plot. Since only straight lines are involved, it is relatively simple to obtain a few key points at frequencies such as $\omega = 1/T$ and $\omega = 1$, and then draw straight lines with the proper slopes through these points. This is the procedure used to obtain the $|G(j\omega)|$ curve in Fig. 10–11.1. For the frequency range above $\omega = (1/T)$ there are two straight-line segments with slopes of −6 db/octave and the resulting curve has a slope of −12 db/octave. The net effect of the fixed gain term K is to shift the result of the remaining parts up or down, depending on whether K is greater or less than unity.

One last point is made with this example. When the transfer function of Eq. 10–11.1 is converted to the Laplace Transform domain, the result is

$$G(S) = K/s\,(1 + Ts) \tag{10–11.8}$$

It is evident from this equation that a system pole exists at $s = 0$, and another real pole exists at $s = -(1/T)$. When these poles are related to their counterparts in the Bode plot, it is seen that the pole at the origin causes the simple straight line with slope of −6 db/octave, while the pole at $s = -(1/T)$ leads to the break frequency and the two-line-segment approximate curve.

Example 10–11.2. As a second example, we consider the transfer function

$$G(j\omega) = 1/[1 + j\omega/10^5 + (j\omega/10^5)^2] \tag{10–11.9}$$

This transfer function introduces a new type of term that corresponds to a complex pole pair in the s-plane. Conversion to decibels leads to

$$(\text{db}) = -20 \log_{10} |[1 + j\omega/10^5 + (j\omega/10^5)^2]| \tag{10–11.10}$$

$$= -10 \log_{10} \{[1 - (\omega/10^5)^2]^2 + (\omega/10^5)^2\} \tag{10–11.11}$$

An approximation technique is used again by considering ranges of frequency for which $(\omega/10^5)$ is alternately less than, and then greater than, unity. When $(\omega/10^5)$ is much less than unity, the approximation used neglects all terms in the argument of the logarithm except the unity term. This leads to simply zero decibels gain for this interval. For

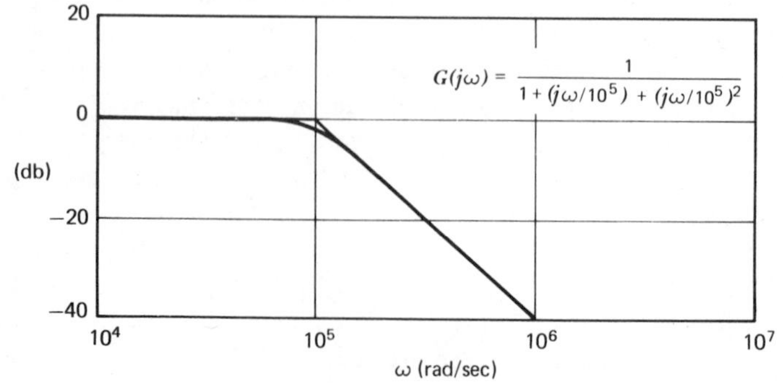

Fig. 10–11.2

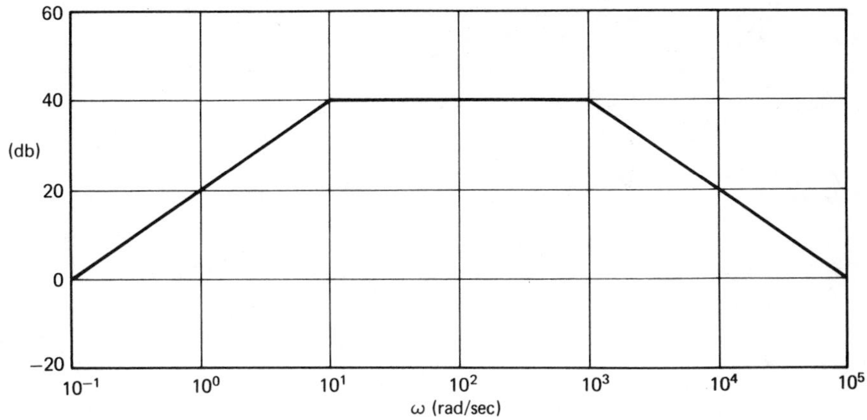

Fig. 10-11.3

$(\omega/10^5)$ much greater than unity, all terms except $(\omega/10^5)^4$ are neglected. The result is again a straight line, this time with slope of -40 *db*/decade or -12 *db*/octave. Figure 10-11.2 illustrates the resulting Bode plot. We again have a break frequency; this time, however, a so-called *double break* is involved with the lines changing from zero slope to -40 *db*/decade, which is twice what the slope was in the previous example. Additionally, there is a region around the break frequency where it is necessary to incorporate more corrections than was done before. Normalized correction curves for such terms are available, but the ready availability of digital computers makes the use of such correction curves of questionable value.

Example 10-11.3. A final example of Bode magnitude curves is presented in

$$G(j\omega) = 10^2(j\omega/10) \,/\, [(1 + j\omega/10)(1 + j\omega/10^3)] \qquad (10\text{-}11.12)$$

There are four resulting decibel terms in this case, but all are of a type previously considered. The resulting Bode plot is shown in Fig. 10-11.3, where the various factors are also shown.

10-12. BODE PHASE PLOTS

As mentioned in Section 10-1, a frequency response transfer function is complex and possesses a phase part as well as a magnitude or gain part. The previous section has considered the magnitude part, and it is now time to consider some aspects of the phase-vs.-frequency plots.

For such simple transfer functions as $(j\omega)^n$ and $(1/j\omega)^n$, the phase is independent of frequency and is simply $90n^\circ$ and $-90n^\circ$, respectively. However, for terms of the type of $(1 + j\omega T)$ and $1/(1 + j\omega T)$ there is an actual variation of phase with frequency given by $\tan^{-1}(\omega T)$ and $-\tan^{-1}(\omega T)$. As frequency varies from very small values, much less than the break frequency, to much larger than this, the phase varies from approximately

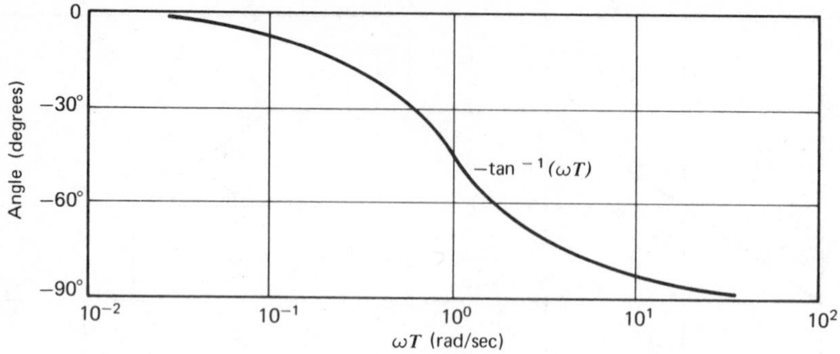

Fig. 10-12.1

zero to ±90°. At the break frequency, the phase is ±45°. There is a smooth transition between the limiting values. Figure 10-12.1 illustrates a typical phase curve.

While it is possible to use straight-line approximations for phase curves, at least three line segments are required, and it is questionable whether there are real benefits to be derived from such approximations. This is especially true in view of the availability of digital computer programs for making such plots.

Example 10-12.1. The transfer function of Ex. 10-11.1 is to be reconsidered, and the phase curve plotted. For this case, it should be pointed out that the fixed gain K does not contribute to the phase curve, whereas the $1/j\omega$ term contributes −90° of phase independent of frequency. The $1/(1+j\omega T)$ term contributes a phase that varies from approximately zero for $\omega \ll 1/T$, to −90° for frequencies much greater than $1/T$. The phase terms add algebraically to the give the total phase curve of Fig. 10-12.2.

Example 10-12.2. The transfer function of Ex. 10-11.3 is to be considered. The phase curve for this case is made up of three terms. The $(j\omega/10)$ term in the numerator contri-

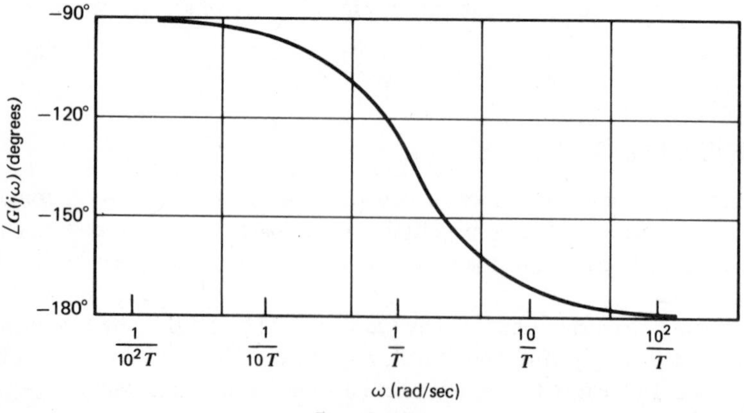

Fig. 10-12.2

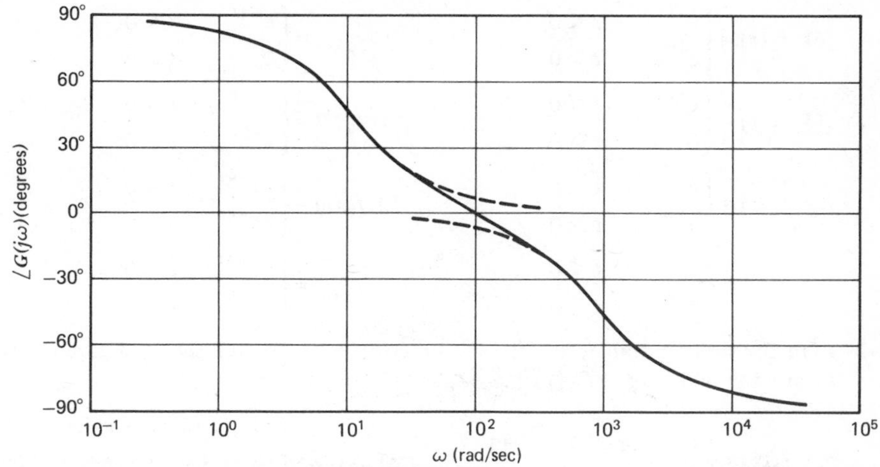

Fig. 10-12.3

butes $+90°$ independent of frequency, whereas each denominator term contributes a phase part that varies from 0 to $-90°$. The resultant of these three parts is shown in Fig. 10-12.3.

CONCLUSION

Addition of the Fourier transform to our collection of tools for system analysis makes it easily possible to consider the frequency-spectrum reshaping accomplished by a given system. It provides a key link between the complex frequency domain of the Laplace s-plane, and commonly encountered gain and phase characteristics of systems. The use of so-called *Bode gain and phase plots* further enhance our ability to understand the frequency domain nature of our systems by use of simple plotting methods.

PROBLEMS

10-1 Find the Fourier transform for each of the following functions. If a transform does not exist, explain why it does not in terms of the Dirichlet conditions, and in terms of the two-sided Laplace transforms. First, sketch the magnitude and the angle of $F_2(j\omega)$ vs. ω. Second, sketch the even part and the j times the odd part of $F_2(j\omega)$.

(a) $f_2(t) = \begin{cases} \epsilon^{+2t} & t<0 \\ \epsilon^{-2t} & t>0 \end{cases}$

(b) $f_2(t) = \begin{cases} \epsilon^{+2t} & t<0 \\ -\epsilon^{-2t} & t>0 \end{cases}$

(c) $f_2(t) = \begin{cases} \epsilon^{-2t} & t<0 \\ \epsilon^{+2t} & t>0 \end{cases}$

(d) $f_2(t) = \begin{cases} -\epsilon^{+2t} & t<0 \\ \epsilon^{-2t} & t>0 \end{cases}$

(e) $f_2(t) = \begin{cases} \epsilon^{+4t} & t<0 \\ \epsilon^{+2t} & t>0 \end{cases}$

(f) $f_2(t) = \begin{cases} \epsilon^{+4t} & t<0 \\ 0 & t>0 \end{cases}$

(g) $f_2(t) = \begin{cases} 0 & t<0 \\ \epsilon^{+4t} & t>0 \end{cases}$

(h) $f_2(t) = \begin{cases} 0 & t<0 \\ \epsilon^{-4t} & t>0 \end{cases}$

(i) $f_2(t) = \begin{cases} \epsilon^{-4t} & t<0 \\ 0 & t>0 \end{cases}$

(j) $f_2(t) = \begin{cases} \epsilon^{-4t} & t<0 \\ \epsilon^{-2t} & t>0 \end{cases}$

(k) $f_2(t) = \begin{cases} \epsilon^{+4t} & t<0 \\ -\epsilon^{-4t} & t>0 \end{cases}$

(l) $f_2(t) = \begin{cases} -\epsilon^{+4t} & t<0 \\ \epsilon^{-4t} & t>0 \end{cases}$

10-2 Find the Fourier transform for each function given. Sketch the even part and the j times the odd part of $F_2(j\omega)$ vs. ω.

(a) $f_2(t) = \begin{cases} 1 & \text{for } -1<t<+1 \\ 0 & \text{elsewhere} \end{cases}$

(b) $f_2(t) = \begin{cases} 1 & \text{for } -2<t<+2 \\ 0 & \text{elsewhere} \end{cases}$

(c) $f_2(t) = \begin{cases} 1 & \text{for } -3<t<-1 \\ 1 & \text{for } \ 1<t<3 \\ 0 & \text{elsewhere} \end{cases}$

(d) $f_2(t) = \begin{cases} -1 & \text{for } -3<t<-1 \\ +1 & \text{for } \ 1<t<3 \\ 0 & \text{elsewhere} \end{cases}$

(e) $f_2(t) = \begin{cases} -1 & \text{for } -2<t<0 \\ +1 & \text{for } \ 0<t<2 \\ 0 & \text{elsewhere} \end{cases}$

(f) $f_2(t) = \begin{cases} +1 & \text{for } -2<t<0 \\ -1 & \text{for } \ 0<t<2 \\ 0 & \text{elsewhere} \end{cases}$

10-3 The function in Problem 10-2(b) can be built up by using the function in (a) two times. The function in (a) can be shifted to the right one unit for the first part and then shifted to the left one unit for the second part. When these two parts are added, the function of (b) results. Find the $F_2(j\omega)$ for (b), by finding the transforms for the two parts and adding.

10-4 Repeat Problem 10-3, except build up the function of (c) from that of (a) to find the $F_2(j\omega)$ for (c).

10-5 Repeat Problem 10-4, except when the function of (a) is shifted to the left, the sign is changed to a negative value in order to build up the function for (d).

10-6 Repeat Problem 10-3, except when the function of (a) is shifted to the left, the sign is changed to a negative value in order to build up the function for (e).

10-7 Find $F_2(j\omega)$ for each of the functions shown.

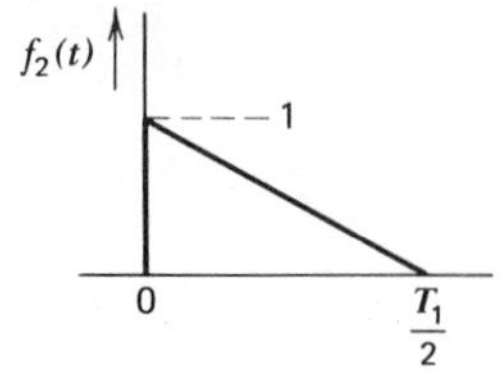

(a)

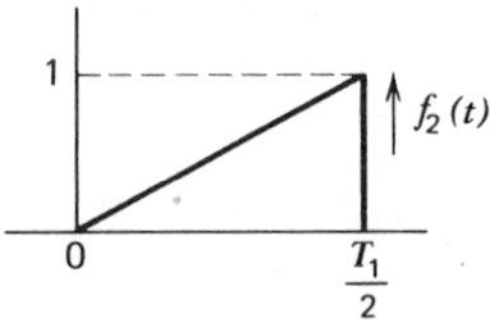

(b)

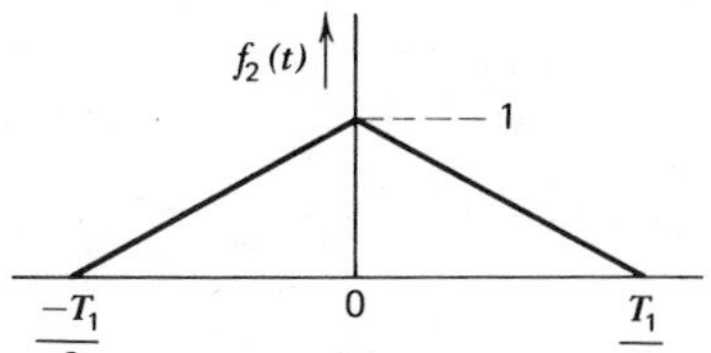

(c)

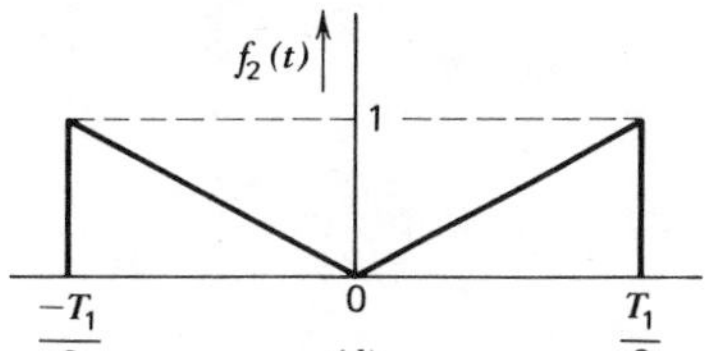

(d)

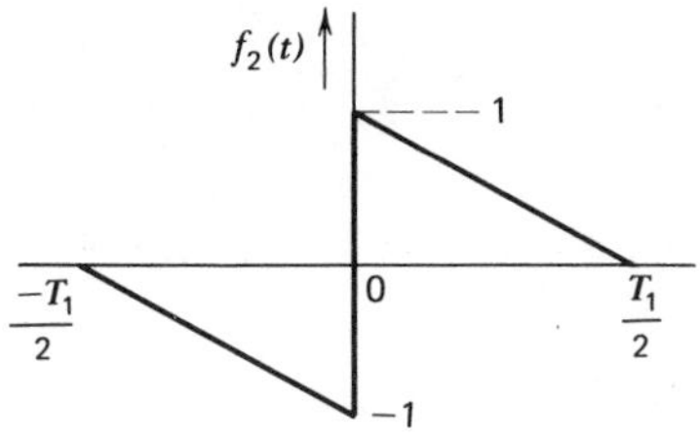

(e)

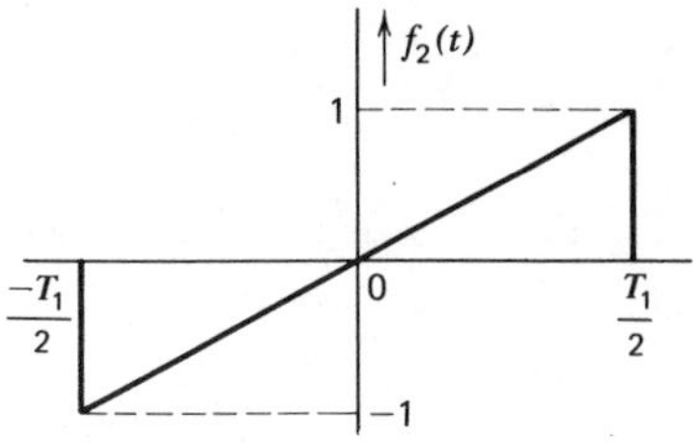

(f)

10–8 Start with the $f_2(t)$ of Problem 10–7(a), and show how this $f_2(t)$ can be used to build up the $f_2(t)$ of (c). From the $F_2(j\omega)$ for (a) find the $F_2(j\omega)$ for (c).

10–9 Repeat Problem 10–8, except build up the $f_2(t)$ of (e) by using the $f_2(t)$ of (a).

10–10 Repeat Problem 10–8, except build up the $f_2(t)$ of (d) by using the $f_2(t)$ of (b).

10–11 Repeat Problem 10–8, except build up the $f_2(t)$ of (f) by using the $f_2(t)$ of (b).

10–12 Start with the $f_2(t)$ and $F_2(j\omega)$ of Problem 10–7(c). Let T_1 approach infinity, and find $F_2(j\omega)$ in the limit. This result should be the $F_2(j\omega)$ of Eq. 10–5.7. Why?

10–13 Repeat Problem 10–12, except use the $f_2(t)$ of Problem 10–7(a), and check the result with the $F_2(j\omega)$ of Eq. 10–5.18.

10–14 Repeat Problem 10–12, except use the $f_2(t)$ of Problem 10–7(e), and check the result with the $F_2(j\omega)$ of Eq. 10–5.22.

10–15 Verify the 2nd, 3rd, and 4th entries under the first family of functions in Table 10–5.1.

10–16 Repeat Problem 10–15, except use the second family of functions.

10–17 Repeat Problem 10–15, except use the third family of functions.

10–18 Given the $f_2(t)$ of Problem 10–2(c) and the corresponding $F_2(j\omega)$: recover the $f_2(t)$ from this $F_2(j\omega)$ by using the inverse Fourier transform integral in a manner similar to that of Example 10–6.1.

10–19 Repeat Problem 10–18, using the $f_2(t)$ of Problem 10–2(d).

10–20 Repeat Problem 10–18, using the $f_2(t)$ of Problem 10–2(e).

10–21 Repeat Problem 10–18, using the $f_2(t)$ of Problem 10–2(f).

10–22 Find the inverse Fourier transforms for the following functions by any method you wish.

(a) $F_2(j\omega) = \dfrac{8}{\omega^2 + 15 - j2\omega}$ (c) $F_2(j\omega) = \dfrac{6}{\omega^2 + 9}$

(b) $F_2(j\omega) = \dfrac{-3j\omega}{\omega^2 + 8 - j2\omega}$ (d) $F_2(j\omega) = \dfrac{2j\omega}{\omega^2 + 9}$

(e) $F_2(j\omega) = \dfrac{8}{\omega^2 + 15 + j2\omega}$

(f) $F_2(j\omega) = \dfrac{3j\omega}{\omega^2 + 8 - j2\omega}$

(g) $F_2(j\omega) = \dfrac{-2j\omega}{\omega^2 + 25}$

(h) $F_2(j\omega) = \dfrac{10}{\omega^2 + 25}$

(i) $F_2(j\omega) = \dfrac{-2j\omega + 4}{\omega^2 + 12 + j4\omega}$

(j) $F_2(j\omega) = \dfrac{11}{\omega^2 + 24 + j5\omega}$

(k) $F_2(j\omega) = \dfrac{-2j\omega}{\omega^2 + 64}$

(l) $F_2(j\omega) = \dfrac{16}{\omega^2 + 64}$

10–23 For the following Fourier transforms, find the corresponding $f_2(t)$, and sketch.

(a) $F_2(j\omega) = \begin{cases} 1 & \text{for } -1 < \omega < +1 \\ 0 & \text{elsewhere} \end{cases}$

(b) $F_2(j\omega) = \begin{cases} 1 & \text{for } -2 < \omega < +2 \\ 0 & \text{elsewhere} \end{cases}$

(c) $F_2(j\omega) = \begin{cases} 1 & \text{for } -3 < \omega < -1 \\ 1 & \text{for } \; 1 < \omega < 3 \\ 0 & \text{elsewhere} \end{cases}$

(d) $jF_2(j\omega) = \begin{cases} -1 & \text{for } -3 < \omega < -1 \\ +1 & \text{for } \; 1 < \omega < 3 \\ 0 & \text{elsewhere} \end{cases}$

(e) $jF_2(j\omega) = \begin{cases} -1 & \text{for } -2 < \omega < 0 \\ +1 & \text{for } \; 0 < \omega < 2 \\ 0 & \text{elsewhere} \end{cases}$

(f) $jF_2(j\omega) = \begin{cases} 1 & \text{for } -2 < \omega < 0 \\ -1 & \text{for } \; 0 < \omega < 2 \\ 0 & \text{elsewhere} \end{cases}$

10–24 Verify the first three entries in Table 10–7.1.

10–25 Verify the 4th and 5th entries in Table 10–7.1.

10–26 Verify the 6th and 7th entries in Table 10–7.1.

10–27 Verify the last two entries in Table 10–7.1.

10–28 The voltages listed below are applied to the circuit of Fig. 10–8.2. Use the inverse Fourier transform integral to find the output $i = c_2(t)$ in a manner similar to that of Example 10–8.1.

(a) $e_s = 10\epsilon^{-6t} u_{-1}(t)$
(b) $e_s = 10\epsilon^{-8t} u_{-1}(t)$
(c) $e_s = 10\epsilon^{-12t} u_{-1}(t)$
(d) $e_s = 10\epsilon^{-4t} u_{-1}(t)$
(e) $e_s = 1$, for all t.

10–29 A system has a $g(t)$ and $r_2(t)$ as given. Solve for $c_2(t)$ using methods similar to those of Examples 10–8.2 though 10–8.5.

(a) $r_2(t) = \begin{cases} \epsilon^{+5t} & t < 0 \\ \epsilon^{-3t} & t > 0 \end{cases}$

$g(t) = \epsilon^{-2t} \qquad t > 0$

(b) $r_2(t) = \begin{cases} \epsilon^{-3t} & t < 0 \\ \epsilon^{+5t} & t > 0 \end{cases}$

$g(t) = \epsilon^{-2t} \qquad t > 0$

(c) $r_2(t) = \begin{cases} \epsilon^{+6t} & t < 0 \\ \epsilon^{-3t} & t > 0 \end{cases}$

$g(t) = \epsilon^{-3t} \qquad t > 0$

(d) $r_2(t) = \begin{cases} \epsilon^{-3t} & t < 0 \\ \epsilon^{+6t} & t > 0 \end{cases}$

$g(t) = \epsilon^{-3t} \qquad t > 0$

(e) $r_2(t) = \begin{cases} \epsilon^{+4t} & t<0 \\ \epsilon^{-6t} & t>0 \end{cases}$

$g(t) = \epsilon^{-4t} \quad t>0$

(f) $r_2(t) = \begin{cases} \epsilon^{+6t} & t<0 \\ \epsilon^{+2t} & t>0 \end{cases}$

$g(t) = \epsilon^{-4t} \quad t>0$

(g) $r_2(t) = \begin{cases} \epsilon^{+6t} & t<0 \\ \epsilon^{-6t} & t>0 \end{cases}$

$g(t) = \epsilon^{-4t} \quad t>0$

(h) $r_2(t) = \begin{cases} \epsilon^{+6t} & t<0 \\ \epsilon^{-6t} & t>0 \end{cases}$

$g(t) = \epsilon^{+1t} u_{-1}(t)$

(i) $r_2(t) = t\, Sgn(t)$

$g(t) = \epsilon^{-1t} \quad t>0$

(j) $r_2(t) = t\, Sgn(t)$

$g(t) = 4\epsilon^{-8t} \quad t>0$

(k) $r_2(t) = t^2\, Sgn(t)$

$g(t) = \epsilon^{-1t} \quad t>0$

(l) $r_2(t) = -Sgn(t)$

$g(t) = \epsilon^{-6t} \quad t>0$

10–30 Work each part of Problem 10–29 as assigned, except find $c_2(t)$ by using convolution.

10–31 In developing a Bode plot without calibrated log–log or semi-log graph paper, it is desired to determine the frequency corresponding to the mid- and $\frac{3}{4}$-distance points between ends of a decade frequency interval on a $\log_{10}$ scale. Use the ratio method of Eq. 10–10.5 to determine the frequencies required.

10–32 Use the ration relationship of Eq. 10–10.11 to determine $|G_1|$, $|G_2|$, $|G_3|$, ω_c, as shown.

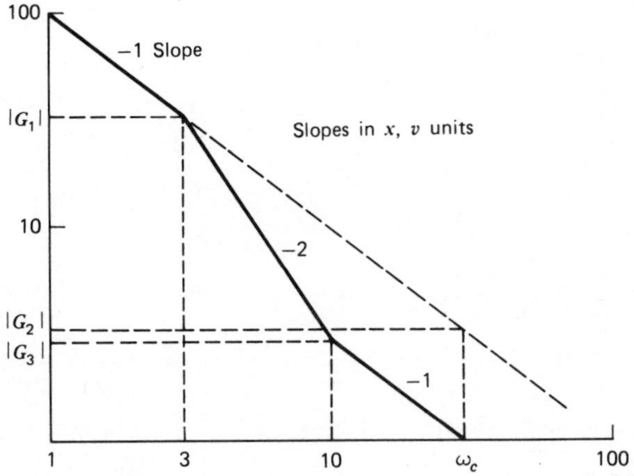

10–33 Use both log–log and semi-log graph paper to construct Bode magnitude plots for the following transfer functions. Consider the number of cycles of log coordinates needed and the flexibility in choosing scales on the two types of paper.

(a) $G(j\omega) = 10\,(1 + j\omega/4)\,/\,(1 + j\omega/40)$

(b) $G(j\omega) = 20/[(j\omega)\,(1 + j\omega/20)]$

10–34 The development of Bode magnitude plots in the text used the forms $(1 + j\omega T)$ and $1/(1 + j\omega T)$. Consider instead the forms $(\alpha + j\omega)$ and $1/(\alpha + j\omega)$. What difference is there between the curves for these forms and those used previously? Is there any advantage in using one form or the other?

10–35 Construct Bode Magnitude plots for the following transfer functions, using straight-line approximations and then corrections.

(a) $G(j\omega) = (1/\sqrt{10}) / (1 + j\omega/2)$
(b) $G(j\omega) = 300 / (3 + j\omega)(1 + j3\omega)$
(c) $G(j\omega) = 10^6 / [(j\omega)(10^4 + j\omega)]$
(d) $G(j\omega) = (5 \times 10^8)(j\omega + 100) / [(j\omega + 10)(j\omega + 10^3)(j\omega + 5 \times 10^3)]$
(e) $G(j\omega) = (100)(j\omega) / [(j\omega)^2 + (j\omega)(6) + 100]$
(f) $G(j\omega) = (j\omega/10)^2 /(1 + j\omega/10)^2$
(g) $G(j\omega) = \epsilon^{-j\omega/2} / (1 + j\omega/2)$
(h) $G(j\omega) = -(j\omega - 4) / (j\omega + 4)$

10–36 How would a highly resonant system with a lightly damped s-plane pole pair carry over into the frequency response, and what feature of the Bode magnitude plot would indicate this resonance?

10–37 In some cases the power series expansion about $\omega T = 0$ for $\tan^{-1}(\omega T)$ is used when considering the phase of $G(j\omega)$. Use the appropriate power series forms for $\omega T < 1$, and then for $\omega T > 1$, and plot the error from the correct value vs. log frequency when using only the linear term.

10–38 Plot phase curves for the transfer functions of Problem 10–35.

Chapter **11**

Relationships Among the Fourier Series, the Fourier Transform, and the Laplace Transform

11-1. INTRODUCTION

There is no one way of presenting the relationship among the Laplace transform, the two-sided Laplace transform, the Fourier transform, and the Fourier series. Some books start this discussion with the Fourier series; then they introduce the Fourier transform, and finally the Laplace transform. Although we have obviously approached this subject from the opposite direction, it is helpful to an overall perspective to examine this alternate approach.

11-2. THE FOURIER TRANSFORM FROM THE FOURIER SERIES

The Fourier series is a method of describing either a function that is periodic with a period T_1, or a function that is defined in a limited region of t as given by

$$t_0 < t < t_0 + T_1 \tag{11-2.1}$$

and there is no concern about matching the function outside this region. In this discussion, we assume the function is periodic and exists for all time; hence, we use the subscript on $f_2(t)$ to be consistent with the earlier notation. An $f_2(t)$ is periodic with a period T_1, if

$$f_2(t) = f_2(t + T_1) \tag{11-2.2}$$

We start from the form of the Fourier series that is most commonly covered in introductory courses as

$$f_2(t) = \frac{a_0}{2} + \sum_{n=1}^{\infty} (a_n \cos n\omega_1 t + b_n \sin n\omega_1 t) \tag{11-2.3}$$

where ω_1 is the angular frequency of the fundamental

$$\omega_1 = 2\pi f_1 = \frac{2\pi}{T_1}$$

and $n = 1, 2, 3, \ldots$.

The following mathematical relationships are helpful

$$\int_t^{t+T_1} \sin m\omega_1 t \sin n\omega_1 t \, dt = \begin{cases} \dfrac{T_1}{2} & m = n \neq 0 \\ 0 & m \neq n \end{cases} \tag{11-2.4}$$

$$\int_t^{t+T_1} \sin m\omega_1 t \cos n\omega_1 t \, dt = 0 \qquad \text{for all } m \text{ and } n \tag{11-2.5}$$

$$\int_t^{t+T_1} \cos m\omega_1 t \cos n\omega_1 t \, dt = \begin{cases} \dfrac{T_1}{2} & m = n \neq 0 \\ 0 & m \neq n \\ T_1 & m = n = 0 \end{cases} \tag{11-2.6}$$

where m and n are any positive integer or zero.

If Eq. 11-2.3 is multiplied by $\cos n\omega_1 t$, and the result integrated over one complete period, all the terms on the right will integrate to zero except the nth term involving the $\cos n\omega_1 t$ term.

$$\int_{t}^{t+T_1} f_2(t) \cos n\omega_1 t \, dt = \cdots + 0 + 0 + \int_{t}^{t+T_1} a_n \cos n\omega_1 t \cos n\omega_1 t \, dt + 0 + 0 \cdots \tag{11-2.7}$$

With the use of Eq. 11-2.6, this becomes

$$\int_{t}^{t+T_1} f_2(t) \cos n\omega_1 t \, dt = \frac{a_n T_1}{2} \qquad n = 0, 1, 2, 3, \ldots \tag{11-2.8}$$

which can be solved for a_n as

$$a_n = \frac{2}{T_1} \int_{t}^{t+T_1} f_2(t) \cos n\omega_1 t \, dt \qquad n = 0, 1, 2, \ldots \tag{11-2.9}$$

These last two equations are also good for $n = 0$, because the Fourier series of Eq. 11-2.3 is written with $a_0/2$.

In a similar manner, Eq. 11-2.3 can be multiplied by $\sin n\omega_1 t$ and integrated over one complete period. The result is solved for b_n as

$$b_n = \frac{2}{T_1} \int_{t}^{t+T_1} f_2(t) \sin n\omega_1 t \, dt \qquad n = 1, 2, 3, \ldots \tag{11-2.10}$$

For the above development to be correct, the $f_2(t)$ must satisfy the Dirichlet conditions (for the Fourier series), which are summarized briefly as: the $f_2(t)$ must be single valued, and sectionally continuous with a finite number of finite discontinuities in any finite interval. These conditions are more severe than are sometimes needed and can be violated under certain conditions. A train of equally spaced impulses is one example.

It is convenient to use the Fourier series in the exponential form. The expansions of the $\cos n\omega_1 t$ and $\sin n\omega_1 t$ terms are given by

$$\begin{aligned} \cos n\omega_1 t &= \frac{\epsilon^{+jn\omega_1 t} + \epsilon^{-jn\omega_1 t}}{2} \\ \sin n\omega_1 t &= \frac{\epsilon^{+jn\omega_1 t} - \epsilon^{-jn\omega_1 t}}{2j} \end{aligned} \tag{11-2.11}$$

and substituted into Eq. 11-2.3 as

$$f_2(t) = \frac{a_0}{2} + \sum_{n=1}^{\infty} \frac{a_n}{2} \left(\epsilon^{+jn\omega_1 t} + \epsilon^{-jn\omega_1 t}\right) + \sum_{n=1}^{\infty} \frac{b_n}{2j} \left(\epsilon^{+jn\omega_1 t} - \epsilon^{-jn\omega_1 t}\right) \tag{11-2.12}$$

These terms are regrouped as

$$f_2(t) = \frac{a_0}{2} + \frac{1}{2} \sum_{n=1}^{\infty} (a_n - jb_n)\, \epsilon^{+jn\omega_1 t} + \frac{1}{2} \sum_{n=1}^{\infty} (a_n + jb_n)\, \epsilon^{-jn\omega_1 t} \tag{11-2.13}$$

This equation can be written in a more compact form by introducing a_{-n} and b_{-n}. We obtain a_{-n} from a_n by taking the Eq. 11-2.9 and replacing n by $-n$.

$$a_{-n} = \frac{2}{T_1}\int_t^{t+T_1} f_2(t)\cos(-n)\omega_1 t\, dt = \frac{2}{T_1}\int_t^{t+T_1} f_2(t)\cos n\omega_1 t\, dt = a_n \quad (11\text{-}2.14)$$

The term b_{-n} is found from Eq. 11-2.10 in a similar manner:

$$b_{-n} = \frac{2}{T_1}\int_t^{t+T_1} f_2(t)\sin(-n)\omega_1 t\, dt = -\frac{2}{T_1}\int_t^{t+T_1} f_2(t)\sin n\omega_1 t\, dt = -b_n \quad (11\text{-}2.15)$$

With this notation, Eq. 11-2.13 can be written as

$$f_2(t) = \frac{a_0}{2} + \frac{1}{2}\sum_{n=1}^{\infty}(a_n - jb_n)\,\epsilon^{+jn\omega_1 t} + \frac{1}{2}\sum_{n=-1}^{-\infty}(a_n - jb_n)\,\epsilon^{+jn\omega_1 t} \quad (11\text{-}2.16)$$

This can be summed as

$$f_2(t) = \sum_{n=-\infty}^{+\infty}\left(\frac{a_n - jb_n}{2}\right)\epsilon^{+jn\omega_1 t} \quad (11\text{-}2.17)$$

The a_0 term is absorbed in this equation when n takes on the value of zero. For notational convenience, $F_2(jn\omega_1)$ is defined as

$$F_2(jn\omega_1) = \frac{a_n - jb_n}{2} \qquad n = 0, \pm 1, \pm 2, \ldots \quad (11\text{-}2.18)$$

Equation 11-2.17 is put in the form

$$f_2(t) = \sum_{n=-\infty}^{\infty} F_2(jn\omega_1)\,\epsilon^{+jn\omega_1 t} \quad (11\text{-}2.19)$$

This equation will be discussed shortly.

The a_n and b_n in Eq. 11-2.18 are replaced with their integral equivalents in Eqs. 11-2.9 and 10, respectively, as

$$\begin{aligned} F_2(jn\omega_1) &= \frac{a_n - jb_n}{2} \\ &= \frac{1}{2}\left[\frac{2}{T_1}\int_t^{t+T_1} f_2(t)\cos n\omega_1 t\, dt - j\frac{2}{T_1}\int_t^{t+T_1} f_2(t)\sin n\omega_1 t\, dt\right] \\ &= \frac{1}{T_1}\int_t^{t+T_1} f_2(t)(\cos n\omega_1 t - j\sin n\omega_1 t)\, dt \\ &= \frac{1}{T_1}\int_t^{t+T_1} f_2(t)\,\epsilon^{-jn\omega_1 t}\, dt \end{aligned} \quad (11\text{-}2.20)$$

Table 11-2.1

Function	Direct transform	Inverse transform
One-sided Laplace transform	$F(s) = \int_0^{\infty} f(t)\, \epsilon^{-st}\, dt$	$f(t) = \frac{1}{2\pi j} \int_{c-j\infty}^{c+j\infty} F(s)\, \epsilon^{st}\, ds$
Two-sided Laplace transform	$F_2(s) = \int_{-\infty}^{+\infty} f_2(t)\, \epsilon^{-st}\, dt$	$f_2(t) = \frac{1}{2\pi j} \int_{c-j\infty}^{c+j\infty} F_2(s)\, \epsilon^{st}\, ds$
Fourier transform	$F_2(j\omega) = \int_{-\infty}^{+\infty} f_2(t)\, \epsilon^{-j\omega t}\, dt$	$f_2(t) = \frac{1}{2\pi} \int_{-\infty}^{+\infty} F_2(j\omega)\, \epsilon^{+j\omega t}\, d\omega$
Fourier series transform	$F_2(jn\omega_1) = \frac{1}{T_1} \int_t^{t+T_1} f_2(t)\, \epsilon^{-jn\omega_1 t}\, dt$	$f_2(t) = \sum_{n=-\infty}^{+\infty} F_2(jn\omega_1)\, \epsilon^{+jn\omega_1 t}$

These last two equations are now discussed. In Eq. 11-2.20, we start with a periodic function $f_2(t)$, which is multiplied by $\epsilon^{-jn\omega_1 t}$, and integrate over one complete period; a frequency representation of the $f_2(t)$ is thus obtained. We call this the direct Fourier series transform. The integration is only over one complete period because the $f_2(t)$ function in any period contains all the information in the entire $f_2(t)$.

Equation 11-2.19 starts with the frequency representation $F_2(jn\omega_1)$, and from this recreates the original time function. We call this the inverse Fourier series transform. Even though an infinite number of frequencies are involved, they are a countable infinity. That is, if the $f_2(t)$ equation is written out, we can point to one term and say this is the fundamental, and to another term and say this is the 2nd harmonic, etc. Therefore, a summation is all that is necessary.

Table 11-2.1 summarizes the one-sided Laplace, the two-sided Laplace, the Fourier, and the Fourier series transforms. It is informative to study this table and to observe the similarities and differences among these transforms.

Example 11-2.1. The first example is the rectangular pulse shown in Fig. 11-2.1. The pulse has a height K, a width t_0, and a period T_1.

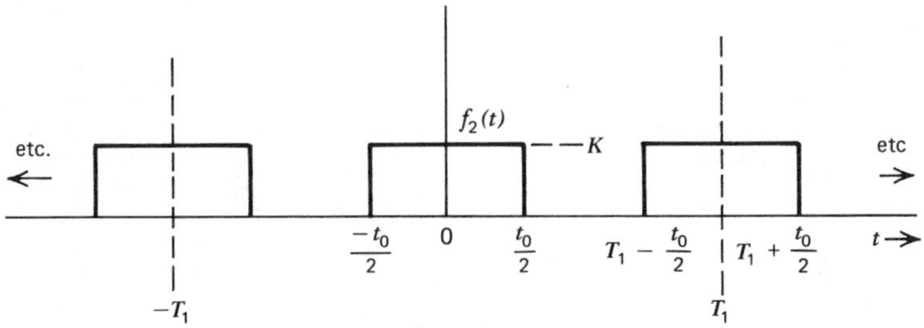

Fig. 11-2.1

In the direct Fourier series transform of Eq. 11-2.20, the integration is from t to $t + T_1$. For convenience, we let $t = -T_1/2$, and the equation becomes:

$$F_2(jn\omega_1) = \frac{1}{T_1}\int_{-T_1/2}^{+T_1/2} f_2(t)\,\epsilon^{-jn\omega_1 t}\,dt \tag{11-2.21}$$

For this example, the equation reduces to

$$F_2(jn\omega_1) = \frac{1}{T_1}\int_{-t_0/2}^{+t_0/2} K\,\epsilon^{-jn\omega_1 t}\,dt$$

$$= \frac{K\epsilon^{-jn\omega_1 t}}{T_1(-jn\omega_1)}\Bigg|_{-t_0/2}^{+t_0/2} = \frac{2K}{T_1}\left[\frac{\sin(n\omega_1 t_0/2)}{n\omega_1}\right] \tag{11-2.22}$$

and is put in the form $\sin x/x$

$$F_2(jn\omega_1) = \frac{Kt_0}{T_1}\left[\frac{\sin(n\omega_1 t_0/2)}{n\omega_1 t_0/2}\right] \tag{11-2.23}$$

This equation has meaning only when n is an integer; however, if $n\omega_1$ is treated as a continuous variable, an envelope can be plotted that helps in visualizing the equation. This is done for Eq. 11-2.23 in Fig. 11-2.2.

The first point to the right of $n\omega_1 = 0$ where the envelope has a value of zero occurs when

$$\frac{n\omega_1 t_0}{2} = \pi \qquad \text{or} \qquad n\omega_1 = \frac{2\pi}{t_0} \tag{11-2.24}$$

The other envelope crossings are determined in a similar manner.

Example 11-2.2. This example is a continuation of the preceding example. The $f_2(t)$ of Fig. 11-2.1 is redrawn in Fig. 11-2.3(a) with a change in scale. In (b), the shape of the individual pulses remains the same, but every other pulse is removed. The new period is

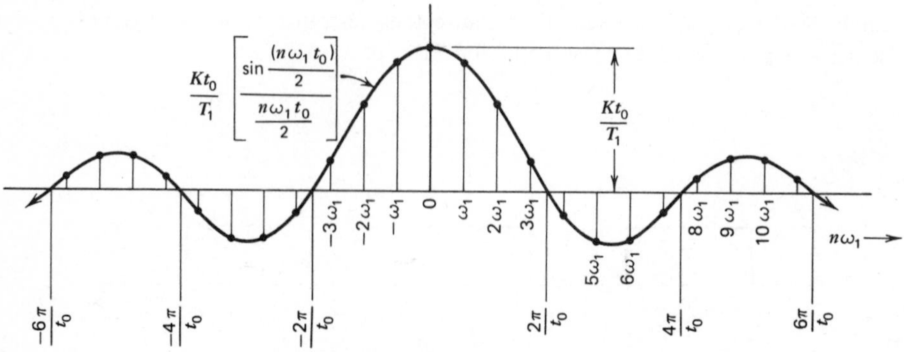

Fig. 11-2.2

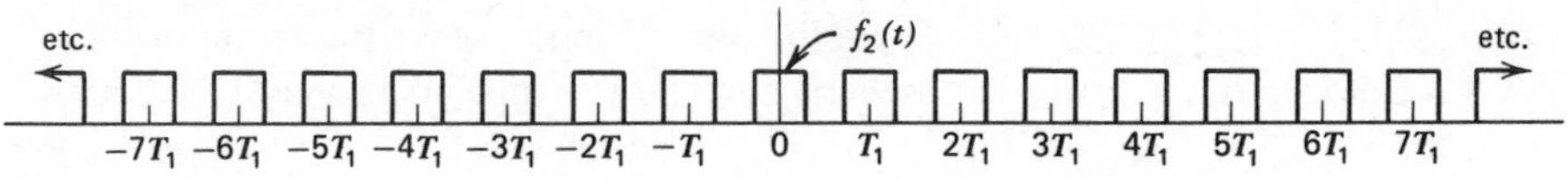

$f_2(t)$ is the same wave shape as that shown in Fig. 11–2–1

(a)

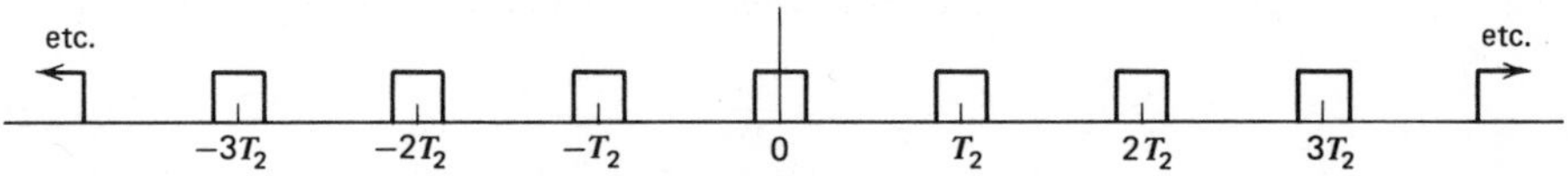

Each pulse has the same shape as the pulse in Part *a*.

(b)

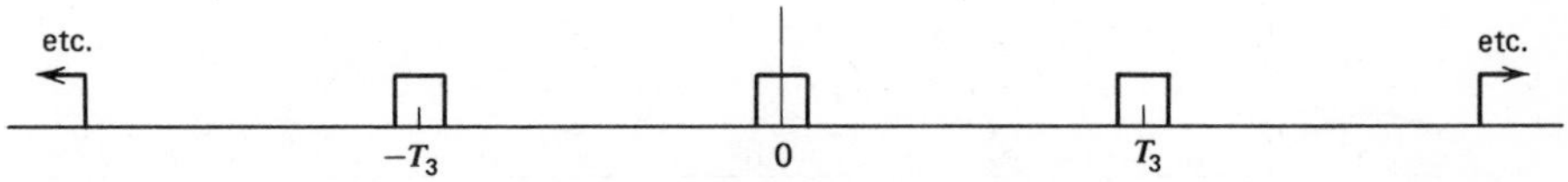

Each pulse has the same shape as the pulse in Part *a*

(c)

The pulse has the same shape as the pulse in Part *a*.

(d)

Fig. 11-2.3

labeled T_2, where

$$T_2 = 2T_1 \tag{11-2.25}$$

In (c), the shape of the individual pulses remains the same, but every other pulse is removed from (b). The new period is labeled T_3, where

$$T_3 = 2T_2 = 4T_1 \tag{11-2.26}$$

If every other pulse is removed indefinitely, finally only one pulse at the origin will be left, as suggested in (d), and thus this $f_2(t)$ has a period

$$T = \infty \tag{11-2.27}$$

Next we wish to see what is happening to the Fourier series transforms for this sequence of $f_2(t)$ functions. The transform for the $f_2(t)$ in Fig. 11-2.3(a) is the same as that found in the previous example and is shown again in Fig. 11-2.4(a). The earlier figure (Fig. 11-2.2) should be referred to for more detail.

In Fig. 11-2.4(a), (b), (c), and (d) are the direct Fourier series transform of the $f_2(t)$ of Fig. 11-2.3(a), (b), (c), and (d), respectively. The envelope takes on a value of zero as determined only by t_0. Since the pulse width remains the same, the envelope crossings of each of the four curves remains the same.

The maximum height of the envelope is Kt_0/T. Each time the period T is doubled, the height of the envelope is divided by two. In Fig. 11-2.4(d), the period is infinite and the maximum height of the envelope is zero.

Each time the period T is doubled, the corresponding ω is cut in half, as given by

$$\omega = \frac{2\pi}{T} \tag{11-2.28}$$

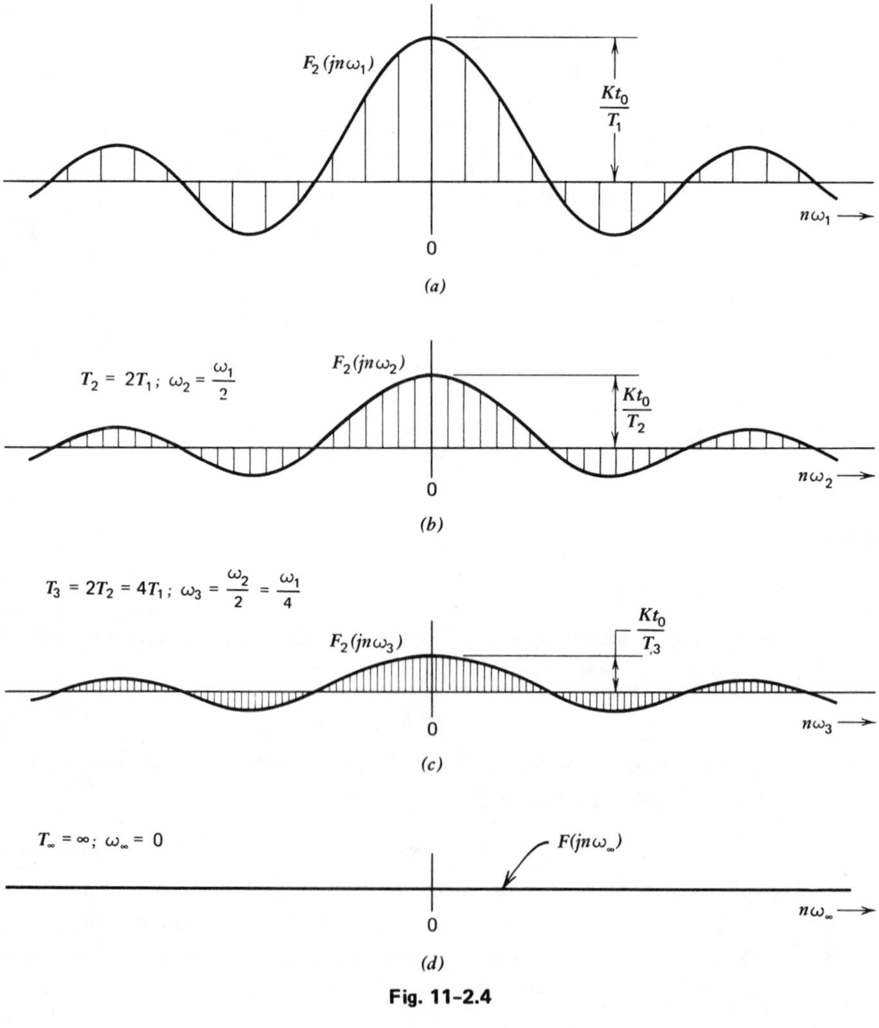

Fig. 11-2.4

Part (b) of the figure has twice as many spectral lines as does (a), and (c) has twice as many lines as (b). As the period T becomes larger, the height of the envelope becomes smaller and the spacing between spectral lines becomes smaller. In the limit, as $T \to \infty$, the height of the envelope and the spacing between spectral lines is zero.

For the preceding discussion, we have used the notation T_1 as representing a fixed period, and we changed the length of the period defining $T_2 = 2T_1, T_3 = 4T_1, \ldots$. However, it is now more convenient simply to use T_1 as representing any period, so that for the rest of this discussion, T_1 is a variable and hence ω_1 is also a variable.

Figure 11-2.4(d) is certainly not very useful. Even though the height of the $F_2(jn\omega_1)$ curve is zero everywhere, it is still desirable to see the relative difference among these zeros. The thing that causes the difficulty is the T_1, which goes to infinity in the denominator of the equation for $F_2(jn\omega_1)$. To bring this curve up out of the mud so that its shape can be examined, we multiply the $F_2(jn\omega_1)$ by T_1, as

$$T_1\, F_2(jn\omega_1) = \frac{T_1}{T_1} \int_t^{t+T_1} f_2(t)\, \epsilon^{-jn\omega_1 t}\, dt = \int_t^{t+T_1} f_2(t)\, \epsilon^{-jn\omega_1 t}\, dt \tag{11-2.29}$$

The arguments that are usually presented are: as

$$T_1 \longrightarrow \infty \tag{11-2.30}$$

the ω_1 approaches a differential

$$\omega_1 \longrightarrow d\omega \tag{11-2.31}$$

and $n\omega_1$ becomes a continuous variable

$$n\omega_1 \longrightarrow \omega \tag{11-2.32}$$

Since T_1 is infinite, the limits of integration must be from $-\infty$ to $+\infty$. With these changes, the integral of Eq. 11-2.29 becomes

$$T_1\, F_2(jn\omega_1) = \int_{-\infty}^{+\infty} f_2(t)\, \epsilon^{-j\omega t}\, dt = F_2(j\omega) \tag{11-2.33}$$

which is precisely the Fourier transform that we have been using.

Since the direct Fourier series transform has been modified, we must also modify the inverse Fourier series transform of Eq. 11-2.19 in an appropriate manner

$$f_2(t) = \frac{1}{T_1} \sum_{n=-\infty}^{+\infty} [T_1\, F_2(jn\omega_1)]\, \epsilon^{+jn\omega_1 t} \tag{11-2.34}$$

The $1/T_1$ out in front can be replaced by $\omega_1/2\pi$. The limiting properties of Eqs. 11-2.30, 31, 32 and 33 are applied to this equation. In the limit, the summation becomes an integral. With these changes, Eq. 11-2.34 is written as

$$f_2(t) = \frac{1}{2\pi} \int_{-\infty}^{+\infty} F_2(j\omega)\, \epsilon^{+j\omega t}\, d\omega \tag{11-2.35}$$

The result is that Eqs. 11-2.33 and 35 reduce to the Fourier transform equations we have been using.

Up to this point, we have presented this discussion without editorial comment. Essentially what this argument says is, if you place enough points close together, the result is a continuous curve. What this development does is to use a countable infinity to approximate an uncountable infinity. As T_1 gets larger, in the development that accompanies Fig. 11-2.4, we can still number the spectral points on the envelope and say here is *Point No. 1*, here is *Point No. 2*, and so on. However, between *Points 1* and *2*, there are still an uncountable infinity of points.

When we were discussing two-sided Laplace transforms and Fourier transforms that exist in the limit, we said there was some *arm waving* involved. If a lecturer had to wave his arms for the previous development, for this present discussion he would not only have to wave them, he would also have to jump up and down and pound on the chalk board at the same time. For these reasons, we started the discussion throughout this book from the point of view of the Laplace transform. The present discussion does reinforce some ideas mentioned before. The spectral points on the envelope of the direct Fourier series transforms are just magnitudes or numbers. When this function is multiplied by T_1, and T_1 goes to infinity, the Fourier transform of a periodic function becomes a train of impulses.

Another way to say this is that the envelope of a direct Fourier series transform of a periodic function cannot be compared to a frequency spectrum obtained from the Fourier transform of an aperiodic function. If the second function were plotted to the same scale as the first function, the height would be zero, or perhaps it is better to say the function would have a differential height.

11-3. FROM THE FOURIER TRANSFORM TO THE LAPLACE TRANSFORM

As mentioned earlier, many textbooks start with the Fourier series and move to the Fourier transform and finally to the Laplace transform. Since we have started discussing this line of reasoning, we should finish by showing how to obtain the Laplace transform from the Fourier transform. A simple example will clarify the idea.

Example 11-3.1. Suppose we wish to find the Fourier transform of

$$f_2(t) = \begin{cases} \epsilon^{+4t} & t<0 \\ \epsilon^{+2t} & t>0 \end{cases} \tag{11-3.1}$$

An examination of $f_2(t)$ shows that this function does not satisfy the Dirichlet conditions. However, why should we stop if we can introduce a mathematical artifice that will allow us to continue. We proceed by multiplying $f_2(t)$ by a converging factor $\epsilon^{-\sigma t}$, where σ is of a value whereby the new function

$$\epsilon^{-\sigma t} f_2(t) \tag{11-3.2}$$

does satisfy the Dirichlet conditions. For the $f_2(t)$ of this example, σ can take on any value within

$$2<\sigma<4 \tag{11-3.3}$$

The σ is given a specific value of 3, and the new function

$$\epsilon^{-\sigma t} f_2(t) = \epsilon^{-3t} f_2(t) = \begin{cases} \epsilon^{+1t} & t<0 \\ \epsilon^{-1t} & t>0 \end{cases} \tag{11-3.4}$$

does satisfy the Dirichlet condition and a Fourier transform can be found.

This Example is Generalized. Given that an $f_2(t)$ does not satisfy the Dirichlet conditions, but that $\epsilon^{-\sigma t} f_2(t)$, with the proper choice of σ, does.

The conventional Fourier transform integral

$$F_2(j\omega) = \int_{-\infty}^{+\infty} f_2(t)\, \epsilon^{-j\omega t}\, dt \tag{11-3.5}$$

does not exist, but the integral defined for the new function does.

$$\int_{-\infty}^{+\infty} f_2(t)\, \epsilon^{-\sigma t}\, \epsilon^{-j\omega t}\, dt = \int_{-\infty}^{+\infty} f_2(t)\, \epsilon^{-(\sigma+j\omega)t}\, dt \tag{11-3.6}$$

If we replace the $(\sigma + j\omega)$ term in the exponent with s, that is

$$s = \sigma + j\omega \tag{11-3.7}$$

the resulting equation then becomes a function of s, which is

$$\int_{-\infty}^{+\infty} f_2(t)\, \epsilon^{-st}\, dt = F_2(s) \tag{11-3.8}$$

This equation is the two-sided Laplace transform with which we have been working.

If an $f_2(t)$ is such that a strip of convergence exists for $F_2(s)$, then when this $f_2(t)$ is multiplied by the converging factor $\epsilon^{-\sigma t}$, where σ is within the strip of convergence, the new time function $\epsilon^{-\sigma t} f_2(t)$ satisfies the Dirichlet conditions. These are essentially two different ways of expressing the same idea.

Finally, if the time function is defined only for positive time, the one-sided Laplace transform results.

11-4. THE FOURIER SERIES OBTAINED FROM THE FOURIER TRANSFORM

We don't necessarily propose that the following is the best procedure for obtaining a Fourier series, but this method is included to complete the discussion, by way of examples.

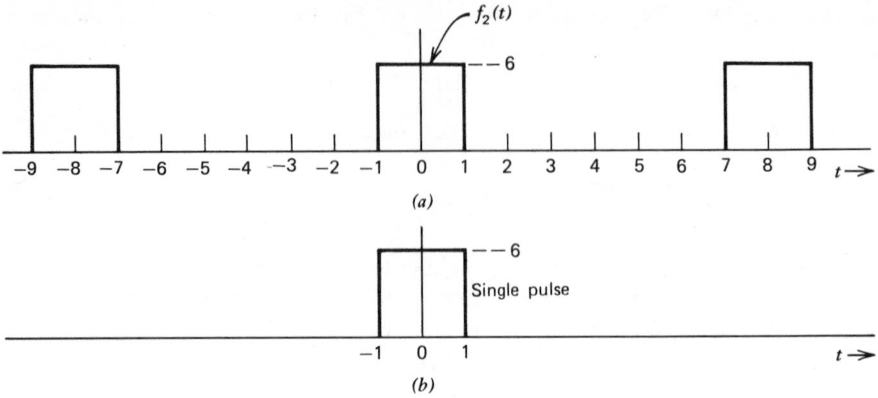

Fig. 11-4.1

Example 11-4.1. We wish to find the Fourier series of the wave shape shown in Fig. 11-4.1(a). We start by throwing away everything about the periodic function except the shape of the function in the period from $-T_1/2$ to $+T_1/2$, as in (b). Next, the Fourier transform is found, by methods that are already familiar, as

$$F_2(j\omega) = 12\left[\frac{\sin\,\omega}{\omega}\right] \tag{11-4.1}$$

which is plotted in Fig. 11-4.2(a).

The following steps are taken to make the transition, from the continuous spectrum $F_2(j\omega)$ for the aperiodic pulse, to the envelope and line spectrum $F_2(jn\omega_1)$ for the periodic wave shape. The area under the curve of the aperiodic pulse is equal to the height of $F_2(j\omega)$ at $\omega = 0$, which from Eq. 11-4.1 is equal to 12. This correspondence between area and height is Property 11-6.14, developed later in this chapter.

The height of the envelope of $F_2(jn\omega_1)$ at $n\omega_1 = 0$ is equal to the average value (the d-c component) of the periodic wave shape. Since the area under the single pulse is equal to 12, and the period T_1 is equal to 8, this average value is equal to 12/8, or 1.5. From this, the envelope for the $F_2(jn\omega_1)$ can be scaled as in Fig. 11-4.2(b).

To complete Fig. 11-4.2(b), the location of the spectral lines must be found. We know these are spaced ω_1 units apart, and for this example

$$\omega_1 = \frac{2\pi}{T_1} = \frac{2\pi}{8} = \frac{\pi}{4} \tag{11-4.2}$$

The height of the spectral line can be calculated from

$$F_2(jn\omega_1) = 1.5\left[\frac{\sin\,(n\pi/4)}{n\pi/4}\right] \tag{11-4.3}$$

for as many values of n as are desired. The corresponding frequency terms are combined in pairs, and the usual Fourier series results. This is left as a problem at the end of the chapter.

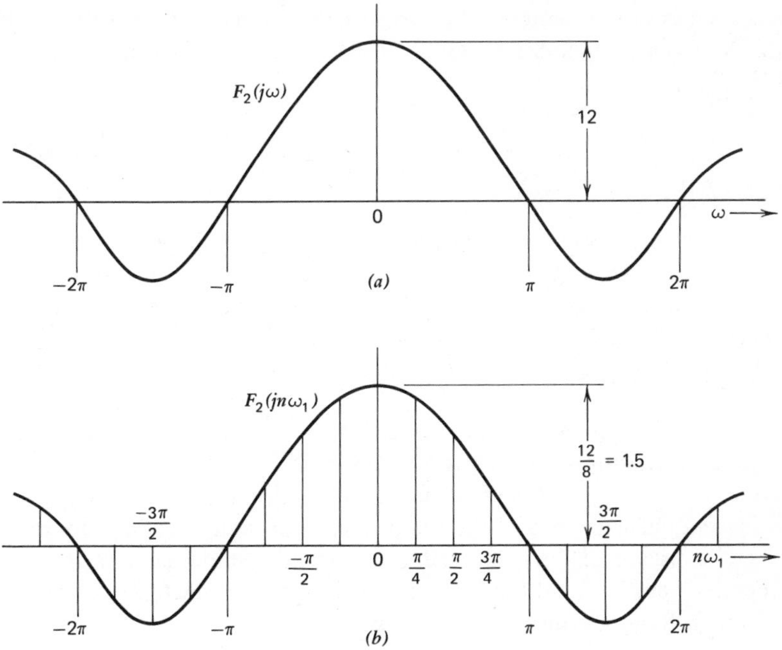

Fig. 11-4.2

Example 11-4.2. We wish to find the Fourier series for the wave shape shown in Fig. 11-4.3. The wave shape in the fundamental period is the same as that in the previous example, but the period T_1 has been changed.

The height of the envelope of $F_2(jn\omega_1)$ at $n\omega_1 = 0$ is equal to 12/3, or 4. The new ω_1 is found to be

$$\omega_1 = \frac{2\pi}{T_1} = \frac{2\pi}{3} \tag{11-4.4}$$

The height of the spectral lines can be calculated from

$$F_2(jn\omega_1) = 4\left[\frac{\sin(2n\pi/3)}{2n\pi/3}\right] \tag{11-4.5}$$

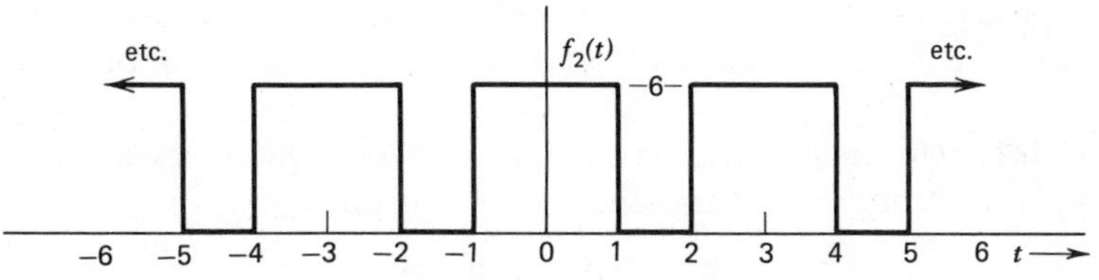

Fig. 11-4.3

for as many values of n as desired. The corresponding frequency terms are combined in pairs, and the usual Fourier Series results. This is left as a problem at the end of the chapter.

COMMENTARY

The last two examples were deliberately chosen to be as simple as possible and yet demonstrate the general idea. Since the $f_2(t)$'s used were even functions, the resulting Fourier series were made up of only cosine terms, or as far as $F_2(jn\omega_1)$ is concerned, all the b's were zero. An $f_2(t)$ that is neither an even nor an odd function, would have an $F_2(jn\omega_1)$ that contains an even part and an odd part, but the same general procedure could be followed.

Again it is not proposed that this is necessarily the best way to obtain a Fourier series. It does show, however, that it is possible to go from the Fourier transform to the Fourier series in a very smooth and systematic way. Also, this development can add to a designer's overall understanding of the Fourier series. If he looks at a periodic $f_2(t)$, and then looks at the shape of the pulse in the fundamental period, he may be able to visualize its Fourier transform and the Fourier series for the periodic $f_2(t)$, out of knowledge he already has, without pursuing the computation in detail.

11-5. PROPERTIES OF FOURIER SERIES, FOURIER TRANSFORM, AND LAPLACE TRANSFORM

A number of the properties of the one-sided Laplace transform are developed with a certain amount of detail in Chapter 4. It is beyond the scope of this book to be that complete about the properties of the two-sided Laplace transform, the Fourier transform, and the Fourier series transform; but still it seems reasonable to make the discussion as general as possible. To do this, the notation shown in Table 11-5.1 is introduced. The subscipt in $f_2(t)$ is dropped to leave $f(t)$, as the subscript in $\mathcal{L}_2[f(t)]$ indicates a two-sided function. The subscript in $\mathcal{F}_s[f(t)]$ indicates a Fourier series transform. The notation $\mathcal{G}[f(t)]$ is a general symbol that represents any or all of the other four transforms. In other words, a property stated in terms of $\mathcal{G}[f(t)]$ holds for all four transforms.

The numbering on the properties coincides as far as possible with the numbering in Chapter 4. Finally, a minimum number of proofs are included, because many of these are extensions of those in Chapter 4 and are left for problems at the end of the chapter.

Property 11-5.1. The transform of a product of a constant times a function is the product of the constant times the transform of the function.

$$\mathcal{G}[Kf(t)] = K\,\mathcal{G}[f(t)] \qquad (11\text{-}5.1)$$

Property 11-5.2. The transform of a sum of time functions is the sum of the transform of the individual time functions, provided that each transform exists.

$$\mathcal{G}[f_1(t) + f_2(t) + \cdots + f_n(t)] = \mathcal{G}[f_1(t)] + \mathcal{G}[f_2(t)] + \cdots + \mathcal{G}[f_n(t)] \quad (11\text{-}5.2)$$

Property 11-5.3. The Transform of the Derivative of a Function

$$\mathcal{L}_2[f'(t)] = s\,\mathcal{L}_2[f(t)] \quad (11\text{-}5.3)$$

$$\mathcal{F}[f'(t)] = j\omega\,\mathcal{F}[f(t)] \quad (11\text{-}5.4)$$

$$\mathcal{F}_s[f'(t)] = jn\omega_1\,\mathcal{F}_s[f(t)] \quad (11\text{-}5.5)$$

Since the time functions associated with each of the three equations exist for all t, there is no possibility of any initial condition entering into these equations as is the case for the one-sided transform.

Property 11-5.4. The Transform of the Integral of a Function. Each of the following equations is qualified in the following discussion.

$$\mathcal{L}_2 \int_{-\infty}^{t} f(t)\,dt = \frac{\mathcal{L}_2[f(t)]}{s} \quad (11\text{-}5.6)$$

$$\mathcal{F}\left[\int_{-\infty}^{t} f(t)\,dt\right] = F(j0)\,\pi\,u_0(j\omega) + \frac{F(j\omega)}{j\omega} \quad (11\text{-}5.7)$$

$$\mathcal{F}_s\left[\int_{-\infty}^{t} f(t)\,dt\right] = \frac{\mathcal{F}_s[f(t)]}{jn\omega_1} \quad (11\text{-}5.8)$$

The Two-Sided Laplace Transform. The right side of Equation 11-5.6 can be thought of as $1/s$ multiplying $\mathcal{L}_2[f(t)]$. The $1/s$ converges to the right of the $j\omega$ axis. If the $\mathcal{L}_2[f(t)]$ has a strip of convergence that is entirely to the left of the $j\omega$-axis, the two-sided Laplace transform of the integral does not converge anywhere and the transform does not exist. If the strip of convergence of $\mathcal{L}_2[f(t)]$ includes, or is to the right of, the $j\omega$-axis, the two-sided Laplace transform of the integral exists and is given by Eq. 11-5.6.

The Fourier Transform. If you picture going from the two-sided Laplace transform of Eq. 11-5.6 to the Fourier transform of Eq. 11-5.7 by replacing s with $j\omega$, certain aspects of the situation become clear. If the $\mathcal{L}_2[f(t)]$ has an s in the numerator to cancel the s in the denominator, and its strip of convergence includes the $j\omega$-axis, there is no problem in replacing s with $j\omega$. However, if the s is not canceled, Eq. 11-5.6 converges to the right of the $j\omega$-axis, and hence the s cannot be replaced with $j\omega$ except through some limiting action. This explains the $F(j0)\,\pi\,u_0(j\omega)$ term in Eq. 11-5.7.

The Fourier Series Transform. The discussion of Eq. 11-5.8 is left as a problem.

Property 11-5.5. The Transform of ϵ^{-at} Times a Function. We assume that the $\mathcal{G}[f(t)]$ exists; then the following equations are for the transform of $[\epsilon^{-at} f(t)]$ in terms of the transform of $f(t)$.

$$\mathcal{L}_2[\epsilon^{-at} f(t)] = F(s + a) \tag{11-5.9}$$

$$\mathcal{F}[\epsilon^{-at} f(t)] = F(j\omega + a) \tag{11-5.10}$$

$$\mathcal{F}_s[\epsilon^{-at} f(t)] = F_s(jn\omega_1 + a) \tag{11-5.11}$$

A discussion of these equations is left to the problems.

The use of Eq. 11-5.9 makes it possible to extend the significance of the 2nd and 3rd family of functions in Table 10-5.1 to where the line of convergence of the two-sided Laplace transform can be any line parallel to the $j\omega$-axis in the complex plane. This idea is covered in the problems.

Properties 11-5.6 and 7. The Transform of $f^n(t)$. In Chapter 4, the Laplace transform of the second and the nth derivatives were considered separately. Since the second of these is more general and includes the first, only the nth derivative is considered here. We assume that the $\mathcal{G}[f(t)]$ exists; then

$$\mathcal{L}_2[f^n(t)] = s^n \mathcal{L}_2[f(t)] \tag{11-5.12}$$

$$\mathcal{F}[f^n(t)] = (j\omega)^n \mathcal{F}[f(t)] \tag{11-5.13}$$

$$\mathcal{F}_s[f^n(t)] = (jn\omega_1)^n \mathcal{F}_s[f(t)] \tag{11-5.14}$$

A discussion of these equations is left to the problems.

Property 11-5.8. The Transform of the nth Integral of a Function. We assume that $\mathcal{G}[f(t)]$ exists; then

$$\mathcal{L}_2[f^{-n}(t)] = \frac{\mathcal{L}_2[f(t)]}{s^n} \tag{11-5.15}$$

$$\mathcal{F}[f^{-n}(t)] = (\textit{see statements below}) \tag{11-5.16}$$

$$\mathcal{F}_s[f^{-n}(t)] = \frac{\mathcal{F}_s[f(t)]}{(jn\omega_1)^n} \tag{11-5.17}$$

The Two-Sided Laplace Transform. If the transform for the first integral exists, so does the transform for the nth integral (see property 11-5.4).

The Fourier Transform. Again picture going from Eq. 11-5.15 to Eq. 11-5.16 by replacing s with $j\omega$. The Fourier transform can only exist through some limiting action and is somewhat complex. The details are left as a problem.

The Fourier Series Transform. The discussion of Eq. 11-5.17 is left as a problem.

Table 11-5.1

Name of the transform	Symbol
One-sided Laplace transform	$\mathcal{L}[f(t)]$
Two-sided Laplace transform	$\mathcal{L}_2[f(t)]$
Fourier transform	$\mathcal{F}[f(t)]$
Fourier Series transform	$\mathcal{F}_s[f(t)]$
A general symbol that indicates any or all of the above four transforms	$\mathcal{G}[f(t)]$

Property 11-5.9. The Dead-Time Delay Operator. We assume that $\mathcal{G}[f(t)]$ exists. The new time function $f_T(t)$ is the old time function shifted by T seconds to the right if T is positive, and to the left if T is negative.

$$\mathcal{L}_2[f_T(t)] = \epsilon^{-Ts} F(s) \tag{11-5.18}$$

$$\mathcal{F}[f_T(t)] = \epsilon^{-j\omega T} F(j\omega) \tag{11-5.19}$$

$$\mathcal{F}_s[f_T(t)] = \epsilon^{-jn\omega_1 T} F_s(jn\omega_1) \tag{11-5.20}$$

It should be noted that

$$|\epsilon^{-j\omega T} F(j\omega)| = |F(j\omega)| \tag{11-5.21}$$

This equation can be expressed in words by saying that no matter how far a time function is shifted in time, the magnitude spectrum remains unchanged and only the phase spectrum is affected.

Property 11-5.10. The Convolution Integral. The convolution integral is independent of transform methods and is an alternate method of finding a system output.

Property 11-5.11. Convolution in the Frequency Domain. Property 4-12.11 is first extended to the two-sided Laplace transform. If $f_1(t)$ and $f_2(t)$ both have two-sided Laplace transforms, as given by

$$\begin{aligned} F_1(s) &= \mathcal{L}_2[f_1(t)] \\ F_2(s) &= \mathcal{L}_2[f_2(t)] \end{aligned} \tag{11-5.22}$$

the two-sided Laplace transform of the product is given by

$$\mathcal{L}_2[f_1(t) f_2(t)] = \frac{1}{2\pi j} \int_{c-j\infty}^{c+j\infty} F_1(s-p) F_2(p)\, dp \tag{11-5.23}$$

Since the integration is in the complex plane, we will not use this integral.

For the Fourier transform, this integral is real. To show that this is so, the following changes of variables are used

$$s = j\omega \qquad p = ju$$

The c is set to zero, and the limits of integration become

$$\begin{aligned} p &= -j\infty \longrightarrow u = -\infty \\ p &= +j\infty \longrightarrow u = +\infty \end{aligned} \tag{11-5.24}$$

The differential is

$$dp = j\,du \tag{11-5.25}$$

Upon substitution, the frequency convolution becomes

$$\mathfrak{F}[f_1(t)\,f_2(t)] = \frac{1}{2\pi}\int_{-\infty}^{+\infty} F_1[j\,(\omega - u)]\,F_2(ju)\,du \tag{11-5.26}$$

Except for the $1/2\pi$ out in front, the frequency convolution is identical to the time convolution studied earlier.

Property 11-5.12. The Transform of a Change of Scale. The Two-Sided Laplace Transform. Equation 4-12.39 can be extended to the two-sided Laplace transform as follows. Given

$$\mathcal{L}_2[f(t)] = F(s) \tag{11-5.27}$$

then for a real positive constant a

$$\mathcal{L}_2[f(at)] = \frac{1}{a}F\left(\frac{s}{a}\right) \tag{11-5.28}$$

If a is negative, this becomes

$$\mathcal{L}_2[f(at)] = \frac{1}{-a}F\left(\frac{s}{a}\right) \tag{11-5.29}$$

The last two equations are combined as

$$\mathcal{L}_2[f(at)] = \frac{1}{|a|}F\left(\frac{s}{a}\right) \tag{11-5.30}$$

The Fourier Transform. Equation 11-5.30 becomes

$$\mathfrak{F}[f(at)] = \frac{1}{|a|}F\left(\frac{j\omega}{a}\right) \tag{11-5.31}$$

The Fourier Series Transform. Equation 11-5.30 becomes

$$\mathfrak{F}_s[f(at)] = \frac{1}{|a|}F_s\left(\frac{jn\omega_1}{a}\right) \tag{11-5.32}$$

A Discussion of the Scaling Property. These equations show that as a time function spreads out, the corresponding frequency spectrum contracts, and conversely that as a time function contracts, the frequency spectrum spreads out.

The Reversal Property. A special case of the scaling property is when a takes on a value of minus one.

$$\mathcal{L}_2\,[f(-t)] = F(-s) \tag{11-5.33}$$

$$\mathfrak{F}\,[f(-t)] = F(-j\omega) \tag{11-5.34}$$

$$\mathfrak{F}_s\,[f(-t)] = F_s(-jn\omega_1) \tag{11-5.35}$$

Expressed in words, the last two equations say that if the time functions are folded around the (t = 0) axis, the corresponding frequency functions are folded around the ($\omega = 0$) axis.

Property 11-5.13. The Transform of $[(-t)^n\,f(t)]$. The Two-sided Laplace Transform. Equation 4-12.44 can be generalized to the two-sided Laplace transform as

$$\frac{d^n\,F(s)}{ds^n} = \mathcal{L}_2\,[(-t)^n\,f(t)] \tag{11-5.36}$$

The Fourier Transform. In this case the differentiation is with respect to ω, which leads to

$$(j)^n\,\frac{d^n\,F(j\omega)}{d\omega^n} = \mathfrak{F}\,[(t)^n\,f(t)] \tag{11-5.37}$$

The Fourier Series Transform. A discussion of this is left as a problem.

11-6. ADDITIONAL PROPERTIES OF THE FOURIER TRANSFORM

We center our interest on the Fourier transform for additional properties.

Property 11-6.14. The Height of the $F(j\omega)$ Curve at $\omega = 0$. We start with the equation for the Fourier transform

$$F(j\omega) = \int_{-\infty}^{+\infty} f(t)\,\epsilon^{-j\omega t}\,dt \tag{11-6.1}$$

and we let $\omega = 0$, giving

$$F(j0) = \int_{-\infty}^{+\infty} f(t)\,dt \tag{11-6.2}$$

Expressed in words, this equation says that the height of the frequency spectrum at $\omega = 0$ is equal to the area under the $f(t)$ curve.

Property 11-6.15. The Height of the $f(t)$ Curve at $t = 0$. We start with the equation

$$f(t) = \frac{1}{2\pi} \int_{-\infty}^{+\infty} F(j\,\omega)\, \epsilon^{+j\omega t}\, d\omega \tag{11-6.3}$$

and let t take on the value of zero, giving

$$f(0) = \frac{1}{2\pi} \int_{-\infty}^{+\infty} F(j\omega)\, d\omega$$

Expressed in words, this equation says that the height of the time function at $t = 0$ is equal to $1/2\pi$ times the area under the $F(j\omega)$ curve.

11-7. EVEN AND ODD FUNCTIONS

When the two-sided Laplace transforms or the Fourier transforms are used, the concept of even and odd functions becomes very useful.

If $f(t)$ is an even function, by definition

$$f_e(t) = f_e(-t) \tag{11-7.1}$$

If $f(t)$ is an odd function, by definition

$$f_o(t) = -f_o(-t) \tag{11-7.2}$$

An $f(t)$ that is neither even nor odd can be divided into an even part and an odd part, as by

$$f(t) = f_e(t) + f_o(t) \tag{11-7.3}$$

The $f(-t)$ can be written as

$$f(-t) = f_e(-t) + f_o(-t) = f_e(t) - f_o(t) \tag{11-7.4}$$

Equations 11-7.3 and 11-7.4 can be combined and solved for $f_e(t)$ and $f_o(t)$, as

$$f_e(t) = \frac{f(t) + f(-t)}{2} \qquad f_o(t) = \frac{f(t) - f(-t)}{2} \tag{11-7.5}$$

THE DIRECT FOURIER TRANSFORM

With these thoughts in mind, the Fourier transform

$$F(j\omega) = \int_{-\infty}^{+\infty} f(t)\, \epsilon^{-j\omega t}\, dt \tag{11-7.6}$$

is rewritten with $f(t)$ and $\epsilon^{-j\omega t}$ expanded as

$$f(t) = f_e(t) + f_o(t) \qquad \epsilon^{-j\omega t} = \cos \omega t - j \sin \omega t$$

$$F(j\omega) = \int_{-\infty}^{+\infty} [f_e(t) + f_o(t)]\, [\cos \omega t - j \sin \omega t]\, dt \tag{11-7.7}$$

This equation is written as

$$F(j\omega) = \int_{-\infty}^{+\infty} f_e(t) \cos \omega t\, dt + \int_{-\infty}^{+\infty} f_o(t) \cos \omega t\, dt$$
$$-j \int_{-\infty}^{+\infty} f_e(t) \sin \omega t\, dt - j \int_{-\infty}^{+\infty} f_o(t) \sin \omega t\, dt \tag{11-7.8}$$

In the first term, both $f_e(t)$ and $\cos \omega t$ are even functions of t; then the product $f_e(t) \cos \omega t$ is also an even function of t and the integral from $-\infty$ to 0 is equal to the integral from 0 to $+\infty$. Therefore, this first term can be written as

$$\int_{-\infty}^{+\infty} f_e(t) \cos \omega t\, dt = 2 \int_0^{+\infty} f_e(t) \cos \omega t\, dt \tag{11-7.9}$$

In the second term, $f_o(t)$ is an odd function, and $\cos \omega t$ an even function of t; then the product $f_o(t) \cos \omega t$ is an odd function of t, and the integral from $-\infty$ to 0 is equal in magnitude to, but has the opposite sign from, the integral from 0 to $+\infty$. Therefore, this term is equal to zero.

By a similar reasoning process, the next two terms are

$$\int_{-\infty}^{+\infty} f_e(t) \sin \omega t\, dt = 0 \qquad \int_{-\infty}^{+\infty} f_o(t) \sin \omega t\, dt = 2 \int_0^{\infty} f_o(t) \sin \omega t\, dt \tag{11-7.10}$$

Equation 11-7.8 can be written as

$$F(j\omega) = 2 \int_0^{\infty} f_e(t) \cos \omega t\, dt - j2 \int_0^{\infty} f_o(t) \sin \omega t\, dt \tag{11-7.11}$$

These integrations are performed with respect to time, but t does not appear in the results, which are a function of ω. Since $\cos \omega t$ is an even function of ω, the first term of Eq. 11-7.11 is the even part of $F(j\omega)$, or

$$F_e(j\omega) = 2 \int_0^{\infty} f_e(t) \cos \omega t\, dt \tag{11-7.12}$$

Similarly, the odd part of $F(j\omega)$ is

$$F_o(j\omega) = -j2 \int_0^{\infty} f_o(t) \sin \omega t\, dt \tag{11-7.13}$$

The $F(j\omega)$ can be written as

$$F(j\omega) = F_e(j\omega) + F_o(j\omega) \tag{11-7.14}$$

Since $f(t)$ is real, the $F_e(j\omega)$ is real and $F_o(j\omega)$ is imaginary. If t is replaced with x, where x is complex, then $f(x)$ will be complex. Therefore, in this book we are talking about a special case, but it is a very important special case.

THE INVERSE FOURIER TRANSFORM

Next let us look at the inverse Fourier transform

$$f(t) = \frac{1}{2\pi}\int_{-\infty}^{+\infty} F(j\omega)\, \epsilon^{+j\omega t}\, d\omega \tag{11-7.15}$$

This equation is rewritten with $F(j\omega)$ and $\epsilon^{+j\omega t}$ expanded as

$$F(j\omega) = F_e(j\omega) + F_o(j\omega) \qquad \epsilon^{+j\omega t} = \cos\omega t + j\sin\omega t$$

$$f(t) = \frac{1}{2\pi}\int_{-\infty}^{+\infty} [F_e(j\omega) + F_o(j\omega)]\, [\cos\omega t + j\sin\omega t]\, d\omega \tag{11-7.16}$$

This equation can be expanded into four terms analogous to Eq. 11-7.8, two of which integrate to zero, with the remaining two written as

$$f(t) = \frac{1}{2\pi}\left[2\int_{0}^{\infty} F_e(j\omega)\cos\omega t\, d\omega + 2j\int_{0}^{\infty} F_o(j\omega)\sin\omega t\, d\omega\right] \tag{11-7.17}$$

These integrations are performed with respect to ω, but ω does not appear in the results, which are functions of t. Since $\cos\omega t$ is an even function of t, and $\sin\omega t$ is an odd function of t, these two integrals lead to

$$f_e(t) = \frac{2}{2\pi}\int_{0}^{\infty} F_e(j\omega)\cos\omega t\, d\omega \tag{11-7.18}$$

$$f_o(t) = \frac{2j}{2\pi}\int_{0}^{\infty} F_o(j\omega)\sin\omega t\, d\omega \tag{11-7.19}$$

The $f(t)$ can be written as

$$f(t) = f_e(t) + f_o(t)$$

As we look over these equations, we are able to see additional properties of the Fourier transform.

Property 11-7.16. The Fourier Transform of an Even Function is an Even Function. If $f(t)$ is an even function, Eq. 11-7.11 reduces to

$$F_e(j\omega) = 2\int_0^\infty f_e(t)\cos\omega t\,dt \qquad (11\text{-}7.20)$$

which is the even part of $F(j\omega)$.

If $F(j\omega)$ is an even function, Eq. 11-7.17 reduces to

$$f_e(t) = \frac{2}{2\pi}\int_0^\infty F_e(j\omega)\cos\omega t\,d\omega \qquad (11\text{-}7.21)$$

which is the even part of $f(t)$.

Except for the $1/2\pi$ in front of the inverse transform, these last two equations have exactly the same form with only an exchange of ω and t. As soon as a Fourier transform for an even function is determined, the result can be used in the opposite direction. In Example 10-4.5, the time function

$$f(t) = \begin{cases} 1 & -T/2 < t < +T/2 \\ 0 & \text{elsewhere} \end{cases} \qquad (11\text{-}7.22)$$

was found to have the direct Fourier transform

$$F(j\omega) = T\left[\frac{\sin(\omega T/2)}{\omega T/2}\right] \qquad (11\text{-}7.23)$$

From this, the inverse Fourier Transform of

$$F(j\omega) = \begin{cases} 1 & -\Omega/2 < \omega < +\Omega/2 \\ 0 & \text{elsewhere} \end{cases} \qquad (11\text{-}7.24)$$

can be determined by multiplying Eq. 11-7.23 by $1/2\pi$ and with the appropriate exchange of notation becomes

$$f(t) = \frac{\Omega}{2\pi}\left[\frac{\sin(\Omega t/2)}{\Omega t/2}\right] \qquad (11\text{-}7.25)$$

Property 11-7.17. The Fourier Transform of an Odd Function is an Odd Function. If $f(t)$ is an odd function, Eq. 11-7.11 reduces to

$$F_o(j\omega) = -j2\int_0^\infty f_o(t)\sin\omega t\,dt \qquad (11\text{-}7.26)$$

This shows that the odd part of $F(j\omega)$ is imaginary.

If $F(j\omega)$ is an odd function, Eq. 11-7.17 reduces to

$$f_o(t) = \frac{2j}{2\pi} \int_0^\infty F_o(j\omega) \sin \omega t \, d\omega \tag{11-7.27}$$

The symmetry between Eqs. 11-7.26 and 11-7.27 is not as complete as the symmetry between the equations for $f_e(t)$ and $F_e(j\omega)$. The main reason for this is that $f_o(t)$ is real and $F_o(j\omega)$ is imaginary. If $j F_o(j\omega)$ is considered, the product is real. Equations 11-7.26 and 27 are rewritten featuring this $[j F_o(j\omega)]$ product as

$$[j F_o(j\omega)] = 2 \int_0^\infty f_o(t) \sin \omega t \, dt \tag{11-7.28}$$

$$f_o(t) = \frac{2}{2\pi} \int_0^\infty [j F_o(j\omega)] \sin \omega t \, d\omega \tag{11-7.29}$$

The symmetry between the two equations is placed in evidence and can be used to extend a Fourier transform table, as suggested for the real parts.

Property 11-7.18. The Conditions on $F(j\omega)$ Whereby the Inverse Transform is Real. Based on the last two properties, the inverse Fourier transform of some $F(j\omega)$ will yield a real $f(t)$ if the following two conditions are met.

1 The even part of $F(j\omega)$ is real.
2 The odd part of $F(j\omega)$ is imaginary.

A SPECIAL CASE INVOLVING EVEN AND ODD FUNCTIONS

Before leaving the subject of even and odd functions, there is a special case that should be examined. This situation is an $f(t)$ function that is zero for negative times, as given by

$$f(t) = \begin{cases} 0 & t < 0 \\ f_b(t) & t > 0 \end{cases} \tag{10-7.30}$$

The sketch in Fig. 11-7.1(a) suggests such a function; (b) shows the resulting $f(-t)$ curve, (c) the $f_e(t)$, and (d) the $f_o(t)$.

The $f(t)$ contains both an even part and an odd part.

$$f(t) = f_e(t) + f_o(t) \tag{11-7.31}$$

and the $F(j\omega)$ has the form

$$F(j\omega) = F_e(j\omega) + F_o(j\omega) \tag{11-7.32}$$

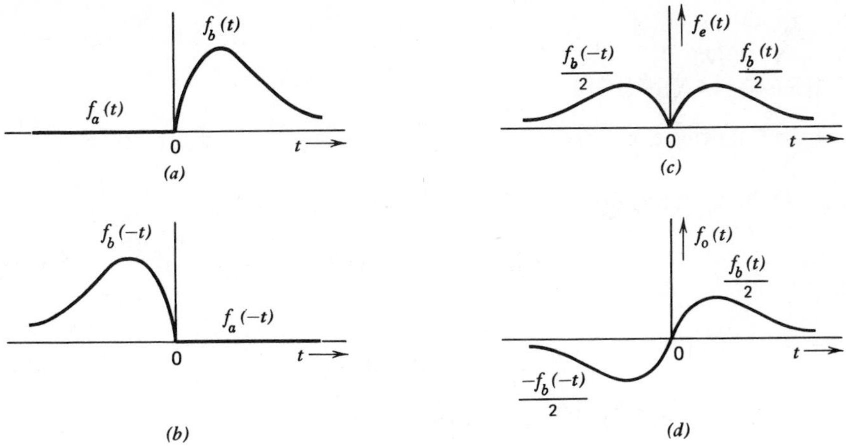

Fig. 11-7.1

where

$$F_e(j\omega) = 2\int_0^\infty \frac{f_b(t)}{2} \cos \omega t \, dt \tag{11-7.33}$$

and

$$F_o(j\omega) = -2j\int_0^\infty \frac{f_b(t)}{2} \sin \omega t \, dt \tag{11-7.34}$$

The inverse transform of these functions yields

$$f_e(t) = \frac{2}{2\pi}\int_0^\infty F_e(j\omega) \cos \omega t \, d\omega \tag{11-7.35}$$

and

$$f_o(t) = \frac{2j}{2\pi}\int_0^\infty F_o(j\omega) \sin \omega t \, d\omega \tag{11-7.36}$$

When $t > 0$

$$f_e(t) = f_o(t) \tag{11-7.37}$$

and when $t < 0$

$$f_e(t) = -f_o(t) \tag{11-7.38}$$

The constraints of these last two equations places restrictions on the $F_e(j\omega)$ and $F_o(j\omega)$, which we shall refer to later.

11-8. ADDITIONAL APPLICATIONS OF THE FOURIER TRANSFORM

Each of the following examples uses one or more of the properties just discussed.

Example 11-8.1. In Example 11-5.1, the Fourier transform of

$$f(t) = 1 \qquad \text{for all } t \tag{11-8.1}$$

was found to be

$$F(j\omega) = 2\pi\, u_0(j\omega) \tag{11-8.2}$$

From this result, we wish to find the Fourier transform of

$$f(t) = \cos \omega_0 t \qquad \text{for all } t \tag{11-8.3}$$

This is done as follows

$$\mathfrak{F}[1 \cos \omega_0 t] = \mathfrak{F}\left[1\left(\frac{\epsilon^{+j\omega_0 t} + \epsilon^{-j\omega_0 t}}{2}\right)\right]$$

$$= 1/2\,\mathfrak{F}[1\epsilon^{+j\omega_0 t}] + 1/2\,\mathfrak{F}[1\epsilon^{-j\omega_0 t}]$$

$$= \pi\{u_0[j(\omega - \omega_0)] + u_0[j(\omega + \omega_0)]\} \tag{11-8.4}$$

The $\mathfrak{F}[1]$ of Eq. 11-8.2 is an example of a Fourier transform that exists in the limit, and, since this was used to find $\mathfrak{F}[\cos \omega_0 t]$, Eq. 11-8.4 is also a transform that exists in the limit, because the $\cos \omega_0 t$ violates the Dirichlet conditions.

Incidentally, Properties 11-5.1, 11-5.2, and 11-5.5 were used in the above development.

Example 11-8.2. In a similar manner, the $\mathfrak{F}[\sin \omega_0 t]$ is found as

$$\mathfrak{F}[\sin \omega_0 t] = \mathfrak{F}\left[\frac{\epsilon^{j\omega_0 t} - \epsilon^{-j\omega_0 t}}{2j}\right] = -j\pi\{u_0[j(\omega - \omega_0)] - u_0[j(\omega + \omega_0)]\} \tag{11-8.5}$$

Again $\sin \omega_0 t$ violates the Dirichlet conditions.

Example 11-8.3. When the transforms just developed are combined with the Fourier series, the Fourier transform for any periodic wave can be found. For example, the square wave shape shown in Fig. 11-8.1(a) has the Fourier series

$$f(t) = 0.5 + 0.638 \cos 15.7t - 0.213 \cos 47.1t + 0.128 \cos 78.5t + \cdots \tag{11-8.6}$$

and the corresponding Fourier transform is

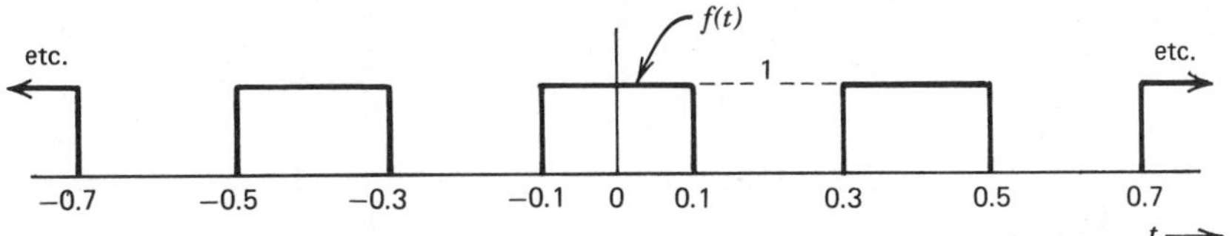

(a)

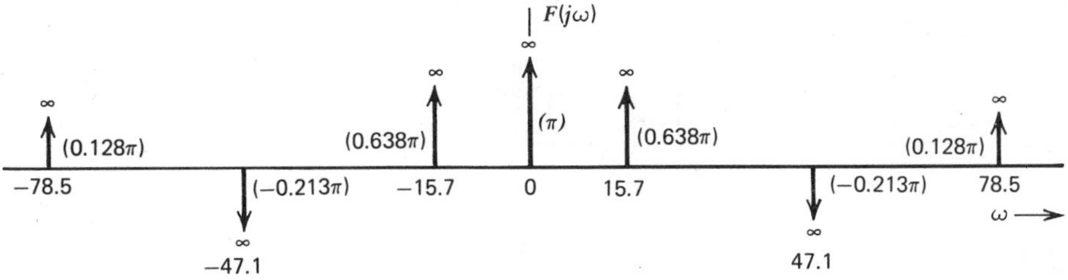

(b)

Fig. 11-8.1

$$F(j\omega) = 2\pi \left\{ 0.5u_0(j\omega) + \frac{0.638}{2}(u_0[j(\omega - 15.7)] + u_0[j(\omega + 15.7)]) - \frac{0.213}{2}(u_0[j(\omega - 47.1)] + u_0[j(\omega + 47.1)]) + \frac{0.128}{2}(u_0[j(\omega - 78.5)] + u_0[j(\omega + 78.5]) + \cdots \right\} \tag{11-8.7}$$

This Fourier transform is shown in Fig. 11-8.1(b).

Example 11-8.4. We wish to find the Fourier transform for an infinite train of unit impulses spaced T seconds apart, as shown in Fig. 11-8.2(a). The symbol $u_T(t)$ is used to represent this infinite train of impulses.

$$u_T(t) = \sum_{n=-\infty}^{+\infty} u_0(t - nT) \tag{11-8.8}$$

One way to proceed is to approximate the train of impulses with the train of pulses shown in Fig. 11-8.2(b). The area under each pulse is unity independent of δ. The Fourier series of this train of pulses can be taken in the conventional manner and a limiting process can be performed on the result by letting $\delta \to 0$. As $\delta \to 0$, the train of pulses in (b) approaches the train of impulses in (a), and the following Fourier series in the limit is obtained.

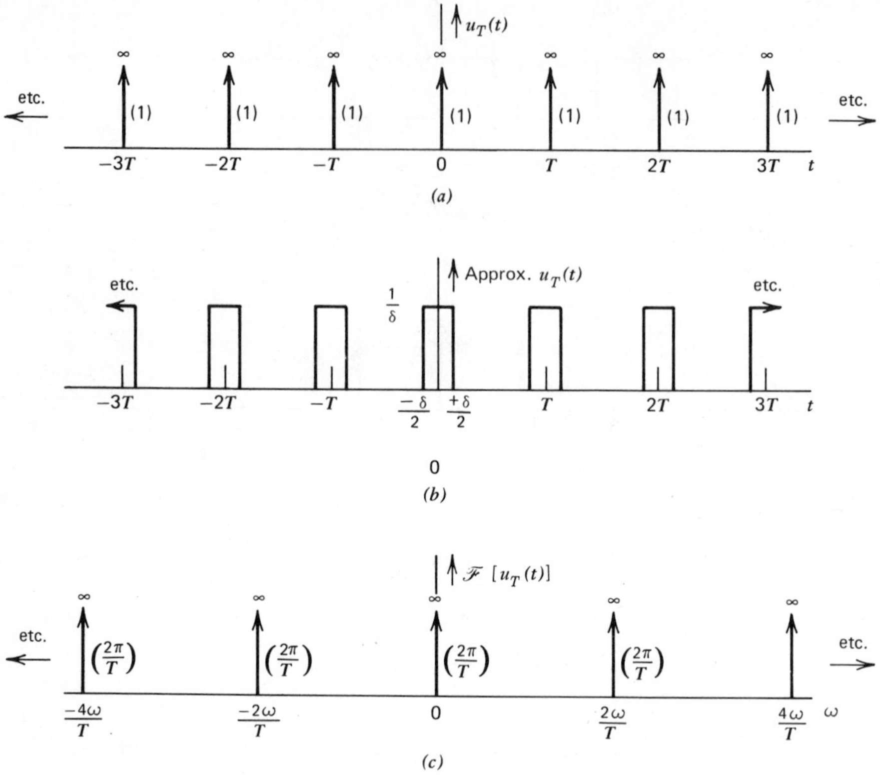

Fig. 11-8.2

$$u_T(t) = \frac{1}{T} + \sum_{n=1}^{\infty} \frac{2}{T} \cos n \frac{2\pi}{T} t \tag{11-8.9}$$

One of the properties usually listed in classical mathematics for a Fourier series is that the coefficient of the nth term approaches zero as n approaches infinity. This property is violated here because this Fourier series is obtained through a limiting process.

The Fourier transform of the d-c component yields an impulse at the origin, and the Fourier transform of each of the cosine terms yields a pair of impulses located a distance $+n2\pi/T$ and $-n2\pi/T$ along the ω-axis. The result is that the Fourier transform of the train of impulses in the time domain is another train of impulses in the frequency domain, as given by

$$\mathcal{F}\,[u_T(t)] = \frac{2\pi}{T} \sum_{n=-\infty}^{+\infty} u_0\left[j\left(\omega - \frac{2\pi n}{T}\right)\right] \tag{11-8.10}$$

This train of impulses is shown in Fig. 11-8.2(c).

As the impulses in the time domain are spaced farther apart (that is, T is increased) the impulses in the frequency domain decrease in weight and move closer together. As the impulses in the time domain move closer together, just the opposite happens.

FREQUENCY CONVOLUTION

The next three examples show the usefulness of the frequency convolution.

Example 11-8.5. We are given a time signal $f(t)$ that has a Fourier transform suggested by the spectrum in Fig. 11-8.3(a). A signal of this type is known as a band limited signal because it has no spectral components outside $\pm\omega_1$. This $f(t)$ is used to form the envelope of a modulated signal, as given by

$$f(t) \cos \omega_0 t \tag{11-8.11}$$

where ω_0 is the radian frequency of the carrier. The spectrum of $\cos \omega_0 t$ is shown in (b).

We want to find the spectrum of the modulated signal. Since the modulated signal is the product of two time functions, we can convolve the spectrum in Fig. 11-8.3(a) with that in (b) to obtain the desired spectrum as shown in (c).

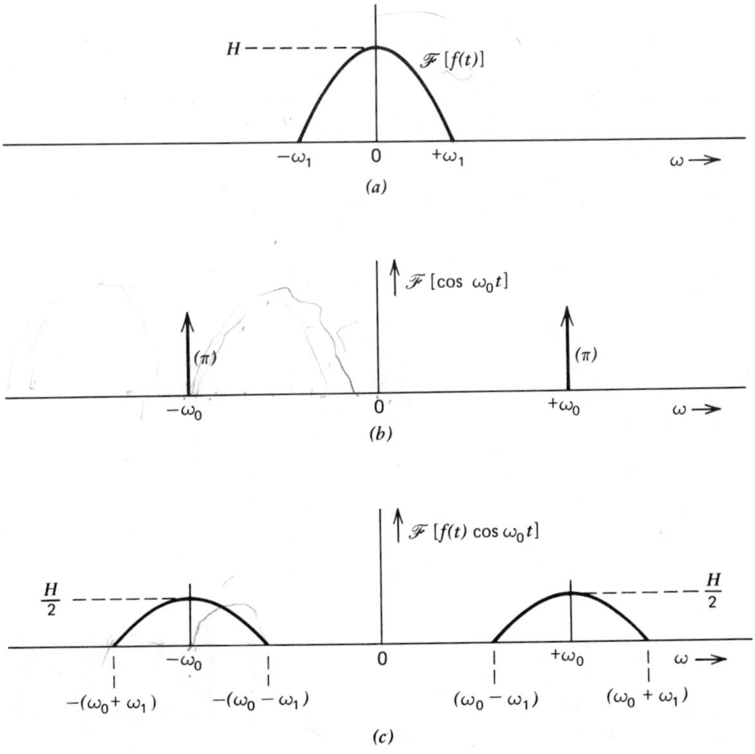

Fig. 11-8.3

The area under each impulse in (b) is equal to π, but the convolution formula has a $1/2\pi$ out in front; hence each lobe in (c) looks exactly like that in (a) except that it is divided by 2 and shifted upward and downward in frequency by an amount equal to ω_0. If $\omega_0 < \omega_1$, these two lobes overlap, and information in the original $f(t)$ becomes scrambled. Therefore, it is impossible to recover the original $f(t)$ by any demodulation procedure.

Example 11-8.6. Next we wish to explore the Fourier transform of

$$f(t) = \begin{cases} \cos \omega_0 t & -\dfrac{T}{2} < t < \dfrac{T}{2} \\ 0 & \text{for all other } t \end{cases} \tag{11-8.12}$$

and to see how the Fourier transform builds up as $T \to \infty$. We note that this $f(t)$ can be thought of as the time product of the $f(t)$ of Ex. 10-4.5 and $\cos \omega t$. For identification purposes we refer to this earlier time function as a gate function, which is defined as

$$f^G(t) = \begin{cases} 1 & -\dfrac{T}{2} < t < \dfrac{T}{2} \\ 0 & \text{for all other } t \end{cases} \tag{11-8.13}$$

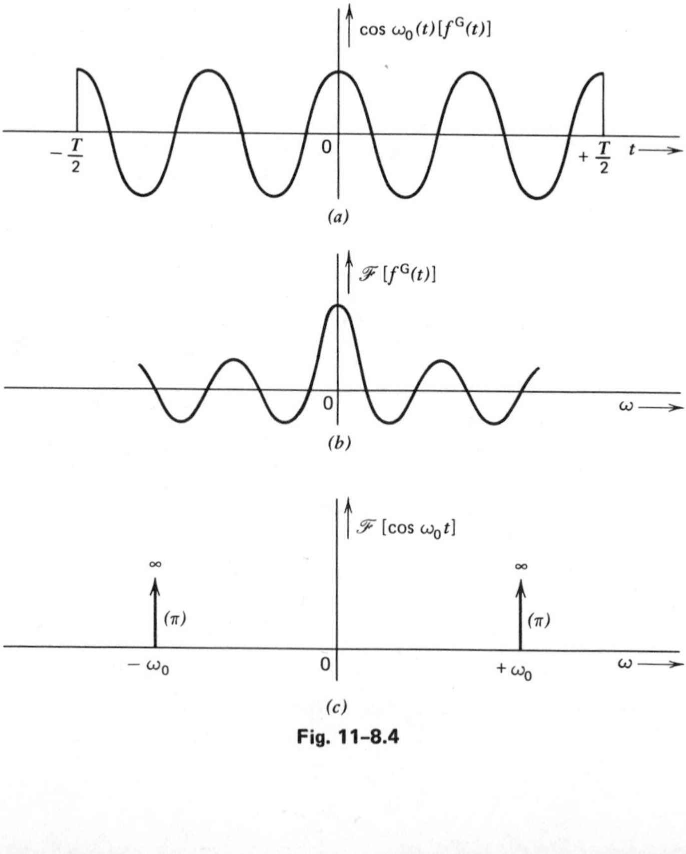

Fig. 11-8.4

The Fourier transform for the gate function is found in Eq. 10–4.21 to be

$$\mathcal{F}\,[f^G(t)] = T\left[\frac{\sin(\omega T/2)}{\omega T/2}\right] \tag{11-8.14}$$

To begin, we choose a T whereby 4 complete periods of the $\cos \omega_0 t$ curve are passed by the gate function, as shown in Fig. 11–8.4(a). In (b) is shown the Fourier transform for the gate function, and in (c), the transform of $\cos \omega_0 t$. The convolutions of these two frequency functions can be visualized easily. The $[f^G(t)]$ curve is multiplied by 1/2 and is shifted to the left by ω_0 due to one of the impulses, and is also shifted to the right by ω_0 due to the other impulse and the two components added. This is not done in the figure but is left for the reader to visualize.

Suppose that T is doubled. The height of the curve in (b) will double, and each crossing occurs at 1/2 the previous ω. The curve in (c) remains the same, but each component in the convolution of (b) and (c) will have twice the height it had before but will ripple at a rate of twice as often.

Suppose that T is doubled again and again and again. It is easy to visualize how the convolution is made up of two pulses, one centered at $-\omega_0$ and the other at $+\omega_0$. As $T \to \infty$, these pulses approach the impulses that are the transform of $\cos \omega_0 t$.

Example 11–8.7. We are given a band limited time signal $f(t)$ that has a Fourier transform suggested by the spectrum of Fig. 11–8.5(a). This $f(t)$ is multiplied by $u_T(t)$, the infinite train of unit-impulses spaced T seconds apart as defined by

$$u_T(t) = \sum_{n=-\infty}^{+\infty} u_0(t - nT) \tag{11-8.15}$$

We wish to find the Fourier transform of the product that can be found by the convolution of the two frequency spectrums. The Fourier transform of $u_T(t)$ was developed in Eq. 11–8.10 as an infinite train of impulses in the frequency domain, and is suggested in Fig. 11–8.5(b). The frequency convolution of the curve in (a) with that of (b) is divided by 2π, as suggested in (c). This is the Fourier transform of the product of the two time functions.

To discuss this example, we examine the product $f(t)\,u_T(t)$. The $u_T(t)$ term is zero at all times, except when an impulse occurs. The value of any particular impulse $u_0(t - nT)$ is modified by the value of $f(t)$ at this time, as given by $f(nT)\,u_0(t - nT)$. Therefore, the product of $f(t)\,u_T(t)$ is also a train of impulses whose values carry information contained in $f(t)$ at the times when the impulses occur. We can say that the $f(t)$ is sampled, and the sampling rate is determined by the period of the impulse train.

If the frequency content of the $f(t)$ signal is limited to the band of frequencies $-\omega_1 < \omega < +\omega_1$, as suggested in Fig. 11–8.5(a), and the sampling rate is high enough, the spectrum of $\mathcal{F}\,[f(t)\,u_T(t)]$, shown in Fig. 11–8.5(c), reproduces the original spectrum over and over again an infinite number of times. However, each reproduction is a faithful portrayal of all the frequencies in the original $f(t)$ signal.

If the sampling frequency is decreased, these separate reproductions of the spectrum of $f(t)$ move closer and closer together. When the sampling frequency is equal to twice

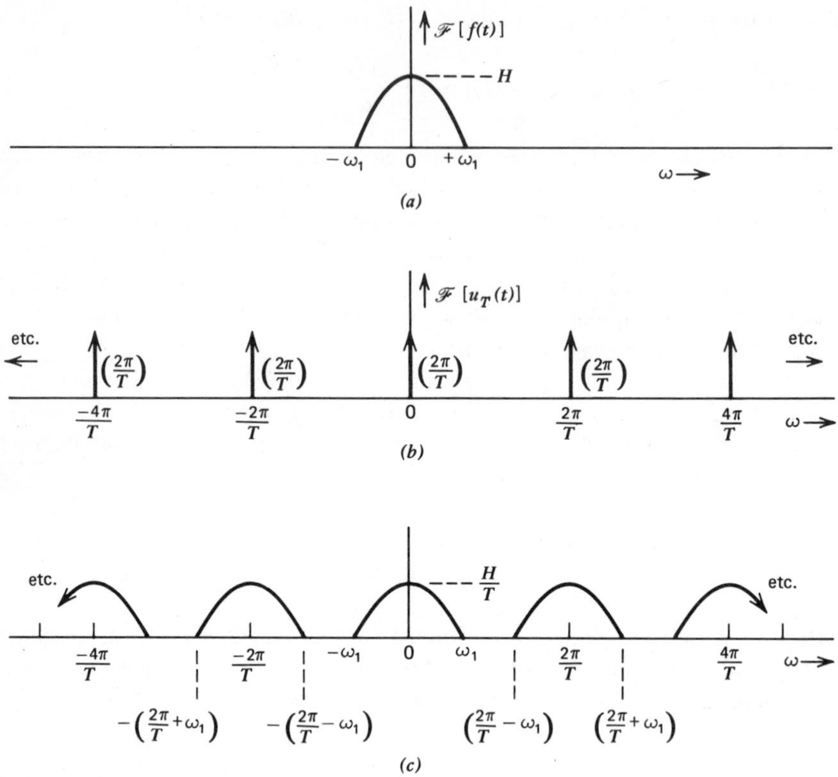

Fig. 11-8.5

the highest frequencies in $f(t)$, these separate reproductions just touch. If the sampling rate is reduced still more, these reproductions overlap, and adjacent reproductions get scrambled together. Therefore, the sampling frequency must be at least twice the highest frequency contained in $f(t)$ in order to be able to regain the information contained in $f(t)$ from the sampled signal. This minimum sampling frequency is called the Nyquist frequency.

11-9. THE IDEAL FILTER

As discussed previously, a spectrum $R(j\omega)$ is applied to a system, and as the signal passes through the system, the spectrum is modified by the filtering action of the system, to produce the output spectrum. If a system is built for the specific purpose of being a filter, then it is designed to pass certain frequencies and to attenuate other frequencies. Examples of filters are mechanical, electrical, pneumatical, and acoustical.

If a person were to design a low-pass filter (as one example), it would seem logical to attempt to design a filter that would pass without attenuation all angular frequencies below a certain cut-off angular frequency ω_c, not allowing to pass any angular frequencies above this value. We use this concept to define the following transfer function $G(j\omega)$ for the ideal low-pass filter.

$$G(j\omega) = \begin{cases} 1 & -\omega_c < \omega < \omega_c \\ 0 & \text{elsewhere} \end{cases} \tag{11-9.1}$$

To explore this situation more fully we assume a unit impulse is applied to the input of this system; that is

$$r(t) = u_0(t) \tag{11-9.2}$$

The Fourier transform of the input is

$$R(j\omega) = 1 \qquad \text{for all } \omega \tag{11-9.3}$$

The $C(j\omega)$ is

$$C(j\omega) = R(j\omega)\,G(j\omega) = \begin{cases} 1 & -\omega_c < \omega < \omega_c \\ 0 & \text{elsewhere} \end{cases} \tag{11-9.4}$$

This function is identical to Eq. 11-7.24, except for a change in notation, and the inverse Fourier transform of the output can be found from Eq. 11-7.25, as

$$c(t) = \frac{\omega_c}{\pi}\left[\frac{\sin \omega_c t}{\omega_c t}\right] \tag{11-9.5}$$

A plot of the input and the output are shown in Fig. 11-9.1(a) and (b), respectively.

It doesn't take a very observant person to react rather quickly to these results. An impulse is applied to this ideal filter at $t = 0$, and the Fourier transform tells us that the result is an output that begins before $t = 0$. How can this be? We know that in the real world this cannot happen! When we dreamed up the characteristics of the ideal filter as

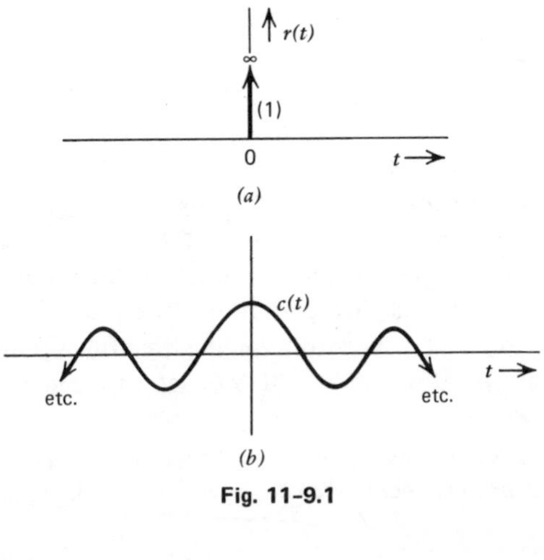

Fig. 11-9.1

given in Eq. 11-9.1, we were unknowingly proposing a system that cannot be built. We say that such a system is not a physically realizable system.

THE CAUSALITY CONDITION

The subject of physical realizability is a very broad subject, with a vast literature of its own. Here we are talking about only one aspect of the subject. We use an heuristic argument and say that a physically realizable system cannot have an output response before the input is applied. This particular type of system is said to be a causal system, that is, it meets the causality condition. The ideal filter just discussed is a non-causal system.

THE PALEY-WIENER CRITERION

If an impulse is applied to a physically realizable system at $t = 0$, the impulse response of this system must be zero before $t = 0$. This is the time domain method of talking about a system that meets the causality condition. Next, we wish to examine how this same condition can be expressed in the frequency domain.

Toward the end of Section 11-7, we discussed the Fourier transform of a time signal that was zero for negative time, as defined by Eq. 11-7.30. Equations 11-7.35, 36, 37 and 38 were proposed at that time, and these equations impose constraints on the $G_e(j\omega)$ and $G_o(j\omega)$. These constraints have been interpreted into the well-known Paley–Wiener[1] criterion, as given by

$$\int_{-\infty}^{+\infty} \frac{|\ln |G(j\omega)||}{1+\omega^2}\,d\omega < \infty \tag{11-9.6}$$

If this condition is satisfied, the corresponding system is a causal system.

Let us examine this equation and attempt to develop a feeling for its significance. Suppose a $G(j\omega)$ has a value of zero at some $j\omega = j\omega_1$ as given by

$$|G(j\omega_1)| = 0 \tag{11-9.7}$$

then

$$|\ln |G(j\omega_1)|| = |\ln 0| = \infty \tag{11-9.8}$$

However, it is still possible for the integral of Eq. 11-9.7 to be finite. Next, let us suppose that the $|G(j\omega)|$ is zero over some band of frequencies, no matter how small. The integral of Eq. 11-9.6 will then be infinite. We see from this that the $|G(j\omega)|$ may be zero at some set of discrete frequencies, but it cannot be zero over a band of frequencies, no matter how narrow. The ideal low-pass filter of Eq. 11-9.1 certainly violates the Paley–Wiener criterion, and hence the corresponding time function is non-causal.

To continue our examination of Eq. 11-9.6, let us assume that the shape of the

[1] Raymond E. A. C. Paley and Norbart Wiener, "Fourier Transforms in the Complex Domain," *Americal Mathematical Society Colloquium Publication 19*, New York, 1934.

$|G(j\omega)|$ curve behaves according to some such function as

$$|G(j\omega)| = C\,\epsilon^{-Y^2} \tag{11-9.9}$$

When this is substituted into Eq. 11-9.6, the result is infinite. If this line of reasoning is pursued farther, it can be shown that if $|G(j\omega)|$ falls to zero faster than a function of exponential order, the Paley-Wiener criterion cannot be met.

THE APPROXIMATION PROBLEM

Many times in this book we are very close to an important concept that is known as *the approximation problem*. It doesn't seem appropriate to go into this problem in great depth in a book of this sort, but a few comments are in order. The ideal filter is excellent for a discussion of this concept.

A designer has, for example, the job of building a low-pass filter. He speculates as to how to do this and defines the filtering characteristic of the ideal filter. As he explores the situation, he discovers that it is physically impossible to build a filter with the characteristics he desires. He cannot stop and say that he is through, because the low-pass filter is still needed. One way he can proceed is to approximate the ideal filter characteristics in such a way that the physical realizability conditions are met by these modified characteristics. Then he designs a filter that will achieve these modified objectives. There is a very extensive literature on the approximation problem, because it is such an extremely important subject to the designer.

CONCLUSION, CHAPTERS 9, 10, AND 11

We have discussed the relationships among the one-sided Laplace transform, the two-sided Laplace transform, the Fourier transform, and the Fourier series transform. Since this coverage is intended to be only an introduction, many of the finer points and the more mathematical developments are by-passed. A physical interpretation has been given when possible, and simple examples have been used to demonstrate the broad principles involved.

There are many subjects so very close to those discussed that it is tempting to put them all in. Such things include the energy density spectrum, correlation functions, statistical methods, noise, signal-to-noise ratio, various forms of modulation, and so on.

The material covered can also lead toward many different related fields of study, among those of circuit theory, control system theory, and communication theory. If a person becomes interested in any one of these, he should consult a book written for the specific area.

PROBLEMS

11-1 Find a_n and b_n using Eqs. 11-2.9 and 10, respectively, and write out the Fourier series in the form of Eq. 11-2.3 using four non-zero terms.

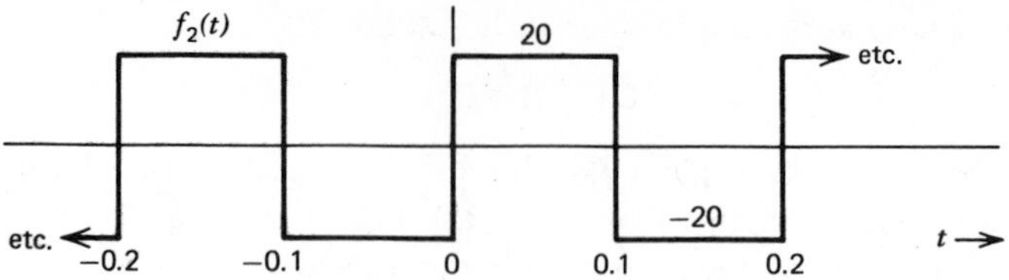

(a)

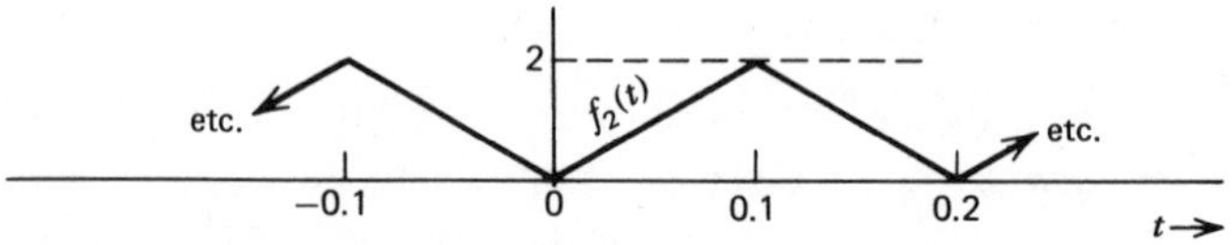

(b)

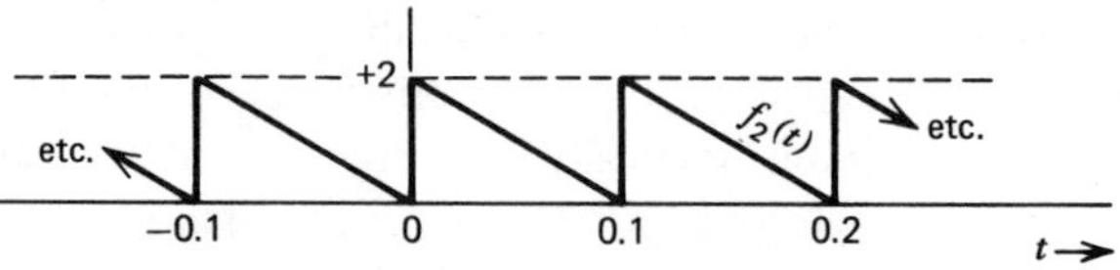

(c)

11-2 Use the time functions of Problem 11-1, except find $F_2(jn\omega_1)$, using Eq. 11-2.20. Write $f_2(t)$ as given in Eq. 11-2.17 out to eight non-zero terms, then combine terms and write the result in the form of Eq. 11-2.3.

11-3 In Example 11-2.1, t_0 and T_1 take on the specific values as given. Sketch a curve similar to the one in Fig. 11-2.2, and use the curve to explain why certain harmonics are zero.

(a) $t_0 = 0.1$ and $T_1 = 0.2$
(b) $t_0 = 0.1$ and $T_1 = 0.4$
(c) $t_0 = 0.1$ and $T_1 = 0.8$

11-4 Use the $f_2(t)$ functions given in Problem 10-1. If $f_2(t)$ does not satisfy the Dirichlet conditions, see if you can find a converging factor whereby $\epsilon^{-\sigma t} f_2(t)$ does. If no such factor exists, explain why. If such a factor does exist, find the Fourier transform of $[\epsilon^{-\sigma t} f_2(t)]$.

11-5 Complete Example 11-4.1 by writing the Fourier series in the form of Eq. 11-2.3.

11-6 Repeat Problem 11-5 for Example 11-4.2.

11-7 Repeat Problem 11-5 for the functions of Problem 11-3.

11-8 Repeat Problem 11-3 for the wave shape shown.

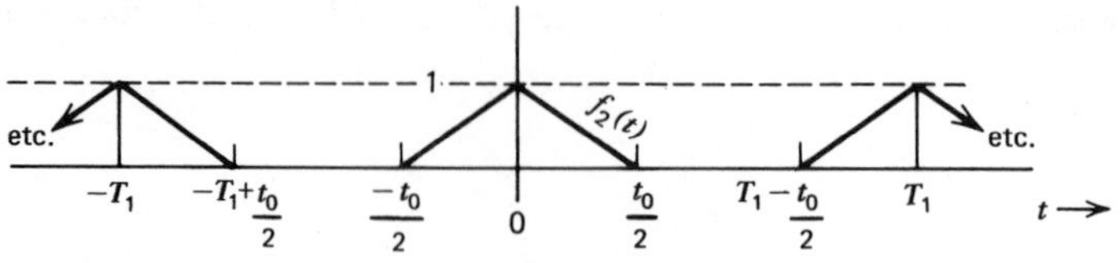

11-9 Repeat Problem 11-5 for the functions of Problem 11-8.

11-10 Develop a proof for the properties as assigned for the two-sided Laplace transform, the Fourier transform, and the Fourier series transforms. Discuss any limitation the property may have.

(a) Property 11-5.1.
(b) Property 11-5.2.
(c) Property 11-5.3.
(d) Property 11-5.4.
(e) Property 11-5.5.
(f) Property 11-5.7.
(g) Property 11-5.9.
(h) Property 11-5.12.
(i) Property 11-5.13.

11-11 Use the $f_2(t)$ functions given in Problem 10-1. Find the $\int_{-\infty}^{t} f_2(t)\,dt$ by three different methods: (1) use calculus; (2) use Property 11-5.4 for the two-sided Laplace transform; and (3) use Property 11-5.4 for the Fourier transform. If any of these functions does not exist, explain why.

11-12 The proof of Property 11-5.8 for the Fourier transform is complex. Complete the proof in the following manner. Given $\mathcal{L}_2[f(t)] = F_2(s)$ and $\mathcal{L}_2[f^{-n}(t)] = F_2(s)/s^n$: Make the expansion

$$\frac{F_2(s)}{s^n} = \frac{C_n}{s^n} + \frac{C_{n-1}}{s^{n-1}} + \cdots$$

then take the inverse Laplace transform. Finally, use the $F_2(j\omega)$ of the 1st Family of Functions given in Table 10-5.1 to find the Fourier transform. Discuss the limitation of the method.

11-13 Note that the function in Problem 11-1(a) is the derivative of the function in (b). Verify that Property 11-5.3, for the Fourier series transform, holds for this case.

11-14 Find the Fourier transforms and sketch for the Fourier series given in:

(a) Problem 2-35.
(b) Problem 2-36.
(c) Problem 2-37.
(d) Problem 2-38.

11-15 Find the Fourier series for the function given in Problem 11-1(a) in the following manner. The derivative of this function leads to two trains of impulses. Use Eq. 11-8.9 for each train, and then integrate the result one time.

11-16 Repeat Problem 11-15, except find the Fourier series for the function given in Problem 11-1(b) by taking the derivative two times and integrating the result two times.

11-17 Repeat Problem 11-15, except perform the necessary steps to find the Fourier series for the function of Problem 11-1(c).

11-18 Given an $f_2(t)$

$$f_2(t) = \begin{cases} 0 & t < 0 \\ 1\epsilon^{-2t} & t > 0 \end{cases}$$

find the even and odd parts of this $f_2(t)$.

11-19 Repeat Problem 11-18 for the function shown.

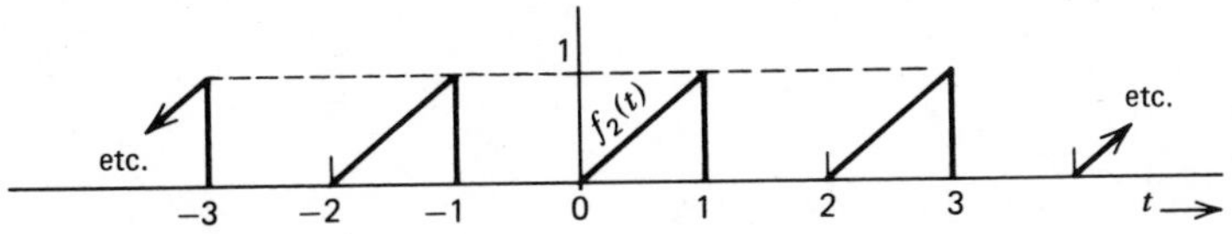

11-20 For each of the three values of T, draw a rough sketch for the convolution of the two frequency functions, as discussed in Example 11-8.6.

11-21 Draw rough sketches of the Fourier transform of

$$[\cos \omega_0 t]\, u_{-1}(t)$$

by use of the convolution of the two frequency spectrums.

11-22 Repeat Problem 11-21 for

$$[\sin \omega_0 t]\, u_{-1}(t)$$

Chapter **12**

Signal-Flow Graphs

12-1. INTRODUCTION

Circuit and system designers usually develop the habit of thinking in terms of input signals and output signals, and the manner in which these signals are transmitted through a system or a portion of a system. Signal flow is thus a natural concept that can be used in analyzing the behavior of a system.

The signal-flow graph is a representation of a system in a topological form that features the interconnective aspects of the system. Signal-flow graphs and block diagrams are very closely related and have essentially the same purpose.

Historically, the block diagram representation was developed and used before the advent of the signal-flow graph. However, the signal-flow graph was developed with a more uniform notation and with a graphical representation that is more convenient to draw and easier to visualize. Also the workers in the field of signal-flow graphs developed more elegant methods for manipulating and reducing the graphs than were available for block diagrams. Although it is true that these methods of reduction can be applied to block diagrams, it seems more fitting to study them in their original setting.

The block diagram shown in Fig. 12-1.1 is similar to those developed and used in Chapters 7 and 8. It is based on notation originally developed in control system theory.

The block may represent an overall system or only a single component or element, depending on the detail desired. Signal $r(t)$ is the reference input signal, $g(t)$ is the impulse response of the element or elements being considered, and $c(t)$ is the controlled quantity, all represented as time functions. $R(s)$, $G(s)$, and $C(s)$ are their respective Laplace transforms. (It is also possible to use two-sided Laplace transforms, $R_2(s)$ and $C_2(s)$.)

As discussed several times in this book, the output $c(t)$ of the block can be found in two different ways. The transform of the output is

$$C(s) = R(s)\,G(s) \tag{12-1.1}$$

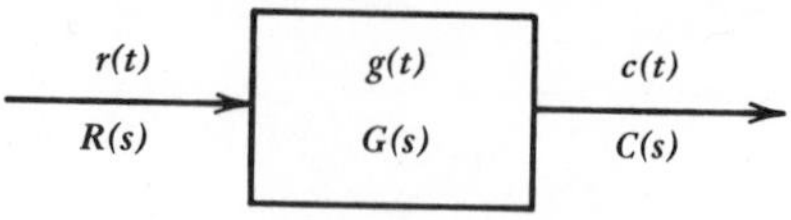

Fig. 12-1.1

and $c(t)$ is the inverse transform of $C(s)$. The $c(t)$ can also be found by use of the convolution integral, as

$$c(t) = \int_{-\infty}^{\infty} r(t - \tau)\, g(\tau)\, d\tau \tag{12-1.2}$$

The convenience of manipulating the transformed quantities makes this method very attractive so far as the signal-flow diagram is concerned, and for the most part transformed quantities are used in this chapter.

For a starting point in developing signal-flow graphs, we assume a system that can be described by a set of linear differential equations with constant coefficients. When these are transformed, the result is a set of simultaneous algebraic equations that describe the system. These equations can be arranged in a particular form, as in

$$\begin{aligned}
X_1(s) &= T_{11}(s)\,X_1(s) + T_{12}(s)\,X_2(s) + \cdots + T_{1n}(s)\,X_n(s) \\
&\quad + B_{11}(s)\,U_1(s) + \cdots + B_{1m}(s)\,U_m(s) \\
X_2(s) &= T_{21}(s)\,X_1(s) + T_{22}(s)\,X_2(s) + \cdots + T_{2n}(s)\,X_n(s) \\
&\quad + B_{21}(s)\,U_1(s) + \cdots + B_{2m}(s)\,U_m(s) \\
&\;\;\vdots \\
X_n(s) &= T_{n1}(s)\,X_1(s) + T_{n2}(s)\,X_2(s) + \cdots + T_{nn}(s)\,X_n(s) \\
&\quad + B_{n1}(s)\,U_1(s) + \cdots + B_{nm}(s)\,U_m(s)
\end{aligned} \tag{12-1.3}$$

The reason for this form is not at all clear at this time, but will be developed as the discussion proceeds. The symbol $U_i(s)$ is used to represent the transform of the ith input, and the notation suggests there are m such inputs. The inputs are sometimes referred to as independent variables. The symbol $X_j(s)$ is used to represent the transform of the jth dependent variable, and the notation suggests there are n such variables. In signal-flow graphs, the independent and the dependent variables are represented by *nodes*. *Branches* with transfer functions connect these nodes and show signal-flow paths.

Two dependent *nodes* and the *branch* connecting the nodes are shown in Fig. 12-1.2. The arrow on the branch indicates the direction of signal flow, with the transfer function placed adjacent to the branch. In the double subscript on the transfer function, the first subscript represents the output, and the second the input.

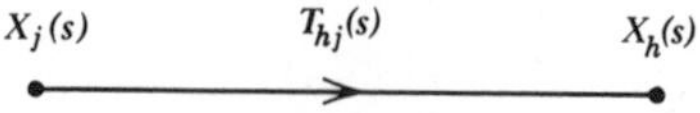

Fig. 12-1.2

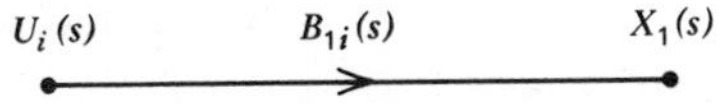

Fig. 12-1.3

$$T_{hj}(s) = \frac{X_h(s)}{X_j(s)} \tag{12-1.4}$$

When multiplied through, Eq. 12-1.4 becomes

$$X_h(s) = T_{hj}(s)\, X_j(s) \tag{12-1.5}$$

A physical interpretation can be given to this last equation as follows. As the signal $X_j(s)$ travels through the transfer function, it is modified into a new signal whose Laplace Transform is the product $T_{hj}(s)\, X_j(s)$. For this reason, the $T_{hj}(s)$ in signal-flow terminology is referred to as a *transmittance*.

One independent node, one dependent node, and the corresponding transmittance are shown in Fig. 12-1.3. Some authors do not distinguish between the transmittances of Figs. 12-1.2 and 12-1.3, and use the symbol T for both, but we find it convenient to make this distinction. The transform of the signal at the $X_1(s)$ node is given by

$$X_1(s) = B_{1i}(s)\, U_i(s) \tag{12-1.6}$$

Although we take up the basic details and definitions in the next section, it can be seen from the present discussion that the signal-flow graph uses *nodes* to represent both dependent and independent *variables*, *directed branches* to represent signal *transmission paths*, with the *transfer function* written beside the branch to *describe* the *effect* of passing a signal through a given branch in the direction of the arrow.

With these thoughts in mind, we return to examine Eqs. 12-1.3. The first equation expresses the dependent variable $X_1(s)$ as a linear combination of signals transmitted from all independent and dependent variable nodes. The transmittances $T_{12}(s)$ and $B_{13}(s)$ follow the notation suggested in Figs. 12-1.2 and 12-1.3, respectively. The second of Eqs. 12-1.3 expresses the dependent variable $X_2(s)$ in a similar manner. We take up the details of drawing and interpreting signal-flow graphs in the next section.

In summary, signal-flow graphs are graphical structures depicting the interrelations of dependent and independent variables in an actual system or, more correctly, in a mathematical model of an actual system. In reality the term *signal* is an engineering addition to associate a pure mathematical tool with a field of application. It should be pointed out that signal-flow graphs are not unique, and that any one of several forms of the signal-flow graph can represent a system or model, just as there are many ways to perform an elimination of dependent variables in solving Eq. 12-1.3 for a specific dependent variable.

12-2. BASIC DETAILS AND DEFINITIONS

As pointed out in Section 12-1, simple nodes, directed branches, and transmittances are fundamental tools used in signal-flow graphs. Even the simple one-branch, two-node graphs of Figs. 12-1.2 and 12-1.3 can serve to help define the terms *source node* and *sink node*. *Source nodes* represent *independent input* signals or variables and, as such, *have no branches coming into them*. However, they may supply signals to several nodes. Node $U_i(s)$ in Fig. 12-1.3 is a source node and has only branches leaving it. *Sink nodes*, on the other hand, are totally dependent on other variables or signals and *have only entering branches*. Node $X_1(s)$ in Fig. 12-1.3 is a sink node that also makes it evident that sink nodes represent system output variables.

We need to discuss two more rules before we will be ready to proceed with examples.

The Incoming Branch Rule. The value of the variable represented by a node is equal to the sum of all the signals entering that node.

The Outgoing Branch Rule. The value of the variable represented by a node is transmitted along all branches leaving that node.

Example 12-2.1. Fig. 12-2.1 is used for the following discussion. The $U_1(s)$ and $U_2(s)$ are source nodes, and the $X_3(s)$, $X_4(s)$, and $X_5(s)$ are sink nodes. Each of the signals $U_1(s)$ and $U_2(s)$ is transmitted along two branches. The value of the $X_1(s)$ node is equal to the sum of the two incoming signals, hence the equation

$$X_1(s) = B_{11}(s)\,U_1(s) + B_{12}(s)\,U_2(s) \tag{12-2.1}$$

The value of the $X_1(s)$ signal is transmitted along the two branches leaving the $X_1(s)$ node, hence the equations

$$\begin{aligned} X_3(s) &= T_{31}(s)\,X_1(s) \\ X_4(s) &= T_{41}(s)\,X_1(s) \end{aligned} \tag{12-2.2}$$

The following equations are obtained in a similar manner.

$$\begin{aligned} X_2(s) &= B_{21}(s)\,U_1(s) + B_{22}(s)\,U_2(s) \\ X_5(s) &= T_{52}(s)\,X_2(s) \end{aligned} \tag{12-2.3}$$

Here in Section 12-2, we are working from a given signal-flow graph and obtain the equations the graph represents. In the next, Section 12-3, we develop the graph from given equations.

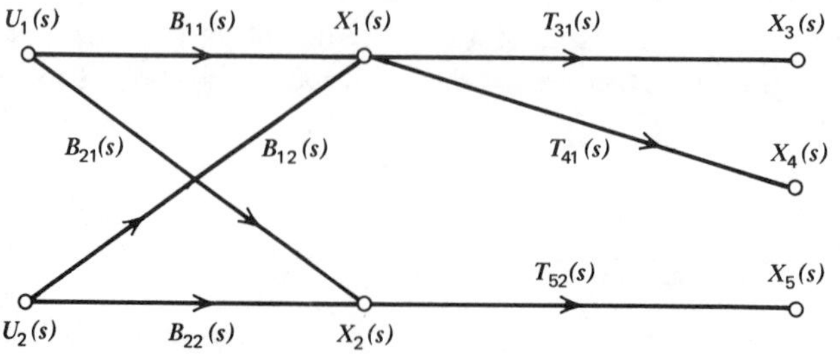

Fig. 12-2.1

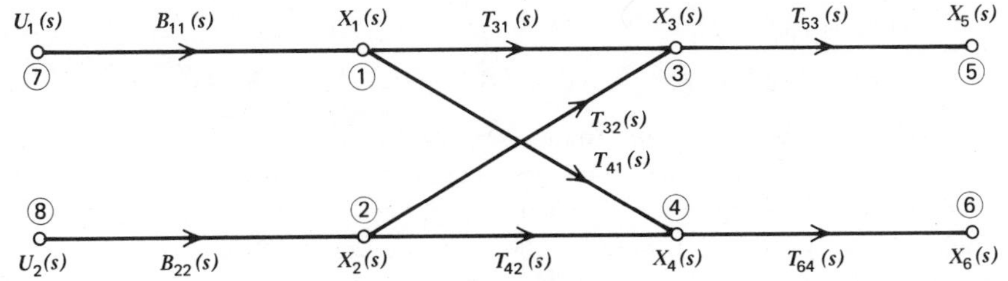

Fig. 12-2.2

Example 12-2.2. As a second example, discussed in more complete detail, we use the signal-flow graph of Fig. 12-2.2. The variable represented by each node is labeled. In addition to this, the nodes are numbered for convenient reference. The node numbering is usually done in an arbitrary manner, with no real significance to the way it is done.

The corresponding set of equations is

$$\begin{aligned} X_1(s) &= B_{11}(s)\,U_1(s) \\ X_2(s) &= B_{22}(s)\,U_2(s) \\ X_3(s) &= T_{31}(s)\,X_1(s) + T_{32}(s)\,X_2(s) \\ X_4(s) &= T_{41}(s)\,X_1(s) + T_{42}(s)\,X_2(s) \\ X_5(s) &= T_{53}(s)\,X_3(s) \\ X_6(s) &= T_{64}(s)\,X_4(s) \end{aligned} \tag{12-2.4}$$

The resulting form of these equations should be compared with the form of Eqs. 12-1.3. The same form results, with a number of coefficients being zero.

In general, a node may have several branches entering and several branches leaving as depicted for the node representing dependent signal $X_k(s)$ in Fig. 12-2.3. For the case

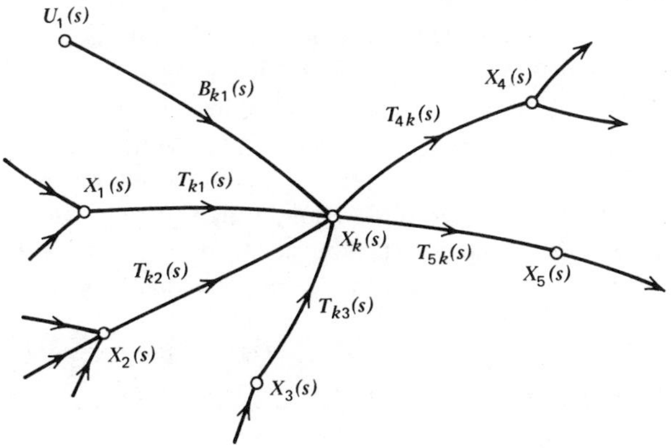

Fig. 12-2.3

considered, $X_k(s)$ is given by

$$X_k(s) = T_{k1}(s)X_1(s) + T_{k2}(s)X_2(s) + T_{k3}(s)X_3(s) + B_{k1}(s)U_1(s) \qquad (12\text{-}2.5)$$

The various outgoing branches from the $X_k(s)$ node do not enter the equation for a node signal value unless one or more outgoing branches return directly to their originating node, forming so-called *self-loops*. In some sense, nodes behave as if they could supply signals to an arbitrary number of outgoing paths without *loading* effects.

To further illustrate the form of the system equations resulting for the signal-flow graph of Fig. 12-2.2, the equations are arranged in the form of Eqs. 12-2.6, with matrices used to make the systematic arrangement more evident.

$$\begin{bmatrix} 1 & 0 & 0 & 0 & 0 & 0 \\ 0 & 1 & 0 & 0 & 0 & 0 \\ -T_{31} & -T_{32} & 1 & 0 & 0 & 0 \\ -T_{41} & -T_{42} & 0 & 1 & 0 & 0 \\ 0 & 0 & -T_{53} & 0 & 1 & 0 \\ 0 & 0 & 0 & -T_{64} & 0 & 1 \end{bmatrix} \begin{bmatrix} X_1 \\ X_2 \\ X_3 \\ X_4 \\ X_5 \\ X_6 \end{bmatrix} = \begin{bmatrix} B_{11} & 0 \\ 0 & B_{22} \\ 0 & 0 \\ 0 & 0 \\ 0 & 0 \\ 0 & 0 \end{bmatrix} \begin{bmatrix} U_1 \\ U_2 \end{bmatrix} \qquad (12\text{-}2.6)$$

A general notation that can represent any signal-flow graph is

$$[T(s)]\,X(s)] = [B(s)]\,U(s)] \qquad (12\text{-}2.7)$$

We shall return to this representation several times.

The $[T(s)]$ matrix of Eq. 12-2.6 is a very special type, and is said to be in *lower triangular form*. This name results from the fact that all elements above the principal diagonal are zeros.

If a matrix is in the triangular form, the value of the $X(s)$'s can be found without inverting the matrix. In lower triangular form, $X_1(s)$ can be found from the first equation; with $X_1(s)$ known, $X_2(s)$ can be found from the second equation; with $X_1(s)$ and $X_2(s)$ known, $X_3(s)$ can be found from the third equation, and so on.

The reason the $[T(s)]$ for the graph of Fig. 12-2.2 is in triangular form is because of the simplicity of the original signal-flow graph and the special nature of this graph; namely, it is what is called a *cascade graph*. Such graphs have no loops. A loop exists when you can travel along the branches in the direction of the arrows and return to the starting node. As the name implies, cascade signal-flow graphs contain only direct *series* paths from inputs toward outputs. Such cascade graphs can be reduced to simple input-output relations by direct sequential substitution, using equations such as Eqs. 12-2.4. In this case, the equations for X_1 and X_2 can be substituted into the equations for X_3 and X_4, which in turn can be substituted into the X_5 and X_6 equations, to give

$$\begin{aligned} X_5(s) &= B_{11}(s)T_{31}(s)T_{53}(s)U_1(s) + B_{22}(s)T_{32}(s)T_{53}(s)U_2(s) \\ X_6(s) &= B_{11}(s)T_{41}(s)T_{64}(s)U_1(s) + B_{22}(s)T_{42}(s)T_{64}(s)U_2(s) \end{aligned} \qquad (12\text{-}2.8)$$

These same results can be determined, tracing *paths* of signal flow through the graph, by moving in the direction of signal flow along a set of branches in sequence and multiplying the transmittances encountered in tracing these paths. For example, the equation for $X_5(s)$ in Eqs. 12-2.8 can be obtained by starting at the $U_1(s)$ node and moving along a path through the transmittances $B_{11}(s)$, $T_{31}(s)$, and $T_{53}(s)$. Then, starting again at the $U_2(s)$ node and moving along a path through the transmittances $B_{22}(s)$, $T_{32}(s)$ and $T_{53}(s)$, etc., we obtain the second term. These two results are added, using the principal of superposition to obtain $X_5(s)$. The equation for $X_6(s)$ can be obtained in a similar manner.

Not all signal-flow graphs are cascade graphs, because many graphs contain a loop or a set of loops. As stated earlier, a loop exists when one can travel along the branches in the direction of the signal flow and encounter a node a second time.

When loops exist in a graph, the computation of outputs is modified from the simple sequential substitution procedure outlined above, and the signal-flow graph is then called a *feedback* graph.

Example 12-2.3. Figure 12-2.4 is an example of the feedback graph. There are two loops in this graph. Transmittance T_{11} is the gain for what is called a *self-loop*, while $T_{41}T_{24}T_{32}T_{13}$ is the path gain or transmission for a second loop. The reader should note that this signal-flow graph is *not* the same as that of Fig. 12-2.2. Equations of the form of Eqs. 12-2.4 and 6 for this graph are

$$\begin{aligned}
X_1(s) &= T_{11}(s)X_1(s) + T_{13}(s)X_3(s) + B_{11}(s)U_1(s)\\
X_2(s) &= T_{24}(s)X_4(s) + B_{22}(s)U_2(s)\\
X_3(s) &= T_{32}(s)X_2(s)\\
X_4(s) &= T_{41}(s)X_1(s)\\
X_5(s) &= T_{53}(s)X_3(s)\\
X_6(s) &= T_{64}(s)X_4(s)
\end{aligned} \tag{12-2.9}$$

$$\begin{bmatrix} (1-T_{11}) & 0 & -T_{13} & 0 & 0 & 0\\ 0 & 1 & 0 & -T_{24} & 0 & 0\\ 0 & -T_{32} & 1 & 0 & 0 & 0\\ -T_{41} & 0 & 0 & 1 & 0 & 0\\ 0 & 0 & -T_{53} & 0 & 1 & 0\\ 0 & 0 & 0 & -T_{64} & 0 & 1 \end{bmatrix} \begin{bmatrix} X_1\\ X_2\\ X_3\\ X_4\\ X_5\\ X_6 \end{bmatrix} = \begin{bmatrix} B_{11} & 0\\ 0 & B_{22}\\ 0 & 0\\ 0 & 0\\ 0 & 0\\ 0 & 0 \end{bmatrix} \begin{bmatrix} U_1\\ U_2 \end{bmatrix} \tag{12-2.10}$$

The difference in form of these equations from those of Eqs. 12-2.4 and 6 should be noted. More than just a direct substitution is required to eliminate X_1 from the remaining equations. The matrix of system coefficients does not appear in a triangular form. Also, it is not possible to rearrange these equations so that $[T(s)]$ is in triangular form. The self-loop with loop gain T_{11} makes itself evident by appearing in the 1-1 diagonal element of the coefficient matrix of Eq. 12-2.10. While the solution of the above equations for the output signals $X_5(s)$, $X_6(s)$ is still elementary, the presence of loops does cause some change in the solution algorithm.

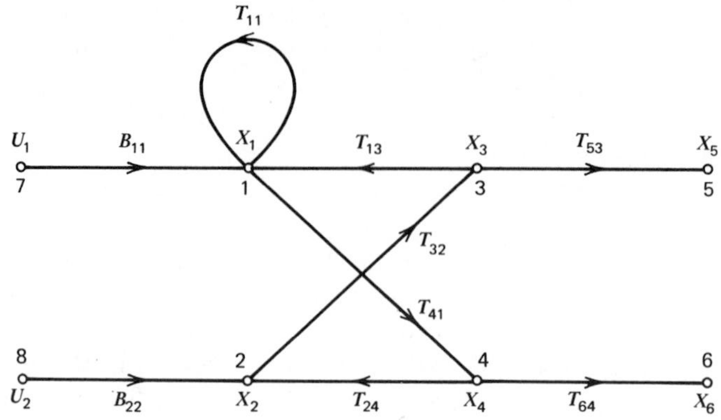

Fig. 12-2.4

We have been very careful in presenting the meaning of the term *loop*; now let us be equally careful about how we will use the term *path*. A *path* is a route along which one can trace the direction of signal flow from an input node to an output node *without* involving any loops in the process. For example, let us return to Fig. 12-2.4 and determine the path or paths from input U_1 to output X_5. We start at U_1 and have only one choice so we proceed to X_1. There are two transmittances leaving X_1, but if we choose the T_{11} branch we return to X_1 and are involved in a loop. Therefore, we follow branch T_{41} to X_4, and again have two choices. Branch T_{64} does not achieve our objective of reaching X_5, so we cannot choose this path; instead we move along T_{24} to X_2, and then along T_{32} to X_3. Leaving X_3 we again have two choices. The branch T_{13} takes us back to X_1 and involves a loop which we do not want; this means we travel along T_{53} and finally complete our mission. An inspection of the graph reveals only one *path* from U_1 to X_5.

To be completely explicit, we use small p to indicate the path itself, and capital P to indicate the product of the transmittances encountered in traveling the path. In the example just above, p is defined by the sequence of numbered nodes as

$$p = (7, 1, 4, 2, 3, 5) \tag{12-2.11}$$

whereas P is

$$P = B_{11}T_{41}T_{24}T_{32}T_{53} \tag{12-2.12}$$

If only the branches and nodes involved in the two loops of Fig. 12-2.4 are retained in a new graph, the result is called a *loop subgraph*. It contains a subset of the original set of branches and nodes. The concept of loops in signal-flow graphs is extremely important as will become more evident when the various methods for determining overall transmissions and other properties are considered later. Fig. 12-2.5(a) depicts this loop subgraph. We use small l to indicate the loop itself, and capital L to indicate the product of the transmittances encountered in traversing the loop. For example in Fig. 12-2.5(a), l_2 is defined as a sequence of numbered nodes, as

$$l_2 = (1, 4, 2, 3, 1) \tag{12-2.13}$$

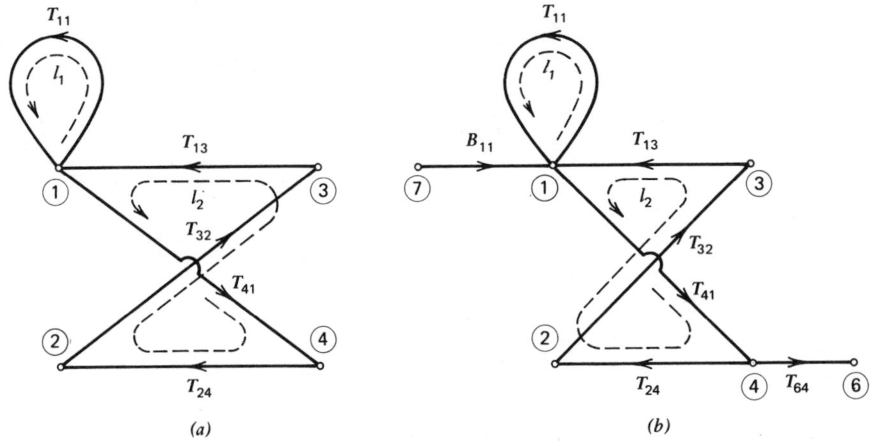

Fig. 12-2.5

whereas

$$L_2 = T_{41}T_{24}T_{32}T_{31} \tag{12-2.14}$$

One last basic concept to be discussed in this section is that of *touching* and *non-touching paths* and *loops*. Whenever two or more paths, two or more loops, or paths and loops, contain a *common branch or node*, they are said to touch. For example, Fig. 12-2.5(a) has two loops, l_1 and l_2, that share node 1 in common. Thus, the loops in this subgraph touch. Similarly in Fig. 12-2.5(b), which is another subgraph of Fig. 12-2.4, the path from node 7 to node 6 via transmissions $B_{11}T_{41}T_{64}$ touches both loop l_1 and loop l_2.

Example 12-2.4. The signal-flow graph of Fig. 12-2.6 is considered next. There are three input-output paths defined by

$$\begin{aligned} p &= (1, 6) \\ p &= (1, 2, 3, 6) \\ p &= (1, 4, 5, 6) \end{aligned} \tag{12-2.15}$$

and two loops

$$\begin{aligned} l_1 &= (2, 3, 2) \\ l_2 &= (4, 5, 4) \end{aligned} \tag{12-2.16}$$

Loops l_1 and l_2 do not touch. The path p_1 does not touch either l_1 or l_2. Path p_2 touches l_1 but not l_2, and path p_3 touches l_2 but not l_1.

The preceding definitions and concepts provide an adequate basis for proceeding to the problem of actually constructing signal-flow graphs for physical systems, for reducing them to obtain overall input-output relations, and for considering other special uses of such graphs.

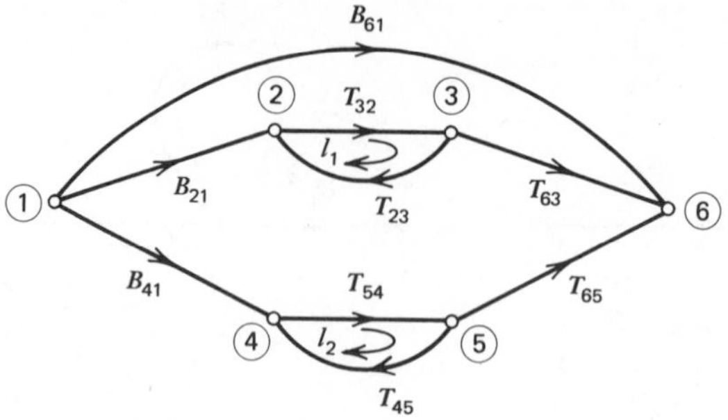

Fig. 12-2.6

12-3. THE CONSTRUCTION OF SIGNAL-FLOW GRAPHS

Example 12-3.1. Before we start with an actual system, let us first consider a set of equations and their signal-flow graph. The equations to be considered are:

$$\begin{aligned} 8 &= 2X_1 + 4X_2 + 6X_3 \\ 0 &= 2X_1 + 2X_2 + 4X_3 \\ 0 &= 2X_1 + 6X_2 + 2X_3 \end{aligned} \tag{12-3.1}$$

Due to the algebraic nature of these equations, no Laplace transformation is needed. Let us do the wrong thing first to understand why we set up the signal-flow diagrams the way we do. Each of the three equations is solved for X_1 as

$$\begin{aligned} X_1 &= 4 - 2X_2 - 3X_3 \\ X_1 &= -X_2 - 2X_3 \\ X_1 &= -3X_2 - X_3 \end{aligned} \tag{12-3.2}$$

The signal-flow graph for these equations is shown in Fig. 12-3.1(a), (b), and (c), respectively. After these three figures are drawn, the question is, what do we do next? We cannot just superimpose these three graphs, because we would get useless results.

What we need to do is to solve each of the three equations for a *different* variable, and to show all three equations on the same graph. This means that we have a choice among a number of different graphs to represent the same set of equations. As one possible graph, we solve the first equation for X_1, the second for X_2, and the third for X_3, as

$$\begin{aligned} X_1 &= 4 - 2X_2 - 3X_3 \\ X_2 &= -X_1 - 2X_3 \\ X_3 &= -X_1 - 3X_2 \end{aligned} \tag{12-3.3}$$

In Fig. 12-3.2(a), we locate and label the nodes and show the branches that yield the first of Eqs. 12-3.3. In (b), we add the branches that yield the second of Eqs. 12-3.3, and

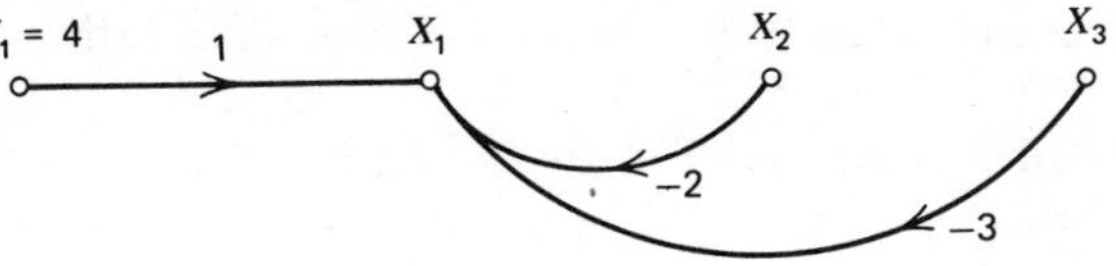

(a)

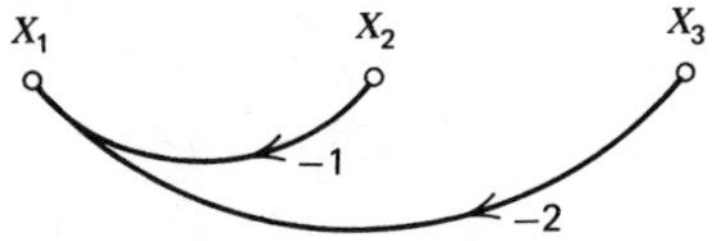

(b)

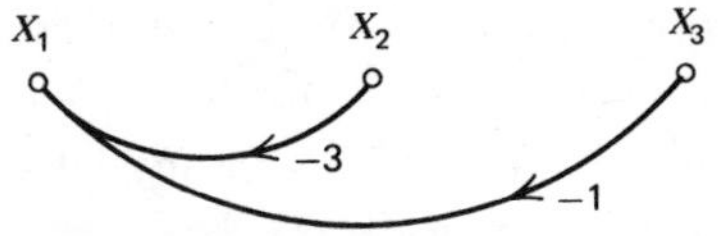

(c)

Fig. 12-3.1

(a)

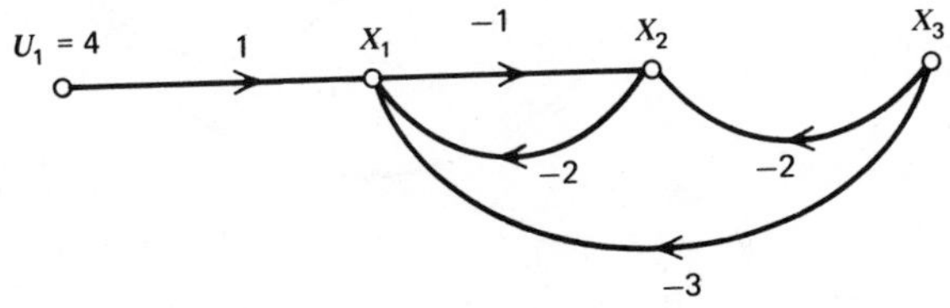

(b)

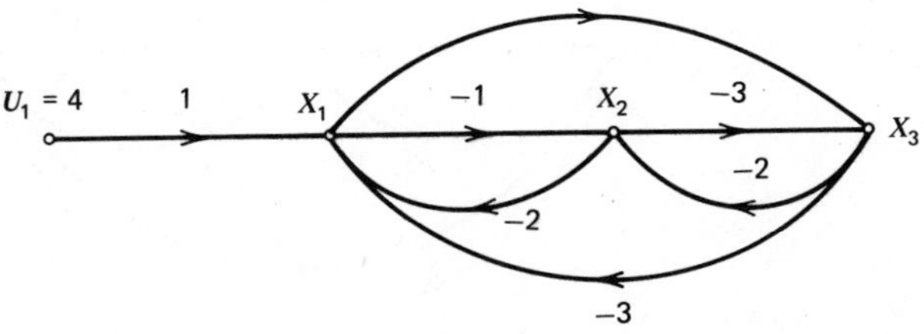

(c)

Fig. 12-3.2

in (c), we add the branches that yield the third of Eqs. 12-3.3, thus obtaining the completed graph.

As a second possible graph, we solve the first of Eqs. 12-3.1 for X_2, the second for X_3, and the third for X_1, as

$$\begin{aligned} X_2 &= 2 - \frac{X_1}{2} - \frac{3}{2}X_3 \\ X_3 &= -\tfrac{1}{2}X_1 - \tfrac{1}{2}X_2 \\ X_1 &= -3X_2 - X_3 \end{aligned} \tag{12-3.4}$$

We develop the signal-flow graph in Fig. 12-3.3(a), (b), and (c), by adding the branches necessary to describe each of the three equations of Eqs. 12-3.4.

A little thought will reveal that if we have a set of equations similar to Eqs. 12-3.1 with the number of X's equal to n, there would be $n!$ signal-flow graphs we could draw.

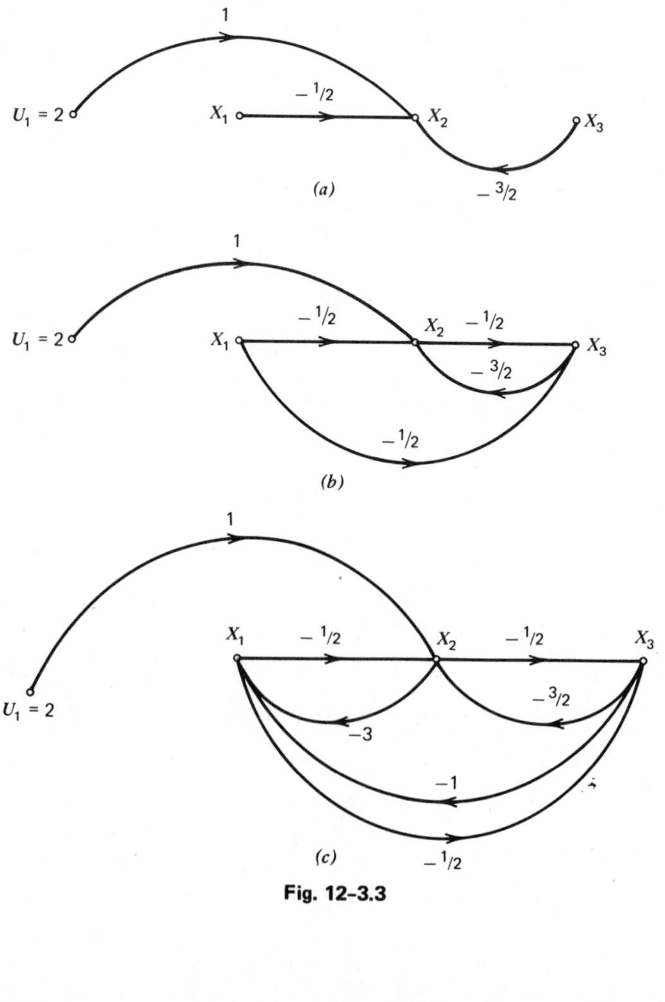

Fig. 12-3.3

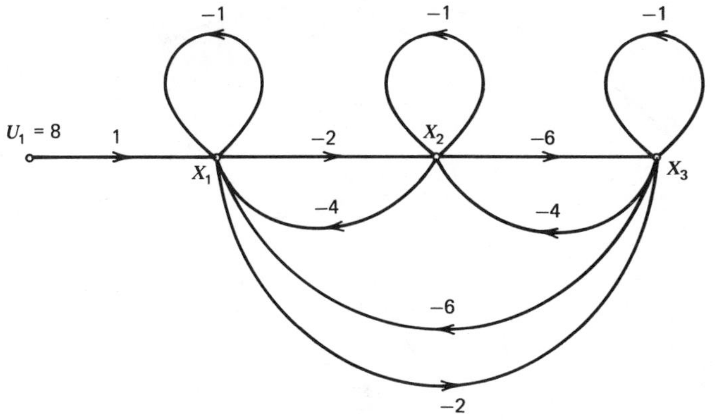

Fig. 12-3.4

using the method just outlined. In addition to these, however, there are many other types of graphs that can be drawn. We again write Eqs. 12-3.1 in a different form to suggest this idea.

$$\begin{aligned} X_1 &= 8 - X_1 - 4X_2 - 6X_3 \\ X_2 &= -2X_1 - X_2 - 4X_3 \\ X_3 &= -2X_1 - 6X_2 - X_3 \end{aligned} \tag{12-3.5}$$

The signal-flow graph for these equations is shown in Fig. 12-3.4.

The point being made at this time is that any set of equations (or system) can have a large number of signal-flow graphs associated with it. Each of these graphs is an adequate representation of the system, even though one graph may display a given characteristic of a system better than the others. An appreciation of this idea can only be developed after considerable experience.

Before we leave this discussion, let us return to Eqs. 12-1.3, written for the special case in which $n = 3$ and $m = 2$. To make this notation more compact, the equations are written without the Laplace transform notation, although this is implied.

$$\begin{aligned} X_1 &= T_{11}X_1 + T_{12}X_2 + T_{13}X_3 + B_{11}U_1 + B_{12}U_2 \\ X_2 &= T_{21}X_1 + T_{22}X_2 + T_{23}X_3 + B_{21}U_1 + B_{22}U_2 \\ X_3 &= T_{31}X_1 + T_{32}X_2 + T_{33}X_3 + B_{31}U_1 + B_{32}U_2 \end{aligned} \tag{12-3.6}$$

The corresponding signal-flow graph is shown in Fig. 12-3.5.

In the following Section 12-4 we define and learn how to combine parallel branches. With this thought in mind, we say that for $n = 3$ and $m = 2$ the graph of Fig. 12-3.5 is a *complete graph*. That is, no matter what additional branch is added, it will be in parallel with, and can be combined with, an existing branch. This is why we started the entire

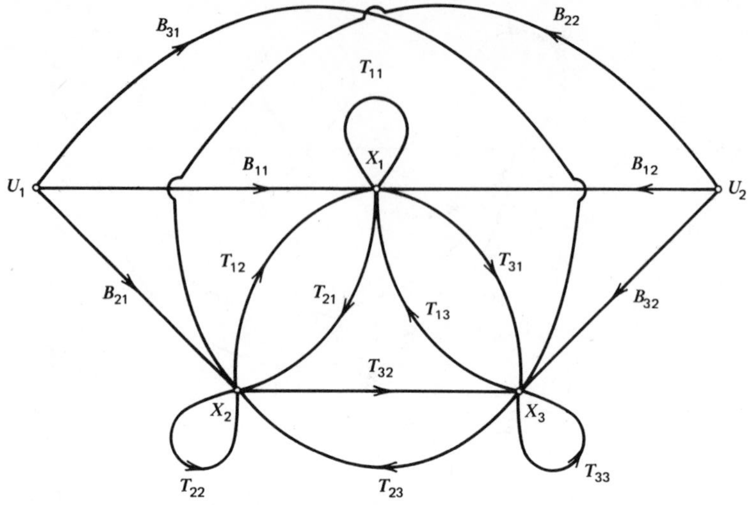

Fig. 12-3.5

discussion with Eqs. 12-1.3. For given values of n and m, any graph that can be drawn can be obtained from a complete graph by removing some of its branches.

Example 12-3.2. As indicated previously, a signal-flow graph is really a picture of a set of algebraic relations. Thus, one method for constructing signal-flow graphs for specific physical systems is to write an independent set of differential equations, transform them, and then draw the signal-flow graph from these equations.

The electro-mechanical system of a d'Arsonval meter movement, depicted in Fig. 12-3.6, is used as a first example in obtaining a signal-flow graph from a physical system. In reality this device is typical of any armature-controlled motor with inertia, viscous friction, and spring load. The basic linearized equations describing system operation are

$$i_a R + L\frac{di_a}{dt} + K_g\frac{d\theta}{dt} = e_a \qquad (12\text{-}3.7)$$

$$J\frac{d^2\theta}{dt^2} = -B\frac{d\theta}{dt} - K_s\theta + K_T i_a \qquad (12\text{-}3.8)$$

In considering these equations, cause-effect relations lead to the conclusion that the signal-flow graph should contain $E_a(s)$ as an independent input node, $\Theta(s)$ as an output node, with any or all of the variables $I_a(s)$, $s\,\Theta(s)$, $s^2\,\Theta(s)$ as additional dependent variable nodes. Alternately, slightly modified variables, such as developed and accelerating torque and generated voltage in the armature, could be used as variables. These would certainly lend more physical feel for the problem. Transforming the system equations, neglecting initial conditions as done to develop transfer functions, and grouping terms to use $I_a(s)$,

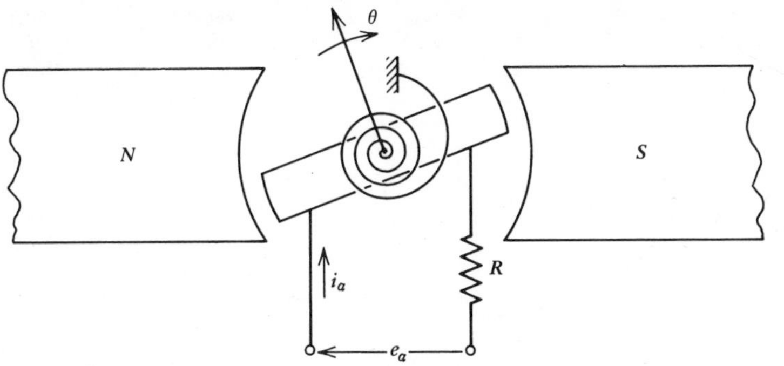

Fig. 12-3.6

$s^2\,\Theta(s)$, $s\,\Theta(s)$, and $\Theta(s)$ as dependent variables, leads to

$$(R + Ls)\,I_a(s) + K_g\,[s\,\Theta(s)] = E_a(s) \tag{12-3.9}$$

$$J[s^2\,\Theta(s)] = -B[s\,\Theta(s)] - K_s[\Theta(s)] + K_T\,I_a(s) \tag{12-3.10}$$

These equations do not appear in the form used in Section 12-2. There are a number of rearrangements that can be used, if it is desired to achieve such a form. For example

$$\begin{aligned} I_a(s) &= \frac{-K_g}{(R + Ls)}\,[s\,\Theta(s)] + \frac{1}{(R + Ls)}\,E_a(s) \\ [s^2\,\Theta(s)] &= (K_T/J)\,I_a(s) - (B/J)\,[s\,\Theta(s)] - (K_s/J)\,[\Theta(s)] \\ [s\,\Theta(s)] &= (1/s)\,[s^2\,\Theta(s)] \\ \Theta(s) &= (1/s)\,[s\,\Theta(s)] \end{aligned} \tag{12-3.11}$$

can be used where the last two relations are identities that are needed to relate some of the dependent variables that are used. The form is exactly that of Eq. 12-1.3. The development of the corresponding signal-flow graph is shown in Fig. 12-3.7. Actual graph construction starts by drawing all of the nodes in some arbitrary configuration. The above equations are then implemented graphically, using the result mentioned in Section 12-2.

Figure 12-3.7(b) shows the result of forming $I_a(s)$ as required by the first of Eqs. 12-3.11, and (c) adds the construction required for the second of Eqs. 12-3.11. The remaining two integration operations are filled in to give the completed signal-flow graph in (d).

Some items worth pointing out in Fig. 12-3.7(d) are: (a) there are three loops, each of which touches the other two loops; (b) there is only one path from the input, $E_a(s)$, to the output, $\Theta(s)$, and this path touches all loops; (c) the output, $\Theta(s)$, has been shown as a sink node by adding a unity transmission branch from node 4 to node 5; (d) initial conditions have been ignored.

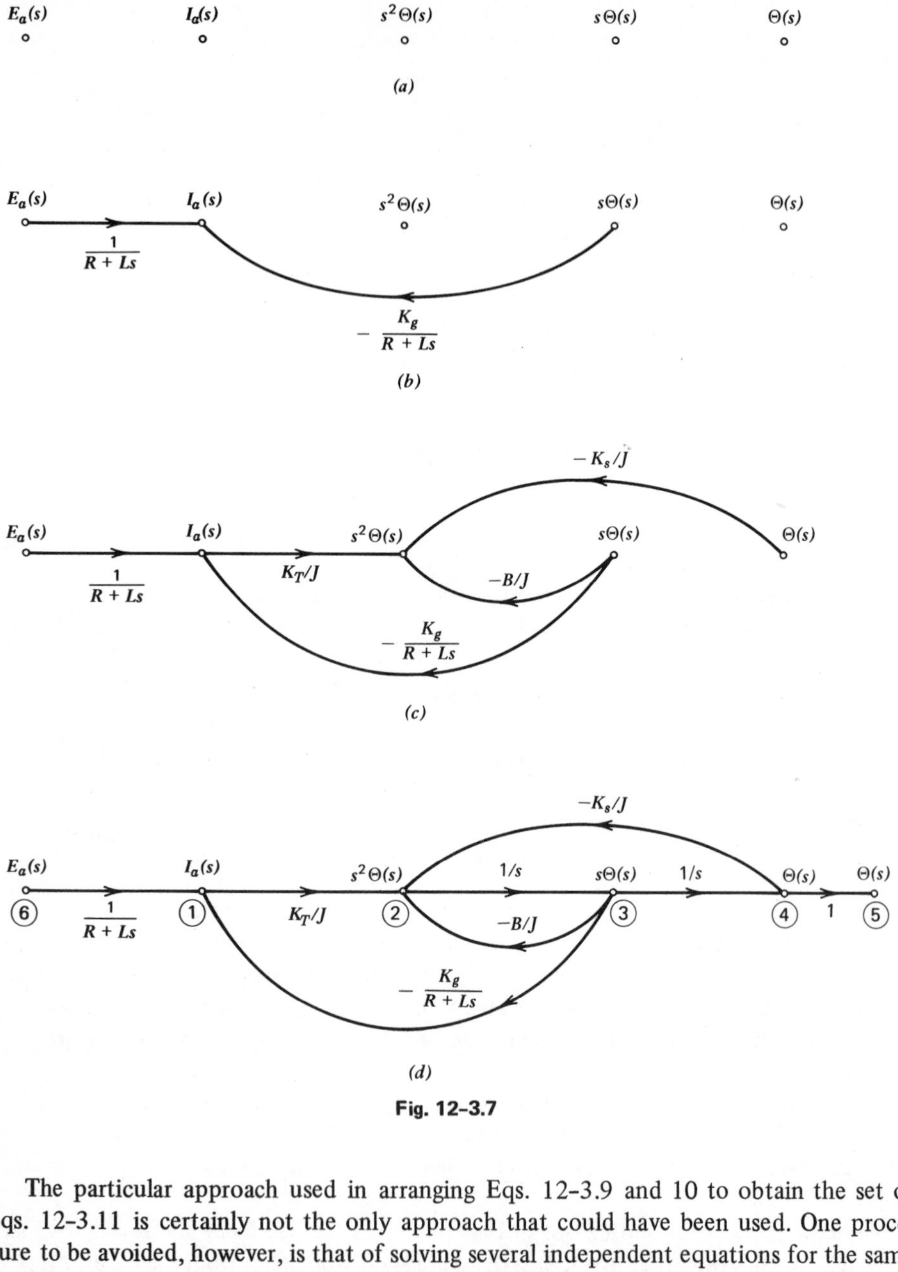

Fig. 12-3.7

The particular approach used in arranging Eqs. 12-3.9 and 10 to obtain the set of Eqs. 12-3.11 is certainly not the only approach that could have been used. One procedure to be avoided, however, is that of solving several independent equations for the same variable, as was explained in Example 12-3.1. If Eqs. 12-3.9 and 10 were both solved for $I_a(s)$, the result would be an impasse, with two equations for the same thing.

To again demonstrate that signal-flow graphs are not unique, the system of this example is reformulated, using I_a, $(s\,I_a)$, T_{dev}, T_{acc}, $[s^2\,\Theta(s)]$, $[s\,\Theta(s)]$, and $\Theta(s)$ as de-

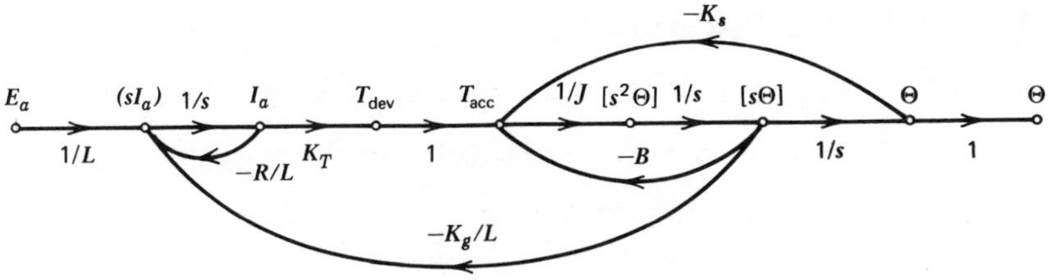

Fig. 12-3.8

pendent variables, with the augmented and revised set of equations given as

$$
\begin{aligned}
L(sI_a) &= -RI_a - K_g[s\,\Theta(s)] + E_a(s) \\
I_a &= 1/s\,(sI_a) \\
T_{\text{dev}} &= K_T I_a \\
T_{\text{acc}} &= T_{\text{dev}} - B[s\,\Theta(s)] - K_s\,\Theta(s) \\
[s^2\,\Theta(s)] &= 1/J\,(T_{\text{acc}}) \\
[s\,\Theta(s)] &= 1/s\,[s^2\,\Theta(s)] \\
\Theta(s) &= 1/s\,[s\,\Theta(s)]
\end{aligned}
\tag{12-3.12}
$$

These equations are then used to form the signal-flow graph of Fig. 12-3.8.

Example 12-3.3. A signal-flow graph is to be developed for the circuit of Fig. 12-3.9. One set of differential equations for solving this circuit by using loop currents is

$$
\begin{aligned}
i_1R_1 + Ld(i_1 - i_2)/dt + (1/C_1)\int_0^t i_1\,dt + v_{C1}(0) &= e_s \\
i_2R + Ld(i_2 - i_1)/dt + (1/C_2)\int_0^t i_2\,dt + v_{C2}(0) &= 0
\end{aligned}
\tag{12-3.13}
$$

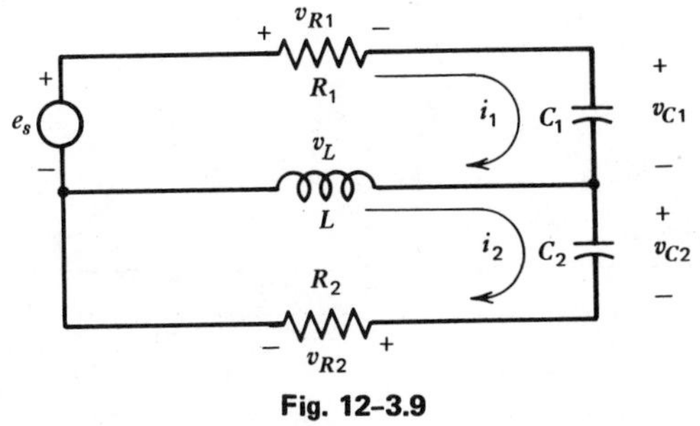

Fig. 12-3.9

which is then Laplace transformed to obtain

$$
\begin{aligned}
&I_1(s)\,[R_1 + Ls + (1/C_1 s)] - I_2(s)\,(Ls) = E_s(s) - v_{C1}(0)/s + (L)\,[i_1(0) - i_2(0)] \\
&-I_1(s)\,(Ls) + I_2(s)\,[R_2 + Ls + (1/C_2 s)] = (L)\,[i_2(0) - i_1(0)] - v_{C2}(0)/s
\end{aligned}
\qquad (12\text{-}3.14)
$$

The simplest approach at this point is to solve the first of Eqs. 12-3.14 for $I_1(s)$, and the second for $I_2(s)$. Nodes for the various transformed variables and initial condition inputs are then drawn and interconnected, as required by the revised or rearranged equations. Figure 12-3.10(a) shows the result of interconnecting terms to satisfy the $I_1(s)$ equation, while (b) adds the second equation and interconnects the two relationships where common variables appear. It might be noted that auxillary nodes have been used to simplify the branch transmittances where complex terms would normally appear several times.

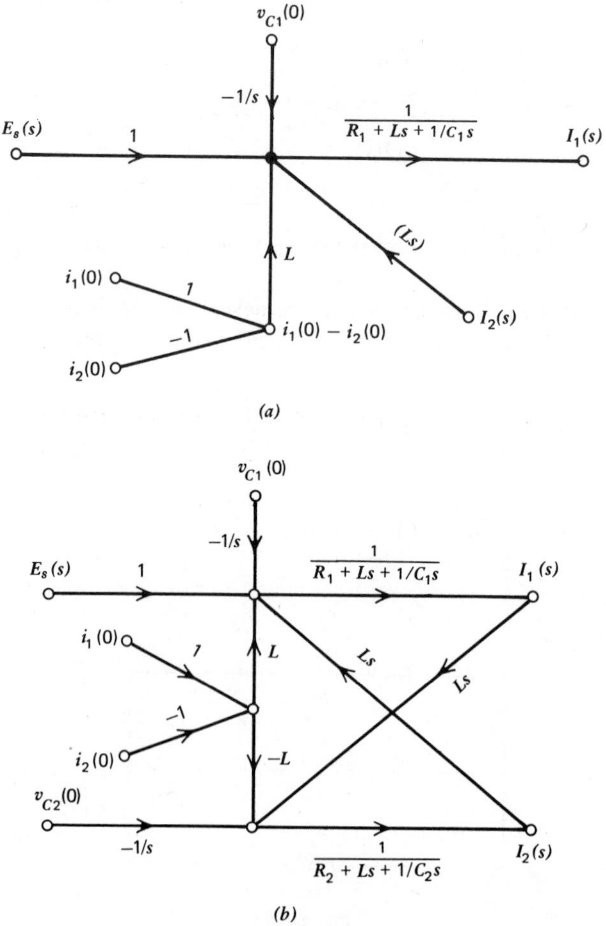

Fig. 12-3.10

The above procedure is not as closely related to the physical circuit as it might be. It is possible to write Kirchhoff's Voltage Law for the two loops in terms of the voltages across the various circuit elements, and then to relate these in terms of element volt-ampere relationships and the circuit topology. Equations 12-3.15, 16 and 17 provide the needed relations.

$$v_{R1} + v_{C1} + v_L = e_s$$
$$-v_L + v_{C2} + v_{R2} = 0 \qquad (12\text{-}3.15)$$

$$v_{R1} = i_1 R_1 \qquad v_C = (1/C_1) \int_0^t i_1\, dt + v_C(0)$$

$$v_{R2} = i_2 R_2 \qquad v_C = (1/C_2) \int_0^t i_2\, dt + v_C(0) \qquad (12\text{-}3.16)$$

$$v_L = L\, d(i_1 - i_2)/dt$$

$$V_{R1}(s) = I_1(s) R_1 \qquad V_{C1}(s) = I_1(s)/(C_1 s) + v_{C1}(0)/s$$
$$V_{R2}(s) = I_2(s) R_2 \qquad V_{C2}(s) = I_2(s)/(C_2 s) + v_{C2}(0)/s \qquad (12\text{-}3.17)$$
$$V_L(s) = (Ls)\,[I_1(s) - I_2(s)] - L\,[i_1(0) - i_2(0)]$$

In order to develop a signal-flow diagram using these relationships, the following procedure can be used:

1 Solve the first of Eqs. 12-3.15 for one of the dependent voltages, such as $V_{R1}(s)$.
2 Generate $I_1(s)$ from this $V_{R1}(s)$, using the appropriate volt-ampere relation of Eqs. 12-3.17.
3 Generate $V_{C1}(s)$ and $V_L(s)$, using the appropriate volt-ampere relationships; Fig. 12-3.11(a) shows the results to this point.
4 Solve the second loop equation of Eqs. 12-3.15 for a second dependent voltage, such as V_{R2} or $V_{R2}(s)$.
5 Generate $I_2(s)$ from this $V_{R2}(s)$, using the appropriate volt-ampere relationship from Eqs. 12-3.17.
6 Generate the remaining dependent voltage $V_{C2}(s)$ from $I_2(s)$.
7 Interçonnect the previously generated parts as required by the occurrence of common variables.

The completed signal-flow graph is shown in Fig. 12-3.11(b). It should be noted that independent inputs and initial conditions are added as needed by the implementation of the various equations.

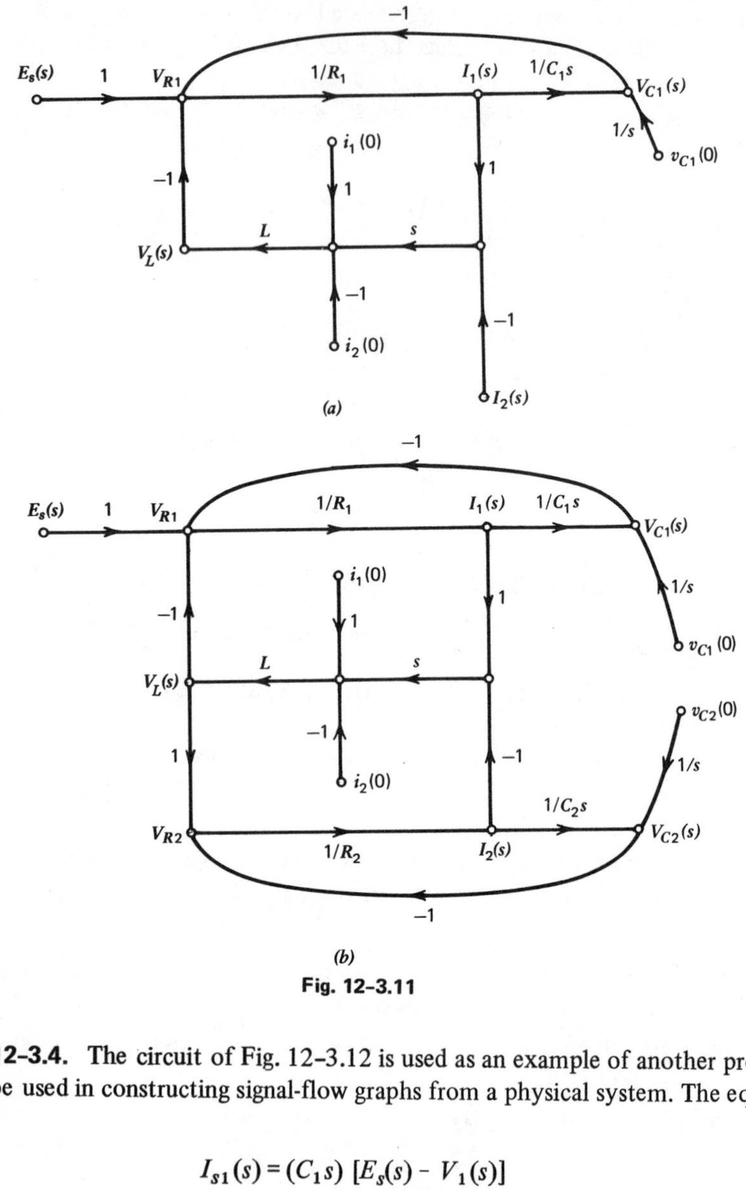

Fig. 12-3.11

Example 12-3.4. The circuit of Fig. 12-3.12 is used as an example of another procedure that can be used in constructing signal-flow graphs from a physical system. The equations used are

$$
\begin{aligned}
I_{s1}(s) &= (C_1 s)\,[E_s(s) - V_1(s)] \\
V_1(s) &= [I_{s1}(s) - I_{12}(s)]\,R_1 \\
I_{12}(s) &= (C_2 s)\,[V_1(s) - V_2(s)] \\
V_2(s) &= [I_{12}(s) - I_{23}(s)]\,R_2 \\
I_{23}(s) &= (C_3 s)\,[V_2(s) - V_3(s)] \\
V_3(s) &= [I_{23}(s) - I_{3o}(s)]\,R_3
\end{aligned}
\qquad (12\text{-}3.18)
$$

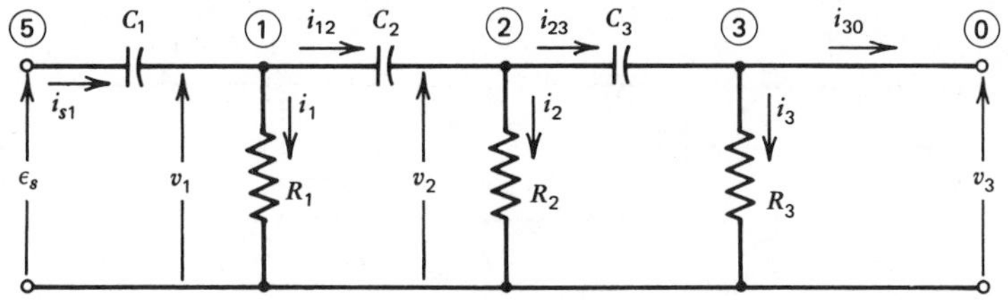

Fig. 12-3.12

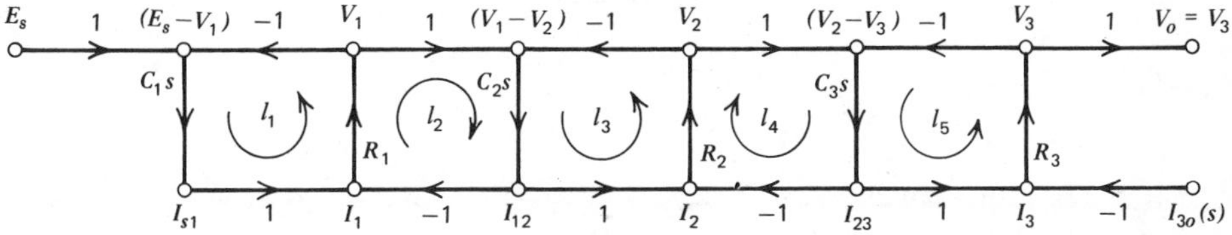

Fig. 12-3.13

Figure 12-3.13 is the signal-flow graph derived from these equations, as described previously. Several facts should be noted in considering this example. The particular selection of Kirchhoff's Laws and the Ohm's Law equations should be noted as appropriate for ladder networks. There is only one path from E_s as input to V_3 as output; however, I_{3o} appears as a second input representing output loading. There are five loops in the graph, and the path from E_s to V_3 touches all loops. The path from $I_{3o}(s)$ to V_3 touches only one of the loops. However, not all loops touch each other as, for example, l_1 does not touch l_3, l_4, or l_5.

Example 12-3.5. As a fourth example for obtaining a signal-flow graph from a physical system, a thermal system is considered. A source of heat is used to heat material in a vessel. There are heat losses to the environment proportional to the difference between material temperature and ambient temperature. Uniform material temperature is presumed with the system depicted in Fig. 12-3.14; no change of material state is presumed:

T_m = material temperature

T_a = ambient temperature

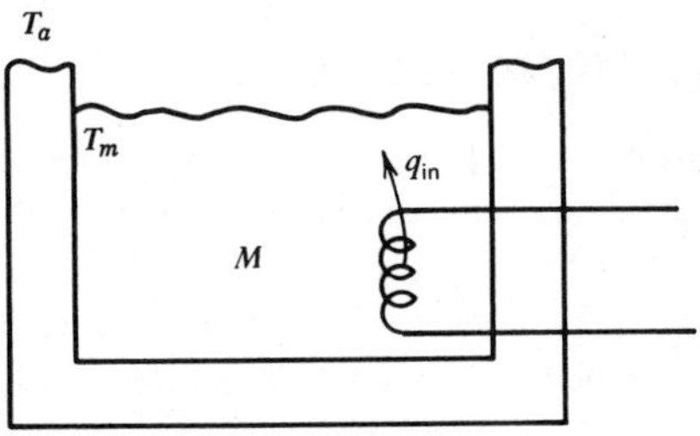

Fig. 12-3.14

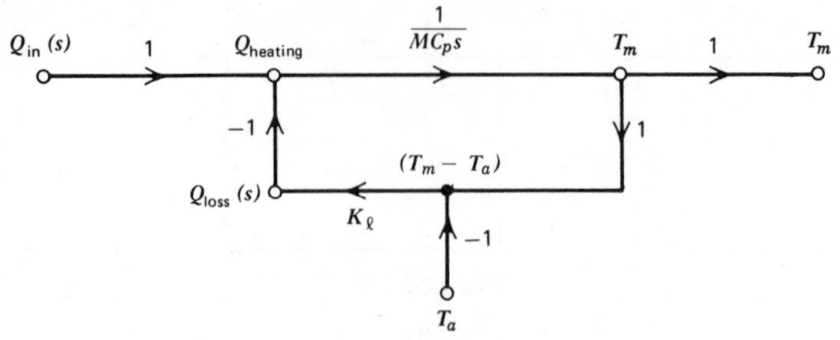

Fig. 12-3.15

q_{in} = heat rate input

q_ℓ = heat rate loss

K_ℓ = heat loss coefficient

M = mass material

$q_{heating}$ = net heat rate to raise material temp.

C_p = specific heat

The basic heat flow equations describing the system are

$$MC_p \frac{dT_m}{dt} = q_{in} - q_{loss} = q_{heating} \qquad (12\text{-}3.19)$$

$$q_{loss} = K_\ell (T_m - T_a)$$

The signal-flow graph is constructed by again using the equations and cause-effect relations. As shown in Fig. 12-3.15, the signal-flow graph is particularly simple, with one loop and two inputs.

The procedures for constructing signal-flow graphs described in the examples of this section are but a few of many techniques. Certain types of models for systems lead to standard forms for some signal-flow graphs. However, these standard forms are just something else to put in tables, or remember, and are not deemed worth presenting at this point. The basic cause-effect relationship is still fundamental.

12-4. REDUCTION OF SIGNAL-FLOW GRAPHS

The *reduction of a signal-flow graph*, intuitively, is to make a graph simpler or to bring it into a specified form. Generally, the ultimate use of a signal-flow graph is to obtain overall input-output transmittances.

Before we start the subject of reducing the graph, let us review the subject of solving the system equations by using matrix methods. To do this, we first rewrite Eq. 12-2.7 as

$$[T]X] = [B]U] \qquad (12\text{-}4.1)$$

To emphasize the general case with m inputs and n of the X-variables, this equation is written as

$$[T]_{nn}\, X]_{n1} = [B]_{nm}\, U]_{m1} \tag{12-4.2}$$

The inverse of the $[T]_{nn}$ matrix is found, and both sides of the equation are pre-multiplied by the inverse, as

$$[T]_{nn}^{-1}\, [T]_{nn}\, X]_{n1} = [T]_{nn}^{-1}\, [B]_{nm}\, U]_{m1} \tag{12-4.3}$$

which can be written as

$$X]_{n1} = [[T]^{-1}\, [B]]_{nm}\, U]_{m1} \tag{12-4.4}$$

The $[T]^{-1}\,[B]$ product is written this way to emphasize that there are $(n \times m)$ elements in the solution.

Let us suppose that the system only has one input ($m = 1$), and of the n X-variables we only want to find one, say X_i. If we follow the usual matrix solution, we find not only X_i but all of the rest of the n variables as outputs. When the corresponding signal-flow graph is reduced by methods to be discussed shortly, only one input-output transmittance needs to be found, as suggested in Fig. 12-4.1(a). There are two concepts to be pointed out here. First, when a signal-flow graph is reduced, this is an alternate method to inverting a matrix; and second, when matrix methods are used, not only is X_i determined, but all of the rest of the X_n variables are found. In other words the matrix methods give more answers than necessary. This is one of the advantages of signal-flow graphs not mentioned previously.

Figure 12-4.1(b) indicates a reduced system with two inputs but only one output variable required, whereas (c) suggests one input with two outputs required. The general case is indicated in (d), with m inputs and all n of the X-variables required to be outputs. There are $(n \times m)$ transmittances required, corresponding to the $(n \times m)$ elements in the solution matrix of Eq. 12-4.4.

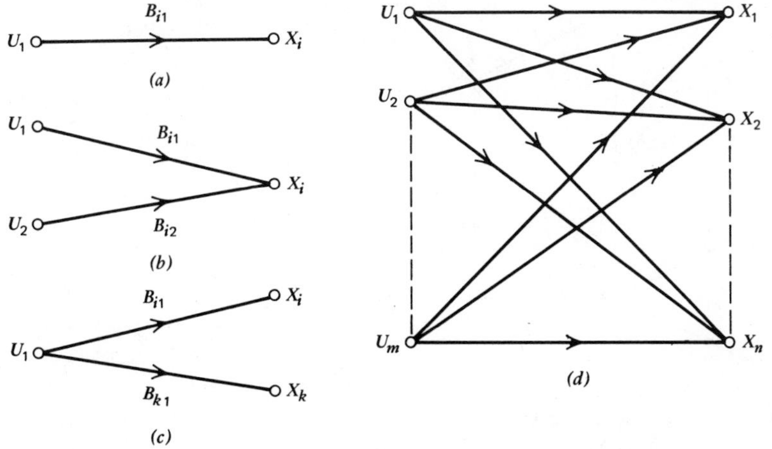

Fig. 12-4.1

In reducing a signal-flow graph, two procedures are frequently used. One is a step-by-step reduction of parts of the graph, while the second is a direct procedure for writing the overall transmission using what is called *Mason's Rule*. A few aspects of the step-by-step procedure are considered first.

STEP-BY-STEP REDUCTION OF SIGNAL-FLOW GRAPHS

Certain rather obvious but basic elementary equivalencies exist in signal-flow graphs and serve as the basis for the step-by-step reduction process. The most useful of these are presented in Fig. 12–4.2, with the algebraic justification for the signal-flow graph modifications for each.

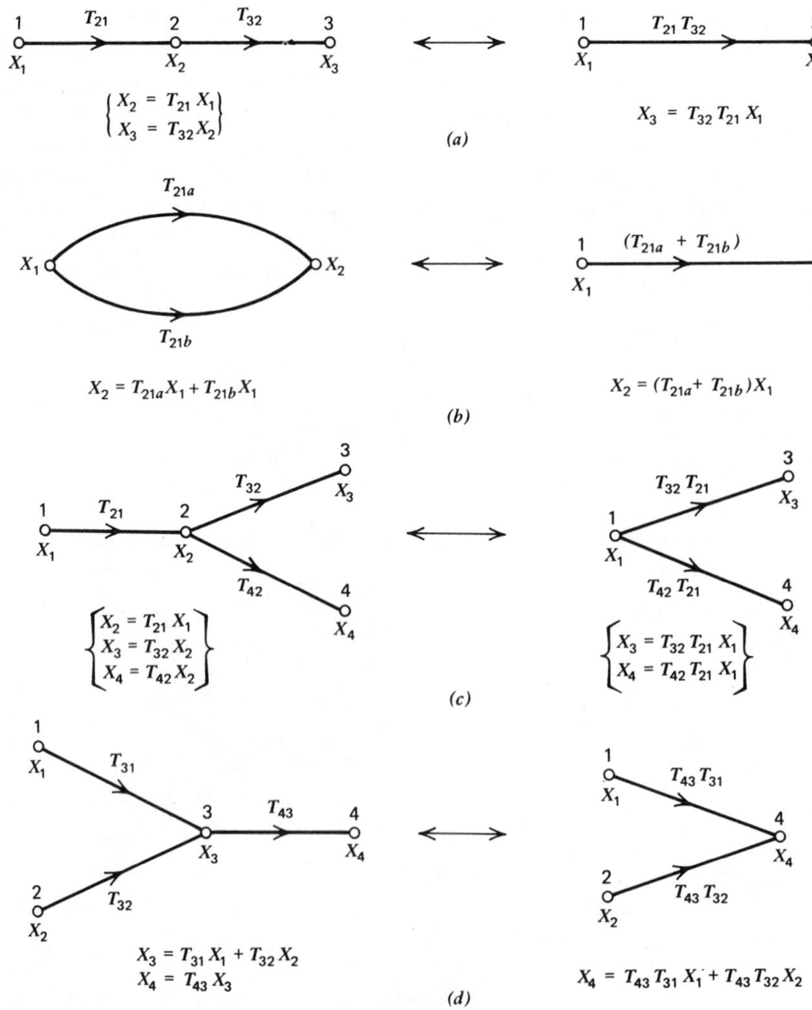

Fig. 12-4.2

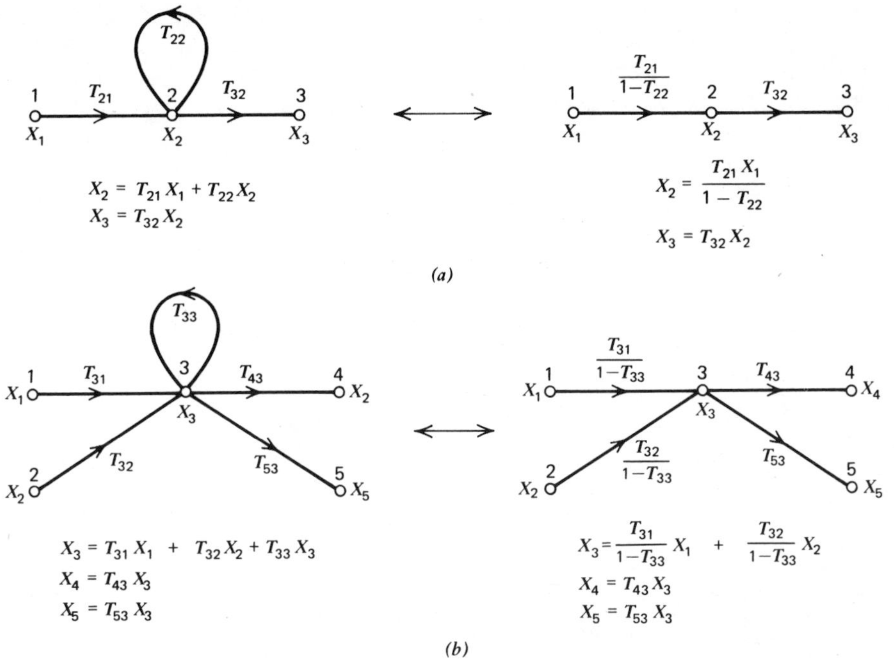

Fig. 12-4.3

Before proceeding further, one additional equivalency is of such fundamental importance as to warrant separate consideration. This is the self-loop mentioned previously. Figure 12-4.3(a) shows this equivalency for a somewhat special case, while (b) depicts a somewhat more general case. The self-loop can be eliminated in a straightforward manner, dividing all incoming transmittances to the node with the self-loop by $(1 - T_{jj})$. This is apparent from observing the effect of collecting X_j terms in the equations to get $(1 - T_{jj})X_j$, and then dividing by the coefficient. There is *no change* in transmittances *for branches leaving* node X_j. Further reductions of the signal-flow graphs of Fig. 12-4.3 are possible by using the equivalencies of Fig. 12-4.2.

The application of the above equivalencies in the step-by-step reduction process is illustrated in the following examples.

Example 12-4.1. We are to solve the three simultaneous equations of Eqs. 12-3.1 for X_1 by reducing the signal-flow graph of Fig. 12-3.2. The graph is drawn again as Fig. 12-4.4(a), featuring X_1 as the output. We can remove either node 4 or node 3 first. We choose to remove node 4, and an auxiliary subgraph, shown in (b), is drawn as an aid in the process. Node 4 has two branches incoming from nodes 2 and 3, and two branches outgoing to nodes 2 and 3. In (c) is shown the subgraph of (b), with node 4 removed. The transmittances shown in (c) are obtained by multiplying the transmittances in (b) as the appropriate paths are traversed. The transmittances in (c) are placed on the graph in (d), as shown by the solid lines. The branches of the original graph not involved in removing node 4 are shown by dashed lines.

Obviously a person can go from the graph in Fig. 12-4.4(a) to that in (d) in one step; the graphs in (b) and (c) are included to make the step clear. The graph in (e) is obtained from (d) by combining parallel paths, and the graph in (f) is obtained from (e) by removing the self-loops. Node 3 can be removed in (f) by multiplying $(-\frac{1}{5})(-\frac{7}{2})$, and the resulting self-loop is removed, as shown in (g). From the final graph in (h), X_1 is found to be $X_1 = -\frac{20}{3}$.

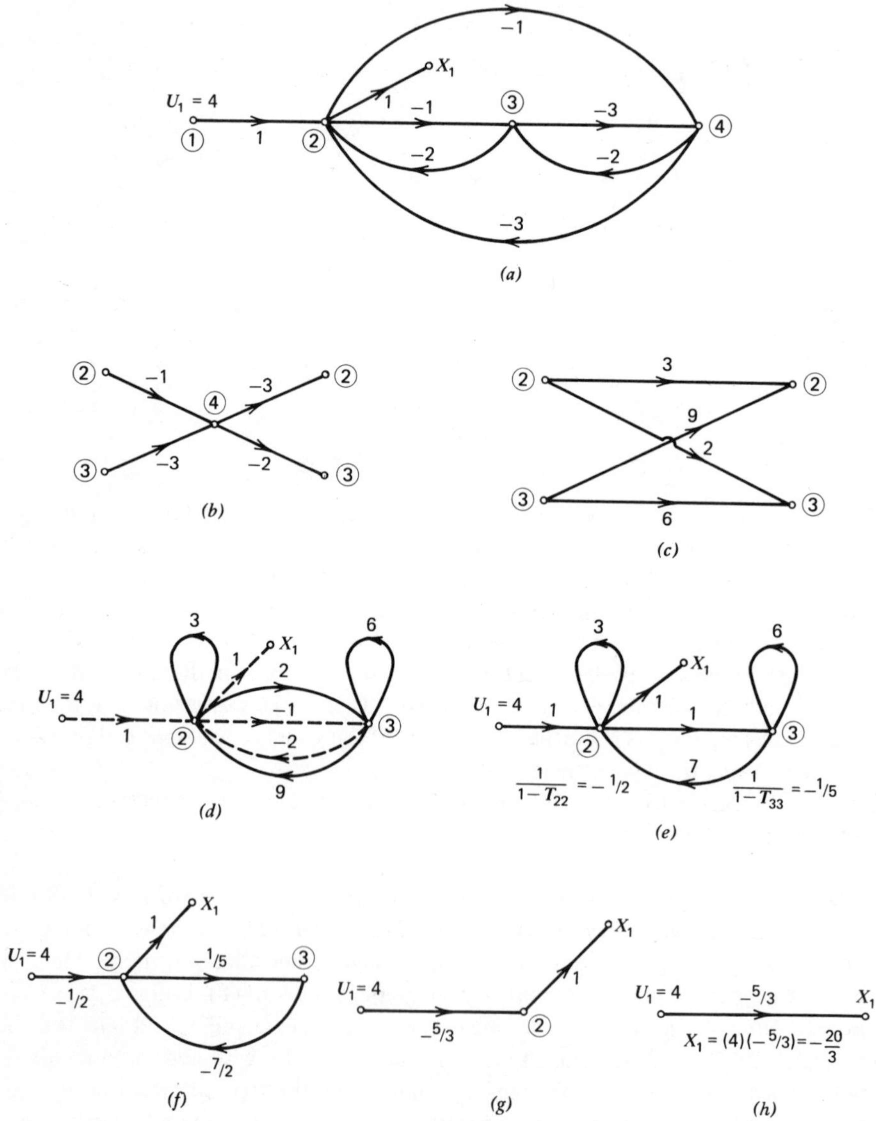

Fig. 12-4.4

Example 12-4.2. The d'Arsonval meter movement problem of Example 12-3.2 is reconsidered at this point. It is desired to reduce the signal-flow graph to obtain the $\Theta(s)/E_a(s)$ transfer function. The original signal-flow graph is repeated as Fig. 12-4.5(a), except T_{dev} has been eliminated by using the equivalencies of Figs. 12-4.2 and 3. In applying the step-by-step reduction, the procedure is certainly not unique.

As a first step, node $I_a(s)$ is eliminated by applying the equivalency (d) from Fig. 12-4.2. The result is shown in Fig. 12-4.5(b). The equivalency in Fig. 12-4.2(d) is next used to obtain Fig. 12-4.5(c). The parallel branches from $\dot{\theta}$ to T_{accel} are combined by using the equivalency in Fig. 12-4.2(b) to give Fig. 12-4.5(d). The feedback loop from T_{accel} to $\dot{\theta}$ is converted to a self-loop at $\dot{\theta}$ by the equivalency in Fig. 12-4.2(d). Finally, the self-loop is reduced by the equivalency in Fig. 12-4.3(a) to give the final result of Fig. 12-4.5(f).

Example 12-4.3. For a third example of the step-by-step reduction process, Example 12-3.4 is reconsidered. Figure 12-4.6 depicts the reduction process in stages. The complete reduction is not detailed, only the first steps. The algebra becomes messy, although conceptually the procedure is straightforward. The final form of this reduction process is considered later in the direct method of using Mason's Rule.

DIRECT REDUCTION OF SIGNAL-FLOW GRAPHS

Perhaps the most useful part of signal-flow graph work is the formula of S. J. Mason for overall transmission. This formula is stated, without proof, in

$$\frac{X_{out}(s)}{U_{in}(s)} = \frac{\sum_{i=1}^{m} P_i(s)\,\Delta_i(s)}{\Delta(s)} \tag{12-4.5}$$

In this formula special symbols are used

$\Delta(s)$ = graph determinant

$P_i(s)$ = ith input to output path transmittance

$\Delta_i(s)$ = ith path cofactor

and these require some discussion.

The graph determinant, $\Delta(s)$, is considered first. As discussed several times before, the signal-flow graph is a representation of a set of equations given in matrix form, as

$$[T(s)]X(s)] = [B(s)]\,U(s)] \tag{12-4.6}$$

and the $\Delta(s)$ can be obtained from these equations, as

$$\Delta(s) = |T(s)| \tag{12-4.7}$$

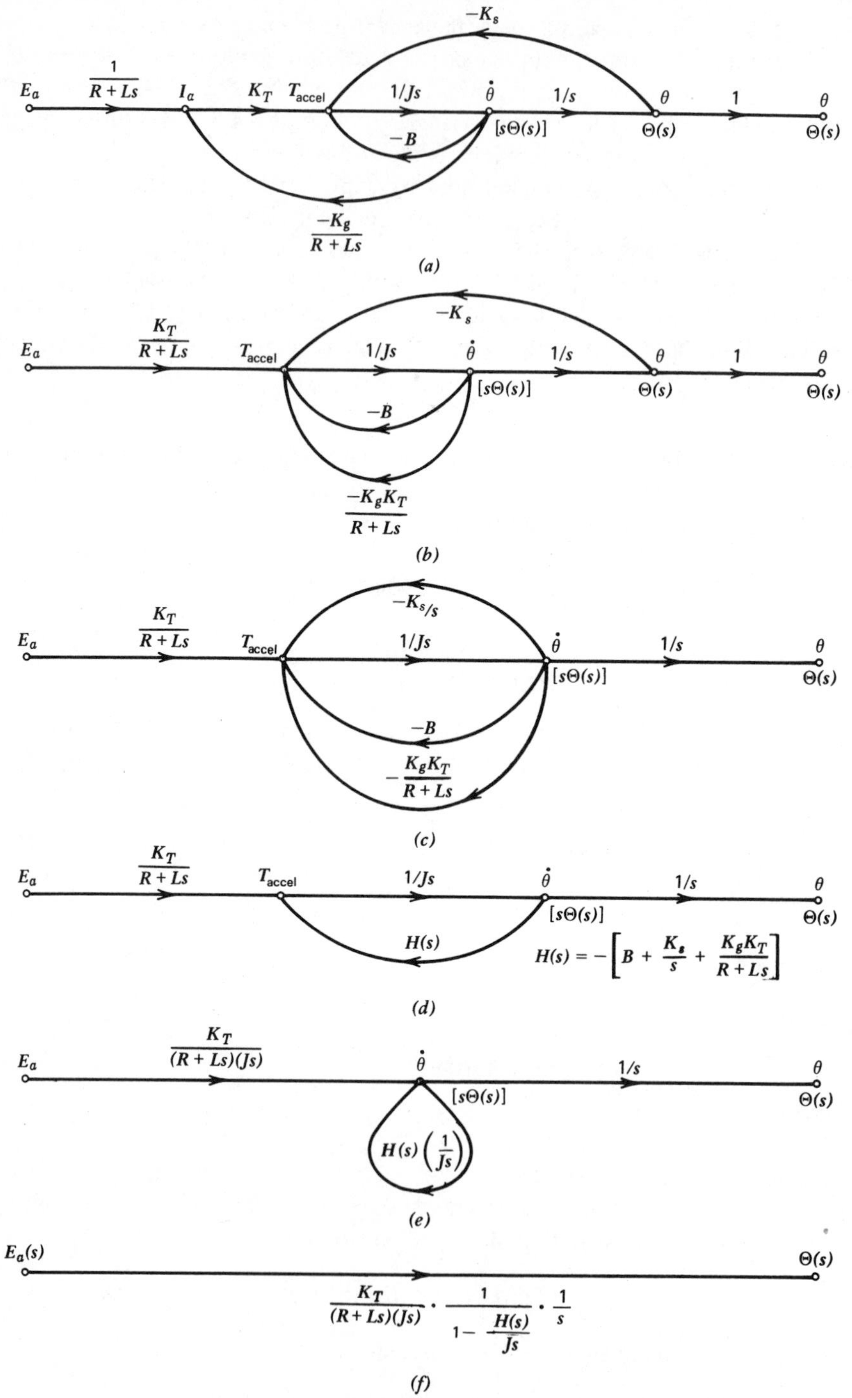

$-K_s$
$\frac{1}{R+Ls}$
E_a
I_a
K_T
T_{accel}
$1/Js$
$\dot{\theta}$
$[s\Theta(s)]$
$1/s$
θ
$\Theta(s)$
1
$-B$
$\frac{-K_g}{R+Ls}$
(a)
$\frac{K_T}{R+Ls}$
$\frac{-K_g K_T}{R+Ls}$
(b)
$-K_{s/s}$
$-\frac{K_g K_T}{R+Ls}$
(c)
$H(s)$
$H(s) = -\left[B + \frac{K_s}{s} + \frac{K_g K_T}{R+Ls}\right]$
(d)
$\frac{K_T}{(R+Ls)(Js)}$
$H(s)\left(\frac{1}{Js}\right)$
(e)
$E_a(s)$
$\frac{K_T}{(R+Ls)(Js)} \cdot \frac{1}{1-\frac{H(s)}{Js}} \cdot \frac{1}{s}$
(f)

Fig. 12-4.5

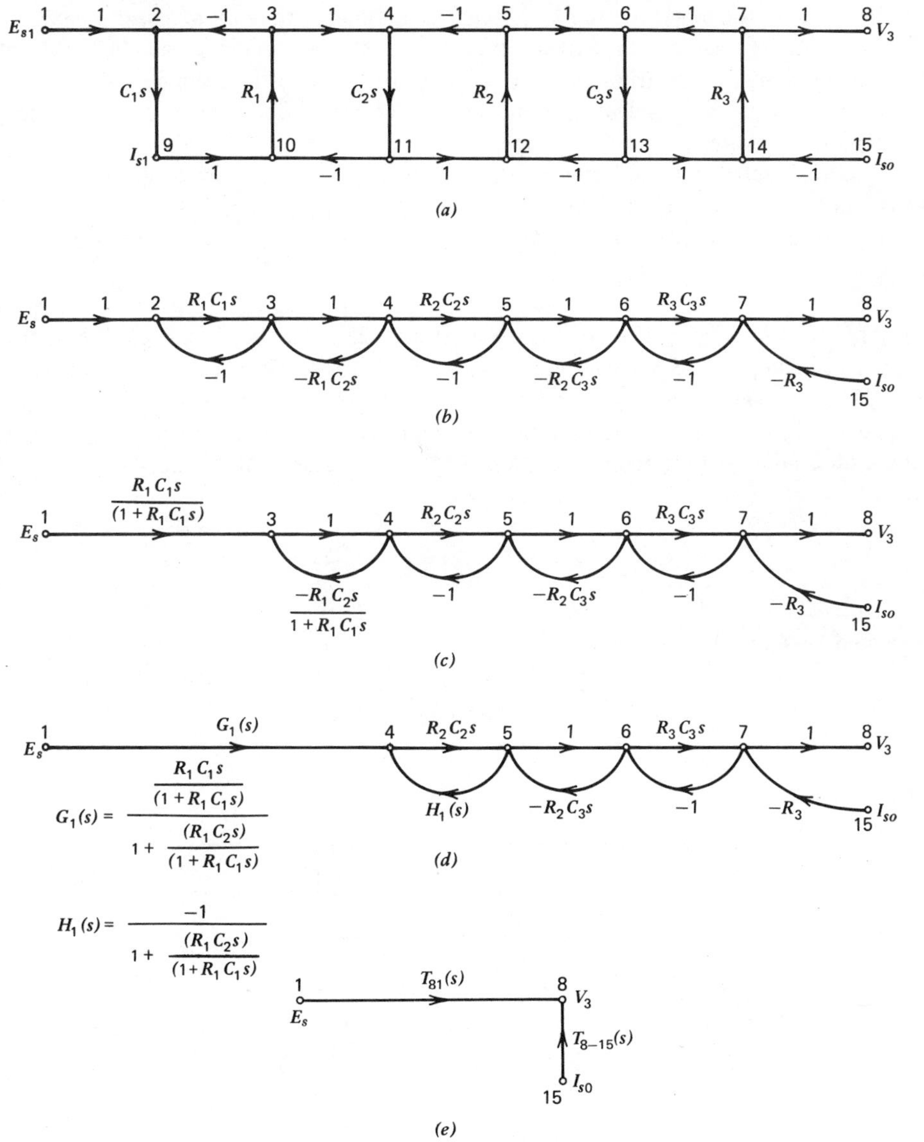

Fig. 12-4.6

The $\Delta(s)$ can also be determined directly from the signal-flow graph, by the formula

$$\Delta(s) = 1 - \sum_{i=1}^{n} L_i + \sum_{k=1}^{r} (L_i L_j)_k^{\text{N.T.}} - \sum_{\ell=1}^{s} (L_i L_j L_k)_\ell^{\text{N.T.}} + \cdots \qquad (12\text{-}4.8)$$

The L terms are loop transmittances, and the first summation includes all loops in the signal-flow graph. Terms of the form $(L_i L_j)_k^{\text{N.T.}}$ in the second summation are products of two loop gains for loops that *do not touch* according to the *non-touching concept* of Section 12-2. The summation extends over all possible non-touching pairs in the signal-flow graph. Similarly, terms of the form $(L_iL_jL_k)_{\ell}^{\text{N.T.}}$ in the third summation are products of *three non-touching loop* transmittances. The summation extends over all possible combinations of three non-touching loops.

The graph determinant formula continues with summations involving alternating signs and combinations of four, five, etc., non-touching loops, until no more such combinations exist. It is important to note that $\Delta(s)$ *depends only upon the loops* in the signal-flow graph.

Example 12-4.4. Figure 12-3.15, which illustrates the signal-flow graph for the thermal system of Example 12-3.5 is considered again. Figure 12-4.7 illustrates the original graph and the *loop subgraph* (having only paths in loops). In this case, it is readily apparent that only one loop exists in the graph. Thus it is impossible to have any combination of two, three, etc., non-touching loops. Thus the graph determinant, $\Delta(s)$, consists of only the parts

$$\Delta(s) = 1 - L_1(s) = 1 + \frac{K_\ell}{MC_ps} \tag{12-4.9}$$

Example 12-4.5. As a more involved example, attention is directed to the electrical network of Example 12-4.3 and Fig. 12-4.6(a). There are five loops in this signal-flow graph and the loop subgraph is shown in Fig. 12-4.8(a).

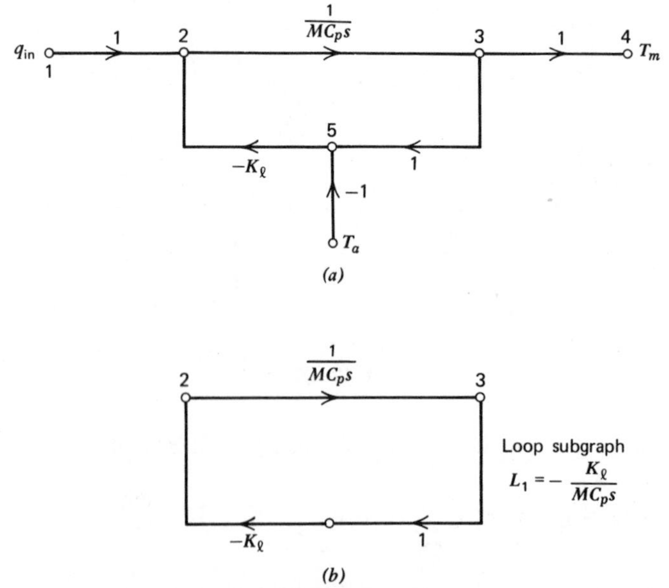

Fig. 12-4.7

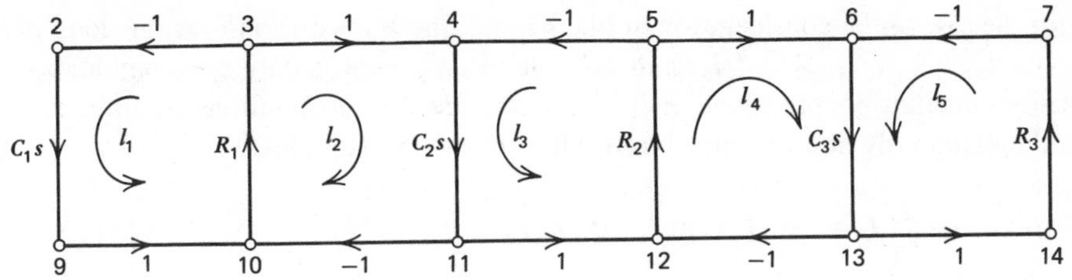

(a)

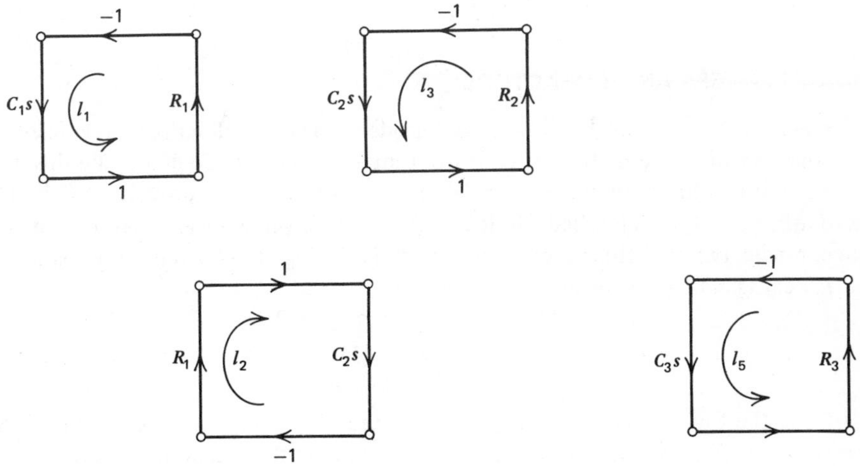

(b) Two of the nontouching loop pair subgraphs

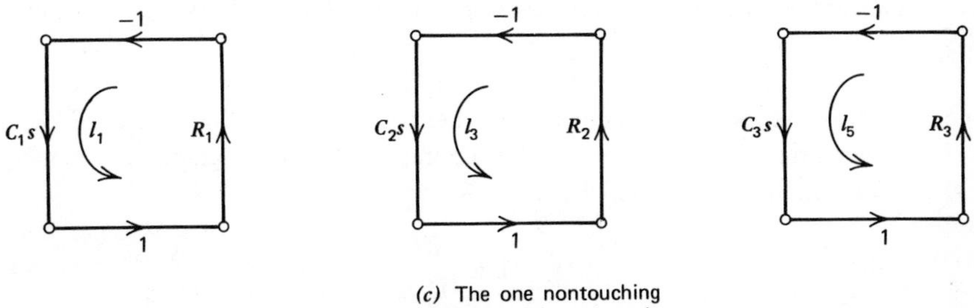

(c) The one nontouching loop triplet subgraph

Fig. 12–4.8

For the case under consideration in Fig. 12-4.8, there are six non-touching loop pairs, namely $l_1 l_3$, $l_1 l_4$, $l_1 l_5$, $l_2 l_4$, $l_2 l_5$, and $l_3 l_5$. Similarly, there is only one combination of three non-touching loops, which is $l_1 l_3 l_5$. There are no combinations of four, five, or more simultaneously non-touching loops. Thus the graph determinant is

$$\Delta(s) = 1 - \sum_{i=1}^{5} L_i + \sum_{k=1}^{6} (L_i L_j)_k^{\text{N.T.}} - L_1 L_3 L_5 \tag{12-4.10}$$

$$\begin{aligned}\Delta(s) = 1 &+ (R_1 C_1 s) + (R_1 C_2 s) + (R_2 C_2 s) + (R_2 C_3 s) + (R_3 C_3 s) + (R_1 C_1 R_2 C_2 s^2) \\ &+ (R_1 C_1 R_2 C_3 s^2) + (R_1 C_1 R_3 C_3 s^2) + (R_1 C_2 R_2 C_3 s^2) + (R_1 C_2 R_3 C_3 s^2) \\ &+ (R_2 C_2 R_3 C_3 s^2) + (R_1 C_1 R_2 C_2 R_3 C_3 s^3)\end{aligned} \tag{12-4.11}$$

PATH TRANSMITTANCES AND COFACTORS

The significance of P_i in Eq. 12-4.5 is considered next. As described previously, this term is the transmittance of the ith path from input to output. To determine this quantity, a path from the input under consideration is traced in the direction of the path arrows until the output is reached. No looping is allowed, which means that no node may be encountered twice in this process. For example, in Fig. 12-4.7, only one path from q_{in} to T_m exists, which results in

$$P_1 = (1) \frac{1}{MC_p s} (1) \tag{12-4.12}$$

Similarly, in the Fig. 12-4.6(a) signal-flow graph used in Example 12-4.3, only one path exists from E_s to V_3. One path also exists from I_{3o} to I_3 in this signal-flow graph.

Associated with each path transmission in Eq. 12-4.5 is a path cofactor, $\Delta_i(s)$. This path cofactor is obtained by *deleting from* $\Delta(s)$ (or setting equal to zero) *all terms containing loops that touch path* P_i. In case all loops touch a given path, the path cofactor is unity. For the case of the thermal system of Example 12-4.4 the only path from q_{in} to T_m touches the only loop and so the cofactor, Δ_1, for this path is simply 1.0.

Several examples are used to illustrate the complete use of Mason's formula, Eq. 12-4.5.

Example 12-4.6. As described above, the thermal system of Examples 12-3.5 and 12-4.4 is rather simple and is used again. In Example 12-4.4 the graph determinant for this system was found to be Eq. 12-4.13.

$$\Delta(s) = 1 + \frac{K_\ell}{MC_p s} \tag{12-4.13}$$

As described above, only one path from q_{in} to T_m exists and has a gain $1/(MC_p s)$. Since this path touches the only loop, its cofactor is 1.0, as found by deleting loop L_1 from

$\Delta(s)$. The contribution of q_{in} to T_m can now be written by Mason's rule as

$$\left.\frac{T_m(s)}{q_{\text{in}}(s)}\right|_{T_a=0} = \frac{P_1\Delta_1}{\Delta} = \frac{(1/MC_p s)(1)}{1 + K_\ell/MC_p s} \tag{12-4.14}$$

$$= \frac{1}{MC_p s + K_\ell} \tag{12-4.15}$$

However, $q_{\text{in}}(s)$ is not the only independent input contributing to $T_m(s)$ in this system. The ambient temperature is also an input, and the contribution of $T_a(s)$ must be considered. By using *superposition of effects* in this *linear model*, the total response can be obtained.

For the T_a-to-T_m signal transmission, only one path exists, and it touches the only loop. Thus the T_a contribution is

$$\left.\frac{T_m(s)}{T_a(s)}\right|_{q_{\text{in}}=0} = \frac{(K_\ell/MC_p s)(1)}{1 + K_\ell/MC_p s} = \frac{K_\ell}{MC_p s + K_\ell} \tag{12-4.16}$$

It should be noted that considering paths from a different input node does not change the graph determinant, $\Delta(s)$, which was used in Eq. 12-4.13. The total output is obtained by solving Eqs. 12-4.15 and 12-4.16 for $T_m(s)$ and adding the two contributions. The reduced signal-flow graph is shown in Fig. 12-4.9.

Example 12-4.7. As another example of the application of Mason's Rule, the ladder network problem of Fig. 12-4.6(a) is completed. The contribution to $V_3(s)$ from $E_s(s)$ is considered first. As mentioned above, there is only the one path involved with gain

$$P_1(s) = R_1C_1R_2C_2R_3C_3s^3 \tag{12-4.17}$$

and since all loops touch this path, all loops are deleted from $\Delta(s)$ to obtain

$$\Delta_1 = 1.0 \tag{12-4.18}$$

The graph determinant, $\Delta(s)$, is given by Eq. 12-4.11. Thus,

$$\left.\frac{V_3(s)}{E_s(s)}\right|_{I_{3o}=0} = \frac{(R_1C_1R_2C_2R_3C_3s^3)(1)}{\Delta(s)} \tag{12-4.19}$$

This would be the voltage gain for the circuit of Fig. 12-3.12 with no load on the output, or $I_{3o} = 0$, and the input driven by a voltage source.

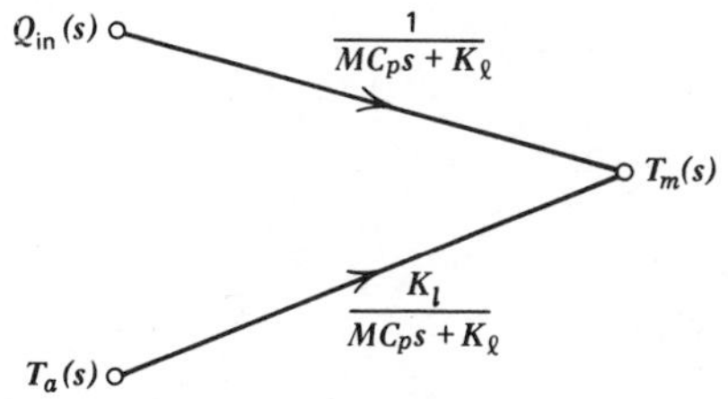

Fig. 12-4.9

It is of interest to consider the effect of de-energizing the voltage driving source, E_s, and driving the network with the current source, I_{3o}. The output could be taken as either V_3 or I_{s1}, depending on the transmission of interest. To continue the example, V_3 is taken as the output. The graph determinant, $\Delta(s)$, is the same for this case as in Eq. 12-4.11. However, the one path from $I_{3o}(s)$ to $V_3(s)$ in Fig. 12-4.6(a) does not touch all loops. A path cofactor of other than unity is obtained. The path cofactor is obtained from the $\Delta(s)$ of Eq. 12-4.11 by deleting all terms that involve loop l_5 as defined in Fig. 12-4.8, since the path under consideration touches only l_5. This leads to

$$P_1(s) = -R_3 \qquad (12\text{-}4.20)$$

$$\Delta_1(s) = (1 + R_1 C_1 s + R_1 C_2 s + R_2 C_2 s + R_2 C_3 s + R_1 C_1 R_2 C_2 s^2$$
$$+ R_1 C_1 R_2 C_3 s^2 + R_1 C_2 R_2 C_3 s^2) \quad (12\text{-}4.21)$$

and to the desired transfer function

$$\frac{V_3(s)}{I_{3o}(s)} = \frac{(-R_3)\,\Delta_1(s)}{\Delta(s)} \qquad (12\text{-}4.22)$$

in which $\Delta_1(s)$ is given by Eq. 12-4.21 and $\Delta(s)$ by Eq. 12-4.11.

The ease of formulating the various transfer functions, once the signal-flow graph is drawn, is a particularly attractive feature of this approach. It might be pointed out that two other transfer functions could be of interest in the signal-flow graph of Fig. 12-4.6. These are $I_{s1}(s)/E_s(s)$ and $I_{s1}(s)/I_{3o}(s)$ and both use the same graph determinant as above. So long as the terminal loading is not changed, the graph determinant will not change. The labor involved in using Mason's Rule here should be compared with the step-by-step reduction started in Example 12-4.3.

Example 12-4.8. As another example of applying Mason's Rule, the d'Arsonval meter movement of Examples 12-3.2 and 12-4.2 is reconsidered. Referring to Fig. 12-4.5(a), it is evident that there are three loops and that they all touch. The graph determinant is therefore

$$\Delta(s) = 1 + \frac{B}{Js} + \frac{K_s}{Js^2} + \frac{K_g K_T}{(R+Ls)\,(Js)} \qquad (12\text{-}4.23)$$

The only path from $E_a(s)$ to $\Theta(s)$ has the gain

$$P_1(s) = \frac{K_T}{(R+Ls)\,(Js^2)} \qquad (12\text{-}4.24)$$

and, since all loops touch this path, its cofactor is

$$\Delta_1(s) = 1.0 \qquad (12\text{-}4.25)$$

The overall transfer function is now

$$\frac{\Theta(s)}{E_a(s)} = \frac{\{K_T/(R+Ls)\,(Js^2)]\}\,(1)}{1 + (1/Js)\{B + [K_s/s] + [K_g K_T/(R+Ls)]\}} \qquad (12\text{-}4.26)$$

which checks the expression found in Example 12-4.2 and Fig. 12-4.5 by the step-by-step reduction procedure.

The step-by-step reduction procedure and the use of Mason's Rule for direct reduction as illustrated in this section are the main tools used in obtaining the desired information from signal-flow graphs. The remaining sections describe some special techniques that aid in obtaining even more information from signal-flow graphs.

12-5. THE AUGMENTED GRAPH

We apply the term *augmented graph* to a concept proposed by Mason.[1] Suppose we wish to find a transmittance from some input node U_i to some output node X_k in a system as suggested in Fig. 12-5.1(a). Although this transmittance is the solution we desire, we indicate it in (b) and augment the graph by adding another transmittance $1/B_{ki}$ from the output node back to the input node. We have now formed a loop that has a loop transmittance, as shown by

$$(B_{ki})\left(\frac{1}{B_{ki}}\right) = 1 \tag{12-5.1}$$

If we find the system determinant of the augmented graph denoted by $\Delta^A(s)$, we find

$$\Delta^A(s) = 1 - 1 = 0 \tag{12-5.2}$$

We use this equation to determine B_{ki} as indicated in Example 12-5.1.

Example 12-5.1. The signal-flow diagram to be used is shown in Fig. 12-5.2(a). This diagram is developed in a later section although the diagram is simplified and the notation is changed for our present purpose.

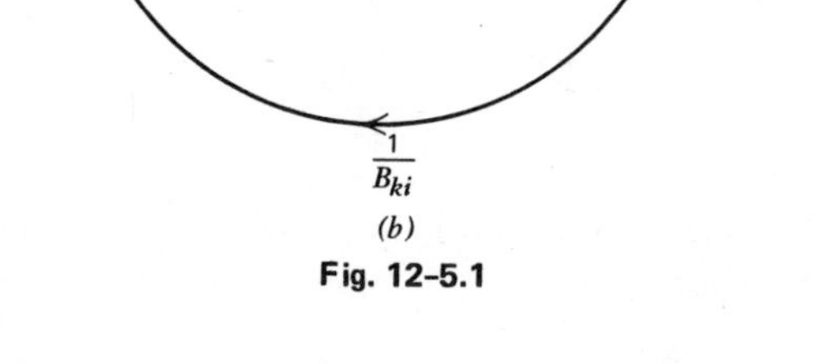

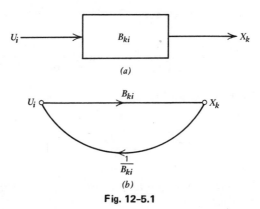

Fig. 12-5.1

[1] S. J. Mason and H. J. Zimmerman, *Electronic Circuits, Signals and Systems*, John Wiley & Sons, New York, 1960, pp. 116.

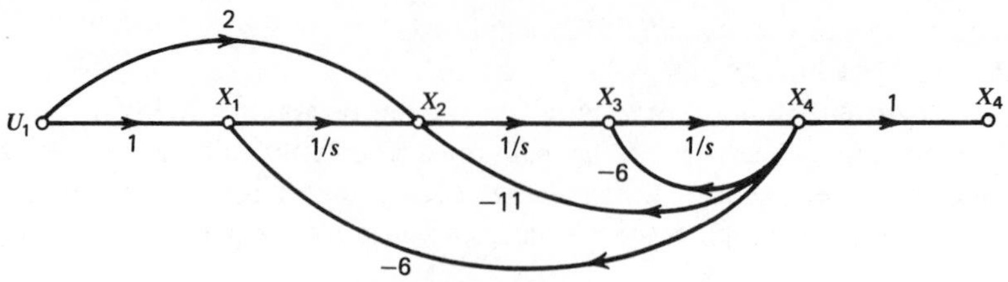

Fig. 12-5.2(a)

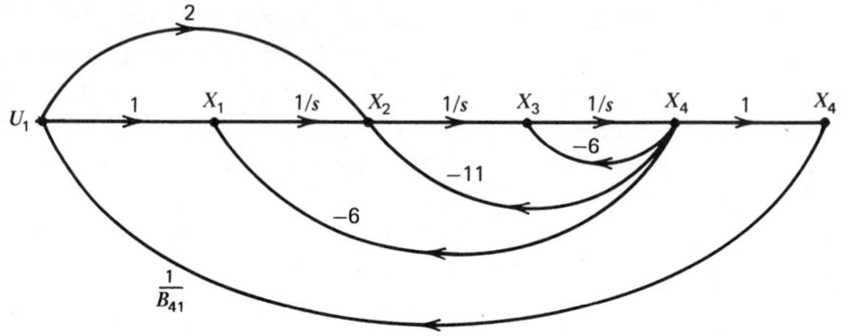

(b)

Fig. 12-5.2(b)

The system transmittance B_{41} is found in the usual way in (a), where

$$L_1 = \frac{-6}{s} \qquad L_2 = \frac{-11}{s^2} \qquad L_3 = \frac{-6}{s^3}$$

$$P_1(s) = \frac{1}{s^3} \qquad \Delta_1(s) = 1 \qquad P_2(s) = \frac{2}{s^2} \qquad \Delta_2(s) = 1$$

$$B_{41} = \frac{X_4}{U_1} = \frac{2/s^2 + 1/s^3}{1 + 6/s + 11/s^2 + 6/s^3} = \frac{2s+1}{s^3 + 6s^2 + 11s + 6}$$

The augmented graph is drawn in Fig. 12-5.2(b). For this graph we have the same three loops as in (a), repeated as

$$L_1 = \frac{-6}{s} \qquad L_2 = \frac{-11}{s^2} \qquad L_3 = \frac{-6}{s^3} \tag{12-5.3}$$

but in addition to this, there are two new loops which result from augmenting the graph. These new loop transmittances are

$$\frac{1}{B_{41}s^3} \qquad \frac{2}{B_{41}s^2} \tag{12-5.4}$$

The determinant for the augmented graph is

$$\Delta^A(s) = 0 = 1 + \frac{6}{s} + \frac{11}{s^2} + \frac{6}{s^3} - \frac{1}{B_{41}s^3} - \frac{2}{B_{41}s^2} \tag{12-5.5}$$

which can be solved for B_{41}, as

$$B_{41} = \frac{2s + 1}{s^3 + 6s^2 + 11s + 6} \tag{12-5.6}$$

This checks the previous result.

The augmented graph transforms what would be paths in the usual Mason's Rule approach into loops. The transmittance of each of the loops of this type is multiplied by the reciprocal of the transmittance to be determined. All the steps that were necessary before are also necessary now, except the procedure is more systematized. The concept of non-touching loops, etc., can be extended to include the paths by the use of the augmented graph. The main advantage is when the graph is reduced using a computer. Using the augmented graph approach, a program need only be written to determine loops. The loop transmittances that contain the reciprocal of the desired transmittance are placed in the numerator and the others are placed in the denominator. Thus, the reciprocal of the desired transmittance serves as a sort of flag to separate the loops into two sets.

12-6. SIGNAL-FLOW GRAPHS AND SIMULATION DIAGRAMS

The idea of simulation involves analog or digital computer models of physical systems. If signal-flow diagrams are developed using analog computer type of operations—that is, summing, integration, constant gains, etc.—then the major difference between signal-flow graphs of this sort and analog computer program diagrams is primarily a matter of notation. One particular approach is taken up in the examples that follow. The basic problem considered is the development of signal-flow diagrams to realize a transfer function in the form of Eq. 12-6.1, using only the common analog computer type of operations.

$$\frac{Y(s)}{U(s)} = G(s) = \frac{b_m s^m + \cdots + b_0}{s^n + a_{n-1}s^{n-1} + \cdots + a_0} \qquad m \leqslant n \tag{12-6.1}$$

Example 12-6.1. A technique called *direct programing* is used to develop a simulation diagram for the transfer function

$$\frac{Y(s)}{U(s)} = \frac{2s + 1}{s^3 + 6s^2 + 11s + 6} \tag{12-6.2}$$

The first step requires dividing numerator and denominator by the highest power of s in the denominator. This results in

$$\frac{Y(s)}{U(s)} = \frac{2/s^2 + 1/s^3}{1 + 6/s + 11/s^2 + 6/s^3} \tag{12-6.3}$$

The development of the simulation diagram proceeds by *realizing* the denominator of the transfer function as a signal-flow graph determinant, $\Delta(s)$, and the numerator as a sum of path transmissions times path cofactors.

To simplify matters, since many arbitrary choices must be made, all loops are forced to touch so that the denominator, $\Delta(s)$, consists of only the terms

$$\Delta(s) = 1 - \sum_{i=1}^{n} L_i \tag{12-6.4}$$

Thus, non-touching loop pairs, triplets, etc., are eliminated by choice or design. Similarily by choice, all input-output paths are made to touch all loops, thereby forcing all path cofactors to be unity. The numerator terms are, then, just path gains.

Inspection of the denominator of Eq. 12-6.3, subject to the constraints of Eq. 12-6.4, reveals that there must be three loops with gains $-6/s$, $-11/s^2$, and $-6/s^3$. To realize these with a minimum number of integrators, $1/s$ terms, apparently three integrators will have to be cascaded to obtain the $1/s^3$ needed. Therefore, to start the simulation signal-flow graph, three integrators are cascaded, as shown in Fig. 12-6.1(a). Unit-gain isolating branches have been added for a reason that is mentioned later. Since this transfer function describes a third-order system, three internal new variables are required. These variables are labeled x_1, x_2 and x_3.

The three required loops can now be closed in several ways, one of which is that shown in Fig. 12-6.1(b). It should be noted that all loops touch as mentioned previously. At this point the use of the unity gain branches is considered. The $(1/s)$ operation is merely an integration, and if its output is x_j, then its input must be $\dot{x}_j$. Several branches may enter the $\dot{x}_j$ node, but *only* the $(1/s)$ branch may lead from $\dot{x}_j$ to x_j together with perhaps an initial-condition input. In order to label the output of each integrator with a different x_j-variable, and feed this into a following $\dot{x}_{j-1}$ node that has other inputs, it is necessary to separate the x_j and $\dot{x}_{j-1}$ nodes by some path. The unity gain path is used for this purpose. Figure 12-6.2 illustrates the problem. In Fig. 12-6.2(a), the variable $\dot{x}_1$ is *not* the variable x_2 from Fig. 12-6.1(b). However, as in Fig. 12-6.2(b) it could be called an x_2, but then the input to the left hand integrator in Fig. 12-6.2 is not $\dot{x}_2$ because two paths combine to form x_1. Thus, if the variable labeling is desired as in Fig. 12-6.1(b), the unit-gain isolating paths are needed.

To complete the simulation diagram, the numerator of Eq. 12-6.3 must be realized. To force all input-output paths to touch all loops, as mentioned, these paths are forced through the x_1 node that is shared by all loops. A unity gain path from x_1 to y is added as the only allowed path from the loop subgraph to the output node, y. The two paths' gains of $(2/s^2)$ and $(1/s^3)$ can now be added, as shown in Fig. 12-6.1(c), to connect the input, u, to the output, y. Again it should be noted that these paths touch all loops and therefore have unity path cofactors.

In some cases it is desired to introduce initial conditions in the simulation diagram. This can be done by adding suitable inputs to the variable that is the output of each integrator operation. For example, from Laplace transforms

$$\mathcal{L}\left\{\frac{dx}{dt}\right\} = sX(s) - x(0^+) \tag{12-6.5}$$

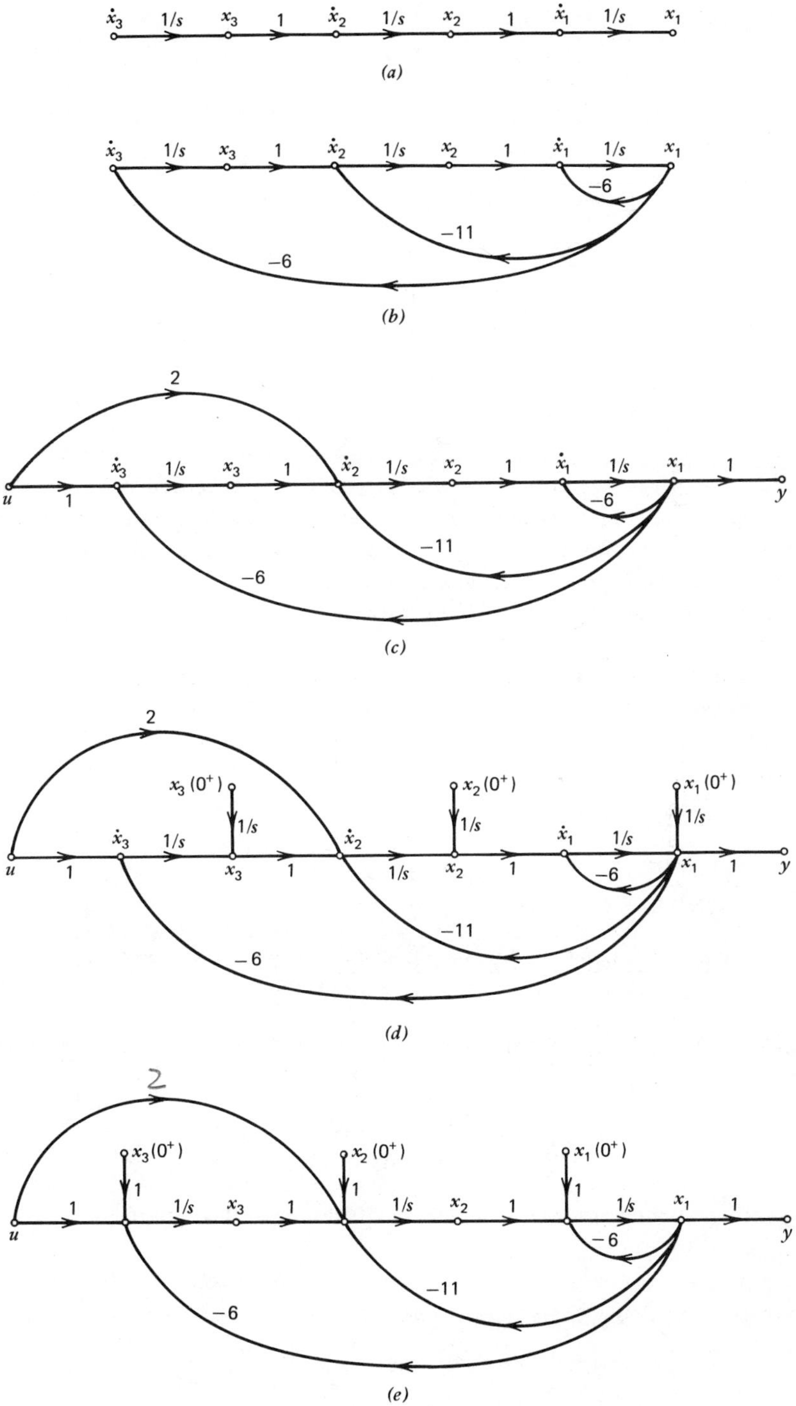

$\dot{x}_3$ 1/s x_3 1 $\dot{x}_2$ 1/s x_2 1 $\dot{x}_1$ 1/s x_1
(a)
$\dot{x}_3$ 1/s x_3 1 $\dot{x}_2$ 1/s x_2 1 $\dot{x}_1$ 1/s x_1
−6
−11
−6
(b)
2
u 1 $\dot{x}_3$ 1/s x_3 1 $\dot{x}_2$ 1/s x_2 1 $\dot{x}_1$ 1/s x_1 1 y
−6
−11
−6
(c)
2
$x_3(0^+)$ 1/s
$x_2(0^+)$ 1/s
$x_1(0^+)$ 1/s
u 1 $\dot{x}_3$ 1/s x_3 1 $\dot{x}_2$ 1/s x_2 1 $\dot{x}_1$ 1/s x_1 1 y
−6
−11
−6
(d)
2
$x_3(0^+)$ 1
$x_2(0^+)$ 1
$x_1(0^+)$ 1
u 1 1/s x_3 1 1/s x_2 1 1/s x_1 1 y
−6
−11
−6
(e)

Fig. 12-6.1

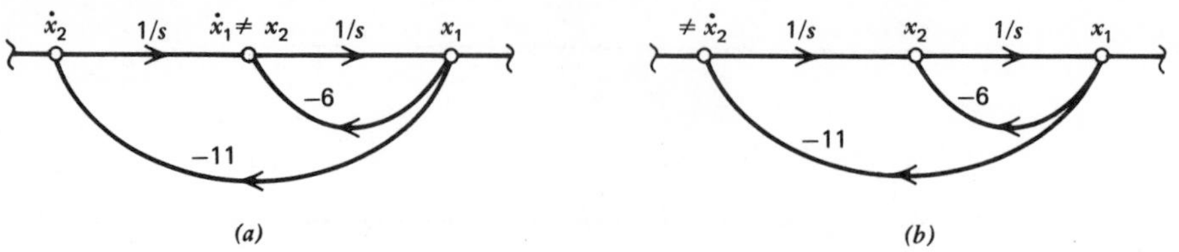

Fig. 12-6.2

$$X(s) = \frac{1}{s}\,\mathcal{L}\left\{\frac{dx}{dt}\right\} + \frac{x(0^+)}{s} \tag{12-6.6}$$

$$X(s) = \frac{1}{s}\left[\mathcal{L}\left\{\frac{dx}{dt}\right\} + x(0^+)\right] \tag{12-6.7}$$

This indicates that each $X_j(s)$ node should have an extra input that is $x_j(0+)/s$. As indicated in Fig. 12-6.1(d), these inputs can be added as impulse inputs through an extra integral, or $(1/s)$, operation. Also, as in Fig. 12-6.1(e), the initial conditions can be added by impulse inputs to the basic integrators, Eq. 12-6.7.

The problem of relating the $x_j(0+)$ initial conditions to $y(0)$, $\dot{y}(0)$, $\ddot{y}(0)$, etc., is another matter that must be considered. The original transfer function Eq. 12-6.2 by definition assumes zero initial conditions. However, most likely the transfer function replaces a differential equation relationship of the form

$$\dddot{y} + 6\ddot{y} + 11\dot{y} + 6y = 2\dot{u} + u \tag{12-6.8}$$

This equation transforms to give

$$Y(s) = \frac{(2s+1)\,U(s)}{s^3 + 6s^2 + 11s + 6} + \frac{y(0^+)s^2 + [6y(0^+) + \dot{y}(0^+)]s + [11y(0^+) + 6\dot{y}(0^+) + \ddot{y}(0^+) - 2u(0^+)]}{s^3 + 6s^2 + 11s + 6} \tag{12-6.9}$$

From either the (d) or the (e) signal-flow graph of Fig. 12-6.1, the initial condition transmission is

$$Y(s)_{ic} = \frac{x_1(0^+)s^2 + x_2(0^+)s + x_3(0^+)}{s^3 + 6s^2 + 11s + 6} \tag{12-6.10}$$

Term-by-term matching of power of s coefficients results in the required initial conditions

$$\begin{aligned} x_1(0^+) &= y(0^+) \\ x_2(0^+) &= 6y(0^+) + \dot{y}(0^+) \\ x_3(0^+) &= 11y(0^+) + 6\dot{y}(0^+) + \ddot{y}(0^+) - 2u(0^+) \end{aligned} \tag{12-6.11}$$

An alternate approach is to note that

$$\begin{aligned}
y(t) &= x_1(t) \\
\dot{y}(t) &= \dot{x}_1(t) = -6x_1(t) + x_2(t) \\
\ddot{y}(t) &= \ddot{x}_1(t) = -6\dot{x}_1(t) + \dot{x}_2(t) \\
&= -6[-6x_1(t) + x_2(t)] + [-11x_1(t) + x_3(t) + 2u(t)] \\
&= 25x_1(t) - 6x_2(t) + x_3(t) + 2u(t)
\end{aligned} \tag{12-6.12}$$

In matrix form for $t = 0^+$

$$\begin{bmatrix} y(0^+) \\ \dot{y}(0^+) \\ \ddot{y}(0^+) \end{bmatrix} = \begin{bmatrix} 1 & 0 & 0 \\ -6 & 1 & 0 \\ 25 & -6 & 1 \end{bmatrix} \begin{bmatrix} x_1(0^+) \\ x_2(0^+) \\ x_3(0^+) \end{bmatrix} + \begin{bmatrix} 0 \\ 0 \\ 2 \end{bmatrix} u(0+) \tag{12-6.13}$$

which gives

$$\begin{bmatrix} x_1(0^+) \\ x_2(0^+) \\ x_3(0^+) \end{bmatrix} = \begin{bmatrix} 1 & 0 & 0 \\ 6 & 1 & 0 \\ 11 & 6 & 1 \end{bmatrix} \begin{bmatrix} y(0^+) \\ \dot{y}(0^+) \\ \ddot{y}(0^+) \end{bmatrix} - \begin{bmatrix} 0 \\ 0 \\ 2 \end{bmatrix} u(0+) \tag{12-6.14}$$

as found previously.

The technique used in this example is but one of several methods referred to as *direct programing*. A second version of this approach is presented in the following example in order to illustrate that such schemes are certainly not unique.

Example 12-6.2. As in Example 12-6.1, the transfer function

$$\frac{Y(s)}{U(s)} = \frac{2s + 1}{s^3 + 6s^2 + 11s + 6} \tag{12-6.15}$$

is to be programed in a simulation diagram. The steps to be followed parallel those of the previous example, except that all loops and input-output paths are forced to pass through the x_3 node. The results are shown in Fig. 12-6.3, with the details left for the

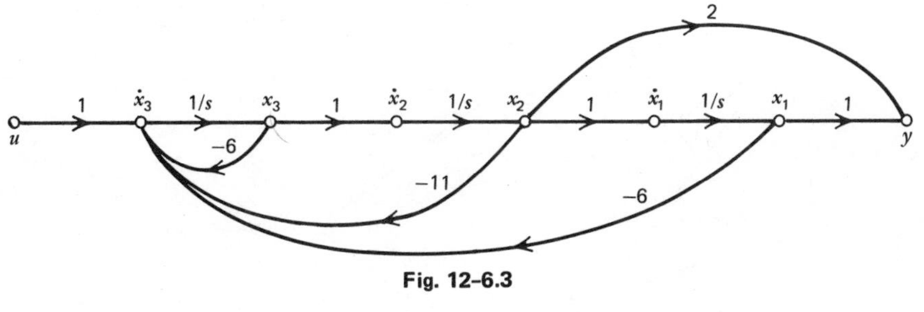

Fig. 12-6.3

reader to consider. Initial conditions can again be filled in as in Example 12-6.1, if desired, but new relations to $y(0), \dot{y}(0), \ddot{y}(0)$ exist.

While other versions of direct programing exist, the two presented here are most typical and easiest to remember.

CONCLUSION

The signal-flow graph is an interesting and useful tool. Only the surface of its total usefulness has been explored in this chapter. Nevertheless it is hoped that sufficient detail has been presented to spur the reader's interest to look further and also to use the techniques that have been introduced.

PROBLEMS

12-1 Given the signal flow graph as shown:

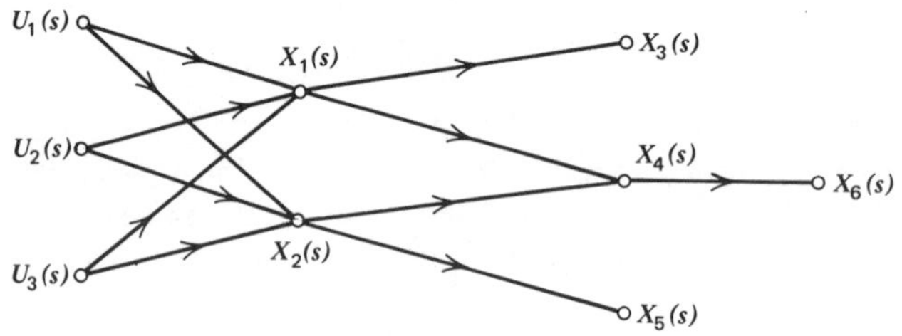

(a) Label each transmittance with its proper symbol.

(b) Write the set of equations represented by this graph, then put them in matrix form. Are these equations in triangular form?

(c) Solve these equations for the values of the sink nodes by direct substitution, then check the results by tracing input–output paths and multiplying the transmittances encountered.

12-2 Repeat Problem 12-1 for the following signal-flow graph:

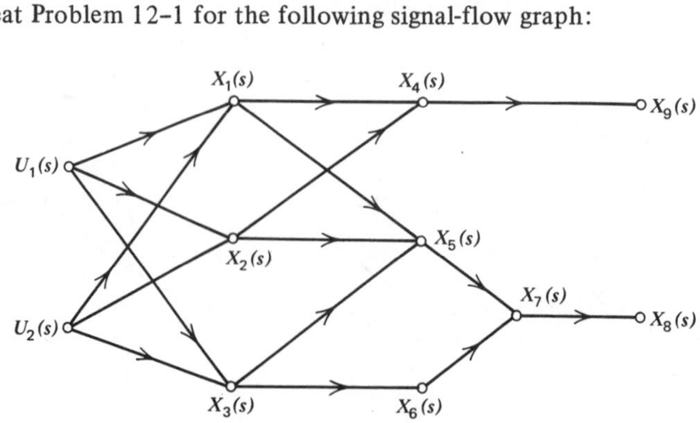

12-3 Repeat Problem 12-1 for the following signal-flow graph:

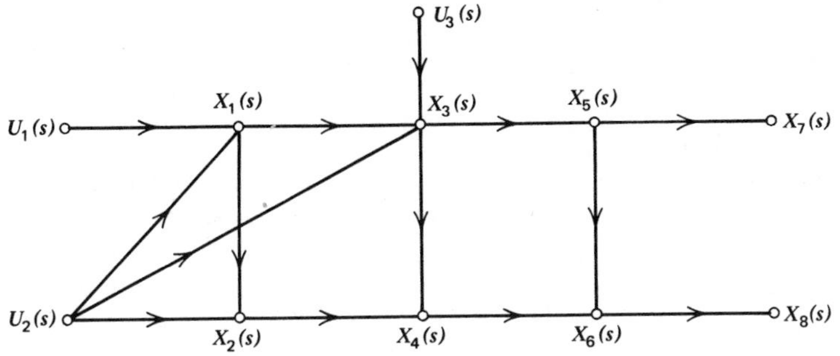

12-4 Start with the figure of Problem 12-1, and add: (1) self loops at $X_1(s)$ and $X_2(s)$; (2) branches directed from $X_3(s)$ to $X_4(s)$, from $X_4(s)$ to $X_5(s)$, from $X_4(s)$ to $X_1(s)$, and from $X_5(s)$ to $X_1(s)$.

(a) Label each transmittance with its proper label.
(b) Write the set of equations represented by this graph, then put them in matrix form.
(c) See if you can rearrange these equations and put them in triangular form.
(d) Enumerate all loops (l) and write out each of the loop transmittance product (L).
(e) Enumerate all the input to sink paths (p), and write out all the path transmittance products (P).

12-6 Repeat Problem 12-4, except start with the graph of Problem 12-2 and add the following branches: (1) self loops at $X_1(s)$, $X_5(s)$ and $X_6(s)$; and (2) branches from $X_4(s)$ to $X_5(s)$, from $X_5(s)$ to $X_6(s)$, and from $X_6(s)$ to $X_2(s)$.

12-7 Given the following equations:

(1) $5 = 2X_1 + 3X_2 + 4X_3$ (2) $6 = 3X_1 + 5X_2 + 4X_3$
(3) $10 = 4X_1 + 6X_2 + 3X_3$

Solve the first equation for X_1, the second equation for X_2, and the third equation for X_3. Draw the resulting signal-flow diagram.

12-8 Use the equations of Problem 12-7. Solve the first equation for X_2, the second equation for X_3 and the third equation for X_1. Draw the resulting signal-flow diagram.

12-9 Start with the equations of Problem 12-7, but write them in the following manner:

(1) $X_1 = 5 - X_1 - 3X_2 - 4X_3$ (2) $X_2 = 6 - 3X_1 - 4X_2 - 4X_3$
(3) $X_3 = 10 - 4X_1 - 6X_2 - 2X_3$

Draw the resulting signal-flow diagram.

12-10 Repeat Problem 12-7 for the following equations:

(4) $4 = 5X_1 - 3X_2 - X_3$ (5) $2 = -2X_1 + 3X_2 - X_3$
(6) $1 = -3X_1 - 4X_2 + 6X_3$

12-11 Repeat Problem 12-8, except use the equations of Problem 12-10.

12-12 Start with the equations of Problem 12-10, but write them in the following manner:

(4) $X_1 = 4 - 4X_1 + 3X_2 + X_3$ (5) $X_2 = 2 + 2X_1 - 2X_2 + X_3$
(6) $X_3 = 1 + 3X_1 + 4X_2 - 5X_3$

Draw the resulting signal-flow diagram.

12-13 Given the circuit as shown: write the window current equations. Finish the problem in a manner analogous to that of Problem 12-7.

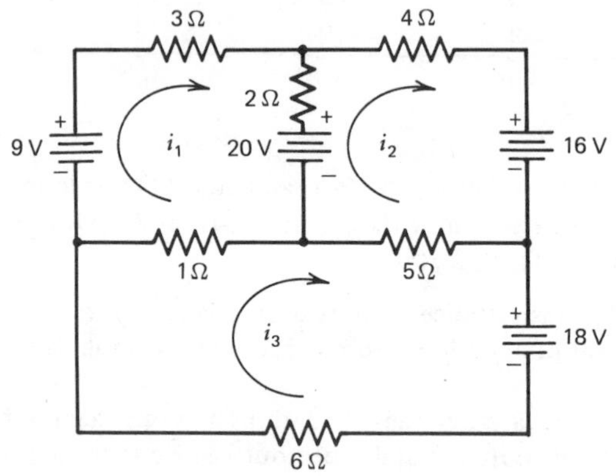

12-14 Use the circuit of Problem 12-13, except use the loop currents as shown on the accompanying sketch. Finish the problem in a manner analogous to that of Problem 12-7.

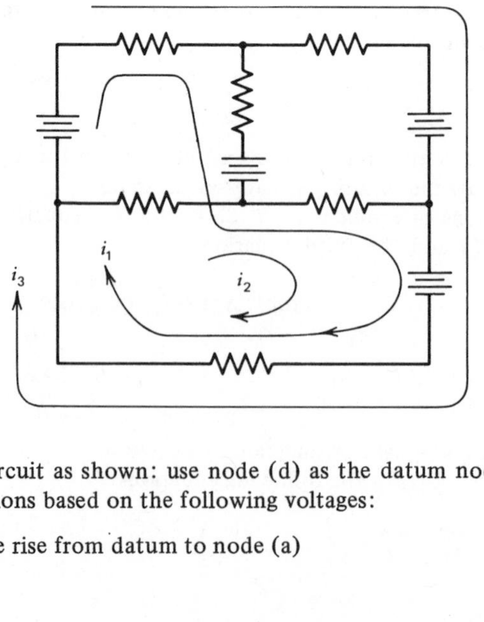

12-15 Given the circuit as shown: use node (d) as the datum node and write node-to-datum equations based on the following voltages:

v_1 = voltage rise from datum to node (a)

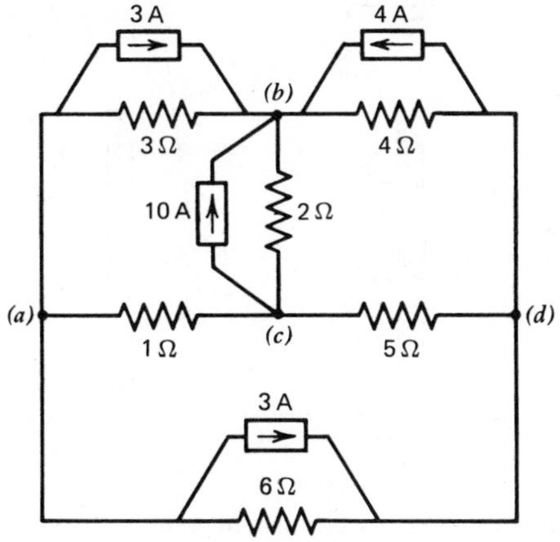

v_2 = voltage rise from datum to node (b)

v_3 = voltage rise from datum to node (c)

Finish the problem in a manner analogous to that of Problem 12–7.

12–16 Use the circuit of Problem 12–15, except use node (c) as datum, with the voltages defined as:

v_1 = voltage rise from datum to node (a)

v_2 = voltage rise from datum to node (b)

v_3 = voltage rise from datum to node (d)

Finish the problem in a manner analogous to that of Problem 12–7.

12–17 The low pass filter shown is to be solved for various items as indicated.

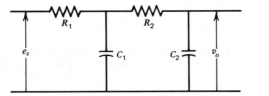

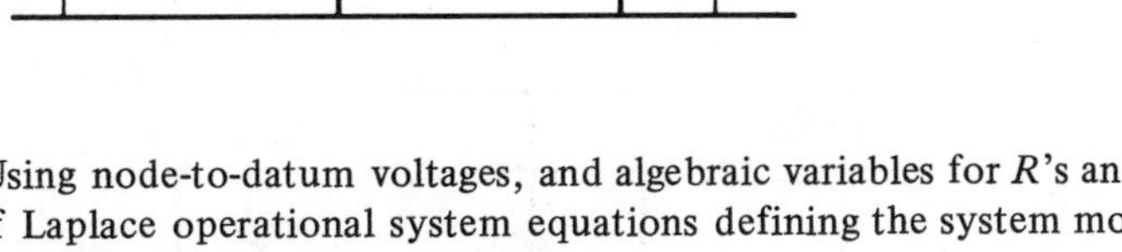

(a) Using node-to-datum voltages, and algebraic variables for R's and C's, write a set of Laplace operational system equations defining the system model, and construct a signal-flow diagram for the filter.

(b) Use step-by-step reduction of the signal-flow diagram to find $V_o(s)/E_s(s)$.

(c) Find the loops in the signal-flow diagram, and from these the graph determinant.

(d) Find the cofactor and the path transmittance of the path from $E_s(s)$ to $V_o(s)$.

(e) Use Mason's Rule to find $V_o(s)/E_s(s)$.

(f) Locate the poles and zeros of the transfer function $V_o(s)/E_s(s)$.

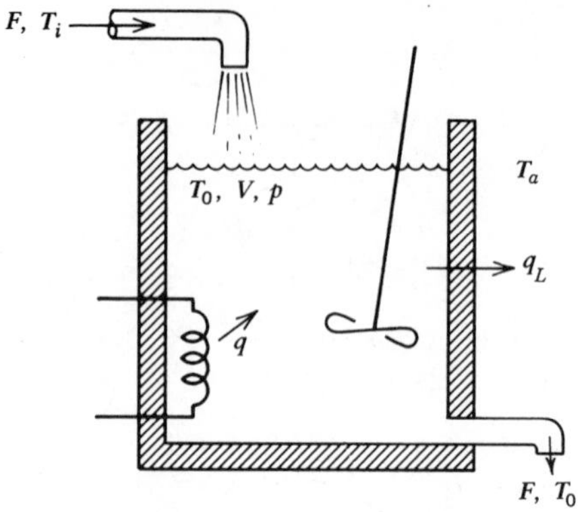

12-18 A typical simplified thermal system of the general type used before, in Chapter 8, is shown, where

F = input and output constant weight flow rate
V = fluid volume in tank
ρ = fluid weight/unit volume
T_o = tank and output fluid temperature
T_a = air temperature
T_i = input fluid temperature
q_L = heat loss rate, $K_L(T_o - T_a)$
q = independent heat input rate

(a) Write a set of differential equations for this system, Laplace transform them, and construct a signal-flow graph for the system. What quantities are independent inputs (source nodes)?
(b) Identify the loops in the graph, and determine $\Delta(s)$.
(c) Use step-by-step reduction to find $T_o(s)$.
(d) Use Mason's Rule to find $T_o(s)$.

12-19 Construct a signal-flow graph for the twin-tee network shown. Determine $V_o(s)/E_s(s)$ by using Mason's Rule. Check the result by a non-signal-flow graph method.

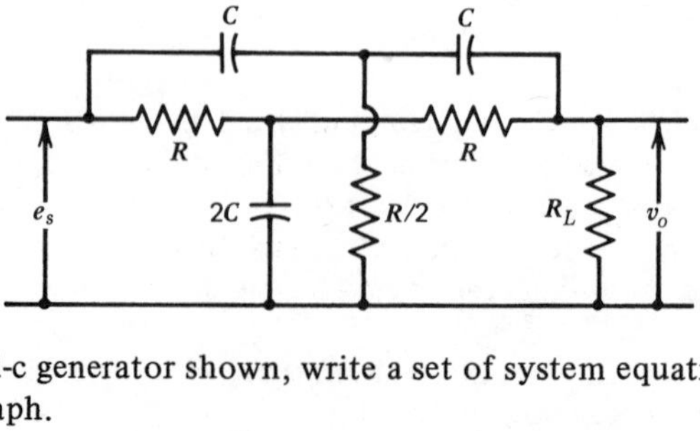

12-20 (a) For the d-c generator shown, write a set of system equations, and construct a signal-flow graph.
(b) Determine the system poles and zeros for $V_o(s)/E_f(s)$.

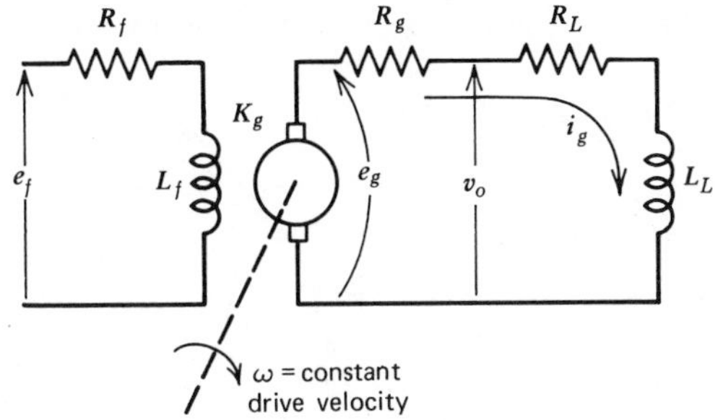

(c) Use the augmented graph technique of Section 12-5 to find the transmittance, $I_g(s)/E_f(s)$.
(d) Repeat (c) for the transmittance, $V_o(s)/E_f(s)$.

12-21 (a) Determine the graph determinant for the signal-flow graph shown.

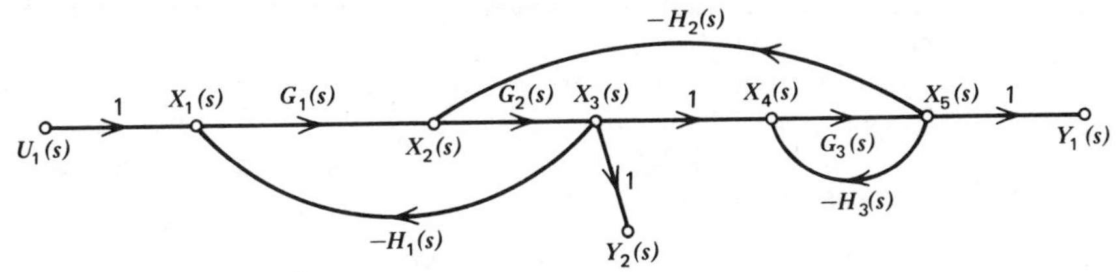

(b) Determine the path cofactor for the transmittance, $Y_1(s)/U_1(s)$.
(c) Repeat (b) for $Y_2(s)/U_1(s)$.

12-22 The d-c motor-generator system shown is to be studied by use of signal-flow graphs.

(a) Write a set of differential equations for the system, and Laplace transform them.
(b) Construct a signal-flow graph for the system with $\Omega_m(s)$ and $I_a(s)$ as outputs.
(c) Find $\Omega_m(s)/E_f(s)$ and $I_a(s)/E_f(s)$.
(d) Determine $i_a(t)$ and $\omega_m(t)$, if $e_f(t)$ is a step function of E_{f0} volts.

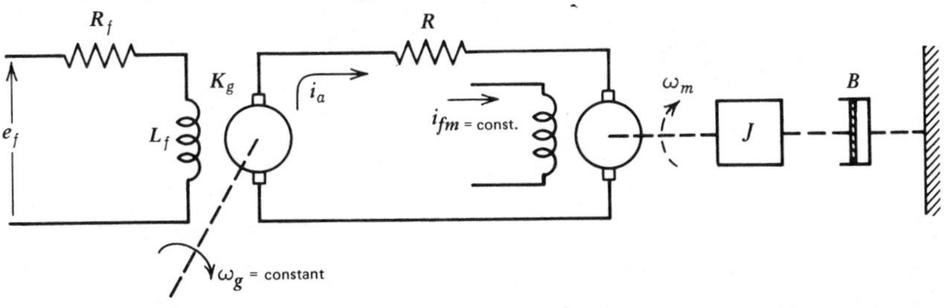

12-23 (a) Use the augmented graph of Section 12-5 to find $Y_1(s)/U_1(s)$ for the system shown for Problem 12-21.
(b) Repeat (a) for $Y_2(s)/U_1(s)$.

12-24 (a) Determine the graph determinant for the signal-flow graph shown.

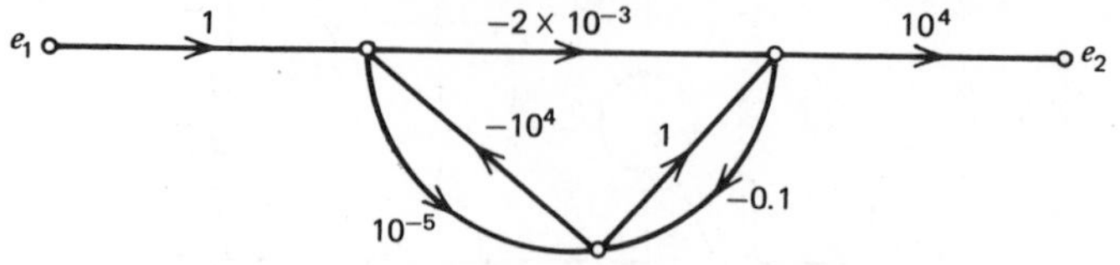

(b) Why can $e_1(t)$ and $e_2(t)$, etc., be used in this graph rather than $E_1(s)$, etc.?
(c) What is the transmittance of $E_2(s)/E_1(s)$?

12-25 A simplified model of an economic system related to the national personal income is to be based upon the following assumptions:

1. Actual national income is subtracted from the desired national income to obtain an error that is used to determine government spending by passing the error through the lag transfer function $K_{gs}/(1+s)$ (time is in years).
2. Funds available for business to use in producing national income consist of government spending, consumer spending, and private business investment (considered an independent input).
3. Consumer spending is 80% of after-tax actual national income.
4. Taxes are 20% of actual national income.
5. Actual national income is determined by passing funds available for producing income through the transfer function, $K_{fi}/(1+0.5s)$.

(a) Draw a signal-flow graph to model this system.
(b) Determine the transmittance from changes in desired national income to actual national income.
(c) Determine the transmittance from private business investment to actual national income.
(d) Discuss the effect of the *gains* K_{gs} and K_{fi} on the system behavior.
(e) What effects result from changes in the tax rate?

12-26 (a) Determine the loops and the graph determinant for the signal-flow graph shown.

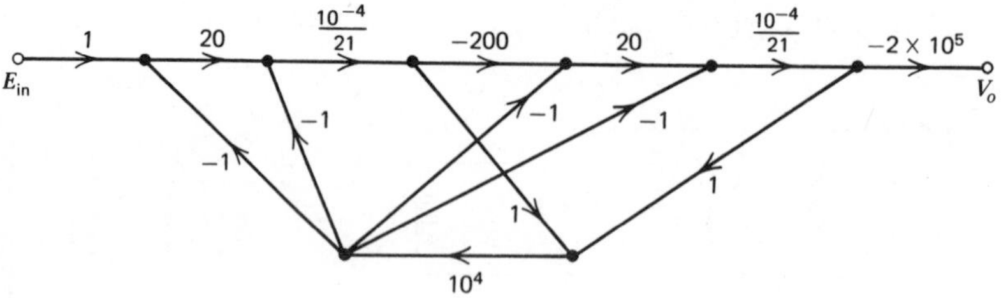

(b) Use step-by-step reduction to find $V_o(s)/E_{in}(s)$.
(c) Use Mason's Rule to find $V_o(s)/E_{in}(s)$.

Chapter 13

State Variables: I

13-1. INTRODUCTION

Some of the most useful system analysis and design approaches ever to be developed have become known as state-variable or state-space techniques. The development of these techniques has occurred mainly since 1955 as the result originally of work in control systems. In reality, the basic mathematical tools used in the state-variable approach have been in existence for many years, and so the main new contribution has been that of associating the mathematical techniques with the physical problems and interpreting the results.

The development of state-variable techniques evolved mainly in efforts to obtain more general techniques of formulating system problems, of gaining insight into the behavior of systems, and of finding mathematical formulations that better lent themselves to computer solution methods. State-variable methods do indeed lend themselves to the gaining of insight into system dynamic behavior, as shown later in this chapter and in the next where *modes* are discussed. The state equations have also provided convenient models for computer solution, although more general formulations involving sparse tableau equations have been developed.[1] The particular formulation of a given circuit or system problem depends to a great extent on the objective of a given study, be it design, simple analysis, optimization, fast numerical solution, or whatever. In general, the view of systems provided by state variables is believed to be of sufficient importance to warrant at least introductory study, though many system and circuit problems can now be approached through very general computer programs that require only a minimal input of data and so can handle a great variety of system studies.

To effectively use state variable techniques, it is essential to have some reasonable background in matrix and vector space theory. The presentation here and in Chapter 14, however, requires only a minimum background. For more depth, the reader is referred to Appendix D and to texts or outlines on matrix and vector space theory.

[1] G. D. Hachtel, R. K. Brayton, and F. G. Gustavson, "The Sparse Tableau Approach to Network Analysis and Design," *IEEE Trans, Circuit Theory, CT-18,* Jan., 1971, pp. 101-13.

The concept of system state and methods of selecting system state variables are considered in this chapter, and the solution of the basic state-vector differential equation and the interpretation of response modes are considered in Chapter 14.

13-2. THE CONCEPT OF SYSTEM STATE

Most introductions to the study of the dynamics of various physical systems, or for that matter the study of ordinary differential equations, consider the fact that a set of so-called *initial conditions* together with present and future forcing functions are sufficient to determine the present and future response of a system. In essence this involves a *system state* philosophy. The set of numbers taken as the initial conditions represent the state of the system as it evolved from its past history. In general, the *set of numbers* referred to above changes as times other than $t = 0$ are considered; however, at any particular time—for example, $t = t_0$—complete information concerning the effects of past history ($t < t_0$) on the system are contained in the set of numbers.

Thus, associated with the *internal condition* of a dynamic system is at least one set of numbers that change with time and describe this condition. Such a set of n numbers for an nth-order system is generally arranged as an $(n \times 1)$ column matrix or column vector and called a system *state vector* or, since the numbers vary with time, the term *state variables* is also used to describe the set of components of a state vector. It has become standard notation in the literature to use the symbols $x_j(t)$ $(j = 1, 2, 3, \cdots, n)$ to represent system state variables, and

$$\mathbf{x}(t) = \begin{bmatrix} x_1(t) \\ x_2(t) \\ \vdots \\ x_n(t) \end{bmatrix} \tag{13-2.1}$$

to represent a system state vector.

When using vector representations, an underlying vector space is implied. Almost intuitively, a geometrical representation with two rectangular coordinates is used to represent a vector with two components. In reality, an underlying set of two linearily independent two-component vectors exists to serve as a *basis* for the so-called *two-space*, and a given set of components for a general 2-vector merely expresses the components of this new vector *along* the original basis vectors. More specifically, the components of a vector express the unique linear combination of the basis vectors needed to form the desired vector.

Several terms used above perhaps need clarification. A column vector is merely an ordered set of quantities arranged as a column of elements, as in Eq. 13-2.1. If, in a set of such vectors, every sum of two vectors is still a vector of the set, then the set of vectors is said to be *closed under addition.* If the product of a scalar and a vector is still a vector of the set, then the set of vectors is said to be *closed under scalar multiplication.* Any set of n-component vectors that is closed under addition and scalar multiplication is called a *vector space.* A vector space is really a set of vectors with certain properties.

Example 13-2.1. The set of all n-component vectors with complex number values for the components constitutes an n-dimensional vector space over the field of complex numbers. This vector space, together with real-number vector spaces, is of primary interest in this chapter. Note that the two closed properties mentioned above are satisfied in the cases of this example.

Linear (and, for that matter, non-linear) combinations of quantities are quite common forms. For example

$$\mathbf{x}_3 = -3.5\mathbf{x}_1 + 2.0\mathbf{x}_2 \tag{13-2.2}$$

represents the vector $\mathbf{x}_3$ as a linear combination of the two vectors $\mathbf{x}_1$ and $\mathbf{x}_2$. A slight extension of the linear combination concept leads to the idea of linear independence. If, for a set of vectors $\mathbf{x}_j$, the *only* solution to the equation

$$a_1\mathbf{x}_1 + a_2\mathbf{x}_2 + \cdots + a_m\mathbf{x}_m = \mathbf{0} \tag{13-2.3}$$

where all a_i's are scalar constants, is that all a_i's must be zero; then the set of vectors are linearly independent. Alternatively, none of the vectors $\mathbf{x}_j$ can be expressed as a linear combination of the remaining vectors.

Example 13-2.2. To illustrate the above concepts, the set of vectors

$$\mathbf{x}_1 = \begin{bmatrix} 1 \\ 0 \end{bmatrix} \quad \mathbf{x}_2 = \begin{bmatrix} 0 \\ 1 \end{bmatrix} \quad \mathbf{x}_3 = \begin{bmatrix} 2 \\ -1 \end{bmatrix} \quad \mathbf{x}_4 = \begin{bmatrix} -1 \\ -1 \end{bmatrix} \tag{13-2.4}$$

are considered. It is apparent that

$$\mathbf{x}_3 = 2\mathbf{x}_1 - \mathbf{x}_2 \tag{13-2.5}$$

$$\mathbf{x}_4 = -\mathbf{x}_1 - \mathbf{x}_2 \tag{13-2.6}$$

so that both $\mathbf{x}_3$ and $\mathbf{x}_4$ can be expressed as linear combinations of $\mathbf{x}_1$ and $\mathbf{x}_2$. Alternately

$$2\mathbf{x}_1 - \mathbf{x}_2 - \mathbf{x}_3 = \mathbf{0} \tag{13-2.7}$$

$$\mathbf{x}_1 + \mathbf{x}_2 + \mathbf{x}_4 = \mathbf{0} \tag{13-2.8}$$

so that the set of vectors $\mathbf{x}_1$, $\mathbf{x}_2$, $\mathbf{x}_3$ is not linearly independent (a set of a_i's not all zero exists for a relation of the type of Eq. 13-2.3), and also the sets $\mathbf{x}_1$, $\mathbf{x}_2$, $\mathbf{x}_4$, and $\mathbf{x}_1$, $\mathbf{x}_2$, $\mathbf{x}_3$, $\mathbf{x}_4$, are not linearly independent. However, a set consisting of any two of the vectors of Eq. 13-2.4 is linearly independent.

The aim of this discussion is to show that there is some minimum number of linearily independent vectors in any vector space. Such a minimum, linearily independent, set of vectors is called a *basis* for the space. Any other vector can then be expressed as a unique linear combination of the vectors in the *basis set.* Vectors $\mathbf{x}_1$ and $\mathbf{x}_2$ in Eqs. 13-2.4 are the most common basis set for two-dimensional space. Extensions of these to n-vectors with n-components results in the most common *basis* for n-dimensional space. This so-called *E-basis* will be used extensively in this chapter. It might be pointed out that normally n vectors each of n-components are required as a basis for n space; however, a limit of n-components is not necessary. For example, two three-component vectors can serve as the basis of a two-dimensional space embedded in the three-dimensional space. The two-dimensional space would consist of all vectors in the *plane* formed by all possible linear combinations of the two vectors.

Example 13-2.3. If the vectors

$$\mathbf{x}_1 = \begin{bmatrix} 1 \\ 0 \\ 0 \end{bmatrix} \qquad \mathbf{x}_2 = \begin{bmatrix} 0 \\ 1 \\ 0 \end{bmatrix} \tag{13-2.9}$$

are considered, they can be seen to serve as a basis for a two-dimensional space or plane. However, this *subspace* is part of the higher dimensional three-space. Nevertheless, $\mathbf{x}_1$ and $\mathbf{x}_2$ form a basis for a two-space. The vectors

$$\mathbf{x}_3 = \begin{bmatrix} 1 \\ 1 \\ 0 \end{bmatrix} \qquad \mathbf{x}_4 = \begin{bmatrix} 0 \\ -1 \\ 0 \end{bmatrix} \tag{13-2.10}$$

form another basis for the same two-space, as spanned by vectors $\mathbf{x}_1$ and $\mathbf{x}_2$, so that it is evident that *there are many different bases for a given vector space.*

Returning to the discussion of the state of a system, it can be said that the state vector really describes the system's internal condition *in some state space.* The particular state vector will depend on the underlying set of basis vectors used. Various different state vector representations are developed in the following sections, indicating that state variables for a given system can be selected in many different ways to suit various purposes.

It is helpful in describing the state of a system to consider Fig. 13-2.1. In (a) are shown specific inputs, u_i, outputs, y_i, and internal system state variables, x_j, indicating that there

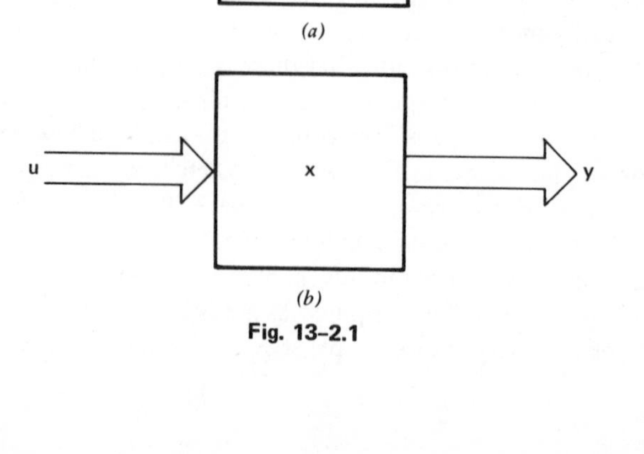

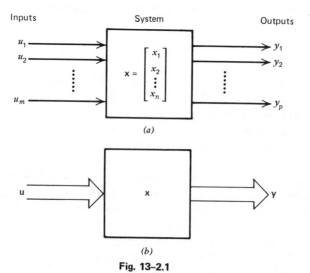

Fig. 13–2.1

is a difference between output variables and state variables. In general, output variables are linear combinations of the state variables and perhaps of the inputs. In (b) is shown a common representation of a multiple-input-multiple-output system using vector representation.

In summary, a minimum set of system variables, in which the values of the variables at some time, t_0, together with the system inputs for $t \geqslant t_0$, uniquely determine the value of the variables for $t \geqslant t_0$, is called a set of state variables. This can be stated concisely by

$$\mathbf{x}(t) = \mathbf{f}[\mathbf{x}(t_0), \mathbf{u}(t), t] \qquad t \geqslant t_0 \tag{13-2.11}$$

where the $\mathbf{f}[\cdot\cdot\cdot]$ symbol implies a vector of functions, and $\mathbf{x}(t)$ is a state vector. Additionally, the system output vector $\mathbf{y}(t)$ must also be writable as

$$\mathbf{y}(t) = \mathbf{g}[\mathbf{x}(t_0), \mathbf{u}(t), t] \qquad t \geqslant t_0 \tag{13-2.12}$$

13-3. SELECTION OF STATE VARIABLES

In accordance with the discussion of Section 13-2 regarding initial conditions and state variables, several ways to select suitable sets of state variables come to mind. In mechanical systems, knowledge of the initial positions and velocities of all masses is frequently used, together with force or position inputs, to determine dynamic response. Apparently, using the positions and velocities of all masses as a set of state variables is a possibility. Equivalently, these quantities allow the computation of the kinetic energy and the potential energy. System energy must be related to the system state. In fact, a development of state variables based on energy functions can be made; however, the reader is referred to the literature[2] for this approach.

In the case of electrical circuits, it is known that knowledge of the initial voltages across capacitors and the currents through inductors, together with the inputs, suffice to determine the responses for present and future time. Therefore, a selection of such currents and voltages gives a set of suitable state variables. It should be pointed out that the system-stored energy is also determined by these variables.

In many systems such physical variables provide a natural choice for a state vector. In many cases, however, some different selection is easier to obtain and also to use for analysis and design. For example, sets of state variables particularily easy to obtain are the outputs of all integrators in an analog computer simulation diagram, or in a signal-flow graph simulation diagram. Initial values for the outputs of all integrators in such a simulation, plus the forcing functions or inputs, result in a unique response. Thus, the requirements for a state vector are met. These integrator outputs may or may not correspond directly to physical system variables. To make the above discussion more lucid, the following subsections and examples are presented.

[2] D. G. Schultz and J. L. Melsa, *State Functions and Linear Control Systems*, McGraw-Hill Book Company, New York, 1967, Chapter 3.

13-3.1. PHYSICAL VARIABLES AS STATE VARIABLES

As mentioned above for mechanical systems, positions and velocities of masses are useful state variables.

Example 13-3.1. Figure 13-3.1 depicts the first example to be used for illustrating the selection of state variables.
Normally two positions, z_1 and z_2, are selected, and Newton's Second Law is used to write

$$M_1\ddot{z}_1 = -B_1\dot{z}_1 - K_1z_1 + B_2(\dot{z}_2 - \dot{z}_1) + K_2(z_2 - z_1) + f_1(t) \qquad (13\text{-}3.1)$$

$$M_2\ddot{z}_2 = -K_3z_2 - B_2(\dot{z}_2 - \dot{z}_1) - K_2(z_2 - z_1) + f_2(t) \qquad (13\text{-}3.2)$$

With the selection and identification

$$\begin{aligned} x_1 &= z_1 & x_2 &= \dot{z}_1 \\ x_3 &= z_2 & x_4 &= \dot{z}_2 \end{aligned} \qquad (13\text{-}3.3)$$

a state-variable formulation is posssible. Equations 13-3.3 can be introduced into Eqs. 13-3.1 and 2 to give

$$M_1\dot{x}_2 = -B_1x_2 - K_1x_1 + B_2(x_4 - x_2) + K_2(x_3 - x_1) + f_1(t) \qquad (13\text{-}3.4)$$

$$M_2\dot{x}_4 = -K_3x_3 - B_2(x_4 - x_2) - K_2(x_3 - x_1) + f_2(t) \qquad (13\text{-}3.5)$$

in state-variable form. Additionally, it is evident that

$$\dot{x}_1 = x_2 \qquad \dot{x}_3 = x_4 \qquad (13\text{-}3.6)$$

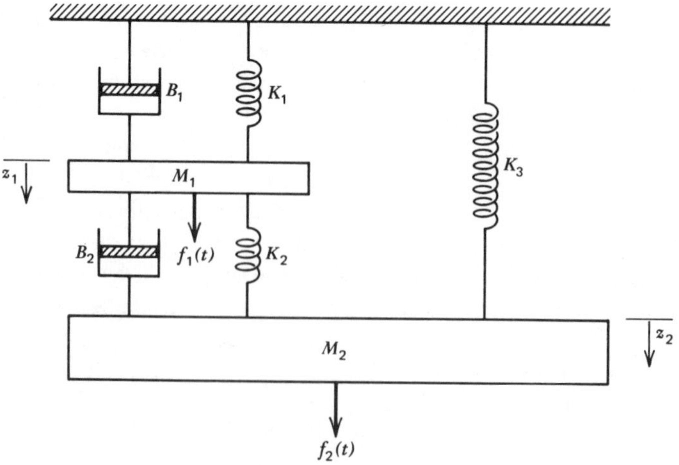

Fig. 13-3.1

A rearrangement, together with the use of matrices and vectors, results in

$$\begin{bmatrix} \dot{x}_1 \\ \dot{x}_2 \\ \dot{x}_3 \\ \dot{x}_4 \end{bmatrix} = \begin{bmatrix} 0 & 1 & 0 & 0 \\ \frac{-(K_1+K_2)}{M_1} & \frac{-(B_1+B_2)}{M_1} & \frac{K_2}{M_1} & \frac{B_2}{M_1} \\ 0 & 0 & 0 & 1 \\ \frac{K_2}{M_2} & \frac{B_2}{M_2} & \frac{-(K_2+K_3)}{M_2} & \frac{-B_2}{M_2} \end{bmatrix} \begin{bmatrix} x_1 \\ x_2 \\ x_3 \\ x_4 \end{bmatrix} + \begin{bmatrix} 0 \\ \frac{f_1(t)}{M_1} \\ 0 \\ \frac{f_2(t)}{M_2} \end{bmatrix} \tag{13-3.7}$$

This equation is now in standard state variable form, and the system state vector **x** and its derivative $\dot{\mathbf{x}}$ take the form

$$\dot{\mathbf{x}} = [A]\mathbf{x} + [B]\mathbf{u} \tag{13-3.8}$$

where $[A]$ is the system matrix, $[B]$ is the input matrix not explicitly written out, and **u** is the input vector. The matrix $[B]$ and the vector **u** can be written as

$$[B] = \begin{bmatrix} 0 & 0 \\ \frac{1}{M_1} & 0 \\ 0 & 0 \\ 0 & \frac{1}{M_2} \end{bmatrix} \qquad \mathbf{u} = \begin{bmatrix} f_1(t) \\ f_2(t) \end{bmatrix} \tag{13-3.9}$$

Notable in Eq. 13-3.7 is the use of four first-order differential equations to describe the system rather than the original two second-order equations. Such sets of first-order differential equations are a basic part of the state variable approach. It is not possible to use the form of Eq. 13-3.7 in the non-linear system case, as the components of the state vector will no longer enter the differential equations in linear combinations. An $[A]$ matrix can no longer be separated out. For non-linear systems, the vector-matrix state differential equations are written as

$$\left.\begin{aligned} \dot{x}_1 &= h_1[\mathbf{x}, \mathbf{u}, t] \\ \dot{x}_2 &= h_2[\mathbf{x}, \mathbf{u}, t] \\ &\cdots\cdots \\ \dot{x}_n &= h_n[\mathbf{x}, \mathbf{u}, t] \end{aligned}\right\} \tag{13-3.10}$$

in which the $h_j[\ldots]$ are, in general, non-linear functions.

As indicated previously, the selection of the state variables for Eq. 13-3.3 is certainly not the only choice that can be made; however, the consideration of other choices is deferred until later.

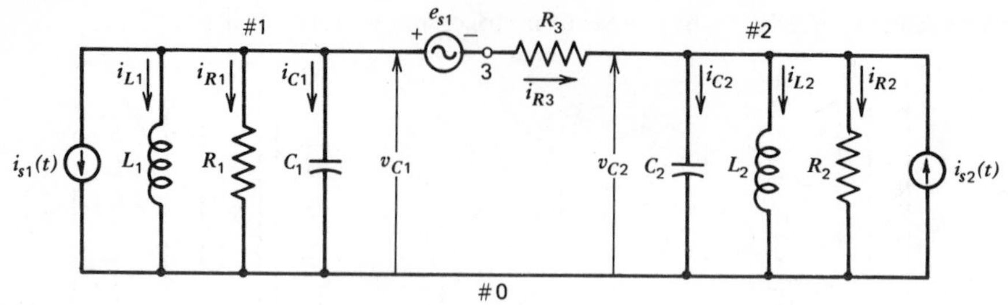

Fig. 13-3.2

Example 13-3.2. Capacitor voltages and inductor currents provide a suitable set of state variables for the network of Figure 13-3.2.
As an arbitrary step, the association of state variables with the above mentioned physical system variables is made as

$$\left.\begin{aligned} x_1 &= v_{C1} & x_3 &= i_{L1} \\ x_2 &= v_{C2} & x_4 &= i_{L2} \end{aligned}\right\} \tag{13-3.11}$$

Basic Ohm's and Kirchoff's Law relations are now used to formulate the vector-matrix state differential equations.

Before considering a general procedure for development of equations, a step-by-step procedure is used. It is evident from Fig. 13-3.2 that

$$L_1 \frac{di_{L1}}{dt} = v_{L1} = v_{C1} \qquad C_1 \frac{dv_{C1}}{dt} = i_{C1} \tag{13-3.12}$$

$$L_2 \frac{di_{L2}}{dt} = v_{L2} = v_{C2} \qquad C_2 \frac{dv_{C2}}{dt} = i_{C2} \tag{13-3.13}$$

$$-(i_{L1} + i_{R1} + i_{R3} + i_{s1}) = i_{C1} \tag{13-3.14}$$

$$-i_{L2} - i_{R2} + i_{R3} + i_{s2} = i_{C2} \tag{13-3.15}$$

$$i_{R3} = \frac{(v_{C1} - v_{C2} - e_{s1})}{R_3} \tag{13-3.16}$$

Introducing the state variables of Eqs. 13-3.11 gives the following equations:

$$\dot{x}_1 = \frac{1}{C_1}\left\{-\frac{x_1}{R_1} - \frac{x_1}{R_3} + \frac{x_2}{R_3} - x_3 - i_{s1} + \frac{e_{s1}}{R_3}\right\} \tag{13-3.17}$$

$$\dot{x}_2 = \frac{1}{C_2}\left\{-\frac{x_2}{R_2} - x_4 + \frac{x_1}{R_3} - \frac{x_2}{R_3} - \frac{e_{s1}}{R_3} + i_{s2}\right\} \tag{13-3.18}$$

$$\dot{x}_3 = \frac{x_1}{L_1} \tag{13-3.19}$$

$$\dot{x}_4 = \frac{x_2}{L_2} \tag{13-3.20}$$

In the vector-matrix form of Eq. 13-3.8 the result is

$$\begin{bmatrix} \dot{v}_{C1} \\ \dot{v}_{C2} \\ \dot{i}_{L1} \\ \dot{i}_{L2} \end{bmatrix} = \begin{bmatrix} \dot{x}_1 \\ \dot{x}_2 \\ \dot{x}_3 \\ \dot{x}_4 \end{bmatrix} = \begin{bmatrix} -\frac{1}{C_1}\left(\frac{1}{R_1}+\frac{1}{R_3}\right) & \left(\frac{1}{C_1 R_3}\right) & -\frac{1}{C_1} & 0 \\ \left(\frac{1}{R_3 C_2}\right) & -\frac{1}{C_2}\left(\frac{1}{R_2}+\frac{1}{R_3}\right) & 0 & -\frac{1}{C_2} \\ \left(\frac{1}{L_1}\right) & 0 & 0 & 0 \\ 0 & \left(\frac{1}{L_2}\right) & 0 & 0 \end{bmatrix} \begin{bmatrix} x_1 \\ x_2 \\ x_3 \\ x_4 \end{bmatrix} +$$

$$\begin{bmatrix} \frac{e_{s1}}{R_3 C_1} - \frac{i_{s1}}{C_1} \\ \frac{i_{s2}}{C_2} - \frac{e_{s1}}{R_3 C_2} \\ 0 \\ 0 \end{bmatrix} \qquad (13\text{-}3.21)$$

Again the input matrix $[B]$ and control vector $\mathbf{u}$ can be written as

$$[B] = \begin{bmatrix} \frac{1}{R_3 C_1} & -\frac{1}{C_1} & 0 \\ -\frac{1}{R_3 C_2} & 0 & \frac{1}{C_2} \\ 0 & 0 & 0 \\ 0 & 0 & 0 \end{bmatrix} \qquad \mathbf{u} = \begin{bmatrix} e_{s1} \\ i_{s1} \\ i_{s2} \end{bmatrix} \qquad (13\text{-}3.22)$$

The above procedure resulted in the desired set of first-order differential equations by mere chance (or by devious selection of the circuit used in the example). If two or more capacitor currents enter into a given Kirchhoffs' Voltage Law equation, then equations of the form of Eq. 13-3.14, 15, or 16 will have several first derivatives, and there will be a need to solve equations simultaneously to get only one first derivative in each, as in Eqs. 13-3.21.

General procedures for developing the $[A]$ and $[B]$ matrices that avoid the simultaneous equation problem mentioned above are available in the literature(3,4) with one scheme described in Section 13-4.

[3] P. M. DeRusso, R. J. Roy, and C. M. Close, *State Variables for Engineers*, John Wiley & Sons, Inc., New York, 1965, pp. 330-332.

[4] David J. Comer, *Computer Analysis of Circuits*, International Textbook Company, Scranton, Pa., 1971, pp. 94-176.

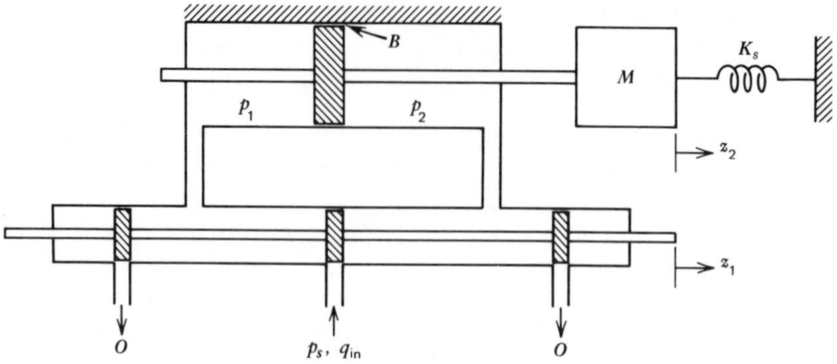

Fig. 13-3.3

Example 13-3.3. To diversify the study of the use of physical system variables as state variables, the hydraulic system of Fig. 13-3.3 is considered. A signal-flow graph is developed to show explicitly the relationships involved among physical variables and to aid in selecting state variables.

As described in Chapter 8, the input flow rate is a function of the load pressure drop, $(p_1 - p_2) = p_L$, and the pilot valve displacement, z_1. For incremental changes, the linear relationship

$$q_{in} = K_1 z_1 - K_2 p_L \tag{13-3.23}$$

is used. The input flow rate is also equal to the sum of: (1) piston displacement rate; (2) leakage flow rate; and (3) compression flow rate, as described, respectively, by

$$q_p = A_p \dot{z}_2 \tag{13-3.24}$$

$$q_l = K_l p_L \tag{13-3.25}$$

$$q_C = K_C \dot{p}_L \tag{13-3.26}$$

Mechanical system relations result in

$$M\ddot{z}_2 = A_p p_L - B\dot{z}_2 - K_s z_2 \tag{13-3.27}$$

One form for the signal-flow graph is given in Fig. 13-3.4, which, it must be admitted, has been drawn in a special way so as to use only integral operations.

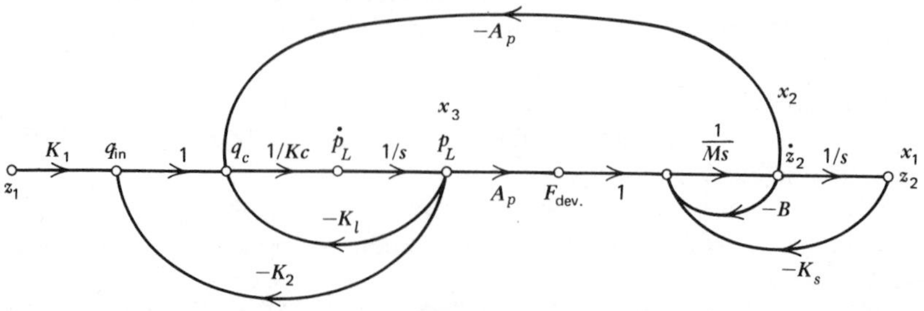

Fig. 13-3.4

In order to select a suitable set of state variables, the output of each integral operation can be used. The establishment of initial conditions for each integral operation, together with the input, is sufficient to determine the future state. Thus, let the state variables be defined as in

$$\left.\begin{array}{ll} x_1 = z_2 & x_3 = p_L \\ x_2 = \dot{z}_2 & u_1 = z_1 \end{array}\right\} \qquad (13\text{-}3.28)$$

The state differential equations are then given by

$$\dot{x}_1 = x_2$$

$$\dot{x}_2 = \frac{1}{M}(A_p x_3 - B x_2 - K_s x_1)$$

$$\dot{x}_3 = \frac{1}{K_C}(-A_p x_2 - K_l x_3 - K_2 x_3 + K_1 u_1) \qquad (13\text{-}3.29)$$

which are derived by inspection of the signal-flow graph of Fig. 13-3.4. Alternately, the state differential equations could have been written from the original physical equations; however, there would have been some doubt about the validity of any set of variables as being state variables. It should be pointed out that the state variables selected are directly related to physical system quantities—which is not always the case.

Example 13-3.4. As a last example of the use of physical variables as state variables, the speed control system of Figure 13-3.5(a) is considered. As described in Chapter 7, the dynamics of the system can be represented by the block diagram of Fig. 13-3.5(b), which also indicates a suitable choice for state variables as being the outputs of the two integral operations.

$$x_1 = \omega \qquad x_2 = \Delta i_f \qquad u_1 = e_{\text{in}} \qquad (13\text{-}3.30)$$

By inspection of the block diagram, the state vector-matrix differential equations can be written as

$$\dot{x}_1 = \frac{1}{J}(K_T x_2 - B x_1)$$

$$\dot{x}_2 = \frac{1}{(L_f + M)}[-R_f x_2 - K_A K_{\text{tach}} x_1 + K_A u] \qquad (13\text{-}3.31)$$

The $[A]$ and $[B]$ matrices are as given in

$$[A] = \begin{bmatrix} \dfrac{-B}{J} & \dfrac{K_T}{J} \\ \dfrac{-K_A K_{\text{tach}}}{L_f + M} & \dfrac{-R_f}{L_f + M} \end{bmatrix} \qquad [B] = \begin{bmatrix} 0 \\ \dfrac{K_A}{L_f + M} \end{bmatrix} = \mathbf{b} \qquad (13\text{-}3.32)$$

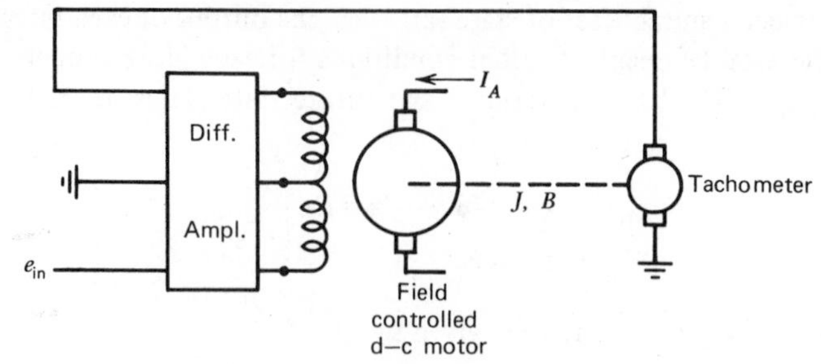

Fig. 13-3.5(a)

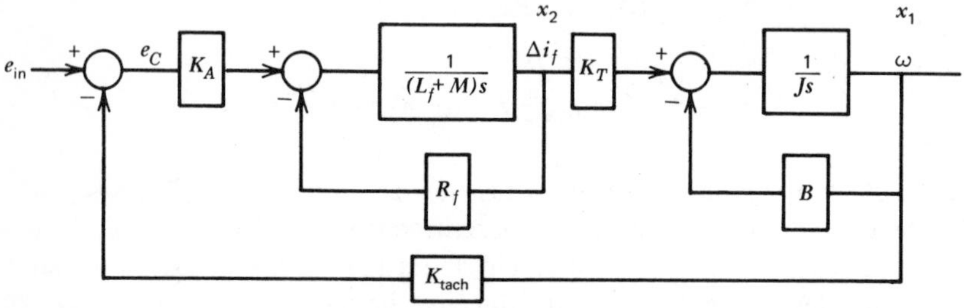

Fig. 13-3.5(b)

Again it should be pointed out that, if x_j is a state variable at the output of an integral operation, then $\dot{x}_j$ is the variable at the input to the integrator and can be found by summing all incoming signals.

An item that has been neglected in the previous examples is the relationship between the state variables and the system outputs. When physical variables are chosen as state variables, it may very well happen that all state variables are desired outputs, so that Eq. 13-2.12 may appear as

$$\mathbf{y} = [I]\mathbf{x} \tag{13-3.33}$$

where $[I]$ is the identity matrix. However, in general, the desired system output vector may be formed from only a few state variables, or may be some linear combination of the state variables and inputs, so that $\mathbf{y}$ may be written

$$\mathbf{y} = [C]\mathbf{x} + [D]\mathbf{u} \tag{13-3.34}$$

For the control system of this example, perhaps only the output speed, ω, and the error voltage, e_c are desired. Then the output equation 13-3.34 can be written as

$$\begin{bmatrix} y_1 \\ y_2 \end{bmatrix} = \begin{bmatrix} 1 & 0 \\ -K_{tach} & 0 \end{bmatrix} \begin{bmatrix} \omega \\ \Delta i_f \end{bmatrix} + \begin{bmatrix} 0 \\ 1 \end{bmatrix} e_{in} \tag{13-3.35}$$

or

$$\mathbf{y} = \begin{bmatrix} 1 & 0 \\ -K_{tach} & 0 \end{bmatrix} \mathbf{x} + \begin{bmatrix} 0 \\ 1 \end{bmatrix} u \tag{13-3.36}$$

13-3.2. SIMULATION DIAGRAMS AND STATE VARIABLE SELECTION

Applying the signal-flow simulation graph techniques developed in Section 12-6 to the selection of state variables is quite useful. Again the outputs of integrators are particularly easy state variable selections. In this case the transfer function used as a starting point provides only the overall input-output relationship, and the status of state variables as truly *internal* variables becomes apparent. Also, the fact that the state variables need not directly be physical system variables becomes apparent. Certainly, if the simulation diagram is implemented on an analog computer, the integrator outputs are real voltages, but these voltages may not be analogous to any real physical quantity of any significance.

Example 13-3.5. To illustrate state variables as related to *direct programming*, the transfer function

$$\frac{Y(s)}{U(s)} = \frac{s^2 + 3s + 2}{s^4 + 16s^3 + 91s^2 + 236s + 280} \tag{13-3.37}$$

is used. Division of numerator and denominator by s^4 and developing the signal-flow graph as in Section 12-6 results in Fig. 13-3.6. The state variables have been labelled as x_1 through x_4 in the diagram. By observing each $\dot{x}_j$ node, it is now possible to write

$$\left.\begin{aligned} \dot{x}_1 &= -16x_1 + x_2 \\ \dot{x}_2 &= -91x_1 + x_3 + u \\ \dot{x}_3 &= -236x_1 + x_4 + 3u \\ \dot{x}_4 &= -280x_1 + 2u \\ y &= x_1 \end{aligned}\right\} \tag{13-3.38}$$

or

$$\dot{\mathbf{x}} = \begin{bmatrix} -16 & 1 & 0 & 0 \\ -91 & 0 & 1 & 0 \\ -236 & 0 & 0 & 1 \\ -280 & 0 & 0 & 0 \end{bmatrix} \mathbf{x} + \begin{bmatrix} 0 \\ 1 \\ 3 \\ 2 \end{bmatrix} u \tag{13-3.39}$$

$$y = [1 \quad 0 \quad 0 \quad 0]\, \mathbf{x} \tag{13-3.40}$$

Determination of initial conditions for the state vector $\mathbf{x}(0)$ can be accomplished by considering that $y(0)$, $\dot{y}(0)$, $\ddot{y}(0)$, and $\dddot{y}(0)$ may be specified along with possible impulses

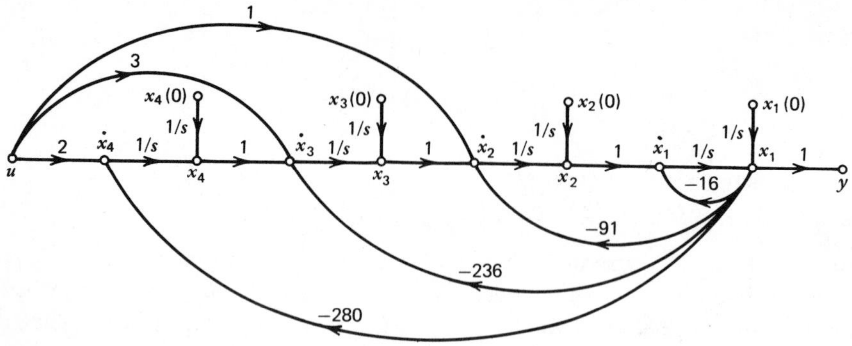

Fig. 13-3.6

in u and its derivatives, $\dot{u}$ and $\ddot{u}$, and that $\mathbf{x}$, y, and u are related by Eqs. 13-3.38. For example, the relations of

$$y(0) = x_1(0) \tag{13-3.41}$$

$$\begin{aligned} \dot{y}(t) &= \dot{x}_1(t) = -16x_1(t) + x_2(t) \\ \dot{y}(0) &= -16x_1(0) + x_2(0) \qquad x_2(0) = \dot{y}(0) + 16y(0) \end{aligned} \tag{13-3.42}$$

$$\begin{aligned} \ddot{y}(t) &= -16\dot{x}_1(t) + \dot{x}_2(t) = -16[-16x_1 + x_2] + [-91x_1 + x_3 + u] \\ \ddot{y}(0) &= 165y(0) - 16x_2(0) + x_3(0) + u(0) \end{aligned} \tag{13-3.43}$$

are typical of such computations. A matrix formulation can be done to systematize the process.

Example 13-3.6. An alternate direct programming formulation of Eq. 13-3.37 is shown in Fig. 13-3.7, and the state variable differential equations are

$$\dot{\mathbf{x}} = \begin{bmatrix} 0 & 1 & 0 & 0 \\ 0 & 0 & 1 & 0 \\ 0 & 0 & 0 & 1 \\ -280 & -236 & -91 & -16 \end{bmatrix} \mathbf{x} + \begin{bmatrix} 0 \\ 0 \\ 0 \\ 1 \end{bmatrix} u \tag{13-3.44}$$

$$y = [2 \quad 3 \quad 1 \quad 0]\ \mathbf{x} \tag{13-3.45}$$

The particular formulation or selection of x_j's used here results in a set of state variables called *phase variables*, as related to a rather old concept of phase space in mechanics. It should be pointed out that the state vector $\mathbf{x}$ in this example is generally *different* from the state vector $\mathbf{x}$ in the previous example. Perhaps different symbols should be used, but standard practice makes it desirable to keep $\mathbf{x}$ as the state vector, except in certain special cases.

Examples 13-3.5 and 6 start from a given transfer function and develop a state vari-

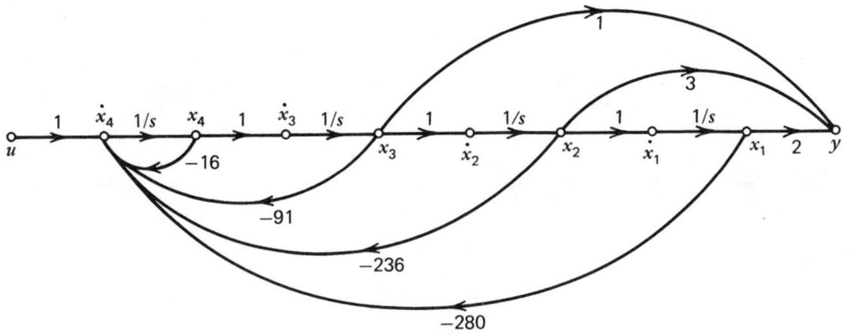

Fig. 13-3.7

able representation. It is interesting to consider the reverse process, for a single input u and a single output, y. If the state equations are given by

$$\dot{\mathbf{x}} = \mathbf{A}\mathbf{x} + \mathbf{B}u$$
$$y = \mathbf{C}\mathbf{x} + \mathbf{D}u \tag{13-3.46}$$

where the brackets around the matrices **A, B, C**, and **D** have been dropped for simplicity, the Laplace transform term-by-term, or element-by-element, for these equations, assuming zero initial conditions (as normally assumed in transfer functions), results in

$$\left.\begin{aligned} s\mathbf{X}(s) &= \mathbf{A}\mathbf{X}(s) + \mathbf{B}U(s) \\ Y(s) &= \mathbf{C}\mathbf{X}(s) + \mathbf{D}U(s) \end{aligned}\right\} \tag{13-3.47}$$

$$\mathbf{X}(s) = [sI - A]^{-1}\mathbf{B}U(s) \tag{13-3.48}$$

$$Y(s) = \{\mathbf{C}[sI - A]^{-1}\mathbf{B} + \mathbf{D}\}\,U(s) \tag{13-3.49}$$

In order to solve the first of Eqs. 13-3.47 for $\mathbf{X}(s)$, as in Eq. 13-3.48, it is necessary to introduce an identity matrix as $s\mathbf{X}(s) = sI\mathbf{X}(s)$, to transpose $\mathbf{A}\mathbf{X}(s)$, to factor $\mathbf{X}(s)$ out on the right from $[sI\mathbf{X}(s) - A\mathbf{X}(s)]$, and to multiply on the left by the inverse, $(sI - A)^{-1}$. A direct substitution of Eq. 13-3.48 into the second of Eqs. 13-3.47, and factoring $U(s)$ out on the right, results in Eq. 13-3.49. Division of both sides of Eq. 13-3.49 by $U(s)$ gives the transfer function. A simple second-order example (to simplify the inverse) is used to illustrate the above results.

$$\frac{Y(s)}{U(s)} = G(s) = \{\mathbf{C}[sI - A]^{-1}\mathbf{B} + \mathbf{D}\} \tag{13-3.50}$$

Example 13-3.7. Using the direct programming approach of Example 13-3.6, the transfer function of

$$\frac{Y(s)}{U(s)} = G(s) = \frac{s^2 + 3s + 2}{s^2 + 4s + 8} \tag{13-3.51}$$

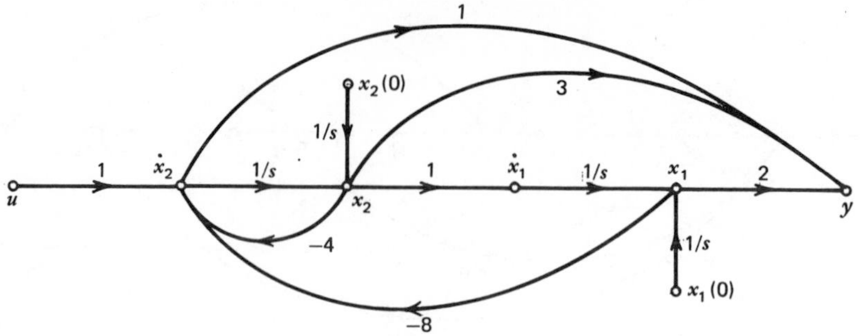

Fig. 13-3.8

is simulated as in Fig. 13-3.8. The direct path from $\dot{x}_2$ to y is an extra feature, due to the fact that both numerator and denominator of $G(s)$ are of equal degree. This extra path touches all loops, and has a path gain of 1.0 and a path cofactor of 1.0. The graph determinant is given by

$$\Delta(s) = 1 + \frac{4}{s} + \frac{8}{s^2} \tag{13-3.52}$$

so that the extra path contributes the part of $G(s)$, given by

$$G_1(s) = \frac{(1.0)(1.0)}{1 + (4/s) + (8/s^2)} = \frac{s^2}{s^2 + 4s + 8} \tag{13-3.53}$$

The reader should consider what happens if the graph form of Example 13-3.5 is used here.

Returning to the state variable formulation results in

$$\dot{\mathbf{x}} = \begin{bmatrix} 0 & 1 \\ -8 & -4 \end{bmatrix} \mathbf{x} + \begin{bmatrix} 0 \\ 1 \end{bmatrix} u \tag{13-3.54}$$

$$y = 2x_1 + 3x_2 + \dot{x}_2 = [-6 \; - 1]\; \mathbf{x} + [1]\; u \tag{13-3.55}$$

Extra effort is now required to determine the $[C]$ matrix, as indicated in Eq. 13-3.55. The matrix $[sI - A]^{-1}$ is determined, as in

$$[sI - A] = s\begin{bmatrix} 1 & 0 \\ 0 & 1 \end{bmatrix} - \begin{bmatrix} 0 & 1 \\ -8 & -4 \end{bmatrix} = \begin{bmatrix} s & -1 \\ 8 & (s+4) \end{bmatrix} \tag{13-3.56}$$

$$[sI - A]^{-1} = \frac{1}{s(s+4) + 8} \begin{bmatrix} (s+4) & 1 \\ -8 & s \end{bmatrix} \tag{13-3.57}$$

$$[sI - A]^{-1}\mathbf{B} = \begin{bmatrix} \dfrac{(s+4)}{s^2 + 4s + 8} & \dfrac{1}{s^2 + 4s + 8} \\ \dfrac{-8}{s^2 + 4s + 8} & \dfrac{s}{s^2 + 4s + 8} \end{bmatrix} \begin{bmatrix} 0 \\ 1 \end{bmatrix} = \begin{bmatrix} \dfrac{1}{s^2 + 4s + 8} \\ \dfrac{s}{s^2 + 4s + 8} \end{bmatrix} \tag{13-3.58}$$

$$\mathbf{C}[sI - A]^{-1}\mathbf{B} = \frac{-6}{s^2 + 4s + 8} - \frac{s}{s^2 + 4s + 8} = \frac{-(s+6)}{s^2 + 4s + 8} \tag{13-3.59}$$

followed by $G(s)$, as in

$$G(s) = \mathbf{C}[sI - A]^{-1}\mathbf{B} + \mathbf{D} = 1 - \frac{(s+6)}{s^2 + 4s + 8} = \frac{s^2 + 3s + 2}{s^2 + 4s + 8} \tag{13-3.60}$$

A complete cycle has been traced resulting in a check of the original formulation.

Example 13-3.8. In some cases, *iterative* or *cascade programming* is a convenient simulation tool for selecting state variables. To illustrate this technique, the $G(s)$ of Eq. 13-3.37 and Example 13-3.5 is considered.

$$G(s) = \frac{s^2 + 3s + 2}{s^4 + 16s^3 + 91s^2 + 236s + 280} = \frac{(s+1)(s+2)}{[(s+2)^2 + (2)^2](s+5)(s+7)} \tag{13-3.61}$$

This transfer function is now arbitrarily split into three parts to be cascaded.

$$G_1(s) = \frac{1}{s+7} \qquad G_2(s) = \frac{(s+1)}{(s+5)} \tag{13-3.62}$$

$$G_3(s) = \frac{(s+2)}{(s+2)^2 + (2)^2} \tag{13-3.63}$$

The $G_1(s)$ and $G_2(s)$ parts can be easily programed, as in Fig. 13-3.9(a), while it is convenient to program $G_3(s)$ as in (b). The programming of $G_3(s)$ is based on manipulating

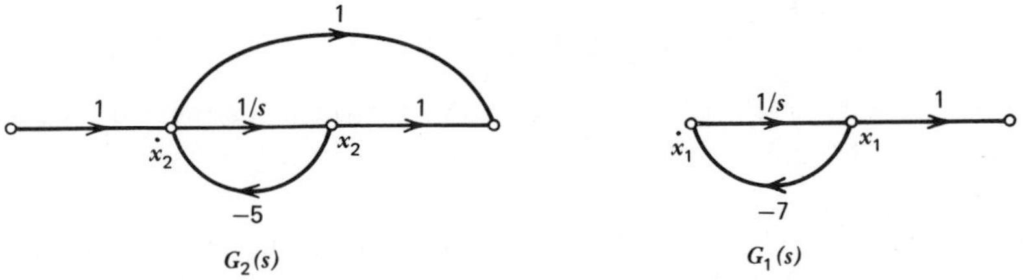

Fig. 13-3.9(a)

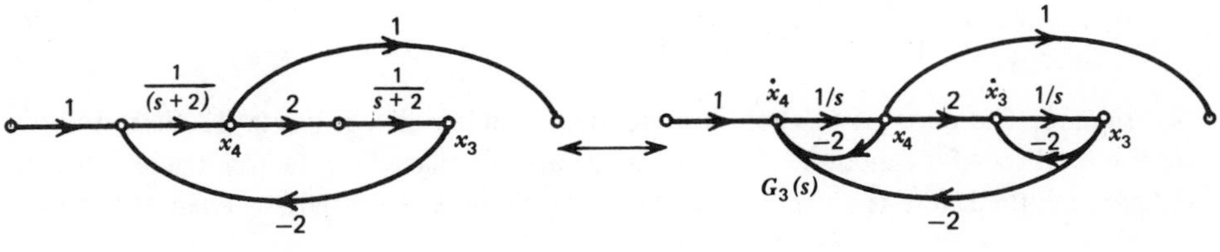

Fig. 13-3.9(b)

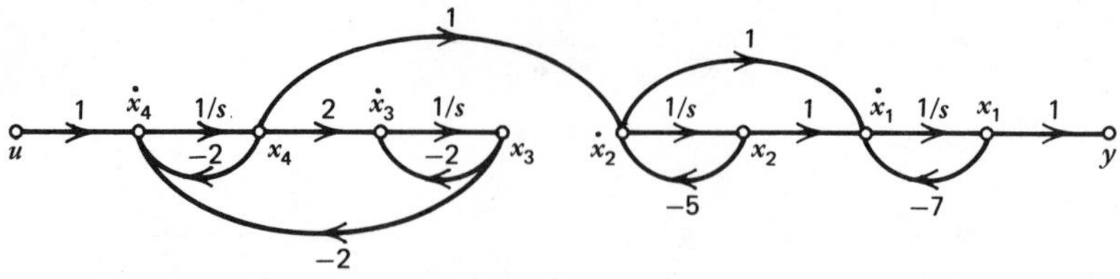

Fig. 13-3.9(c)

$G_3(s)$ as in

$$G_3(s) = \frac{1/(s+2)}{1 + [(2)^2/(s+2)^2]} = \frac{(P_1)(\Delta_1)}{\Delta} \tag{13-3.64}$$

and Mason's transmission rule.

Cascading the various parts of the transfer function results in Fig. 13-3.9(c) and a new set of state variables. State equations are now given by

$$\left.\begin{aligned} \dot{x}_1 &= -7x_1 + x_2 + \dot{x}_2 = -7x_1 - 4x_2 + x_4 \\ \dot{x}_2 &= -5x_2 + x_4 \\ \dot{x}_3 &= -2x_3 + 2x_4 \\ \dot{x}_4 &= -2x_3 - 2x_4 + u \\ y &= x_1 \end{aligned}\right\} \tag{13-3.65}$$

$$\dot{\mathbf{x}} = \begin{bmatrix} -7 & -4 & 0 & 1 \\ 0 & -5 & 0 & 1 \\ 0 & 0 & -2 & 2 \\ 0 & 0 & -2 & -2 \end{bmatrix} \mathbf{x} + \begin{bmatrix} 0 \\ 0 \\ 0 \\ 1 \end{bmatrix} u \qquad y = [0 \;\; 0 \;\; 0 \;\; 1]\,\mathbf{x} \tag{13-3.66}$$

There is still some extra effort in obtaining $\dot{x}_1$, due to the equal powers of s in numerator and denominator of $G_2(s)$.

Example 13-3.9. *Parallel programming* is the last of the continuous system simulation techniques to be related to state variable selection. As a first example, the transfer function of

$$\frac{Y(s)}{U(s)} = G(s) = \frac{35(s+1)}{s(s+5)(s+7)} = \frac{1}{s} + \frac{14}{s+5} - \frac{15}{s+7} \tag{13-3.67}$$

is considered. The parts of $G(s)$ are now connected in parallel, as in Fig. 13-3.10, to provide the total transfer function. The gains are arbitrarily placed on the input side. Because the approach used here is so special and important, the state variables taken at the out-

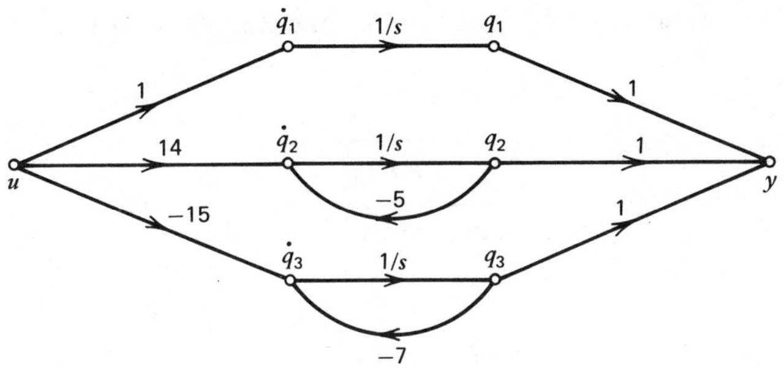

Fig. 13-3.10

puts of each integrator are designated as q_j's. The state equations are given as

$$\left.\begin{aligned} \dot{q}_1 &= u \\ \dot{q}_2 &= -5q_2 + 14u \\ \dot{q}_3 &= -7q_3 - 15u \\ y &= q_1 + q_2 + q_3 \end{aligned}\right\} \qquad (13\text{-}3.68)$$

$$\dot{\mathbf{q}} = \begin{bmatrix} 0 & 0 & 0 \\ 0 & -5 & 0 \\ 0 & 0 & -7 \end{bmatrix} \mathbf{q} + \begin{bmatrix} 1 \\ 14 \\ -15 \end{bmatrix} u \qquad (13\text{-}3.69)$$

$$y = [1 \quad 1 \quad 1]\, \mathbf{q}$$

Only *diagonal* entries appear in the **A** system state matrix, leading to the designation of such an **A** matrix as a **Λ** matrix. The importance of this case lies in the fact that the state differential equations are *decoupled.* Each can be solved independently of the others. For example, if $u(t)$ is zero, then

$$\left.\begin{aligned} q_1(t) &= q_1(0+) \\ q_2(t) &= q_2(0+)\, \epsilon^{-5t} \\ q_3(t) &= q_3(0+)\, \epsilon^{-7t} \end{aligned}\right\} \qquad (13\text{-}3.70)$$

as each of Eqs. 13-3.68 is a simple first-order differential in q_i. This special case is extensively investigated in Section 14-5.

It is not always possible or desirable to completely do a partial fraction expansion or a diagonalization of the **A** matrix. Whenever complex or multiple poles exist, special forms are desirable. To illustrate these cases. Examples 13-3.10 and 11 are used.

Example 13-3.10. Equation 13-3.37 involves a transfer function with complex poles as specifically shown in

$$\frac{Y(s)}{U(s)} = G(s) = \frac{(s+1)(s+2)}{(s+5)(s+7)[(s+2)^2+(2)^2]} \tag{13-3.71}$$

This transfer function can be expanded so as to keep the complex poles at $s = -2 \pm j2$ paired together, as in Eq. 13-3.72 or 73.

$$G(s) = \frac{(1/377)(21s - 34)}{(s+2)^2+(2)^2} + \frac{6/13}{s+5} - \frac{15/29}{s+7} \tag{13-3.72}$$

$$G(s) = \frac{(21/377)(s+2)}{(s+2)^2+(2)^2} - \frac{(38/377)(2)}{(s+2)^2+(2)^2} + \frac{6/13}{s+5} - \frac{15/29}{s+7} \tag{13-3.73}$$

The form used in Fig. 13-3.9(b) for realizing the complex pole terms is used again, together with standard parallel programming of the remaining parts, to obtain Fig. 13-3.11(a). To realize the two terms involving complex poles, the two paths through the separate complex pole part are used. In one case, through q_4 the path gain is $21/[377(s+2)]$, with cofactor 1.0, while the other path gain is $(-38/377)[2/(s+2)^2]$, with cofactor 1.0, as shown in simplified form in Fig. 13-3.11(b).

By using the simulation diagram it is a simple matter to select the integrator outputs as state variables and write

$$\dot{\mathbf{q}} = \left[\begin{array}{cc|cc} -5 & 0 & 0 & 0 \\ 0 & -7 & 0 & 0 \\ \hline 0 & 0 & -2 & 2 \\ 0 & 0 & -2 & -2 \end{array}\right] \mathbf{q} + \begin{bmatrix} 1 \\ 1 \\ 0 \\ 1 \end{bmatrix} u \tag{13-3.74}$$

$$y = \left[\tfrac{6}{13} \quad \tfrac{-15}{29} \quad \tfrac{-38}{377} \quad \tfrac{21}{377}\right] \mathbf{q}$$

Choosing state variables in the above manner leads to a nearly diagonal $[A]$ matrix, except that the complex pole terms enter in two-by-two blocks. If poles are at $s = \alpha \pm j\beta$, the typical block will be in the form

$$\begin{bmatrix} \alpha & \beta \\ -\beta & \alpha \end{bmatrix}$$

as shown in the lower right hand corner of the $[A]$ matrix in Eq. 13-3.74. Other simple pole terms will still appear along the diagonal as do -5 and -7 above.

The advantage of using the above formulation, over a strict diagonalization, is that the resulting time domain terms are real time functions of the form $\epsilon^{\alpha t} \cos \beta t$ and $\epsilon^{\alpha t} \sin \beta t$ rather than complex terms such as $\epsilon^{(\alpha \pm j\beta)t}$. Such complex terms do little toward simplifying visualization of the responses as considered in Sections 14-4 and 14-5.

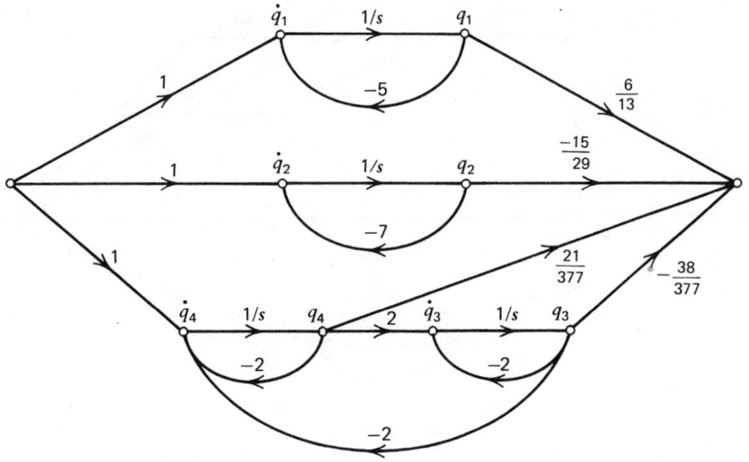

(a)

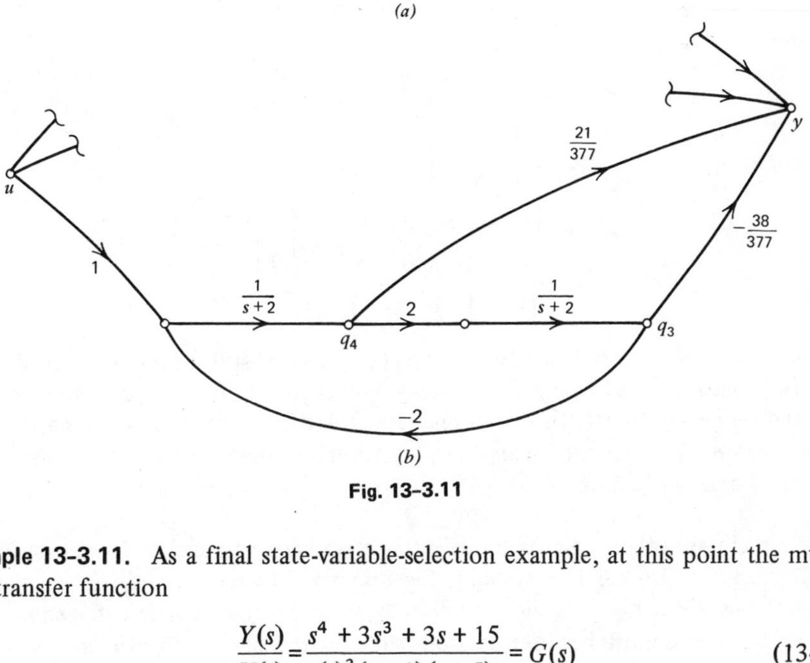

(b)

Fig. 13-3.11

Example 13-3.11. As a final state-variable-selection example, at this point the multiple pole transfer function

$$\frac{Y(s)}{U(s)} = \frac{s^4 + 3s^3 + 3s + 15}{(s)^3(s+3)(s+5)} = G(s) \tag{13-3.75}$$

is considered.

The partial fraction expansion of this $G(s)$ leads to

$$G(s) = \frac{1}{s^3} - \frac{\frac{1}{3}}{s^2} + \frac{\frac{1}{9}}{s} - \frac{\frac{1}{9}}{s+3} + \frac{1}{s+5} \tag{13-3.76}$$

In order to realize this form for the transfer function with a minimum number of integrators, the signal-flow graph of Fig. 13-3.12 is used. The state equations can now

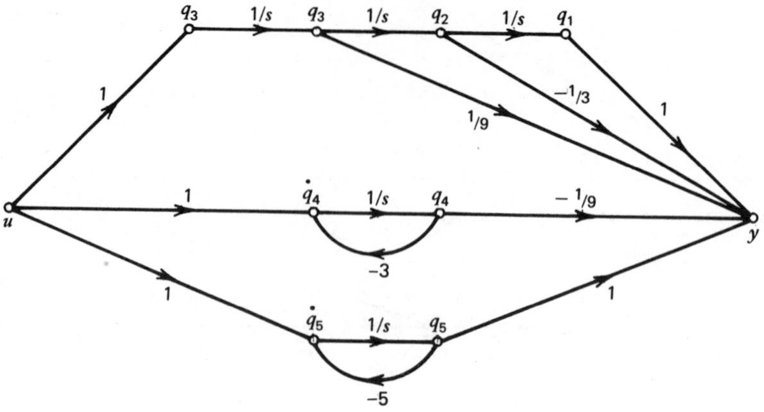

Fig. 13-3.12

be written

$$\dot{\mathbf{q}} = \left[\begin{array}{ccc|cc} 0 & 1 & 0 & 0 & 0 \\ 0 & 0 & 1 & 0 & 0 \\ 0 & 0 & 0 & 0 & 0 \\ \hline 0 & 0 & 0 & -3 & 0 \\ 0 & 0 & 0 & 0 & -5 \end{array}\right] \mathbf{q} + \begin{bmatrix} 0 \\ 0 \\ 1 \\ 1 \\ 1 \end{bmatrix} u \tag{13-3.77}$$

$$y = [1 \ \ -\tfrac{1}{3} \ \ \tfrac{1}{9} \ \ -\tfrac{1}{9} \ \ 1]\,\mathbf{q}$$

The important aspects of this formulation are: (1) the poles still appear along the diagonal of the $[A]$ matrix, with the triple pole at $s = 0$ blocked together, and the simple poles at $s = -3$ and $s = -5$ separate; (2) a row of ones appears right above the poles at $s = 0$. Such a form is standard in matrix theory and is called a *Jordan canonical form.* This form will be encountered again in Section 14-5.

While there are other schemes for selecting state variables, such as in multiple-input-multiple-output and non-linear systems, the discussion presented so far is adequate for many purposes. Certainly programming directly from the system differential equations is desirable, but can almost be inferred from some of the simulation diagrams presented. Some additional material on state variable selection for discrete and multi-variable systems is presented in Chapter 14.

13-4. INDEPENDENCE OF EQUATIONS, THE 2*b* SIGNAL-FLOW GRAPH, AND STATE VARIABLE SELECTION

As mentioned in Section 13-3, standard methods exist for choosing state variables for networks in a manner that assures independence of the resulting first-order differential equations. As an introduction to the topic of independent equations, a consideration of sets of algebraic equations from the signal-flow graph standpoint is presented.

The existence of a signal-flow graph in itself does not guarantee the independence of the original equations. To demonstrate this, let us fabricate a set of equations whose determinant is zero. We do this in the following example.

Example 13-4.1. We choose the following set of equations.

$$\begin{aligned} u(t) &= 4x_1 + 2x_2 + 1x_3 \\ 0 &= 3x_1 + 4x_2 + 3x_3 \\ 0 &= 7x_1 + 6x_2 + 4x_3 \end{aligned} \tag{13-4.1}$$

Note that if the system determinant is written, the third row is the sum of the first two rows and hence has a value of zero. However, a signal-flow graph can still be drawn. Each equation is solved for the term on the principle diagonal, as

$$\begin{aligned} x_1 &= u(t)/4 - x_2/2 - x_3/4 \\ x_2 &= -3x_1/4 - 3x_3/4 \\ x_3 &= -7x_1/4 - 3x_2/2 \end{aligned} \tag{13-4.2}$$

The signal-flow graph of Fig. 13-4.1 can now be constructed. The system determinant is evaluated using Mason's Rule, as

$$\Delta = 1 - \tfrac{3}{8} - \tfrac{9}{8} + \tfrac{9}{32} - \tfrac{7}{16} + \tfrac{21}{32} = 0 \tag{13-4.3}$$

If we change the $4x_1$ in the first of Eqs. 13-4.1 to $2x_1$, the topology of the resulting signal-flow graph is identical to that of Fig. 13-4.1 and yet the new system determinant is not zero.

This problem of independence among the equations is not unique to signal-flow graphs, but is one that must be faced in writing the system equations, regardless of how the equations are solved. However, the signal-flow graph is not a solution to this problem. This problem has been solved in circuit theory through the use of a *tree* (among other methods). The currents in the link branches of the tree form one independent set and lead to what are known as *loop-current equations.* The voltages across the tree branches

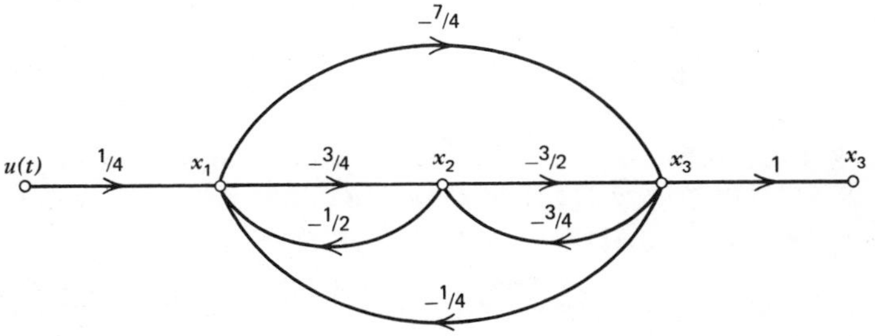

Fig. 13-4.1

form another independent set and lead to what are known as *node-pair voltage equations.* Both of these sets of equations can be derived from a more fundamental set of equations by a method referred to as the $2b$-method.

In our discussion of state variables, we pointed out that one possible physical interpretation of state variables in circuit theory is to use currents in the inductors and voltages on the capacitors. It should be noted that neither loop-current nor node-pair voltage equations are satisfactory for this purpose, but the equations must be written in a mixed form so that both inductor currents and capacitor voltages appear in the equations. One way of doing this is to return to the $2b$ set of equations, and the use of the signal-flow graph as a way of reducing the equations to the state variable form. This idea is explained in the next section.

Finally, these circuit theory ideas of linear independence among equations can be applied to other physical systems. One rather roundabout method that is simple in concept, would be to draw the electric circuit that is the analog of the given physical system, and then apply the circuit theory concept to the resulting circuit. We do not pursue this idea, but limit ourselves to circuit theory application with the thought that these ideas can be extended to all linear physical systems.

13-4.1. THE 2*b* SIGNAL-FLOW DIAGRAM

A network consists of b branches. Each branch has associated with it a current and a voltage, and in this sense there are $2b$ unknowns.[5] When a set of equations is written, this set must contain $2b$ equations in order to solve for the $2b$ unknowns. There are b volt-ampere relationships that exist (one for each branch). A tree is chosen that has n branches, which with the remaining l branches, called *link branches*, gives

$$b = n + l \tag{13-4.4}$$

The number of independent Kirchhoff's Current Law (KCL) equations that can be written is n, and the number of independent Kirchhoff's Voltage Law (KVL) equations is l. Hence, the total number of equations is

$$\begin{aligned} &\text{Volt-ampere relationships} = b \\ &\text{KCL equations (KCLE's)} \ \ = n \\ &\text{KVL equations (KVLE's)} \ \ = \underline{l \qquad\qquad} \\ &\qquad\qquad\qquad\qquad\qquad\ b + n + l = 2b \end{aligned} \tag{13-4.5}$$

If the volt-ampere relationships and the KCLE's are substituted into the KVLE's the result is the loop-current set of equations.

If the volt-ampere relationships and KVLE's are substituted into the KCLE's, the result is the node-pair voltage set of equations.

Alternately, some of the KCLE's and KVLE's can be saved, with the result referred to as *equations in the mixed form*, which has as one application the writing of state variable equations.

[5] E. A. Guillemin, *Introductory Circuit Theory*, John Wiley & Sons, Inc., New York, 1953, pp. 64–79.

The $2b$ signal-flow graph is a geometrical interpretation of the $2b$ equations, from which can be obtained the loop current, the node-pair voltage equations, and the equations in the mixed form. The $2b$ graph is sometimes referred to as *the primitive flow-graph.*[6]

Sometimes in using topology in circuit theory it is convenient to remove the independent sources and leave only their internal impedances. However, for the present discussion, we will count the sources as branches. A voltage source will be a branch with a voltage constraint, and a current source a branch with a current constraint.

The process of setting up the $2b$ graph is first to choose a tree. A tree is chosen wherein all the voltage sources and capacitors are in the tree, and all inductors and current sources are in the links. There are cases in which this presents problems, and that subject, though beyond the scope of this book, will be referred to elsewhere.

All the voltages associated with tree branches are referred to as *an independent set of voltages*, and the currents associated with the link branches as *an independent set of currents.* The currents associated with the tree branches are referred to as *a dependent set of currents*, and the voltages associated with the link branches as *a dependent set of voltages.* The reasons for these designations will be reviewed from the point of view of the signal-flow graph shortly.

Figure 13-4.2 shows how the nodes are distributed in a block diagram form.

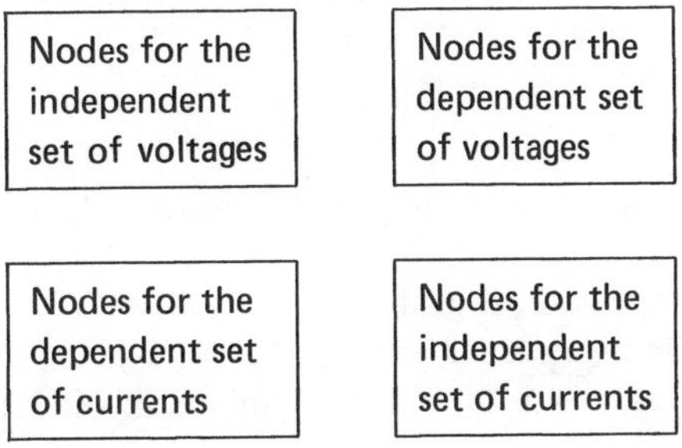

Fig. 13-4.2

Perhaps the best way to continue is through the use of examples. Although the primary purpose of this discussion is to introduce a method of writing state variable equations, the first example is an all-resistive network that introduces a method of setting up the diagram itself.

Example 13-4.2. The circuit of Fig. 13-4.3(a) is used in this example. The tree as shown in (b) is chosen so that the two voltage sources are in the tree, and the current source is a link. The other choices are arbitrary. There are six branches (counting sources), and these are so numbered that the voltage sources are numbered first, then the remain-

[6]S. J. Mason and H. J. Zimmerman, *Electronic Circuits, Signals and Systems*, John Wiley & Sons, Inc., New York, 1960, pp. 140-145.

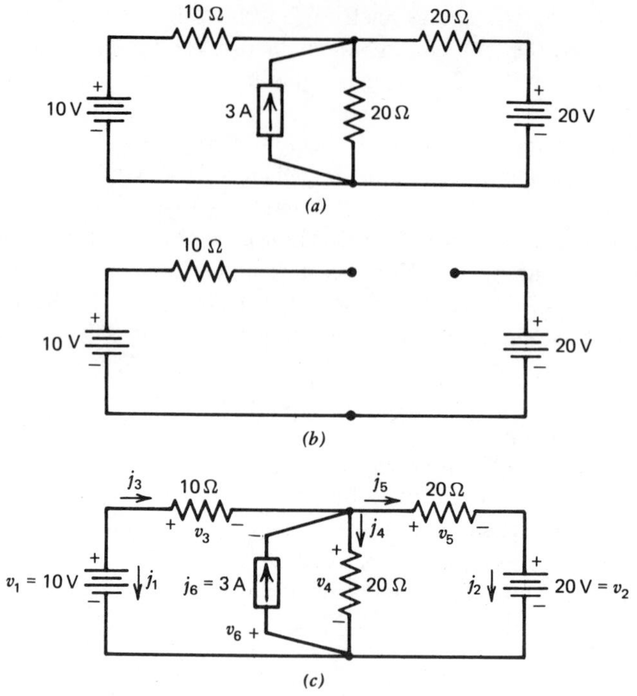

Fig. 13-4.3

ing tree branches, and then finally the current source. The numbering is shown in Fig. 13-4.3(c).

The process of drawing the $2b$ signal-flow graph starts by enumerating the independent and dependent sets of nodes and arranging them as suggested in Fig. 13-4.2.

The three branch voltages form the independent set of voltages, and are

$$\{v_1, v_2, v_3\} \tag{13-4.6}$$

The dependent set of voltages, the voltages associated with the link branches, are

$$\{v_4, v_5, v_6\} \tag{13-4.7}$$

The independent set of currents, the currents in the link branches, are

$$\{j_4, j_5, j_6\} \tag{13-4.8}$$

The dependent set of currents, the currents in the tree branches, are

$$\{j_1, j_2, j_3\} \tag{13-4.9}$$

All four sets of nodes are shown in Fig. 13-4.4. The notation ISBV (and DSBV) stands for *independent* (and *dependent*) *set of branch voltages*.

The next thing to do is write the necessary KVLE's and KCLE's. These give the relationship between the dependent sets and the independent sets. The KVLE's are written

$$\left.\begin{matrix} v_1 \\ v_2 \\ v_3 \end{matrix}\right\}\text{ISBV} \qquad \text{DSBV}\left\{\begin{matrix} v_4 \\ v_5 \\ v_6 \end{matrix}\right.$$

$$\left.\begin{matrix} j_1 \\ j_2 \\ j_3 \end{matrix}\right\}\text{DSBC} \qquad \text{ISBC}\left\{\begin{matrix} j_4 \\ j_5 \\ j_6 \end{matrix}\right.$$

(*a*)

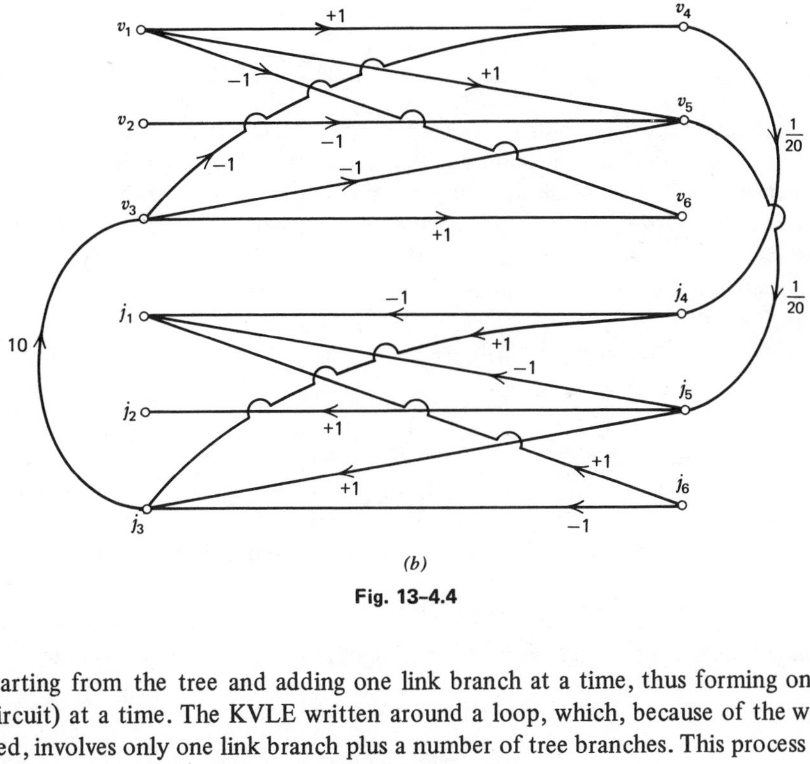

(*b*)

Fig. 13-4.4

by starting from the tree and adding one link branch at a time, thus forming one loop (or circuit) at a time. The KVLE written around a loop, which, because of the way it is formed, involves only one link branch plus a number of tree branches. This process is continued in the numerical sequence for each link branch.

By taking the voltage across the link branch with a positive sign in each loop equation, the set of KVLE's can be represented in a particularly simple form, as developed below for the example being considered. In expanded form, they are

$$\begin{aligned} -v_1 + v_3 + v_4 &= 0 \\ -v_1 + v_2 + v_3 + v_5 &= 0 \\ v_1 - v_3 + v_6 &= 0 \end{aligned} \qquad (13\text{-}4.10)$$

It is also useful to consider the matrix representation for these equations, which shows particularly well the systematic features involved. For this example, the result is

$$\begin{bmatrix} -1 & 0 & 1 & 1 & 0 & 0 \\ -1 & 1 & 1 & 0 & 1 & 0 \\ 1 & 0 & -1 & 0 & 0 & 1 \end{bmatrix} \begin{bmatrix} v_1 \\ v_2 \\ v_3 \\ \hline v_4 \\ v_5 \\ v_6 \end{bmatrix} = \begin{bmatrix} 0 \\ 0 \\ 0 \end{bmatrix} \tag{13-4.11}$$

or in symbolic form

$$[\beta]\mathbf{v} = [\beta_T \,\vdots\, \beta_L] \begin{bmatrix} v_T \\ \hline v_L \end{bmatrix} = \mathbf{0} \tag{13-4.12}$$

where $[\beta]$ is called the *tieset matrix*, $[\beta_T]$ is the tree part, or partition, of $[\beta]$, and $[\beta_L]$ is the link partition, which is always an identity matrix with only diagonal unity elements. This structure of $[\beta]$ is the result of numbering the tree branches with the lower numbers, and link branches with higher numbers, and then forming the loops by dropping one link at a time into the circuit to form loops. The matrix formulation is particularly useful for computer circuit analysis methods; however, space does not permit development of this area at this point.

It is possible to expand Eq. 13-4.12 as

$$[\beta_T]\mathbf{v}_T + [\beta_L]\mathbf{v}_L = \mathbf{0} \tag{13-4.13}$$

Since $[\beta_L]$ is an identity matrix

$$[\beta_L]\mathbf{v}_L = \mathbf{v}_L \tag{13-4.14}$$

and Eq. 13-4.13 can be solved for the link voltages $\mathbf{v}_L$ in terms of tree voltages $\mathbf{v}_T$, as

$$\mathbf{v}_L = [-\beta_T]\mathbf{v}_T \tag{13-4.15}$$

In expanded form, for the example being considered, the resulting equations are

$$\begin{aligned} v_4 &= v_1 - v_3 \\ v_5 &= v_1 - v_2 - v_3 \\ v_6 &= -v_1 + v_3 \end{aligned} \tag{13-4.16}$$

These equations can now be added to the signal-flow graph of Fig. 13-4.4(b).

The KCLE's are now written by shorting in succession all tree branch voltages except one, and writing the KCLE for each resulting circuit. Due to the way in which the equation is written, it involves only one tree branch current plus a number of link branch currents. All other tree branches are shorted to one node or the other of the one energized tree branch. For this example, the resulting KCLE's are

$$\begin{aligned} j_1 + j_4 + j_5 - j_6 &= 0 \\ j_2 - j_5 &= 0 \\ j_3 - j_4 - j_5 + j_6 &= 0 \end{aligned} \tag{13-4.17}$$

As with the KVLE's (3-4.11) these KCLE's can be systematically arranged using matrices as

$$\left[\begin{array}{ccc|ccc} 1 & 0 & 0 & 1 & 1 & -1 \\ 0 & 1 & 0 & 0 & -1 & 0 \\ 0 & 0 & 1 & -1 & -1 & 1 \end{array}\right] \begin{bmatrix} j_1 \\ j_2 \\ j_3 \\ \hline j_4 \\ j_5 \\ j_6 \end{bmatrix} = \begin{bmatrix} 0 \\ 0 \\ 0 \end{bmatrix} \tag{13-4.18}$$

or in symbolic form as

$$[\alpha]\mathbf{j} = [\alpha_T \,|\, \alpha_L] \begin{bmatrix} j_T \\ \hline j_L \end{bmatrix} = \mathbf{0} \tag{13-4.19}$$

where $[\alpha]$ is called the *cutset matrix*, $[\alpha_T]$ is the tree partition of $[\alpha]$, and $[\alpha_L]$ is the link partition. The $[\alpha_T]$ is an identity matrix in this case, so that Eq. 13-4.19 can be written

$$[\alpha_T]\mathbf{j}_T + [\alpha_L]\mathbf{j}_L = \mathbf{0} \tag{13-4.20}$$

and the tree branch currents can be solved for, in terms of the link branch currents, as

$$\mathbf{j}_T = -[\alpha_L]\mathbf{j}_L \tag{13-4.21}$$

In expanded form, the equations for this example are

$$\begin{aligned} j_1 &= -j_4 - j_5 + j_6 \\ j_2 &= j_5 \\ j_3 &= j_4 + j_5 - j_6 \end{aligned} \tag{13-4.22}$$

These equations are added to the signal-flow graph of Fig. 13-4.4(b).

The only thing needed to complete the graph are the volt-ampere relationships. The equations are

$$\begin{gathered} j_4 = v_4/20 \qquad j_5 = v_5/20 \\ v_3 = 10j_3 \end{gathered} \tag{13-4.23}$$

While these equations can also be expressed in matrix form, the fact that some of the branches are ideal voltage or current sources prevents us from writing a volt-ampere type of equation for such branches. As care must therefore be exercised for these branches, the matrix approach is not developed here. The above equations are now added to Fig. 13-4.4(b), completing the graph.

An overall discussion of the final graph is in order. All the transmittances at the top of the graph flow from left to right, and are a graphical expression of the KVLE's. All the transmittances at the bottom flow from right to left, and are a graphical expression of the KCLE's. All the transmittances on the right flow down, and since the output is current and the input is voltage, each of these transmittances is an admittance. All the transmittances on the left flow up, and since the output is a voltage and the input is a current, each of these transmittances is an impedance. The sum of the transmittances on the left and on the right are the volt-ampere relationships.

As with all signal-flow graphs, nodes with all branches leaving are source nodes. In this graph, these are nodes v_1, v_2, and j_6, which correspond to voltage sources and current source in the original problem. Nodes that have all branches entering are sink or output nodes. As this graph is presently drawn, nodes j_1, j_2, and v_6 are output nodes, and these correspond to the currents in the voltage sources and the voltage across the current source. In the next example, these nodes are omitted to make the diagram simpler and because they do not add significantly to the discussion.

An inspection of this graph, Fig. 13-4.4, can now be used to indicate the meaning of the terms *independent set of voltages* and *independent set of currents*. Suppose the independent set of voltages is known. This information flows to the right, and all dependent voltages can be found. As soon as the dependent voltages are known, this information flows down on the right, and the independent set of currents can be found. Finally, this information flows to the left, and the dependent set of currents are determined. Therefore, out of the $2b$ unknowns, if the n independent voltages are known, all other quantities can be found.

Similarly, if the independent set of currents is known, this information flows to the left, and all dependent currents can be found. As soon as the dependent currents are known, this information flows up at the left, and the independent voltages can be found. Finally, this information flows to the right, and the dependent voltages are found. Therefore, out of the $2b$ unknowns, if the l independent currents are known, all other quantities can be found.

If all the nodes except the source nodes and the nodes representing the independent set of currents are removed by the node removal method, the resulting graph describes the loop-current equations for this circuit for the given tree. If all the nodes except the source nodes and the nodes representing the independent set of voltages are removed by the node removal method, the resulting graph describes the node-pair voltage equations for this circuit for the given tree.

13-4.2. STATE VARIABLES AND THE 2*b* METHOD

Example 13-4.3. As a continuation of the discussion at the end of the last section, the circuit of Fig. 13-4.5(a) is used to show how these same ideas can be extended to the writing of state variable equations. The tree in (b) is chosen so that the voltage source

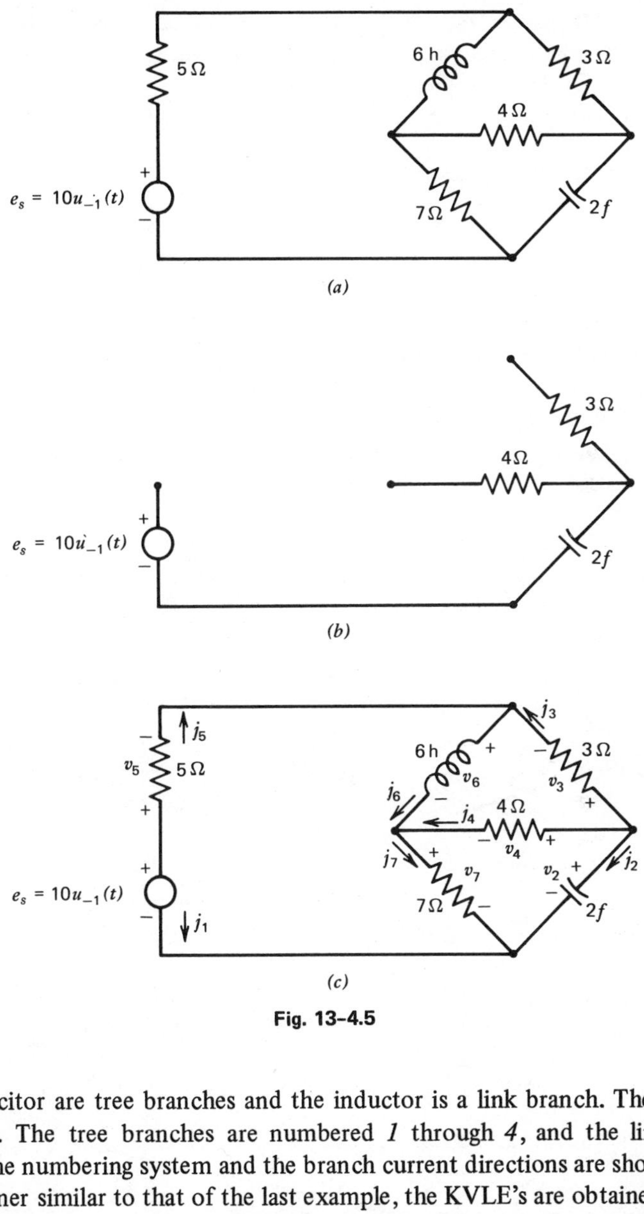

Fig. 13-4.5

and the capacitor are tree branches and the inductor is a link branch. The other choices are arbitrary. The tree branches are numbered *1* through *4*, and the link branches *5* through *7*. The numbering system and the branch current directions are shown in (c).

In a manner similar to that of the last example, the KVLE's are obtained by dropping the link branches in place, one at a time, in numerical sequence, writing the KVLE's around the resulting loop, and solving these equations for the dependent voltages, as

$$\begin{aligned} v_5 &= v_1 - v_2 + v_3 \\ v_6 &= -v_3 + v_4 \\ v_7 &= v_2 - v_4 \end{aligned} \qquad (13\text{-}4.24)$$

The KCLE's are obtained by shorting all but one tree branch voltage, writing the resulting KCLE taking each tree branch one at a time in numerical sequence, and solving for the resulting equations for the dependent current, as

$$
\begin{aligned}
j_1 &= -j_5 \\
j_2 &= \ \ j_5 - j_7 \\
j_3 &= -j_5 + j_6 \\
j_4 &= -j_6 + j_7
\end{aligned}
\tag{13-4.25}
$$

The first of Eqs. 13-4.25 involves the current in the voltage source, and we ignore it to make the resulting diagram simpler.

The $2b$ signal-flow graph is shown in Fig. 13-4.6. This graph is similar to the one for the last example, except that storage elements are present.

The state variables are the voltage on the capacitor, v_2, and the current in the inductor, j_6. These are arbitrarily denoted by

$$
\begin{aligned}
x_1 &= v_2 \\
x_2 &= j_6
\end{aligned}
\tag{13-4.26}
$$

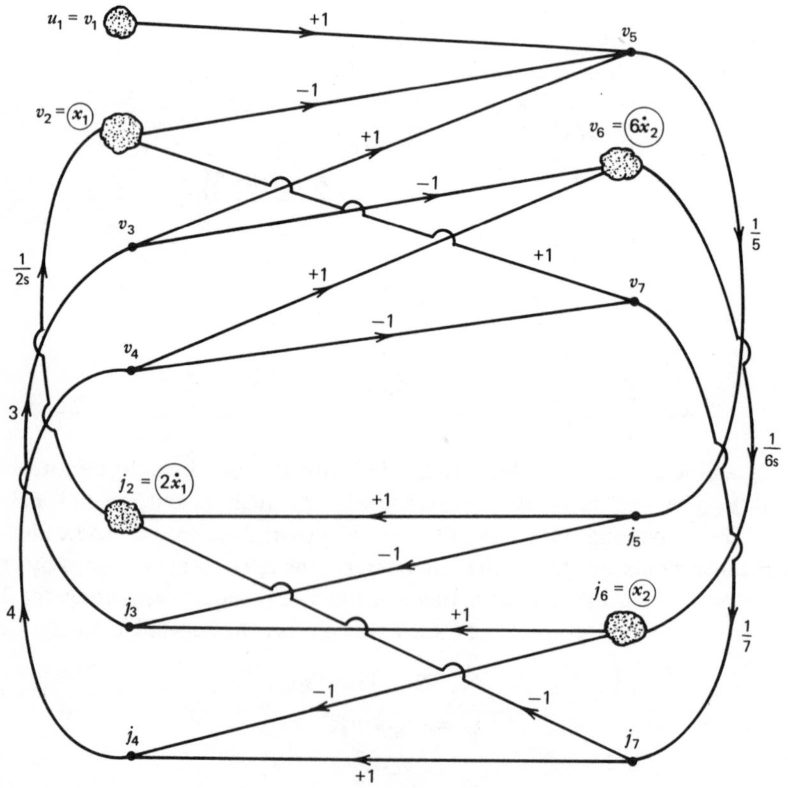

Fig. 13-4.6

The derivative of the state variables are also needed. We note that

$$j_2 = 2\frac{dv_2}{dt} = 2\dot{x}_1 \tag{13-4.27}$$

and that

$$v_6 = 6\frac{dj_6}{dt} = 6\dot{x}_2 \tag{13-4.28}$$

The constants 2 and 6 could be moved to another location in the signal-flow diagrams; the only rule that need be followed is to not change any loop transmittance. It is simpler, however, to leave these constants where they are.

The source node and the nodes associated with the state variables and their derivatives are enlarged in Fig. 13-4.6, and the state variable designation associated with each node is shown circled.

The easiest way to proceed conceptually (although extremely laborious in the carrying out of the actual steps) would be to remove all the nodes that are not enlarged, one at a time. The result would then be a graph from which the state variable equations could be written.

A more elegant method is to split the $x_1, 2\dot{x}_1, x_2$, and $6\dot{x}_2$ nodes, and to reduce the resulting graph between one input half-node to the associated output half-nodes, by the use of Mason's Rule. The resulting graph is rather difficult to visualize, and the same result is obtained by the following procedure.

USE OF THE DIMINISHED GRAPH

To use this method, all source nodes, state variable nodes, and their associated branches are removed from the original graph, and the resulting graph is called the *diminished graph*. For the graph of Fig. 13-4.6, the diminished graph is shown in Fig. 13-4.7(a). (Copies of this graph might be made, because it will be used over and over.)

The final reduced graph is shown in Fig. 13-4.8. This graph has 8 branches, but two of these transmittances ($1/2s$ and $1/6s$) are already known from the original graph of Fig. 13-4.6. Therefore, 6 branch transmittances need to be determined. The method of determining four of these is shown in Fig. 13-4.7.

The transmittance from $u_1 = v_1$ to $2\dot{x}_1$ is determined by adding these two nodes, and the appropriate branches that connect these nodes, to the diminished graph shown in Fig. 13-4.7(b). This graph is reduced by using Mason's Rule. It should be noticed that one of the two loops is dangling and does not enter into the transmittance, which is found as

$$\frac{2\dot{x}_1}{u_1} = \frac{1/5(1)}{1 + 3/5} = 1/8 \tag{13-4.29}$$

The transmittance from x_1 to $2\dot{x}_1$ is determined by adding these two nodes, and the appropriate branches that connect these nodes, to the diminished graph shown in Fig.

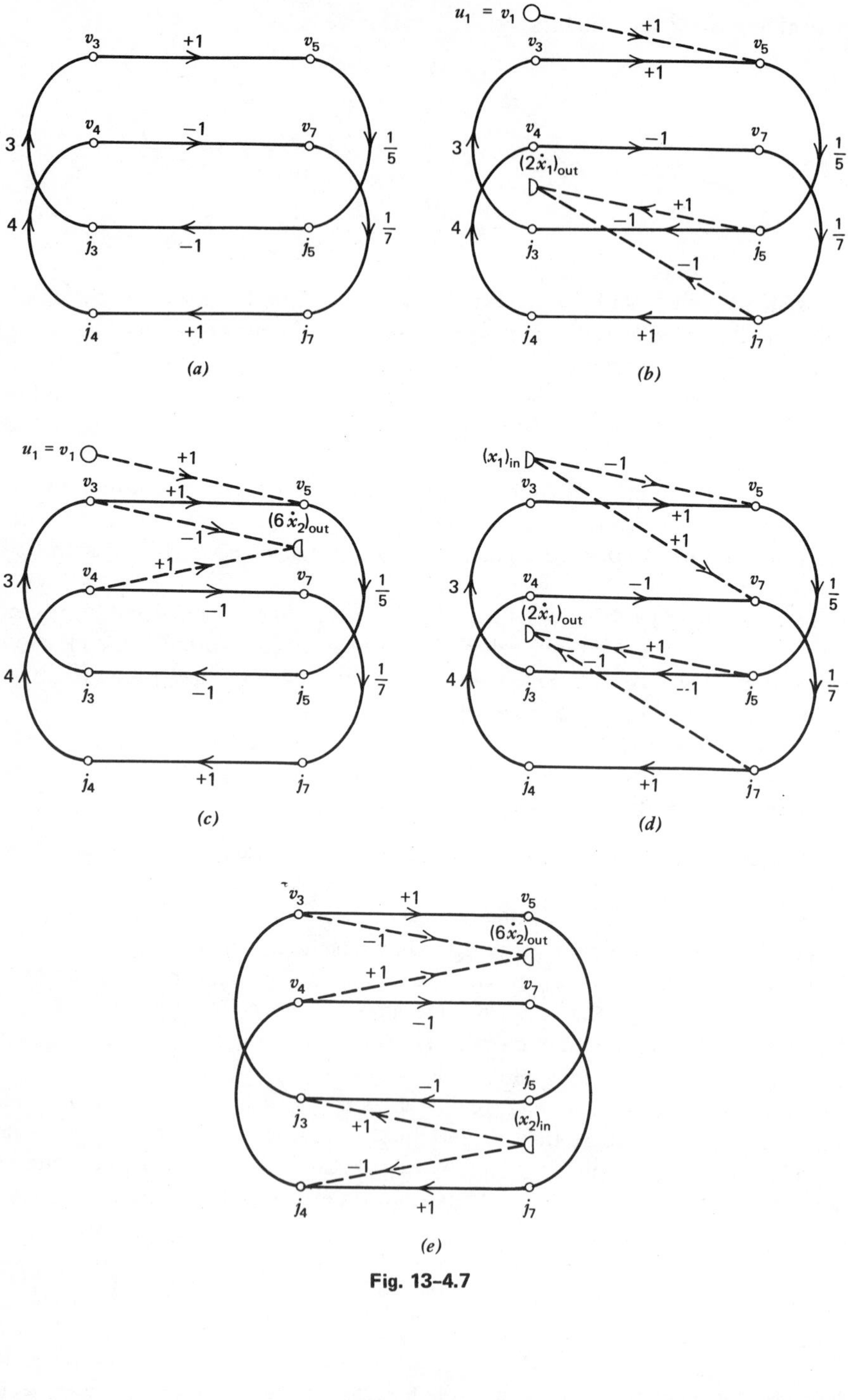

v3
+1
v5
v4
−1
v7
3
4
1/5
1/7
j3
j5
j4
j7
(a)
u1 = v1
(2ẋ1)out
(b)
(6ẋ2)out
(c)
(x1)in
(d)
(x2)in
(e)

Fig. 13-4.7

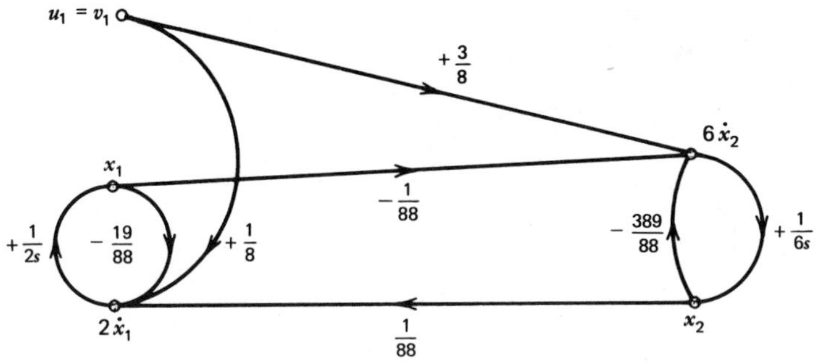

Fig. 13-4.8

13-4.7(d). This graph is reduced by using Mason's Rule, as

$$\frac{2\dot{x}_1}{x_1} = \frac{-\frac{1}{5}(1+\frac{4}{7}) - \frac{1}{7}(1+\frac{3}{5})}{1+\frac{3}{5}+\frac{4}{7}+(-\frac{3}{5})(-\frac{4}{7})} = -\frac{19}{88} \tag{13-4.30}$$

The other transmittances are found in a similar manner, and the results are added to the graph in Fig. 13-4.8.

The following equations are written from the graph.

$$\begin{aligned} 2\dot{x}_1 &= -\tfrac{19}{88}x_1 + \tfrac{1}{88}x_2 + \tfrac{1}{8}u_1 \\ 6\dot{x}_2 &= -\tfrac{1}{88}x_1 - \tfrac{389}{88}x_2 + \tfrac{3}{8}u_1 \end{aligned} \tag{13-4.31}$$

The first equation is divided by two, and the second equation by six, and the final equations are written as

$$\dot{\mathbf{x}} = \begin{bmatrix} -\frac{19}{176} & \frac{1}{176} \\ -\frac{1}{528} & -\frac{389}{528} \end{bmatrix} \mathbf{x} + \begin{bmatrix} \frac{1}{16} \\ \frac{1}{16} \end{bmatrix} u_1 \tag{13-4.32}$$

Again our present subject is not how to solve these equations, but rather how to write them. Therefore, let us assume that we have solved Eqs. 13-4.32 somehow, and x_1 and x_2 are now known. The remaining problem is how to find the value of all the remaining branch currents and branch voltages.

Let us examine Fig. 13-4.6 again to see what the problem is. The values of the enlarged nodes are known, but the values of the other nodes have yet to be determined. In the previous example, we suggested that if either the set of independent branch voltages, or the indeoendent set of branch currents were known, all other quantities could be found. There are four independent branch voltages, and only two of these are known, which of course leaves two voltages as unknowns. There are three independent branch currents, but only one of these is known, again leaving two unknowns to be determined. What we need to do is to determine either v_3 and v_4, or j_5 and j_7, and all the other quantities follow immediately. To continue, we go after the equations for v_3 and v_4.

We again consider the graph of Fig. 13-4.6, with the following observations. The values of the x_1, x_2, $2\dot{x}_1$, and $6\dot{x}_2$ nodes are known. These nodes can now be split, with

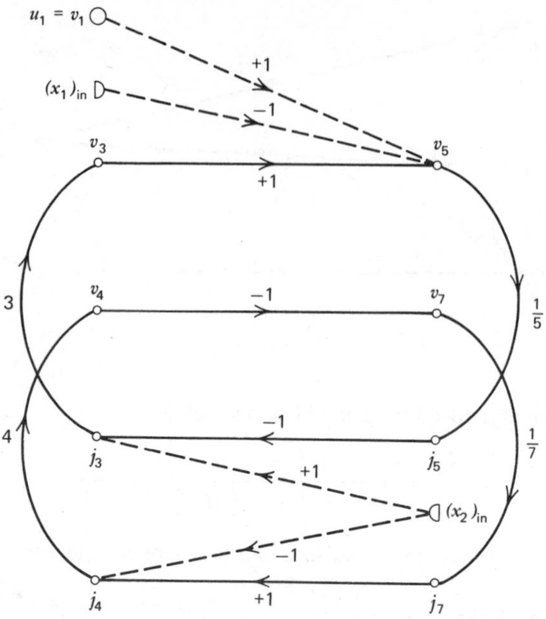

Fig. 13-4.9

the appropriate value being fed into the system at each input node. In other words, these nodes now act as source nodes. The output halves of the split nodes act as system outputs, and will have the proper value if all the other nodes have their proper values. Therefore, the output halves of the split nodes and their associated branches can be ignored. What we are really doing is returning to the diminished graph of Fig. 13-4.7(a) and adding u_1, x_1, and x_2 as inputs. The resulting graph is shown in Fig. 13-4.9. For v_3 and v_4, by reducing the graph using Mason's Rule,

$$v_3 = \frac{(3/5)(1)}{1 + (3/5)} x_1 + \frac{3(1)}{1 + (3/5)} x_2 + \frac{-(3/5)(1)u_1}{1 + (3/5)}$$
$$v_4 = \frac{-4(1)x_2}{1 + (4/7)} \tag{13-4.33}$$

which can be rewritten as

$$\begin{bmatrix} v_3 \\ v_4 \end{bmatrix} = \begin{bmatrix} \frac{3}{8} & \frac{15}{8} \\ 0 & -\frac{28}{11} \end{bmatrix} \mathbf{x} + \begin{bmatrix} -\frac{3}{8} \\ 0 \end{bmatrix} u_1 \tag{13-4.34}$$

As stated earlier, now that the independent set of branch voltages is known, all the other quantities can be determined.

COMMENTARY

The $2b$ signal-flow graph is a straightforward procedure from which the state variable equations can be set up. The preceding example is certainly not all-inclusive, but is presented to show the basic concept of the procedure. Since the graph is based on a tree, the problem of linear independence among the equations is eliminated.

One important disadvantage of the $2b$ signal-flow graph in general is the large number of loops that exists in these graphs. When the system determinant is evaluated using Mason's Rule, a large majority of the terms cancel. It is possible to show how these non-canceling terms can be picked out ahead of time, thus saving much effort; however, this development is beyond the scope of this text.

Another problem not discussed in the previous example is how to set up state variable equations for circuits that contain tiesets of capacitors and/or cutsets of inductors. This problem is not unique to the $2b$ method.

PROBLEMS

Unless otherwise stated, all vectors are specified with components relative to the standard E-basis.

13-1 (a) The ordered triples of numbers $(-1,2,1)$, $(0,2,0)$, and $(1,0,0)$ are to be written as vectors $\mathbf{x}_1$, $\mathbf{x}_2$, and $\mathbf{x}_3$, respectively. What must be true, if the vectors are linearly independent?

(b) Consider $[\mathbf{x}_1\ \mathbf{x}_2\ \mathbf{x}_3]\mathbf{c} = \mathbf{0}$, where $\mathbf{c}$ is a vector of three constants, and $\mathbf{0}$ is a vector of three zero elements (Null vector). How can this vector-matrix equation be used to determine the linear dependence or independence of $\mathbf{x}_1$, $\mathbf{x}_2$, and $\mathbf{x}_3$ for this case?

(c) Can the three given $\mathbf{x}$ vectors be a basis for three-dimensional space? *Explain.*

13-2 A common and essential concept in vector space work is that of orthogonality, which in a restricted sense involves perpendicularity. More generally, the orthogonality of two real n-dimensional vectors, $\mathbf{x}_1$ and $\mathbf{x}_2$, requires that $\mathbf{x}_1^T\mathbf{x}_2 = 0$. For complex elements the concept involves $\mathbf{x}_1^{*T}\mathbf{x}_2 = 0$, where $\mathbf{x}_1^{*T}$ is the complex conjugate transpose of $\mathbf{x}_1$. This latter expression is frequently termed the inner product, written as $(\mathbf{x}_1, \mathbf{x}_2)$.

(a) Are any of the three vectors of Problem 13-1(a) orthogonal to any other vectors in this set?

(b) The quantity $\mathbf{x}^{*T}\mathbf{x}$ or $(\mathbf{x}, \mathbf{x})$ is commonly used to define the square of *distance* or *length* when dealing with vector spaces. Alternately, $(\mathbf{x}, \mathbf{x})^{1/2} = \|x\|$ is defined as the norm of the vector $\mathbf{x}$. If $\mathbf{x}_1 = [1, -2, 3]^T$, $\mathbf{x}_2 = [0, 1, 0, -1]^T$, and $\mathbf{x}_3 = [1 - j1, 1, 2 + j1]^T$, find $\|\mathbf{x}_1\|$, $\|\mathbf{x}_2\|$, $\|\mathbf{x}_3\|$, $\mathbf{x}_1/\|\mathbf{x}_1\|$, $\mathbf{x}_2/\|x_2\|$, $\mathbf{x}_3/\|\mathbf{x}_3\|$, $(\mathbf{x}_1, \mathbf{x}_3)$.

(c) What must be true of the vector $\mathbf{x}/\|\mathbf{x}\|$?

13-3 (a) Can the vectors $\mathbf{x}_1 = [-1, 2, 1]^T$ and $\mathbf{x}_2 = [0, 0, 1]^T$ form a basis for a 2-space? *Explain.*

(b) If $\mathbf{x}_1$ and $\mathbf{x}_2$ from (a) can be a basis, sketch the space spanned, and show the two vectors.

(c) Generalize any conclusions you can from (a) and (b).

13-4 (a) For $\mathbf{x}_1 = [\sqrt{3}, 0]^T$, $\mathbf{x}_2 = [\sqrt{3}, 1]^T$, find $(\mathbf{x}_1, \mathbf{x}_2)$, $(\mathbf{x}_2, \mathbf{x}_1)$, $(\mathbf{x}_1/\|\mathbf{x}_1\|, \mathbf{x}_2)$, $(\mathbf{x}_2, \mathbf{x}_1/\|\mathbf{x}_1\|)$, $(\mathbf{x}_1, \mathbf{x}_2/\|\mathbf{x}_2\|)$, $(\mathbf{x}_2/\|\mathbf{x}_2\|, \mathbf{x}_1)$.

(b) Explain the significance of the quantities in (a), in terms of *projections*. Use sketches of 2-space as appropriate.

13-5 An orthonormal basis set of vectors is a set of mutually orthogonal, unit-length vectors, such as the E-basis set of Section 13-2. However, the E-basis is certainly not the only possible orthonormal basis set.

(a) In 2-space, determine a second vector that can be used with $\mathbf{x}_1 = [1/\sqrt{2}, -1/\sqrt{2}]^T$ as an orthonormal basis.

(b) Can any other vectors be used in combination with the specified $\mathbf{x}_1$ to form an orthonormal basis? *Explain.*

13-6 The *projection* of a vector $\mathbf{x}_2$ along the direction of a vector $\mathbf{x}_1$ can be found as $(\mathbf{x}_2, \mathbf{x}_1/\|\mathbf{x}_1\|)$.

(a) If a vector of length equal to that of the projection of $\mathbf{x}_2$, along the direction of $\mathbf{x}_1$ is subtracted from $\mathbf{x}_2$, what must be true of the resulting vector? *Verify* your conclusion with an example.

(b) If $\mathbf{x}_1 = [-1, \sqrt{3}]^T$, and $\mathbf{x}_2 = [0, 2]^T$, find the projection of $\mathbf{x}_2$ along the direction of $\mathbf{x}_1$, subtract a vector, of length equal to the projection of $\mathbf{x}_2$ along the $\mathbf{x}_1$ direction, from $\mathbf{x}_2$ to yield a new vector $\mathbf{x}_3$. Form $(\mathbf{x}_2, \mathbf{x}_3)$.

13-7 Find a reference describing the Gram–Schmidt orthogonalizing process, and apply it to obtain an orthonormal basis for 2-space from $\mathbf{x}_1 = [1, 2]^T$, $\mathbf{x}_2 = [1, 1/2]^T$.

13-8 A time-varying vector is given by $\mathbf{x}(t) = \epsilon^{-5t}[1, -1, -1]^T$, $t \geqslant 0$. Describe how this vector changes with time. Find $\|\mathbf{x}(t)\|$.

13-9 If $\mathbf{x}(t) = [\epsilon^{-t} \cos 2t, \epsilon^{-t} \sin 2t, 1]^T$; $t \geqslant 0$, find $\|\mathbf{x}(t)\|$ and describe how this vector changes with time.

13-10 If $\mathbf{x}_1(t) = [t\epsilon^{-t}, \epsilon^{-t}, 0]^T$; $t \geqslant 0$, $\mathbf{x}_2(t) = [\epsilon^{-t}, -t\epsilon^{-t}, 1]^T$; $t \geqslant 0$, are $\mathbf{x}_1(t)$ and $\mathbf{x}_2(t)$ orthogonal?

13-11 The network shown is to be modeled by vector-matrix state differential equations. Use capacitor voltages as state variables and write the state equations in $\dot{\mathbf{x}} = [A]\,\mathbf{x} + [B]\,\mathbf{u}$ form.

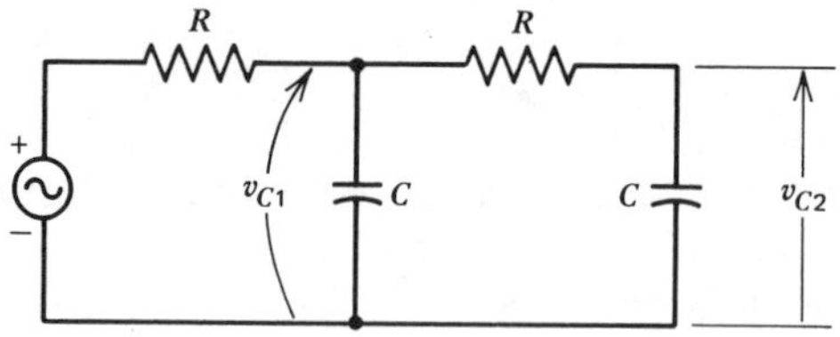

13-12 Use position and velocity of mass as state variables, and develop the state differential equations for the system shown.

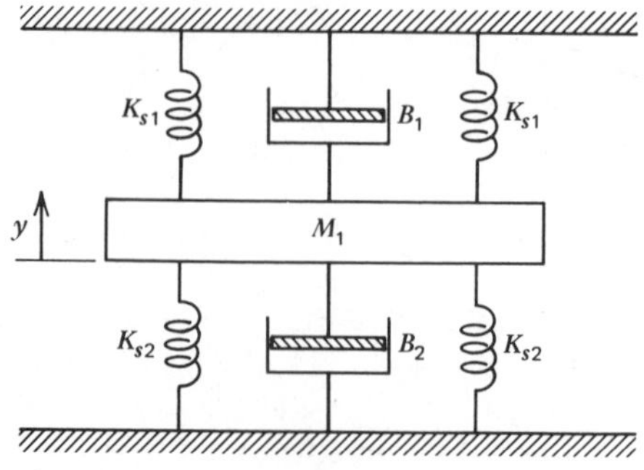

13–13 Repeat Problem 13–12 for the system shown.

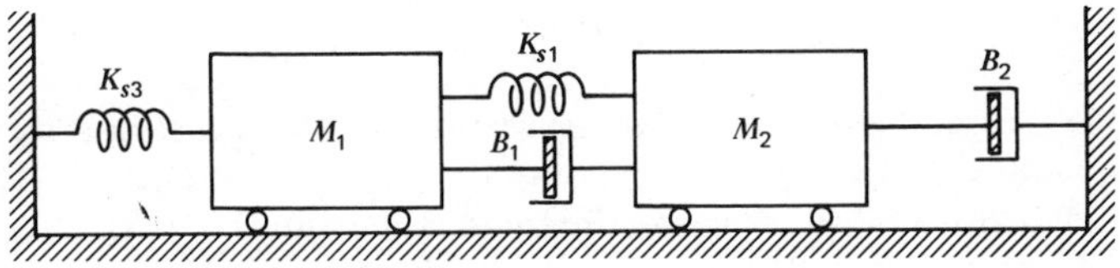

13–14 Repeat Problem 13–12 for the system shown.

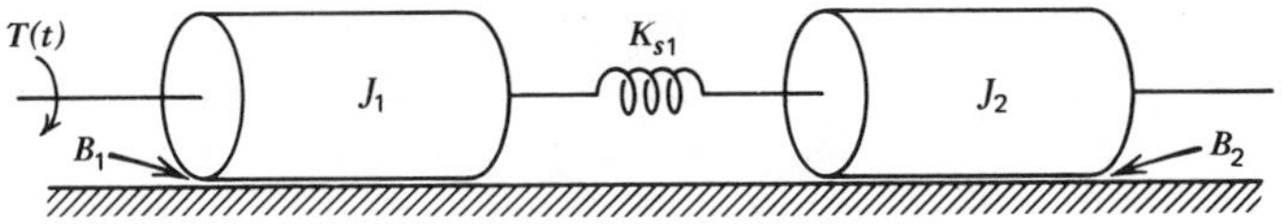

13–15 Develop a signal-flow diagram for the fluidic system shown, using a minimum number of integrators ($1/s$ transmittances). Write a set of state differential equations, using the outputs of ($1/s$) branches as state variables.

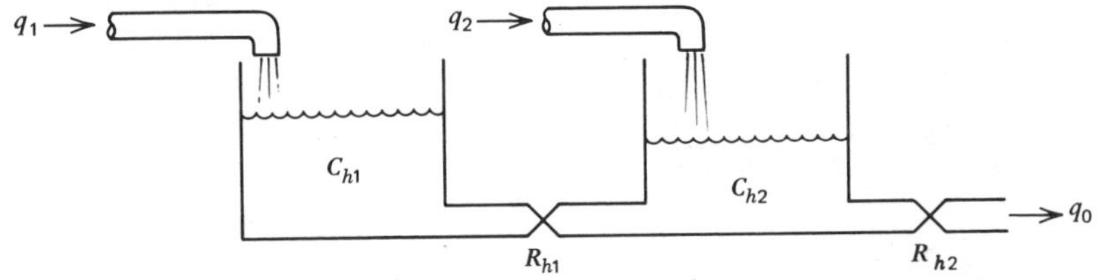

13–16 Repeat Problem 13–15 for the thermal system shown. Consider that: (a) a heater supplies q_{in} Joule/min; (b) mass M_1 is at a uniform temperature T_1; (c) heat flow from mass M_1 to M_2 occurs through a thermal resistance R_{t1} according to $(T_1 - T_2)/R_{t1}$ Joule/min; (d) mass M_2 is at a uniform temperature T_2; (e) heat flows from mass M_2 to the ambient environment through a thermal resistance R_{t2} according to $(T_2 - T_a)/R_{t2}$ Joule/min.

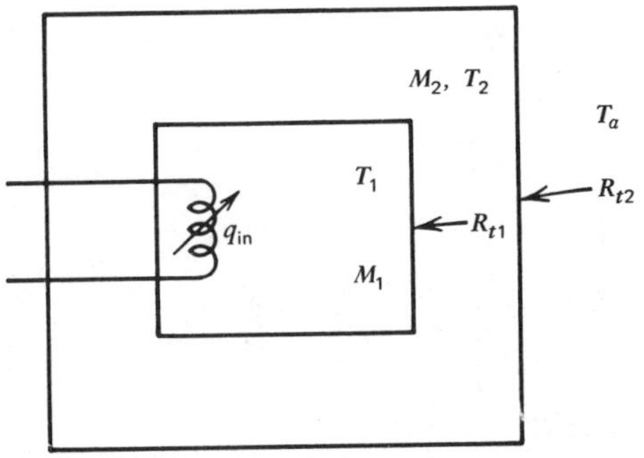

13–17 (a) Develop a signal-flow graph simulation for the transfer function

$$\frac{Y(s)}{U(s)} = \frac{(s+2)}{(s^2+s+2)(s+1)}$$

Select a set of state variables, and write the vector-matrix state differential equations $\dot{\mathbf{x}} = [A]\mathbf{x} + [B]\mathbf{u}$ and output equations $\mathbf{y} = [C]\mathbf{x} + [D]\mathbf{u}$.

(b) Use Eq. 13–3.50 to redevelop the transfer function $Y(s)/U(s)$.

(c) Use Mason's Rule and signal-flow graph theory to verify $Y(s)/U(s)$.

13–18 A system is modeled by the block diagram shown. Select a set of state variables, and formulate the corresponding state differential equations.

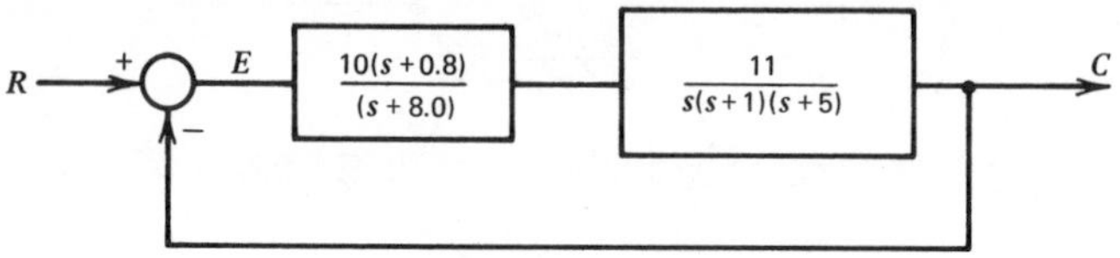

13–19 A certain system has the overall transfer function

$$\frac{Y(s)}{U(s)} = \frac{147}{(s+12)(s^2+3.2s+12.25)}$$

Use a parallel programming approach to determine a set of state variables, and develop the state differential equations.

13–20 (a) Determine a tree for the network shown, with all capacitor and source branch voltages in the tree. Write the cutset and tieset equations.

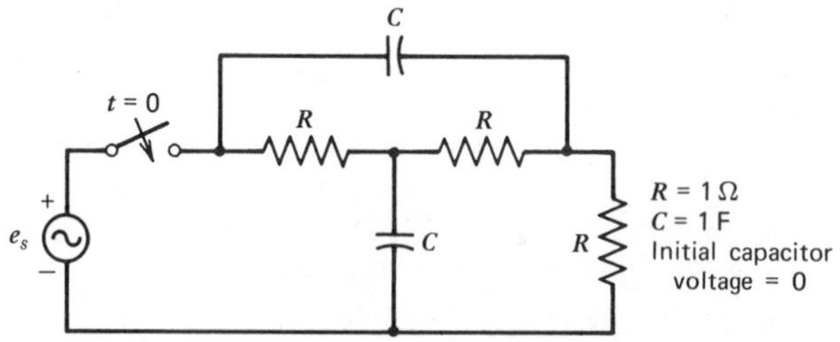

(b) Use the procedure of Section 13–4.2 to develop a set of state differential equations.

13–21 Repeat Problem 13–20 for the network shown.

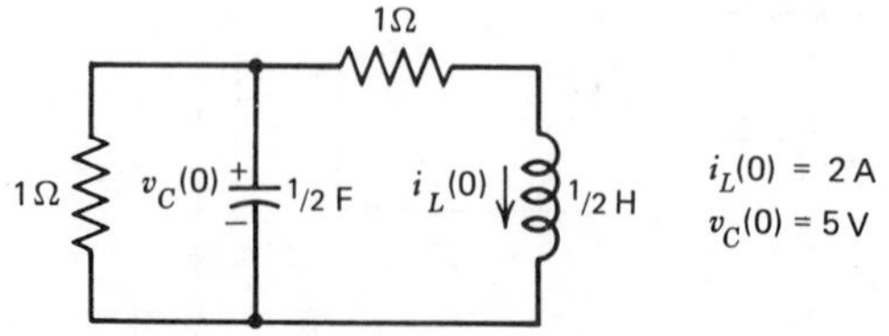

13–22 Develop a state variable differential equation model for the system with the block diagram representation shown.

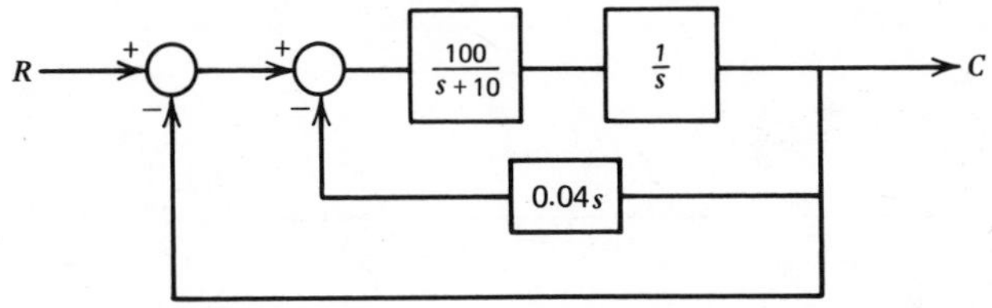

Chapter 14

State Variables: II

14-1. INTRODUCTION

The previous chapter presented an introduction to the selection and formulation of state variable equations. A continuation of these concepts to include the solution of the state differential equations, system trajectories, eigenvalues and eigenvectors, multiple input-output systems, discrete systems, and system controllability and observability is presented in this chapter.

14-2. SOLUTION OF THE STATE VECTOR-MATRIX DIFFERENTIAL EQUATION

Many pages have been devoted to formulating the linear system dynamic model in terms of the vector-matrix differential equation

$$\dot{\mathbf{x}} = \mathbf{A}\,\mathbf{x} + \mathbf{B}\,\mathbf{u} \tag{14-2.1}$$

However, only the brief discussion in Example 13-3.9 mentioned anything about determining $\mathbf{x}(t)$ or solving Eq. 14-2.1. Certainly it would be possible to redevelop a single nth-order differential equation and apply the usual differential equation theory or Laplace transform theory. However, why do this after all of the effort devoted to formulation in terms of state variables? As indicated in Eqs. 13-3.47, it is possible to use Laplace transformation directly, element by element, in Eq. 14-2.1. The procedure gives

$$s\,\mathbf{X}(s) - \mathbf{x}(0) = \mathbf{A}\,\mathbf{X}(s) + \mathbf{B}\,\mathbf{U}(s) \tag{14-2.2}$$

$$[sI - A]\;\mathbf{X}(s) = \mathbf{x}(0) + \mathbf{B}\,\mathbf{U}(s) \tag{14-2.3}$$

$$\mathbf{X}(s) = [sI - A]^{-1}\,\mathbf{x}(0) + [sI - A]^{-1}\,\mathbf{B}\,\mathbf{U}(s) \tag{14-2.4}$$

Again, it should be pointed out that the order of matrices in a product is important.

If element-by-element inverse Laplace transforms are taken in Eq. 14-2.4, the result can be written as the sum of the initial condition part and the forcing function part. As indicated in

$$\mathbf{x}(t) = \mathcal{L}^{-1}\{[sI - A]^{-1}\}\,\mathbf{x}(0) + \mathcal{L}^{-1}\{[sI - A]^{-1}\,\mathbf{B}\,\mathbf{U}(s)\} \qquad (14\text{-}2.5)$$

the term $\mathcal{L}^{-1}[sI - A]^{-1}$ is important. In fact, it is so important, that it is designated by the symbol $[\phi(t)]$ and is called the *state transition matrix* or *fundamental matrix*. This definition is emphasized by

$$[\Phi(s)] = [sI - A]^{-1} \qquad (14\text{-}2.6)$$

$$[\phi(t)] = \mathcal{L}^{-1}\{[sI - A]^{-1}\} \qquad (14\text{-}2.7)$$

Example 14-2.1. The system of Eq. 13-3.54 is used to illustrate the application of Eqs. 14-2.6 and 7, with results from Eqs. 13-3.56 and 57 also utilized.

$$\dot{\mathbf{x}} = \begin{bmatrix} 0 & 1 \\ -8 & -4 \end{bmatrix}\mathbf{x} + \begin{bmatrix} 0 \\ 1 \end{bmatrix} u \qquad (14\text{-}2.8)$$

$$[\Phi(s)] = [sI - A]^{-1} = \begin{bmatrix} \dfrac{(s+4)}{(s+2)^2 + 2^2} & \dfrac{1}{(s+2)^2 + (2)^2} \\[2ex] \dfrac{-8}{(s+2)^2 + 2^2} & \dfrac{s}{(s+2)^2 + (2)^2} \end{bmatrix} \qquad (14\text{-}2.9)$$

$$[\phi(t)] = \mathcal{L}^{-1}\{\Phi(s)\} = \begin{bmatrix} \epsilon^{-2t}(\cos 2t + \sin 2t) & (\tfrac{1}{2}\epsilon^{-2t}\sin 2t) \\ -(4\epsilon^{-2t}\sin 2t) & \epsilon^{-2t}(\cos 2t - \sin 2t) \end{bmatrix} \qquad (14\text{-}2.10)$$

In case $u(t) = 0$ with only initial conditions exciting the system, the results of Eq. 14-2.5 and 14-2.10 can be combined to give

$$\mathbf{x}(t) = [\phi(t)]\,\mathbf{x}(0) \qquad t \geqslant 0. \qquad (14\text{-}2.11)$$

$$x_1(t) = x_1(0)\,\epsilon^{-2t}(\cos 2t + \sin 2t) + \tfrac{1}{2}x_2(0)\,\epsilon^{-2t}\sin 2t \qquad (14\text{-}2.12)$$

$$x_2(t) = -4x_1(0)\,\epsilon^{-2t}\sin 2t + x_2(0)\,\epsilon^{-2t}(\cos 2t - \sin 2t) \qquad (14\text{-}2.13)$$

The matrix $[\phi(t)]$ has allowed the state of the system for $t \geqslant 0$ to be determined from the state at $t = 0$. Thus, $[\phi(t)]$ really describes the *transition of the state* of the system and thereby deserves the name *state transition matrix*.

Rather than obtain $[\phi(t)]$ analytically by forming $[sI - A]$ and obtaining the matrix inverse, it is generally easier to use simulation signal-flow graphs and Mason's Rule as follows. If individual equations are written for the initial condition part of Eq. 14-2.4, the results in

$$\begin{bmatrix} X_1(s) \\ X_2(s) \end{bmatrix} = \begin{bmatrix} \Phi_{11}(s) & \Phi_{12}(s) \\ \Phi_{21}(s) & \Phi_{22}(s) \end{bmatrix}\begin{bmatrix} x_1(0) \\ x_2(0) \end{bmatrix} \qquad (14\text{-}2.14)$$

$$X_1(s) = \Phi_{11}(s)\,x_1(0) + \Phi_{12}(s)\,x_2(0)$$
$$X_2(s) = \Phi_{21}(s)\,x_1(0) + \Phi_{22}(s)\,x_2(0) \tag{14-2.15}$$

are obtained. The $x_1(0)$ and $x_2(0)$ are taken as impulses for this system. If $x_2(0)$ is made zero, then Eqs. 14-2.15 reduce to

$$\left.\begin{aligned} X_1(s) &= \Phi_{11}(s)\,x_1(0) \\ X_2(s) &= \Phi_{21}(s)\,x_1(0) \end{aligned}\right\} \quad \begin{aligned} U(s) &= 0 \\ x_2(0) &= 0 \end{aligned} \tag{14-2.16}$$

These states can be determined directly from the signal-flow graph of Fig. 13-3.8, which is repeated as Fig. 14-2.1 except for omission of the branches to y.

To determine $\Phi_{11}(s)$ the transmission from $x_1(0)$ to $X_1(s)$ is computed with $x_2(0) = 0$. The graph determinant, $\Delta(s)$, path transmission, $P_1(s)$, and path cofactor, $\Delta_1(s)$ are given in

$$\Delta(s) = 1 + \frac{4}{s} + \frac{8}{s^2} \qquad P_1(s) = \frac{1}{s} \qquad \Delta_1(s) = 1 + \frac{4}{s} \tag{14-2.17}$$

$$\Phi_{11}(s) = \frac{X_1(s)}{x_1(0)} = \frac{(1/s)(1+4/s)}{1+4/s+8/s^2} = \frac{(s+4)}{(s^2+4s+8)} \tag{14-2.18}$$

Similarly the remaining elements of $[\Phi(s)]$ are found as follows:

$$\Phi_{21}(s) = \left.\frac{X_2(s)}{x_1(0)}\right|_{x_2(0)=0} = \frac{(-8/s^2)(1)}{1+4/s+8/s^2} = \frac{-8}{s^2+4s+8} \tag{14-2.19}$$

$$\Phi_{12}(s) = \left.\frac{X_1(s)}{x_2(0)}\right|_{x_1(0)=0} = \frac{(1/s^2)(1)}{\Delta(s)} = \frac{1}{s^2+4s+8} \tag{14-2.20}$$

$$\Phi_{22}(s) = \left.\frac{X_1(s)}{x_2(0)}\right|_{x_1(0)=0} = \frac{(1/s)(1)}{\Delta(s)} = \frac{s}{s^2+4s+8} \tag{14-2.21}$$

The procedure above is general, and can be described in general terms. To find $\Phi_{ij}(s)$, determine the transmission from the jth initial condition as the input to the ith state-variable node as the output with all other initial conditions and inputs set equal to zero. Notice that the graph determinant, $\Delta(s)$, need be computed only once.

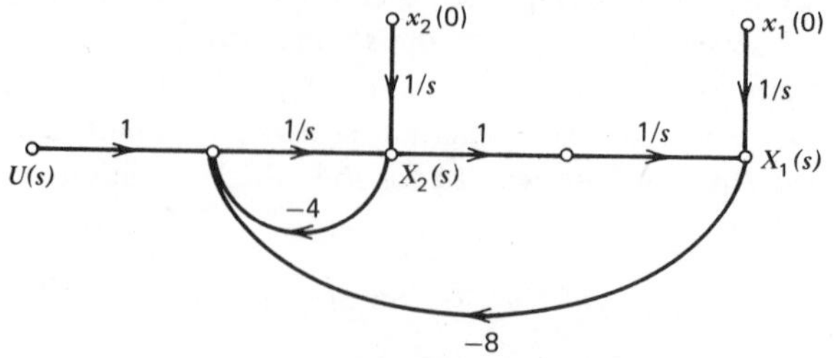

Fig. 14-2.1

Even more information can be obtained from an extension of the above procedure to the U inputs, and the forcing function part of the Laplace transform of the response given by $[sI - A]^{-1}$ **B U**(s) in Eq. 14-2.4. By considering all initial conditions zero, and energizing the u's one at a time, elements of $[sI - A]^{-1}$ **B** can be found.

In the example being considered, namely Fig. 14-2.1, the forcing function response matrix has only two elements.

$$[sI - A]^{-1}\,\mathbf{B} = [H(s)] = \begin{bmatrix} H_{11}(s) \\ H_{21}(s) \end{bmatrix} \qquad (14\text{-}2.22)$$

To find the $H_{ij}(s)$ elements, the following procedure is used, together with Fig. 14-2.1.

$$H_{11}(s) = \left.\frac{X_1(s)}{U(s)}\right|_{\mathbf{x}(0)=0} = \frac{(1/s^2)(1)}{\Delta(s)} = \frac{1}{s^2 + 4s + 8} \qquad (14.2.23)$$

$$H_{21}(s) = \left.\frac{X_2(s)}{U(s)}\right|_{\mathbf{x}(0)=0} = \frac{(1/s)(1)}{\Delta(s)} = \frac{s}{s^2 + 4s + 8} \qquad (14\text{-}2.24)$$

The results above can be checked by computing $[\Phi(s)]$ **B**. It is left to the reader to generalize this result.

The use of Laplace transformation to solve the state differential equations has led to referencing the initial state to $t = 0$. However, the basic philosophy of state variables is that knowledge of the state vector at anytime, t_0, together with the input $\mathbf{u}(t)$ for $t \geqslant t_0$, suffices to uniquely determine $\mathbf{x}(t)$, $t \geqslant t_0$. By using matrix techniques to solve Eq. 14-2.1, it is possible to show that the state transition matrix $[\phi(t)]$ is really a function of t and t_0.

However, it is not too difficult to arrive at this conclusion theoretically. Since $[\phi(t_0)]$ determines the state transition from the state $\mathbf{x}(0)$ at $t = 0$, to the state $\mathbf{x}(t_0)$ at $t = t_0$, it is reasonable that a state transition matrix in turn takes the system from state $\mathbf{x}(t_0)$ at t_0, to state $\mathbf{x}(t_1)$ at t_1. Since no basic change in dynamics is involved for a linear, constant coefficient system, as the time application of impulse inputs is changed, only an elapsed time $(t_1 - t_0)$ is involved in the last transition step. This leads to a state transition matrix $[\phi(t_1 - t_0)]$ to accomplish the transition from t_0 to t_1. In fact, starting the system from state $\mathbf{x}(t_0)$ is equivalent to applying a set of impulses at t_0 as initial condition inputs in the system simulation diagram. In general, $\mathbf{x}(t)$ is given by

$$\mathbf{x}(t) = [\phi(t - t_0)]\,\mathbf{x}(t_0) \qquad \mathbf{u} = 0. \qquad (14\text{-}2.25)$$

For time varying systems a state transition matrix still exists, this is a function of not only the elapsed time $(t - t_0)$, but also specifically of t_0, since the shape of the impulse response changes with the time of application of the impulses. For such cases, Eq. 14-2.26 expresses the result symbolically, but $[\phi(t, t_0)]$ may be impossible to find analytically in closed form.

$$\mathbf{x}(t) = [\phi(t, t_0)]\,\mathbf{x}(t_0) \qquad (14\text{-}2.26)$$

Several additional very important properties are possessed by $\phi(t)$. The first of these is established by using Eq. 14-2.25, with $t = t_0$. From this it is determined that

$$\mathbf{x}(t_0) = [\phi(t_0 - t_0)]\ \mathbf{x}(t_0) \qquad \mathbf{x}(t_0) = [\phi(0)]\ \mathbf{x}(t_0) \tag{14-2.27}$$

so that

$$[\phi(0)] = [I] \tag{14-2.28}$$

A second property is established by using Eq. 14-2.25 to obtain

$$\mathbf{x}(t_1) = [\phi(t_1 - t_0)]\mathbf{x}(t_0) \qquad \mathbf{x}(t_2) = [\phi(t_2 - t_1)]\ \mathbf{x}(t_1) \tag{14-2.29}$$

$$\mathbf{x}(t_2) = [\phi(t_2 - t_0)]\ \mathbf{x}(t_0) \tag{14-2.30}$$

so that

$$[\phi(t_2 - t_0)] = [\phi(t_2 - t_1)]\ [\phi(t_1 - t_0)] \tag{14-2.31}$$

If t_2 is set equal to t_0 in Eq. 14-2.31, the result is Eq. 14-2.32 and 33.

$$[\phi(t_0 - t_0)] = [\phi(0)] = [I] = [\phi(t_0 - t_1)]\ [\phi(t_1 - t_0)] \tag{14-2.32}$$

$$[\phi(t_0 - t_1)] = [\phi(t_1 - t_0)]^{-1} \tag{14-2.33}$$

The last result is a very interesting one which greatly simplifies finding the matrix inverse. As a special case, consider $t_0 = 0$, and $[\phi(t)]^{-1} = [\phi(-t)]$. All that is required to obtain the inverse is to change the sign of the argument, t.

For example, from Eq. 14-2.10 is obtained

$$[\phi(t)]^{-1} = [\phi(-t)] = \begin{bmatrix} \epsilon^{2t}(\cos 2t - \sin 2t) & -\frac{1}{2}\epsilon^{2t}\sin 2t \\ 4\epsilon^{2t}\sin 2t & \epsilon^{2t}(\cos 2t + \sin 2t) \end{bmatrix} \tag{14-2.34}$$

The reader might be energetic enough to derive this inverse by other usual methods.

Example 14-2.2. To illustrate the above properties, the system of Fig. 14-2.2 is used. For this case

$$\dot{\mathbf{x}} = \begin{bmatrix} 0 & 1 \\ 0 & -2 \end{bmatrix}\mathbf{x} + \begin{bmatrix} 0 \\ 1 \end{bmatrix}\mathbf{u}_1 \tag{14-2.35}$$

$$[\Phi(s)] = \begin{bmatrix} \dfrac{1}{s} & \dfrac{1}{s(s+2)} \\ 0 & \dfrac{1}{s+2} \end{bmatrix} \tag{14-2.36}$$

$$[\phi(t)] = \begin{bmatrix} 1 & \frac{1}{2}(1 - \epsilon^{-2t}) \\ 0 & \epsilon^{-2t} \end{bmatrix} \tag{14-2.37}$$

$$[\phi(0)] = \begin{bmatrix} 1 & 0 \\ 0 & 1 \end{bmatrix} = [I] \tag{14-2.38}$$

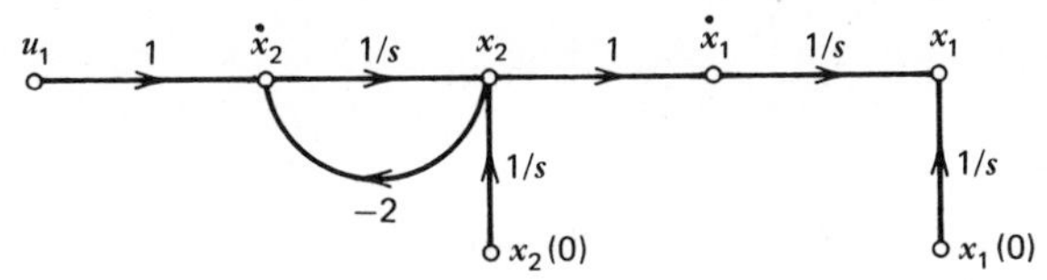

Fig. 14-2.2

$$[\phi(t_2 - t_1)]\ [\phi(t_1 - 0)] = \begin{bmatrix} 1 & \frac{1}{2}(1 - \epsilon^{-2(t_2 - t_1)}) \\ 0 & \epsilon^{-2(t_2 - t_1)} \end{bmatrix} \begin{bmatrix} 1 & \frac{1}{2}(1 - \epsilon^{-2t_1}) \\ 0 & \epsilon^{-2t_1} \end{bmatrix} \tag{14-2.39}$$

$$= \begin{bmatrix} 1 & \frac{1}{2}[1 - \epsilon^{-2t_1} + \epsilon^{-2t_1}(1 - \epsilon^{-2t_2 - t_1)})] \\ 0 & (\epsilon^{-2t_1})(\epsilon^{-2(t_2 - t_1)}) \end{bmatrix} \tag{14-2.40}$$

$$= \begin{bmatrix} 1 & \frac{1}{2}(1 - \epsilon^{-2t_2}) \\ 0 & \epsilon^{-2t_2} \end{bmatrix} = [\phi(t_2 - 0)] \tag{14-2.41}$$

$$[\phi(t)]^{-1} = \left(\frac{1}{\epsilon^{-2t}}\right) \begin{bmatrix} \epsilon^{-2t} & -\frac{1}{2}(1 - \epsilon^{-2t}) \\ 0 & 1 \end{bmatrix} \tag{14-2.42}$$

$$= \begin{bmatrix} 1 & \frac{1}{2}(1 - \epsilon^{2t}) \\ 0 & \epsilon^{2t} \end{bmatrix} = [\phi(-t)] \tag{14-2.43}$$

Except for a brief discussion of the forced part of the system state response in Example 14-2.1 and Eqs. 14-2.23 and 24, little has been said about it. This has been possible because of the superposition property of linear systems. From the Laplace transform Eq. 14-2.4, the forced state response component is given by

$$\mathbf{X}(s) = [H(s)]\ \mathbf{U}(s) \tag{14-2.44}$$

As developed for general Laplace transforms, the inverse of a product of transforms leads to a convolution in the time domain. The same is true in this case, since element-by-element multiplication gives sums of products of transforms such as $\Sigma_{j=1}^{m} H_{ij}(s)\,\mathbf{U}_j(s)$. Thus, the time domain forced part of the solution can be written as

$$\mathbf{x}(t) = \int_0^t [h(t - \tau)]\ \mathbf{u}(\tau)\,d\tau = \int_0^t [\phi(t - \tau)]\ [B]\ \mathbf{u}(\tau)\,d\tau \tag{14-2.45}$$

when the system is started at $t = 0$. If the system is started at $t = t_0$, the result for the forced part of the response is

$$\mathbf{x}(t) = \int_{t_0}^t [\phi(t - \tau)]\ [B]\ \mathbf{u}(\tau)\,d\tau \tag{14-2.46}$$

The complete solution consists of both the forced response and the natural response, as given by

$$\mathbf{x}(t) = [\phi(t - t_0)]\ \mathbf{x}(t_0) + \int_{t_0}^t [\phi(t - \tau)]\ [B]\ \mathbf{u}(\tau)\,d\tau \tag{14-2.47}$$

Example 14-2.3. To illustrate the use of Eqs. 14-2.44 and 47, the system of Fig. 14-2.2 is reconsidered. Using transforms, the following result is obtained for $u_1(t)$ as a unit-step input.

$$\mathbf{X}(s) = [\Phi(s)]\ \mathbf{x}(0) + [\Phi(s) B]\ U_1(s) \tag{14-2.48}$$

$$\mathbf{X}(s) = \begin{bmatrix} \frac{1}{s} & \frac{1}{s(s+2)} \\ 0 & \frac{1}{(s+2)} \end{bmatrix} \mathbf{x}(0) + \begin{bmatrix} \frac{1}{s(s+2)} \\ \frac{1}{(s+2)} \end{bmatrix} \left(\frac{1}{s}\right) \tag{14-2.49}$$

$$\mathbf{x}(t) = \begin{bmatrix} 1 & \frac{1}{2}(1-\epsilon^{-2t}) \\ 0 & \epsilon^{-2t} \end{bmatrix} \mathbf{x}(0) + \begin{bmatrix} \left(\frac{t}{2} - \frac{1}{4} + \frac{1}{4}\epsilon^{-2t}\right) \\ \frac{1}{2}(1-\epsilon^{-2t}) \end{bmatrix} \tag{14-2.50}$$

Using the convolution approach of Eq. 14-2.47 with $t = 0$ leads to the results below. Only the method for the forced part is different from what has been done.

$$\int_0^t [\phi(t-\tau)]\ [B]\ u(\tau)\, d\tau = \int_0^t \begin{bmatrix} 1 & \frac{1}{2}(1-\epsilon^{-2(t-\tau)}) \\ 0 & \epsilon^{-2(t-\tau)} \end{bmatrix} \begin{bmatrix} 0 \\ 1 \end{bmatrix} (1)\, d\tau \tag{14-2.51}$$

$$= \int_0^t \begin{bmatrix} \frac{1}{2}(1-\epsilon^{-2(t-\tau)}) \\ \epsilon^{-2(t-\tau)} \end{bmatrix} d\tau \tag{14-2.52}$$

$$= \begin{bmatrix} \frac{1}{2}\left(t - \frac{\epsilon^{-2t}\epsilon^{2\tau}}{2}\right)\Big|_0^t \\ \frac{\epsilon^{-2t}\epsilon^{2\tau}}{2}\Big|_0^t \end{bmatrix} = \begin{bmatrix} \left(\frac{t}{2} - \frac{1}{4} + \frac{\epsilon^{-2t}}{4}\right) \\ \frac{1}{2}(1-\epsilon^{-2t}) \end{bmatrix} \tag{14-2.53}$$

It is of interest to use one of the properties of the state transition matrix to simplify the use of Eqs. 14-2.46 and 47. By Eq. 14-2.31, $[\phi(t-\tau)]$ may be written as

$$[\phi(t-\tau)] = [\phi(t-0)]\ [\phi(0-\tau)] = [\phi(t)]\ [\phi(-\tau)] \tag{14-2.54}$$

so that

$$\int_{t_0}^t [\phi(t-\tau)B]\ \mathbf{u}(\tau)\, d\tau = [\phi(t)] \int_{t_0}^t [\phi(-\tau)B]\ \mathbf{u}(\tau)\, d\tau \tag{14-2.55}$$

Using this result in Eq. 14-2.51 gives

$$\begin{bmatrix} 1 & \frac{1}{2}(1-\epsilon^{-2t}) \\ 0 & \epsilon^{-2t} \end{bmatrix} \int_0^t \begin{bmatrix} \frac{1}{2}(1-\epsilon^{2\tau}) \\ \epsilon^{2t} \end{bmatrix} d\tau = [\phi(t)] \begin{bmatrix} \frac{t}{2} - \frac{\epsilon^{2t}}{4} + \frac{1}{4} \\ \frac{1}{2}(\epsilon^{2t}-1) \end{bmatrix} \quad 14\text{-}2.56)$$

$$= \begin{bmatrix} \left(\frac{t}{2} - \frac{1}{4} + \frac{1}{4}\epsilon^{-2t}\right) \\ \frac{1}{2}(1-\epsilon^{-2t}) \end{bmatrix} \tag{14-2.57}$$

Development of the solution of the vector-matrix differential equation for the state of a system in this section has been based mainly on Laplace transforms and a minimum of matrix operations—namely multiplications, additions, and inverses. An approach based on functions of matrices and classical differential equation theory is also possible. The following section provides an introduction to this approach.

14.3. MATRIX FORMULATION FOR THE SOLUTION OF STATE EQUATIONS

In the introductory study of differential equations, the first-order linear DE

$$\dot{x} = ax \tag{14-3.1}$$

is a common starting point. Its solution is obtained as

$$x(t) = x(0)\,\epsilon^{at} = x(t_0)\,\epsilon^{a(t-t_0)} \tag{14-3.2}$$

which indicates the dependence of $x(t)$ on $x(0)$ or on $x(t_0)$, together with the time durations $(t - 0)$ or $(t - t_0)$.

It is quite tempting to consider the extension of this form to the matrix use indicated in

$$\dot{\mathbf{x}}(t) = [A]\ \mathbf{x}(t) \tag{14-3.3}$$

which is a state differential equation with no forcing function. Such an extension leads to

$$\mathbf{x}(t) = \epsilon^{[A]t}\,\mathbf{x}(0) = \epsilon^{[A](t-t_0)}\,\mathbf{x}(t_0) \tag{14-3.4}$$

in which a function of a matrix, $\epsilon^{[A]t}$, is encountered. Such a thing needs definition, and so further reference is made to the definition of the scalar function ϵ^{at}. The infinite series of

$$\epsilon^{at} = 1 + (at) + (at)^2/2! + (at)^3/3! + \cdots \tag{14-3.5}$$

is used here. It can be shown that a similar series using $[A]$ instead of a is an appropriate definition of $\epsilon^{[A]t}$. This is given in

$$\epsilon^{[A]t} = [I_n] + [A]t + \frac{[A]^2 t^2}{2!} + \frac{[A]^3 t^3}{3!} + \cdots \tag{14-3.6}$$

where $[A]^n$ is the matrix $[A]$ multiplied by itself n times, and $[I_n]$ is the $n \times n$ identity matrix. Thus, $\epsilon^{[A]t}$ is a function of a matrix which also results in a matrix.

To verify that Eq. 14-3.4 is a solution of Eq. 14-3.3, the following steps are used.

$$\dot{\mathbf{x}} = \frac{d\mathbf{x}}{dt} \stackrel{?}{=} \frac{d}{dt}\left[\epsilon^{[A](t-t_0)}\mathbf{x}(t_0)\right] = \frac{d}{dt}\left[\epsilon^{[A](t-t_0)}\right]\mathbf{x}(t_0) \tag{14-3.7}$$

$$\frac{d\,\epsilon^{[A](t-t_0)}}{dt} = \frac{d}{dt}\left\{I + \mathbf{A}(t - t_0) + \frac{\mathbf{A}^2(t - t_0)^2}{2!} + \cdots\right\} \tag{14-3.8}$$

$$= \mathbf{A} + \mathbf{A}^2(t - t_0) + \frac{\mathbf{A}^3(t - t_0)^2}{2!} + \cdots \tag{14-3.9}$$

$$= \mathbf{A}\left[I + A\,(t - t_0) + \frac{A^2(t - t_0)^2}{2!} + \cdots\right] \tag{14-3.10}$$

$$= [A]\ \epsilon^{[A](t-t_0)} = \epsilon^{[A](t-t_0)}[A] \tag{14-3.11}$$

Direct substitution into Eq. 14-3.3, as in

$$\dot{\mathbf{x}}(t) \equiv [A]\ \epsilon^{[A](t-t_0)}\ \mathbf{x}(t_0) = [A]\ \mathbf{x} \tag{14-3.12}$$

completes the verification. Thus, two forms have now been shown to be solutions of Eq. 14-3.3. From Section 14-2 the Laplace transform solution is obtained, and from Eq. 14-3.4 a time domain solution results.

$$\mathbf{x}(t) = \mathcal{L}^{-1}\ \{[sI - A]^{-1}\}\ \mathbf{x}(0) = \epsilon^{[A]t}\ \mathbf{x}(0) \tag{14-3.13}$$

By identifying corresponding factors

$$\mathcal{L}^{-1}\ \{[sI - A]^{-1}\} = [\phi(t)] = \epsilon^{[A]t} \tag{14-3.14}$$

$$[\phi(t - t_0)] = \epsilon^{[A](t-t_0)} \tag{14-3.15}$$

is obtained.

Several other properties of the matrix $\epsilon^{[A]t}$ are interesting, as shown in

$$\epsilon^{-[A]t_1}\ \epsilon^{[A]t_2} = \left[I - At_1 + \frac{A^2 t_1^2}{2!} + \cdots\right]\left[I + At_2 + \frac{A^2 t_2^2}{2!} + \cdots\right]$$

$$= I + \mathbf{A}t_2 + \frac{\mathbf{A}^2 t_2^2}{2!} - \mathbf{A}t_1 - \mathbf{A}^2 t_1 t_2 - \frac{\mathbf{A}^3 t_1 t_2^2}{2!}$$

$$+ \frac{\mathbf{A}^2 t_2^2}{2!} + \frac{\mathbf{A}^3 t_1^2 t_2}{2!} + \frac{\mathbf{A}^4 t_1^2 t_2^2}{(2!)^2} + \cdots \tag{14-3.16}$$

$$= I + \mathbf{A}(t_2 - t_1) + \frac{\mathbf{A}^2(t_2 - t_1)^2}{2!} + \frac{\mathbf{A}^3(t_2 - t_1)^3}{3!} + \cdots \tag{14-3.17}$$

$$= \epsilon^{[A](t_2 - t_1)} \tag{14-3.18}$$

For $t_2 = t_1$ in Eq. 14-3.17

$$\epsilon^{[A](-t_1)}\ \epsilon^{[A]t_1} = [I] \tag{14-3.19}$$

With the additional result of

$$\epsilon^{-[A]t_1} = [\epsilon^{[A]t_1}]^{-1} \tag{14-3.20}$$

It also follows that

$$\epsilon^{[A](0)} = [I] \tag{14-3.21}$$

Equations 14–3.16 through 21 also verify the properties of $[\phi(t)]$ heuristically developed in Section 14–2.

In order to obtain a single matrix expression for $\epsilon^{[A]t}$ at this point, the infinite series of Eq. 14–3.6 has to be summed, or Eq. 14–3.14 has to be used. There is an alternate matrix procedure that is also useful. To show this, however, some preliminary development is necessary. It is known that

$$[sI - A]^{-1} = \frac{\text{Adjoint } [sI - A]}{|sI - A|} \tag{14-3.22}$$

for the inverse applies as well as

$$|sI - A|\,[sI - A]^{-1} = \text{Adjoint } [sI - A] \tag{14-3.23}$$

$$|sI - A|\,[I] = P(s)\,[I] = \{\text{Adjoint } [sI - A]\}\,[sI - A] \tag{14-3.24}$$

developed from this by post-multiplying by $[sI - A]$. In this development the system characteristic polynomial $P(s) = |sI - A|$ is introduced. This polynomial will have n roots designated as s_i, $i = 1, 2, \ldots, n$ so that $P(s) = 0$ for any root, s_i. Now, if the complex scalar s is replaced with the matrix $[A]$ in Eq. 14–3.24, the result is

$$P([A])\,[I] = \{\text{Adjoint } ([A]\,I - [A])\}\,([A]\,I - [A]) = [0] \tag{14-3.25}$$

The substitution of a matrix for a scalar appears to be a rather drastic step. However, it should be pointed out that the scalar s is multiplied by an identity matrix, so that both $s\mathbf{I}$ and $\mathbf{AI}$ are $(n \times n)$ matrices. Equation 14–3.25 is a zero or null matrix on the right, because $[AI] - [A] = [A] - [A] = [0]$. The former $P(s)$ polynomial is now a matrix polynomial $P[A]$, as indicated in

$$P(s) = a_n s^n + a_{n-1} s^{n-1} + \cdots + a_o \tag{14-3.26}$$

$$P([A]) = a_n [A]^n + a_{n-1} [A]^{n-1} + \cdots + a_o [I] \tag{14-3.27}$$

An identity matrix is used for $[A]^0$, so that all terms in $P([A])$ are $(n \times n)$ matrices. From Eq. 14–3.25 the main result is

$$P([A]) = [0] \tag{14-3.28}$$

in which $P(\,)$ is the form for the system characteristic polynomial for the system described by $\dot{\mathbf{x}} = \mathbf{Ax}$. The result of Eq. 14–3.28 is called the *Cayley-Hamilton Theorem*, which states that a matrix satisfies it's own characteristic equation.

Example 14–3.1. To illustrate some of the previous results, the system described by

$$\dot{\mathbf{x}} = \begin{bmatrix} -7 & 1 \\ -12 & 0 \end{bmatrix} \mathbf{x} \tag{14-3.29}$$

is considered. For this system the following equations result.

$$[sI - A] = \begin{bmatrix} (s+7) & -1 \\ 12 & s \end{bmatrix} \tag{14-3.30}$$

$$P(s) = |sI - A| = s(s+7) + 12 = (s+3)(s+4) \quad (14\text{-}3.31)$$

$$P([A]) = \begin{bmatrix} -7 & 1 \\ -12 & 0 \end{bmatrix}^2 + 7\begin{bmatrix} -7 & 1 \\ -12 & 0 \end{bmatrix} + 12\begin{bmatrix} 1 & 0 \\ 0 & 1 \end{bmatrix} \quad (14\text{-}3.32)$$

$$= \begin{bmatrix} 37 & -7 \\ 84 & -12 \end{bmatrix} + \begin{bmatrix} -49 & 7 \\ -84 & 0 \end{bmatrix} + \begin{bmatrix} 12 & 0 \\ 0 & 12 \end{bmatrix} = \begin{bmatrix} 0 & 0 \\ 0 & 0 \end{bmatrix} \quad (14\text{-}3.33)$$

$$[sI - A]^{-1} = \Phi(s) = \frac{\text{adj}\,(sI - A)}{|sI - A|} = \begin{bmatrix} \dfrac{s}{(s+3)(s+4)} & \dfrac{1}{(s+3)(s+4)} \\ \dfrac{-12}{(s+3)(s+4)} & \dfrac{(s+7)}{(s+3)(s+4)} \end{bmatrix} \quad (14\text{-}3.34)$$

$$\epsilon^{[A]t} = \phi(t) = \begin{bmatrix} (-3\epsilon^{-3t} + 4\epsilon^{-4t}) & (\epsilon^{-3t} - \epsilon^{-4t}) \\ -12(\epsilon^{-3t} - \epsilon^{-4t}) & (4\epsilon^{-3t} - 3\epsilon^{-4t}) \end{bmatrix} \quad (14\text{-}3.35)$$

One of the interesting uses of the Cayley-Hamilton Theorem is to evaluate powers of matrices. From Eq. 14-3.33, it can be seen that $[A]^2$ can be determined in terms of $[A]$, as in

$$[A]^2 = [0] - 7[A] - 12[I] \quad (14\text{-}3.36)$$

In fact, because of the Cayley-Hamilton Theorem, $[A]^n$ can be found in terms of $[A]^{n-1}$, $[A]^{n-2}, \ldots [A]^0$, as shown above for $n = 2$. Thus, any polynomial in $[A]$ can be reduced to an expression involving only $[A]^{n-1}$ and lower powers.

Example 14-3.2. If the arbitrary matrix polynomial of

$$F([A]) = [A]^4 + 2[A]^2 + [I] \quad (14\text{-}3.37)$$

is used with $[A]$ from Example 14-3.1, the expression for $F([A])$ can be expressed as follows. From Eq. 14-3.36, $[A]^2$ is found in terms of $[A]$ and $[I]$, and $[A]^3$ and $[A]^4$ are now found in

$$\left.\begin{aligned} [A]^3 &= [A]\,[A]^2 = [A]\,\{-7[A] - 12[I]\} = -7[A]^2 - 12[A] \\ &= -7\{-7[A] - 12[I]\} - 12[A] = 37[A] + 84[I] \end{aligned}\right\} \quad (14\text{-}3.38)$$

$$\left.\begin{aligned} [A]^4 &= [A]\,[A]^3 = 37[A]^2 + 84[A] = 37\{-7[A] - 12[I]\} + 84[A] \\ &= -175[A] - 444[I] \end{aligned}\right\} \quad (14\text{-}3.39)$$

Finally $F([A])$ can be written as

$$F([A]) = -175[A] - 444[I] + 2\{-7[A] - 12[I]\} + [I] = -189[A] - 467[I] \quad (14\text{-}3.40)$$

To return to the development of an alternate way of evaluating a function of a matrix, the following procedure is used. If the matrix function $F([A])$ can be expanded in a convergent power series in $[A]$, it can be reduced to at most a function of $[A]^{n-1}$ and lower powers of $[A]$, by using the above procedure. Again n is the order of this $[A]$

matrix. If the functional form $F(\ )$ is used with the scalar variable s, and a division by the characteristic polynomial $P(s)$ is carried out, the results of

$$\frac{F(s)}{P(s)} = W(s) + \frac{R(s)}{P(s)} \qquad F(s) = W(s)\,P(s) + R(s) \tag{14-3.41}$$

are obtained. In this equation, the series expansion of $F(s)$ is assumed, and $W(s)$ consists of the whole power of s part of the division by $P(s)$ and $R(s)$ is the remainder, which must have s^{n-1} as the highest power of s in it. Replacing s with $[A]$ in the right-hand relation of Eq. 14-3.41, and using the fact that $P([A]) = [0]$, gives the result

$$F([A]) = W([A])\,P([A]) + R([A]) = [0] + R([A]) \tag{14-3.42}$$

To use this result, the problem is to evaluate the $R([A])$ remainder part which is known to contain powers of $[A]$ no higher than $(n - 1)$ as shown in

$$R([A]) = k_{n-1}\,[A]^{n-1} + \cdots + k_1\,[A] + k_0\,[I] \tag{14-3.43}$$

The remainder functional form must be the same in both the scalar Eq. 14-3.41 and the matrix Eq. 14-3.42. For s equal to the roots, s_i, of $P(s)$, Eq. 14-3.41 degenerates to

$$F(s_i) = 0 + R(s_i) = k_{n-1}\,s_i^{n-1} + \cdots + k_1 s_i + k_0 \tag{14-3.44}$$

Thus, the k_j's can be found from the scalar form and then used in the matrix form. To find all k_j coefficients, n equations must be obtained from Eq. 14-3.44. Examples 14-3.3, 4, and 5 illustrate how to develop and use these equations.

Example 14-3.3. The matrix polynomial $F([A])$ of Example 14-3.2 and Eq. 14-3.37 is reconsidered.

$$F([A]) = [A]^4 + 2[A]^2 + [I] \tag{14-3.45}$$

$$F(s) = s^4 + 2s^2 + 1 \tag{14-3.46}$$

$$[A] = \begin{bmatrix} -7 & 1 \\ -12 & 0 \end{bmatrix} \qquad P(s) = s^2 + 7s + 12 \tag{14-3.47}$$

In this case $F(s)/P(s)$ can actually be divided out in the form of Eq. 14-3.41, as in

$$\frac{F(s)}{P(s)} = (s^2 - 7s - 39) + \frac{-189s - 467}{P(s)} \tag{14-3.48}$$

in which the remainder, $R(s)$, is $(-189s - 467)$. By direct comparison with Eq. 14-3.44, the coefficients needed for $R([A])$ are $k_1 = -189$ and $k_0 = -467$. Thus, the reduced expression for $F([A])$ from Eqs. 14-3.42 and 43 becomes

$$F([A]) = -189[A] - 467[I] \tag{14-3.49}$$

As an alternate method of finding k_1 and k_0, the following procedure, as described above in Eq. 14-3.44, is used with $s_1 = -3$, $s_2 = -4$.

$$F(s_1) = R(s_1) = F(-3) = -3k_1 + k_0 \tag{14-3.50}$$

$$\left.\begin{aligned} F(s_1) &= (-3)^4 + 2(-3)^2 + 1 = -3k_1 + k_0 \\ F(s_2) &= (-4)^4 + 2(-4)^2 + 1 = -4k_1 + k_0 \end{aligned}\right\} \tag{14-3.51}$$

Equations 14-3.51 are now solved for k_1 and k_0 to obtain $k_1 = -189$ and $k_0 = -467$ as above.

Example 14-3.4. As a more typical application of the above procedure for evaluating functions of matrices, the exponential $\epsilon^{[A]t}$ from Example 14-3.1 is considered. The process is started in

$$F([A]) = \epsilon^{[A]t} = R([A]) \tag{14-3.52}$$

$$F(s_i) = \epsilon^{s_i t} = R(s_i) = k_1 s_i + k_0 \tag{14-3.53}$$

which, with $s_1 = -3$ and $s_2 = -4$ substituted, leads to

$$\left.\begin{aligned} \epsilon^{-3t} &= k_1(-3) + k_0 \\ \epsilon^{-4t} &= k_1(-4) + k_0 \end{aligned}\right\} \tag{14-3.54}$$

These equations can be solved for $k_1 = (\epsilon^{-3t} - \epsilon^{-4t})$ and $k_0 = (4\epsilon^{-3t} - 3\epsilon^{-4t})$. When substituted into $R[A]$ in Eq. 14-3.52, the result is

$$\epsilon^{[A]t} = k_1[A] + k_0[I] = (\epsilon^{-3t} - \epsilon^{-4t})\begin{bmatrix} -7 & 1 \\ -12 & 0 \end{bmatrix} + (4\epsilon^{-3t} - 3\epsilon^{-4t})\begin{bmatrix} 1 & 0 \\ 0 & 1 \end{bmatrix} \tag{14-3.55}$$

$$= \begin{bmatrix} (-3\epsilon^{-3t} + 4\epsilon^{-4t}) & (\epsilon^{-3t} - \epsilon^{-4t}) \\ -12(\epsilon^{-3t} - \epsilon^{-4t}) & (4\epsilon^{-3t} - 3\epsilon^{-4t}) \end{bmatrix} \tag{14-3.56}$$

which checks Eq. 14-3.35.

Example 14-3.5. The procedure used when multiple roots of $P(s)$ exist is illustrated by considering the $[A]$ matrix of

$$[A] = \begin{bmatrix} 0 & 1 \\ -1 & -2 \end{bmatrix} \tag{14-3.57}$$

and again the matrix exponential, $\epsilon^{[A]t}$. In this case $P(s)$ has a double root at $s_1 = s_2 = -1$, as shown in

$$P(s) = |sI - A| = \begin{vmatrix} s & -1 \\ 1 & s+2 \end{vmatrix} = s^2 + 2s + 1 = (s+1)^2 \tag{14-3.58}$$

$$F([A]) = \epsilon^{[A]t} = R([A]) = k_1[A] + k_0[I] \tag{14-3.59}$$

$$F(s_i) = \epsilon^{s_i t} = R(s_i) = k_1 s_i + k_0 \tag{14-3.60}$$

Thus, only one equation results if ($s_1 = -1$) is substituted into Eq. 14-3.60. A second equation can be obtained by differentiating the $F(s)$ equation, as in

$$\left.\frac{dF(s)}{ds}\right|_{s_i} = t\,\epsilon^{s_i t} = \left.\frac{dR(s)}{ds}\right|_{s_i} = k_1 \tag{14-3.61}$$

The validity of this operation depends on the fact that for a double pole, $dP(s)/ds|_{s_i} = 0$. This is shown in

$$P(s) = (s - s_i)^2 Q(s) \tag{14-3.62}$$

$$\left.\frac{dP(s)}{ds}\right|_{s_i} = \left.\left\{2(s - s_i)\,Q(s) + (s - s_i)^2 \frac{dQ(s)}{ds}\right\}\right|_{s=s_i} = 0 \tag{14-3.63}$$

Also $dF(s)/ds$ is shown in

$$F(s) = W(s)\,P(s) + R(s) \tag{14-3.64}$$

$$\left.\frac{dF(s)}{ds}\right|_{s_i} = \left\{\frac{dW}{ds}P(s) + W(s)\frac{dP}{ds} + \frac{dR}{ds}\right\}_{s_i} = \left.\frac{dR}{ds}\right|_{s_i} \tag{14-3.65}$$

For higher order poles more derivatives can be used.

From Eqs. 14-3.60 and 61 the two equations for k_1 and k_0 are obtained, leading to the solutions for $k_1 = t\,\epsilon^{-t}$ and $k_0 = (t\,\epsilon^{-t} + \epsilon^{-t})$.

$$F(s_1) = \epsilon^{-t} = k_1(-1) + k_0 \qquad \left.\frac{dF(s)}{ds}\right|_{s_1} = t\,\epsilon^{-t} = k_1 \tag{14-3.66}$$

When these results are substituted into Eq. 14-3.59, the final result can be obtained.

$$\epsilon^{[A]t} = (t\,\epsilon^{-t})\begin{bmatrix} 0 & 1 \\ -1 & -2 \end{bmatrix} + (t\,\epsilon^{-t} + \epsilon^{-t})\begin{bmatrix} 1 & 0 \\ 0 & 1 \end{bmatrix} \tag{14-3.67}$$

$$= \begin{bmatrix} (t+1) & t \\ -t & (1-t) \end{bmatrix} \epsilon^{-t} \tag{14-3.68}$$

Example 14-3.6. As a final example of the matrix function evaluation, the matrix function

$$F([A]) = \cos([A]\;t) \tag{14-3.69}$$

is considered. This function can be defined by an infinite matrix series of the form used for $\cos\theta$. However, a single matrix is desired for $F([A])$. To be specific, $[A]$ is given as in

$$[A] = \begin{bmatrix} -6 & 1 \\ -8 & 0 \end{bmatrix} \tag{14-3.70}$$

For this case the steps are detailed in

$$P(s) = |sI - A| = s^2 + 6s + 8 = (s+2)(s+4) \tag{14-3.71}$$

$$F([A]) = \cos([A]t) = k_1[A] + k_0 \tag{14-3.72}$$

$$F(s_i) = \cos(s_i t) = k_1 s_i + k_0 \tag{14-3.73}$$

$$\begin{aligned}\cos(-2t) &= -2k_1 + k_0\\ \cos(-4t) &= -4k_1 + k_0\end{aligned} \tag{14-3.74}$$

$$k_1 = \tfrac{1}{2}(\cos 2t - \cos 4t) \qquad k_0 = (2\cos 2t - \cos 4t) \tag{14-3.75}$$

$$\cos([A]t) = k_1\begin{bmatrix}-6 & 1\\ -8 & 0\end{bmatrix} + k_0\begin{bmatrix}1 & 0\\ 0 & 1\end{bmatrix} = \begin{bmatrix}(k_0 - 6k_1) & k_1\\ -8k_1 & k_0\end{bmatrix} \tag{14-3.76}$$

$$= \begin{bmatrix}(-\cos 2t + 2\cos 4t) & \tfrac{1}{2}(\cos 2t - \cos 4t)\\ -4(\cos 2t - \cos 4t) & (2\cos 2t - \cos 4t)\end{bmatrix} \tag{14-3.77}$$

As the final item considered in this section, the forced state response for the system

$$\dot{\mathbf{x}} = \mathbf{Ax} + \mathbf{Bu} \tag{14-3.78}$$

is considered. Rather than to proceed to develop a completely new matrix solution, the results of Eqs. 14-2.47 and 14-3.15 are used. Substituting $\epsilon^{[A]t}$ for $\phi(t)$ leads directly to

$$\mathbf{x}(t) = \epsilon^{[A](t-t_0)}\mathbf{x}(t_0) + \int_{t_0}^{t} \epsilon^{[A](t-\tau)}\mathbf{Bu}(\tau)\,d\tau \tag{14-3.79}$$

as the desired result. This form for the solution has very little to offer over previous techniques, so far as computation is concerned, and so is included only for completeness.

14-4. THE CONCEPT OF SYSTEM TRAJECTORIES

In Sections 14-2 and 14-3, straight forward techniques were developed to obtain the solution of the state differential equations. No attempt was made to interpret these solutions. In Section 13-2, some discussion centered on the concept of a vector space, including the set of *basis vectors* as well as coordinates *along* the basis vectors. Such a vector space is fundamental to the state variable modeling of a system. In fact, the components of the state vector, $\mathbf{x}(t)$, are the components of the state of the system along some set of basis vectors. It is convenient to use the rectangular coordinate E-basis set described in Section 13-2, with state variable $x_i(t)$ as the component along $\mathbf{E}_i$. Thus, at any given time, t_1, the *state of a system*, $\mathbf{x}(t_1)$, can be represented as a *point in* this n-dimensional *state space.* Since the components of the state vector change with time, the state point moves in state space, tracing a locus called the *system trajectory.* System trajectories provide extensive insight into system dynamic behavior, including stability characteristics. The distance of the state point from equilibrium points, where $\dot{\mathbf{x}} = \mathbf{0}$, or where all rates of change are zero, is an important factor in stability analysis.

To illustrate some of the aspects of system trajectories, several examples are presented. For simplicity, only two- and three-dimensional space is treated, although mathematically there is no problem in extending the concepts to n-dimensional space.

Example 14-4.1. A very simplified model for decay of iodine and xenon in a nuclear reactor after shutdown[1] is given by

$$\left.\begin{aligned} \frac{dI}{dt} &= -0.104I \\ \frac{dXe}{dt} &= -0.0753Xe + 0.104I \end{aligned}\right\} \qquad (14\text{-}4.1)$$

in which I (not an identity matrix here) and Xe are the concentrations of iodine and xenon, respectively. To assign state variables, let $x_1 = I$, and $x_2 = Xe$ so that the above system can be written as

$$\begin{bmatrix} \dot{x}_1 \\ \dot{x}_2 \end{bmatrix} = \begin{bmatrix} -0.104 & 0 \\ 0.104 & -0.0753 \end{bmatrix} \begin{bmatrix} x_1 \\ x_2 \end{bmatrix} \qquad (14\text{-}4.2)$$

for which the solution is obtained, by any of the previously described techniques, as

$$\mathbf{x}(t) = \begin{bmatrix} \epsilon^{-0.104t} & 0 \\ -3.62(\epsilon^{-0.104t} - \epsilon^{-0.0753t}) & (\epsilon^{-0.0753t}) \end{bmatrix} \mathbf{x}(0) \qquad (14\text{-}4.2a)$$

Figure 14-4.1 shows a two-dimensional state space, together with several trajectories for the system when started from various initial conditions. It is apparent from the figure exactly how the state of the system approaches the equilibrium state at the origin of the state space. Time points can be added along the various trajectories, if it is desired to be more quantitative. Only the parts in the first quadrant would have physical significance, since negative numbers of atoms/unit volume are rather difficult to obtain.

The case considered in Example 14-4.1 is one in which only negative real poles are present in $[\Phi(s)]$, thereby resulting in exponentially decaying *modes* in the response. In general, each simple s-plane pole in $[\Phi(s)]$ contributes a response *mode* $K\epsilon^{s_i t}$ to the response. In Example 14-4.1, state variable $x_1(t)$ contains only the $\epsilon^{-0.104t}$ mode, regardless of the initial state vector. State variable $x_2(t)$, on the other hand, contains both mode $\epsilon^{-0.104t}$ and mode $\epsilon^{-0.0753t}$, with their particular weighting being dependent on the initial state, $\mathbf{x}(0)$.

In general, with an arbitrary selection of state variables for a system, the result in each component of the state vector will be a linear combination of several modes. However, the parallel programming approach described in Section 13-3 leads to a separation of the modes. Special attention is directed to mode separation in Section 14-5.

[1] Electronics Associates Inc., *Study of Xenon Poisoning in a Nuclear Reactor*, General Purpose Analog Computation Applications Study No. 13.4.3a.

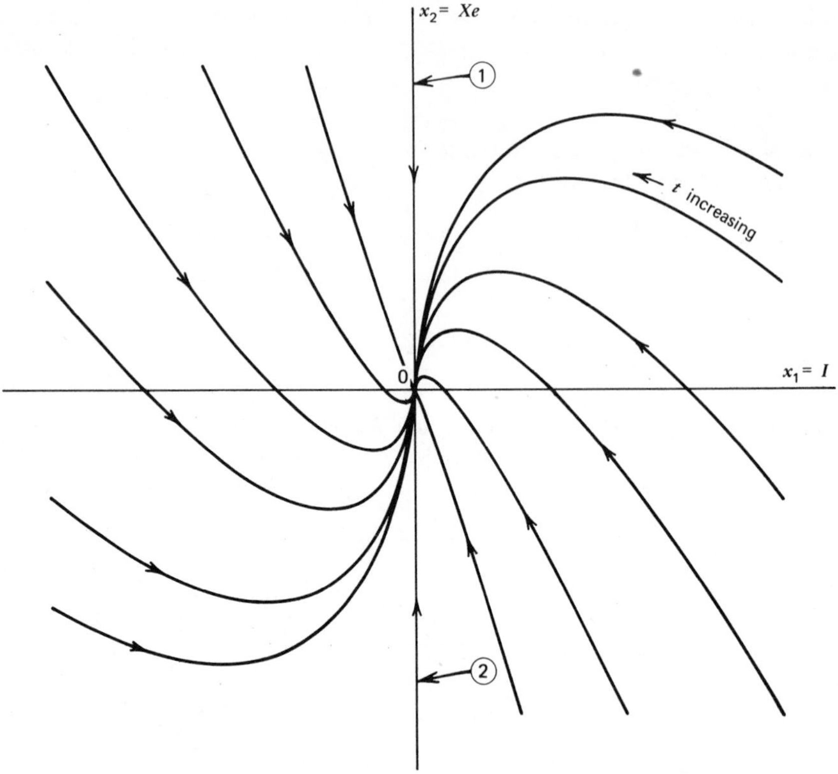

Fig. 14-4.1

Example 14-4.2. While trajectories of the type shown in Fig. 14-4.1 are typical for systems with real poles, there is some difference for systems with complex poles. As an illustration of this, the system of Example 14-2.1 is reconsidered.

$$\dot{\mathbf{x}} = \begin{bmatrix} 0 & 1 \\ -8 & -4 \end{bmatrix} \mathbf{x} + \begin{bmatrix} 0 \\ 1 \end{bmatrix} u \tag{14-4.3}$$

While the analytical solution was obtained in the earlier example (Eqs. 14-2.12 and 13) for $u = 0$, the trajectories for the system can also be obtained very easily using an analog or digital computer and an x-y plotter. Figure 14-4.2 shows a simulation signal-flow

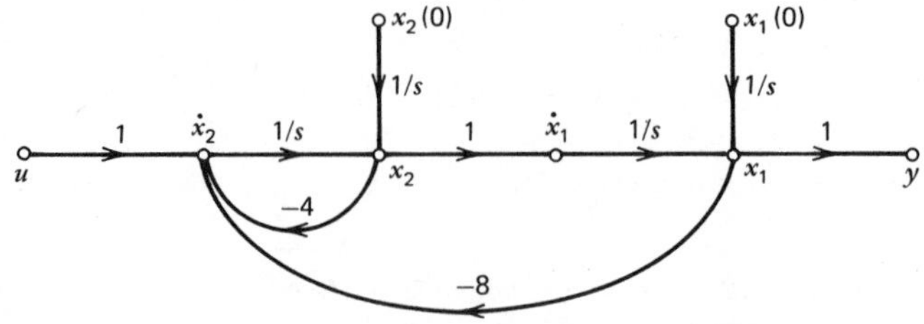

Fig. 14-4.2

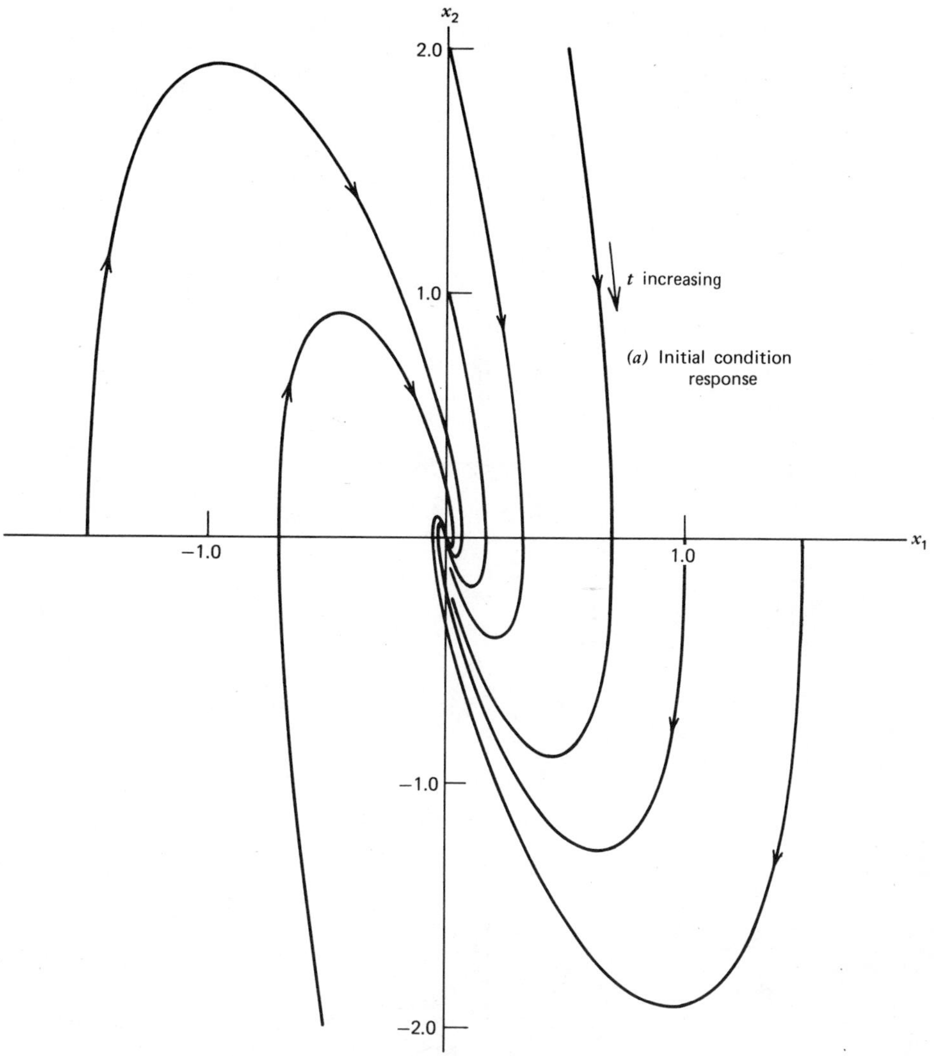

Fig. 14-4.3(a)

graph for the system. Figure 14-4.3(a) shows the resulting trajectories that are recorded with only initial condition inputs, while in (b) is shown the trajectory for the step input only. The spiral nature of the trajectories is typical of the response due to complex pole modes of a system. Also the shift in the equilibrium point for the step input should be noted. For this case the new equilibrium point is found when $\dot{\mathbf{x}} = \mathbf{0}$ in Eq. 14-4.3, which leads to $x_{2ss} = 0, x_{1ss} = \frac{1}{8}$, as given in

$$x_{2ss} = 0$$

$$-8x_1 - 4x_2 + u_1 = 0 \qquad x_{1ss} = \tfrac{1}{8} \tag{14-4.4}$$

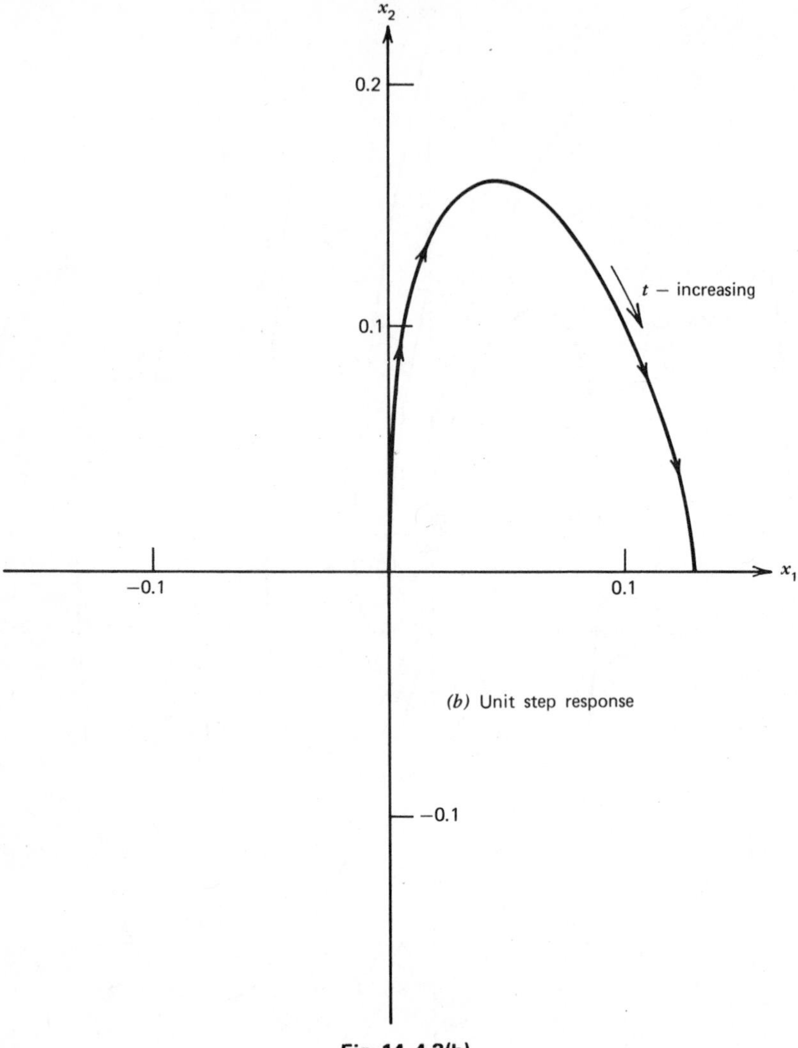

(b) Unit step response

Fig. 14-4.3(b)

Example 14-4.3. To illustrate the fact that different sets of state variables can be chosen to represent a given system, Example 14-4.2 is reformulated, as shown in Fig. 14-4.4. For this case $\mathbf{z}$ is used as the state vector to emphasize the difference from the $\mathbf{x}$ vector of Example 14-4.2. The new state equations are given by

$$\dot{\mathbf{z}} = \begin{bmatrix} -4 & 1 \\ -8 & 0 \end{bmatrix} \mathbf{z} + \begin{bmatrix} 0 \\ 1 \end{bmatrix} u \tag{14-4.5}$$

It should be noted that the only difference in the present formulation from that of Example 14-4.2 is in the form of direct programming that has been used to realize the transfer

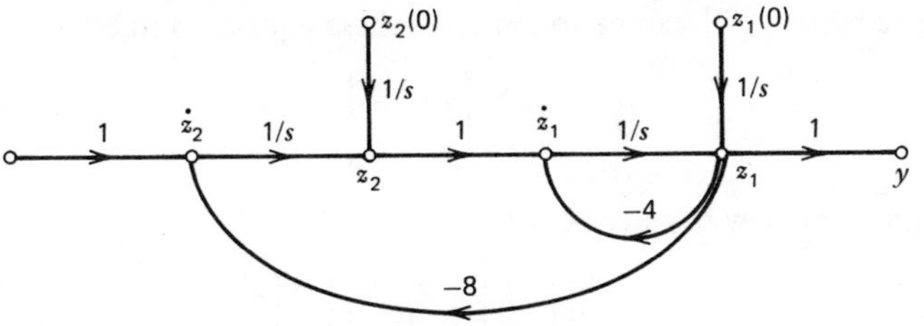

Fig. 14-4.4

function $G(s)$ in

$$\frac{Y(s)}{U(s)} = \frac{1}{s^2 + 4s + 8} = G(s) \tag{14-4.6}$$

Since the state of a system is a fundamental property represented by a vector, all different state coordinates and basis vector representations must be related by a linear transformation of the form given by

$$\mathbf{x} = [T]\mathbf{z} \tag{14-4.7}$$

It can be shown that the above transformation is equivalent to using a different set of basis vectors in representing a fundamental vector. This concept can be developed as follows. The vector $\mathbf{x}$ was used to represent the system with respect to the E-basis. The columns of $[T]$ are then taken as a new set of basis vectors called the T-basis expressed relative to the E-basis. Next the elements of $\mathbf{z}$ are considered as components of the state vector relative to the T-basis[2], so that Eq. 14-4.7 may be rewritten as

$$\mathbf{x}_e = (z_1)\,\mathbf{T}_{1_e} + (z_2)\,\mathbf{T}_{2_e} \tag{14-4.8}$$

in which the $\mathbf{T}_i$ vectors are columns of $[T]$. Additional e subscripts are used on $\mathbf{x}_e$, $\mathbf{T}_{1_e}$, and $\mathbf{T}_{2_e}$ to indicate the basis used for reference. The only remaining problem is to determine the matrix $[T]$.

To find $[T]$, substitution of Eq. 14-4.7 into Eq. 14-4.3 leads to

$$\mathbf{T}\dot{\mathbf{z}} = \begin{bmatrix} 0 & 1 \\ -8 & -4 \end{bmatrix} \mathbf{T}\mathbf{z} + \begin{bmatrix} 0 \\ 1 \end{bmatrix} u \tag{14-4.9}$$

$$\dot{\mathbf{z}} = \mathbf{T}^{-1} \begin{bmatrix} 0 & 1 \\ -8 & -4 \end{bmatrix} \mathbf{T}\mathbf{z} + \mathbf{T}^{-1} \begin{bmatrix} 0 \\ 1 \end{bmatrix} u \tag{14-4.10}$$

These results have to give Eq. 14-4.5 so that the relations of

$$\mathbf{T}^{-1} \begin{bmatrix} 0 & 1 \\ -8 & -4 \end{bmatrix} \mathbf{T} = \begin{bmatrix} -4 & 1 \\ -8 & 0 \end{bmatrix} \qquad \mathbf{T}^{-1} \begin{bmatrix} 0 \\ 1 \end{bmatrix} = \begin{bmatrix} 0 \\ 1 \end{bmatrix} \tag{14-4.11}$$

$$\begin{bmatrix} 0 & 1 \\ -8 & -4 \end{bmatrix} \mathbf{T} = \mathbf{T} \begin{bmatrix} -4 & 1 \\ -8 & 0 \end{bmatrix} \qquad \begin{bmatrix} 0 \\ 1 \end{bmatrix} = \mathbf{T} \begin{bmatrix} 0 \\ 1 \end{bmatrix} \tag{14-4.12}$$

[2] Frank Ayres, *Theory and Problems of MATRICES*, Schaum Publishing Company, New York, 1962, pp. 88-89.

are required. The matrix $[T]$ can be found from these equations to be

$$[T] = \begin{bmatrix} 1 & 0 \\ -4 & 1 \end{bmatrix} \tag{14-4.13}$$

which in turn gives the new basis vectors as

$$\mathbf{T}_{1_e} = \begin{bmatrix} 1 \\ -4 \end{bmatrix} \quad \mathbf{T}_{2_e} = \begin{bmatrix} 0 \\ 1 \end{bmatrix} \tag{14-4.14}$$

Figure 14-4.5 shows the relationship between the two basis vector sets. The solutions of Eq. 14-4.5 for $\mathbf{z}(t)$ gives the state vector components relative to the T-basis of Eq. 14-4.14.

Equation 14-4.5 can be solved for the $\mathbf{z}(t)$, for $u = 0$, to give

$$\mathbf{z}(t) = \begin{bmatrix} \epsilon^{-2t}(\cos 2t - \sin 2t) & \frac{1}{2}\epsilon^{-2t}\sin 2t \\ -4\epsilon^{-2t}\sin 2t & \epsilon^{-2t}(\cos 2t + \sin 2t) \end{bmatrix} \mathbf{z}(0) \tag{14-4.15}$$

To verify the relationships between $\mathbf{x}$ and $\mathbf{z}$ discussed above, let the initial state $\mathbf{x}(0) = [1 \quad 0]^T$. According to Eq. 14-4.7 the initial vector $\mathbf{z}(0)$ should be $\mathbf{z}(0) = [T]^{-1}\mathbf{x}(0) = [1 \quad 4]^T$. In Fig. 14-4.5 it can be seen that the point $(1, 0)$ can be obtained from $(1)\mathbf{T}_1 + (4)\mathbf{T}_2$. For the $\mathbf{z}(0)$ given above, the resulting $\mathbf{z}(t)$ is found to be

$$\mathbf{z}(t) = \begin{bmatrix} \epsilon^{-2t}(\cos 2t + \sin 2t) \\ 4\epsilon^{-2t}\cos 2t \end{bmatrix} \tag{14-4.16}$$

Using Eq. 14-4.7 again should give $\mathbf{x}(t)$.

$$\mathbf{x}(t) = \begin{bmatrix} 1 & 0 \\ -4 & 1 \end{bmatrix} \mathbf{z}(t) = \begin{bmatrix} \epsilon^{-2t}(\cos 2t + \sin 2t) \\ -4\epsilon^{-2t}\sin 2t \end{bmatrix} \tag{14-4.17}$$

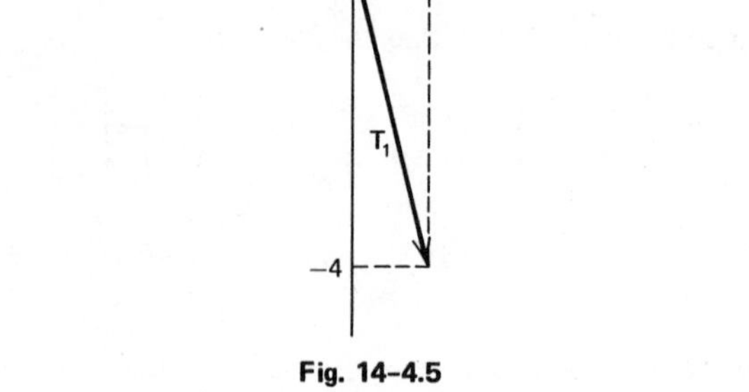

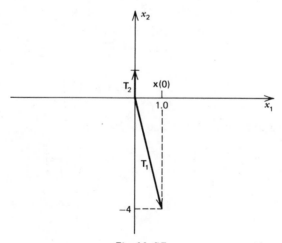

Fig. 14-4.5

This result does check with the solution of the **x** equations, 14–2.12 and 13, for the specified **x**(0).

Various ways for presenting the system trajectories can be used. If the **z** representation is used, but trajectories in the **x** space of Fig. 14–4.5 are desired, then the x–y plotter inputs must be formed from $\mathbf{x} = [T]\mathbf{z}$, or $x_1 = z_1$, $x_2 = -4z_1 + z_2$. These linear combinations are formed with an adder. The x-axis of the plotter is then driven with the x_1 signal, and the y-axis with the x_2 signal. These trajectories should be indistinguishable from those obtained from the **x** state variables of Fig. 14–4.3.

Instead of the above presentation, the **z** coordinates can be plotted on a rectangular or E-basis, by simply letting the x-axis be driven by z_1 and the y-axis by z_2. The results are a skewed version of the spirals obtained from the **x** state variables. Fig. 14–4.6 shows typical trajectories for the **z** state variables plotted on the E-basis.

The above examples demonstrate that the various choices of state variables are really

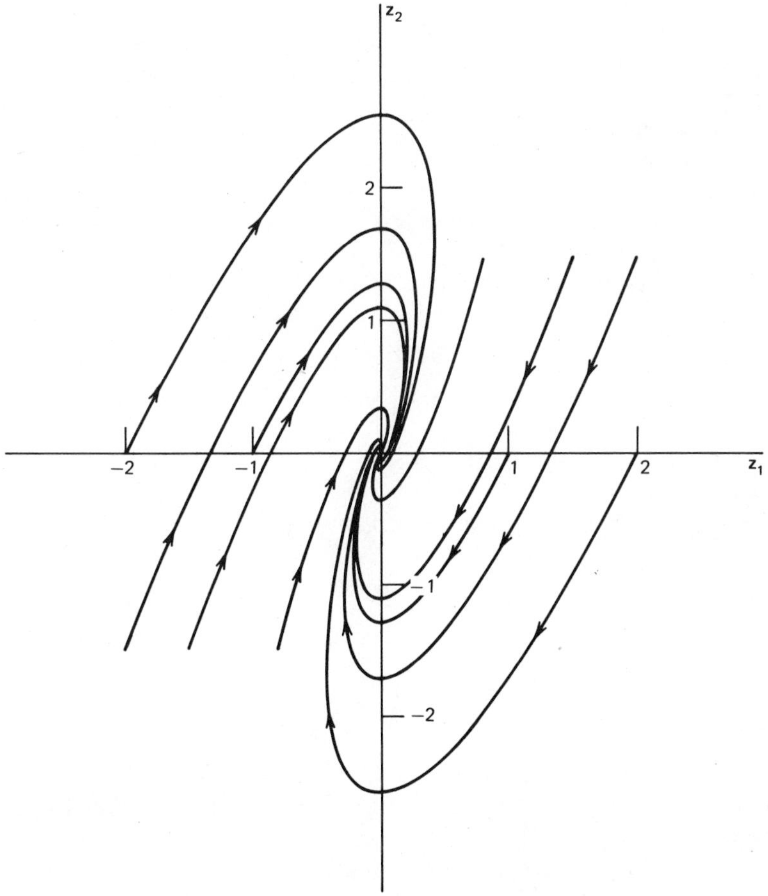

Fig. 14–4.6

pictures of a system in terms of different sets of basis vectors or from different points of view. Section 14-5 continues this discussion by describing a very special picture, or basis set.

14-5. NORMAL COORDINATES AND EIGENVECTORS

Several times in previous sections, reference has been made to the methods for selecting state variables so as to decouple the state equations or to *isolate the modes.* In fact, the parallel programming scheme of Section 13-3 does precisely this when distinct or simple poles, s_i, exist in $[\Phi(s)]$. The resulting form for the state differential equations in the time domain can thus be written as

$$\dot{\mathbf{q}} = \begin{bmatrix} s_1 & 0 & 0 & \cdots\cdots \\ 0 & s_2 & 0 & \cdots\cdots \\ 0 & 0 & s_3 & 0 & \cdots \\ \cdots & \cdots & \cdots & \cdots & \cdots \end{bmatrix} \mathbf{q} + \begin{bmatrix} b_{11} & b_{12} & \cdots \\ b_{21} & b_{22} & \cdots \\ b_{31} & b_{32} & \\ \cdots & \cdots & \cdots \end{bmatrix} \mathbf{u} \tag{14-5.1}$$

as described in Example 13-3.9. Because this is such a special representation, q is used instead of **x**. Most significant in this result is the form for the $[A]$ matrix, which is a diagonal matrix specially designated as $[\Lambda]$.

In this special case, the state equation is $\dot{\mathbf{q}} = \mathbf{\Lambda q} + \mathbf{B}_n \mathrm{U}$, where $\mathbf{B}_n$ is just a special **B** matrix used with the **q** state vector.

If the state transition matrix, $[\phi(t)]$, is computed for the system, as described in Eq. 14-5.1, the result is

$$[\phi(t)]_\Lambda = \begin{bmatrix} \epsilon^{s_1 t} & 0 & 0 & 0 & \cdot \\ 0 & \epsilon^{s_2 t} & 0 & 0 & \cdot \\ 0 & 0 & \epsilon^{s_3 t} & 0 & \cdot \\ \cdots & \cdots & \cdots & \cdots & \cdots \end{bmatrix} \tag{14-5.2}$$

Use of this matrix in the general solution equation for $\mathbf{q}(t)$ results in

$$\mathbf{q}(t) = \begin{bmatrix} \epsilon^{s_1 t} & 0 & 0 & \cdot \\ 0 & \epsilon^{s_2 t} & 0 & \cdot \\ 0 & 0 & \epsilon^{s_3 t} & \cdot \\ \cdots & \cdots & \cdots & \cdots \end{bmatrix} \mathbf{q}(0) + \int_0^t \begin{bmatrix} \epsilon^{s_1(t-\tau)} & 0 & 0 \\ 0 & \epsilon^{s_2(t-\tau)} & 0 \\ \cdots & \cdots & \cdots \end{bmatrix} \mathbf{B}_n\, \mathrm{U}(\tau)\, d\tau \tag{14-5.3}$$

which shows that any given component of $\mathbf{q}(t)$ will have a response determined by only one system pole and the inputs. Thus, only one system mode appears in each component of $\mathbf{q}(t)$. The modes are separated.

As described in Section 14-4, the various sets of state variables are related by some linear transformation, which for this case is represented by

$$\mathbf{x} = [M]\mathbf{q} = [\mathbf{M}_1 \mathbf{M}_2 \cdots]\mathbf{q} \tag{14-5.4}$$

A transformation matrix $[M]$ is used in this case rather than $[T]$, as in Eq. 14-4.7, because of the special form of $\mathbf{\Lambda}$ involved. This matrix is called the *modal matrix*, and its columns become *modal*, *characteristic*, or *eigen-vectors* for the system. These vectors serve as a new set of basis vectors, as described in Section 14-4, and the elements of the state vector $\mathbf{q}$ are components of the state vector along these modal vectors. Putting this result together with the fact that each component of $\mathbf{q}(t)$ involves only one system mode results in this picture of what is called the *normal form* representation of the system state. If the state of a system represented by, say, $\mathbf{x}(0)$ is started along one of the eigenvectors, $\mathbf{M}_i$, only the mode $(\epsilon^{s_i t})$ will exist in the response, and the trajectory will follow only that vector as t increases ($\mathbf{u}$ is assumed zero here). Examples that follow will show this in detail. It might be pointed out that such responses are shown in Fig. 14-4.1 by trajectories labeled (1) and (2).

Several mathematical relations are still needed to complete the picture of mode separation. In particular, the relations between the $\mathbf{x}$ equations and $\mathbf{q}$ equations are needed, together with some way for determining $[M]$. The following equations show the required steps.

$$\begin{aligned} \dot{\mathbf{x}} &= [A]\mathbf{x} + [B]\mathbf{u} \\ \mathbf{y} &= [C]\mathbf{x} + [D]\mathbf{u} \end{aligned} \tag{14-5.5}$$

$$\begin{aligned} \dot{\mathbf{q}} &= [M^{-1}AM]\mathbf{q} + [M^{-1}B]\mathbf{u} = \mathbf{\Lambda}\mathbf{q} + \mathbf{B}_n\mathbf{u} \\ \mathbf{y} &= [CM]\mathbf{q} + [D]\mathbf{u} \end{aligned} \tag{14-5.6}$$

$$\mathbf{M}^{-1}[A]\mathbf{M} = \mathbf{\Lambda} \tag{14-5.7}$$

$$[A]\ [\mathbf{M}_1\mathbf{M}_2\mathbf{M}_3 \cdots] = \mathbf{M\Lambda} = [s_1\mathbf{M}_1 s_2\mathbf{M}_2\ s_3\mathbf{M}_3 \cdots] \tag{14-5.8}$$

From Eq. 14-5.8 are obtained n relations of the form of

$$[A]\mathbf{M}_j = s_j\mathbf{M}_j \qquad j = 1, 2, \ldots, n \tag{14-5.9}$$

$$[s_jI - A]\mathbf{M}_j = \mathbf{0} \qquad j = 1, 2, \ldots, n \tag{14-5.10}$$

Such equations are identical to those used in matrix work when studying *eigenvalues* and *eigenvectors*, where s_j are called the eigenvalues, and $\mathbf{M}_j$ the eigenvectors. However, the s_j are nothing more than the roots of the system polynomial, $P(s) = |sI - A| = 0$, which in turn appear as poles in $[\Phi(s)]$. The eigenvectors, $\mathbf{M}_j$, can be found by solving Eqs. 14-5.10. An arbitrary choice of one component always occurs in each $\mathbf{M}_j$, as will be shown in Examples 14-5.1 and 2.

Example 14-5.1. The system represented by Eqs. 14-5.11 is considered from the modal or normal response point of view in this example.

$$\begin{aligned} \dot{\mathbf{x}} &= \begin{bmatrix} 0 & 1 \\ -12 & -7 \end{bmatrix} \mathbf{x} + \begin{bmatrix} 0 \\ 1 \end{bmatrix} u \\ y &= [1\ \ 0]\mathbf{x} \end{aligned} \tag{14-5.11}$$

For this system various quantities required for the **x** and **q** representations are given by

$$P(s) = |sI - A| = s^2 + 7s + 12 = (s+3)(s+4) \tag{14-5.12}$$

$$[\phi(t)]_{\mathbf{x}} = \begin{bmatrix} (4\epsilon^{-3t} - 3\epsilon^{-4t}) & (\epsilon^{-3t} - \epsilon^{-4t}) \\ -12(\epsilon^{-3t} - \epsilon^{-4t}) & -3\epsilon^{-3t} + 4\epsilon^{-4t}) \end{bmatrix} \tag{14-5.13}$$

$$[\Lambda] = \begin{bmatrix} -3 & 0 \\ 0 & -4 \end{bmatrix} \qquad \begin{matrix} s_1 = -3 \\ s_2 = -4 \end{matrix} \tag{14-5.14}$$

$$[s_1 I - A]\mathbf{M}_1 = \begin{bmatrix} -3 & -1 \\ 12 & 4 \end{bmatrix} \begin{bmatrix} m_{11} \\ m_{21} \end{bmatrix} = \begin{bmatrix} 0 \\ 0 \end{bmatrix} = \mathbf{0} \tag{14-5.15}$$

$$[s_2 I - A]\mathbf{M}_2 = \begin{bmatrix} -4 & -1 \\ 12 & 3 \end{bmatrix} \begin{bmatrix} m_{12} \\ m_{22} \end{bmatrix} = \begin{bmatrix} 0 \\ 0 \end{bmatrix} = \mathbf{0} \tag{14-5.16}$$

$$3m_{11} + m_{21} = 0 \tag{14-5.17}$$

$$4m_{12} + m_{22} = 0 \tag{14-5.18}$$

From Eqs. 14-5.17 and 18 it is apparent that one element of $[M]$ can be chosen arbitrarily in each column. This really results from the fact that the equations for the m_{ij} elements of each column are not independent. To proceed, $m_{11} = 1$ and $m_{12} = 1$ are used so that

$$\mathbf{M}_1 = \begin{bmatrix} 1 \\ -3 \end{bmatrix} \qquad \mathbf{M}_2 = \begin{bmatrix} 1 \\ -4 \end{bmatrix} \tag{14-5.19}$$

$$\mathbf{M} = \begin{bmatrix} 1 & 1 \\ -3 & -4 \end{bmatrix} \qquad \mathbf{M}^{-1} = \begin{bmatrix} 4 & 1 \\ -3 & -1 \end{bmatrix} \tag{14-5.20}$$

$$\mathbf{M}^{-1}\mathbf{B} = \begin{bmatrix} 4 & 1 \\ -3 & -1 \end{bmatrix} \begin{bmatrix} 0 \\ 1 \end{bmatrix} = \begin{bmatrix} 1 \\ -1 \end{bmatrix} = \mathbf{B}_n \tag{14-5.21}$$

$$\mathbf{CM} = [1 \quad 0] \begin{bmatrix} 1 & 1 \\ -3 & -4 \end{bmatrix} = [1 \quad 1] = \mathbf{C}_n \tag{14-5.22}$$

can be computed. Additionally, the state transition matrix for the **q** state variables can be computed as Eq. 14-5.23.

$$[\phi(t)]_{\mathbf{q}} = \mathcal{L}^{-1}\{[sI - \Lambda]^{-1}\} = \begin{bmatrix} \epsilon^{-3t} & 0 \\ 0 & \epsilon^{-4t} \end{bmatrix} \tag{14-5.23}$$

Location of the eigenvectors, $\mathbf{M}_1$ and $\mathbf{M}_2$, in the **x** state space is shown in Fig. 14-5.1. As mentioned previously, if the state of the system is started anywhere along the eigenvectors, only one mode is excited, and the trajectory remains on the vector. For example, if $\mathbf{x}(0) = [-1 \quad 3]^T$ is used with $u = 0$, then

$$\mathbf{x}(t) = [\phi(t)]_{\mathbf{x}}\,\mathbf{x}(0) = \begin{bmatrix} -\epsilon^{-3t} \\ 3\epsilon^{-3t} \end{bmatrix} \tag{14-5.24}$$

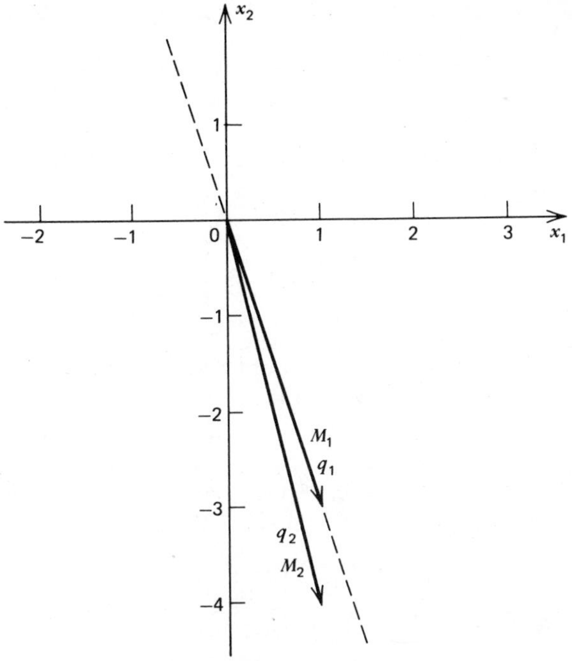

Fig. 14-5.1

$$\mathbf{q}(0) = \mathbf{M}^{-1}\mathbf{x}(0) = \begin{bmatrix} 4 & 1 \\ -3 & -1 \end{bmatrix}\begin{bmatrix} -1 \\ 3 \end{bmatrix} = \begin{bmatrix} -1 \\ 0 \end{bmatrix} \tag{14-5.25}$$

$$\mathbf{q}(t) = [\phi(t)]_{\mathbf{q}}\, \mathbf{q}(0) = \begin{bmatrix} -\epsilon^{-3t} \\ 0 \end{bmatrix} \tag{14-5.26}$$

$$\mathbf{x}(t) = \mathbf{M}\mathbf{q}(t) = \begin{bmatrix} x_1(t) \\ x_2(t) \end{bmatrix} = \begin{bmatrix} q_1(t) + q_2(t) \\ -3q_1(t) - 4q_2(t) \end{bmatrix} \tag{14-5.27}$$

verify the excitation of only the ϵ^{-3t} mode as well as the relationship between **x** and **q**.

For the case when complex poles exist in $[\Phi(s)]$, it may not be desirable to separate the modes due to $s_1 = \alpha + j\beta$ and $s_2 = \alpha - j\beta$. The individual complex exponentials, $\epsilon^{(\alpha \pm j\beta)t}$, do not have much appeal. Instead, pairing such poles as done in Example 13-3.10 and Fig. 13-3.11 is much more useful. The responses are then of the form $\epsilon^{\alpha t} \sin \beta t$ and $\epsilon^{\alpha t} \cos \beta t$. Instead of using the transformation $\mathbf{x} = [M]\mathbf{q}$ as before to obtain $[\Lambda]$, an additional change is required to pair the complex poles. This change is given, for a third-order system, in

$$\mathbf{x} = [M]\,[R]\,\mathbf{z} = [M]\left[\begin{array}{cc|c} \frac{1}{2} & -j/2 & 0 \\ \frac{1}{2} & j/2 & 0 \\ \hline 0 & 0 & 1 \end{array}\right]\mathbf{z} \tag{14-5.28}$$

$$\mathbf{z} = \mathbf{R}^{-1}\mathbf{M}^{-1}\mathbf{x} = \left[\begin{array}{cc|c} 1 & 1 & 0 \\ j & -j & 0 \\ \hline 0 & 0 & 1 \end{array}\right][M]^{-1}\mathbf{x} \tag{14-5.29}$$

The arrangement of $[R]$ is such as to place the effects of the complex poles $(\alpha \pm j\beta)$ in the approximate $\mathbf{\Lambda}$ matrix, as shown in

$$\hat{\mathbf{\Lambda}} = \begin{bmatrix} \alpha & \beta & 0 \\ -\beta & \alpha & 0 \\ 0 & 0 & s_3 \end{bmatrix} \tag{14-5.30}$$

The $[R]$ and $\hat{\mathbf{\Lambda}}$ matrices can be extended in order and the position of the (α, β) block moved to the desired diagonal location, but the reader is referred to the literature for this discussion.[3]

Example 14-5.2. As a simple illustration of the above procedure, the system of Example 14-2.1 is reconsidered.

$$\dot{\mathbf{x}} = \begin{bmatrix} 0 & 1 \\ -8 & -4 \end{bmatrix}\mathbf{x} + \begin{bmatrix} 0 \\ 1 \end{bmatrix} u \tag{14-5.31}$$

$$[\phi(t)]_{\mathbf{x}} = \begin{bmatrix} \epsilon^{-2t}(\cos 2t + \sin 2t) & \frac{1}{2}\epsilon^{-2t}\sin 2t \\ -4\epsilon^{-2t}\sin 2t & \epsilon^{-2t}(\cos 2t - \sin 2t) \end{bmatrix} \qquad \begin{aligned} s_1 &= -2 + j2 \\ s_2 &= -2 - j2 \end{aligned} \tag{14-5.32}$$

From

$$[s_1 I - A]\mathbf{M}_1 = \begin{bmatrix} -2 + j2 & -1 \\ 8 & 2 + j2 \end{bmatrix}\mathbf{M}_1 = \mathbf{0} \tag{14-5.33}$$

$$[s_2 I - A]\mathbf{M}_2 = \begin{bmatrix} -2 - j2 & -1 \\ 8 & 2 - j2 \end{bmatrix}\mathbf{M}_2 = \mathbf{0} \tag{14-5.34}$$

suitable solutions for $\mathbf{M}_1$ and $\mathbf{M}_2$ are given, as the columns of $\mathbf{M}$ in

$$[M] = \begin{bmatrix} 1 & 1 \\ -2 + j2 & -2 - j2 \end{bmatrix} \qquad \mathbf{M}^{-1} = \begin{bmatrix} \frac{1}{2}(1 - j) & -j\frac{1}{4} \\ \frac{1}{2}(1 + j) & j\frac{1}{4} \end{bmatrix} \tag{14-5.35}$$

Using $[R]$ and $[R]^{-1}$ from Eqs. 14-5.28 and 29 and $[A]$ from Eq. 14-5.31 gives $\hat{\Lambda}$, as in

$$\hat{\mathbf{\Lambda}} = \mathbf{R}^{-1}\mathbf{M}^{-1}\mathbf{AMR} \tag{14-5.36}$$

$$\mathbf{MR} = \begin{bmatrix} 1 & 1 \\ -2 + j2 & -2 - j2 \end{bmatrix}\begin{bmatrix} \frac{1}{2} & -j/2 \\ \frac{1}{2} & j/2 \end{bmatrix} = \begin{bmatrix} 1 & 0 \\ -2 & 2 \end{bmatrix} \tag{14-5.37}$$

[3] K. Ogata, *State Space Analysis of Control Systems*, Prentice Hall, Inc., Englewood Cliffs, N.J., 1967, p. 144.

$$\mathbf{R}^{-1}\mathbf{M}^{-1} = (\mathbf{MR})^{-1} = \begin{bmatrix} 1 & 0 \\ 1 & \frac{1}{2} \end{bmatrix} \tag{14-5.38}$$

$$\hat{\mathbf{\Lambda}} = \begin{bmatrix} 1 & 0 \\ 1 & \frac{1}{2} \end{bmatrix} \begin{bmatrix} 0 & 1 \\ -8 & -4 \end{bmatrix} \begin{bmatrix} 1 & 0 \\ -2 & 2 \end{bmatrix} = \begin{bmatrix} -2 & 2 \\ -2 & -2 \end{bmatrix} \tag{14-5.39}$$

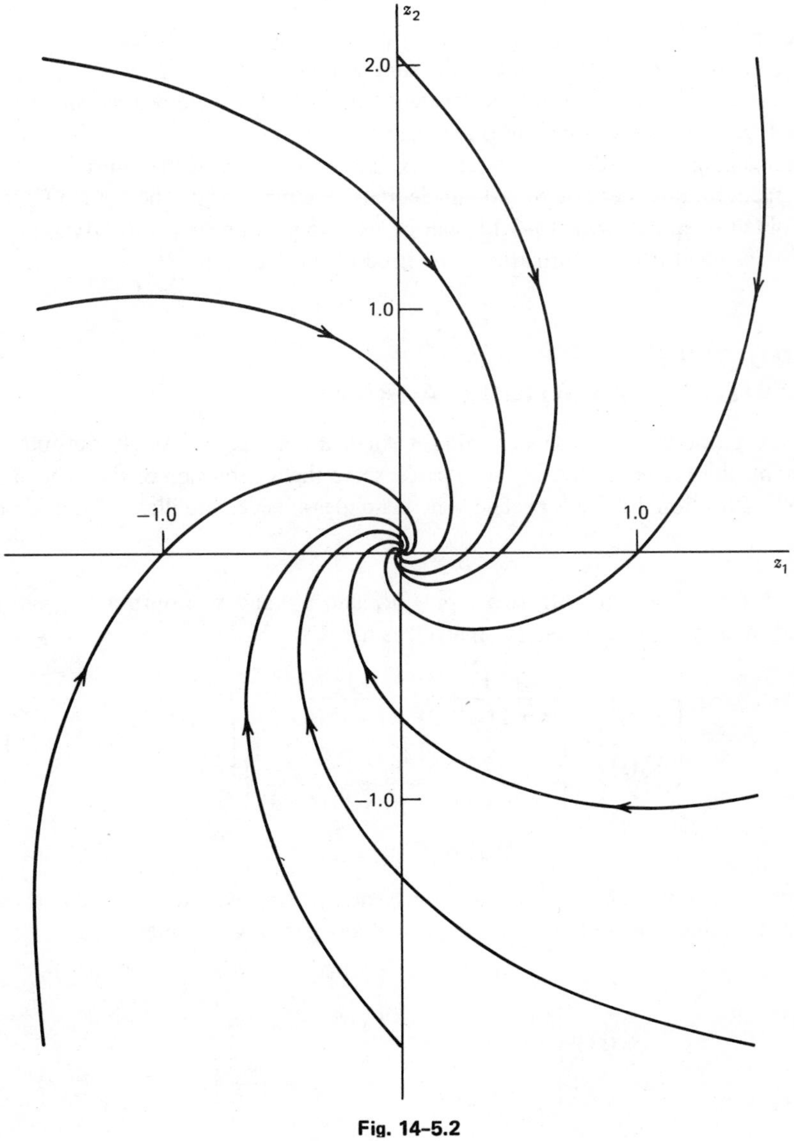

Fig. 14-5.2

The system is now represented by

$$\dot{\mathbf{z}} = \begin{bmatrix} -2 & 2 \\ -2 & -2 \end{bmatrix} \mathbf{z} + [R^{-1}M^{-1}]\ \mathbf{B}\,\mathbf{u} \qquad (14\text{–}5.40)$$

with $\mathbf{z}$ and the original $\mathbf{x}$ state vectors related by Eq. 14–5.28. For this system the state transition matrix is found to be

$$[\phi(t)]_{\mathbf{z}} = \epsilon^{-2t} \begin{bmatrix} \cos 2t & \sin 2t \\ -\sin 2t & \cos 2t \end{bmatrix} \qquad (14\text{–}5.41)$$

It can be seen that initial conditions, with one component of $\mathbf{z}(0)$ equal to zero, will cause the components of $\mathbf{z}(t)$ to contain the simple modes $\epsilon^{-2t}\cos 2t$, or $\epsilon^{-2t}\sin 2t$.

A plot showing trajectories for this system, with z_1 and z_2 along a rectangular basis, is given in Fig. 14–5.2. The trajectories are logarithmic spirals.

One remaining case needs consideration in this section. This is the multiple-pole case. However, space simply does not permit an adequate treatment of it. The Jordan Canonical form, as obtained in Example 13–3.11, can be used to get as near a Λ matrix as possible. The reader desiring further information is referred to the literature.[4]

14–6. MULTIPLE INPUT-OUTPUT SIMULATION DIAGRAMS

Most of the previous discussion has utilized single-inputs, u, and single outputs, y. In many systems this is inadequate. For such cases, some slight extension of the programming techniques of Section 13–3.2 is needed. The main ideas needed are illustrated in Example 14–6.1.

Example 14–6.1. A system with two inputs, u_1 and u_2, and two outputs, y_1 and y_2, is considered to have a transfer function matrix, as in

$$\begin{bmatrix} Y_1(s) \\ Y_2(s) \end{bmatrix} = \begin{bmatrix} \dfrac{1}{(s+1)(s+2)} & \dfrac{(3s+5)}{(s+1)(s+2)} \\ \dfrac{2}{(s+1)} & \dfrac{4}{(s+1)(s+2)} \end{bmatrix} \begin{bmatrix} U_1(s) \\ U_2(s) \end{bmatrix} \qquad (14\text{–}6.1)$$

$$\mathbf{Y}(s) = [G(s)]\,\mathbf{U}(s) \qquad (14\text{–}6.2)$$

In order to realize a simulation diagram for this system, a version of parallel programming is used. Each element of $[G(s)]$ is expanded in a partial fraction expansion, as

$$G(s) = \begin{bmatrix} \left(\dfrac{1}{s+1} - \dfrac{1}{s+2}\right) & \left(\dfrac{2}{s+1} + \dfrac{1}{s+2}\right) \\ \left(\dfrac{2}{s+1}\right) & \left(\dfrac{4}{s+1} - \dfrac{4}{s+2}\right) \end{bmatrix} \qquad (14\text{–}6.3)$$

[4]S. C. Gupta, *Transform and State Variable Methods in Linear Systems*, John Wiley & Sons, Inc., New York, pp. 247–71, 1966.

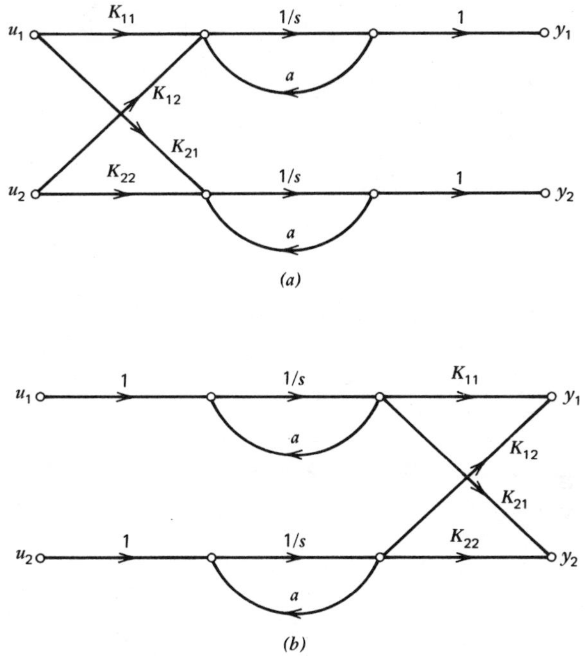

Fig. 14-6.1

The resulting $[G(s)]$ matrix is then divided into separate coefficient matrices for each pole as shown in

$$G(s) = \left(\frac{1}{s+1}\right)\begin{bmatrix} 1 & 2 \\ 2 & 4 \end{bmatrix} + \left(\frac{1}{s+2}\right)\begin{bmatrix} -1 & 1 \\ 0 & -4 \end{bmatrix} \tag{14-6.4}$$

Attention is now devoted to a typical pole and its coefficient matrix. In general, the result will be as shown in Fig. 14-6.1(a) or (b). It appears that two realizations of the $1/(s - a)$ factor are required. However, it may happen that the ratios K_{11}/K_{21} and K_{12}/K_{22} are the same, or in other words the rows of the $[K]$ coefficient matrix are proportional. In such a case only one realization of the $1/(s - a)$ factor is required. This is the case for the $1/(s + 1)$ term in Eq. 14-6.4, and Fig. 14-6.2(a) shows the resulting simulation diagram for this pole while Fig. 14-6.2(b) shows the complete simulation diagram. It should be noted that the proportionality factor 2.0 between rows appears on the path to y_2.

The above procedure can be stated in more general terms. The requirement for proportionality of rows in the coefficient matrix is changed to consideration of what is called the rank of the matrix. Rank is defined as the order of the largest non-zero determinant that can be formed from the coefficient matrix. A given pole has to be realized a number of times equal to the rank of its coefficient matrix. Thus the *minimum number of integrators* (and the number of state variables) *is equal to the sum of the ranks of all coefficient matrices.*

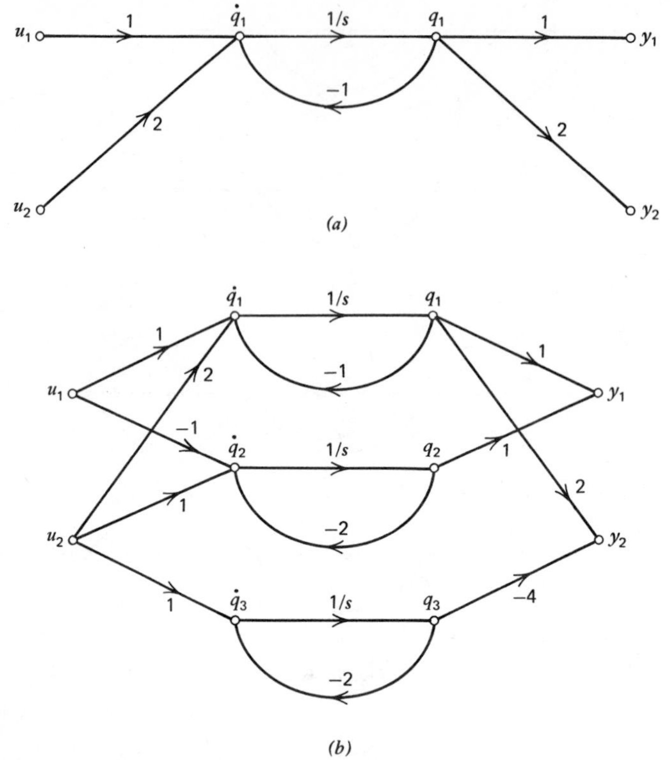

Fig. 14-6.2

Example 14-6.2. As a further example of the multiple input-output case, the one-pole, three-input-output case of Eq. 14-6.5 is considered.

$$\mathbf{Y}(s) = \left(\frac{1}{s+1}\right)\begin{bmatrix} 1 & 1 & 1 \\ -1 & 0 & 0 \\ 0 & 1 & 1 \end{bmatrix}\mathbf{U}(s) = G(s)\,[K_{-1}]\,\mathbf{U}(s) \qquad (14\text{-}6.5)$$

It is fairly obvious that *row 3* of the coefficient matrix is the sum of *rows 1* and *2*, and so $|K_{-1}|$ is zero. There are several (2×2) non-zero determinants in $[K_{-1}]$. Thus the rank of $[K_{-1}]$ is 2, and so two realizations of the pole at $s = -1$ are required. In reality, any two rows of $[K_{-1}]$ are independent. Thus, picking *rows 1* and *2*, the simulation diagram for these rows is developed, as shown in Figure 14-6.3(a). The realization of y_3 or the third row of $[G(s)]$ can now be carried out by making use of the fact that it is the sum of *rows 1* and *2*. Signals that form y_1 and y_2 are added at the node for y_3, as indicated by the dotted lines in Fig. 14-6.3(b). It might be noted that the u_1 contribution to y_3 cancels, due to signals traveling two paths.

While there are other realizations of multiple input-output transfer function matrices, the above examples at least illustrate a typical and useful approach to use in obtaining a set of state variables. It also shows the true order of the system.

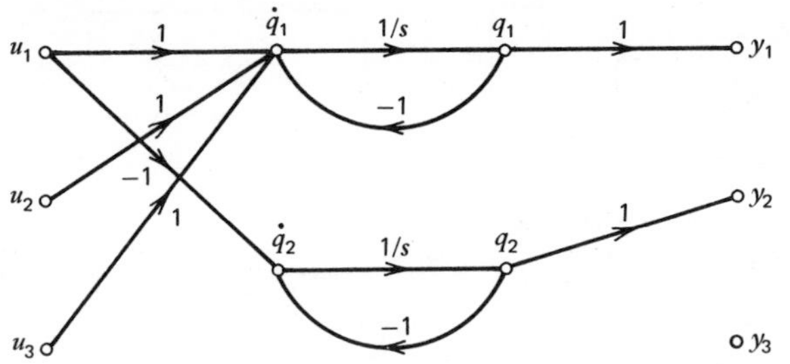

(a)

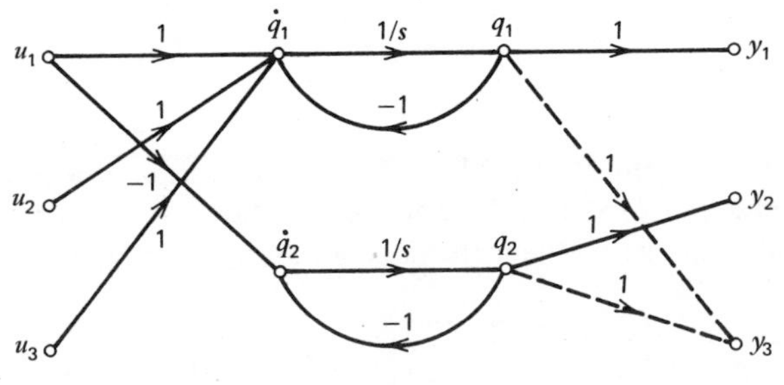

(b)

Fig. 14-6.3

14-7. STATE VARIABLES FOR DISCRETE SYSTEMS AND SAMPLED-DATA SYSTEMS

Perhaps one of the most useful aspects of the state variable approach is the unification possible in the formulation and solution of systems containing sampled signals as well as continuous signals. Space limitations permit only an introduction to open and closed loops with samplers followed by zero-order holds and also purely discrete systems.

The simplest cases to use in extending the continuous state variable techniques to sampled-data systems are those of the type shown in Fig. 14-7.1. While there are several ways to formulate the state variable approach to the solution of the system of Fig. 14-7.1, only two are considered in the following examples.

One method treats the variable m at the output of the sampler as an input to the system rather than as a state variable. This is possible since it is explicitly determined as a piecewise constant function of the true input $u(t)$ and the type of sampler.

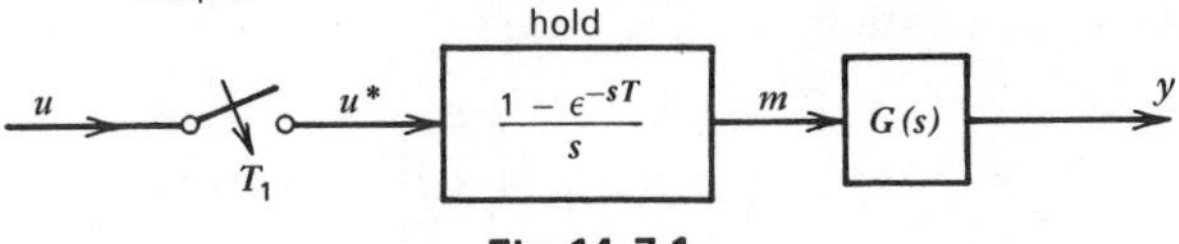

Fig. 14-7.1

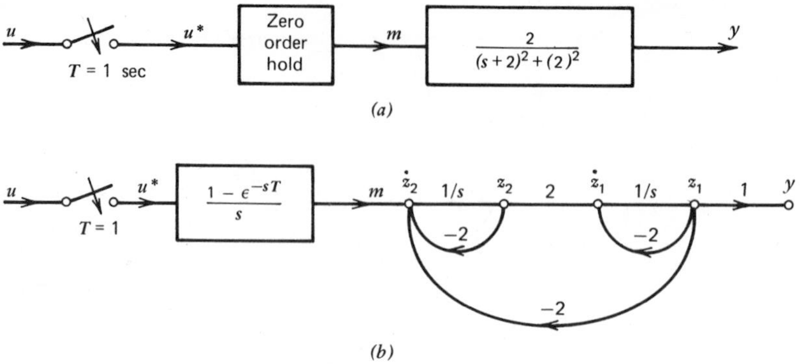

Fig. 14-7.2

Example 14-7.1. The system of Fig. 14-7.2 is used to illustrate the procedure described above. In the formulation used in this example the state equations are written as Eqs. 14-7.1. In this method, none of the state variables are discontinuous for $m(t)$ functions

$$\begin{bmatrix} \dot{z}_1 \\ \dot{z}_2 \end{bmatrix} = \begin{bmatrix} -2 & 2 \\ -2 & -2 \end{bmatrix} \begin{bmatrix} z_1 \\ z_2 \end{bmatrix} + \begin{bmatrix} 0 \\ 1 \end{bmatrix} m(t) \tag{14-7.1}$$

that possess only simple jumps. The state of the system during a sampling interval $(kT \leqslant t < \overline{k+1}T)$ can be computed from the state transition matrix, the initial state of the system just after sampler operation, $\mathbf{z}(kT^+)$, and the forced response due to a constant forcing function, $m(kT^+)$. The required relation is expressed by

$$\mathbf{z}(t) = \phi(t - kT)\,\mathbf{z}(kT^+) + \left\{ \int_{kT}^{t} \phi(t-\tau)\, B\, d\tau \right\} m(kT^+) \qquad kT \leqslant t < (k+1)T \tag{14-7.2}$$

There is no real difficulty in obtaining the response *between* sampling instants.

The solution proceeds sequentially from $t = 0$ by evaluating the sampling interval response for $0 \leqslant t < T$. At $t = T$ any discontinuities in $\mathbf{z}(T)$ and $m(T)$ are effected. These values are used as initial conditions and input respectively for $T \leqslant t < 2T$. This sequence is repeated for as long a time interval as desired.

For $u(t)$ equal to a unit impulse and $\mathbf{z}(0) = \mathbf{0}$ the response of the system of this example is developed in the following equations. Values are computed at the midpoint of each period as well as at the sampling instants.

$$\left[\phi\left(\frac{T}{2}\right)\right] = \left[\phi\left(\frac{1}{2}\right)\right] = \epsilon^{-1} \begin{bmatrix} \cos(1 \text{ rad}) & \sin 1 \\ -\sin 1 & \cos 1 \end{bmatrix} = \begin{bmatrix} 0.1988 & 0.3096 \\ -0.3096 & 0.1988 \end{bmatrix} \tag{14-7.3}$$

From a basic property of transition matrices we have

$$\left[\phi\left(\frac{kT}{2}\right)\right] = \left[\phi\left(\frac{T}{2}\right)\right]^k \tag{14-7.4}$$

Additionally, the defining relationship

$$\left[H\left(\frac{T}{2}\right)\right] = \int_0^{T/2} [\phi(\lambda)]\, B\, d\lambda \tag{14-7.5}$$

is used. It is also desirable, by a change of variable $(\overline{k+1}\,T - \tau) = \lambda$, to show the relation

$$\int_{kT}^{(k+1)T} [\phi(\overline{k+1}\,T - \tau]\, B\, d\tau = \int_0^{T} [\phi(\lambda)]\, B\, d\lambda = [H(T)] \tag{14-7.6}$$

For the specific case

$$[H(T/2)] = \int_0^{T/2} \epsilon^{-2\lambda} \begin{bmatrix} \cos 2\lambda & \sin 2\lambda \\ -\sin 2\lambda & \cos 2\lambda \end{bmatrix} \begin{bmatrix} 0 \\ 1 \end{bmatrix} d\lambda \tag{14-7.7}$$

$$= \begin{bmatrix} \displaystyle\int_0^{T/2} \epsilon^{-2\lambda} \sin 2\lambda\, d\lambda \\ \displaystyle\int_0^{T/2} \epsilon^{-2\lambda} \cos 2\lambda\, d\lambda \end{bmatrix} = \frac{1}{4} \begin{bmatrix} 1 - \epsilon^{-T} (\sin T + \cos T) \\ 1 - \epsilon^{-T} (\cos T - \sin T) \end{bmatrix} \tag{14-7.8}$$

$$= [H(1/2)] = \begin{bmatrix} 0.1229 \\ 0.2777 \end{bmatrix} \tag{14-7.9}$$

$$m(t) = \begin{cases} 1 & 0 \leqslant t < T \\ 0 & T \leqslant t \end{cases} \tag{14-7.10}$$

$$\left.\begin{aligned} \mathbf{z}(T/2) &= [\phi(T/2)]\, \mathbf{z}(0)^{\nearrow 0} + [H(T/2)]\, m(0^{+}) \\ \mathbf{z}(T) &= [\phi(T/2)]\, \mathbf{z}(T/2) + [H(T/2)]\, m(T/2) \\ \mathbf{z}(3T/2) &= [\phi(T/2)]\, \mathbf{z}(T) + [H(T/2)]\, m(T^{+}) \\ \mathbf{z}(2T) &= [\phi(T/2)]\, \mathbf{z}(3T/2) + [H(T/2)]\, m(3T/2) \end{aligned}\right\} \tag{14-7.11}$$

$$\begin{aligned} &\mathbf{z}(1/2) = \begin{bmatrix} 0.1229 \\ 0.2777 \end{bmatrix} \quad \mathbf{z}(1) = \begin{bmatrix} 0.2333 \\ 0.2949 \end{bmatrix} \quad \mathbf{z}(3/2) = \begin{bmatrix} 0.1377 \\ -0.0136 \end{bmatrix} \\ &\mathbf{z}(2) = \begin{bmatrix} 0.02316 \\ -0.04533 \end{bmatrix} \quad \mathbf{z}(5/2) = \begin{bmatrix} -0.00943 \\ -0.01618 \end{bmatrix} \quad \mathbf{z}(3) = \begin{bmatrix} -0.00687 \\ -0.0003 \end{bmatrix} \end{aligned} \tag{14-7.12}$$

The procedure used above can be applied where the sampling is not periodic, thereby being a very powerful technique. Digital computer programs can be easily written to evaluate the above matrices.

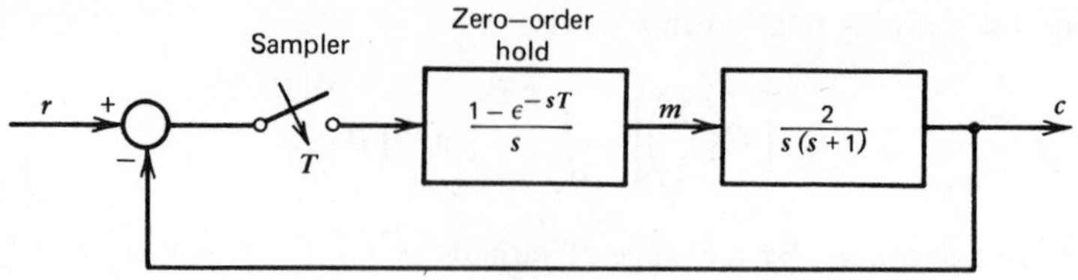

Fig. 14-7.3(a)

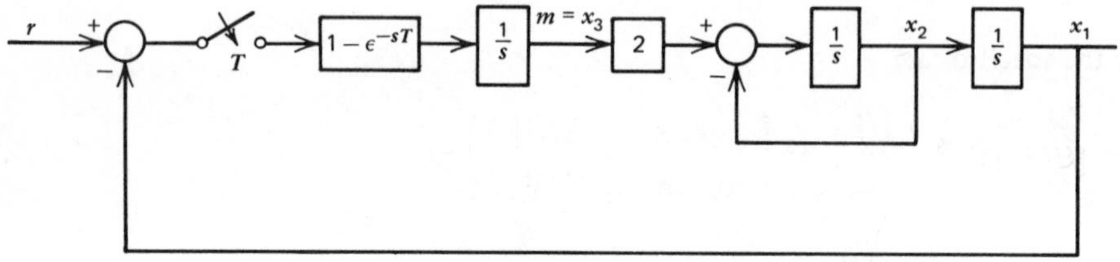

Fig. 14-7.3(b)

Example 14-7.2. An alternate formulation for sampled-data systems is illustrated by considering the system of Fig. 14-7.3. The hold output, m, can be included as a component of the state vector. However, the discontinuities in it have to be taken care of by a special jump transition matrix. This formulation is described as follows.

$$\begin{bmatrix} \dot{x}_1 \\ \dot{x}_2 \\ \dot{x}_3 \end{bmatrix} = \begin{bmatrix} 0 & 1 & 0 \\ 0 & -1 & 2 \\ 0 & 0 & 0 \end{bmatrix} \mathbf{x} \tag{14-7.13}$$

$$[\Phi(s)] = \begin{bmatrix} s & -1 & 0 \\ 0 & (s+1) & -2 \\ 0 & 0 & s \end{bmatrix}^{-1} = \frac{1}{s^2(s+1)} \begin{bmatrix} s(s+1) & s & 2 \\ 0 & s^2 & 2s \\ 0 & 0 & s(s+1) \end{bmatrix} \tag{14-7.14}$$

$$= \begin{bmatrix} \dfrac{1}{s} & \dfrac{1}{s(s+1)} & \dfrac{2}{s^2(s+1)} \\ 0 & \dfrac{1}{s+1} & \dfrac{2}{s(s+1)} \\ 0 & 0 & \dfrac{1}{s} \end{bmatrix} \tag{14-7.15}$$

$$[\phi(t)] = \begin{bmatrix} 1 & (1-\epsilon^{-t}) & 2(t-1+\epsilon^{-t}) \\ 0 & \epsilon^{-t} & 2(1-\epsilon^{-t}) \\ 0 & 0 & 1 \end{bmatrix} \tag{14-7.16}$$

$$\mathbf{x}(kT^+) = \begin{bmatrix} 1 & 0 & 0 \\ 0 & 1 & 0 \\ -1 & 0 & 0 \end{bmatrix} \mathbf{x}(kT^-) + \begin{bmatrix} 0 \\ 0 \\ 1 \end{bmatrix} r(kT^+) \tag{14-7.17}$$

$$= [L]\,\mathbf{x}(kT^-) + [N]\,r(kT^+) \tag{14-7.18}$$

Matrices $[L]$ and $[N]$ account for the jumps in state component x_3 or m_1 at sampling instants. Cyclical application of the state transition matrix and the jump matrices permit a development of the solution, $\mathbf{x}(t)$, for as long a time as desired. For the specific system of this example the following equations illustrate the required procedure with points obtained only at the sampling instants. However, no real difficulty is involved if intersample points are desired. A unit step function input, $r(t)$, is used, as well as zero initial conditions.

$$\left.\begin{aligned} \mathbf{x}(0^+) &= [L]\,\mathbf{x}(0^-) + [N]\,r(0^+) \\ &= \begin{bmatrix} 1 & 0 & 0 \\ 0 & 1 & 0 \\ -1 & 0 & 0 \end{bmatrix} \begin{bmatrix} 0 \\ 0 \\ 0 \end{bmatrix} + \begin{bmatrix} 0 \\ 0 \\ 1 \end{bmatrix} (1) = \begin{bmatrix} 0 \\ 0 \\ 1 \end{bmatrix} \end{aligned}\right\} \tag{14-7.19}$$

$$\phi(T) = \phi(1) = \begin{bmatrix} 1.0 & 0.63212 & 0.73576 \\ 0 & 0.36799 & 1.26424 \\ 0 & 0.0 & 1.0 \end{bmatrix} \tag{14-7.20}$$

$$\mathbf{x}(T^-) = [\phi(1)]\,\mathbf{x}(0^+) = \begin{bmatrix} 2(T - 1 + \epsilon^{-T}) \\ 2(1 - \epsilon^{-T}) \\ 1.0 \end{bmatrix} = \begin{bmatrix} 0.73576 \\ 1.26424 \\ 1.0 \end{bmatrix} \tag{14-7.21}$$

$$\mathbf{x}(T^+) = [L]\,\mathbf{x}(1^-) + [N]\,r(1^+) = \begin{bmatrix} 0.73576 \\ 1.26424 \\ 0.26424 \end{bmatrix} \tag{14-7.22}$$

$$\mathbf{x}(2T^-) = [\phi(1)]\,\mathbf{x}(T^+) = \begin{bmatrix} 1.72932 \\ 0.79914 \\ 0.26424 \end{bmatrix} \tag{14-7.23}$$

$$\mathbf{x}(2T^+) = [L]\,\mathbf{x}(1^-) + [N]\,r(2T^+) = \begin{bmatrix} 1.72932 \\ 0.79914 \\ -0.72932 \end{bmatrix} \tag{14-7.24}$$

$$\mathbf{x}(3T^-) = [\phi(1)]\,\mathbf{x}(2T^+) = \begin{bmatrix} 1.69787 \\ -0.62805 \\ -0.72932 \end{bmatrix} \tag{14-7.25}$$

$$\mathbf{x}(3T^+) = \begin{bmatrix} 1.69787 \\ -0.62805 \\ -0.69787 \end{bmatrix} \quad \mathbf{x}(4T^-) = \begin{bmatrix} 0.78741 \\ -1.11331 \\ -0.69787 \end{bmatrix} \quad \mathbf{x}(4T^+) = \begin{bmatrix} 0.78741 \\ -1.11331 \\ 0.21259 \end{bmatrix} \qquad (14\text{–}7.26)$$

A general recursion formula for the solution can be developed, if it is desired to program the solution on a digital computer.

The previous examples indicate two specific ways of formulating state variable solutions of sampled-data systems. These can be extended without major modification to more complex systems. However, there is another class of systems in which only discrete data exists, or systems that can be adequately described by difference equations rather than differential equations. For such systems a state variable representation can be developed from the difference equations by use of *unit delays*.

Example 14-7.3. A discrete system is described by the difference equation 14-7.27.

$$y(\overline{k+2}\,T) + 5y(\overline{k+1}\,T) + 6y(kT) = 4v(kT) \qquad (14\text{–}7.27)$$

In order to use the unit delay and a state variable representation, use is made of the relationship of Fig. 14-7.4. A simulation block diagram for the system can be developed as shown in Fig. 14-7.5. Discrete state variables can be obtained by assigning *one* to the output of each unit delay. Thus, for this system, two state variables are needed, with the resulting discrete state equations being given by

$$\left.\begin{aligned} x_1(\overline{k+1}\,T) &= x_2(kT) \\ x_2(\overline{k+1}\,T) &= -6x_1(kT) - 5x_2(kT) + 4v(kT) \end{aligned}\right\} \qquad (14\text{–}7.28)$$

$$y(kT) = x_1(kT) \qquad (14\text{–}7.29)$$

$$\mathbf{x}(\overline{k+1}\,T) = \begin{bmatrix} 0 & 1 \\ -6 & -5 \end{bmatrix} \mathbf{x}(kT) + \begin{bmatrix} 0 \\ 4 \end{bmatrix} v(kT) \qquad (14\text{–}7.30)$$

$$y(kT) = [1 \quad 0]\, \mathbf{x}(kT) \qquad (14\text{–}7.31)$$

The state transition matrix for this case can be developed by methods analogous to those for continuous systems. There is a discrete transformation method called the z-transform method, which can be used in a manner analogous to the Laplace transform method. However, this technique has not been presented, and so only the Cayley-Hamilton Theorem method will be used.

The state transition matrix can be developed from Eq. 14-7.30 with $v(kT) = 0$, together with the fact that the function that needs evaluation as $f([A])$ in the Cayley-

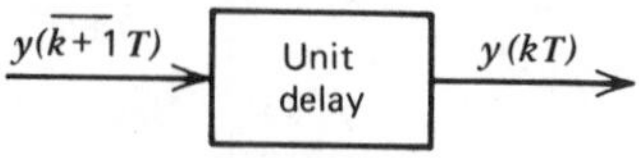

Fig. 14-7.4

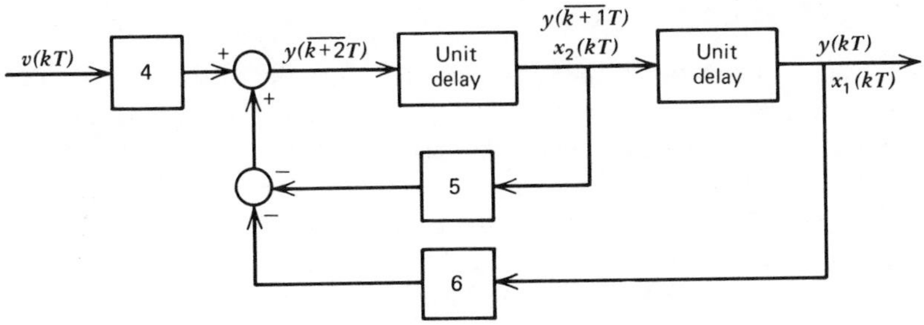

Fig. 14-7.5

Hamilton Theorem turns out to be $[A]^n$. This latter fact is obtained by repeated use of Eq. 14-7.30 to find that the transition from $\mathbf{x}(0)$ to $\mathbf{x}(nT)$ is given by

$$\mathbf{x}(nT) = [A]^n\, \mathbf{x}(0) \tag{14-7.32}$$

when $v(kT) = 0$.

For the $[A]$ matrix used, the eigenvalues are found to be -2 and -3, by using $|sI - A| = 0$. For the (2×2) matrix $[A]$ used, the following equations result.

$$F([A]) = k_1 [A] + k_0 [I] = [A]^n \tag{14-7.33}$$

$$(-2)^n = -2k_1 + k_0 \tag{14-7.34}$$

$$(-3)^n = -3k_1 + k_0 \tag{14-7.35}$$

$$k_1 = (-2)^n - (-3)^n \tag{14-7.36}$$

$$k_0 = 3(-2)^n - 2(-3)^n \tag{14-7.37}$$

Substituting back into Eq. 14-7.33 provides the required transition matrix $[A]^n$.

$$[\phi(nT)] = [A]^n = \begin{bmatrix} [3(-2)^n - 2(-3)^n] & [(-2)^n - (-3)^n] \\ 6[(-3)^n - (-2)^n] & [(-2)^{n+1} - (-3)^{n+1}] \end{bmatrix} \tag{14-7.38}$$

It should be noted that, while the transition matrix $[\phi(nT)]$ indicates a dependence upon the sampling interval T, the actual matrix contains only the number of the sampling interval n. This results from the fact that the original discrete state equation 14-7.30 incorporated the T dependency into the $[A]$ matrix used.

Nearly all the previous developments for continuous systems can be paralleled for discrete systems. However, space does not permit such a development here, and the reader is referred to the literature for additional material.[5,6]

[5] P. M. DeRusso, R. J. Roy, and C. M. Close, *State Variables for Engineers*, John Wiley & Sons, Inc., New York, 1965, Chapter 6.
[6] K. Ogata, *State Space Analysis of Control Systems*, Prentice-Hall, Inc., Englewood Cliffs, N.J., 1967.

14-8. CONTROLLABILITY

The use of transfer functions to model systems indicates the existence of various input-output relationships, with at least some of the inputs affecting each output. As developed in the last two chapters, one also needs to consider the internal state of the system. It is possible for certain state variables to vary, independent of input variables, in systems or in linearized models of non-linear systems. While such a situation is not common, it nevertheless needs to be recognized. The occurrence of such uncontrolled states means that we cannot force the system state to move from one arbitrary state point in state space to some other point.

In general, the study of controllability considers the conditions required for changing the state of a system from some arbitrary state point $\mathbf{x}_0$, at time t_0, to some other point such as the origin at some later finite time t_f, by application of a control vector over the time interval (t_0, t_f). The initial time is commonly taken as $t_0 = 0$. Requirements for controllability of linear systems can be determined by use of Eq. 14-3.79 and the above statement for changing the state of the system. The result is expressed in

$$\mathbf{x}(t_f) = \mathbf{0} = \epsilon^{\mathbf{A}t_f}\,\mathbf{x}_0 + \int_0^{t_f} \epsilon^{\mathbf{A}(t_f-\tau)}\,\mathbf{B}\,\mathbf{u}(\tau)\,d\tau \tag{14-8.1}$$

$$\mathbf{x}_0 = -\int_0^{t_f} \epsilon^{-\mathbf{A}\tau}\,\mathbf{B}\,\mathbf{u}(\tau)\,d\tau \tag{14-8.2}$$

In Eq. 14-8.2, the initial state is an arbitrary vector, and it is desired that some control vector, $\mathbf{u}(\tau)$, exist to satisfy the equation, if the system is to be controllable. It is assumed in the above equations that the state vector is of the dimension of n, that the matrix $\mathbf{A}$ is $(n \times n)$, that the matrix $\mathbf{B}$ is $(n \times m)$, and that the control vector $\mathbf{u}(\tau)$ is $m \times 1$.

Equation 14-8.2 can be rearranged to simplify the determination of the required conditions for the existence of a unique solution. The matrix exponential $e^{-\mathbf{A}\tau}$ can be expressed as

$$\begin{aligned}\epsilon^{-\mathbf{A}\tau} &= k_0(\tau)\,I + k_1(\tau)\,\mathbf{A} + \cdots + k_{n-1}(\tau)\,\mathbf{A}^{n-1} \\ &= \sum_{j=0}^{n-1} k_j(\tau)\,\mathbf{A}^j\end{aligned} \tag{14-8.3}$$

Use of this equation and a rearrangement of terms allows Eq. 14-8.2 to be rewritten as

$$\mathbf{x}_0 = -\sum_{j=0}^{n-1} \mathbf{A}^j\mathbf{B}\left\{\int_0^{t_f} k_j(\tau)\,\mathbf{u}(\tau)\,d\tau\right\} \tag{14-8.4}$$

It is convenient to replace the integrals with new vectors, $\boldsymbol{\gamma}_j$, as in

$$\boldsymbol{\gamma}_j = \int_0^{t_f} k_j(\tau)\,\mathbf{u}(\tau)\,d\tau \qquad j = 0, 1, 2 \ldots n-1 \tag{14-8.5}$$

This now allows the right side of Eq. 14–8.4 to be rewritten in matrix form as

$$\mathbf{x}_0 = -[B \mid AB \mid A^2B \mid \cdots \mid A^{n-1}B] \begin{bmatrix} \boldsymbol{\gamma}_0 \\ \boldsymbol{\gamma}_1 \\ \cdot \\ \cdot \\ \cdot \\ \boldsymbol{\gamma}_{n-1} \end{bmatrix} \tag{14–8.6}$$

It is helpful to consider the dimensions of the various matrices in Eq. 14–8.6. The $[A^jB]$ array is $(n \times mn)$, while the total $\boldsymbol{\gamma}$ vector is $(mn \times 1)$. It should also be noted that the control vector $\mathbf{u}(\tau)$ is related to the $\boldsymbol{\gamma}_j$ vectors.

Equation 14–8.6 can now be viewed as an expansion of the arbitrary vector $\mathbf{x}_0$ in terms of vectors represented by the columns of the $[A^jB]$ matrix, with the expansion coefficients related to the components of $\mathbf{u}(\tau)$. For a unique expansion to exist, this matrix must have a rank of n. In other terms there must be exactly n independent column vectors in the $[A^jB]$ matrix. Since this matrix has n rows, its rank cannot exceed n, and it could be less than n.

If there are fewer than n independent columns, then a unique expansion of $\mathbf{x}_0$ as required in Eq. 14–8.6 is not possible. It is possible to prove that the linear system being considered is controllable if, and only if, the rank of the matrix

$$[B \mid AB \mid A^2B \mid \cdots \mid A^{n-1}B]$$

is n or the same as the dimension of the state vector.

If the system under consideration has only distinct, simple eigenvalues, it is instructive to consider the controllability concept for the normalized system of Section 14–5 for which the state vector matrix differential equation is

$$\dot{\mathbf{q}} = \boldsymbol{\Lambda}\mathbf{q} + \mathbf{B}_n\mathbf{u} \tag{14–8.7}$$

For this case it is required that the input matrix $\mathbf{B}_n$ have no rows containing all zeros. Otherwise there would be certain modes that could not be affected by any input vector components. This would prevent total control of the state as required to achieve arbitrary changes in state.

Example 14–8.1. Controllability of the system of

$$\dot{\mathbf{x}} = \begin{bmatrix} 0 & 1 \\ -12 & -7 \end{bmatrix} \mathbf{x} + \begin{bmatrix} 0 \\ 1 \end{bmatrix} \mathbf{u} \tag{14–8.8}$$

is considered as a first example. For this case the $[A^jB]$ matrix is given by

$$[A^jB] = [\mathbf{b} \mid A\mathbf{b}] = \begin{bmatrix} 0 & 1 \\ 1 & -7 \end{bmatrix} \tag{14–8.9}$$

In this case the two columns of this matrix are independent. The determinant is not zero, and the rank is *two*, thereby assuring controllability of this system.

The normal form for this system is given as

$$\dot{\mathbf{q}} = \begin{bmatrix} -3 & 0 \\ 0 & -4 \end{bmatrix} \mathbf{q} + \begin{bmatrix} 1 \\ -1 \end{bmatrix} u(t) \tag{14-8.10}$$

(see Example 14-5.1). In this case there are no all-zero rows in $[B_n] = \mathbf{b}_n$, and the system is controllable.

Example 14-8.2. As a second example of controllability, the system of

$$\dot{\mathbf{x}} = \begin{bmatrix} 0 & 1 \\ -12 & -7 \end{bmatrix} \mathbf{x} + \begin{bmatrix} 1 \\ -3 \end{bmatrix} u \tag{14-8.11}$$

is considered. It should be noted that the modes of this system are the same as those of the previous example but that the input matrix is different.

For this case the matrix of interest is that of

$$[A^j B] = [\mathbf{b} \,|\, A\mathbf{b}] = \begin{bmatrix} 1 & -3 \\ -3 & 9 \end{bmatrix} \tag{14-8.12}$$

The columns of this matrix are not independent, as the second column is (−3) times the first column. The system is not controllable. From a two-dimensional state space standpoint, it is apparent that the two columns of $[A^j B]$, viewed as vectors, do not span the total *two* space but only a *one* subspace or line. Thus, not all $\mathbf{x}_0$ points can be expressed as a linear combination of these two vectors, and the system is not controllable.

It might be helpful to consider the complete expansion of Eq. 14-8.6 for this system.

$$\mathbf{x}_0 = -\begin{bmatrix} 1 & -3 \\ -3 & 9 \end{bmatrix} \begin{bmatrix} \gamma_0(u) \\ \gamma_1(u) \end{bmatrix} \tag{14-8.13}$$

$$\begin{aligned} \gamma_0 &= \int_0^{t_f} k_0(\tau)\, u(\tau)\, d\tau \\ \gamma_1 &= \int_0^{t_f} k_1(\tau)\, u(\tau)\, d\tau \end{aligned} \tag{14-8.14}$$

From the total γ vector, which involves the control input $u(t)$, and the non-independence of the columns of $[A^j B]$, it is apparent that there is no control, $u(t)$, that can be used to express all possible $\mathbf{x}_0$ initial state vectors in the manner required for controllability.

The normal form for this system is given by

$$\dot{\mathbf{q}} = \begin{bmatrix} -3 & 0 \\ 0 & -4 \end{bmatrix} \mathbf{q} + \begin{bmatrix} 1 \\ 0 \end{bmatrix} u \tag{14-8.15}$$

The input vector $\mathbf{b}$ has a row (or single element, in this case) that is zero. Thus, the mode due to the system eigenvalue or s-plane pole at −4 is totally unaffected by the input. This part of the state of the system depends only on the initial component of the system state along this eigenvector. Again, we must conclude that the system is not controllable.

Example 14-8.3. A further development of the two previous examples in terms of transfer functions is helpful. It is necessary to specify the output equation, and to simplify matters we will take the output vector to be simply the state vectors in the various cases. In other terms, the output vector, $\mathbf{y}(t)$, is given by

$$\mathbf{y} = [I]\,\mathbf{x} \qquad \mathbf{y} = [I]\,\mathbf{q} \tag{14-8.16}$$

From Eq. 14-2.4 for $\mathbf{X}(s)$ and the Laplace transforms of the first of Eqs. 14-8.16, it is possible to write the transfer function from the input $U(s)$ to the output $\mathbf{Y}(s)$ as

$$\mathbf{Y}(s) = [I]\,[sI - A]^{-1}\,[B]\,U(s) = \frac{1}{(s+3)(s+4)}\begin{bmatrix} 1 \\ s \end{bmatrix} U(s) \tag{14-8.17}$$

The initial condition inputs have been ignored in this expression, as is customary in transfer function analysis.

There are no problems in developing the above transfer functions with a specific transfer between the input and each output component.

For the system of Eq. 14-8.11 a transfer function development as done above leads to

$$\mathbf{Y}(s) = I[sI - A]^{-1}\,[B]\,U(s) = \frac{1}{(s+3)(s+4)}\begin{bmatrix} (s+4) \\ -3(s+4) \end{bmatrix} U(s) \tag{14-8.18}$$

$$= \frac{1}{(s+3)}\begin{bmatrix} 1 \\ -3 \end{bmatrix} U(s) \tag{14-8.19}$$

In this case it is apparent from this analysis that the usual cancellation of poles and zeros leads to a transfer function without the system eigenvalue or s-plane pole at -4. We have no control over this lost mode from the input. Such a situation is the result of either an uncontrollable system, or an unobservable one, as described in the next section. For the example being considered the problem arises because the system is uncontrollable.

14-9. OBSERVABILITY

In a manner somewhat analogous to the relationship between control inputs and system states, it is also of interest to consider the relationship between system states and system outputs. Such a study leads to the concept of observability. More specifically, if an observation of the output vector, $\mathbf{y}(t)$, of a system, for a time interval from some t_0 to some later time t_f, together with system input, $u(t)$, for this period, allows all arbitrary initial states, $\mathbf{x}_0$, at t_0, to be determined, then the system is said to be *observable*. In general, this indicates that all system modes must contribute to the output.

The simplest approach in considering the requirements for a system to be observable is to use the normal form for the system, as in Section 14-5. This requires a representation of the system as in

$$\begin{aligned} \dot{\mathbf{q}} &= \boldsymbol{\Lambda}\mathbf{q} + \mathbf{B}_n\mathbf{u} \\ \mathbf{y} &= \mathbf{C}_n\mathbf{q} + \mathbf{D}\mathbf{u} \end{aligned} \tag{14-9.1}$$

where **q** is of dimension of n, **u** is of dimension m, and **y** is of dimension r.

For all modes of **q** to affect **y**, it can be seen that the matrix $\mathbf{C}_n$ can have no *columns* of all zeros. If such a column of zeros were to occur, there would be no way for the corresponding component of the **q** vector to contribute to the output vector, **y**. Thus, an observation of **y** could never determine the existence of the missing mode in the system state vector **q**. The system would then be said to be unobservable.

It is possible to develop a general test for the observability of linear systems in terms of the **A** and **C** matrices, much as was done in the case of controllability using the **A** and **B** matrices. However, the details are somewhat more tedious, and only the final result is presented here. The reader is referred to the literature for the development.[7] The indicated test states that a system is observable if, and only if, the matrix

$$[C^T \mid A^T C^T \mid A^{2^T} C^T \mid \cdots \mid A^{n-1^T} C^T]$$

has n linearly independent columns, or it is rank n. This condition is related to the existence of n linearly independent rows in the set of matrices $\{C, CA, CA^2, \ldots, CA^{n-1}\}$, and the ability to uniquely express arbitrary initial states $\mathbf{x}_0$ in terms of such rows when $\mathbf{y}(t)$ is known.

Example 14-9.1. To illustrate the above observability concepts, the system of

$$\begin{aligned} \dot{\mathbf{x}} &= \begin{bmatrix} 0 & 1 \\ -35 & -12 \end{bmatrix} \mathbf{x} + \begin{bmatrix} 0 \\ 1 \end{bmatrix} u \\ \mathbf{y} &= \begin{bmatrix} 1 & 0 \\ 0 & 1 \end{bmatrix} \mathbf{x} \end{aligned} \tag{14-9.2}$$

is used. For this case, n is *two*, and the matrix $[C^T A^{j^T}]$ becomes

$$[C^T \mid A^T C^T] = \left[\begin{array}{cc|cc} 1 & 0 & 0 & -35 \\ 0 & 1 & 1 & -12 \end{array}\right] \tag{14-9.3}$$

From this equation it can be seen that any pair of columns from the set $\{1, 2, 4\}$ are independent. Alternately, the rank of the subject matrix is *two*, and so the system is said to be observable.

Example 14-9.2. A system defined by the mathematical model of Eqs. 14-9.2 is used again, with the exception that the output matrix **C** is given by

$$[C] = \left(\frac{1}{2}\right) \begin{bmatrix} -5 & -1 \\ 5 & 1 \end{bmatrix} \tag{14-9.4}$$

[7] M. Athans and P. Falb, *Optimal Control*, McGraw-Hill Book Co., New York, 1966, pp. 207–11.

For this case the $[C^T A^{i^T}]$ matrix is given by

$$[C^T \mid A^T C^T] = \begin{bmatrix} -5/2 & 5/2 & 35/2 & -35/2 \\ -1/2 & 1/2 & 7/2 & -7/2 \end{bmatrix} \tag{14-9.5}$$

It is apparent from this matrix that all columns are proportional thereby indicating only one independent column and a rank of one for the matrix. This system is not observable.

Consideration of the normal form for the system under study indicates that the system equations are those of

$$\dot{\mathbf{q}} = \begin{bmatrix} -5 & 0 \\ 0 & -7 \end{bmatrix} \mathbf{q} + \begin{bmatrix} 1/2 \\ -1/2 \end{bmatrix} u$$
$$\mathbf{y} = \begin{bmatrix} 0 & 1 \\ 0 & 1 \end{bmatrix} \mathbf{q} \tag{14-9.6}$$

It can be seen that the first column of all zeros in the $\mathbf{C}_n$ matrix prevents the mode, due to the eigenvalue of -5, from appearing in the output vector, $\mathbf{y}$. Therefore observations of $\mathbf{y}$ can never give any indication of the presence of one of the modes of the system state.

Example 14-9.3. The transfer function matrix from system input to system output for the model used in Example 14-9.2 is considered in this example. For this case we have the development

$$\mathbf{Y}(s) = [C]\,[sI - A]^{-1}\,[B]\,U(s) \tag{14-9.7}$$

$$= \frac{1}{2} \begin{bmatrix} -5 & -1 \\ 5 & 1 \end{bmatrix} \begin{bmatrix} s & -1 \\ 35 & (s+12) \end{bmatrix}^{-1} \begin{bmatrix} 0 \\ 1 \end{bmatrix} U(s) \tag{14-9.8}$$

$$= \frac{1}{2(s+5)(s+7)} \begin{bmatrix} -(s+5) \\ (s+5) \end{bmatrix} \tag{14-9.9}$$

$$= \frac{1}{2(s+7)} \begin{bmatrix} -1 \\ 1 \end{bmatrix} \tag{14-9.10}$$

Again it is apparent that the cancellation of a pole and zero in the transfer function has lead to the loss of a system mode so far as the input-output relationship is concerned. However, the mode still exists internally in the system state vector and affects the internal dynamics of the system.

The general significance of the examples pertaining to transfer functions is that internal system modes may be lost if only transfer functions are used to model systems. Fortunately, most systems of practical significance are both controllable and observable, and the transfer function approach includes all of the system internal modes. Specifically, transfer functions contain only the controllable and observable states or modes of the system being considered. Based upon the controllability and observability concepts introduced in the last two sections, it is possible to divide system modes into the following four categories: (1) controllable but not observable; (2) controllable and observable; (3) not controllable but observable; and (4) not controllable and not observable.

CONCLUSION

An attempt has been made in previous sections to provide an introductory but still useful presentation of basic state variable topics. We have omitted such topics as stability considerations, numerical solution methods, optimality, and many others. Only the surface has been touched and, hopefully, the reader will pursue other sources of more advanced material. Integration of the state variable material with previous topics has been attempted in order to build a broader and more useful set of mathematical tools for the reader.

PROBLEMS

14-1 A certain system is represented by the signal-flow diagram shown.

(a) For the state variables x_1 and x_2 determine the state transition matrix $[\Phi(s)]$ using $[sI - A]^{-1}$.

(b) Find $[\Phi(s)]$, element-by-element, using signal-flow graph methods.

(c) Find $[\phi(t)]$, using $\mathcal{L}^{-1}\{[\Phi(s)]\}$.

(d) Find the forced part of the state vector response, $u(t)$ being a unit-step function.

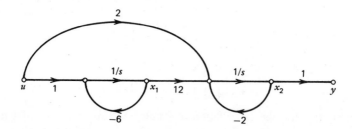

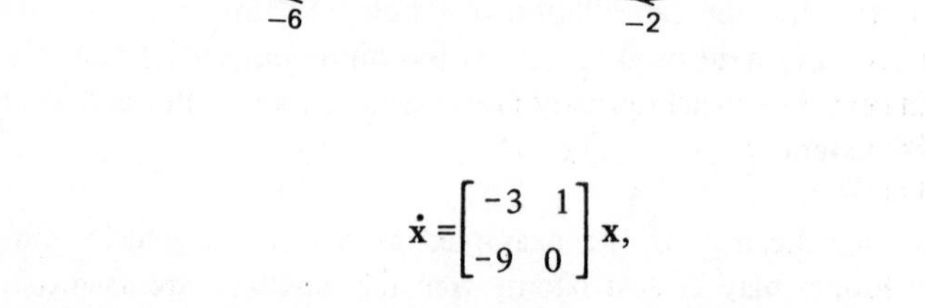

14-2 (a) If

$$\dot{\mathbf{x}} = \begin{bmatrix} -3 & 1 \\ -9 & 0 \end{bmatrix} \mathbf{x},$$

determine the closed form matrix for the state transition matrix, $\epsilon^{[A]t}$, using the Cayley–Hamilton theorem and an appropriate method of determining the coefficients in $f([A]) = k_0[I] + k_1[A]$.

(b) Repeat (a), but use $\mathcal{L}^{-1}\{[sI - A]^{-1}\}$ to find $\epsilon^{[A]t}$.

14-3 (a) If $[A]$ in $\dot{\mathbf{x}} = [A]\mathbf{x}$ is given as

$$[A] = \begin{bmatrix} 0 & 1 & 0 \\ 0 & 0 & 1 \\ -15 & -23 & -9 \end{bmatrix}$$

determine $[sI - A]^{-1} = [\Phi(s)]$.

(b) Find $\mathcal{L}^{-1}\{[\Phi(s)]\} = \epsilon^{[A]t}$ by Laplace transform inversion.

(c) Construct a signal-flow graph representing the system state variables as described by $\dot{\mathbf{x}} = [A]\mathbf{x}$.

(d) Use the signal-flow graph of (c) to find $[\Phi(s)]$, element by element.

(e) Find $\epsilon^{[A]t}$ by the Cayley–Hamilton method.

14-4 If

$$[A] = \begin{bmatrix} 0 & 1 & 0 \\ 0 & 0 & 1 \\ -3 & -3 & -1 \end{bmatrix},$$

find a closed form matrix evaluation of $\cos([A]t)$.

14-5 Determine the eigenvalues and a set of eigenvectors for the system of Problem 14-1.

14-6 (a) Use parallel programming to obtain a signal-flow simulation diagram for the the system described by the transfer function

$$\frac{Y(s)}{U(s)} = \frac{3(s+2)}{(s+1)(s+3)}.$$

Select the appropriate *normal* coordinate state variables, $\mathbf{q}$, and write the state differential equations. Verify that the state variables are normal by showing that $[A] = [\Lambda]$. What is the input matrix $[B_n]$?

(b) Determine $\epsilon^{[\Lambda]t}$.

(c) If $U(s) = 1/(s+1)$, and the initial state of the system, $q(0)$, is zero, determine the total response.

14-7 (a) Determine a set of eigenvectors for the system.

$$\dot{\mathbf{x}} = \begin{bmatrix} 0 & 1 \\ -1.5 & -2.5 \end{bmatrix} \mathbf{x}$$

(b) Sketch the eigenvectors in two-dimensional E-space.

(c) Choose $\mathbf{x}(0)$ along one of the eigenvectors, and determine $\mathbf{x}(t)$.

(d) Repeat (c) for $\mathbf{x}(0)$ along the other eigenvector.

14-8 The system shown is to be studied.

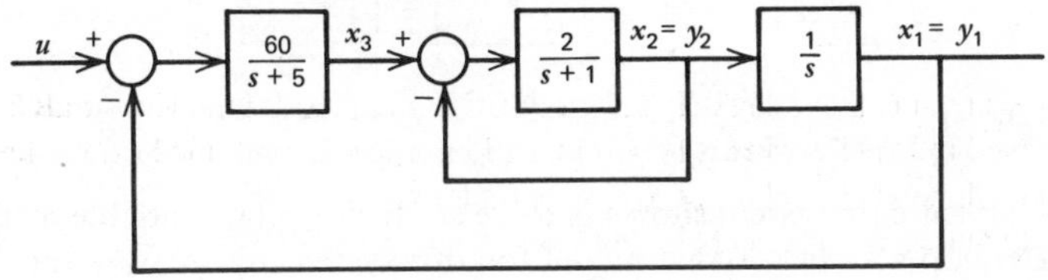

(a) Write a set of state differential equations using x_1, x_2, and x_3. Write the system output equation using y_1 and y_2 as outputs.
(b) Determine the system eigenvalues.
(c) Determine a set of system eigenvectors; use the eigenvectors to form a modal matrix, $[M]$.
(d) Use $[M]$ to transform $[A]$ into $[\Lambda]$, $[B]$ into $[B_n]$, and $[C]$ into $[C_n]$. Do not use the $[R]$ matrix of Eq. 14–5.28.

14-9 Repeat Problem 14-8, except use the transformation to $[\hat{\Lambda}]$ of Eq. 14–5.30.

14-10 A two-input, two-output system has the transfer function matrix given. Determine a state variable representation for the system, using simulation signal-flow graphs and the theory of Section 14–6.

$$\begin{bmatrix} \dfrac{1}{s+2} & \dfrac{1/2}{s+2} \\ \dfrac{3s+8}{(s+2)(s+4)} & \dfrac{(2.5s+6)}{(s+2)(s+4)} \end{bmatrix} \mathbf{U}(s)$$

14-11 A multiple input–output system is modeled by

$$\mathbf{Y}(s) = \begin{bmatrix} \dfrac{s^2 - s - 25}{s(s^2+25)} & \dfrac{s^2+s+32}{(s+7)} \\ 0 & \dfrac{7}{s(s+7)} \\ \dfrac{s^2+s+25}{s(s^2+25)} & \dfrac{s^2+s+25}{s(s^2+25)} \end{bmatrix} \mathbf{U}(s)$$

(a) Determine the system modes.
(b) Determine the rank of the coefficient matrix for each mode. What minimum number of state variables are needed to model the system.
(c) Develop a simulation diagram for the system using a minimum number of state variables, and write the state differential equations and output equations.

14-12 Develop a state variable model for the system shown, and obtain the solution for at least three sampling periods. Assume the sampler starts at $t = 0$ and that only the initial condition $x_2(0^+) = 1.0$ drives the system.

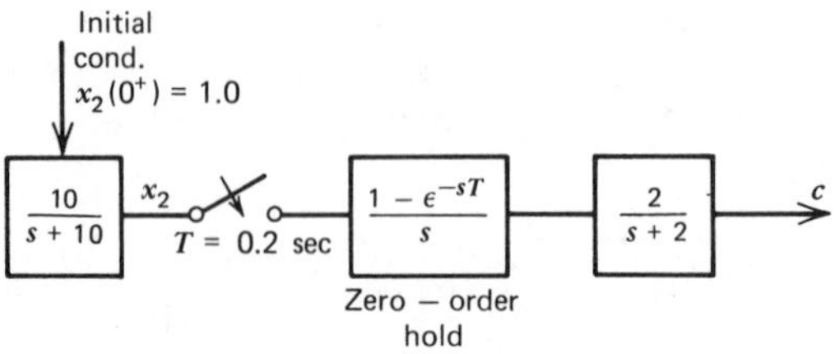

14-13 The sampled-data system shown is to be studied by state variable methods.
(a) Develop a state variable model for this system by treating the output of the zero-order hold as a piecewise constant variable determined by $c(t)$. With this approach, $e(t) = r(t) - c(kT)$, $kT \leqslant t < (\overline{k+1}\,T)$, and the system runs *open loop* between sampling intervals.

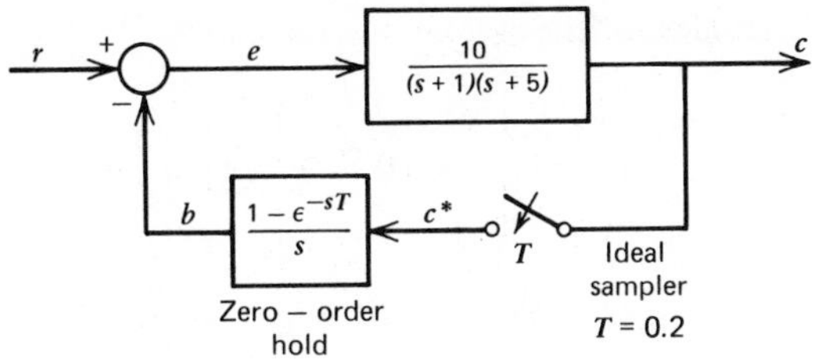

(b) Develop a computer program for solving for $c(t)$, and obtain the solution for at least five sampling periods. $r(t) = u_{-1}(t)$

(c) Reformulate the system, treating the output of the zero-order hold as a third state variable.

14–14 A discrete system is represented by the difference equation

$$y(k+2) + 11y(k+1) + 28y(k) = u(k).$$

Develop a discrete state equation model for the system, and obtain the solution for the system state for at least five steps, when $u(k) = 1.0$, $k = 0\ 1, 2, \ldots$, and $y(1) = y(0) = 0.0$.

14–15 Develop a state equation model for the system represented by the difference equation

$$y(k+3) + 3y(k+2) + 3y(k+1) + y(k) = 2u(k+1) + u(k)$$

and obtain a solution for at least five steps, when $y(2) = y(1) = y(0) = 0.0$; $u(k) = \epsilon^{-2k}$; $k = 0, 1, 2, \ldots$

14–16 Determine whether the system model

$$\dot{\mathbf{x}} = \begin{bmatrix} -5 & -4 \\ 0 & -1 \end{bmatrix} \mathbf{x} + \begin{bmatrix} 3 \\ 0 \end{bmatrix} u$$

is controllable. If not controllable, what mode(s) pose the problem?

14–17 A parallel programming model for a system is shown. Describe the modes relative to controllability and observability, and develop the transfer function $Y(s)/U(s)$.

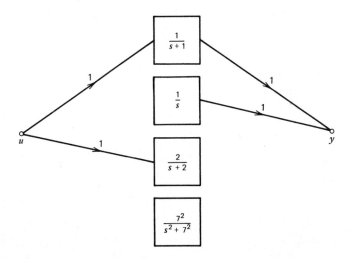

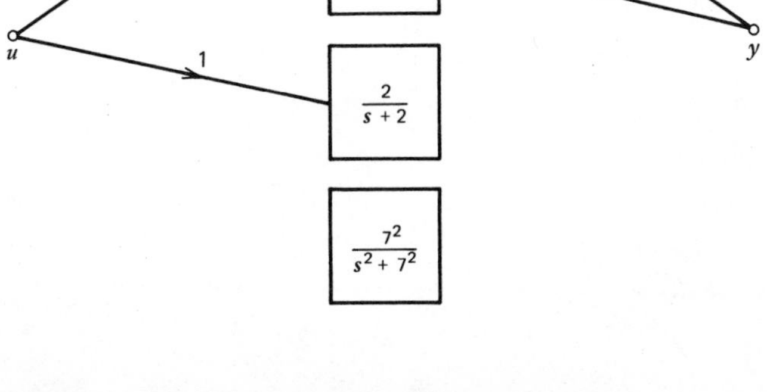

14-18 Determine the modes for the system described by

$$\dot{\mathbf{x}} = \begin{bmatrix} -1 & 0 & 0 \\ 0 & 0 & 0 \\ 0 & -2 & -2 \end{bmatrix} \mathbf{x} + \begin{bmatrix} 1 \\ 0 \\ 2 \end{bmatrix} u$$

$$y = [1 \quad 1 \quad 0]\, \mathbf{x}$$

Is this system controllable? Is it observable? Transform the system to normal coordinates, and discuss the controllable and observable aspects of the system.

Appendix **A**

The Impulse Family of Functions

A-1. INTRODUCTION

The impulse family of functions serves a useful purpose in many fields of applied science and supplies coherence to many concepts that otherwise would seem somewhat fragmented.

A-2. A FAMILY OF FUNCTIONS

THE $u_0(t)$ FUNCTION

Figure A-2.1(a) shows a rectangle with a height of $1/\delta$ and a base of δ, so that the area under the curve is unity. If δ becomes smaller, the base of the rectangle becomes smaller and in the limit as $\delta \to 0$ the height increases without bound, but the area still equals unity and the result is called a unit impulse. If the area were K, the result would be called an impulse of value K, or of weight K. Because it is impossible to show literally an infinite height and a zero width, (b) represents the unit impulse occurring at $t = 0$. The arrow pointing up, with the ∞ mark above, indicates that the function goes to infinity, and the unity symbol inside brackets indicates the area under the curve to be unity. The symbol $u_0(t)$ is used to represent this function. In (c) is indicated the impulse of weight K, occurring at time T; and the symbol $K u_0(t - T)$ represents this function.

The wave shape in Fig. A-2.1(a) is not the only one that could be used to introduce the impulse function. Any wave shape whose area remains at unity as the base approaches zero would do equally well.

There are a number of situations in which the use of the impulse may be confusing. Since the impulse function is the result of a limiting process, one way to eliminate con-

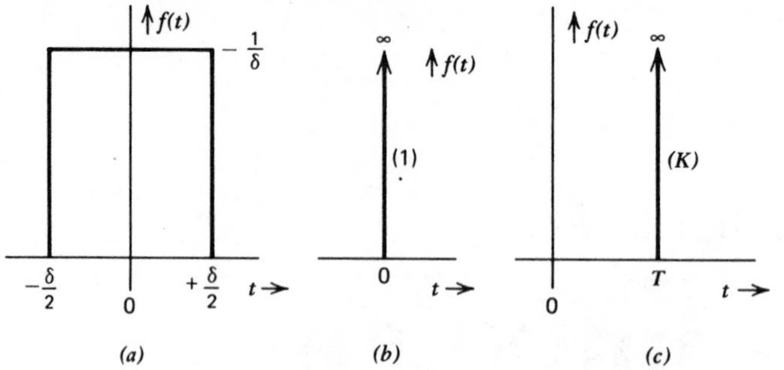

Fig. A-2.1

fusion is to let δ take on some arbitrary small value. After this, letting δ approach zero should clarify the use of the impulse. If it should still be confusing, let δ remain arbitrarily small and do not bother with the limiting process.

The point of view here is that the impulse $u_0(t)$ occurs at $t = 0$. At $t = 0-$ (the limiting value of t, as t approaches zero through negative values) the impulse has not occurred. At $t = 0+$ (the limiting value of t, as t approaches zero through positive values) the impulse has already occurred.

If the statements in the last paragraph need clarifying, follow the suggestion just made and replace the impulse in Fig. A-2.1(b) with the pulse in (a); then let δ become arbitrarily small, but not take on the limiting value of 0. The statements of the last paragraph now read, at $t = -\delta/2$ the pulse has not occurred, the pulse occurs between $t = -\delta/2$ and $t = +\delta/2$, and at $t = +\delta/2$ the pulse has occurred.

THE $u_{-1}(t)$ FUNCTION

The unit impulse is shown again in Fig. A-2.2(a), and the integral of $u_0(t)$ is in (b). To visualize this, think about moving a point across the curve in (a), and picture the integral as being the area under the curve to the left of the point as the point moves from $(-\infty)$ to $(+\infty)$. The area under the curve from $(-\infty)$ to $(0-)$ is zero. At $t = 0$ the impulse is passed, contributing an area of unity, and from $(0+)$ to $(+\infty)$ no additional area is added. This function is called the unit-step function and is denoted by $u_{-1}(t)$. This notation indicates that the $u_0(t)$ function has been integrated one time. The step function $u_{-1}(t)$

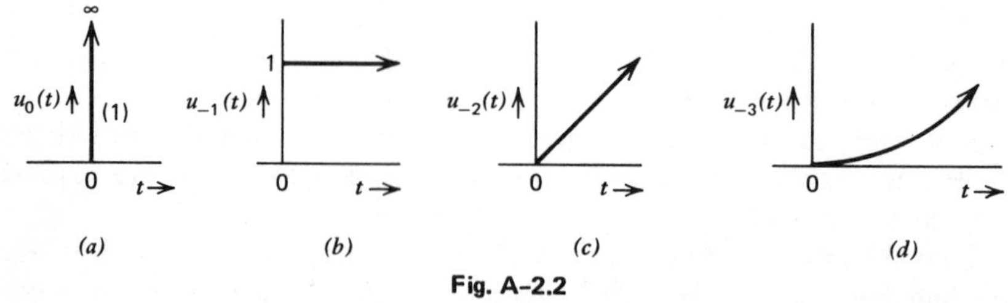

Fig. A-2.2

can be defined as

$$u_{-1}(t) = \begin{cases} 0 & t<0 \\ 1 & t>0 \end{cases} \tag{A-2.1}$$

If a step of height K occurs at $t = T$, the function becomes

$$Ku_{-1}(t - T) = \begin{cases} 0 & t<T \\ K & t>T \end{cases} \tag{A-2.2}$$

USES OF THE UNIT-STEP FUNCTION

The unit-step function is used often in a descriptive way. For example, a constant voltage source applied to a circuit at $t = 0$ can be expressed as

$$e_s = Eu_{-1}(t) \tag{A-2.3}$$

which can also be written as

$$e_s = \begin{cases} 0 & t<0 \\ E & t>0 \end{cases} \tag{A-2.4}$$

A similar example would be in the impressing of a constant current source onto a circuit at $t = 0$. This is written as

$$i_s = Iu_{-1}(t) \tag{A-2.5}$$

As an additional example, a force might be applied to a mechanical system. The function being applied need not be a constant, but can be any function of time. For example, an applied sinusoidal forcing function is given by

$$f(t) = F_m \sin(\omega t + \lambda)\, u_{-1}(t) \tag{A-2.6}$$

which is a shorthand way of expressing

$$f(t) = \begin{cases} 0 & t<0 \\ F_m \sin(\omega t + \lambda) & t>0 \end{cases} \tag{A-2.7}$$

where λ is the point in the cycle of the sinusoid at which $t = 0$ occurs.

THE $u_{-2}(t)$ FUNCTION

The next question to ask is, what is the integral of $u_{-1}(t)$? We again visualize the result by using the graphical interpretation of integration. As we move a point from $(-\infty)$ to (0) across the curve of Fig. A-2.2(b), the area under the curve is zero, and hence the integral of $u_{-1}(t)$ is zero. As we move the point across the $u_{-1}(t)$ curve from $(0+)$ to $(+\infty)$ the area to the left of the point is equal to t. The resulting function is shown in (c) and is called the *unit-ramp function.* The symbol for this function is $u_{-2}(t)$, which indicates that the $u_0(t)$ function has been integrated two times. The $u_{-2}(t)$ is a shorthand way of expressing

$$u_{-2}(t) = \begin{cases} 0 & t < 0 \\ t & t > 0 \end{cases} \tag{A-2.8}$$

which can also be written as

$$u_{-2}(t) = t u_{-1}(t) \tag{A-2.9}$$

THE $u_{-3}(t)$ FUNCTION

To continue the discussion, we inquire about the integral of the $u_{-2}(t)$. To be different, here we use an analytical approach. The first tendency would be to write such an equation as

$$u_{-3}(t) = \int_{-\infty}^{t} u_{-2}(t)\, dt \tag{A-2.10}$$

To interpret this equation, let us use the definitions of $u_{-2}(t)$ as given by Eqs. A-2.8, and rewrite Eq. A-2.10 for $t < 0$ as

$$u_{-3}(t) = \int_{-\infty}^{t} 0\, dt = 0 \tag{A-2.11}$$

and for $t > 0$ as

$$u_{-3}(t) = \int_{-\infty}^{0} 0\, dt + \int_{0}^{t} t\, dt = \frac{t^2}{2} \tag{A-2.12}$$

Hence, $u_{-3}(t)$ can be written as

$$u_{-3}(t) = \frac{t^2}{2} u_{-1}(t) \tag{A-2.13}$$

THE $u_{-n}(t)$ FUNCTION

Each member of this family of functions can be obtained by integrating the function just before it. By inspection, the next function is

$$u_{-4}(t) = \frac{t^3}{3 \cdot 2} u_{-1}(t) \tag{A-2.14}$$

The general term is

$$u_{-n}(t) = \frac{t^{n-1}}{(n-1)!} u_{-1}(t) \tag{A-2.15}$$

DIFFERENTIATING THE IMPULSE FAMILY OF FUNCTIONS

If we can move from one member of this family to the next member by integration, we certainly can go in the opposite direction by the use of differentiation. As examples: $u_{-4}(t)$ is the derivative of $u_{-5}(t)$; the $u_0(t)$ is the derivative of the $u_{-1}(t)$; etc.

If we begin at $u_0(t)$ and integrate, we generate the $u_{-n}(t)$ part of this infinite family of functions. Likewise we could begin at $u_0(t)$ and differentiate to obtain the remainder of these functions. The first derivative of $u_0(t)$ is denoted by $u_{+1}(t)$, and the second derivative by $u_{+2}(t)$. The nth derivative of $u_0(t)$ is denoted by $u_{+n}(t)$. We will not attempt to give a physical interpretation to these functions here, but this is done in Chapter 4.

CONCLUSION

If we lived in a perfect world, everyone would agree on one set of symbols to denote this family of functions. As you might guess, this is not the case. Many use the symbol $u(t)$ to have the same meaning as $u_{-1}(t)$, and the symbol $\delta(t)$ to indicate $u_0(t)$. If these authors differentiate $\delta(t)$, or integrate $u(t)$, other symbols have to be invented. The advantage of the $u_{+n}(t)$, $u_0(t)$, $u_{-n}(t)$ notation is that the symbol itself describes the function being used.

The impulse function was first proposed by the physicist Dirac and is quite often referred to as the *Dirac δ-function.* Until recently, this function has been used primarily by people interested in applications. The validity of $\delta(t)$, and the manner of its use, had been disputed by mathematicians. It can be argued that the impulse function is not a function in the true meaning of the word, and some mathematicians have called this impulse family a *family of fiction.* Recently, a concept called *the theory of distributions* has been developed by L. Schwarz and others, and their theory, beyond the scope of this book, places a rigorous foundation under the use of the impulse function.

Appendix B

Hyperbolic Functions

B-1. INTRODUCTION

In this book, the hyperbolic functions are used to help show the relationships among the solutions of systems that are overdamped, critically damped, and oscillatory.

B-2. HYPERBOLIC FUNCTIONS VS. CIRCULAR FUNCTIONS

Hyperbolic functions are related to the hyperbola in a manner somewhat similar to the way circular functions are related to the circle. The geometry of hyperbolic functions is not of interest here, but the analytic similarities and differences between the hyperbolic and the circular functions are useful. Euler's relation states

$$\epsilon^{j\beta t} = \cos \beta t + j \sin \beta t \qquad \epsilon^{-j\beta t} = \cos \beta t - j \sin \beta t \tag{B-2.1}$$

which yield the relationships

$$\cos \beta t = \frac{\epsilon^{+j\beta t} + \epsilon^{-j\beta t}}{2} \qquad \sin \beta t = \frac{\epsilon^{+j\beta t} - \epsilon^{-j\beta t}}{2j} \tag{B-2.2}$$

In a manner similar to Eqs. B-2.2, the hyperbolic cosine and sine functions can be defined as

$$\cosh bt = \frac{\epsilon^{bt} + \epsilon^{-bt}}{2} \qquad \sinh bt = \frac{\epsilon^{bt} - \epsilon^{-bt}}{2} \tag{B-2.3}$$

A few other hyperbolic functions are given by

$$\tanh bt = \frac{\sinh bt}{\cosh bt} \qquad \coth bt = \frac{1}{\tanh bt}$$

$$\operatorname{sech} bt = \frac{1}{\cosh bt} \qquad \operatorname{csch} bt = \frac{1}{\sinh bt}$$

Equations B-2.3 then yield the relationships

$$\epsilon^{bt} = \cosh bt + \sinh bt \qquad \epsilon^{-bt} = \cosh bt - \sinh bt \tag{B-2.4}$$

Often in circuit work, b is imaginary; thus $b = j\beta$, where β is real. In such a situation, the hyperbolic functions become

$$\cosh bt = \frac{\epsilon^{bt} + \epsilon^{-bt}}{2} = \cosh j\beta t = \frac{\epsilon^{j\beta t} + \epsilon^{-j\beta t}}{2} = \cos \beta t \tag{B-2.5}$$

$$\sinh bt = \frac{\epsilon^{+bt} - \epsilon^{-bt}}{2} = \sinh j\beta t = \frac{\epsilon^{j\beta t} - \epsilon^{-j\beta t}}{2} = j\left[\frac{\epsilon^{j\beta t} - \epsilon^{-j\beta t}}{2j}\right] = j \sin \beta t \tag{B-2.6}$$

Similarly, β may be imaginary, or $\beta = +jb$, where b is real, and the circular functions become

$$\cos \beta t = \frac{\epsilon^{j\beta t} + \epsilon^{j\beta t}}{2} = \cos jbt = \frac{\epsilon^{j(jb)t} + \epsilon^{-j(jb)t}}{2} = \frac{\epsilon^{-bt} + \epsilon^{bt}}{2} = \cosh bt \tag{B-2.7}$$

$$\sin \beta t = \frac{\epsilon^{+j\beta t} - \epsilon^{j\beta t}}{2j} = \sin jbt = \frac{\epsilon^{j(jb)t} - \epsilon^{-j(jb)t}}{2j} = j\left[\frac{\epsilon^{bt} - \epsilon^{-bt}}{2}\right] = j \sinh bt \tag{B-2.8}$$

By use of the preceding equations, a complete set of identities for hyperbolic functions may be derived from known circular functions. In the following example, the circular function identity for the sum of two angles is the starting point.

$$\sin(\beta_1 t + \beta_2 t) = \sin \beta_1 t \cos \beta_2 t + \sin \beta_2 t \cos \beta_1 t \tag{B-2.9}$$

Suppose that both β_1 and β_2 become imaginary, or

$$\beta_1 = jb_1 \qquad \beta_2 = jb_2$$

The equation becomes

$$\sin j(b_1 t + b_2 t) = \sin jb_1 t \cos jb_2 t + \sin jb_2 t \cos jb_1 t$$

or

$$j \sinh(b_1 t + b_2 t) = j \sinh b_1 t \cosh b_2 t + j \sinh b_2 t \cosh b_1 t$$

and, finally, the corresponding hyperbolic identity is

$$\sinh(b_1 t + b_2 t) = \sinh b_1 t \cosh b_2 t + \sinh b_2 t \cosh b_1 t \tag{B-2.10}$$

As a second example, the circular function identity is the starting point:

$$\cos(\beta_1 t + \beta_2 t) = \cos \beta_1 t \cos \beta_2 t - \sin \beta_1 t \sin \beta_2 t \tag{B-2.11}$$

Again suppose that β_1 and β_2 become imaginary. This equation then becomes

$$\cos j(b_1 t + b_2 t) = \cos jb_1 t \cos jb_2 t - \sin jb_1 t \sin jb_2 t$$

$$\cosh (b_1 t + b_2 t) = \cosh b_1 t \cosh b_2 t - j^2 \sinh b_1 t \sinh b_2 t$$

and finally the corresponding hyperbolic identity is

$$\cosh (b_1 t + b_2 t) = \cosh b_1 t \cosh b_2 t + \sinh b_1 t \sinh b_2 t \qquad \text{(B-2.12)}$$

The derivatives for sinh bt and cosh bt may be obtained from the exponential definitions

$$\frac{d}{dt} [\cosh bt] = \frac{d}{dt} \left[\frac{\epsilon^{bt} + \epsilon^{-bt}}{2} \right] = b \left[\frac{\epsilon^{bt} - \epsilon^{-bt}}{2} \right] = b \sinh bt \qquad \text{(B-2.13)}$$

$$\frac{d}{dt} [\sinh bt] = \frac{d}{dt} \left[\frac{\epsilon^{bt} - \epsilon^{-bt}}{2} \right] = b \left[\frac{\epsilon^{bt} + \epsilon^{-bt}}{2} \right] = b \cosh bt \qquad \text{(B-2.14)}$$

When other identities are needed, they are derived in a similar manner.

Appendix **C**

Independence Among Circuit Variables

C-1. INTRODUCTION

An area of concern in systems theory is the writing of a set of equations that can be solved to yield the desired responses. In circuit theory, however, the methods of writing such equations are well developed[1]; and through the use of analogues, these methods can be extended to other systems.

C-2. LINEAR INDEPENDENCE

The circuit of Fig. C-2.1(a) contains six elements, two voltage sources, and one current source. Each element is called a *branch*, and a point where two or more branches are connected is called a *node*. There are two possible methods to pursue, and they give the same result as will be shown. One method counts sources as if they were branches; using this method, the circuit in (a) contains 9 branches and 6 nodes. The other method removes the sources, leaving only their internal impedances, which for a voltage source is a short and for a current source is an open; with the sources removed, circuit (a) reduces to (b) and has 6 branches and 4 nodes. We begin the discussion with circuits of the second type.

To make the presentation more general, we picture a circuit with b branches and with n_t nodes.[2] The branch currents are indicated by the symbol j. When the numbering and the direction of the j's are chosen, the circuit is said to be oriented. The v's and j's are

[1] For a more detailed discussion, see E. A. Guillemin, *Introductory Circuit Theory*, John Wiley & Sons, Inc., New York, 1953, pp. 5-63.

[2] The notation b, n_t, and some of the other symbols used here are not universally accepted; several other systems of notation appear in the literature.

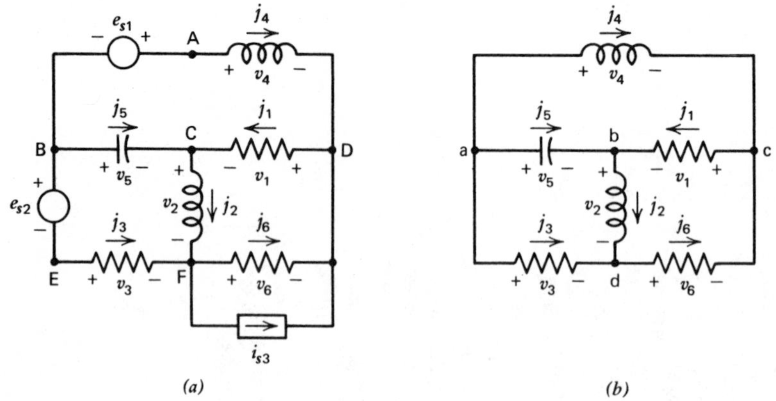

Fig. C-2.1

related through the volt-ampere relationship for the type of element involved. The circuit has b unknown currents and b unknown voltages, or a total of $2b$ unknowns. Many of these unknowns are interrelated, and if a proper subset of these $2b$ unknowns is determined, the remaining unknowns can be found from this subset. If, for example, the b branch currents are determined, the b branch voltages can be found by using the volt-ampere relationships.

Next we explore the concept that if a proper subset of the b branch currents is known, the remaining branch currents can be found. The first group is said to be an *independent set*, and the second a *dependent set.*

C-3. THE INDEPENDENT SET OF BRANCH CURRENTS

The circuit of Fig. C-2.1(b) is used. Suppose somehow we can determine the current j_4. Since j_4 is known, as far as determining any of the remaining branch currents is concerned, j_4 can be replaced with a current source as in Fig. C-3.1(a), and j_4 becomes one member of an independent set. Having j_4 known, does not in itself determine any of the remaining branch currents, and we are free to select any one of the remaining 5 currents as the next known current. The current j_6 is so selected and is replaced with a current source as in (b). With j_4 and j_6 fixed, j_1 is determined and cannot be selected as a member of the independent set. However, any one of the other three currents can be selected, and j_5 is chosen and is replaced with a current source as in (c).

The fact that the three current sources in (c) control the remaining three branch currents is shown in (d), where each of the current sources is set equal to zero, leaving only its internal impedance, which is infinite. Of the six branch currents, j_4, j_5, and j_6 form one independent set, and j_1, j_2, and j_3 form the corresponding dependent set.

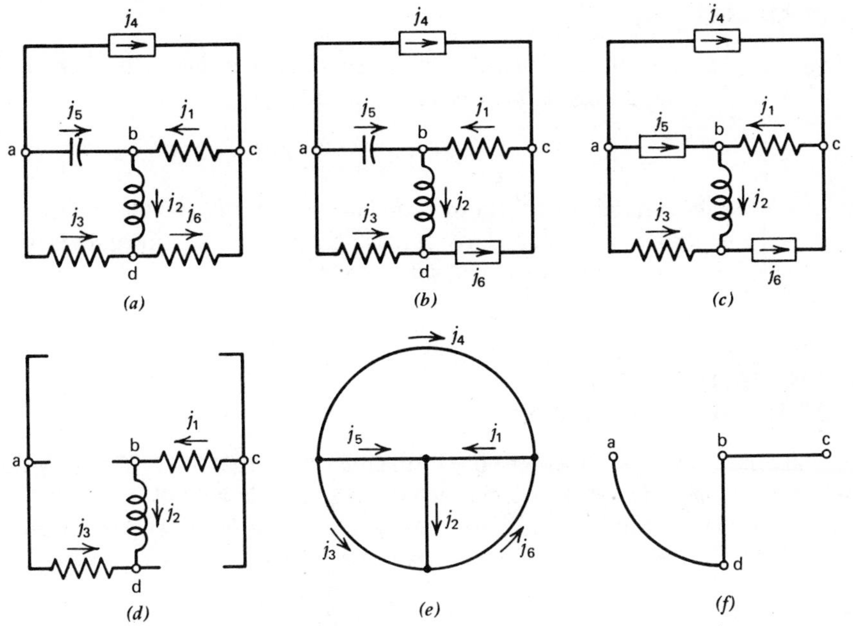

Fig. C-3.1

C-4. GRAPH OF A CIRCUIT

The preceding discussion is not concerned with the nature of the elements, but only with the manner in which they are connected. To focus attention on the interconnections, each element can be shown by a line, and the result is called *the graph of the circuit*. The graph and the orientation of the circuit of Fig. C-2.1(b) are shown in Fig. C-3.1(e), and the graph of (d) is shown in (f) which is called *the tree of the graph.*

THE TREE OF THE GRAPH

The graph in Fig. C-3.1(f) is said to be a *tree*, and its branches *tree branches*, because of the treelike properties of the graph and because the branches form no closed loops on circuits.

The tree can be constructed in the following manner. Branches from the original graph are added until all the nodes are connected with no closed loops. Since this procedure is not unique, a graph can have more than one tree. For example the graph in Fig. C-3.1(e) has 16 distinct trees.

The number of tree branches for a connected graph with n_t nodes is

$$n = n_t - 1 \tag{C-4.1}$$

The first branch added connects two nodes, and each additional branch connects only one additional node.

LINK BRANCHES

The branches that are not tree branches are called *link branches*, and there are l of these.[3] Since all b branches are either tree or link branches

$$b = n + l \qquad l = b - n = b - n_t + 1 \tag{C-4.2}$$

Of the b branch currents, l of them form an independent set, and the remaining n of them form the dependent set. Not just any l branch currents can be chosen, because they must be the link branch currents associated with some tree.

C-5. THE INDEPENDENT SET OF BRANCH VOLTAGES

Suppose in some manner we determine one of the branch voltages of Fig. C-2.1(b). So far as the other voltages are concerned, this voltage can be replaced with a voltage source. The voltage v_1 is used and is replaced with a source as in Fig. C-5.1(a). The voltage v_2 is chosen next and is replaced with a voltage source as in (b). With v_1 and v_2 fixed, v_6 is

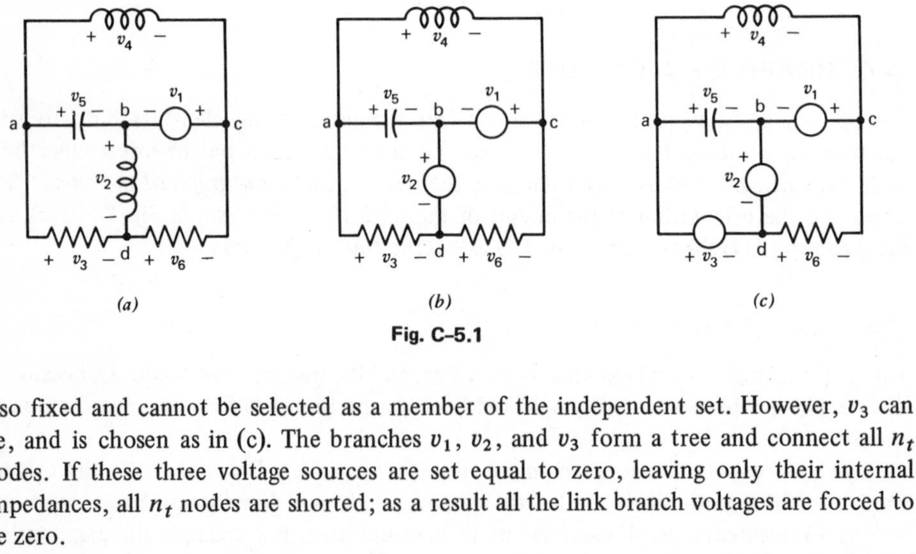

Fig. C-5.1

also fixed and cannot be selected as a member of the independent set. However, v_3 can be, and is chosen as in (c). The branches v_1, v_2, and v_3 form a tree and connect all n_t nodes. If these three voltage sources are set equal to zero, leaving only their internal impedances, all n_t nodes are shorted; as a result all the link branch voltages are forced to be zero.

Of the b branch voltages, n of them form an independent set, and the remaining l of them form the dependent set. Not just any n branch voltages can be selected, because they must be the voltages associated with the branches of some tree.

[3] The set of link branches is sometimes referred to as a *cotree*.

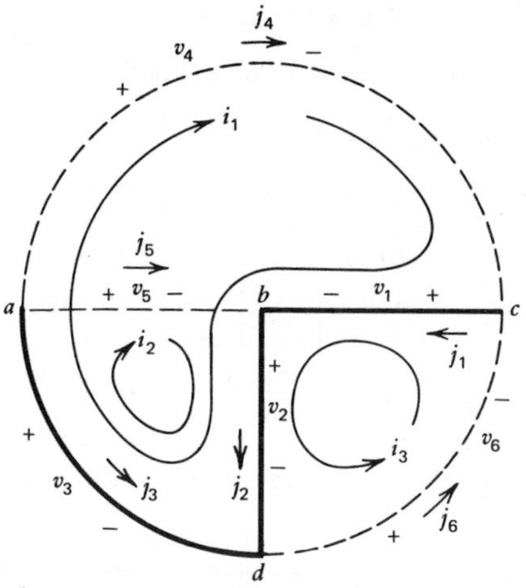

Fig. C-6.1

C-6. THE CIRCUIT-SET SCHEDULE

The same circuit is used again, and its graph is shown in Fig. C-6.1. The tree branches are shown as solid lines, and the link branches as dashed lines.

The link branch currents can be interpreted as being loop currents in the following way. Each link branch is added to the tree separately. For example, add the j_4 branch to the tree of Fig. C-6.1. The tree and this j_4 branch form one closed path, around which the current j_4 can exist. This loop current is i_1, shown in Fig. C-6.1. The link branch j_4 is removed, link branch j_5 is added to the tree, thus forming only one closed path, and the loop current i_2 is determined. In a similar manner the branch j_6 and the tree determine i_3. The direction of the j currents determine the direction of the loop currents, as shown by

$$j_4 = i_1 \qquad j_5 = i_2 \qquad j_6 = i_3 \tag{C-6.1}$$

The relationship among the loop and branch currents is shown in a compact form in a *circuit-set schedule* (sometimes called a *tie-set schedule*). Figure C-6.2 is such a schedule for the graph and tree of Fig. C-6.1. We develop this schedule by rows. If the loop current coincides in direction with a branch current, a (+1) is entered in the schedule. If the loop current is in the opposite direction, a (−1) is entered. If no loop current exists in a specific branch, a 0 is used. For example, *row 1* shows that i_1 is in the same direction as the branch currents j_1, j_2 and j_4, that it is in the opposite direction of j_3, and that it does not exist in the j_5 and j_6 branches. The other rows are developed in a similar manner.

The columns of the schedule show the relationships among the branch currents and the loop currents, which in turn show how the tree branch currents can be determined from the link branch currents. The last three columns yield the equations already given as

Circuit-Set Schedule

Loop paths defined by	Branches defined by					
	$j1$	$j2$	$j3$	$j4$	$j5$	$j6$
i_1	+1	+1	−1	+1	0	0
i_2	0	+1	−1	0	+1	0
i_3	+1	+1	0	0	0	+1

Fig. C-6.2

Eqs. C-6.1. The following equations came from the first three columns.

$$j_1 = i_1 + i_3 = j_4 + j_6$$
$$j_2 = i_1 + i_2 + i_3 = j_4 + j_5 + j_6 \quad \text{(C-6.2)}$$
$$j_3 = -i_1 - i_2 = -j_4 - j_5$$

Since these equations can be obtained from the graph of Fig. C-6.1, this demonstrates that the schedule of Fig. C-6.2 can also be developed by columns.

Each row of the schedule describes the branches' encounter around a closed path, and a Kirchhoff's Voltage Law equation can be written around the loops as

$$v_1 + v_2 - v_3 + v_4 = 0$$
$$v_2 - v_3 + v_5 = 0 \quad \text{(C-6.3)}$$
$$v_1 + v_2 + v_6 = 0$$

The equations can be solved for the link branch voltages, in terms of the tree branch voltages, as

$$v_4 = -v_1 - v_2 + v_3$$
$$v_5 = -v_2 + v_3 \quad \text{(C-6.4)}$$
$$v_6 = -v_1 - v_2$$

Equations C-6.2 show how the dependent branch currents can be found in terms of the independent branch currents, and Eqs. C-6.4 show how the dependent branch voltages can be found in terms of the independent branch voltages.

The information in the Fig. C-6.2 schedule is shown in matrix form as

$$[\beta] = \left[\begin{array}{ccc|ccc} 1 & 1 & -1 & 1 & 0 & 0 \\ 0 & 1 & -1 & 0 & 1 & 0 \\ 1 & 1 & 0 & 0 & 0 & 1 \end{array}\right] = [\beta_T \mid \beta_L] \quad \text{(C-6.5)}$$

The $[\beta_T]$ is the tree portion, and $[\beta_L]$ the link portion, of $[\beta]$. Based on the manner in which the matrix is developed, $[\beta_L]$ is a unit-matrix.

C-7. THE CUT-SET SCHEDULE

Equations C-6.1 are used to emphasize that the link branch currents form an independent set, by replacing the j's with i's. Similarly, the v's of the tree-branch voltages can be replaced with e's, as given by

$$v_1 = e_1 \qquad v_2 = e_2 \qquad v_3 = e_3 \tag{C-7.1}$$

Figure C-7.1 is a schedule developed for the graph of Fig. C-6.1. If any one of the three tree branch voltages is non-zero, and the other two voltages are set equal to zero, the four nodes are divided into two subsets; one subset is at the (+) end of the non-zero voltage, and the other is at the (−) end. The nodes at the plus end are identified in the schedule.

Cut-Set Schedule

Cut-sets defined by	Branches defined by						Nodes at the (+) end of a cut-set
	v_1	v_2	v_3	v_4	v_5	v_6	
e_1	+1	0	0	−1	0	−1	c
e_2	0	+1	0	−1	−1	−1	b, c
e_3	0	0	+1	+1	+1	0	a

Fig. C-7.1

The term *cut-set* can be given the following physical interpretation. A group of branches exist between the two subsets of nodes; and if these branches are *cut*, the circuit is divided into two parts. Each such set of branches is a cut set based on some tree.

The rows of the schedule are an expression of Kirchhoff's Current Law between the two subsets of nodes or the currents in the corresponding cut-set. For example, the e_2 row results from shorting e_1 and e_3. Nodes b and c are at the (+) end of e_2, and nodes a and d are at the (−) end. The current equation for the branch currents going from the (+) end of e_2 toward the (−) end of e_2 is

$$j_2 - j_4 - j_5 - j_6 = 0 \tag{C-7.2}$$

The other two rows of the schedule come from

$$j_1 - j_4 - j_6 = 0 \qquad j_3 + j_4 + j_5 = 0 \tag{C-7.3}$$

Equations C-7.2 and 3 can be solved for the tree branch currents in terms of the link branch currents, as

$$j_1 = j_4 + j_6 \qquad j_2 = j_4 + j_5 + j_6 \qquad j_3 = -j_4 - j_5 \tag{C-7.4}$$

The first three columns of the cut-set schedule yield Eqs. C-7.1, and the last three yield

$$\begin{aligned} v_4 &= -e_1 - e_2 + e_3 = -v_1 - v_2 + v_3 \\ v_5 &= -e_2 + e_3 = -v_2 + v_3 \\ v_6 &= -e_1 - e_2 = -v_1 - v_2 \end{aligned} \tag{C-7.5}$$

Since these equations can be obtained from the graph of Fig. C-6.1, this demonstrates that the schedule of Fig. C-7.1 can also be developed by columns. The information in the Fig. C-7.1 schedule can be written in matrix form as

$$[\alpha] = \left[\begin{array}{ccc|ccc} 1 & 0 & 0 & -1 & 0 & -1 \\ 0 & 1 & 0 & -1 & -1 & -1 \\ 0 & 0 & 1 & 1 & 1 & 0 \end{array}\right] = [\alpha_T \mid \alpha_L] \tag{C-7.6}$$

The $[\alpha_T]$ is the tree portion and is a unit matrix. The $[\alpha_L]$ is the link portion.

Equations C-6.4 from the rows of $[\beta]$ are equal to Eqs. C-7.5 from the columns of $[\alpha]$. Equations C-6.2 from the columns of $[\beta]$ are equal to Eqs. C-7.4 from the rows of $[\alpha]$. Therefore the rows (columns) of $[\beta]$ contain the same information as the columns (rows) of $[\alpha]$. These equations are a specific example of a general property given by

$$[\beta_T] = -[\alpha_L]^t \tag{C-7.7}$$

Expressed in words: if the branch numbering and the choice of the tree are the same for $[\beta]$ and $[\alpha]$, the tree portion of $[\beta]$ is the negative transpose of the link portion of $[\alpha]$.

C-8. THE 2*b* EQUATIONS

As discussed earlier, a circuit has a total of $2b$ unknowns, and $2b$ equations must be written to solve for the unknowns. As discussed in Section C-6, l independent KVLE's can be written around the l loops, and as discussed in Section C-7, n independent KCLE's can be written for the n cut-sets. In addition, there are the b volt-ampere relationships. These equations add up to $2b$, as

$$l + n + b = 2b \tag{C-8.1}$$

These are the $2b$ equations from which the $2b$ unknowns can be found; they may be solved in many different ways. If the VAR's and the KCLE's are substituted into the KVLE's, the result is the l loop current equations. If the VAR's and the KVLE's are substituted into the KCLE's, the result is the n node pair voltage equations. Alternately, the two procedures just discussed can be blended by saving some of the KCLE's and some of the KVLE's, with the results referred to as equations in the mixed form. One application of equations in the mixed form is the writing of state variable equations.

C-9. THE WINDOW (OR MESH) CURRENT EQUATIONS

If the graph of a circuit is flat,[4] the graph has the aspect of a window consisting of many small panes. A little reflection reveals that any branch current can be determined from a

[4] A flat graph is one that can be drawn of a flat surface or on a sphere without any of the lines crossing.

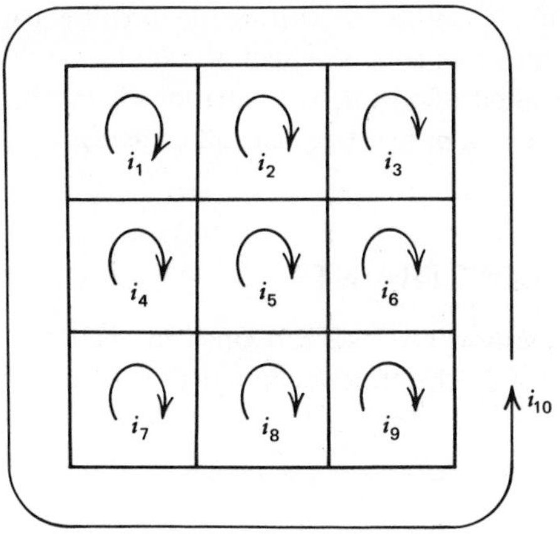

Fig. C-9.1

set of currents that exist in the branches surrounding the pane. The currents i_1 through i_9 in Fig. C-9.1 are the choices usually made for window (or mesh) currents. These currents have the property that not more than two currents exist in any one branch, and these will be in opposite directions. With these thoughts in mind, the current i_{10} is added. To generalize this, a flat graph with l loops has $(l + 1)$ possible window currents. Since any of these window currents can be chosen as reference, there are $(l + 1)$ possible sets of window current equations that can be used. For the example being considered $l + 1 = 10$. The use of one of the $(l + 1)$ loops as reference, parallels the choosing of one of the $n_t = (n + 1)$ nodes as reference, as discussed in the next section.

Window currents cannot be used on a non-flat graph; however, loop currents can be used on either a flat graph or a non-flat graph. Therefore, when general properties of circuits are being explored, loop currents should be used.

C-10. NODE-TO-DATUM VOLTAGES

If one of the n_t nodes of a graph is chosen as reference (or datum), the voltages from the other n nodes to this datum also form a sufficient set from which all b branch voltages can be determined. Therefore there are $(n + 1) = n_t$ possible sets of node-to-datum equations that can be written. Each branch exists between two nodes. Each branch voltage can be found from the two node-to-datum voltages associated with these two nodes. If one end of a branch is connected to datum, this branch voltage can be found from one node-to-datum voltage.

The node-to-datum voltages can be used on a non-flat graph as well as on a flat graph, and in this sense they are as general as a set of tree-branch voltages.

C-11. STATE-VARIABLE EQUATIONS

One method of setting up the state-variable equations is to place all voltage sources and capacitors in tree branches and all current sources and inductors in link branches. Therefore in part, the choice of the method of writing the circuit equations is dictated by the location of these sources and elements. For such situations, neither the window-current nor the node-to-datum method of writing the equations is satisfactory. However, the $2b$ method based on the choice of a proper tree can still be used.

C-12. SOURCES AS CONSTRAINTS

We are now back to a question that was left open in Section C-2. The circuit of Fig. C-2.1(b) has $n_t = 4$, and $b = 6$. Therefore the number of the voltage variables to be determined is

$$n = n_t - 1 = 4 - 1 = 3 \qquad \text{(C-12.1)}$$

and the number of the current variables is

$$l = b - n = 6 - 3 = 3 \qquad \text{(C-12.2)}$$

The circuit of (a) has $n_t = 6$, and $b = 9$, when the sources are counted as branches. The subscript A is added to both n_A and l_A for the moment. These quantities are determined by

$$n_A = n_t - 1 = 6 - 1 = 5 \qquad \text{(C-12.3)}$$

$$l_A = b - n_A = 9 - 5 = 4 \qquad \text{(C-12.4)}$$

Of the five n_A, two are constrained to take on specific values by the voltage sources, and the number left to be determined is

$$n = n_A - 2 = 5 - 2 = 3 \qquad \text{(C-12.5)}$$

Of the four l_A, one is constrained by the current source, and the number left to be determined is

$$l = l_A - 1 = 4 - 1 = 3 \qquad \text{(C-12.6)}$$

Therefore the sources may or may not be counted as branches, because the net result is the same.

Appendix **D**

Basic Matrix Theory

D-1. INTRODUCTION

Orderly schemes for handling all sorts of mathematical and technical problems are essential to the development of most present-day systems theory. This is particularly true when dealing with high-order systems. The use of rectangular arrays of elements in rows and columns is the basis for one such orderly scheme. Such an array is called a *matrix*. An ordered pair of real numbers designates the position of an element in a matrix, the first number designating the row, and the second number the column. Such an array is

$$\mathbf{M} = \begin{bmatrix} m_{11} & m_{12} & m_{13} \\ m_{21} & m_{22} & m_{23} \end{bmatrix} = [m_{ij}] \qquad \text{(D-1.1)}$$

with i the row index ranging from 1 to 2, and j the column index ranging from 1 to 3. Brackets generally surround the matrix symbol, except when specific symbols have been defined otherwise. It can be seen that a matrix contains a number of elements equal to the number of rows times the number of columns.

The array of elements constituting a matrix may represent anything from the coefficients of a set of linear algebraic equations, to experimentally measured data elements. After defining suitable operations involving matrices, it will be more apparent why the matrix is an indispensible mathematical tool.

D-2. SPECIAL MATRICES

A number of special matrices need to be defined before proceeding very far with matrix operations. One common type of matrix, having an equal number of rows and columns and hence row and column indices with identical sets of integers, is called a *square matrix*.

An example of this is

$$\mathbf{M} = \begin{bmatrix} 5 & 8 & 1 \\ -2 & 0 & 0 \\ 2 & -1 & 9 \end{bmatrix} \tag{D-2.1}$$

A special case of the square matrix is one with all unity elements along the main diagonal of the array, and with zero elements for all other entries. Such an array is called an *identity* matrix. An example of this, with four rows and columns, and the **I** symbol generally used for such matrices, is

$$\mathbf{I} = \begin{bmatrix} 1 & 0 & 0 & 0 \\ 0 & 1 & 0 & 0 \\ 0 & 0 & 1 & 0 \\ 0 & 0 & 0 & 1 \end{bmatrix} \tag{D-2.2}$$

A square matrix can be used to define a single numerical quantity, called a *determinant*, inherent in the array of elements. The determinant is symbolized by vertical lines enclosing an array, as in

$$|M| = \begin{vmatrix} 5 & 8 & 1 \\ -3 & 0 & 0 \\ 2 & -1 & 9 \end{vmatrix} \tag{D-2.3}$$

and can be extracted from the matrix in accord with a well-defined algorithm. In a series of operations, elements are taken in specific order, from specific positions, and formed into a product, which is given a specific positive or negative sign preparatory to a summing of the products to get the desired determinant. Space does not permit complete exposition of the general rule for determinants, but the schemes commonly used for 2-by-2 and 3-by-3 arrays are

$$|M| = \begin{vmatrix} 5 & 7 \\ 1 & -2 \end{vmatrix} = (5)(-2) - (7)(1) = -17 \tag{D-2.4}$$

$$|M| = \begin{vmatrix} 5 & 8 & 1 \\ -3 & 0 & 0 \\ 2 & -1 & 9 \end{vmatrix} = (5)(0)(9) + (8)(0)(2) + (1)(-1)(-3) - (1)(0)(2) - (8)(-3)(9) - (5)(-1)(0) = 219 \tag{D-2.5}$$

where the arrows indicate the routes along which elements are selected for the various determinant product terms, and the signs are allocated to the product terms. It should be noted that no two elements in any product term are from the same row and column. It should also be noted that a determinant is the single number derived from a square array, and not the array itself.

A determinant formed by striking out rows and columns so as to leave a square array is said to be a *minor determinant* of the matrix. Thus, two possible minors for the matrix of Eq. D-2.1 are given by

$$|M_1| = \begin{vmatrix} 5 & 8 \\ -3 & 0 \end{vmatrix} \qquad |M_2| = \begin{vmatrix} -3 & 0 \\ 2 & 9 \end{vmatrix} \tag{D-2.6}$$

The largest non-zero minor for a matrix establishes what is called the *rank of the matrix*. This rank is equal to the number of rows (or columns) in the square minor matrix. For the matrix of Eq. D-2.1, the rank is 3, since the largest non-zero determinant matrix composed of elements from this square matrix includes all its elements. There are various systematic schemes for determining the rank of a matrix that should be investigated by the reader interested in such techniques. One additional example of rank is illustrated in

$$\mathbf{M} = \begin{bmatrix} 2 & 3 & 1 \\ 4 & 6 & 2 \end{bmatrix} \tag{D-2.7}$$

where the rank of the matrix **M** is 1, because all its 2-by-2 minor arrays have zero determinant values.

If a new matrix is formed by interchanging the rows and columns of a matrix, the new matrix is said to be the *transpose* of the original matrix. Such an operation involves, in general terms, simply interchanging the row and column subscripts on the elements of the original matrix. Thus, m_{13} will become the (3, 1) element in the transpose of a matrix **M**.

Cases in which a matrix equals its transpose are said to be symmetrical, and such a matrix is called a *symmetrical matrix*. In these cases, $m_{ij} = m_{ji}$.

Some arrays represented by matrices involve only a single row or a single column, which as is natural, are said to be *row* or *column matrices*. Where such matrices are associated with a suitable coordinate system, and suitable operations are defined, the row or column matrix can be designated as a *row* or *column vector*, frequently with, for example, a change in notation as follows: $[X] = X] = \mathbf{x}$ (for a column vector).

D-3. MATRIX ADDITION AND SUBTRACTION

In general, the matrices we will deal with are defined over the field of real, or complex, numbers. In such cases the basic arithmetic operations pose no difficult problem. It would seem somewhat reasonable to define *matrix addition* as the normal operation of adding elements in corresponding positions in two matrices, and this is precisely what is done, although it must be mentioned that the two matrices being added will have to be identical in size. Such matrices are said to be conformable for addition. A similar concept and operation are used for *matrix subtraction.* The general definition of matrix addition and subtraction are illustrated by

$$\mathbf{M} = [m_{ij}] \qquad \mathbf{K} = [k_{pr}] \qquad \mathbf{M} \pm \mathbf{K} = [m_{ij} \pm k_{ij}] \tag{D-3.1}$$

Matrix addition, in a specific example, is shown by

$$\mathbf{A} = \begin{bmatrix} 3 & 2 & 1 \\ 0 & -4 & 7 \end{bmatrix} \quad \mathbf{B} = \begin{bmatrix} 4 & 0 & -1 \\ 0 & 3 & 9 \end{bmatrix} \quad \mathbf{A} + \mathbf{B} = \begin{bmatrix} 7 & 2 & 0 \\ 0 & -1 & 16 \end{bmatrix} \tag{D-3.2}$$

In view of this definition of matrix addition and subtraction, it should be noted that the associative, distributive, and commutative properties of addition hold for matrices.

D-4. MATRIX MULTIPLICATION

Because of the presence of a number of elements in a matrix, more than the usual arithmetic multiplication process is involved in matrix multiplication; some form of indexing must be invoked. For this reason the definition of matrix multiplication is simply stated and then explained. The product of a matrix **A** times a matrix **B** written as **AB**, is defined only when the number of columns of **A** is the same as the number of rows of **B**. The multiplication formula then is illustrated by

$$c_{ij} = \sum_{r=1}^{n} a_{ir} b_{rj} \qquad \begin{matrix} i = 1, 2, \ldots, p \\ j = 1, 2, \ldots, q \end{matrix} \tag{D-4.1}$$

In this case it is presumed that there are n columns in **A** and n rows in **B**, and that there are p rows in **A** and q columns in **B**. The result in a matrix **C** that has p rows and q columns.

As an example, let us consider the matrices given in

$$\mathbf{A} = \begin{bmatrix} 1 & 2 & 4 \\ 0 & 1 & 0 \end{bmatrix} \quad \mathbf{B} = \begin{bmatrix} 1 & 1 & 1 \\ -3 & 0 & 2 \\ 0 & 4 & -5 \end{bmatrix} \tag{D-4.2}$$

with the matrix product as in

$$\begin{gathered} \mathbf{C} = \mathbf{AB} = [c_{ij}] \qquad c_{ij} = \sum_{r=1}^{3} a_{ir} b_{rj} \qquad \begin{matrix} i = 1, 2 \\ j = 1, 2, 3 \end{matrix} \\ \mathbf{C} = \mathbf{AB} = \begin{bmatrix} -5 & 17 & -15 \\ -3 & 0 & 2 \end{bmatrix} \end{gathered} \tag{D-4.3}$$

As an aid in visualizing and remembering the operations involved in matrix multiplication, it can be noted that the elements in a column of the product matrix are obtained, as illustrated in Fig. D-4.1, by (1) placing elements from the corresponding column of the right matrix horizontally above the rows of the left matrix; (2) forming the product of

$$\begin{matrix} & & \mathit{1} & \mathit{-3} & \mathit{0} \end{matrix}$$

$$\mathbf{C} = \mathbf{AB} = \begin{bmatrix} 1 & 2 & 4 \\ 0 & 1 & 0 \end{bmatrix} \begin{bmatrix} \mathit{1} & 1 & 1 \\ \mathit{-3} & 0 & 2 \\ \mathit{0} & 4 & -5 \end{bmatrix}$$

$$c_{11} = (\mathit{1})(1) + (\mathit{-3})(2) + (\mathit{0})(4) = -5$$

$$c_{21} = (\mathit{1})(0) + (\mathit{-3})(1) + (\mathit{0})(0) = -3$$

$$\begin{matrix} & & \mathit{1} & \mathit{0} & \mathit{4} \end{matrix}$$

$$\mathbf{C} = \mathbf{AB} = \begin{bmatrix} 1 & 2 & 4 \\ 0 & 1 & 0 \end{bmatrix} \begin{bmatrix} 1 & \mathit{1} & 1 \\ -3 & \mathit{0} & 2 \\ 0 & \mathit{4} & -5 \end{bmatrix}$$

$$c_{12} = (\mathit{1})(1) + (\mathit{0})(2) + (\mathit{4})(4) = 17$$

$$c_{22} = (\mathit{1})(0) + (\mathit{0})(1) + (\mathit{4})(0) = 0$$

Fig. D-4.1

the shifted elements with those directly below; and (3) adding product terms to get a single new row entry in the product matrix in a position corresponding to the row in left matrix and the column in the right matrix.

In assessing the operation of matrix multiplication it can now be noted that the sequence of the matrices in a product is important; it cannot be reversed—the commutative property does *not* hold. In general, reversing the order of the matrices in a product operation may lead to a situation where the multiplication is not defined, as the number of columns of the left matrix may no longer equal the number of rows of the right matrix. It can be indicated that the operation of matrix multiplication is still distributive and associative. This allows matrix products such as **ABC** to be performed as **(AB)C** or as **A(BC)**. It does require careful attention to the fact that the number of rows of **B** must equal the number of columns of **A**, and the number of rows of **C** must equal the number of columns of **B**. The final product matrix will have a number of rows equal to the number of rows of **A**, and a number of columns equal to the number of columns of **C**. The distributive property will allow expressions such as **A(B + C)** to be written as **AB + AC**. Again the main deviation in properties of matrix multiplication from regular scalar operations is that the sequence of terms in a matrix product is critical; the left matrix (or premultiplying matrix) must remain on the left, and the right matrix (or post-multiplying matrix) must remain on the right.

D-5. INVERSE MATRICES

In general, the operation of division is not defined when dealing with matrices. However, a somewhat analogous operation is available for square matrices, in the form of what are called *inverse matrices*. Such matrices have the property that they yield an identity matrix when pre- or post-multiplied by the square matrix from which an inverse is derived. Thus, if we desire the inverse of the square matrix $\mathbf{M}$, we would have

$$\mathbf{M}\mathbf{M}^{-1} = \mathbf{M}^{-1}\mathbf{M} = \mathbf{I} \tag{D-5.1}$$

where $\mathbf{M}^{-1}$ is the inverse matrix of $\mathbf{M}$.

A great amount of effort has been devoted to the development of computational methods for determining inverse matrices. In general, space does not permit the exposition of such schemes; however, a brief overview of one such scheme is presented in the next section. It perhaps should also be pointed out that not all square matrices have inverses. If the determinant of a square matrix is zero, the matrix will have no inverse. Though this is beyond the scope of this text, so-called *pseudo-inverse matrices* do exist for some non-square matrices.

D-6. ADJOINT MATRICES AND INVERSES

For square matrices it is possible to define a matrix called the *adjoint matrix*, or simply adj $\mathbf{C}$, which must satisfy such a relationship as that in

$$\mathbf{C} \cdot \text{Adj } \mathbf{C} = |C|\mathbf{I} = \text{Adj } \mathbf{C} \cdot \mathbf{C} \tag{D-6.1}$$

By premultiplying the left equality by $\mathbf{C}^{-1}$ and substituting the fact that $\mathbf{C}^{-1}\mathbf{C} = \mathbf{I}$, together with the property that multiplication by an inverse matrix simply reproduces the matrix being multiplied by the identity matrix, we are able to obtain

$$\begin{aligned} \mathbf{C}^{-1}\mathbf{C} \cdot \text{Adj } \mathbf{C} &= \mathbf{C}^{-1}|C|\mathbf{I} \\ \mathbf{I} \cdot \text{Adj } \mathbf{C} &= \mathbf{C}^{-1}|C|\mathbf{I} \\ \text{Adj } \mathbf{C} &= \mathbf{C}^{-1}|C| \\ \mathbf{C}^{-1} &= \text{Adj } \mathbf{C}/|C| \end{aligned} \tag{D-6.2}$$

From Eqs. D-6.2 it can be seen that the inverse could be obtained if we found some way to calculate Adj $\mathbf{C}$. It can be shown that the adjoint matrix is made up of what are called *cofactors* of the original matrix $\mathbf{C}$, which are defined in the following paragraphs.

If the ith row and jth column of matrix $\mathbf{C}$ are removed, and a determinant is formed from the remaining elements, we have what is called one of the *first minors* of $\mathbf{C}$. First minors are then multiplied by $(-1)^{i+j}$, we have the *cofactors* for all row and column positions of the matrix $\mathbf{C}$ corresponding to the row and column positions i and j, respectively. If we use the symbol ψ_{ij} to designate a cofactor of $\mathbf{C}$, obtained by forming the determinant of elements remaining after striking out the ith row and jth column of $\mathbf{C}$ and attaching the appropriate sign, we have Fig. D-6.1, which illustrates the formation of a few cofactors for the indicated 2-by-2 and 3-by-3 matrices.

$$\mathbf{C} = \begin{bmatrix} 3 & -7 \\ 0 & 1 \end{bmatrix} \qquad \begin{aligned} \psi_{11} &= (-1)^{1+1}(1) = 1 \\ \psi_{12} &= (-1)^{1+2}(0) = 0 \\ \psi_{21} &= (-1)^{2+1}(-7) = 7 \\ \psi_{22} &= (-1)^{2+2}(3) = 3 \end{aligned}$$

(*a*)

$$\mathbf{B} = \begin{bmatrix} 1 & 1 & 1 \\ -3 & 0 & 2 \\ 0 & 4 & -5 \end{bmatrix} \qquad \begin{aligned} \beta_{11} &= (-1)^{1+1} \begin{vmatrix} 0 & 2 \\ 4 & -5 \end{vmatrix} \\ \beta_{12} &= (-1)^{1+2} \begin{vmatrix} -3 & 2 \\ 0 & -5 \end{vmatrix} \\ &\cdots\cdots\cdots\cdots \end{aligned}$$

(*b*)

Fig. D-6.1

The *adjoint matrix* is formed by arranging all cofactors in a new matrix according to the reverse of the row and column position index of the cofactor. Thus, Adj $\mathbf{C} = [\psi_{ji}]$. In the case of the 2-by-2 matrix from Fig. D-6.1, we have

$$\text{Adj } \mathbf{C} = \begin{bmatrix} 1 & 7 \\ 0 & 3 \end{bmatrix} \tag{D-6.3}$$

By using the above formulation for the adjoint matrix and the defining relationship of Eq. D-6.2, we are able to verify the original statements concerning inverses, as in

$$\begin{aligned} \mathbf{C}^{-1} = \text{Adj } \mathbf{C}/|C| &= \begin{bmatrix} 1 & 7 \\ 0 & 3 \end{bmatrix} \Big/ \begin{vmatrix} 3 & -7 \\ 0 & 1 \end{vmatrix} \\ &= \frac{1}{3} \begin{bmatrix} 1 & 7 \\ 0 & 3 \end{bmatrix} \end{aligned} \tag{D-6.4}$$

$$\mathbf{C}\mathbf{C}^{-1} = \mathbf{I} = \begin{bmatrix} 3 & -7 \\ 0 & 1 \end{bmatrix} \begin{bmatrix} 1 & 7 \\ 0 & 3 \end{bmatrix} \frac{1}{3} = \begin{bmatrix} 1 & 0 \\ 0 & 1 \end{bmatrix}$$

With this very brief overview of inverses and adjoint matrices we close our discussion of matrix theory and urge the interested reader to pursue the matter further in texts devoted specifically to this subject.

Appendix **E**

Laplace Transform Formulas

LAPLACE TRANSFORM FORMULAS

	$F(s)$	$f(t)$
1	$\frac{1}{s}$	$u_{-1}(t)$
2	$\frac{1}{s^n}$, n is a positive integer	$\frac{t^{n-1}}{(n-1)!}$
3	$\frac{1}{s+a}$	ϵ^{-at}
4	$\frac{1}{(s+a)^n}$, n is a positive integer	$\frac{t^{n-1}\epsilon^{-at}}{(n-1)!}$
5	$\frac{\omega}{s^2+\omega^2}$	$\sin \omega t$
6	$\frac{s}{s^2+\omega^2}$	$\cos \omega t$
7	$\frac{b}{s^2-b^2}$	$\sinh bt$
8	$\frac{s}{s^2-b^2}$	$\cosh bt$
9	$\frac{\omega}{(s+a)^2+\omega^2}$	$\epsilon^{-at} \sin \omega t$
10	$\frac{s+a}{(s+a)^2+\omega^2}$	$\epsilon^{-at} \cos \omega t$
11	$\frac{b}{(s+a)^2-b^2}$	$\epsilon^{-at} \sinh bt$

LAPLACE TRANSFORM FORMULAS (*Continued*)

12	$\dfrac{s+a}{(s+a)^2 - b^2}$	$\epsilon^{-at} \cosh bt$
13	$\dfrac{a_1 s + a_0}{(s+a)^2 + \omega^2}$	$A\epsilon^{-at} \sin(\omega t + \alpha)$ where $A\underline{/\alpha} = \left[\dfrac{a_0 - a_1(a - j\omega)}{\omega}\right]$
14	$\dfrac{a_1 s + a_0}{(s+a)(s+b)}$	$\left(\dfrac{a_0 - a_1 a}{b - a}\right)\epsilon^{-at} + \left(\dfrac{a_0 - ba_1}{a - b}\right)\epsilon^{-bt}$
15	$\dfrac{a_1 s + a_0}{(s+a)^2}$	$[a_1 + (a_0 - a_1 a)t]\epsilon^{-at}$
16	$\dfrac{a_2 s^2 + a_1 s + a_0}{(s+a)(s+b)(s+c)}$	$\dfrac{a_2 a^2 - a_1 a + a_0}{(b-a)(c-a)}\epsilon^{-at} + \dfrac{a_2 b^2 - a_1 b + a_0}{(a-b)(c-b)}\epsilon^{-bt} + \dfrac{a_2 c^2 - a_1 c + a_0}{(a-c)(b-c)}\epsilon^{-ct}$
17	$\dfrac{a_2 s^2 + a_1 s + a_0}{(s+a)^2(s+b)}$	$\left[\dfrac{b(a_1 - 2a_2 a) + a_2 a^2 - a_0}{(b-a)^2}\right]\epsilon^{-at} + \left(\dfrac{a_2 a^2 - a_1 a + a_0}{b - a}\right)t\epsilon^{-at} + \left[\dfrac{a_2 b^2 - a_1 b + a_0}{(b-a)^2}\right]\epsilon^{-bt}$
18	$\dfrac{a_2 s^2 + a_1 s + a_0}{(s+a)^3}$	$a_2\epsilon^{-at} + (a_1 - 2a_2 a)t\epsilon^{-at} + \frac{1}{2}(a_2 a^2 - a_1 a + a_0)t^2\epsilon^{-at}$
19	$\dfrac{a_2 s^2 + a_1 s + a_0}{[(s+a)^2 + \omega^2](s+b)}$	$A\epsilon^{-at}\sin(\omega t + \alpha) + \dfrac{a_0 - a_1 b + a_2 b^2}{(a-b)^2 + \omega^2}\epsilon^{-bt}$ where $A\underline{/\alpha} = \dfrac{a_0 - a_1(a - j\omega) + a_2(a - j\omega)^2}{\omega(b - a + j\omega)}$
20	$\dfrac{a_3 s^3 + a_2 s^2 + a_1 s + a_0}{(s+a)(s+b)(s+c)(s+d)}$	$\dfrac{a_0 - a_1 a + a_2 a^2 - a_3 a^3}{(b-a)(c-a)(d-a)}\epsilon^{-at} + \dfrac{a_0 - a_1 b + a_2 b^2 - a_3 b^3}{(a-b)(c-b)(d-b)}\epsilon^{-bt}$ $+ \dfrac{a_0 - a_1 c + a_2 c^2 - a_3 c^3}{(a-c)(b-c)(d-c)}\epsilon^{-ct} + \dfrac{a_0 - a_1 d + a_2 d^2 - a_3 d^3}{(a-d)(b-d)(c-d)}\epsilon^{-dt}$

LAPLACE TRANSFORM FORMULAS (*Continued*)

	$F(s)$	$f(t)$
21	$\dfrac{a_3s^3+a_2s^2+a_1s+a_0}{(s+a)^2(s+b)(s+c)}$	$\dfrac{a_0-a_1a+a_2a^2-a_3a^3}{(b-a)(c-a)}t\epsilon^{-at}+\left[\dfrac{a_1-2a_2a+3a_3a^2}{(b-a)(c-a)}-\dfrac{(a_0-a_1a+a_2a^2-a_3a^3)(b+c-2a)}{(b-a)^2(c-a)^2}\right]\epsilon^{-at}$ $+\dfrac{a_0-a_1b+a_2b^2-a_3b^3}{(a-b)^2(c-b)}\epsilon^{-bt}+\dfrac{a_0-a_1c+a_2c^2-a_3c^3}{(a-c)^2(b-c)}\epsilon^{-ct}$
22	$\dfrac{a_3s^3+a_2s^2+a_1s+a_0}{(s+a)^2(s+b)^2}$	$\dfrac{(a_0-a_1a+a_2a^2-a_3a^3)}{(b-a)^2}t\epsilon^{-at}+\dfrac{(a_0-a_1b+a_2b^2-a_3b^3)}{(a-b)^2}t\epsilon^{-bt}$ $+\left\{\dfrac{a_1-2a_2a+3a_3a^2}{(b-a)^2}-\dfrac{2(a_0-a_1a+a_2a^2-a_3a^3)}{(b-a)^3}\right\}\epsilon^{-at}$ $+\left\{\dfrac{a_1-2a_2b+3a_3b^2}{(a-b)^2}-\dfrac{2(a_0-a_1b+a_2b^2-a_3b^3)}{(a-b)^3}\right\}\epsilon^{-bt}$
23	$\dfrac{a_3s^3+a_2s^2+a_1s+a_0}{(s+a)^3(s+b)}$	$\dfrac{a_0-a_1a+a_2a^2-a_3a^3}{2(b-a)}t^2\epsilon^{-at}+\dfrac{a_0-a_1b+a_2b^2-a_3b^3}{(a-b)^3}\epsilon^{-bt}$ $+\left[\dfrac{a_1-2a_2a+3a_3a^2}{b-a}-\dfrac{a_0-a_1a+a_2a^2-a_3a^3}{(b-a)^2}\right]t\epsilon^{-at}$ $+\left[\dfrac{a_2-3a_3a}{b-a}-\dfrac{a_1-2a_2a+3a_3a^2}{(b-a)^2}+\dfrac{a_0-a_1a+a_2a^2-a_3a^3}{(b-a)^3}\right]\epsilon^{-at}$
24	$\dfrac{a_3s^3+a_2s^2+a_1s+a_0}{(s+a)^4}$	$\frac{1}{6}(a_0-a_1a+a_2a^2-a_3a^3)t^3\epsilon^{-at}+\frac{1}{2}(a_1-2a_2a+3a_3a^2)t^2\epsilon^{-at}+(a_2-3a_3a)t\epsilon^{-at}+a_3\epsilon^{-at}$
25	$\dfrac{a_3s^3+a_2s^2+a_1s+a_0}{[(s+a)^2+\omega^2](s+b)(s+c)}$	$A\epsilon^{-at}\sin(\omega t+\alpha)+\dfrac{a_0-a_1b+a_2b^2-a_3b^3}{[(a-b)^2+\omega^2](c-b)}\epsilon^{-bt}+\dfrac{a_0-a_1c+a_2c^2-a_3c^3}{[(a-c)^2+\omega^2](b-c)}\epsilon^{-ct}$ where $A\underline{/\alpha}=\dfrac{a_0-a_1(a-j\omega)+a_2(a-j\omega)^2-a_3(a-j\omega)^3}{\omega(b-a+j\omega)(c-a+j\omega)}$

LAPLACE TRANSFORM FORMULAS (*Continued*)

26	$\dfrac{a_3s^3 + a_2s^2 + a_1s + a_0}{[(s+a)^2 + \omega^2](s+b)^2}$	$A\epsilon^{-at}\sin(\omega t + \alpha) + \dfrac{a_0 - a_1b + a_2b^2 - a_3b^3}{[(a-b)^2 + \omega^2]}t\epsilon^{-bt} + \left\{\dfrac{a_1 - 2a_2b + 3a_3b^2}{[(a-b)^2 + \omega^2]} - 2\dfrac{(a_0 - a_1b + a_2b^2 - a_3b^3)(a-b)}{[(a-b)^2 + \omega^2]^2}\right\}$ where $A\underline{/\alpha} = \dfrac{a_0 - a_1(a-j\omega) + a_2(a-j\omega)^2 - a_3(a-j\omega)^3}{\omega(b-a+j\omega)^2}$
27	$\dfrac{a_3s^3 + a_2s^2 + a_1s + a_0}{[(s+a)^2 + \omega^2][(s+b)^2 + \sigma^2]}$	$A\epsilon^{-at}\sin(\omega t + \alpha) + B\epsilon^{-bt}\sin(\sigma t + \beta)$ where $A\underline{/\alpha} = \dfrac{a_0 - a_1(a-j\omega) + a_2(a-j\omega)^2 - a_3(a-j\omega)^3}{\omega[(b-a+j\omega)^2 + \sigma^2]}$ $B\underline{/\beta} = \dfrac{a_0 - a_1(b-j\sigma) + a_2(b-j\sigma)^2 - a_3(b-j\sigma)^3}{\sigma[(a-b+j\sigma)^2 + \omega^2]}$
28	$\dfrac{a_3s^3 + a_2s^2 + a_1s + a_0}{[(s+a)^2 + \omega^2]^2}$	$-\dfrac{At\epsilon^{-at}}{2\omega^2}\cos(\omega t + \alpha) + 2B\epsilon^{-at}\cos(\omega t + \beta)$ where $A\underline{/\alpha} = a_0 - a_1(a-j\omega) + a_2(a-j\omega)^2 + a_3(a-j\omega)^3$ $B\underline{/\beta} = \dfrac{a_1 - 2a_2(a-j\omega) + 3a_3(a-j\omega)^2}{(2j\omega)^2} - \dfrac{2[a_0 - a_1(a-j\omega) + a_2(a-j\omega)^2 - a_3(a-j\omega)^3]}{(2j\omega)^3}$
29	$\dfrac{a_4s^4 + a_3s^3 + a_2s^2 + a_1s + a_0}{(s^2 + \omega^2)(s+a)(s+b)(s+c)}$	$\dfrac{a_0 - a_1a + a_2a^2 - a_3a^3 + a_4a^4}{(a^2 + \omega^2)(b-a)(c-a)}\epsilon^{-at} + \dfrac{a_0 - a_1b + a_2b^2 - a_3b^3 + a_4b^4}{(b^2 + \omega^2)(a-b)(c-b)}\epsilon^{-bt} + \dfrac{a_0 - a_1c + a_2c^2 - a_3c^3 + a_4c^4}{(c^2 + \omega^2)(a-c)(b-c)}\epsilon^{-ct} + A\sin(\omega t + \alpha)$ where $A\underline{/\alpha} = \dfrac{a_0 + a_1(j\omega) + a_2(j\omega)^2 + a_3(j\omega)^3 + a_4(j\omega)^4}{\omega(j\omega + a)(j\omega + b)(j\omega + c)}$

LAPLACE TRANSFORM FORMULAS (*Continued*)

	$F(s)$	$f(t)$
30	$\dfrac{a_4 s^4 + a_3 s^3 + a_2 s^2 + a_1 s + a_0}{(s^2+\omega^2)(s+a)^2(s+b)}$	$\dfrac{a_0 - a_1 a + a_2 a^2 - a_3 a^3 + a_4 a^4}{(a^2+\omega^2)(b-a)} t\epsilon^{-at} + \dfrac{a_0 - a_1 b + a_2 b^2 - a_3 b^3 + a_4 b^4}{(b^2+\omega^2)(a-b)^2} \epsilon^{-bt}$ $+ \left[\dfrac{a_1 - 2a_2 a + 3a_3 a^2 - 4a_4 a^3}{(a^2+\omega^2)(b-a)} - \dfrac{(a_0 - a_1 a + a_2 a^2 - a_3 a^3 + a_4 a^4)(\omega^2 - 2ab + 3a^2)}{(a^2+\omega^2)^2(b-a)^2}\right] \epsilon^{-at}$ $+ A \sin(\omega t + \alpha)$ where $A\underline{/\alpha} = \dfrac{a_0 + a_1(j\omega) + a_2(j\omega)^2 + a_3(j\omega)^3 + a_4(j\omega)^4}{\omega(j\omega+a)^2(j\omega+b)}$
31	$\dfrac{a_4 s^4 + a_3 s^3 + a_2 s^2 + a_1 s + a_0}{(s^2+\omega^2)(s+a)^3}$	$A \sin(\omega t + \alpha) + \dfrac{a_0 - a_1 a + a_2 a^2 - a_3 a^3 + a_4 a^4}{2(a^2+\omega^2)} t^2 \epsilon^{-at}$ $+ \left[\dfrac{a_1 - 2a_2 a + 3a_3 a^2 - 4a_4 a^3}{a^2+\omega^2} + \dfrac{2a(a_0 - a_1 a + a_2 a^2 - a_3 a^3 + a_4 a^4)}{(a^2+\omega^2)^2}\right] t\epsilon^{-at}$ $+ \left[\dfrac{2a_2 - 6a_3 a + 12a_4 a^2}{(a^2+\omega^2)} + \dfrac{4a(a_1 - 2a_2 a + 3a_3 a^2 - 4a_4 a^3)}{(a^2+\omega^2)^2}\right.$ $\left. + \dfrac{(6a^2 - 2\omega^2)(a_0 - a_1 a + a_2 a^2 - a_3 a^3 + a_4 a^4)}{(a^2+\omega^2)^3}\right] \epsilon^{-at}$ where $A\underline{/\alpha} = \dfrac{a_0 + a_1(j\omega) + a_2(j\omega)^2 + a_3(j\omega)^3 + a_4(j\omega)^4}{\omega(j\omega+a)^3}$
32	$\dfrac{a_4 s^4 + a_3 s^3 + a_2 s^2 + a_1 s + a_0}{(s^2+\omega^2)[(s+a)^2+\sigma^2](s+b)}$	$A \sin(\omega t + \alpha) + B\epsilon^{-at} \sin(\sigma t + \beta) + \dfrac{a_0 - a_1 b + a_2 b^2 - a_3 b^3 + a_4 b^4}{(b^2+\omega^2)[(a-b)^2+\sigma^2]} \epsilon^{-bt}$ where $A\underline{/\alpha} = \dfrac{a_0 + a_1(j\omega) + a_2(j\omega)^2 + a_3(j\omega)^3 + a_4(j\omega)^4}{\omega[(j\omega+a)^2+\sigma^2](j\omega+b)}$ $B\underline{/\beta} = \dfrac{a_0 - a_1(a-j\sigma) + a_2(a-j\sigma)^2 - a_3(a-j\sigma)^3 + a_4(a-j\sigma)^4}{\sigma[(a-j\sigma)^2+\omega^2](b-a+j\sigma)}$

Index